THE ROUTLEDGE COMPANION TO TRANSMEDIA STUDIES

Around the globe, people now engage with media content across multiple platforms, following stories, characters, worlds, brands, and other information across a spectrum of media channels. This transmedia phenomenon has led to the burgeoning of transmedia studies in media, cultural studies, and communication departments across the academy. *The Routledge Companion to Transmedia Studies* is the ultimate volume for scholars and students interested in comprehending all the various aspects of transmediality. This collection, which gathers together original articles by a global roster of contributors from a variety of disciplines, sets out to contextualize, problematize, and scrutinize the current status and future directions of transmediality, exploring the industries, arts, practices, cultures, and methodologies of studying convergent media across multiple platforms.

Matthew Freeman is Reader in Multiplatform Media at Bath Spa University, UK. He is Co-Director of Bath Spa's Centre for Media Research and acts as REF Champion for the University's Communication, Culture and Media submission to REF. His research examines cultures of production across the borders of media, industries, cultures, and histories, and he is the author of *Historicising Transmedia Storytelling: Early Twentieth-Century Transmedia Story Worlds*, *Industrial Approaches to Media: A Methodological Gateway to Industry Studies*, the co-author of *Transmedia Archaeology: Storytelling in the Borderlines of Science Fiction, Comics and Pulp Magazines* (with Carlos A. Scolari and Paolo Bertetti), and the co-editor of *Global Convergence Cultures: Transmedia Earth* (with William Proctor).

Renira Rampazzo Gambarato is Senior Lecturer in Media and Communication Studies at Jönköping University, School of Education and Communication, Sweden. She is co-editor of the books *Exploring Transmedia Journalism in the Digital Age* (together with Geane Alzamora) and of *Kulturdialoge Brasilien-Deutschland: Design, Film, Literatur, Medien [Cultural Dialogue Brazil-Germany: Design, Film, Literature, Media]* (together with Geane Alzamora and Simone Malaguti). Her current research revolves around transmedia storytelling analysis and complexity of transmedia experiences. More broadly, her research interests include Peircean semiotics, digital culture, international and intercultural media studies, media education, film analysis, and the design and analysis of multiplatform experiences.

THE ROUTLEDGE COMPANION TO TRANSMEDIA STUDIES

Edited by Matthew Freeman and
Renira Rampazzo Gambarato

Routledge
Taylor & Francis Group
NEW YORK AND LONDON

First published 2019
by Routledge
711 Third Avenue, New York, NY 10017

and by Routledge
2 Park Square, Milton Park, Abingdon, Oxon OX14 4RN

Routledge is an imprint of the Taylor & Francis Group, an informa business

Library of Congress Cataloging in Publication Data
A catalog record for this book has been requested

ISBN: 978-1-138-48343-9 (hbk)
ISBN: 978-1-351-05490-4 (ebk)

Typeset in Bembo
by Out of House Publishing

MIX
Paper from responsible sources
FSC® C013056

Printed and bound in Great Britain by
TJ International Ltd, Padstow, Cornwall

CONTENTS

FIGURES

TABLES

CONTRIBUTORS

Geane Carvalho Alzamora is Professor in the Social Communication Department at the Federal University of Minas Gerais, Brazil. She is researcher of CNPq, Brazil (CNPq Productivity Scholarship, process: 311914/2016) and researcher of Fapemig, Brazil (Process: PPM-00263-15). Her postdoctorate is from Pompeu Fabra University, Spain. She holds a Ph.D. in communication and semiotics from Pontifical Catholic University of São Paulo, Brazil; a MA in communication and semiotics also from Pontifical Catholic University of São Paulo and a BA in journalism from Pontifical Catholic University of Minas Gerais, Brazil.

Sarah Atkinson is Head of Department of Culture, Media and Creative Industries at King's College London and co-editor of *Convergence: The International Journal of Research into New Media Technologies*. Sarah has published two books, an edited volume, and numerous articles on the impacts of digital and networked technologies on film and cinema audiences and film production practices.

Kyle Barrett is a lecturer at the University of Waikato. His research focuses on micro-budget film-making and creative practice in various forms. His Ph.D. thesis explored the Scottish film industry in a "post-film" context. He teaches various papers including World Cinema, Studio Production, Film Production, Contemporary Television, and Writing the Web Series. He has been published in *Twin Peaks: Essays on the Original Series*, *Directory of World Cinema: Scotland*, *AMES Media Journal*, *European Journal of Communication*, and *MeCCSA Journal*. He is currently involved in various documentary projects in both Scotland and New Zealand.

Matthias Berg is a postdoctoral research associate at the ZeMKI, Centre for Media, Communication and Information Research, University of Bremen, Germany. He graduated in 2006 from the University of Bremen with an MA in cultural analysis, economy and musicology. Matthias Berg holds a doctoral degree in media and communication studies from the University of Bremen. His doctoral thesis on "Communicative Mobility: Mediated Networking and Work-related Mobility" was awarded the 2016 dissertation prize of the German Communication Association's Media Sociology Section and of Springer VS. From 2010 until 2015 he has been a research associate in the German Research Foundation's Priority Program 1505 "Mediatized Worlds" working on "The Communicative Construction of Communitization" together with Andreas Hepp and Cindy Roitsch. Currently, Matthias Berg is pursuing a postdoctoral research project on the mediatization and digitalization of rural areas in Germany. His research interests include communication and media studies, with a focus on the interrelations of communication and mobility, mediatization research and interpersonal communication, as well as media and community-building and media generations.

Paolo Bertetti has a Ph.D. in semiotics and psychology of symbolic communication. He teaches "sociology of communication" at the University of Siena, where he is also responsible for the organization of teaching of the master's degree in business communication. Former secretary and then vice-president of the AISS, the Italian Association of Semiotic Studies, he is on the board of directors of FELS (Federacion Latinoamericana de semiotica). His research interests concern narratology, transmedia storytelling, semiotic theory, semiotics of film, and semiotics of text. He is also interested in the genres and imagery of contemporary popular culture. His books include: *Il mito Conan. Identità e metamorfosi di un personaggio seriale tra letteratura, fumetto, cinema e televisione* (2011), *Il discorso audiovisivo. Teorie e strumenti semiotici* (2012; spanish translation: *El relato audiovisual. Teorias y hierramentas semioticas*, 2015), *Lo schermo dell'apparire. La teoria della figuratività nella semiotica generativa* (2013). His latest book, *Transmedia Archaeology* (2014; with C. Scolari and M. Freeman), was published in England by Palgrave Macmillan.

Nicoleta Popa Blanariu teaches comparative literature at the Vasile Alecsandri University, in Romania. Her research interests include comparative literature, intercultural studies, intermediality, semiotics, and performance studies. She has published articles in Romanian, French, and English, and books, book chapters, and translations.

Paul Booth is Associate Professor at DePaul University. He is the author of *Crossing Fandoms* (Palgrave 2016), *Digital Fandom 2.0* (Peter Lang 2016), *Playing Fans* (University of Iowa 2015), *Game Play* (Bloomsbury 2015), *Time on TV* (Peter Lang 2012), and *Digital Fandom* (Peter Lang 2010). He has edited *Wiley Companion to Fandom* (Blackwell 2018), *Seeing Fans* (Bloomsbury 2016, with Lucy Bennett), *Controversies in Digital Ethics* (Bloomsbury 2016, with Amber Davisson), and *Fan Phenomena: Doctor Who* (Intellect 2013). He has published numerous articles on fans, social media, and technology. His research interests include fandom, new technologies and media, popular culture, and cult media. He is currently enjoying a cup of coffee.

Mélanie Bourdaa is Associate Professor at the University of Bordeaux Montaigne in communication and information sciences. She analyzes the new television environment and the cultural convergence. She studies particularly American TV series, the phenomenon of fandom in the digital age, and production's strategies (i.e., transmedia storytelling). She also teaches courses on audience and programming, television and cultures, and transmedia storytelling. Mélanie also ran a MOOC entitled "Understanding Transmedia Storytelling" in France. She coordinated with Benjamin Derhy Kurtz a book entitled *The Rise of Transtexts: Challenges and Opportunities* (Routledge, 2016). She is the co-manager of the Design and Media Lab (Université Bordeaux Montaigne – IDEX). She is the head of the research program MediaNum (Valorisation of Cultural Heritage via Transmedia Storytelling).

Frank Branch is a principal level Information Scientist and Ontologist. In this capacity, he leads all information design and ontology efforts for multibillion dollar commercial information platforms. He has a master's in library and information science from the University of Washington specializing in knowledge representation, ontology, and data curation in non-traditional domains. In addition, he has published articles on ontology in domain information design and has patents related to the use of ontology for complex curation tasks. With over 20 years of experience designing enterprise scale content and information management solutions for Fortune 500 companies he has used ontological methods, principles, and practices to develop information ecosystems for a diverse set of domains including finance, advertising, online gaming, popular culture, modern media, and education.

Paola Brembilla is an adjunct professor in television studies at the Università degli Studi di Padova, Italy. In 2016, she earned her Ph.D. in television studies at the Università di Bologna, with a thesis on the competitive strategies and formal evolutions of American TV series. In 2017, she was a research fellow at the same university, collaborating with the Imperial Fashion Group on a corporate storytelling project. She has extensively studied the economics and logics of entertainment and television industries in the United States: in 2010–2011, she was an exchange student at University of California, Berkeley. In 2014, she was a visiting scholar at University of Wisconsin–Madison. She was a speaker at many international conferences, and a keynote speaker at the 2017 *The Art of TV Series* conference (Universidad de Navarra, Pamplona). She has authored several publications and a book, *Game of Strategy* (with Prof. Edoardo Mollona, Giappichelli 2015). Her forthcoming co-edited collection (with Prof. Ilaria A. De Pascalis), *Narrative Ecosystems: Reading Contemporary Serial Television Universes*, is going to be published by Routledge in 2018.

Axel Bruns is an Australian Research Council Future Fellow and Professor in the Digital Media Research Centre at Queensland University of Technology in Brisbane, Australia. He is the author of *Gatewatching and News Curation: Journalism, Social Media, and the Public Sphere* (2018), *Blogs, Wikipedia, Second Life and Beyond: From Production to Produsage* (2008), and *Gatewatching: Collaborative Online News Production* (2005), and a co-editor of *The Routledge Companion to Social Media and Politics* (2016), *Twitter and Society* (2014), *A Companion to New Media Dynamics* (2012), and *Uses of Blogs* (2006). His current work focusses on the study of user participation in social media spaces such as Twitter, and its implications for our understanding of the contemporary public sphere, drawing especially on innovative new methods for analyzing "big social data." His research blog is at http://snurb.info/, and he tweets at @snurb_dot_info. See http://mappingonlinepublics.net/ for more details on his research into social media.

Eefje Op den Buysch is the founder of the Transmedia Storytelling Lab at the Fontys Academy for Creative Industries, in Tilburg, Netherlands. At the lab, students learn to design and execute a complete transmedia storytelling project in a six-month minor program. Eefje obtained her master's degree of science (cum laude) in informatics–human centered multimedia, on the subject of using future scenarios and narratives for the design of a simulated future newsroom. With her knowledge and experience in digital media and information technologies, she led several (research) projects on transmedia, robotics, and online communication. In 2018, Eefje is managing director of *Robot Love*, an art and technology manifestation that asks if we can learn from robots about love.

Marie-Eve Carignan, Ph.D., is a professor of information and public communication at the Université de Sherbrooke. She is co-director of communication, applied communications, and international strategic communications graduate programs. She holds a doctorate in information sciences and communication from the Institut d'études politiques d'Aix-en-Provence and a doctorate in communication from the joint Ph.D. program of the Université de Montréal, the Université du Québec à Montreal, and Concordia University. Prior to that, she worked for more than ten years on many projects in public relations, corporate communication, and management. She focuses her research on cultural industries, media content analysis, journalistic practices and ethics as well as on strategic communication, including public relations, crisis communication, risk, terrorism, and counterterrorism. She is also a researcher at the CHERPA of Sciences Po Aix-en-Provence, an associate member of the Research Laboratory on Audiences of Culture, and a member of the International Network on the Professionalization of the Communicator, which brings together professional researchers and university scholars from France, Belgium, Morocco, the United States, and Canada.

Christy Dena wrote a Ph.D. on transmedia practice, wrote the definition of "transmedial fictions" for the *The Johns Hopkins Encyclopedia of Digital Textuality*, and has been published in journals such as *Convergence* and *ToDiGRA*. Dena is a senior lecturer, and has been the Department Coordinator of Games, and National Chair of Games at SAE Creative Media Institute. As a creative professional, Dena has attracted digital writing awards for her original projects, and is a member of the International Academy of Television Arts and Sciences (Emmys); and an Editorial Board Member of Carnegie Mellon's Entertainment Technology Press (ETC).

Elizabeth Evans is Associate Professor of Film and Television Studies at the University of Nottingham. Her research examines the social, industrial, and technological factors that shape and are shaped by our experiences of increasingly transmedia screen narratives. Her work covers multiple aspects of television studies, audience studies, media studies, games studies, and transmedia studies, and often involves working with practitioners or audiences on their experiences creating and experiencing screen narratives. She is the author of *Transmedia Television: Audiences, New Media and Daily Life* (Routledge 2011), the first exploration of how audiences were responding to the transmedia expansion of television. She has published numerous book chapters and articles in journals including *Convergence, Critical Studies in Television, The International Journal of Communication Studies, Participations*, and *Pervasive Ubiquitous Computing*. She is currently working on a second monograph, *Understanding Transmedia Engagement* (forthcoming, Routledge) which explores how practitioners and audiences define, value, and manage "engagement" with screen content.

Karin Fast is Senior Lecturer in Media and Communication Studies and one of the Geomedia Research Group coordinators, at Karlstad University, Sweden. Her research interests span topics such as (geo-)mediatization, media work, transmediality, and cultural industries (with a particular focus on the music sector). She is editor of *Geomedia Studies: Spaces and Mobilities in Mediatized Worlds* (Routledge 2018; with A. Jansson, J. Lindell, L. R. Bengtsson, and M. Tesfahuney) and has published her work in journals such as *Journal of Computer-Mediated Communication, International Journal of Cultural Studies, European Journal of Cultural Studies*, and *Media Culture and Society*. Her Ph.D. dissertation, *More Than Meets the Eye: Transmedial Entertainment as a Site of Pleasure, Resistance and Exploitation* (2012), dug beneath the surface of one today's most popular media franchises, Hasbro's *Transformers*, to explore industry/fans power-relations.

Kate Fitzpatrick is Senior Strategist at e3 Media. She is an experienced digital marketing professional with experience on both agency and client side. With a purely digital background that began in online PR and moved through to client side marketing, her main passion is for strategy and driving through creative ideas with clear and measurable results. Kate has worked across entertainment, youth, publishing, and financial services (both D2C and B2B) and has a strong drive to understand new technologies and their role in multi-platform communication.

Matthew Freeman is Reader in Multiplatform Media at Bath Spa University, UK. He is Co-Director of Bath Spa's The Centre for Media Research and acts as REF Champion for the University's Communication, Culture and Media submission to REF. His research examines cultures of production across the borders of media, industries, cultures, and histories, and he is the author of *Historicising Transmedia Storytelling: Early Twentieth-Century Transmedia Story Worlds, Industrial Approaches to Media: A Methodological Gateway to Industry Studies*, the co-author of *Transmedia Archaeology: Storytelling in the Borderlines of Science Fiction, Comics and Pulp Magazines* (with Carlos A. Scolari and Paolo Bertetti), and the co-editor of *Global Convergence Cultures: Transmedia Earth*.

Renira Rampazzo Gambarato is Senior Lecturer in Media and Communication Studies at Jönköping University, School of Education and Communication, Sweden. She is co-editor of the books *Exploring Transmedia Journalism in the Digital Age* (together with Geane Alzamora) and of *Kulturdialoge Brasilien-Deutschland: Design, Film, Literatur, Medien* [Cultural Dialogue Brazil-Germany: Design, Film, Literature, Media] (together with Geane Alzamora and Simone Malaguti). Her current research revolves around transmedia storytelling analysis and complexity of transmedia experiences. More broadly, her research interests include Peircean semiotics, digital culture, international and intercultural media studies, media education, film analysis, and the design and analysis of multiplatform experiences.

Max Giovagnoli is the premier transmedia storyteller and producer in Italy. He leads the School of Visual Arts at Istituto Europeo di Design in Rome and he is author of several television series and novels. As a story architect and transmedia architect, he has worked with Disney, Warner Bros., Universal, M2Pictures, and Lucky Red. He has a Ph.D. in narration and imagery, and his essays on transmedia worlds have been published in Italy, the United States, and the United Kingdom. He has also given presentations for TEDxTransmedia, Romics and Cartoons on the Bay, and he is considered one of the most important European fan culture experts and authors. See www.maxgiovagnoli.com.

Šárka Gmiterková is a Ph.D. candidate at the Department of Film and Audio Visual Culture, Masaryk University Brno, Czech Republic, where she finishes her thesis on prewar versus postwar Czech film stardom. She regularly presents her research outcomes in Czech Film Studies journal *Iluminace*, where she also served as guest editor on the topics of local stardom and film acting. Her work was published internationally in *Journal of Celebrity Studies*, *NECSUS*, and in the edited volume *Popular Cinemas in Central Europe: Film Cultures and Histories*.

Nataly Rios Goico is a creative consultant at Conducttr. She is an experienced interactive storyteller with a degree in systems engineering and an MA in digital media. She recently led the TSL work on Hacked, the interactive experience on cyber warfare for Al Jazeera. She is a classically trained ballerina and an award-winning photographer.

Jeff Gomez, CEO of Starlight Runner, is a leading expert in the fields of brand narrative, storyworld development, creative franchise design, and transmedia storytelling. He specializes in the expansion of entertainment properties, premium brands, and socio-political themes into highly successful multiplatform communications and international campaigns. As a producer accredited by the Producers Guild of America, Jeff also develops the storyworlds of films, TV shows, video games, toys, books, comics, apps, virtual reality projects, and theme park attractions. This deepens engagement and accelerates the development of participative communities, resulting in mass audience approval, brand loyalty, and increased revenues. Jeff's pop culture work has impacted such blockbuster properties as Disney's *Pirates of the Caribbean*, James Cameron's *Avatar*, Hasbro's *Transformers*, Sony Pictures' *Spider-Man* and *Men in Black*, Microsoft's *Halo*, and Nickelodeon's *Teenage Mutant Ninja Turtles*.

Jeff has also developed highly successful transmedia campaigns and participative brand narratives for Coca-Cola (*Happiness Factory*), Pepperidge Farm (*Goldfish*), and Spartan Race. Other current clients include Electronic Arts, Sesame Workshop, Disney Parks & Resorts, and World Vision Canada. Jeff's proprietary transmedia methods have also been applied to educational and geo-political causes, accelerating positive self-organized social movements and increasing resistance to crime, violence, and corruption. Through applications of his Collective Journey and transmedia population activation models, Jeff has helped optimize communications for large NGOs, and address crises in Mexico, Colombia, Australia, and the Middle East North Africa region.

Donna Hancox is a senior lecturer in the School of Creative Practice at Queensland University of Technology. In 2017 she was awarded a Smithsonian Research Fellowship and her research is focused on the role of storytelling in social change and the potential for digital technology to amplify marginalized voices.

Stephen Harrington is an associate professor and head of journalism and professional communication at the Queensland University of Technology, Australia. He is the co-author of *Politics, Media and Democracy in Australia: Public and Producer Perceptions of the Political Public Sphere* (Routledge 2017), sole author of *Australian TV News: New Forms, Functions, and Futures* (Intellect 2013), and editor of *Entertainment Values* (Palgrave 2017).

Colin B. Harvey is a writer and narrative designer working across multiple media and specializing in digital storytelling, transmedia narrative, world-building, and shared storyworlds. He has previously worked for Sony and for Rebellion Developments on their Sniper Elite franchise, as well as helping develop British company To Play For's new storytelling platform. As a freelance journalist he has written for the *Guardian*, *Edge*, *Develop*, *RetroGamer*, and *PopMatters*, amongst other publications. Aside from games, Harvey's original short fiction won the first Pulp Idol award, jointly conferred by *SFX Magazine* and Gollancz Books in 2006. He has written licensed tie-in fiction for franchises such as *Doctor Who*, *Highlander*, and *Judge Dredd*, and comic material for *2000AD* and *Commando*. He also contributed the novella "Dead Kelly" to the collection *Journal of the Plague Year*, published in 2014 by Abaddon Books. Harvey is the author of the academic book *Fantastic Transmedia* (Palgrave 2015), an exploration of cross-media storytelling in science fiction and fantasy franchises such as *Star Wars*, Marvel, *Buffy the Vampire Slayer*, and *Halo*, as well as smaller independent projects. He has written and presented extensively on game narrative and transmedia storytelling in a variety of international contexts. His Ph.D., exploring the interrelationship of storytelling and play in video game media using ideas of affect and memory, was conferred in 2009. He is a visiting professor with the Department of Culture, Media and Creative Industries at King's College London and a visiting professor with the Manchester Writing School at Manchester Metropolitan University.

Dan Hassler-Forest works as Assistant Professor of Media and Cultural Studies at Utrecht University. He has published books and articles on superhero movies, comics, transmedia storytelling, adaptation studies, critical theory, and zombies. His most recent book is *Star Wars and the History of Transmedia Storytelling*.

Andreas Hepp is Professor of Media and Communication Studies at the ZeMKI, Centre for Media, Communication and Information Research, University of Bremen, Germany. His main research areas are media and communication theory, media sociology, mediatization research, datafication of social practices, transnational and transcultural communication, and cross-media research. He is author of numerous books and journal articles. His latest book publications are *The Mediated Construction of Reality* (Polity Press 2016, with Nick Couldry) and *Communicative Figurations* (Palgrave 2017, edited with Andreas Breiter and Uwe Hasebrink). Andreas Hepp is involved in the research network "Communicative Figurations" and was co-applicant and PI of the DFG Priority Research Program 1505 "Mediatized Worlds" (2012–2016) and member of the DFG funded Collaborative Research Centre 597 "Transformations of the State," which ended in 2014.

Matt Hills is Professor of Media and Film at the University of Huddersfield, where he is also Co-Director of the Centre for Participatory Culture. Matt is additionally co-editor on the "Transmedia" book series for Amsterdam University Press, and has written six sole-authored monographs, starting with *Fan Cultures* in 2002 (Routledge) and coming up to date with *Doctor Who: The Unfolding Event—Marketing, Merchandising and Mediatizing a Brand Anniversary* in 2015 (Palgrave Pivot). He also edited *New Dimensions of Doctor Who* (I. B. Tauris 2013), and has published more than a hundred book chapters or journal articles on topics such as media fandom and cult film/television. This includes work on transmedia for the open access journal *Participations* and the 2017 Routledge edited collection, *The Rise of Transtexts: Challenges and Opportunities*. Amongst other projects, Matt is currently working on a follow-up to his first book for Routledge, entitled *Fan Studies*.

Alastair Horne is a doctoral research student based at Bath Spa University and the British Library, exploring how mobile devices are changing the relationship between author, text, and reader. This research comprises three complementary strands: analysis of mobile fictions written for smartphones and tablets; exploration of author–reader relationships expressed through social media; and the creation of a short mobile audio fiction to be experienced while walking through Brompton Cemetery. He brings to his research considerable experience within the publishing industry, having spent more than a decade working in a variety of roles for Cambridge University Press, on company-wide innovation, audience development through social media, and digital projects in primary education. His published articles include "The Future of Publishing: A Report on Innovation and the Future of the Book" (Media Futures 2011), "The Future's Live, the Future's Digital" (*Logos* 2012), and "Publishing: The Last (and Next?) Five Years" (*The Indexer* 2017); he has also written on publishing for *FutureBook*, the London Book Fair, and *Publishing Perspectives*. Alastair has been a fan of Doctor Who since 1980 and is currently midway through a chronological re-watch of the entire series.

Indrek Ibrus is Professor of Media Innovation at Tallinn University (TLU). He is also the head of TLU Center of Excellence in Media Innovation and Digital Culture (MEDIT). His main research areas include media innovation, cross- and transmediality, media production, and policy studies. His main current research projects look at the evolution of audiovisual content metadata standards and the convergence between the audiovisual sector and other sectors such as health care, tourism, and education. He is the editor (together with Carlos A. Scolari) of *Crossmedia Innovations: Texts, Markets, Institutions* (Peter Lang 2012). Prof. Ibrus received his Ph.D. from the London School of Economics and Political Science and MPhil from the University of Oslo.

André Jansson is Professor of Media and Communication Studies and Director of the Geomedia Research Group at Karlstad University, Sweden. His research is oriented toward questions of media use, identity, and power from an interdisciplinary perspective. A main research theme is the relationship between mediatization processes and the production of social space. Jansson's most recent books include *Mediatization and Mobile Lives: A Critical Approach* (Routledge 2018), *Communications/Media/Geographies* (Routledge 2017, with P. C. Adams, J. Cupples, K. Glynn, and S. Moores) and *Cosmopolitanism and the Media: Cartographies of Change* (Palgrave Macmillan 2015, with M. Christensen). He is also the co-editor of *Geomedia Studies: Spaces and Mobilities in Mediatized Worlds* (Routledge 2018, with K. Fast, J. Lindell, L. R. Bengtsson, and M. Tesfahuney) and several other books. His work has been published in journals such as *Communication Theory*, *European Journal of Communication*, *European Journal of Cultural Studies*, *International Journal of Cultural Studies*, *Journal of Consumer Culture*, *New Media & Society*, *Tourist Studies*, and *Urban Studies*. Jansson obtained his Ph.D. in 2001 at Gothenburg University, Sweden.

Henry Jenkins is Provost's Professor of Communication, Journalism, Cinematic Arts and Education at the University of Southern California. He joined USC from the Massachusetts Institute of Technology, where he was Peter de Florez Professor in the Humanities and founder and co-director of the Comparative Media Studies Master's Program. Best known for *Convergence Culture: Where Old and New Media Collide*, Jenkins' most recent books include *Participatory Culture in a Networked*

Society (with danah boyd and Mimi Ito) and *By Any Media Necessary: The New Youth Activism* (with Sangita Shresthova, Liana Gamber-Thompson, Neta Kligler-Vilenchik, and Arley Zimmerman). He blogs twice a week at henryjenkins.org.

Hille van der Kaa is the editor-in-chief of BN DeStem. Previously, she worked as a professor and research leader for Fontys Hogeschool and Tilburg University. Her research interests are transmedia storytelling, data journalism, robot journalism, and the influence of technology and journalism. She took the lead in an NWO-funded research on building an automated newsroom in the Netherlands. In addition, Van der Kaa worked as a journalist, marketing manager, and transmedia scriptwriter.

Joakim Karlsen is an assistant professor at the Faculty of Computer Sciences, Østfold University College where he teaches media production, communication design, and interaction design. He researches emerging media production practices for online- and mobile media platforms, and has recently conducted studies on computational journalism in newsrooms, independent transmedia documentary production, and hackathons organized for media workers to learn more about digital making. From 1994 to 2000 he worked with documentary film and video production as director and editor, and several of the films he has worked on has been screened on national television in Norway.

Helen W. Kennedy is Head of the School of Media at the University of Brighton, UK. Her current research interests are feminist interventions into games culture, experience design, and cultural evaluation. She has published widely in game studies and is Principle Investigator in two significant projects—one an international project aimed at the transformation of games (REFIG.ca) and the other a project that puts new immersive technologies in the hands of circus and street artists (https://blogs.brighton.ac.uk/xrcircus/). Over the past three years she has been researching experiential cinema, street games, and other innovative and immersive forms as an aspect of the wider ludification of contemporary culture.

Anna Kérchy is an associate professor at the English Department of the University of Szeged, Hungary. She holds a Ph.D. in literature from the University of Szeged and a DEA in semiology from Université Paris VII, as well as a Habilitation degree in literature and culture from the University of Debrecen. Her research interests include intermedial cultural representations, the post-semiotics of the embodied subject, interfacings of Victorian and postmodern fantastic imagination, gender studies, women's art, fairy tales, and children's/YA literature. She has authored two monographs: *Alice in Transmedia Wonderland. Curiouser and Curiouser New Forms of a Children's Classic* (McFarland 2016) and *Body-Texts in the Novels of Angela Carter: Writing from a Corporeagraphic Point of View* (Mellen 2008). She (co-)edited five essay collections on postmodern reinterpretations of fairy tales, the literary fantastic, the iconology of law and order, the cultural history of Continental European freak shows, posthumanism in fantastic fiction, as well as an *EJES* special journal issue on feminist interventions into intermedial studies, a *Bookbird* special journal issue on translating and transmediating children's literatures and cultures, and an *Americana* special issue on *Interspecies Encounters in Postmillenial Filmic Fantasies*.

Jenny Kidd is Senior Lecturer in the School of Journalism, Media and Cultural Studies, Cardiff University. Jenny researches and teaches across the fields of digital media, heritage studies, and the creative economy, and is author of *Museums in the New Mediascape: Transmedia, Participation, Ethics* (Routledge 2014) and *Representation* (Routledge 2015). She has published on related themes in (for example) the *International Journal of Heritage Studies, Museum and Society, The Journal of Curatorial Studies*, and the *Museums Journal*.

Lisbeth Klastrup is an associate professor with the Culture and Communication Research Group at the IT University of Copenhagen. She wrote her Ph.D. thesis on virtual worlds and storytelling

(2003). Her research currently centers on the popular and professional uses of social media; mundane memes and transmedial communication. She is co-editor of the *International Handbook of Internet Research* (with Hunsinger and Allen, Springer 2010). Her most recent book is *Sociale Netværksmedier* [Social Network Media] (Samfundslitteratur 2016), already widely popular among Danish university students.

Kevin Moloney is a transmedia scholar who consults with public and private news organizations on transmedia story design, and has taught transmedia storytelling since 2012. He holds a Ph.D. in technology, media, and society from the interdisciplinary Atlas Institute at the University of Colorado. For 20 years he was a regular contributor to the *New York Times* covering the Rocky Mountain region. His images appeared on the *Times* front page 50 times, and on section fronts hundreds more. He has photographed more than 960 stories for the US newspaper of record. Moloney's work has also appeared in *U.S. News & World Report, Fortune, Life, Time, Stern, The Chicago Tribune, The Independent, USA Today, Elle, Marie Claire, Business Week*, the *Christian Science Monitor*, and *National Geographic* publications. He was one of two journalists selected as inaugural recipients of the Ford Environmental Journalism Fellowship. For 21 years Moloney was a lecturer of photojournalism at the University of Colorado Boulder. In 2008 his students secured for him the Robin F. Garland Education award, given by the National Press Photographers Association in honor of achievement in photojournalism education. His students are winners of World Press, POYi, Best of Photojournalism, Hearst, and CPOY competitions. They have been listed on PDN magazine's 30 Emerging Photographers lists, and winners of Alicia Patterson, Getty, Inge Morath, and Aftermath Project grants. Moloney also has extensive international journalism training experience having taught photojournalism workshops in Argentina, Chile, the Falkland Islands, Myanmar, and Venezuela.

Erica Negri graduated in foreign languages in 2006 and was subsequently awarded with a master's degree in film production and screenwriting at the Università Cattolica del Sacro Cuore in Milan, Italy. After working for TV-series production company Lux Vide, in 2008 she joined Cattleya, a leading Italian company in film and TV production as a development coordinator and script editor. In 2012 Erica joined beActive, a film, TV, and transmedia production company, experienced in developing entertainment content across multiple platforms, as development officer. Starting from January 2013 Erica has worked as a freelance story editor, transmedia strategist, and academic lecturer both in Italy and Spain. In 2014 Erica earned her PhD in transmedia storytelling, and her research was published as a book entitled *La Rivoluzione Transmediale* (ed. Lindau) in 2015. After two years spent as a TV program producer at The Walt Disney Company Italia, in 2016 Erica was hired by SKY Italia, where she currently works as a delegate producer overseeing local and international productions, from script to delivery.

Hanna Nolasco is a Brazilian master's student in media and contemporary culture at the Federal University of Bahia (UFBA) and holds a bachelor's degree in journalism from the same institution (2013). She participated in an academic exchange program with Université Paris 8 Vincennes-Saint-Denis, in Paris (France), between 2011 and 2012. She is a member of the research groups Laboratory of Analysis in Telefiction (A-Tevê) and Laboratory of Film Analysis (LAF). Her field of interest is the relationship between music and television, more specifically dealing with music programs and music in telenovelas in Brazil, mainly analyzing Globo Network's products, for it is the biggest TV broadcaster in the country. She has developed previous studies on Globo's music programs comprising the period since its creation, in 1965, until 2013. Her current research deals with the role of music in the narrative construction of the telenovela *Cheias de Charme* [*Sparkling Girls*].

Alison Norrington is Founder and Creative Director of StoryCentral Ltd., also working across areas such as Brand and Creative Director, Strategist, Talent and Content Development Lead, Creative Franchise Design and Story Architect, Writer, Digital Producer, and Media Lecturer. She works with brands, storytellers, filmmakers, and theme parks, and her clients include Walt Disney Imagineering R&D, FOX International, YouTube, Sundance TV, CBS Interactive, McCann, and Coca Cola.

Roberta Pearson is Professor of Film and Television Studies at the University of Nottingham. Among her most recent publications are the co-authored *Star Trek and American Television* (University of California Press 2014), and the co-edited *Many More Lives of the Batman* (BFI 2015) and *Storytelling in the Media Convergence Age: Exploring Screen Narratives* (Palgrave Macmillan 2015). The co-edited collection *Contemporary Transatlantic Television Drama* is forthcoming from Oxford University Press in 2018. She is, in total, the author, co-author, editor, or co-editor of 14 books, and author or co-author of over 80 journal articles and book chapters; among the last are several articles/chapters on the Sherlock Holmes character.

Rebekah Phillips is an information specialist/ISO coordinator, who spends her days managing the information needs of her peers at an international high-tech design and engineering company. Her current position also requires that she manage information for associated Fortune 500 companies. She is passionate about creating information environments that are intuitive and adaptable. Her BS in psychology from Southern Oregon University and MA in library and information science from the University of Washington assist in this ongoing pursuit. By integrating human behavioral tendencies with information management techniques, Rebekah endeavors to make life easier for those seeking information. Her previous work managing the information database of a local retail organization helped hone these skills before moving to the field of technology. Though her career in information management is still young, Rebekah has dedicated herself to research on myriad topics for the last decade. Rebekah has also been published in an academic journal for her work in creating a unique ontological structure.

Dan Popa teaches formal languages and automata at the Vasile Alecsandri University, in Romania. His research interests include formal linguistics, interpretation, translators, and compilers. He has a Ph.D. in adaptable modular language construction and is interested in the syntax and semantics of foreign languages. He has built a didactic computer language, which uses national keywords and, as a former game programmer, is interested in storytelling, video games, and multimedia servers.

Robert Pratten is CEO and founder of Transmedia Storyteller Ltd, creator of Conducttr—the immersive experience platform. His ambition is to make everyone's life an adventure. Robert is a thought-leader in transmedia storytelling and author of the book *Getting Started in Transmedia Storytelling: A Practical Guide for Beginners*. His clients include Kodansha, VISA, British Foreign & Commonwealth Office, UK Ministry of Defence, Al Jazeera, and Canal+. He can be found online as @robpratten.

William Proctor is Senior Lecturer in Media, Culture and Communication at Bournemouth University. He has published on various topics including Batman, James Bond, *Ghostbusters*, *Star Wars*, and One Direction. At present, William is completing work on his debut single-authored monograph, *Reboot Culture: Comics, Film, Transmedia*, to be published by Palgrave Macmillan in 2018. He is co-editor of *Disney's Star Wars: Forces of Production, Promotion and Reception* (University of Iowa Press, forthcoming) and *The Scandinavian Invasion: Critical Studies in Nordic Noir and Beyond* (Peter Lang, forthcoming), both with Richard McCulloch. William is also Director of the World Star Wars Project.

Kate Pullinger is a novelist and digital writer. Her most recent work is a ghost story for smartphones that is personalized for every reader. *Breathe* was commissioned as part of the Ambient Literature research project, in collaboration with Visual Editions and Google Creative Lab Sydney. Her novel for smartphones, *Jellybone*, was published in 2017. Her novel, *The Mistress of Nothing*, won Canada's 2009 Governor General's Literary Award for Fiction. In 2014, she adapted her collaborative work of digital multimedia, Flight Paths: A Networked Novel (2007), co-created with Chris Joseph, as the novel *Landing Gear*, which was longlisted for Canada Reads. Also in 2014, she created the digital war memorial, Letter to an Unknown Soldier, with Neil Bartlett; 22,000 members of the public wrote letters to the soldier. Her project Inanimate Alice has also won numerous prizes; 2016 saw the launch of *Inanimate Alice: Episode Six—The Last Gas Station* as well as a teachers' edition of the first five episodes, and 2018 will see the launch of a virtual reality episode, *Perpetual Nomads*. Kate Pullinger is Director of the Centre for Cultural and Creative Industries at Bath Spa University where her primary research interest is in practice-based research on the intersection of fiction and technology, and hybrid collaborative digital forms.

Raúl Rodríguez Ferrándiz is Full Professor of Semiotics of Mass Communication and Transmedia Production at the University of Alicante, Spain. He has been academic coordinator of the Master of Communication and Creative Industries at the same university. He has published the books *Masks of Lying* (XXXV International Essay Prize Ciudad de Valencia, 2017; Valencia, Pre-Textos, 2018); *The Venal Muse: Production and Consumption of Industrial Culture* (International Essay Prize "Miguel Espinosa", Murcia, Tres Fronteras, 2010) and *Apocalypse Show. Intellectuals, TV and End of the Millennium* (Madrid, Biblioteca Nueva, 2001). He has coordinated the volume *The Controversy on Mass Culture in the Inter-War Period: A Critical Anthology* (University of Valencia, 2012) and two monographs dedicated to transmedia storytelling in *Cuadernos de Información y Comunicación* (2014, with Cristina Peñamarín) and *Mediterranean Journal of Communication* (2017, with Vanesa Saiz). He has published papers in *Critical Studies in Media Communication*, *Semiotica*, *International Journal of Communication*, and *Revista de Occidente*.

Ulrike Rohn is visiting professor and senior researcher at Tallinn University's Baltic, Film, Media, Arts and Communication School (BFM) and its Centre of Excellence in Media Innovation and Digital Culture (MEDIT). Dr Rohn is the president of the European Media Management Association (emma), and she is associate editor of the *Journal of Media Business Studies*. Her research interests include media policy, media branding, media business models, the sharing economy, international media strategies, and the relationship between media and culture. Latter research interests have led to her book publication *Cultural Barriers to the Success of Foreign Media Content: Western Media in China, India, and Japan*, published by Peter Lang in 2010. In her main current research project, Ulrike studies European audiovisual policy measurements from a small market's point of view. Ulrike Rohn received her Ph.D. from Friedrich-Schiller-University Jena in Germany.

Inara Rosas is a Brazilian Ph.D. candidate in media and contemporary culture at the Federal University of Bahia (UFBA). Her current research is about the field of screenplay in Brazilian audiovisual productions, especially concerning the trajectory of João Emanuel Carneiro, screenwriter of feature films, TV series and telenovelas. She holds a master's degree from the institution (2014) with a thesis about narrative strategies in Latin American movies regarding the historical period of dictatorship and that have children as protagonists, and a bachelor's degree in journalism (2009) at Federal University of Paraíba (UFPB). She is a member of the research groups Laboratory of Analysis in Telefiction (A-Tevê), Laboratory of Film Analysis (LAF), and the research network Iberoamerican Observatory of Television Studies (OBITEL) Brazil/UFBA, researching Latin American cinema and television, concerning narrative, stylistic and reception analysis.

Pamela Rutledge, Ph.D., MBA, is Director of the Media Psychology Research Center and a professor of media psychology at Fielding Graduate University, where she is lead faculty for the brand psychology and audience engagement doctoral concentration. Rutledge's research interests include the persuasive impact of multiplatform storytelling on consumer intention and brand development, meaning construction across social technologies and the motivation driving audience engagement. Rutledge consults on a variety of media projects, bridging the gap between theory and practice by harnessing psychological science to inform creative strategy. Current projects include eliciting actionable communication patterns and narratives from social media data and providing psychological analysis to identify and integrate behavioral triggers in campaign and message strategies. Dr Rutledge speaks and consults internationally and has published both academic and popular work, most recently authoring the text *Exploring Positive Psychology: The Science of Happiness and Well-Being*. She has written on the social impact of mobile technology in *Global Mobile*, digital media use in the *Handbook of Rehabilitation Psychology*, and the implementation of proactive narratives in the social space for the US Department of Defense Anti-Terrorist Series. Rutledge is also a frequent expert source for the mass media on technology use, social impact and popular culture.

Carlos A. Scolari has a Ph.D. in applied linguistics and communication languages (Università Cattolica del Sacro Cuore, Milan, Italy) and a degree in social communication (Universidad Nacional de Rosario, Argentina). He is Associate Professor at the Department of Communication of the University Pompeu Fabra – Barcelona. He has lectured about digital interfaces, media ecology, transmedia storytelling, and interactive communication in more than 25 European, Latin American, and Asian countries. Most important publications: *Hacer Clic* (2004), *Hipermediaciones. Elementos para una teoría de la comunicación digital interactiva* (2008), *El fin de los medios masivos. El comienzo de un debate* (with M. Carlón, 2009/2014), *Crossmedia Innovations* (with I. Ibrus, 2012), *Narrativas Transmedia* (2013), *Transmedia Archaeology* (with P. Bertetti and M. Freeman, 2014), *Ecología de los medios* (2015), and *Las leyes de la interfaz* (2018). He is the Principal Investigator of the Horizon 2020 "Transmedia Literacy" research project (2015–2018) and the Spanish research project "Transalfabetismos" (2015–2018).

Peter von Stackelberg has worked professionally as a world-builder, transmedia storyteller, journalist, photographer, 3D computer graphic artist, futurist, and professor. He teaches English, composition, new media storytelling, technology and innovation, project management, and systems thinking, and has worked at Alfred State College of Technology (State University of New York) and Keuka College, both in rural western New York state. Peter is the founder of Jericho Hill Publishing, a small publishing firm specializing in print-on-demand books, e-books, and interactive transmedia fiction and non-fiction. He has worked on a series of webinars about visual storytelling, visual creation of storyworlds, and character creation in partnership with *Digital Art Live* magazine. He has also served as a world-builder and transmedia storytelling consultant to projects like *Trail of the Butterfly*, an animated transmedia production developed by HIERROanimación, and *Voltas*, an animated science fiction project developed by Autobotika. In the 1980s, Peter's work as an investigative journalist was awarded the Citation of Merit in Public Service Journalism by Canada's Governor General.

Lorena Peret Teixeira Tárcia is Professor in the Department of Social Communication at the University Center of Belo Horizonte, Brazil. She holds a Ph.D. in communication from the Federal University of Minas Gerais, Brazil; a MA in education from Pontifical Catholic University of Minas Gerais, Brazil, and a BA in journalism also from Pontifical Catholic University of Minas Gerais, Brazil.

Jan-Noël Thon is an assistant professor in media studies and digital media culture at the Department of Culture, Film and Media of the University of Nottingham, UK, and a principal investigator in the Collaborative Research Center 923 "Threatened Orders—Societies under Stress" at the University of Tübingen, Germany. Recent books include *From Comic Strips to Graphic Novels: Contributions to the Theory and History of Graphic Narrative* (co-edited with Daniel Stein, De Gruyter, 2013/2015), *Storyworlds across Media: Toward a Media-Conscious Narratology* (co-edited with Marie-Laure Ryan, University of Nebraska Press, 2014), *Game Studies: Aktuelle Ansätze der Computerspielforschung* [*Game Studies: Current Approaches to Video Game Research*] (co-edited with Klaus Sachs-Hombach, von Halem, 2015), *Transmedial Narratology and Contemporary Media Culture* (University of Nebraska Press, 2016), *Subjectivity across Media: Interdisciplinary and Transmedial Perspectives* (co-edited with Maike Sarah Reinerth, Routledge, 2017), and *Comicanalyse: Eine Einführung* [*Comics Analysis: An Introduction*] (co-authored with Stephan Packard, Andreas Rauscher, Véronique Sina, Lukas Wilde, and Janina Wildfeuer, Metzler, forthcoming 2019).

Susana Tosca is Associate Professor at the Department of Communication and Arts at Roskilde University, Denmark. Her Ph.D. dissertation, a poetics of hypertext literature, was awarded the *summa cum laude* distinction in 2001. She has worked for many years on electronic literature, the storytelling potential of computer games, transmediality, and complex reception processes, with a side interest in fan activity and the distributed aesthetic formats of the network era. Her last book is the third edition of *Understanding Videogames* (Routledge 2016).

Ethan Tussey (Ph.D., University of California, Santa Barbara, 2012) is an assistant professor of film and media at Georgia State University. His book, *The Procrastination Economy: The Big Business of Downtime* (New York University Press 2018) details the economic and social value of mobile device use in the context of the workplace, the commute, the waiting room, and the living room. His work explores the relationship between the entertainment industry and the digitally empowered public. He has contributed book chapters on creative labor, online sports viewing, connected viewing, and crowdfunding to the anthologies *Saturday Night Live and American TV* (Indiana University Press 2013), *Digital Media Sport: Technology and Power in the Network Society* (Routledge 2013), *Connected Viewing: Selling, Sharing, and Streaming Media in a Digital Era* (Routledge 2013), and *Crowdfunding the Future: Media Industries, Ethics, & Digital Society* (Peter Lang, 2015) He is also the coordinating editor of *In Media Res* and the co-founder of the Atlanta Media Project. He teaches classes on television analysis, media industries, and digital media.

Portia Vann is a Ph.D. candidate in the Digital Media Research Centre at Queensland University of Technology. Her research focuses on digital marketing and communications in the sport industry, as organizations develop transmedia strategies to keep up with changes in the media environment. Specifically, as part of her Ph.D. project she examined how sport organizations create and implement social media strategy at large-scale events, working with digital marketing teams at professional sporting events in Australia.

Mark J. P. Wolf is a professor in the Communication Department at Concordia University Wisconsin. His books include *Abstracting Reality: Art, Communication, and Cognition in the Digital Age* (2000), *The Medium of the Video Game* (2001), *Virtual Morality: Morals, Ethics, and New Media* (2003), *The Video Game Theory Reader* (2003), *The Video Game Explosion: A History from PONG to PlayStation and Beyond* (2007), *The Video Game Theory Reader 2* (2008), *Myst and Riven: The World of the D'ni* (2011), *Before the Crash: Early Video Game History* (2012), the two-volume *Encyclopedia of Video Games: The Culture, Technology, and Art of Gaming* (2012), *Building Imaginary Worlds: The Theory and History of Subcreation* (2012), *The Routledge Companion to Video Game Studies* (2014), *LEGO Studies: Examining the Building Blocks of a Transmedial Phenomenon* (2014), *Video Games Around the World* (2015), the four-volume *Video Games and Gaming Cultures* (2016), *Revisiting Imaginary Worlds: A Subcreation Studies*

Anthology (2017), *Video Games FAQ* (2017), *The World of Mister Rogers' Neighborhood* (2017), *The Routledge Companion to Imaginary Worlds* (2017), and *The Routledge Companion to Media Technology and Obsolescence* (forthcoming 2018). He has been invited to speak in North America, South America, Europe, Asia, and Second Life; has had work published in a variety of journals and he is on several editorial boards as well. He lives in Wisconsin with his wife Diane and his sons Michael, Christian, and Francis.

FOREWORD

Henry Jenkins

As hard as it may be for younger readers to understand, we did not have action figures when I was growing up in the 1960s. Children had been playacting stories, creating their own costumes and props, for hundreds of years, probably longer. There had been movie tie-in products as long as there has been mass media. But the apparatus of the modern movie franchise was still taking shape and transmedia was not yet a concept that could be applied to what was happening. Action figures embody some of these shifts—a film franchise like *Star Wars* may release dozens, perhaps hundreds, of action figures over time, which not only represent core characters from the narrative but expand beyond the narrative to flesh out the fictional world (Gray 2010). The best contemporary action figures have costumes that often evoke particular scenes, they have props which suggest their relationship with other elements in the fictional worlds, they have capacities for action which tell us something about how they interact with the other characters (Jenkins 2017a). Action figures are authoring tools which encourage fans to dig deep into the mythology and then go beyond it. Contemporary action figures are bound up with the logic of the transmedia system.

Rather than action figures, we had Soakies, a particular brand of bubble bath soap which licensed the likenesses of various media characters (from Hanna Barbera to Disney, from DC to Warner Brothers); we used the plastic containers as toys afterward to create our own stories. We lacked enough characters to flesh out any given fictional world so most of those stories involved crossovers. We had Colorforms—reusable vinyl stickers that could be stuck onto a fixed backdrop and used to stage character interactions. We had Viewmaster slides and board games and stuffed toys. And comic books deployed many of these same characters, but with little or no regard to the source material's narrative continuity. So, yes, fictions moved across media but the focus was overwhelmingly on characters and rarely—if at all—about extending the narratives or expanding the worlds. We might say that the characters had amnesia since they did not seem to recall where they had been or what they had done before. There are cases, such as Carl Bark's Duckberg comics, where these other media constructed worlds that were richer than could be found on screen but such examples were few and far between. For the most part, these spin-offs told fans little that they did not already know from the film or television series. Of course, serialization within television was still rare—*Zorro* comes to mind—with most shows following more episodic structures. And film franchises—*Planet of the Apes* being the exception—were also not yet common.

As I grew older, I did find opportunities to consume works that extended the plotlines, revisited dramatic turning points, developed the subjectivity of secondary characters, deepened our understanding of fictional worlds, imagined characters that crossed over from one fictional world to another, and even explore multiple and contradictory interpretations of established stories. All of these were functions performed by fan fiction (see my discussion of different ways to rewrite a television series in *Textual Poachers: Television Fans and Participatory Culture*, Jenkins 1992). Fans understood television programs as serials even when the rest of the world saw them as episodic and their stories often represented extensions that expanded the timeline, explored secondary characters, or explored the fictional world. When I started to encounter works that meet a more contemporary understanding of transmedia storytelling, I read them as professional fan fiction—a term that gets applied with derision by some cultural critics, but I saw it as a compliment, since it implied that the writers of these secondary texts had a much deeper engagement with the storyworld than we had seen before.

By the time I became a parent in the early 1980s, more and more media properties—especially for children—were organized into franchises, many of which involved a play with intertextuality, a pattern which Justin Wyatt discussed in *High Concept* (1994) in terms of "the book, the look, and the hook," Marsha Kinder wrote about in *Playing with Power* (1991) as transmedia, and Roberta Pearson and William Uricchio (1991) or Tony Bennett and Marsha Woollacott (1987) explored in terms of the trajectories of popular heroes. All of these are pieces that I taught my students, that shaped how I thought about the media my son consumed, that informed how I and many others read new developments in popular entertainment.

By 1999, there was more and more discussion of a shift in the ways that stories were being told, a shift marked by a cover story of *Entertainment Weekly* that defined 1999 as the "year that changed the movies" (Gordimer 1999). The shift in question indicated the level of innovation and experimentation taking place in genre films (*The Matrix*, *The Blair Witch Project*, *The Sixth Sense*), independent films (*Magnolia*), and art films (*Run Lola Run*). Jeff Gordimer began:

> You can stop waiting for the future of movies. It's already here. Someday, 1999 will be etched on a microchip as the first real year of 21st-century filmmaking. The year when all the old, boring rules about cinema started to crumble. The year when a new generation of directors—weaned on cyberspace and *Cops, Pac-Man* and *Public Enemy*—snatched the flickering torch from the aging rebels of the 1970s. The year when the whole concept of "making a movie" got turned on its head.
>
> *(Gordimer 1999)*

Some of it was hype, yet many of the films and filmmakers Gordimer identified are still central to the discussion two decades later. Critics explored what films were learning from computer and video games, embracing nonlinear or multi-perspectival structures. Meanwhile, television narratives were growing ever-more complex, as Jason Mittell (2015) has noted, with more and more integration of storylines across episodes or development of ensemble casts of characters. But, something more was taking place, something many of us recognized but did not yet have the words to describe. Janet Murray (1997) talked about the encyclopaedic dimensions of digital narratives, Will Brooker (2001) discussed narrative "overflow," Mimi Ito (2005) discussed the "hypersocial" dimensions of Japanese "media mix," Frank Rose (2008) spoke about "The Art of Immersion," P. David Marshall (2002) described "intertextual commodities," and I expanded Kinder's term, "transmedia," to describe an emerging model where narratives and fictional worlds, not just characters, were extending across media platforms. By the time we started to write about transmedia storytelling, we were already soaking in it—not unlike the bubble bath from my old Soakies. We were playing catch up, responding to experiments taking place in various corners of the media industries—my chapter in *Convergence Culture* (Jenkins 2006) focused on *The Matrix* franchise, but sidebars suggested some other parallel examples such as Haxan's promotion of *The Blair Witch Project*, transnational projects as American comics publishers responded to the growing popularity of manga, the Dawson's Desktop project promoting the CW's *Dawson's Creek*, and "The Beast," an alternate reality game linked to the release of the film *A.I.: Artificial Intelligence*. Here, juxtaposition of cases via sidebars substituted for an effort to develop a more coherent picture of the boundaries of transmedia storytelling.

By this point, I had spent the past decade expanding my access to various media-makers. I had consulted with Brenda Laurel (2001) when she was launching Purple Moon games and a section of her book, *Utopian Entrepreneur*, reflected her engagement with ideas from Marsha Kinder that I had introduced to her and her staff. I had run a Creative Leaders Program for Electronic Arts which brought me into contact with Neil Young, Will Wright, Danny Bilson, Bing Gordon, and others who were thinking about the ways that games might forge new relationships with Hollywood franchises. I drafted my first explanation of transmedia flying back from an intense meeting in Los Angeles between top games and film industry people. There were already people out there—Jeff Gomez, Mike Monello, Jay Bushman, Sean Stewart, Maureen McHugh, among them—who were trying to develop new models for richer storyworlds and cross-media extensions. McLuhan tells us that media are put out before they are thought out and a lot of people were putting out by the turn of the millennium. While I might never have anticipated the degree to which people across the media industries would embrace *Convergence Culture*, I wrote my chapter on transmedia storytelling with a strong awareness that I was entering a conversation which would challenge industry, fans, and academia to rethink how they engaged with contemporary entertainment. I refused to read these developments as purely motivated by commercial instincts, having a front-line perspective on the ways that these storytellers talked about the aesthetic potentials of an expanded canvas.

Little did I know that transmedia storytelling would be a flag that people would rally behind. *Convergence Culture* came out in the midst of the Hollywood writers' strike, which centered around how writers would be compensated for digital content that often extended television series: they saw it as part of the storytelling process and the studios saw it as promotion. Transmedia helped them make a case for the value of their contributions. I have since been told stories that people were passing copies up and down the picket lines and as people went off strike, transmedia production started to become more widespread as language for describing what was going on with *Lost*, *Heroes*, and other contemporary television programs. We heard less about 360-degree promotion, the word being pushed by the studios and the networks. Elsewhere, transmedia became language that was taken up by policy-makers, arts funding agencies, and government film bureaus as a way of justifying support for all kinds of digital content.

From the start, then, our understanding of transmedia storytelling emerged at the intersection between theory and practice and through exchanges between academics and media-makers. Such exchanges often took place on the margins, through informal channels—face-to-face exchanges at conferences and local meet-ups, podcast interviews, blog posts, even exchanges on social media (such as Brian Clark's "East Coast-West Coast Distinction"). Even understood in academic terms, it was a space where contributions by graduate students (Geoffrey Long, Ivan Askwith, Christy Dena, Sam Ford, Jeff Watson, among others) were as important as those of more established scholars. Production handbooks exist on our syllabi at the University of Southern California alongside theoretical tomes; our classes often ask students to combine thinking and making. It was an exchange that has become increasingly transnational (taking different shapes depending on whether the media ecology is shaped by commercial or public service logics). As the term has traveled, the concept of transmedia has expanded from an early focus on popular genres to more diverse media (publishing, music, location-based experiences), from entertainment to documentary and journalism, activism and mobilization, education, religion, diplomacy, sports, and branding (Jenkins 2017b).

Each step along the way, we have had to both expand and sharpen our understanding of what we mean by transmedia, as critics and theorists raced to keep up with innovative new practices that often defy any simple, rigid or static definition. To me, transmedia was about a set of relationships across media, not a single model for how different media might "collide." We need lots of different models for the forms that transmedia might take as different creative teams pursue different functions in relation to different stories for different audiences in different national contexts. When my students ask me whether something is or is not transmedia, I usually ask them in what ways it may be useful to read it as transmedia. So, there may be some circumstances where it makes sense to read fan fiction as part of a larger transmedia system while other times, it is important to maintain clear distinctions between canon and fanon, between continuity and multiplicity.

I am often surprised when I and others get accused of ignoring the larger history of transmedia practices in our early writing: my chapter in *Convergence Culture* draws multiple comparisons with earlier media practices. We were preoccupied with trying to figure out what was new about contemporary transmedia stories and with searching for antecedents. But the introduction of a new media ecology shifts how we understood what came before, and in this case, the availability of the concept of transmedia storytelling has sharpened our understanding of earlier media ecologies, changing how we read *The Wizard of Oz*, Walt Disney, Superman, or the Lone Ranger, to cite a few examples.

All of this brings us to the volume you currently hold in your hands. The editors have assembled an international rogue's gallery of some of the world's top thinkers about transmedia. I love that attempts to develop a field theory of transmedia have given way to efforts to describe different configurations transmedia might take in different contexts or in the service of different goals. We have examples here from many different national media ecologies. The book explores the value of different theoretical traditions for analyzing examples of transmedia practices, further expanding the conceptual tools with which we conduct our discussions. I welcome the shift from discussing interactivity toward more nuanced consideration of participatory practices, bringing fandom studies more fully into the mix. I am excited to see some productive tensions here between perspectives that are grounded in claims of medium specificity (whether in terms of different media sectors or the contributions different media affordances offer). And it is wonderful to see such rich discussions of transmedia's contributions beyond the realm of entertainment. These are just a few of the strengths of the current collection, which amply demonstrates the value of an expansive definition of transmedia, as we seek to continually make sense of a period of profound and prolonged media change.

References

Bennett, Tony, and Janet Woollacott. 1987. *Bond and Beyond: The Political Career of a Popular Hero*. London: Methuen.

Brooker, Will. 2001. "Living on *Dawson's Creek*: Teen Viewers, Cultural Convergence, and Television Overflow." *International Journal of Cultural Studies* 4 (4): 456–472.

Gordimer, Jeff. 1999. "1999: The Year That Changed Movies." *Entertainment Weekly*, November 26. Accessed January 4, 2018. http://ew.com/article/1999/11/26/1999-year-changed-movies/.

Gray, Jonathan. 2010. *Show Sold Separately: Promos, Spoilers, and Other Media Paratexts*. New York: New York University Press.

Ito, Mizuko. 2005. "Technologies of Childhood Imagination: Yugioh, Media Mixes and Everyday Cultural Production." In *Structures of Participation in Digital Culture*, edited by Joe Karaganis and Natalie Jeremijenko, 44–67. Durham, NC: Duke University Press.

Jenkins, Henry. 1992. *Textual Poachers: Television Fans and Participatory Culture*. New York: Routledge.

Jenkins, Henry. 2006. *Convergence Culture: Where Old and New Media Collide*. New York: New York University Press.

Jenkins, Henry. 2017a. "Adaptation, Extension, Transmedia." *Film/Literature Quarterly* 45 (2). www.salisbury.edu/lfq/_issues/first/adaptation_extension_transmedia.html.

Jenkins, Henry. 2017b. "Transmedia Logics and Locations." In *The Rise of Transtexts: Challenges and Opportunities*, edited by Benjamin W. L. Derhy Kurtz and Mélanie Bourdaa, 220–240. London: Routledge.

Kinder, Marsha. 1991. *Playing with Power in Movies, Television, and Video Games: From Muppet Babies to Teenage Mutant Ninja Turtles*. Berkeley: University of California Press.

Laurel, Brenda. 2001. *Utopian Entrepreneur*. Cambridge, MA: MIT Press.

Marshall, P. David. 2002. "The New Intertexutal Commodity." In *The New Media Book*, edited by Dan Harries, 110–127. London: British Film Institute.
Mittell, Jason. 2015. *Complex TV: The Poetics of Contemporary Television Storytelling*. New York: New York University Press.
Murray, Janet. 1997. *Hamlet on the Holodeck: The Future of Narrative in Cyberspace*. Cambridge: Free Press.
Pearson, Roberta E., and William Uricchio (eds.). 1991. *The Many Lives of the Batman: Critical Approaches to a Superhero and His Media*. New York: Routledge.
Rose, Frank. 2008. *The Art of Immersion: How the Digital Generation Is Remaking Hollywood, Madison Avenue and How We Tell Stories*. New York: W.W. Norton and Company.
Wyatt, Justin. 1994. *High Concept: Movies and Marketing in Hollywood*. Austin: University of Texas Press.

ACKNOWLEDGMENTS

The crafting of this book has been highly important in terms of shaping our current and future thinking about all things transmedia, an area of study that we have both been immersed within for a number of years now. The process of devising, compiling, and editing this volume has allowed us to re-understand what "transmedia" really is, what its overarching and underpinning characteristics are, how it works, where it works, and—perhaps most importantly of all—why we need it. Our hope is that the broad and expansive scope of this book allows for at least some of the deepest meanings and the fullest potentials of transmediality to be imagined, and for that we wholeheartedly thank each and every one of our contributors for their excellent and focused chapters. And an even bigger thank you must go to Henry Jenkins, whose generous support of this project—as is documented in his insightful foreword—is enormously valued.

INTRODUCTION

Transmedia Studies—Where Now?

Matthew Freeman and Renira Rampazzo Gambarato

Let's start with a question: what is transmedia? Here, we mean this question not as a lead-in to presenting any kind of rudimentary, oft-cited definition, but rather as a genuine question. The transmedia phenomenon has led to the burgeoning of transmedia studies across media, film, television, cultural, and communication studies across the academy, not to mention the wider creative and cultural industries. *The Routledge Companion to Transmedia Studies* seeks to be the ultimate publication for scholars and students interested in comprehending all of the various aspects of transmediality, be it in terms of media industries and their platforms, digital and mobile communications, advertising and marketing sectors, audience behaviors and cultural practices, or socio-political forms like media activism, identity, literacy, and education. This collection, which gathers together original articles by a global roster of contributors from a variety of disciplines and industry backgrounds, sets out to contextualize, problematize, and scrutinize the current status and future directions of transmediality, exploring the industries, practices, cultures, arts, and methodologies of studying convergent media content across multiple media platforms. Now is the time to offer this ultimate publication about transmedia studies, given the central yet multifaceted ways in which transmediality has come to materialize in the media landscape.

Marsha Kinder (1991) first used the term "transmedia" to describe the multiplatform and multimodal expansion of media content. Henry Jenkins (2006) reintroduced the term within the context of digital change and "transmedia storytelling" has subsequently seen widespread adoption and interrogation. Jenkins' (2007) definition of transmedia storytelling as "a process where integral elements of a fiction get dispersed systematically across multiple channels for the purpose of creating a unified and coordinated entertainment experience" has become one of the dominant ways by which the flow of entertainment across media is now understood, especially in a digital and commercial setting where the correlation between transmedia storytelling and the commerce of entertainment has been reinforced in industry. As *Heroes* creator Tim Kring once asserted, transmedia storytelling is "rather like building your Transformer and putting little rocket ships on the side" (Kushner 2008). By providing audiences with more and more content, it seems, transmediality—an umbrella term most fundamentally describing "the increasingly popular industrial practice of using multiple media technologies to present information … through a range of textual forms" (Evans 2011, 1)—is characteristically understood as a commercial practice, enabling as it does for multiple revenue streams and numerous sites of engagement. Marie-Laure Ryan puts it plainly in her assertion that transmedia storytelling is essentially "a way to get us to consume as many products as possible" (2013, 384).

But commercial transmedia storytelling is not the end of the story for transmediality. In fact, Jenkins' description of transmedia storytelling (of a single narrative that is only truly complete when elements from multiple media forms are brought together into a coherent whole) has arguably rarely materialized in quite the fully integrated, plot-intertwining fashion that Jenkins envisaged. Further, as a mode of practice, transmedia storytelling is still most closely associated with what Benjamin Birkinbine, Rodrigo Gómez, and Janet Wasko refer to as the global media giants—"the huge media conglomerates such as Disney and Time-Warner, [which] take advantage of globalization to expand abroad and diversify" (2017, 15). Outside of the conglomerates, though, transmediality has evolved in other ways, namely into a brand development practice or as a way to support traditional media content through transmedia franchising systems (Johnson 2013), to name its other dominant commercial purposes. But transmediality has equally gained wider relevance as digital screen technologies have multiplied, with the so-called "old media" of film and television now experienced through online transmedia distribution practices (Evans 2015), whereby content becomes integrated with social media and other online platforms. Other terms such as "multiplatform" (Jeffery-Poulter 2003), "crossmedia" (Bechmann Petersen 2006), and "second screening" have joined it (Holt and Sanson 2014), but transmediality remains an important concept for understanding the fundamental shifts that digital media technologies have wrought on the media industries and their audiences. More than this, transmediality has since grown into a distinct subfield of scholarly investigation, one that relates to a range of studies across film, television, social media, gaming, marketing, literature, music, journalism, and beyond.

However, the more that transmediality has broadened its definition and its practical use in recent years, the more that it has arguably become something else entirely. Let's not forget that research has defined transmediality through very different disciplinary lenses, be it in terms of storytelling (Jenkins 2006; Evans 2011; Ryan 2013), marketing (Gray 2010; Grainge and Johnson 2015), journalism (Gambarato and Alzamora 2018), world-building (Wolf 2012); historical culture (Freeman 2016), activism (Scolari, Bertetti, and Freeman 2014), literacy (Scolari 2016), and so on. And these different sets of creative and disciplinary lenses should not be underplayed in our understanding of what transmediality is. Mapping the many faces of transmediality is an important task for researchers, for it hints at its multifaceted formations, functions, values, and roles across the wider media landscape.

And yet an almost inevitable consequence of transmediality being approached via so many different disciplinary lenses is that the very definition of transmediality might remain decidedly in flux, meaning different things to different people at different times. In 2011, Brian Clark argued that the potential for transmediality to be (mis)understood as almost everything means that "transmedia," as term, has possibly outlived its usefulness, insisting that only by refining the definition will scholars secure its long-term viability. Clark, we believe, was absolutely right in his critique, and simply because we live in "a digital media environment … [that] calls for a spread of media" (Brinker 2017, 209), it does not mean that everything is transmedial. Revising, refining, and clarifying our understanding of what does—and therefore what does not—constitute a form of "transmedia" is indeed crucial, both to the future of this avenue of study but more importantly to our collective abilities to make sense of how, why, and when media content flows, expands, and moves across multiple media platforms in particular ways, for particular reasons, and with particular effects.

However, we posit that only by embracing the multiplicities and pluralities of transmediality as a cross-disciplinary phenomenon can one fully grasp its prominence. To paraphrase Christy Dena's point from her chapter in this book, it may well make sense to create a simple definition of transmediality so that people understand and recognize it, but doing so is often at the cost of understanding the complete picture. A diverse and ultimate volume interrogating the status, the breadth, the developments, the themes, and the futures of transmediality is thus a timely opportunity for transmedia scholars to reflect on this subfield's current status and to explore potential new directions for future research. Importantly, each contributor in this book has conducted leading research into a particular area of transmedia studies or has done widespread transmedia practice across the cultural industries. Together, our contributors thereby offer a unique perspective on the practices, cultures, arts, and methodologies of studying media across multiple platforms.

Still, this cross-disciplinary approach based on embracing multiplicities and pluralities raises another notable question. If transmediality indeed means different things, in different parts of the globe (see Freeman and Proctor 2018), to different sets of industries, cultures, arts, and disciplines, then how can one go about classifying such different interpretations and divergent industrial practices as the same phenomenon? Doing this successfully—and responsibly—almost means *re*-understanding transmediality, moving far beyond a set of narrow, discipline-specific definitions based on entertainment or storytelling or marketing alone. In effect, it means articulating a more overarching idea of transmediality, albeit one that still addresses the specificity of its workings in different contexts. As Henry Jenkins insists, "this does not mean that transmedia means everything to all people and thus means nothing to anyone. Rather, it means that we need to be precise about what forms of transmedia we are discussing and what claims we are making about them" (2016). This is where the breadth of this book comes in, and it is our embracing of the multiplicities and pluralities of transmediality that also drives the structure of this book. Looking across specific contexts of different industries, cultures, arts, practices, and methodologies of transmediality in turn, we will now use the remainder of this introductory chapter to outline our overarching conceptual interpretation of what transmediality really means, argued in dialogue with the themes and ideas of the subsequent chapters. From there, we also speculate where transmedia studies could go next. And so now we return to our original question, meant with a sense of genuine reflection: *what is transmedia?*

Industries of Transmediality

In her chapter on transmedia television, Elizabeth Evans claims that "these [digital] platforms, and the way they are being utilized by content creators and owners, are contributing to media culture becoming increasingly and inherently transmedial." Similarly, Carlos A. Scolari argues elsewhere that, as of 2017, we are part of a media landscape where almost all content can in some way, shape or form be considered transmedial, meaning that "soon we will assume that all communication industries will be transmedial—it will be integrated into the DNA of media communication" (2017). Somewhat echoing the earlier sentiments of Clark, then, for Scolari (2017), the prevalence of transmedia across the contemporary media industries means that we no longer need to distinguish transmedia communication from other forms of communication.

But transmedia's prevalence is highly questionable and complex, and it is not particularly accurate to assume that transmediality exists across all creative and cultural industries. Indeed, as digital technologies and mobile devices continue to bring media interfaces into the workings of our daily lives, a salient question to consider is not only *what* is transmedia, but also *where* is transmedia? Jenkins' more recent writings on transmediality have begun to consider ideas of transmedia location, meaning "the context from which transmedia products emerge" (Jenkins 2016). There is thus a question in terms of which industries transmediality is now an active part of, and what specific purposes it holds within and across them. The first section of the book comprises 13 chapters around those industries that we believe represent the most dominant transmedia industries today: Film, Documentary, Television, Telenovelas, Comics, Publishing, Games, Music, Journalism, Sports, Social Platforms, Celebrity, and Attractions.

In terms of a focus on industries as a lens through which to better understand what transmediality really is, then, it is evident from this section's configuration of chapters that transmedia industries necessarily embrace both fictional and non-fictional universes. Renira Rampazzo Gambarato's chapter on transmedia journalism usefully reiterates the importance of characterizing transmediality as, first, multiple media platforms, second, as content expansion, and third, as audience engagement. The transmedia DNA of these characteristics is intertwined with fictional entertainment, as emphasized in Kinder's (1991) and Jenkins' (2003, 2006) original research, as much and as well as it is with non-fiction initiatives, as clearly demonstrated by Freeman's (2016) historicized approach to transmedia studies previously. Transmedia phenomena, as a common ground, involve the richness of multiplatform media—it is, as Jenkins notes in his foreword, about a set of relationships across media. Particular media platforms can emerge and disappear, can be in vogue or be ostracized, can change and evolve. Nevertheless, we could not have transmedia dynamics without the support of multiple media platforms and the industries that align them together. Furthermore, this section posits that beyond the digital domain, transmediality can and should involve a variety of alternative combinations between both online and offline platforms. The Internet and all digital technologies unequivocally play a crucial role in (1) disseminating transmedia content, (2) making content easily available worldwide, (3) reaching a diversified range of audiences, (4) enabling audience engagement, and (5) contributing to a participatory culture, for instance. But the possibilities to enrich the audience experience via offline activities, live events, and analogue initiatives, are immense because they can dramatically contribute to (1) the feeling of immersion, (2) the sense of belonging, and (3) the emotional response of audiences, as discussed in the afterword of this collection. These immersive emotions and behavioral practices are key to definitions of transmediality, as is demonstrated in Helen W. Kennedy's chapter on transmedia games, which shows the fruitfulness of applying "play theory" to understandings of transmediality.

Looking across industries as diverse as journalism and the celebrity scene, moreover, it is clear that such playable online or offline transmedia strategies can contribute to a growth of these industries, with the proliferation of content across media platforms building both new storyworlds and new job roles. Chapters on transmedia sports, by Ethan Tussey, and transmedia social platforms, by Portia Vann, Axel Bruns, and Stephen Harrington, for example, both reinforce the globalism associated with transmediality, and particularly the idea that transmediality is partly a tool for enhancing the democratization of media content everywhere. And yet part of the future conceptual breakthrough for transmedia scholars must be to better understand how said democratization of content gels with the innate commerce of many transmedia production motives, as is demonstrated by Šárka Gmiterková's study of the transmedia Kardashian brand and by Matthew Freeman's look at the Warner Bros. Studio Tour in London as a commercially oriented brand extension of the *Harry Potter* storyworld. With any example of transmediality, where is the line between expansion-as-commerce and expansion-as-democratization—and if or when does that contradictory line become in any way problematic?

Transmediality, in fact, is perhaps best understood as a series of conceptual contradictions, as the chapters in Part I show. Sarah Atkinson, positioning "film [as] arguably the most dominant instantiation of the transmedia storytelling phenomenon," sees a tension between "the franchise and campaign binary"—that is, between notions of content and promotion—while Joakim Karlsen hints at the importance of conceptualizing transmedia documentary as a blend of fiction and non-fiction, experience and participation, all combined into a single package. Karlsen's chapter shows the power of transmediality to embody the full potential of participatory media, and yet also points out the innate contradictions that arise when one begins to conceive of non-fiction as something that is itself participatory. Echoing this emphasis on combined tensions, Paola Brembilla explores transmedia music as a set of narrativized and visualized forms of artwork, cross-marketing, and branding. For Brembilla, transmediality is a "streaming of content" afforded by "synergy networks"—a streaming that builds a greater experience for audiences. Importantly, seeing transmediality—most broadly defined—as a stream of content "allows us to account for its versatility and ability to serve several purposes," thus altogether suggesting that transmediality works to give media content greater "cultural and economic value in the contemporary mediascape."

Conceiving of transmediality as a mode of diversification across the cultural industries makes sense, tying in with William Proctor's assertion in his chapter that transmedia comic books are often a secondary or alternative platform for films and television series. Such an idea also gives credence to Alastair Horne's chapter on transmedia publishing, which outlines some of the challenges for transmedia production. Understanding transmediality as diversification also supports Evans' conception of transmediality as something that is deeply rooted in the past and yet is foregrounded by contemporary media industries as a way to stand content apart in a crowded marketplace. For example,

Evans highlights the usefulness of "analogue" theory—academic concepts that originated before the days when "transmedia" was part of the common vernacular—in understanding transmediality in a digital sphere. Evans' chapter on transmedia television shows how particular media—and particular media industries—are *themselves* transmedial, and indeed have always been transmedial, in terms of operations, consumption habits, aesthetics, and so on. If media industries have long extended content across platforms, and audiences have long been encouraged to migrate across a stream of content, then transmediality is best understood as a conceptual approach to producing media via multiple delivery channels that each have combined commercial/democratic objectives at heart, itself enabling creative and participatory opportunities for sustained intellectual and emotional engagement. Simultaneously, from an industrial standpoint, transmediality becomes a means of adapting and diversifying media content so to best afford this kind of sustained intellectual and emotional engagement—as in Inara Rosas and Hanna Nolasco's chapter, which stresses how "lighter plots and shorter narratives" are key to the successful transmedial expansion of telenovelas in Brazil.

Arts of Transmediality

Part II of the book includes seven chapters on Transmedia Storytelling, World-Building, Characters, Genres, Writing, Photography, and Indie. Thinking about what the art of transmediality actually looks like, these seven chapters highlight a number of overlapping themes. Interestingly, Erica Negri's chapter on transmedia indie positions transmediality as a "situation of narrative chaos … [one] that attempts to conciliate narrative forms of digital technologies." In other words, transmediality is itself a conceptual approach to producing media that is intrinsically messy, born out of messy technological disruptions over time, shaped with often messy objectives at heart, and tailored for messy, fragmentary, hard-to-pin-down audiences. Yet from an artistic standpoint, our contributors' understandings of what transmediality *can be* remain more consistent than divergent.

For starters, besides the story to be told or the message to be delivered, which are both fundamental to the art of transmediality and transmedia storytelling more specifically, one such consistency concerns the role of world-building as a core concept. As Jenkins has pointed out elsewhere, the principle of world-building is inherent to the transmedia logic:

> Most forms of transmedia are structured through a process of world-building. The concept of world-building emerged from fantasy and science fiction but has also been applied to documentary or historical fiction. Worlds are systems with many moving parts (in terms of characters, institutions, locations) that can generate multiple stories with multiple protagonists that are connected to each other through their underlying structures. Part of what drives transmedia consumption is the desire to dig deeper into these worlds, to trace their backstories and understand their underlying systems. Fictional texts imagine and design new worlds; documentaries investigate and map existing worlds.
>
> *(Jenkins 2016)*

Regardless of how much a given story overlaps with the "primary" world, the varied dimensions, plausibility, richness of details of fictional and non-fictional transmedia worlds are designed and represented to be as important, intriguing, and compelling as its characters and plots. This creative equivalence is a central distinction of the concept of world-building in particular and transmedia stories in general. The essence of world-building is the strategy that best provides audiences with more stories sharing the same characters and world dynamics, but moreover, it offers them different yet equally immersive media experiences and emotional reactions.

Moreover, Jenkins' characterization of transmediality as that which provides the desire to dig deeper also extends to other chapters across this section, albeit sometimes with messier consequences in ways that reinforce Negri's contextualization of transmediality as chaos. Roberta Pearson's chapter on transmedia characters, for example, defines the art of transmediality as the creative process of making additions to media texts that cohere—or do not cohere—arguing that "cohesion depends upon points of contact between the addition and the transfiction." Pearson shows how audiences gain pleasure from seeing those additions cohere or not cohere, thus lending further weight to our earlier claim that transmediality is in essence a system of diversification.

However, Donna Hancox's chapter on transmedia writing suggests that it is so much more than this, painting a picture of the contemporary transmedia landscape as that which "re-imagines the intersection of media, genre and form to present an entirely new approach to writing." Hancox shows how transmedia fiction is often quite linear in nature, and yet its multifaceted use of multiple platforms affords arguably the best possible mode of storytelling—a mode that is capable of enhancing characterization, emotional and experiential engagement. This idea of transmediality, not as story-building, but as story-enrichment, links to Mélanie Bourdaa's chapter, which shows how transmedia storytelling opens up new possibilities for articulating fictional time.

Altogether, the chapters in Part II indicate that transmediality—from an artistic point of view—is about creating an adventure, one that seeks to transform the world into a story and the story into a storyworld. It is a means of crafting immersion, it seems—and, specifically, offering storytellers creative, pervasive ways to engage audiences emotionally and experientially. Or to put it another way, the art of transmediality is to build *experiences across and between the borders where multiple media platforms coalesce*—experiences that thrive on connecting, sharing, and responding. As Kate Fitzpatrick, a marketing strategist, discusses in the Afterword, "today, the concept of transmedia itself means creating a journey or experience that uses the most relevant mix of channels and platforms for your intended audience." Similarly, Natalie Rios Gioco, a transmedia consultant also interviewed in the Afterword, suggests that transmediality is about "delivering information by experiencing": it is "a system of cause and effect—a distribution of information (cause) that triggers an integrated, expansive response (effect)."

Indeed, characterizing transmediality as an experiential mode of engagement and causal relationships between content and people allows us to go beyond seeing it as a messy side-product of the fragmented media landscape, and also goes beyond describing it as a means of "allowing for different engagement depths," as Kevin Moloney puts it in his chapter on transmedia photography. Going beyond this description, Moloney's chapter does an excellent job in showing how a photograph—a single media image—is capable of hinting at so much more than it shows, bringing together both actual and imagined narrative moments and spaces that co-exist and extend, in the viewer's mind, at least, beyond the borders of the photograph itself. Moloney goes on to argue that

> for producers and critics of transmedia storytelling in any genre, the critical thinking about photographs must not only be how they interact with other media forms used in a project, but how they are also autonomous stories, capable of rich, immersive narrative, fine detail and visual fact presentation.

In terms of studying the artistry of transmediality, in other words, it is important that we return, somewhat contradictorily, to a medium-specific approach to studying individual platforms in order to better understand the function of specific platforms in and across the media landscape. There is a danger that comes with describing the convergences of contemporary media—namely, that convergence becomes directly associated with blending all forms of different media together into single sites of (digital) media artifacts. For even amidst a time of apparent technological convergence, mobile and online media, second screening, and so on, it is crucial to remember that different media still operate with largely specific sets of affordances, practices, policies, and consumption habits (Smith 2018). Thus in order to understand the artistic transmedia potentials of comparatively new platforms, such as augmented reality (AR), we first need to understand what AR—as an individual platform with distinct affordances—can actually *do*. By way of example, elsewhere Freeman (2018) explores the kinds of transmedia interventions represented by *Priya's Shakti*, a project that uses comic books, exhibitions, AR and street art to call attention to the struggles faced by women in India. Focusing on the artistic value of AR, Freeman explains how users are encouraged not to escape reality by entering a fictional world, but instead to think differently about reality by traversing the line between real and virtual (2018).

James Dalby (2017), echoing these same kinds of important social dimensions, argues elsewhere that the true function of any single piece of transmedia content is not simply to enrich, enhance, or augment its companion pieces, but in fact to give one piece of content (a film, a web series, a comic book, a novel, etc.) a new, previously missing dimension that forever shifts the meaning of that piece of content into something else entirely (Dalby 2017). Transmediality, then, has an important *ontological* function to play: at its best, it has the power to shape—and to re-shape—how we perceive the media and the world around it.

Practices of Transmediality

Part III presents seven chapters focused on Transmedia Adaptation, Developer, Production, Commodification, Franchising, Distribution, and Branding and Marketing. Alongside academic perspectives, this section also features chapters written by renowned transmedia practitioners and pioneers, such as Jeff Gomez (Starlight Runner Entertainment), Robert Pratten (Conducttr), and Max Giovagnoli, all sharing their own experiences and perspectives on critical case studies of transmedia projects led by their companies. As hinted previously, practices of transmediality go beyond traditional media franchises, sequels, or adaptations, leading to "integrated media experiences" (Davidson 2010). In the simplest sense, transmedia integration stands for expansion of content across multiple media and formats typically with some level of audience engagement. Christy Dena's chapter, however, argues that thinking of transmedia practice as simply the creation of extensions does not fully encapsulate what transmedia creatives do, nor is it the "only valid design choice for multi-platform-thinking."

Moving beyond notions of extension-making, then, the chapters in this part emphasize and showcase that there is life outside of commercial understandings of transmedia storytelling, countering the recurrent assumption that transmedia equals marketing. Andrea Phillips (2011) has argued previously that this supposition occurs because of economics: "It's not that there are more marketing campaigns using transmedia than anyone else; it's that the marketing campaigns are much, much more visible. Why? Because they have more money to throw around." Freeman (2016) has demonstrated how advertising is intrinsically connected to the early transmedia initiatives of the twentieth century, but this by no means signifies that practices of transmediality are limited to narrow definitions of advertising, marketing, and branding. Instead, Evans' chapter on transmedia distribution articulates that transmediality is a set of "logics" that all involve "branching out into new online spaces" in order to re-locate and to re-contextualize content, (re-)acquiring new audiences.

More than this, the chapters in this part characterize the practices of transmediality as a careful balance between creativity and strategy, echoing aforementioned ideas that it is essentially a blend of content and promotion, fiction and non-fiction, commerce and democratization, experience and participation. For example, Jeff Gomez shows, via a detailed discussion of how he and his team developed the *Pirates of the Caribbean* films into a multiplatform adventure, that the practice of building experiences across and between the borders of multiple platforms is in fact less to do with platform, but is rather a dual process of (1) narratological analysis and (2) something that is "discerned in the storyteller." This balancing act between creativity and strategy is reinforced further in Peter von Stackelberg's chapter on transmedia franchising, which notes that "commercial pressures will drive the adoption of transmedia [practices] across the various media sectors," which will in turn "drive the need for new creative approaches."

There is therefore the sense that the practices of transmediality are driven by conceptions of "themed storytelling," to borrow Alison Norrington's (2017) term, regardless of industry. Gomez, for instance, stresses the importance of an "essence" when crafting transmedia projects, by which he means a kind of thematic x-factor that runs across all media platforms and links the story to the storyteller in emotional and experiential ways. Robert Pratten, too, uses his chapter on transmedia production to position transmedia projects as "living, breathing worlds" that, by spanning countries, languages, platforms, and time, "more closely imitate real life." Even Anna Kérchy, whose chapter is rooted in the commerce of transmedia commodification, understands this practice as the making of "adventures to collect" that can yield "amazement results" over time.

In that sense, transmedia practices are really about crossing time as much as they are about crossing media, operating as systems of production and distribution that cater for the possibilities of tomorrow as well as for the demands of today. And doing so once again means channeling modes of creativity and strategy simultaneously, as per Max Giovagnoli's chapter on transmedia branding and marketing, which stresses the different ways via which creativity sits at the heart of all good transmedia campaigns. Giovagnoli points out how the strategic addition of games, events, and online promotions for a given transmedia brand all work together to enhance emotional investment and enjoyment.

In effect, chapters on the specifics of Transmedia Production, Transmedia Franchising, Transmedia Branding and Marketing, and so on, all clearly demonstrate that practices of transmediality, while prioritizing different agendas and audiences, are *not* storytelling, or marketing, or branding, or commodification, at least in isolation. Rather, practices of transmediality are defined precisely by the bringing together of all of these diverse practices into a single, innovative media package. What varies is which of these diverse practices are foregrounded at particular times. Transmediality is a "concert" of practices, as Gomez puts it, "weaving a tapestry of story that surrounds, immerses, and interacts with the audience."

Cultures of Transmediality

The fourth part of the book is dedicated to 12 chapters about Transmedia Archeology, Heritage, Fandom and Participation, Paratexts, Politics, Charity, Education, Literacy, Social Change, Identities, Psychology, and Religion. Cultures of transmediality explore diachronic and synchronic developments in the realm of transmediality within a human-centered approach and perspective. In our quest for understanding and advancing transmedia studies, putting people's needs in the forefront seems an appropriate way to improve media and communications and reach a more satisfying transmedia experience. Besides the economic advantages that transmedia practices can potentially bring to culture and society, what would be the hearty reason why we would actually need or want transmedia experiences in our lives? We do not necessarily need transmedia dynamics in our lives, but we can definitely take advantage of its techniques and tools to achieve a more meaningful, emotionally connected, and fulfilling media experience. For instance, Marie-Eve Carignan shows in her chapter how notions of transmediality become useful for understanding both the mediatized representation of religion and also the process via which people make sense of a religion. Despite all of the technological advancements in media we are facing, fundamental human needs, instincts, and motivations have not changed radically. As Pamela Rutledge's chapter on transmedia psychology alludes to, people continue to be driven by social connections, meaningful experiences, and the need to share stories that allow them to be part of something larger than themselves.

Transmedia cultures, indeed, are precisely that: experience-centered, technologically augmented conversations, a sharing between storytellers and audiences, between audiences and other audiences, and between online and offline worlds. This is where the concept of "paratext"—i.e., the promos and online materials that "create texts, manage them, and fill them with many of the meanings that we associate with them" (Gray 2010: 6)—becomes particularly useful to understanding transmediality. Matt Hills' chapter on transmedia paratexts examines this meaning-making process further, arguing that paratexts "have been repositioned as a new terrain for audience struggles." And building on aforementioned ideas that transmediality—in its building of immersive, emotional, experiential, and paratextual spaces that closely imitate real life—chapters in this section explore the intrinsic connections between transmediality, culture, and aspects of daily life. This includes Paolo Bertetti's look at the interlacing of transmedia storytelling and changing historical cultures, and Dan Hassler-Forest's account of how popular transmedia franchises such as *Star Wars* are "made meaningful by their specific association with politics." This same idea of cultural interlacing is reinforced by Jenny Kidd's examination of how museums and heritage sites are embracing the experiential and participatory possibilities of transmediality in ways that open up rich possibilities for "identity and nation building," crafting storyworlds of "liminal spaces between known and unknown, past and present, fact and fiction." André Jansson and Karin Fast, too, suggest that transmediality ultimately describes a media ecology "in which social practices are molded by and negotiated through different media technologies, and interweave with various forms of offline communication." In turn, Paul Booth articulates this same interweaving of transmedial social practices in his chapter on transmedia fandom and participation as the recalibration of "what narrative 'is.'"

What such a recalibration of narrative looks like might mean thinking of transmediality as a widened arsenal of media platforms that can aid people in achieving a goal, as Donna Hancox explains in her chapter on transmedia for social change. Or it might mean thinking about the experience of a particular set of transmedia content *not* as a discrete brand or even a story in the traditional sense, but rather as a much more fluid, ephemeral and value-laden transmedia *ethos* (see also Freeman and Taylor-Ashfield 2018). Matthew Freeman's chapter on transmedia charity explains how "the concept of *ethos* is perhaps more useful for characterizing the way audiences navigate transmedia charity projects, with people following beliefs, values, themes, philosophies and meanings (rather than stories) across media." To understand what transmediality really means, we have to talk about navigation, and in particular the ways in which people move across physical and virtual spheres—and what motivates that process of moving. This means analyzing the behaviors and motivations of a media-crossing audience with much more rigor, an idea for future research that is also reinforced first by Lorena Peret Teixeira Tárcia, whose chapter on transmedia education asserts that "transmedia provides a platform for students to learn how to identify, understand, and engage different audiences in their stories," and second by Carlos A. Scolari, who theorizes transmedia literacy as "informal learning strategies" that "facilitate the exchange of experiences" for different groups of learners.

Once again we are back to emphasizing the multiplicities and pluralities of transmediality, then. André Jansson and Karin Fast argue in their chapter that "identity" should be applied as a theoretical framework for understanding transmediality, given that identity—or identities—act as a "complex and negotiated interface between self and society." This idea is echoed elsewhere by Michael Humphrey (2017), who argues that memory is an important part of the transmedia space and one that shapes "the spirit of the self." One potentially important direction for the future of transmedia studies is for scholars to consider the increasing mediatization of life itself, and to better understand what it means to think of our digital lives as complex, intertwining, transmedial experiences.

Methodologies of Transmediality

The last part brings the following chapters: A Narratological Approach to Transmedial Storyworlds and Transmedial Universes; An Ontological Approach to Transmedia Worlds; An Experience Approach to Transmedia Fictions; A Design Approach to Transmedia Projects; A Management Approach to Transmedia Enterprises; A Micro-Budget Approach to Transmedia in Small Nations; A Genettian Approach to Transmedia (Para)Textuality; A Semiotic Approach to Transmedia Storytelling; A Mythological Approach to Transmedia Storytelling; A Qualitative Network Approach to Transmedia Communication; and A Metrics Model for Measuring Transmedia Engagement. Methodologies for studying transmediality are much needed, especially given the way that transmedia studies involves the analysis of hybrid phenomena. The challenge, as Anne Mette Thorhauge, Kjetil Sandvik, and Tem Frank Andersen (2016, 2) have expressed previously, is that "without grasping the broader media environment in which particular media platforms are part … it [is] difficult to demarcate and frame them as phenomena." Elsewhere, James Dalby (2017) goes as far as suggesting that applying theory from "non-transmedial contexts" to what are specifically transmedia texts is limited, arguing that "existing theory is not redundant as such, but can and must be reconsidered for transmedia environments." Dalby's reasons for such an altogether revisionist view stem from, first, the added sense of immersion that may arise from the vast array of available content within any given transmedia story, and, second, from the way in which the active need to piece this vast array of content together in a way that creates meaning—emotionally and/or experientially—transforms rudimentary notions of "audiences" into "participants" (2017).

In response, this section showcases some of the pertinent and original initiatives that aim to fulfil this gap in research methodologies. The discussion revolves around the ontological (things), the epistemological (knowledge), and the phenomenological (experience) parameters involved in transmedia studies. As our authors discuss, these parameters affect the process of ideation, building, and executing transmedia products as well as consuming, interacting, and participating within them. We argue that the process of experience, the act of personally observing, encountering, or undergoing transmedia experiences, is itself key to characterizing transmedia studies at large: a "possible procedure to address the issue of contemporary complexity through a phenomenological approach to the coeval reality" (Ciancia 2015, 133).

Frank Branch and Rebekah Phillips, for instance, stress the need to analyze transmedia content as "real things" on account of the socially profound ways via which transmediality intersects with everyday life. This perspective is reinforced in Nicoleta Popa Blanariu and Dan Popa's chapter, which stresses the "connection between mythical narrative and transmedia storytelling [as being] the *performative* dimension." Jan-Noël Thon, meanwhile, outlines a "toolbox" of transmedial narratology that provides a better understanding of how the pieces of transmedia universes operate according to "redundancy, expansion, and modification," hereby echoing Pearson's earlier claim that the pleasure of transmediality lies in piecing different elements together. Ascribing a mixed-method approach within transmedia studies thus makes a great deal of sense, combining, for example, "aesthetic/formal analysis with the qualitative investigation of user reception in order to get the full picture of the [transmedia] experience," as Susana Tosca and Lisbeth Klastrup claim in their chapter. Using both qualitative and quantitative approaches is something that Eefje Op den Buysch and Hille van der Kaa's chapter on metrics also demonstrates, while Gambarato stresses the value of embracing what she describes in her chapter on design as the "intricate entanglements between all the constituent elements of [super][sub]systems, that is, the set of components, the environment, and the set of relations." These intricate entanglements define transmediality, clarifying the need from a research point of view for diverse disciplinary perspectives, such as those underpinning Ulrike Rohn and Indrek Ibrus' chapter (business and management), Kyle Barrett's chapter (media industry studies), Geane Carvalho Alzamora's chapter (semiotics), and Matthias Berg and Andreas Hepp's chapter (media communications). Berg and Hepp's chapter, in particular, highlights the methodological process of the transmedia scholar to be one of "networking," an idea that is also reinforced in Raúl Rodríguez-Ferrándiz's chapter on transtextuality, which—given the multifaceted, multi-functional and ever-changing nature of transmedia content—stresses the need to "watch over" and "take care" of that content.

Our attempt, then, to reflect upon the proposed question *what is transmedia?* is concluded by recalling Jenkins' (2016) recent postulation that "transmedia—broadly defined—continues to grow in many different directions as people respond to the challenge and opportunities of communicating systematically across multiple platforms." Transmedia, as a term, is merely a descriptor, one that requires meaningful application to different scenarios. That is why Jenkins (2016) insisted that transmedia be used as an adjective instead of a noun. Yet while it is clear to see already that this book paints an enormously varied picture of transmediality, when looking across industries, arts, practices, cultures, and methodologies, it seems that understandings of transmediality are indeed more consistent than divergent.

Specifically, there is a consistent and clear emphasis on understanding transmediality as *experience via technology*, and relatedly on the creativity of audiences, particularly in the context of strategically motivated, democratically augmented media. We return to our conceptualization of transmediality as the building of experiences across and between the borders where multiple media platforms coalesce, altogether refining our understanding of this phenomenon as specifically a mode of themed storytelling that, by blending content and promotion, fiction and non-fiction, commerce and democratization, experience and participation, affords immersive, emotional experiences that join up with the social world in dynamic ways. And in doing so, it becomes more than the sum of its parts—weaving through industry, art, practice, and culture. All of the chapters in this book show vividly how important transmediality remains to understanding communication and culture at large, and hint at the importance of defining transmediality in sociological terms—by which we mean the role of transmediality in helping us all to better understand how we navigate culture as well as our everyday lives.

References

Bechmann Petersen, Anja. 2006. "Internet and Cross Media Production: Case Studies in Two Major Danish Media Organizations." *Australian Journal of Emerging Technology and Society* 4 (2): 94–107.

Birkinbine, Benjamin, Rodrigo Gómez, and Janet Wasko (eds.). 2017. *Global Media Giants*. New York: Routledge.

Brinker, Felix. 2017. "Transmedia Storytelling in the Marvel Cinematic Universe and the Logics of Convergence-Era Popular Seriality." In *Make Ours Marvel: Media Convergence and a Comics Universe*, edited by Matt Yockey, 207–233. Austin: University of Texas Press.

Ciancia, Mariana. 2015. "Transmedia Design Framework: Design-Oriented Approach to Transmedia Research." *International Journal of Transmedia Literacy* 1 (1): 131–145. Accessed January 13, 2018. doi:10.7358/ijtl-2015-001-cian.

Clark, Brian. 2011. "Brian Clark on Transmedia Business Models (Part Two)." *Confessions of an Aca-Fan: The Official Weblog of Henry Jenkins*, November 8. Accessed August 30, 2016. http://henryjenkins.org/2011/11/brian_clarke_on_ transmedia_bus.html.

Dalby, James. 2017. "Transmedia Audiences: The Modular Body Confirmed." Paper presented at the *Transmedia Earth Conference: Global Convergence Cultures*, EAFIT University, October 11–13.

Davidson, Drew. 2010. *Cross-Media Communications: An Introduction to the Art of Creating Integrated Media Experiences*. Pittsburgh: ETC Press.

Evans, Elizabeth. 2011. *Transmedia Television: Audiences, New Media, and Daily Life*. New York: Routledge.

Evans, Elizabeth. 2015. "Building Digital Estates: Transmedia Television in Industry and Daily Life." Paper presented at the *ECREA TV in the Age of Transnationalisation and Transmediation Conference*, Roehampton University, June 22.

Freeman, Matthew. 2016. *Historicising Transmedia Storytelling: Early Twentieth-Century Transmedia Story Worlds*. New York and London: Routledge.

Freeman, Matthew. 2018. "India: Augmented Reality, Transmedia Reality and *Priya's Shakti*." In *Global Convergence Cultures: Transmedia Earth*, edited by Matthew Freeman and William Proctor, 192–205. New York and London: Routledge.

Freeman, Matthew, and William Proctor (eds.). 2018. *Global Convergence Cultures: Transmedia Earth*. New York and London: Routledge.

Freeman, Matthew, and Charlotte Taylor-Ashfield. 2018. "'I Read Comics from a Feministic Point of View': Conceptualizing the Transmedia Ethos of the Captain Marvel Fan Community." *Journal of Fandom Studies* 5 (3): 317–335.

Gambarato, Renira R., and Geane Alzamora (eds.). 2018. *Exploring Transmedia Journalism in the Digital Age*. Hershey, PA: IGI Global.

Gray, Jonathan. 2010. *Show Sold Separately: Promos, Spoilers, and Other Media Paratexts*. New York: New York University Press.

Grainge, Paul, and Catherine Johnson. 2015. *Promotional Screen Industries*. London and New York: Routledge.

Holt, Jennifer, and Kevin Sanson. 2014. *Connected Viewing: Selling, Streaming, and Sharing Media in the Digital Era*. New York and London: Routledge.

Humphrey, Michael. 2017. "The Transmediated Self's Story: Examining Working Narratives in Social Media Ecosystems." Paper presented at the *Transmedia Earth Conference: Global Convergence Cultures*, EAFIT University, October 11–13.

Jeffery-Poulter, Stephen. 2003. "Creating and Producing Digital Content across Multiple Platforms." *Journal of Media Practice* 3 (3): 155–164.

Jenkins, Henry. 2003. "Transmedia Storytelling: Moving Characters from Books to Films to Video-games Can Make Them Stronger and More Compelling." *Technology Review*. Accessed January 12, 2018. www.technologyreview.com/biotech/13052.

Jenkins, Henry. 2006. *Convergence Culture: Where Old and New Media Collide*. New York: New York University Press.

Jenkins, Henry. 2007. "Transmedia Storytelling 101." *Confessions of an Aca-Fan: The Official Weblog of Henry Jenkins*, March 22. Accessed October 4, 2017. http://henryjenkins.org/blog/2007/03/transmedia_storytelling_101.html.

Jenkins, Henry. 2016. "Transmedia What?" *Immerse*, November 15. Accessed January 13, 2018. https://immerse.news/transmedia-what-15edf6b61daa.

Johnson, Derek. 2013. *Media Franchising: Creative License and Collaboration in the Culture Industries*. New York: New York University Press.

Kinder, Marsha. 1991. *Playing with Power in Movies, Television, and Video Games: From Muppet Babies to Teenage Mutant Ninja Turtles*. Berkeley: University of California Press.

Kushner, David. 2008. "Rebel Alliance: How a Small Band of Sci-Fi Geeks Is Leading Hollywood into a New Era." *Fast Company*, May 1. Accessed September 21, 2017. www.fastcompany.com/798975/rebel-alliance.

Norrington, Alison. 2017. "The Present and Future of Transmedia Storytelling." Paper presented at *Transmedia UK: Sector by Sector*, Bath Spa University, November 24.

Phillips, Andrea. 2011. "Transmedia Is Not Marketing." *Deus Ex Machinatio*, January 24. Accessed January 13, 2018. www.deusexmachinatio.com/blog/2011/1/24/transmedia-is-not-marketing.html.

Ryan, Marie-Laure. 2013. "Transmedial Storytelling and Transfictionality." *Poetics Today* 13 (3): 361–388.

Scolari, Carlos A. 2016. "Transmedia Literacy: Informal Learning Strategies and Media Skills in the New Ecology of Communication." *Telos* 103: 13–23.

Scolari, Carlos A. 2017. "Transmedia is Dead: Long Live Transmedia." Paper presented at the *Transmedia Earth Conference: Global Convergence Cultures*, EAFIT University, October 11–13.

Scolari, Carlos A., Paolo Bertetti, and Matthew Freeman. 2014. *Transmedia Archaeology: Storytelling in the Borderlines of Science Fiction, Comics and Pulp Magazines*. New York: Palgrave Macmillan.

Smith, Anthony. 2018. *Storytelling Industries: Narrative Production in the 21st Century*. Basingstoke: Palgrave Macmillan.

Thorhauge, Anne Mette, Kjetil Sandvik, and Tem Frank Andersen. 2016. "Researching Users Across Media." *MedieKultur* 32 (60): 1–5.

Wolf, Mark J. P. 2012. *Building Imaginary Worlds: The Theory and History of Subcreation*. New York: Routledge.

PART I

Industries of Transmediality

1

TRANSMEDIA FILM

From Embedded Engagement to Embodied Experience

Sarah Atkinson

Transmedia film is arguably the most dominant instantiation of transmedia storytelling phenomena. One only has to look at the history of transmedia storytelling where film is the primary media, and the key film and cinema-centric studies which have shaped the field of transmedia studies to appreciate its influence. From spin-off merchandise, to theme parks, to fan-made media—the film and cinema industry has led the way in the creation and commercialization of narrativizing the peripheral surrounding materials of film titles. Of course this is not a new concept for the film industry when it comes to film marketing, promotion, and additional revenue generation from peripheral products, as Thomas Elsaesser previously contended:

> A film, an object we usually consider to be a self-sufficient work, possessing a narrative with its own mode of closure, is being created rather more like a land-mine: to scatter on impact across as wide a topographical and semantic field as possible.
>
> *(Elsaesser 1998, 156)*

The genealogy of the transmedia storytelling "film" is often traced back to 1999, where two frequently cited examples can be seen to have both formed and in many ways shaped the basis of transmedia studies—defining the principles, practices, and techniques of what has come to be referred to as transmedia (storytelling) (Jenkins 2006). Those films are: *The Blair Witch Project* (1999) and *The Matrix* (1999). The theatrical release of *The Blair Witch Project* was preceded by an extended narrative campaign which at the time was referred to as viral marketing. This included a fake television documentary which was aired on the SciFi channel before the film's theatrical release, online websites, a comic and "missing" person leaflets distributed at the film festival, all which provided context, back-story and texturing of the fictional mythology of the Blair Witch. Subject to critical acclaim, cultural recognition, and notoriety, the film has also already been the focus of many academic studies. *The Matrix* universe, meanwhile, spanned a constellation of media platforms, including a trilogy of films, a comic book, the *Animatrix* series of short films, and computer games including *Enter the Matrix*. These different media were scattered with clues and links developed by the creators and augmented by the many surrounding fan interpretations of the expansive Matrix universe.

In 2006, both of these examples were considered to be exemplars of transmedia storytelling by Jenkins in his chapter "Searching for the Origami Unicorn" (which made reference to the film *Blade Runner*) and which centralized *The Matrix* as an exemplar of transmedia storytelling, identifying some of the key tropes and characteristics of the form with a particular focusing on the idea of

Table 1.1 Theorists have tended to delineate between two types of film-based transmedia

Blair Witch Project	East Coast	IP owned by creator	Centripetal
The Matrix	West Coast	IP not owned	Centrifugal

transmediality as the entertainment form par excellence for the era of collective intelligence. At the time, however, Jenkins did not differentiate between "types" of transmediality, though these examples of *The Matrix* and *The Blair Witch Project* set the two *distinct* trajectories of transmedia film in motion.

Two Transmedia Film Trajectories

These trajectories have followed two quite distinct yet frequently converging pathways which have previously been defined by other practitioners and scholars, using various factors: these have included the geographic, intellectual property, and transmedia structures (see Table 1.1).

As Table 1.1 shows, these different theories can be mapped to the two seminal examples of transmedia film. According to Andrea Phillips, West Coast-style transmedia is "more commonly called Hollywood or franchise transmedia" (2012, 13), which operates at major film-studio level (such as *The Matrix*), in contrast to "East Coast transmedia," which Phillips states "tends to be more interactive, and much more web-centric. It overlaps heavily with the traditions of independent film, theater and interactive art. These projects make heavy use of social media, and are often run once over a set period of time rather than persisting forever" (2012, 13–14), thus implicating *The Blair Witch Project*.

Brian Clark's (2011) definitions, meanwhile, are based on the difference of the treatment of intellectual property (IP) ownership in the geographic-based polarities. For Clark, West Coast "thinks more in terms of franchises … and starts from the perspective that creators won't own the IP" and East Coast "starts from the perspective that creators own the IP."

Jason Mittell's centripetal and centrifugal models focus on the structuring of the transmedia universe. He states: "expansionist approach to transmedia, using paratexts to extend the narrative outward into new locales and arenas through an approach we might term 'centrifugal storytelling'" (Mittell 2014, 264). Here Mittell uses the television series *Lost* which spread its narrative universe across books, website, online videos, and an Alternate Reality Game (ARG) (but is also clearly applicable to *The Matrix* model), in comparison to: "the alternate vector, creating transmedia to fold in on itself in a centripetal fashion" (Mittell 2014, 270). Mittell points to the transmedia strategy of television series *Breaking Bad* which focused on increasing the depth of audience engagement with the key characters through a mobile app, additional webisodes, and a fake advertisement, which are strategies that follow from *The Blair Witch Project* in its deepening levels of engagement with the mythology and the characters involved.

Building on these ideas, I propose my own distinctions between these two pathways that I refer to as franchise and campaign transmedia, but as I will demonstrate through the course of this chapter, these categories are almost always subject to convergence and cross-pollination.

Franchise and Campaign Transmedia

In their purest forms, examples that fit into these two categories *tend* to follow the proscribed characteristics as detailed in Table 1.1—with *The Matrix* strand representing the franchise model and *The Blair Witch Project* strand representing the campaign model.

Franchise transmedia follows the IP model—licenses are sold to creators to adapt and extend the IP across different platforms. The dictionary definition of franchise is: "An authorization granted by a government or company to an individual or group enabling them to carry out specified commercial activities, for example acting as an agent for a company's products" (Oxford Dictionaries online 2017).

Although the *Oxford English Dictionary* also recognizes the use of the term franchise as a "general title or concept used for creating or marketing a series of products, typically films or television shows" (Oxford Dictionaries online 2017).

There are numerous examples of transmedia franchises where film is the *central* component (the Marvel Universe, DC Universe, *Transformers*, *Harry Potter*) but it is rare for a singular film to be the genesis of a transmedia *franchise* (with the exception of *Star Wars*), since these franchises all began life as another media form, i.e., comic book, toy, or novel, and are then adapted into a film. Campaign transmedia film, on the other hand, always originates from the film. The definition of campaign is "an organized course of action to achieve a goal" (Oxford Dictionaries online 2017). *The Blair Witch Project* (1999) is the pre-eminent exemplar of campaign transmedia. Campaign transmedia is the more frequent type of transmedia practice which prevail in cinema—primarily in routinized marketing techniques for the film, which have used online web-centric, social media extensions. The cult, the independent and the underground of commercial cinema where innovations continue to emerge, transmedial tendrils which reach out to audiences through online social networks when films mutate into other forms and real-world spaces.

Table 1.2 shows these parallel trajectories of transmediality, and their points of intersection and blurring which I will now go on to discuss in more detail.

Franchise and Campaign Convergence

The release of the 2001 film *A.I. Artificial Intelligence* directed by Steven Spielberg was preceded by a 12-week ARG campaign entitled "The Beast," which merged some of the characteristics initiated by *The Blair Witch Project*, using the tools of the web, and *The Matrix*, utilizing mainstream film distribution tools such as the poster and the trailer, whilst layering real-world engagement in an ARG.

The ARG was "seeded" by clues in the film's posters in spring 2001. One of the credits in the baseline of the poster was a "Sentient Machine Therapist" named Jeanine Salla, an unfamiliar role to those normally seen in a film's production credits, prompting audiences to search for the name online which revealed Jeanine's (fictional) home page and blog. Thus began the game experience which, as Andrea Phillips stated, "turned a fire hose of content on its audience" (Phillips 2012, 28). This widespread textual scattering that Philips refers to was characteristic of "The Beast's" experimental and pioneering nature; textual dispersal logics became more managed, controlled, and predetermined in subsequent campaigns. It was in the "Why So Serious?" promotional campaign for *The Dark Knight* (2008) that the ARG format evolved sufficiently to cohere a logical narrative pathway in a traditional narrative arc, which led to the events depicted at the start of the feature film. A survey revealed that 63 percent of the respondents saw "viral marketing and the Alternate Reality Game 'Why So Serious' as absolutely integral to the film's narrative" (Brooker 2012, 84). "Why So Serious?" employed a treasure hunt campaign in which a website released the details of over 300 locations of bat graffiti across the globe. Audience members were encouraged to photograph the sites of the graffiti and upload them to the website. For each one that they successfully uploaded, they were rewarded with a frame from the trailer. These engagement techniques represented a blend between the commercial and renegade/underground—tactics associated with campaign transmedia were successfully deployed to promote a franchise-based film, through harnessing social media and web interfaces in new ways. These types of extensive "real-world" campaigns were few and far between; far more common are scaled-down transmedia extensions that are delivered and dispersed via social media platforms.

Table 1.2 Transmedia film timeline: mapping the two trajectories of transmedia and their points of convergence

Year of release	*1999*	*2001*	*2007*	*2008*	*2012*	*2013*	*2014*	*2015*	*2017*	*2019*
Campaign	*Blair Witch Project*	*AI*/"The Beast"	"Why So Serious?"	*Cloverfield*	*Prometheus*	*Body/Mind/Change*	*Interstellar* VR tie-in	*Star Wars* Secret Cinema	*Blade Runner* 2049 VR tie-in	*Star Wars* world @ Disney
Franchise	Matrix		Batman/DC	J.J. Abrams Anthology Frachise	Ridley Scott Alien Franchise	David Cronenberg				
Mode of interaction	Engaged	Embedded	Activated	Engaged	Engaged	Embodied engagement	Immersive	Experiential and immersive	Immersive	Experiential and immersive

Transmedia Film and Social Media

The rise of social media and digital devices from the mid-2000s onwards has significantly invigorated transmedia activity around film releases, the types and tropes of transmedia film storytelling tactics have become more sophisticated, designed, embedded, and nuanced.

An example of this is the additional narrative and textural extensions for the film *Prometheus* (2012) which were designed specifically for mobile interfaces and viewing. These included a number of YouTube videos, a website of the film character Elizabeth Shaw's *Project Genesis*, and another website where audience members engaged as an employee of the fictional Weyland Corporation, as well as synchronized second screen apps to be watched simultaneously with the film (see Atkinson 2014, 84–86). The *Prometheus* web-based campaign was aligned to the cultural language and expectations of its imagined spectator. In May 2012, the campaign used LinkedIn to target key social media "influencers" inviting them to apply for a vacancy on the *Prometheus* project. Information taken from the user's curriculum and profile were used to generate personalized messages. *Prometheus* represents a coming-of-age of the mainstream ARG/transmedia film campaign, where corporate sanitization becomes an implicit theme within the experience. The viral aesthetics which characterized previous campaigns and their "underground" conspiratorial nature, are replaced in *Prometheus* by a veneer of corporate logics and aesthetic of officialdom, and the social media mores of the time.

Before *Prometheus*, the theatrical film release of *Cloverfield* (2008) was preceded by a pre-release online ARG (discussed extensively in Atkinson 2014), where aspects of content were released across YouTube, including fake news videos and various websites of fictional organizations (that are not mentioned in the film), including additional non-film characters.

Although both *Prometheus* and *Cloverfield* have been assigned to the "campaign" stream as indicated in Table 1.2—both films were part of existing franchises, *Prometheus* was part of the *Alien* franchise which originally included *Alien* (1979), *Aliens* (1986), *Alien 3* (1993) and *Alien: Resurrection* (1997). *Prometheus* was positioned as a prequel to the original *Alien* film, although this was vehemently criticized by fans of the franchise as oblique and unrelated.

Cloverfield is part of the J. J Abrams anthology franchise which included the manga *Cloverfield/Kishin* comic book (2008), the original *Cloverfield* film and the subsequent *Cloverfield Lane* (2016). The Cloverfield universe is also referenced outside of this anthology in other aspects of J. J Abrams' work. The Slusho! drink (a key expositional object and "diegetic portal" in the Cloverfield transmedia) also features in Abrams' television show *Alias* and the Targuato logo (a fictional organization of great significance in the *Cloverfield* universe) appears on a skyscraper within a futuristic San Francisco skyline in a *Star Trek* Superbowl advert.

However, neither *Prometheus* nor *Cloverfield* take on the modalities of franchise transmedia as they are described above; rather, they take on the subversive and transgressive characteristics of campaign transmedia. What they also both signal is the maturation of transmedia film form where the recurrence of specific globally translatable genres (sci-fi, horror, and action), and the persistence of a particular set of narrative strategies.

Transmedia Film Strategies

A number of strategies have been established and have now become staples in the lexicon of transmedia film—such as companion "apps" and social media extensions which have become a key facet of film marketing campaigns. There is a notable predisposition for certain narrative mechanics and distinctive aesthetics. In *Beyond the Screen* (Atkinson 2014), I proposed that transmediality in film has taken on a number of different approaches through which to meaningfully create merchandising and other commodified engagement opportunities, as well as to facilitate the spread of content, discourse, and discussion about the film across platforms to both extend audiences and deepen their engagements. These forms that are fictional and perform a narrative function beyond promotion, using what I have previously identified as "diegetic portals" (Atkinson 2014), which are common to both franchise and campaign transmedia. The components of a transmedia film will typically involve one or more of the following strategies:

- the extension and enhancement of the world of the film, its textures, themes, and mythology;
- the extension of the story and plotlines of the film which such as back-story, pre-story, post-story and parallel-story plot extensions;
- revealing and rehearsing the narrative structuring of the film in order to guide the audiences' narrative comprehension.

The Extension and Enhancement of the World of the Film

By way of example of this first strategy, the world of *District 9* (2009) literally spilled out onto the streets through its physical installations of benches, billboards, posters, and bus-stops designating areas as "human" or "non-human," which served to extend the mythology and textures of the film. The campaign also incorporated a number of in-film websites. These included d-9.com, which allowed the viewer to elect their species as human or non-human, to then receive subsequent real-time event updates. The in-film company MNU (multinationalunited.com) also had its own website to which Mnuspreadslies.com was the resistance-blog. Mathsfromouterspace.com originated from MNU and included tests which audience members could take in order to qualify for the fictional initiative. Similarly, *The Hunger Games* (2012) fans were able to register for one of the "districts" (which make up the film's narrative vista) via the Capitol.pn website (the "pn" domain suffix makes reference to the film's fictional universe of Panem). Tumblr was also utilized to develop Capitol Couture, showcasing the fashion and costumes of the characters which are one of the defining aesthetics of the film. Thee video-game-oriented site IGN (Imagine Games Network) featured training activities simulations of those that the characters were subjected to in the films.

The Extension of the Story and Plotlines of the Film

This narrative tactic is used to communicate elements of the back-story (as was the case with *Cloverfield*, 2008), pre-story (as demonstrated in *The Dark Knight*, 2008) and post-story (illustrated by *A.I. Artificial Intelligence*, 2001). "The Beast" was set in the future-distant year of 2142, and all of the associated websites were dated as such. As Phillips states, "The Beast" works as "a study in the long-term consequences of the events of the film" (2012, 214). It also served to extend the mythology, texture and themes of the film; for example a website entitled Ballederma sold artificial companions similar to those represented in the film. In addition to websites, "The Beast" also involved a number of live events including one where audience members were able to call the Statue of Liberty security number in order to talk to a security guard to save a kidnapped teenager. Phillips describes the feeling of agency that this project induced through communicating with a fictional character as emblematic of the ARG experience (2012, 214). Hence the categorization of the audience experience of "The Beast" as an embedded one (see Table 1.2).

To Reveal and Rehearse the Narrative Structuring of the Film

Some campaigns have been used to introduce the complexity of the narrative structure to an audience, and to rehearse them in its navigation and interpretation of the film's language. *Inception* (2010), for example, conveys a complex, multi-layered narrative structuring depicting events that are occurring within dreams, and within the dreams of dreams of different characters. The audience is introduced to this complex narrative structure through the mind-crime game which is initially accessed through the "Pasiv Device" website, which was established in the lead-up to the film's release. It exists at the level of the intradiegesis, in which the manual for the Portable Automated Somnacin IntraVenous (Pasiv) device (the tool by which the operator can enter the dreams of others) can be accessed and interacted with. Produced in the style and language of traditional equipment manuals, the user can browse through various basic monochrome animations which demonstrates its use before being directed to the scene of the crime website, which in turn provided a gaming environment for audience members to further explore the thematic dimensions of the film.

Of course, these three modes are all used to market and raise awareness of the films; they all featured as pre-release strategies, which become redundant and un-sustained after the point of the film's release. This is another key distinction between campaign and franchise transmedia—campaigns are time-bound, whereas franchises endure, sustain, and grow over years.

From Embedded Engagement to Embodied Experience

It was between 2013 and 2015 that a shift in the logics of transmedia engagement in film and cinema took place—for one, the principles of ARG return—but also in terms of the manifestation of embodied experience. This manifests in three key projects:

1. David Cronenberg: Evolution exhibition, and the accompanying *Body/Mind/Change* ARG.
2. The recent wave of virtual reality (VR) as a tool for promotion and engagement.
3. The immersive *Star Wars: Empire Strikes Back* experience by Secret Cinema.

Body/Mind/Change

Body/Mind/Change (*BMC*) was referred to as "a multiplatform Augmented Reality Game." It was developed to promote and accompany the retrospective exhibition of film director David Cronenberg which was launched at Toronto International Film Festival in autumn 2013, and an accompanying online virtual exhibition. The *BMC* experience made specific reference and paid homage to the iconographic prop from Cronenberg's 1999 film, *Existenz*, that is namely the "POD"—the biological-technological-organic gaming console that was inserted into the "bio-ports" situated in the lower backs of the characters of the players which enables neurological access to the different levels of the game. In the case of *BMC*, and its inherent nostalgic turn, the POD is reimagined and a new discourse surrounds its existence, reframed for contemporary, digital audiences, using the language and concepts of the present day. The fictional conceit of this project was seeded in a discourse of the real, consistent with the conventions of the ARG, in the online periodical *Filmmaker Magazine*:

> Earlier this year, without much fanfare, David Cronenberg quietly licensed the fictional technology and science found within his films Shivers, The Brood, Scanners, Videodrome and eXistenZ for a mind-bending eight-figure sum. While it is common for a film's IP to make its way into other mediums, such as books, television or games, it is highly unusual for a film's fictional elements to become actual biotechnology.
>
> (*Weiler 2013*)

Here, there is an embedded reflexive reference to transmedia tropes, in what could be described as a meta-transmedia campaign, which I had previously noted of the "immersive pre-premiere experience" of *Hunted*, a global espionage television series in which an explicit reference is made to "the 1 per cent." The campaign centers on a recruitment campaign launched by the in-film firm Byzantium Security, a clandestine intelligence organization. The posters that appeared throughout New York stated, "We're not for everyone, just the 1 per cent who matter." This targeted "call to action" presents a self-reflexive acknowledgment to its imagined audience, an allusion to the 1 per cent who will interact with the campaign, and become an integral facet and driving force of its success (Atkinson 2014, 34).

The *Hunted* project is imbued by the signatory gesture of Cronenberg as a direct reference to the fictional device created in the 1999 film *Existenz*, thereby illuminating a moment of retrospective intertextual cross-pollination. The director himself is centralized in the new fictional POD universe and becomes subject and host to his own fictional creation.

The conceptualization as POD as an organic game console was reimaged for 2013 audiences using the name as an acronym of "PERSONAL ON DEMAND," in which the campaign is focused on a call for human-hosts to volunteer to be implanted with a "sensory learning and data-mining organism." It is the notions of biosurveillance, content discovery, and mind-share space that are conflated within this 2013 iteration of POD, the audience member eventually received their own unique "POD" generated in response to the questions that they have answered and recipients were encouraged to publish and share photos on social media, which in turn act as marketing devices with which to draw attention to the exhibition. This suggestion of a physiological transgression of the POD (explored in the 1999 film in relation to gaming), is revisited with its re-conception in 2013, through the implication of the extension of the materiality of the cinematic text into the human form. This visceral metaphor of embodiment can be seen to be indicative of the sustained discourses of fear that exists around the technologization of society and the body.

Virtual Reality

A further emergent immersive transmedia extension is VR where the world of the film is actualized in three-dimensional space. One of the first VR installations to accompany a feature-film release was the Oculus Rift experience for *Interstellar* (2014), installed at the AMC Metreon in San Francisco, in November 2014. VR has previously featured in exhibitions relating to film, including for Stanley Kubrick's *2001: A Space Odyssey* VR installation, in which the viewer could explore the circular centrifugal space of Discovery One—the spaceship that features in the film—in 360-degree space (at the Stanley Kubrick Dreaming exhibition at Somerset House in summer 2016). We can now see the VR tie-in fast becoming a staple of feature film promotion and exhibition, providing the opportunity to immerse audience members into the fictional universes of films (at least those of particular genres including horror, sci-fi, and action), and also those associated with franchises. Recent franchise-based VR examples include tie-in experiences for *Assassin's Creed* (2016) and *Spider-Man: Homecoming* (2017) as well as for *John Wick* (2014) with a fully realized VR game for the HTC Vive. Other examples include *Dunkirk* (*Save Every Breath*) (2017), *It: Float* (2017), and *Blade Runner 2049*: *Replicant Pursuit* (2017), the latter a VR experience that was embedded into a wider transmedia campaign. Indeed, the world of *Blade Runner* was re-created as part of Comic-Con and a special edition release of Johnnie Walker Director's Cut whisky is planned as an homage to the original film. These latest transmedia innovations are driven by commercial partnerships which are established to promote new technology platforms alongside associated media content and products—Oculus and Disney being a clear example of such techno-cinematic synergies.

Secret Cinema and *Star Wars*

The third example of contemporary transmedia film concerns Secret Cinema, and in particular its immersive experience for *Star Wars: The Empire Strikes Back* (1980) (see Atkinson and Kennedy 2016 and Pett 2016 for in-depth explorations into this particular case study, and for an overview of the work of this organization). This particular production marked a turning point for Secret Cinema—it was its first immersive screening of a franchise-based film, in which the license allegedly took over a year to negotiate with Disney/LucasFilm (Wilson 2015). The experience spanned five months, running from June to September in London and selling 100,000 tickets (each at £75), generating over £7 million at the box office. Upon purchasing a ticket, audience members were invited to join the rebel alliance through an online interface whereby, after a series of questions, they were given one of the following identities: Galactic Explorer, Mercenary, Governor of the Alliance, a member of the Alliance Starfighter Corps or Creative Council. They were encouraged to engage in the in-world communications channel via a themed fictional website, to communicate with other audience members and to purchase costumes and props from the online store ready for their attendance at the immersive screening.

On the evening of the screening, audiences were directed to the secret location—a disused printing press which had been transformed into the Earth Cargo Airlines (ECA) terminal building—and were subject to an immersive experience which was five hours in length, culminating in the augmented screening of the film. During the experience, audience members could perambulate through the Mos Eisley themed marketplace, interacting with characters from the film including C-3PO and R2-D2. The Death Star was flown over the audience's heads and they were delighted by the iconic light sabre battle sequence. Crucially, this example signified a moment of convergence between the dual trajectories of campaign and franchise transmedia, taking on the characteristics of both.

More recently, the proposed *Star Wars* themed-land was announced. The immersive attraction is due to open at the two US Disney parks in 2019 and is said to be based on a never-before-seen planet, a "remote trading port" in the Outer Rim, which is expected to be revealed in the new *Star Wars* films. It is said that visitors will be assigned identities in a very similar vein to that of the the Secret Cinema experience populated by the smugglers, the bounty hunters, and the rogue adventurers. Whether inspired by the high-profile Secret Cinema *Star Wars* event of 2015 or not, this attraction signals yet another shift in transmedia engagement strategies toward immersive and embodied experiences.

Conclusion

BMC, VR, and Secret Cinema all present examples of "retrospective transmedia expansion," a common trope in many emergent instances of transmediality which I define as the re-contextualization of a past film in a contemporary milieu. This is reflected in wider cinema release trends, which in 2017 have been particularly reliant on remakes, sequels, and prequels to prior box office and cultural successes. Crucially, these three instances are not forms of adaptation or remediation—they are new narrative extensions which are based on the world of the film but ones which provide re-interpretations, and ones which provide an embedded and embodied experience for the audience member. As with the earlier forms, these three cases still exhibit the same characteristics of transmedia storytelling as defined earlier, thus demonstrating the extent to which these are by and large film-centric strategies. We have seen how *BMC* extends both the texture of the filmic universe but also the directorial style, while VR extends the cognitive and spatial perception of the storyworld, and Secret Cinema expands the narrative spaces, story-times, and characters of the original.

However, these three different vectors of contemporary transmedia film also provide a richly illustrative example of the move toward the experiential, the immersive, and the embodied in transmedia phenomena, which illuminates the wider context of the rise of the experiential and discourses of experience design and the experience economy (Pine and Gilmore 2011). They also all problematize the franchise and campaign binary that has been proposed in this chapter in sophisticated ways. Still, the two trajectories of franchise and campaign in transmedia film do continue to co-exist and intertwine in the contemporary media landscape, ensuring that the transmediality of film continues to evolve, extend, and expand in new ways.

References

Atkinson, Sarah. 2014. *Beyond the Screen: Emerging Cinema and Engaging Audiences*. New York: Bloomsbury.

Atkinson, Sarah, and Helen W. Kennedy. 2016. "From Conflict to Revolution: The Secret Aesthetic and Narrative Spatialisation in Immersive Cinema Experience Design." *Participations: Journal of Audience & Reception Studies* 13 (1): 252–279.

Brooker, Will. 2012. *Hunting the Dark Knight: Twenty-First Century Batman*. London: I. B. Tauris.
Clark, Brian. 2011. "Reclaiming Transmedia Storyteller." *Facebook*, May 2. Accessed August 14, 2017. www.facebook.com/notes/brian-clark/reclaiming-transmedia-storyteller/10150246236508993.
Elsaesser, Thomas. 1998. "Fantasy Island: Dream Logic as Production Logic." In *Cinema Futures: Cain, Abel or Cable? The Screen Arts in the Digital Age*, edited by Thomas Elsaesser and Kay Hoffman, 150–165. Amsterdam: Amsterdam University Press.
Jenkins, Henry. 2006. *Convergence Culture: Where Old and New Media Collide*. New York: New York University Press.
Mittell, Jason. 2014. "Strategies of Storytelling on Transmedia Television." In *Storyworlds Across Media: Toward a Cross-conscious Narratology*, edited by Marie-Laure Ryan and Ian Noël-Thon, 253–277. Lincoln: University of Nebraska Press.
Oxford Dictionaries online. 2017. "Definition of Campaign." Accessed November 2, 2011. https://en.oxforddictionaries.com/definition/campaign.
Pett, Emma. 2016. "'Stay Disconnected': Eventising Star Wars for Transmedia Audiences." *Participations: Journal of Audience and Reception Studies* 13 (1): 152–169.
Phillips, Andrea. 2012. *A Creator's Guide to Transmedia Storytelling: How to Captivate and Engage Audiences Across Multiple Platforms*. New York: McGraw-Hill.
Pine, B. Joseph, and James H. Gilmore. 2011. *The Experience Economy: Work is Theatre and Every Business a Stage*. Boston: Harvard Business School Press.
Weiler, Lance. 2013. "Pod Wants to Know You." *Filmmaker Magazine*, April 23. Accessed July 1, 2017. http://filmmakermagazine.com/68303-pod-wants-to-know-you/.
Wilson, Benji. 2015. "How Secret Cinema Built The Empire Strikes Back." *Telegraph*. June 5. Accessed August 14, 2017. www.telegraph.co.uk/film/star-wars--the-empire-strikes-back/secret-cinema/.

2

TRANSMEDIA DOCUMENTARY

Experience and Participatory Approaches to Non-Fiction Transmedia

Joakim Karlsen

Siobhan O'Flynn (2012, 143) defines transmedia documentary to be a narrative that is distributed "across more than one platform, it can be participatory or not, can invite audience-generated content or not, tends to be open and evolving, though not always." She frames making transmedia documentaries first and foremost being about designing user experiences, and identifies how the shift "from film-maker to transmedia producer, curator and collaborator now demands a flexibility and willingness to experiment with the means of communication and a commitment to engage in communication" (O'Flynn 2012, 152). In a later paper, she develops how experience design is human-centered with a holistic model "for studying interaction as a qualitative experience across digital and physical media, with its attention to the possible multiplicities of human involvement" (O'Flynn 2016, 84). It is possible to see how this is in continuity with Henry Jenkins' (2007) definition of transmedia storytelling as "a process where integral elements of a fiction get dispersed systematically across multiple delivery channels for the purpose of creating a unified and coordinated entertainment experience." An understanding of non-fiction transmedia as being about telling stories across multiple platforms framed as a design activity, has so far been common in the relatively sparse literature on the topic. Renira Rampazzo Gambarato (2013) follows this approach by addressing "essential features of the design process behind transmedia projects" with the aim to "support the analytic needs of transmedia designers and the applied research in the interest of the media industry" (81). Relying on some of the practical guides to making transmedia (Pratten 2015; Hayes 2011; Phillips 2012), and more theoretic treatments of transmedia storytelling in general (Scolari 2009; Saldre and Torop 2012; Dena 2010; Jenkins 2008, 2009b, 2009a), she develops a model for transmedia project design that can be applied to both fiction and non-fiction projects. One aspect she considers when developing the project design model for transmedia storytelling, is what kind of user engagement a transmedia project facilitates for. She claims that this is crucial in transmedia design and summarizes the issue:

> An interactive project allows the audience to relate to it somehow, for instance, by pressing a button or control, deciding the path to experiencing it, but not being able to co-create and change the story; a participatory project invites the audience to engage in a way that expresses their creativity in a unique, and surprising manner, allowing them to influence the final result. Participation occurs when the audience can, with respect at least to a certain aspect of the project, influence on the set of components, such as the story. Stories that are mainly interactive can be considered as closed systems, in which the audience can act but cannot interfere with the story. Closed systems presuppose interaction but not participation.

> Besides the interactivity, open systems allow participation, i.e. the audience can influence the result and change the outcome.
>
> *(Gambarato 2013, 87)*

This gives a simplified model of two approaches to audience engagement when designing non-fiction transmedia: (1) interactive projects aimed to give the audience an experience where non-fiction transmedia are designed as closed systems; (2) participatory projects allowing the audience to influence the final result, where non-fiction transmedia are designed as open systems. In their analysis of the Fish Fight non-fiction transmedia project, Gambarato and Medvedev (2015) use distinctions between minimalist and maximalist forms for participation (Carpentier 2016; Carpentier and Dahlgren 2013), where maximalist participation focuses on issues of power, and claim that this is the kind of participation the Fish Fight campaign facilitated for by letting the audience contribute to real policy changes by participating in the project.

In the following I will contribute to this discourse by developing these two distinct approaches to non-fiction transmedia design. The first approach, which I label the *experience approach*, relies on storytelling, experience design, and mainly the making of closed transmedia systems. The second approach, which I label the *participatory approach*, relies on facilitation, participatory design, and the making of open transmedia systems. I do not claim that any of the approaches are generally better than the other, but that the second approach is useful when designing for maximalist kinds of participation. In support for developing and discussing the approaches further, I will use "Project Moken" (Cheng Munthe-Kaas, Jensen, and Wiik 2013) as a case, where the producers have gradually moved from the experience approach to the participatory approach over the course of six years.

Designing Non-Fiction Transmedia

The experience approach is well described in the literature on non-fiction transmedia and it is possible to find guidelines and advice supporting this model (Hayes 2011; Pratten 2015; Phillips 2012). This is an author and story centered approach where experiences are designed, planned, and produced for, in a mostly top-down fashion, in line with how most media productions are done today. Considering examples of non-fiction projects, both interactives and transmedia, as catalogued by the MIT Open Documentary Lab ("Open Documentary Lab at MIT" 2017) and the IDFA film festival ("IDFA Doclab Interactive Canon" 2017), they are mostly of this type. It varies a lot how the audience are engaged by virtual reality (VR), augmented reality (AR), games, performances, installations, and so forth in these projects, but audiences are often given little influence on the final result or the characteristics of the main structure of the experience. They are given the opportunity to participate, but often minimally, by selecting, commenting, sharing, voting, signing petitions, and so forth. Even though experience design, as conceptualized by O'Flynn (2015), could accommodate for more maximalist kinds of participation, for making more open transmedia systems, this quickly threatens to "break" the story and by this challenge the position of the author.

Developing the Participatory Approach

The participatory approach for non-fiction transmedia design relies on facilitation, participatory design, and maximalist kinds of audience participation. This is a departure from approaching the design of non-fiction transmedia as designing stories, toward designing platforms that enable members of the audience to influence the final outcome of the project. It is a bottom-up and open approach, not only aiming to let the audience influence the project after it has been "released," but from the start. Principled participatory design (PD) emphasize how participation should be facilitated for in the design process in order to give all parties a voice (Kensing and Greenbaum 2013) and so that what is made is accountable to the problem that is being addressed (Suchman 2002). The ethos of participatory design can be compared and aligned to the ethos of the participatory documentary movements of the 1970s and 1980s (Tripp 2012) and shares a common rationale.

In a recent study, the main challenges involved in designing interactive documentaries, or I-docs, that prioritizes participation before authorship were explored (Green et al. 2017). The researchers criticize many existing interactive documentary projects as having the same "centralized, authorial production structures and tokenistic forms of participation that have characterized traditional, linear documentaries" (Green et al. 2017, 6318), and claim that while they give the users "executionary agency" these projects don't provide "structural agency" or the "ability to inform the context in which this dialogue occurs, or allow users to initiate their own conversations" (Green et al. 2017). They try to find ways to develop an infrastructure that support "structural participation" and frame this as undertaking "infrastructuring," using participatory design and meta-design (Fischer 2011; Björgvinsson, Ehn, and Hillgren 2010; Dantec and DiSalvo 2013) to seek "sustainable configurations of creative making, interactive artifacts and design, with different stakeholders, at different times" (Green et al. 2017, 6319). From their case study, they report that facilitating for structural participation in the context of making an I-doc is generally hard to do. This is supported in the literature, where infrastructuring projects are characterized by their ongoingness (Karasti and Baker 2008), the need for continuous facilitation (Löwgren and Reimer 2013), continuous matchmaking (Björgvinsson, Ehn, and Hillgren 2012), scaffolding (DiSalvo et al. 2014), and participation is ensured by openness and flexibility (Löwgren and Reimer 2013), under-design (Fischer and Giaccardi 2006), and generativity (Monteiro et al. 2012).

According to Gerhard Fischer and Elisa Giaccardi (2006, 427), meta-design "extends the traditional notion of system design beyond the original development of a system to include a co-adaptive process between users and system, in which the users become co-developers or co-designers." Systems must be flexible, meaning that they cannot be designed prior to use, they must evolve at "the hand of the users" and "be designed for evolution." They take care to delineate between design-time, which is future oriented and mostly about planning, and use-time which is situated action. The goal is to allow openness and evolution of the system at use-time, challenging designers to do under design and to focus on designing the "in-between." An interactive art project that have some of these properties is "The Johnny Cash Project" (Milk 2012), where the users can contribute to the project by painting and submitting single frames to a continuously evolving music video. The video, or the end result of the project, is therefore continuously changing by the hands of the users. However, the project has no way for users to change the structure of the webpage or its functionalities and thereby facilitate for *executional* rather than *structural* agency. An example of structural participation in this case would be if the participants could make their own painting tools and share them on the project's website for others to use.

In summary, the participatory approach, as briefly sketched here, gives an indication of what Nico Carpentier (2011) means when claiming that deep or maximalist kinds of audience participation need to be anchored in and protected by strong organizational structures, having less to do with "new" media technologies than the willingness to use the resources needed to facilitate for it, requiring an ongoing and involved process determined by the issue at hand rather than sticking to a preconceived project design.

Case: Project Moken

Project Moken aim to tell the story of the Moken sea nomads living in the Mergui archipelago of the coast of Myanmar and Thailand, to help them save their rapidly vanishing culture. The following description is based on an interview with Mette Cheng Munthe-Kaas, the main producer of the project (Cheng Munthe-Kaas 2017), analyses of what the project has made (Cheng Munthe-Kaas, Jensen, and Wiik 2013; Truong 2012; "One Tree Can Save a Culture" 2013; "Hold Your Breath" 2017; "The Ocean as a Place" 2017; Wiik 2014; Saab 2016) and project descriptions from early and late phases of the project (Cheng Munthe-Kaas 2012, 2015). The project illustrates the two main approaches to designing non-fiction transmedia; designing experiences and designing for participation.

Designing Experiences

In 2011/2012 the producers were invited by the Norwegian Film Institute (NFI) to create a transmedia project based on their documentary film project about the Moken. At this time, NFI gave non-fiction transmedia increased support (Karlsen 2014, 2016). The project received a grant from NFI where approximately 1 million Norwegian Krones (NOK) (equivalent to US$120,000) was earmarked for producing a webpage, a mobile application, and an online campaign based on the documentary film.

The film *No Word for Worry* (Wiik 2014) follows the Moken protagonist "Hook" in his travels, both to make a living and to find a tree to make a traditional Kabang boat with the help of his father. The narrative relies on the dramaturgy common to many documentary films, of following an individual trying to accomplish a well-defined and articulated goal. In 2016, a "the making of" documentary (Saab 2016) was screened on the Norwegian public broadcaster NRK, giving an impression of how the documentary was produced. In the film, the director starts out saying that he did not want to tell his story about the Moken people, but have them tell their story. However, in the film he relates how he wrote the storyline before shooting the film and how he followed this storyline on location, with adaptions to some unexpected events taking place along the way. We see how he works with the subjects, staging situations in accordance with his script. The main character's journey, with his main goal of securing an appropriate tree, would probably never have taken place without the facilitation of the film crew. This said, the treatment is respectful to the subjects portrayed in the film, based on the director's well-developed relationship with them.

In 2012, the project released a tree campaign online ("One Tree Can Save a Culture" 2013), petitioning people to sign their name with the name becoming part of a growing 3D rendered tree (Figure 2.1). With the help of an interactive advertising agency, they had a model of a tree made and worked with ways to let the tree visualize the growing community. The idea was to let the branches and leaves mirror how the tree was shared and signed on Facebook. After some changes in the Facebook application programming interface (API) in 2012, this was no longer possible, so they ended up placing names more randomly in the tree. This part of the project, costing around 400,000 NOK (equivalent to US$48,000), a substantial sum in the context of independent documentary filmmaking in Norway, has been used as evidence by other independent filmmakers in Norway for why non-fiction transmedia production is unsustainable. In their view, the sparse resources available to independent documentary film producers are needed to make traditional documentary films, and should not go to professional designers and developers (Karlsen 2014).

Around the same time, Project Moken released the Dive Moken iPhone app (Truong 2012), where people were challenged to hold their breath while seeing video footage of Hook taking one of his dives. Besides having a simple interface where the user can time herself, the app features an individual score board, options to share the app with others, and links to the other installments of the project; the tree campaign, the YouTube channel, and the project's webpage (Figure 2.2). The producer told me that the app has had a life of its own in some communities but, as with the tree campaign, the changing Facebook terms of use subverted the ambition to open up for people to compete and challenge each other, by having the scoreboard reflect the results of a subset of Facebook users.

The third interactive installment of the project was two stories published on the project webpage (Figure 2.3) ("Hold Your Breath" 2017; "The Ocean as a Place" 2017). "The Ocean as a Place" tells the story of the main challenges to the environment the Moken depend on to uphold their traditional ways. How the government keep them from harvesting from the sea, and how the sea is almost barren because of overfishing and climate change. In addition, it explains what a Kabang boat is, and why this type of boat is central to the Moken culture. "Hold Your Breath" focuses on the Moken's

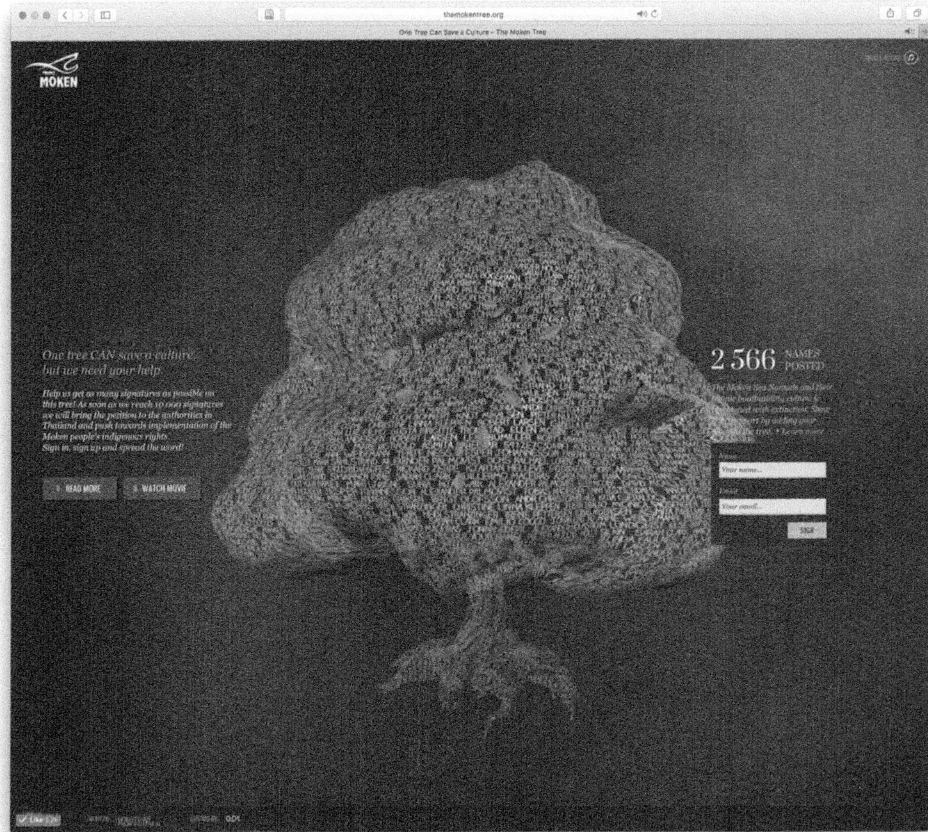

Figure 2.1 The webpage displaying the growing tree, with options to participate by signing the petition.
Source: "One Tree Can Save a Culture" (2013).

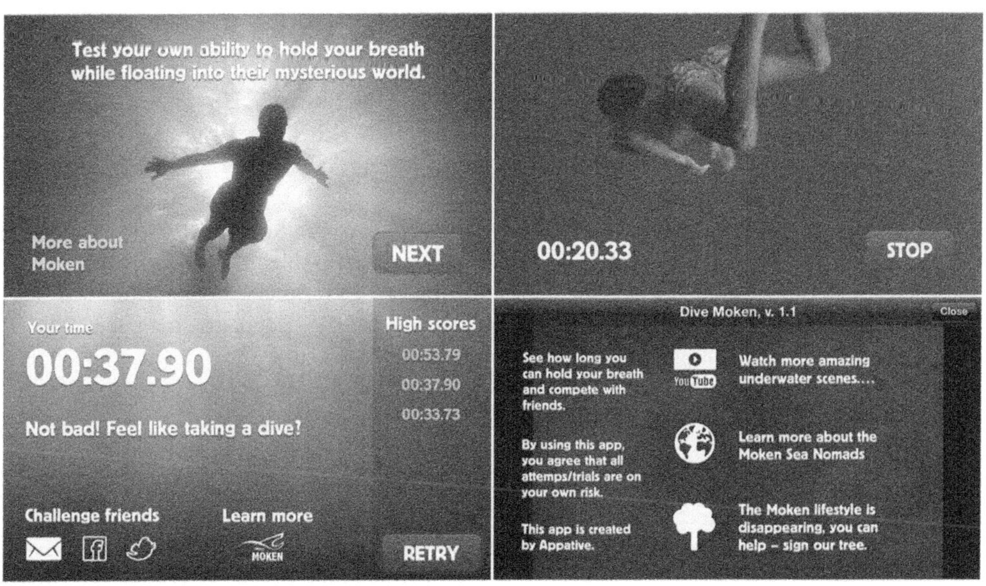

Figure 2.2 Screenshots of the Dive Moken iPhone application.
Source: Andy Truong (2012).

Figure 2.3 The starting pages for the two web stories about the Moken people published on the project's home page.
Source: "The Ocean as a Place" (2017), "Hold Your Breath" (2017).

ability to hold their breath for long stretches of time and focus under water, skills that enable them to harvest the food they need to survive. Both webpages rely on horizontal and vertical scrolling and text combined with images with more information available in pop-ups. In "Hold Your Breath" the vertical scrolling is a metaphor for diving and going deeper and in "The Ocean as a Place" the horizontal scrolling is a metaphor for riding a boat on the surface of the ocean (Cheng Munthe-Kaas 2017). Both pages contain relatively little information, and seem to work more as a way to get people interested in the other installments of the project, than stand-alone web stories. The producers had plans for adding more content to the pages, but ran out of money before being able to do so.

Designing for Participation

When the film had been screened at festivals, in cinemas, and by broadcasters, the transmedia project had run for about two years. Cheng Munthe-Kaas and Wiik were still engaged in the situation of the Moken people and did not end the project. From working with non-fiction transmedia to create attention and engagement around the documentary film, they refocused the project to be about helping the Moken people in any way they could. Cheng Munthe-Kaas (2017) explains this choice:

> It all started with the goal of making the Moken people known to the world, to bring some light to the issue … Now we take initiatives to help them help themselves … In addition to the M.A.P project [Moken against plastic], which aim to give them the opportunity to stay on the ocean, we have a project aiming to build a Kabang boat, where they will build it themselves and we will provide the necessary funds … But now we have spent so much of our private money on this, something we can't do, so we have started campaigns to raise money so they can preserve their culture before it is completely gone.

Several distinct initiatives have been undertaken in this phase of the project, the main ones being Moken Against Plastic ("Project M.A.P.–Moken Against Plastic" 2015) and Moken News ("Moken News" 2017). "Moken Against Plastic" started off as an Indiegogo campaign ("Back on the Waves–Rescuing the Moken Culture" 2017), with the aim of crowdfunding initiatives to help the Moken people live off the ocean again, by clearing away plastic and provide eco-tourism services to the nascent Myanmar tourist industry. The aim of "Moken News" is to help the Moken people tell their story themselves on social media, with the need to provide training, equipment, and help for them to do this. Being filmmakers, the producers use short films to help the campaigns along, but becoming a means to an end rather than an end in itself. Cheng Munthe-Kaas (2017) relates these initiatives to the transmedia project: "I feel that we have started an NGO [non-governmental organization], but it stills continues to be a transmedia project because it still lives on these different platforms." She notes how working with the project has become a lifestyle and something they do in their spare time, after earning a living from other film projects.

In summary, it is possible to see how the issues concerning how to save the Moken way of life has led the producer and director to work with Project Moken as an ongoing, contingent, and involved process, becoming a strategy to both represent an issue and to work with solving it. Documentary filmmaking is part of how they do this, but is not prioritized over other means of communication in their initiative to improve the situation for the Moken.

The Relevance of the Two Approaches

In the following, to assess the relevance of the two approaches to designing non-fiction transmedia, I will briefly consider how the approaches align with how things are usually done in the world of documentary film production. This is an orientation toward the existing cooperative activities and conventions supporting this world, which restricts what is easy to do and what is not. Here I rely on Howard Becker's (1984) analysis of art worlds, but also Bill Nichols' (1991, 2001) classification of documentary modes as historically situated and cooperative practices.

Presenting Project Moken as having two distinct approaches is a simplification and hides the fact that the documentary filmmakers behind the project had a participatory approach throughout the project, with an explicit aim of giving the Moken people a voice and with a continuous concern for being accountable to them. However, it seems that the team behind the project took on an experience design approach when aligning their project to the requirements of NFI, their main funder. When designing the project in accordance with "how things are usually done" in the world of independent documentary film production, Project Moken became top-down, closed, and offered minimalist kinds of participation to the audience. The need to secure the funding offered by NFI gave a top-down and planned project model, supporting the need to make a convincing argument about the likelihood that the project was going to attract mass readership and/or artistic recognition at festivals. This implies that the director and producers had to take most creative decisions at the front end of this part of the project. The top-down approach also aligns well with how films are usually made, by setting up a rather hierarchical project model, where practitioners with different skillsets are hired to do specific jobs. In line with this approach, Project Moken outsourced parts of the production to an interactive advertising agency, something that was seen as unproblematic in itself, with the exception of the costs involved.

In the later stages of the project, the participatory approach has been followed more clearly and has resulted in initiatives that challenge how things are usually done in the world of independent documentary film. The project has been evolving with the issues most pressing for the Moken people, relying on the director's close connection and familiarity with the community. What has been made in the project have become a means to empower the Moken people, with little focus on this content reaching a mass audience, or this being recognized as legitimate outcomes by funders, broadcasters, and peers. Project Moken has become a loosely organized NGO, but still in line with what can be characterized as a non-fiction transmedia project. This move toward a bottom-up, evolving, and community-led approach, has required a continuous battle to find resources, with crowdfunding as the main option so far. The limited resources that are mobilized by crowdfunding are used to help the Moken back onto the sea and to provide a tree for a Kabang boat, not primarily to initiate costly film production. Social media is used as the main distribution platform and instead of hiring media professionals to tell the Moken story, the project tries to train the Moken people to tell their story

Table 2.1 A comparison of the two approaches to designing non-fiction transmedia in Project Moken

Aspects of Project	*Experience Approach*	*Participatory Approach*
Project model	Top-down, planned and hierarchical	Bottom-up and evolving
The role of the director/producer	Storyteller	Facilitator
Securing distribution	Early on	Continuous
Securing funding	Early on	Continuous
Securing viewers/participation	Masses	Communities
Outcomes	Identifiable works of art on multiple platforms	Empowering a community
Skills and learning	Build on existing skills in the media industries	Aim to give the community the skills needed to become fuller participants

themselves. In Table 2.1 the experience approach is compared with the participative approach, based on how this has been undertaken in Project Moken.

When considering Project Moken as a whole, the two approaches can be seen as relying on and feeding into each other. In the first phase of the project, when the Moken experience was designed and produced, the result benefited from the director's and producer's participatory stance in the project, portraying the Moken culture on their terms based on several years of living and working with the Moken. In 2010, a documentary film was produced with funds from the Kon Tiki museum in Oslo, where "Hook" and his brother bring a Kabang boat to Norway, and sail from Stavanger to Oslo, with focus on bridging the traditional boat building cultures of the Moken and the Norwegians. The second and ongoing phase of the project has relied heavily on what was produced in the first phase. The Moken experience, including a feature-length film, an app, a campaign and two interactive web stories, have given a frame of reference and direction for the initiatives taken later.

Conclusion

In conclusion, both the experience and participatory approaches to non-fiction transmedia needs to be developed further, and it is important to understand how they can be combined in productive ways. However, the articulation of the participatory approach challenges the existing logic of non-fiction media production to a larger degree than the experience approach. Non-fiction transmedia in general and documentary in particular have the potential to fulfill the ideals of participatory media with audience engagement, especially via participation, but this will require cooperative activities that up until now have been rare in the context of professional non-fiction media production.

References

"Back on the Waves–Rescuing the Moken Culture." 2017. *Indiegogo*. Accessed September 20, 2017. www.indiegogo.com/projects/1643259.

Becker, Howard Saul. 1984. *Art Worlds*. Berkeley: University of California Press.

Björgvinsson, Erling, Pelle Ehn, and Per-Anders Hillgren. 2012. "Agonistic Participatory Design: Working with Marginalised Social Movements." *CoDesign* 8 (2–3): 127–144. doi:10.1080/15710882.2012.672577.

Carpentier, Nico. 2011. "Contextualising Author-Audience Convergences." *Cultural Studies* 25 (4–5): 517–533. doi:10.1080/09502386.2011.600537.

Carpentier, Nico. 2016. *Media and Participation: A Site of Ideological-Democratic Struggle*. Bristol: Intellect Ltd. www.oapen.org/search?identifier=606390.

Carpentier, Nico, and Peter Dahlgren. 2013. "The Social Relevance of Participatory Theory." *Comunicazioni Sociali* 3: 301–315.

Cheng Munthe-Kaas, Mette. 2012. "Søknad til NFI Transmedia Lab."
Cheng Munthe-Kaas, Mette. 2015. "Project Moken–Moken Against Plastic."
Cheng Munthe-Kaas, Mette. 2017. Producing the Moken Experience, Skype.
Cheng Munthe-Kaas, Mette, Christian Lien Jensen, and Runar Jarle Wiik. 2013. "Project Moken." Accessed June 2, 2017. www.projectmoken.com/.
Dantec, Christopher A. Le, and Carl DiSalvo. 2013. "Infrastructuring and the Formation of Publics in Participatory Design." *Social Studies of Science* 43 (2): 241–264. doi:10.1177/0306312712471581.
Dena, Christy. 2010. "Transmedia Practice: Theorising the Practice of Expressing a Fictional World across Distinct Media and Environments." Ph.D. diss., University of Sydney.
DiSalvo, Carl, Jonathan Lukens, Thomas Lodato, Tom Jenkins, and Tanyoung Kim. 2014. "Making Public Things: How HCI Design Can Express Matters of Concern." In *Proceedings of the 32Nd Annual ACM Conference on Human Factors in Computing Systems*, 2397–2406. CHI '14. New York: ACM. doi:10.1145/2556288.2557359.
Fischer, Gerhard. 2011. "Beyond Interaction: Meta-Design and Cultures of Participation." In *Proceedings of the 23rd Australian Computer-Human Interaction Conference*, 112–121. OzCHI '11. New York: ACM. doi:10.1145/2071536.2071553.
Fischer, Gerhard, and Elisa Giaccardi. 2006. "Meta-Design: A Framework for the Future of End-User Development." In *End User Development*, edited by Henry Lieberman, Fabio Paternò, and Volker Wulf, 427–457. Human-Computer Interaction Series. Dordrecht: Springer. http://link.springer.com/chapter/10.1007/1-4020-5386-X_19.
Gambarato, Renira Rampazzo. 2013. "Transmedia Project Design: Theoretical and Analytical Considerations." *Baltic Screen Media Review* 1 (1): 81–100.
Gambarato, Renira Rampazzo, and Sergei Andreevich Medvedev. 2015. "Fish Fight: Transmedia Storytelling Strategies for Food Policy Change." *International Journal of E-Politics* 6 (3): 43–59. doi:10.4018/IJEP.2015070104.
Green, David Philip, Simon Bowen, Jonathan Hook, and Peter Wright. 2017. "Enabling Polyvocality in Interactive Documentaries Through 'Structural Participation.'" In *Proceedings of the 2017 CHI Conference on Human Factors in Computing Systems*, 6317–6329. CHI '17. New York: ACM. doi:10.1145/3025453.3025606.
Hayes, Gary P. 2011. *How to Write a Transmedia Production Bible*. Sydney: Screen Australia.
"Hold Your Breath." 2017. *Project Moken*. September 19. www.projectmoken.com/experience/?page_id=4.
"IDFA Doclab Interactive Canon." 2017. *IDFA DocLab*. September 26. www.doclab.org/tag/interactive-canon/.
Jenkins, Henry. 2007. "Confessions of an Aca/Fan: Archives: Transmedia Storytelling 101." March 22. http://henryjenkins.org/2007/03/transmedia_storytelling_101.html.
Jenkins, Henry. 2008. *Convergence Culture: Where Old and New Media Collide*. New York: New York University Press.
Jenkins, Henry. 2009a. "Revenge of the Origami Unicorn: The Remaining Four Principles of Transmedia Storytelling." http://henryjenkins.org/2009/12/revenge_of_the_origami_unicorn.html.
Jenkins, Henry. 2009b. "The Revenge of the Origami Unicorn: Seven Principles of Transmedia Storytelling (Well, Two Actually. Five More on Friday)." http://henryjenkins.org/2009/12/the_revenge_of_the_origami_uni.html.
Karasti, Helena, and Karen S. Baker. 2008. "Community Design: Growing One's Own Information Infrastructure." In *Proceedings of the Tenth Anniversary Conference on Participatory Design 2008*, 217–220. PDC '08. Indianapolis: Indiana University. http://dl.acm.org/citation.cfm?id=1795234.1795280.
Karlsen, Joakim. 2014. "Transmediavendingen i Ny Norsk Uavhengig Dokumentarfilm." In *Hvor Går Dokumentaren? Nye Tendenser I Film, Fjernsyn Og På Nett.*, edited by Henrik Grue Bastiansen and Pål Aam, 1st ed., 195–220. Bergen: Fagbokforlaget.
Karlsen, Joakim. 2016. "Aligning Participation with Authorship: Independent Transmedia Documentary Production in Norway." *VIEW Journal of European Television History and Culture* 5 (10): 40–51. doi:10.18146/JETHC111.
Kensing, Finn, and Joan Greenbaum. 2013. "Heritage: Having a Say." In *Routledge International Handbook of Participatory Design*, edited by Jesper Simonsen and Toni Robertson, 21–37. London: Routledge.
Löwgren, Jonas, and Bo Reimer. 2013. *Collaborative Media: Production, Consumption, and Design Interventions*. Kindle Edition. Cambridge, MA: The MIT Press.
Milk, Chris. 2012. "The Johnny Cash Project." www.thejohnnycashproject.com/.
"Moken News." 2017. *Facebook*. September 20. www.facebook.com/MokenNews/.
Monteiro, Eric, Neil Pollock, Ole Hanseth, and Robin Williams. 2012. "From Artefacts to Infrastructures." *Computer Supported Cooperative Work (CSCW)* 22 (4–6): 575–607. doi:10.1007/s10606-012-9167-1.
Nichols, Bill. 1991. *Representing Reality: Issues and Concepts in Documentary*. Bloomington: Indiana University Press.
Nichols, Bill. 2001. *Introduction to Documentary*. Bloomington: Indiana University Press.
"The Ocean as a Place." 2017. *Project Moken*. September 19. Accessed July 12, 2017. www.projectmoken.com/experience/.

O'Flynn, Siobhan. 2012. "Documentary's Metamorphic Form: Webdoc, Interactive, Transmedia, Participatory and beyond." *Studies in Documentary Film* 6 (2): 141–157. doi:10.1386/sdf.6.2.141_1.

O'Flynn, Siobhan. 2015. "Designed Experiences in Interactive Documentaries." In *Contemporary Documentary*, edited by Daniel Marcus and Selmin Kara, 72–86. London and New York: Routledge.

"One Tree Can Save a Culture." 2013. *The Moken Tree*. http://themokentree.org/.

"Open Documentary Lab at MIT." 2017. *Open Documentary Lab at MIT*. September 26. Accessed July 14, 2017. http://opendoclab.mit.edu/.

Phillips, Andrea. 2012. *A Creator's Guide to Transmedia Storytelling: How to Captivate and Engage Audiences across Multiple Platforms*. 1st ed. New York: McGraw-Hill.

Pratten, Robert. 2015. *Getting Started with Transmedia Storytelling*. 2nd ed. London: CreateSpace. Accessed July 14, 2017. www.academia.edu/download/45791832/gettingstartedintransmediastorytell.pdfpdf.

"Project M.A.P.–Moken Against Plastic." 2015. *Project MOKEN*. December 14. Accessed July 4, 2017. http://projectmoken.com/project/project-m-a-p-moken-against-plastic/.

Saab, Anwar. 2016. "Kindred Spirits." *NRK*. Accessed June 2, 2017. https://tv.nrk.no/program/KOID76003314/undervannsfolkets-kamp.

Saldre, Maarja, and Peeter Torop. 2012. "Transmedia Space." In *Crossmedia Innovations: Texts, Markets, Institutions*, edited by Carlos Alberto Scolari and Indrek Ibrus, 25–44. Frankfurt: Peter Lang. http://content.schweitzer-online.de/static/catalog_manager/live/media_files/representation/zd_std_orig__zd_schw_orig/017/988/140/9783631622285_foreword_pdf_1.pdf.

Scolari, Carlos Alberto. 2009. "Transmedia Storytelling: Implicit Consumers, Narrative Worlds, and Branding in Contemporary Media Production." *International Journal of Communication* 3: 586–606. http://dspace.uvic.cat/xmlui/handle/10854/2867.

Suchman, Lucy. 2002. "Located Accountabilities in Technology Production." *Scandinavian Journal of Information Systems* 14 (2). http://aisel.aisnet.org/sjis/vol14/iss2/7.

Tripp, Stephanie. 2012. "From TVTV to YouTube: A Genealogy of Participatory Practices in Video." *Journal of Film and Video* 64 (1): 5–16.

Truong, Andy. 2012. *Dive Moken – Challenge Yourself against One of the Worlds Best Free-Divers!* (version 1.1). Appative AB. Accessed July 12, 2017. https://itunes.apple.com/kz/app/dive-moken-challenge-yourself-against-one-worlds-best/id535114567?mt=8.

Wiik, Runar Jarle. 2014. *No Word for Worry*. Documentary, Adventure, Drama. Accessed July 11, 2017. www.imdb.com/title/tt3583190/.

3

TRANSMEDIA TELEVISION

Flow, Glance, and the BBC

Elizabeth Evans

Transmedia strategies are becoming an increasingly everyday part of the television industry and televisual experiences. Paul Grainge and Catherine Johnson describe transmedia promotion, for instance, as ever more "commonplace" within television (Grainge and Johnson 2015, 124). Digital technologies including the Internet, laptops, smartphones, and tablets have opened up not only new, transmedial ways of watching television content, but also brought new companies into the sector, challenging the boundaries of what "television" is. However, despite the apparent "newness" of these developments, television and transmediality have a long conceptual and empirical history. In particular, some of the foundational theories of television studies continue to be useful for considering how transmedia logics are applied to television.

This chapter will consider two of these theories: Raymond Williams' model of flow and John Ellis' "glance theory." It will explore parallels between the broadcast television phenomena explored through each theory and key characteristics of *transmedia* television. The final section of this chapter will explore how conceptual linkages between television studies and transmedia studies play out within an institutional context. In particular, it will address how notions of transmediality have been at the heart of television since its inception and have merely become more prominent with the embracing of digital platforms. The focus here will be on the British Broadcasting Corporation (BBC), the UK's leading public service broadcaster, and how the Corporation has in fact operated as "transmedia television" since television broadcasting began.

Television, Transmediality, and Time: Liveness and Flow

The first alignment between television and transmediality relates to the importance of time to our understanding of both. Television, especially television broadcasting, is a fundamentally temporal medium. Even as changes such as personal video recorders (PVRs) and video-on-demand (VOD) allow audiences to access television content whenever they choose, the temporal qualities of television persist. Television's liveness, its ability to broadcast events as they happen, is often held up as both a defining characteristic of television broadcasting (Gripsrud 1998, 19; Carroll 2003, 268) and as a legitimating factor that elevates it as "worthy" (Levine 2008). Within the context of transmedia television, practices of transmedia storytelling and transmedia marketing have been used to reinforce live viewing in light of PVRs and VOD services that fragment the audience and, crucially, allow them to skip adverts (see Tussey 2014, 206; Evans 2015, 121–122; Grainge and Johnson 2015, 120). However, it is Raymond Williams' model of television as "flow" that most usefully brings the temporalities of television together with the temporality of transmediality, especially transmedia storytelling.

In his model of "flow," British cultural theorist Williams recounts his experience of US television upon his arrival in Miami after a long transatlantic sea voyage. In a now seminal anecdote, he describes how three different films (one being broadcast in full, the others in the form of trailers) became interspersed with adverts and became one continuous, but incongruous, sequence. As he describes:

> A crime in San Francisco (the subject of the original film) began to operate in an extraordinary counterpoint not only with the deodorant and cereal commercials but with a romance in Paris and the eruption of a prehistoric monster who laid waste to New York.
>
> *(2003 [1974], 92)*

This experience led him to the formulation that:

> it may be even more important to see the true process [of television] as flow: the replacement of a programme series of timed sequential units by a flow series of differently relates units in which the timing, though real, is undeclared, and in which the real internal organisation is something other than the declared organisation.
>
> *(2003 [1974], 93)*

He argues that this flow of related, unrelated, and semi-related content units is a planned part of television's structure. Television, in essence, becomes a collection of different segments of content that are brought together into a whole larger than any individual segment and guided by an ever-present, though potentially invisible, time-based organizational structure.

"Flow" has subsequently become one of the foundational models of television studies, regarded as a defining characteristic of the medium itself (Gripsrud 1998). The organization of television's flow into a schedule functions as a way to structure its endlessness (Ellis 2000, 130–147) and to frame program content for audiences (Weissman 2017). The rise of digital technologies, especially those that allow audiences greater control over their televisual experiences, have been positioned as challenging the centrality of Williams' model to television studies (see, for example, Kompare 2006; Mittell 2011). However, scholars such as Catherine Johnson have argued not only for the continued value of understanding flow in terms of television broadcasting (2013) but also of its application to the newer strategies of programming, marketing, and distribution that are emerging around television (2016).

Key to flow's contribution to understanding transmedia television is its highlighting of the blurred boundaries that exist between different kinds of television content and how audiences need to navigate their way through those blurred boundaries. Whereas the organizational structure for Williams' model is broadcasting, the same concepts can be usefully applied when that organizational structure is transmediality. Most notably, some of the earliest work to consider transmedia television applied the metaphor of flow to their case studies. John Caldwell explores a number of case studies including how events in the television series *Homicide: Life on the Streets* were extended into online videos and viewers were invited to help solve diegetic crimes (2003). Will Brooker, meanwhile, examined online expansions for teen drama *Dawson's Creek* in which viewers were able to uncover additional backstory about the characters and purchase clothing seen on the program. Although neither uses the term "transmedia storytelling" to describe the phenomenon they discuss, their case studies clearly fit the pattern of a narrative deliberately and coherently expanded across different media and as such form examples of nascent digital televisual transmedia storytelling. Notions of flow appear throughout both discussions. For Caldwell, flow operates at the level of an audience who are increasingly moving across different media platforms when he argues that, "programming strategies have shifted from notions of network *program* 'flows' to tactics of *audience/user* 'flows'" (2003, 136, original emphasis). For Brooker, it is how content "overflows" the boundaries of television (2004). As both scholars demonstrate, metaphors of the flow model remain relevant to the transmedia expansion of television.

We can further develop the alignment between transmediality and televisual flow by considering two key characteristics they both share. Both flow and transmediality are concerned with the creation of coherent audience experiences across different pieces of content. Within Williams' theory, the structures of broadcasting move audiences from one piece of content to another. This smooth movement—of audiences and of content—becomes key to ensuring audiences can successfully navigate and "flow" across different media. As Catherine Johnson has argued, the content that sits between television episodes (called "junctions") serve to "bring clarity and order" (2013, 123). Johnson goes on to explore how continuity announcers and idents, those short pieces of content that identify which channel is being watched, mark the points where content changes from program to trailer or adverts and back to program again but also creates coherency through a single brand identity (2013). Within transmedia contexts, overt directions or continuity in authorship and narrative features such as characters serve a similar purpose (Evans 2011a, 28). Instead of keeping the audience watching the same channel, however, the purpose is instead to keep audiences within a transmedia narrative by moving to non-televisual spaces, from the television channel to a website or to even non-audiovisual content. In many ways, by directing audiences away from the television set, onto other screen devices and toward non-televisual forms of content, transmedia television invites audiences to break one flow (that of the television channel) by creating and privileging another (that of the transmedia narrative world). However, both processes actually share the fundamental characteristic of moving audiences between different forms of content in a way that feels coherent and connected.

If the first characteristic shared between televisual flow and transmediality explores how audiences are moved away from television content, the second relates to how they are constantly drawn *back* to television. A frequent use of transmedia content within television is as a bridge between episodes or even seasons of a television series. This particularly occurs in relation to "serial" narratives, in which story arcs are told over a number of weeks with little or no narrative resolution in individual episodes (see Creeber 2004). In addition to "channel flow," we can therefore also consider "serial flow," that televisual storyworlds have necessary temporal gaps within them that take viewers away from the storyworld and toward other content. In these cases, transmedia strategies work to reinforce a sense of televisual flow by maintaining audience interest in a particular series through additional, transmedia content (see Evans 2011a, 37) and ultimately directing them back to the next television episode. Transmedia flows direct audiences away from the television set but also back toward it.

Notions of time and flow are therefore the first parallel between how television and transmediality can be understood. Both ultimately share a number of characteristics, of working to move audiences in particular ways, of working within expanded, serial narrative formats and of creating an overarching coherent experience. Looking to one of the foundational models of television studies, Williams' flow, therefore offers one useful way of understanding transmedia strategies.

The Domesticity of Transmedia Television: Mediating the Glance

If flow connects television and transmediality through the temporal dimensions of both, the second key conceptual framework examined here connects them through the position of televisual technologies within daily life, especially the home. Television has traditionally been understood as a domestic medium, with a consistent strand of television studies positioning television within the social, spatial, technological, and temporal dynamics of the home (see, for example, Morley 1986; Spigel 1992; Silverstone 1994; Gauntlett and Hill 1999). Of the foundational theories concerning television and the domestic, John Ellis' theory of the television "glance" is most valuable for exploring the relationship between television and transmediality. Ellis argues that because television is fundamentally a domestic medium, it:

> does not encourage the same degree of spectator concentration. There is no surrounding darkness, no anonymity of the fellow viewers, no large image, no lack of movement amongst

> the spectators, no rapt attention. TV is not usually the only thing going on, sometimes it is not even the principal thing. TV is treated casually rather than concentratedly.
>
> *(Ellis 1982, 128)*

As a result, he argues, viewers do not "gaze" at the screen as they would do in the cinema, but instead "glance" at the television set as their attention is drawn elsewhere: "the gaze implies a concentration of the spectator's activity into that of looking, the glance implies that no extraordinary effort if being invested in the activity of looking" (1982, 137). Whilst much work has debated Ellis' model, arguing instead that television viewing can be focused and attentive (see, for example, Bacon-Smith 1992, 128; Caldwell 1995, 25–27; Brooker 2005; Evans 2011a, 143), the increasingly transmedial nature of the television industry, the content it produces and the context in which it is consumed makes his argument ever more relevant.

The connection between Ellis's glance theory and notions of transmediality emerges when considering the ways in which digital technologies, most notably portable screen devices, have been integrated into the daily practices of both the television industry and television audiences. On the one hand, transmedia distribution strategies mean that television has left the home via portable screen devices. Whilst Anna McCarthy (2001) has identified a history of television in public spaces, the development of portable screen devices has amplified the convergence between television and non-domestic viewing. Television broadcasters have increasingly developed video-on-demand (VOD) strategies that allow access to television content via laptops, tablets, and smartphones. Companies such as Netflix and Amazon now operate as part of the television industry and equally have inherently transmedial approaches to distribution. As a result, television is not just available at home through the television set; it has expanded onto buses and trains, into cafes and waiting rooms. These spaces are perhaps even more distraction-filled than the home, filled with the general public and other calls on our attention such as the end of our journey or the start of an appointment (Evans 2011a, 142–143). Notions of "glance" therefore become key to television's transmedia expansion outside of the home.

On the other hand, the intersection between the television glance and transmediality has also been drawn out by more recent scholarship that considers how digital technologies have been integrated into the home itself. Key to this is the emergence of "second screen" practices. In such practices, portable media devices (usually a smartphone or tablet) are used alongside the television screen to access online material that may or may not be related to televisual content. Companion apps such as those for *The X Factor* (ITV 2004–) provide gaming opportunities or additional behind the scenes material (see Evans 2015). They therefore function as "transmedia" in a broad sense. In some cases, they may fit patterns of transmedia storytelling, in others they may function more as transmedia marketing or branding, and in others as transmedia distribution. They continue, however, to demonstrate how fundamental transmedia practices have come to be for the television industry and its audiences. In particular, how transmedia strategies facilitate a form of "mediated glance" (Evans 2015, 124).

Hye Jin Lee and Mark Andrejevic argue that the second screen environment "has encouraged the [television] industry to examine potential ways to capitalize on people's propensity to simultaneously watching and browse (or watching and connect/socialize)" (2014, 42). This is often done through the creation of companion apps. These apps are designed to run "live" alongside the broadcast of content ranging from one-hour dramas to reality shows. They facilitate community discussions, include quizzes and puzzles, or provide extra information about the content appearing on the television screen (see also Tussey, 2014). Lee and Andrejevic identify how such strategies serve to reinforce the television industry's parameters within this more transmedial environment. They argue that second screen companion apps are "designed to fold television viewing into the monitored embrace of a digital enclosure: a commercial surround in which one's activities are recorded, stored and mined for marketing purposes" (Lee and Andrejevic 2014, 53). For an increasingly transmedial television industry, creating their own spaces within the app culture of smartphones and tablets allows them to marshal, manage, and monitor audience behavior away from the television set.

Other scholars, however, present a less ideal alignment between the television industry's aims to use second screen devices to "enclose" digital behavior and the realities of audience's second screen behavior. Sheryl Wilson conducted focus groups with audiences around second screen use and found little alignment between such use and television content (2016, 183). Elizabeth Evans, Tim Coughlan, and Vicky Coughlan (2017) equally challenge the framing of second screen activities within the concerted strategies of the television industry. Their aim was to use monitoring technology including cameras an IP (internet protocol) trackers to map the kinds of behaviors implied in Lee and Andrejevic and Tussey's work, in which companion apps work to enclose television-related transmedia behavior. However, they argue that "what actually emerged was relatively little such behaviour" (Evans, Coughlan, and Coughlan 2017, 8). Instead, they found that second screen activities consisted of "disconnected, ephemeral and forgettable multiplatform experiences" (Evans, Coughlan, and Coughlan 2017, 8).

The kinds of experiences that these second screen strategies for television, which extend audience experiences with content transmedially, encourage for their audiences can be usefully understood by returning to Ellis' glance theory. In particular, Ellis' model is useful for its recognition that television has always competed for viewer attention. Within a transmedia televisual culture, that competition now not only consists of the mundane activities of domestic life, but also mediated spaces (Evans 2015, 124) that offer more media content on a different device. As such, the emergence of second screen activities is simply an extension of television's pre-existing conditions of use. Just as Ellis frames television as an inherently distracted behavior, multitasking is equally (if not more so) inherently distracted. Dan Hassoun describes second screen activities as "simultaneous media use," a label that immediately connects them with notions of attention and distraction. He argues that much scholarly attention on converged or transmedia behavior presumes "that users engage with this content one medium at a time" (2012, 273–274). Ideas of transmedia storytelling in which audiences move from one piece of content to another present a similar impression. However, Hassoun instead argues that it is necessary to recognize how "the rise of 'interactive' media has only multiplied the potentials for distraction and split attention in various environments" (2012, 274). Transmedia content and distribution strategies such as companion apps allow broadcasters to keep that distraction within spaces that they ultimately control and manage through what can be understood as "mediated glances" (Evans 2015, 124). Just as notions of television's flow can be applied and adapted to fit the context of transmedia television, so too can an understanding of how television sits, as it always has done, within fragmented and distracted viewer attention. The core ways of understanding broadcast television also offer clear insight into understanding transmedia television.

Case Study: The BBC as Transmedia Television

The above discussion demonstrated how transmedia television can be usefully examined within the frameworks already provided by television studies. More broadly, whilst television has undergone significant changes as a result of digital technologies and transmedia strategies, those changes should not be taken as a radical departure from the way television has always functioned. To a certain extent, the television industry has indeed turned to transmedia strategies as a means to respond to perceived changes in audience behavior. Technological developments such as PVRs, online viewing, and portable screen devices have contributed to a shifting of the previously stable ground on which television stood and an undermining of its core distribution technology of broadcasting.

However, as the above discussion has indicated, the changes wrought by an increasingly transmedial television culture have equally re-emphasized some of the foundational principles of television as both a media form and as a field of scholarship. The relevance of transmediality to television (and vice versa) becomes even more apparent when looking at a case study of UK public service broadcaster the BBC. Examining the corporation's history and current strategies demonstrates how television and transmediality have a long, intertwined relationship.

The BBC was initially launched as a radio broadcaster in 1922. However, only a year later the Corporation had moved into publishing with listings magazine *The Radio Times* and within a decade of its founding took responsibility for developing television services and content, becoming the focal point around which early television experiments coalesced (see Aldridge 2012, 70). From its first decade, then, the BBC began acting not as a radio broadcaster, but as a transmedia institution, a strategy that has only continued since the 1930s. The Corporation's adoption of a text-based television information service, Ceefax, in the early 1970s meant that BBC audiences accessed a combination of audiovisual and text-based materials long before the Internet. The BBC Micro program, an early home computer system, placed it within the computing sector. Throughout its history, BBC programs have been expanded through ancillary markets, merchandise, and early forms of transmedia storytelling (Evans 2011a, 23; Jacobs and Thomas 2017, 12–13). Online services became part of the Corporation's activities as early as 1994 (see Naylor et al. 2000, 140) and the requirement to develop "digital Britain" was officially integrated into its public service remit from 2006 onwards. As an organization, the BBC has always worked across different media and worked to move audiences between different media, often taking a leading role in helping audiences to embrace new media platforms. It is this process of movement, of not simply producing content on different platforms but structuring audience experience across those different platforms, that demonstrates how the "television" of the BBC has always been transmedial.

The BBC's digital strategies further demonstrate how BBC "television" content is often inherently transmedial. James Bennett, for instance, has explored the Corporation's coverage of D-Day commemorations in 2004. This coverage was not only transmedial, but originated with the new media team and then expanded *onto* television (Bennett 2008, 282), undermining any sense of digital platforms simply serving television. The Corporation's strategies for children's content has made use of apps such as Go CBBC and CBeebies Storytime to develop safe spaces for extending children's engagement with television content into gaming spaces. Paul Grainge and Catherine Johnson have explored how the BBC, alongside digital agency Red Bee Media, have employed strategies of transmedia promotion for major natural history series such as *Planet Earth Live* (Grainge and Johnson 2015, 123–134). From the perspective of transmedia distribution, the BBC iPlayer, which allows viewers to access television and radio content for at least 30 days, has come to represent the gold standard of television catch-up services. Across the corporation's activities, strategies that blur the boundaries between media forms and that work to move audiences between the television, their computer, and mobile device or between televisual, prose-based, and gaming content are apparent. Two particular examples of such transmedia strategies—the *Doctor Who Adventure Games* and BBC Three—illuminate both the BBC's role as transmedia television and the continued relevance of flow and glance to understanding this role.

The first example, the *Doctor Who Adventure Games*, points to how transmedia storytelling serves as a way to craft and manage both televisual and transmedia "flows." *Doctor Who* has been one of the BBC's core transmedia brands. Since its first broadcast in 1963, the series has served as a key case study for the licensed expansion of television content and, more recently, transmedia storytelling and branding (see Perryman 2008; Evans 2011a). Here I shall focus on a collection of video games created alongside the first season featuring the Eleventh Doctor (Matt Smith) in 2010. The games could be downloaded for free via the official *Doctor Who* webpage. They were positioned as additional episodes of the television series and, although developed by independent games company Sumo Interactive, were written by several members of the television production team. Four games were released in June 2010 with an additional game launched in October 2011 (see Evans 2014). Most notably, the games were framed in relation to the television episodes in a way that served to move audiences between the television set and the computer, and between viewing the episodes and playing a game, but maintained a sense of coherency. The games not only expanded the content of the television series transmedially, giving the audience new adventures featuring the Doctor and his companions, Amy and Rory, but also involved the crafting of an ideal transmedia user flow from the television series to the games. This in turn served as a bridge in the gap between seasons of the television program.

The games are placed within the official canon of the series. Victoria Jaye, then Head of Multiplatform Commissioning for the BBC, discussed the games as "very much conceived as part of the TV series as four interactive episodes … it's still very much of the *Doctor* Who canon" (Evans 2011b, 111). The official *Doctor Who: The Encyclopedia* positions each game within the story arc of the corresponding television series (Russell 2011, 400), reinforcing the serial flow of the series as encompassing television and gaming. The transmedia flow of *Doctor Who* also formed a core part of the BBC's continuity and marketing strategy. At the end of the final episode of Matt Smith's first series as the Doctor in 2010, titled "The Pandorica Opens" (tx 26/06/2010), the continuity announcer not only directed audiences to other television-based *Doctor Who* content, but also to a series of books. Immediately afterwards, a trailer ran for the *Adventure Games*. On the one hand, this moment uses the relationship between components of the television broadcast flow to initiate the audience's movement onto a different platform and a different form of media content. The transmedia flow of the series was positioned as usurping the broadcast flow of the television channel. On the other hand, the trailer itself explicitly and aesthetically used the notion of transmedia flow. It began with a photographic image of Matt Smith as the Doctor from the television series. That image then mutated and transformed into his videographic likeness from the video game. A visual sense of flow from televisual space to digital space was therefore constructed at the same time as the advertisement worked to encourage the audience's movement from the television set to the computer.

A more recent, and dramatic, strategy enacted by the BBC further demonstrates how intertwined transmediality and television are. In March 2014, the BBC announced that it would be transforming its youth-oriented channel, BBC Three, into an online-only channel. The move was framed in terms that spoke to the variety and multiplicity of transmedia content that moving away from the broadcast schedule would allow. Damien Kavanagh, the controller of BBC Three, announced that the new channel would be able to function across different media forms, producing:

> a range of content which we know young audiences consume and we want to innovate in—short form video, image led storytelling, votes on reactive topics or blog posts from contributors that will make people laugh and think and deliver a richer experience around our content.
>
> *(Kavanagh 2014)*

Kavanagh's reference to audience participation and storytelling through different media forms speaks to models of transmedia storytelling. However, more central to the move were attempts by the BBC to utilize transmedia distribution for the channel by creating content that would live across different media spaces and platforms.

The conversion of BBC Three to online-only is, currently, the digital epitome of the BBC as transmedia television. On the one hand, the move places digital platforms such as the Internet on an equal footing with broadcasting. As a consequence, audiences are moved across different media platforms but always within the "enclosure" of the BBC. As it has been throughout its history, the BBC continues to operate transmedially. On the other hand, it also again reiterates the translation of the foundational theories of television studies into transmedia strategies. Although it may seem that becoming an on-demand service removes the relevance of channel flows, Catherine Johnson has argued that such services are not at all divorced from notions of flow (2017). Such a situation is evident in looking at how elements of flow are retained within BBC Three's home on the iPlayer. Each episode has an individual page that not only allows the viewer to watch that episode but also directs them toward other, related content from BBC Three and the BBC more widely. These hyperlinks function in a similar way to how continuity announcers direct viewers to stay tuned in the junctions after broadcast episodes. At the end of each episode, the following episode in the series is lined up for automatic playback, literally flowing from one episode to the next. Episodes of individual series are still released weekly, reiterating the ebb and flow of serialized television broadcasting. Each of these elements again demonstrate the value in retaining core models of television as the medium becomes more and more transmedial.

Conclusion

Television (and television studies) and ideas of transmediality have a closely entwined relationship. Two of the foundational models of television studies offer applicable frameworks for understanding how television functions as transmedia. Raymond Williams' model of televisual "flow" is relevant for how audiences are moved between the different segments of transmedia content, how they are directed to remain within that content's parameters. John Ellis' "glance" theory speaks to how distraction can be kept within spaces owned and controlled by television broadcasters via transmedia strategies. A case study of the BBC, one of the oldest television institutions, further reiterates the intersection between television and transmediality both at an operational level in the practices of television broadcasters, but also at a strategic level through transmedia storytelling or distribution. This chapter has therefore demonstrated how, as the BBC and other broadcasters seek out ways to address perceived changes in audience behavior, they are in fact working within a long tradition of industrial practice that has long seen the scope of television *as* transmediality.

References

Aldridge, Mark. 2012. *The Birth of British Television: A History*. Basingstoke: Palgrave Macmillan.

Bacon-Smith, Camille. 1992. *Enterprising Women: Television Fandom and the Creation of Popular Myth*. Philadelphia: University of Pennsylvania Press.

Bennett, James. 2008. "Interfacing the Nation: Remediating Public Service Broadcasting in the Digital Television Era." *Convergence* 14 (3): 271–288.

Brooker, Will. 2004. "Living on *Dawson's Creek*: Teen Viewer's, Cultural Convergence and Television Overflow." In *The Television Studies Reader*, edited by Robert C. Allen and Annette Hill, 569–580. London: Routledge.

Brooker, Will. 2005. "Everything will Flow." *Flow* 1 (12), March 18. Accessed December 14, 2017. www.flowjournal.org/2005/03/everything-will-flow/.

Caldwell, John T. 1995. *Televisuality: Style, Crisis and Authority in American Television*. New Brunswick: Rutgers University Press.

Caldwell, John T. 2003. "Second-shift Media Aesthetics: Programming, Interactivity and User Flows." In *New Media: Theories and Practices of Digitextuality*, edited by Anna Everett and John T. Caldwell, 127–144. London: Routledge.

Carroll, Noel. 2003. *Engaging the Moving Image*. New Haven: Yale University Press.

Ellis, John. 1982. *Visible Fictions: Cinema, Television, Video*. London: Routledge.

Ellis, John. 2000. *Seeing Things: Television in the Age of Uncertainty*. London: I. B. Tauris.

Evans, Elizabeth. 2011a. *Transmedia Television: Audiences, New Media and Daily Life*. London: Routledge.

Evans, Elizabeth. 2011b. "The Evolving Ecosystem: An Interview with Victoria Jaye." In *Ephemeral Media: Transitory Screen Culture from Television to YouTube*, edited by Paul Grainge, 105–121. London: BFI Palgrave Macmillan.

Evans, Elizabeth. 2014. "Learning with the Doctor: Pedagogic Strategies in Transmedia *Doctor Who*." In *New Dimensions of Doctor Who: Adventures in Space, Time and Television*, edited by Matt Hills, 134–154. London: I. B. Tauris.

Evans, Elizabeth. 2015. "Layering Engagement: The Temporal Dynamics of Transmedia Television." *Storyworlds: A Journal of Narrative Studies* 7 (2): 111–128.

Evans, Elizabeth, Tim Coughlan, and Vicky Coughlan. 2017. "Building Digital Estates: Multiscreening, Technology Management and Ephemeral Television." *Critical Studies in Television: The International Journal of Television Studies* 12 (2): 191–205.

Gauntlett, David, and Annette Hill. 1999. *TV Living: Television, Culture and Everyday Life*. London and New York: Routledge.

Grainge, Paul, and Catherine Johnson. 2015. *Promotional Screen Industries*. London: Routledge.

Gripsrud, Jostein. 1998. "Television, Broadcasting and Flow: Key Metaphors in TV Theory." In *The Television Studies Book*, edited by Christine Geraghty and David Lusted, 17–32. London: Hodder Arnold.

Hassoun, Dan. 2012. "Tracing Attentions: Toward and Analysis of Simultaneous Media Use." *Television & New Media* 15 (4): 271–288.
Jacobs, Jason, and Deborah J. Thomas. 2017. "From Public-Private Virtue to Cultural-Corporate Amoeba: Mr Blobby and the Rise of BBC Worldwide during the 1990s." *Historical Journal of Film, Radio and Television.* doi:10.1080/01439685.2017.1285149.
Johnson, Catherine. 2013. "The Continuity of Continuity: Flow and the Changing Experience of Watching Broadcast Television." *Key Words: A Journal of Cultural Materialism* 11.
Johnson, Catherine. 2016. "From 'Multiplatform' to 'Online TV': Shifting Paradigms for Understanding Linear and Non-Linear Television." Paper presented to the ECREA conference, Prague, November 10.
Johnson, Catherine. 2017. "Beyond Catch Up: VOD Interfaces, ITV Hub and the Repositioning of Television Online." *Critical Studies in Television: The International Journal of Television Studies* 12 (2): 121–138.
Kavanagh, Damien. 2014. "Blazing a Trail for New BBC Three." *About the BBC Blog,* December 10. Accessed September 15, 2017. www.bbc.co.uk/blogs/aboutthebbc/entries/fa632091-9a8c-304e-a479-e054dc368c47.
Kompare, Derek. 2006. "Publishing Flow: DVD Box Sets and the Reconception of Television." *Television & New Media* 7 (4): 335–360.
Lee, Hye Jin, and Mark Andrejevic. 2014. "Second-screen Theory: From the Democratic Surround to the Digital Enclosure." In *Connected Viewing: Selling, Streaming and Sharing Media in the Digital Age,* edited by Jennifer Holt and Kevin Sanson, 40–61. London and New York: Routledge.
Levine, Elana. 2008. "Distinguishing Liveness: The Changing Meanings of Television Liveness." *Media, Culture and Society* 30 (3): 393–409.
McCarthy, Anna. 2001. *Ambient Television: Visual Culture and Public Space.* Durham, NC: Duke University Press.
Mittell, Jason. 2011. "TiVoing Childhood: Time Shifting a Generation's Concept of Television." In *Flow TV: Television in the Age of Media Convergence,* edited by Michael Kackman, Marine Binfield, Matthew Thomas Payne, Allison Perlman, and Bryan Sebok, 46–54. New York: Routledge.
Morley, David. 1986. *Family Television: Cultural Power and Domestic Leisure.* London and New York: Routledge.
Naylor, Richard, Stephan Driver, and James Cornford. 2000. "The BBC Goes Online: Public Service Broadcasting in the New Media Age." In *Web.Studies: Rewiring Media Studies for the Digital Age,* edited by David Gauntlett, 135–148. London: Hodder Arnold.
Perryman, Neil. 2008. "*Doctor Who* and the Convergence of Media." *Convergence* 14 (1): 21–39.
Russell, Gary. 2011. *Doctor Who Encyclopedia (New Edition).* London: Penguin Books.
Silverstone, Roger. 1994. *Television and Everyday Life.* London and New York: Routledge.
Spigel, Lynn. 1992. *Make Room for TV: Television and Family Ideal in Postwar America.* Chicago: University of Chicago Press.
Tussey, Ethan. 2014. "Myths of the Internet Comedy Club: Andy Samberg and Emergent Practices of Digital Production and Labor." In *Saturday Night Live,* edited by Nick Marx, Ron Becker, and Matt Sienkiewicz, 202–218. Indiana: Indiana University Press.
Weissman, Elke. 2017. "Watching CSI: A Study of British Audiences Watching Channel 5 and 5 USA." *Critical Studies in Television: The International Journal of Television Studies* 12 (2): 174–190.
Williams, Raymond. 2003 [1974]. *Television: Technology and Cultural Form.* London: Routledge.
Wilson, Sheryl. 2016. "In the Living Room: Second Screens and TV Audiences." *Television & New Media* 17 (2): 174–191.

4

TRANSMEDIA TELENOVELAS

The Brazilian Experience

Inara Rosas and Hanna Nolasco

Television language is a fusion of previous expressions of radio, cinema, comics, and other synchronic visual media, such as video clips and computer graphics (Balogh 2005). Television in Brazil was officially inaugurated on September 18, 1950, by TV Tupi, a broadcaster from São Paulo. In its first phase, TV Tupi experimented with this mixture of languages until it was able to develop its own formats and genres. In the 1950s, the *teleteatro*—a filmed theater—was the main fictional genre on Brazilian television, exploring dramaturgic and literary classics with renowned actors of the national theater.

Teledramaturgy in Brazil appeared as soon as television was inaugurated, but the 1970s consolidated telenovela as the great Brazilian television product. So much so that its historical development is intertwined with the very history of the expansion of television in the country. Teledramaturgy became part of the daily life of Brazilians as a product representative of national identity, and telenovelas became the most profitable and popular programs on Brazilian television. Telenovelas began to win international prizes, gain notoriety, and to be exported globally (Hamburger 2005).

With high ratings, telenovelas increased the competition among broadcasters, culminating in the leadership of Rede Globo [Globo Network]. In the beginning, telenovelas did not have a stable time slot in the programming grid in Brazil. Globo Network stood out because it was the only broadcaster to maintain a regular teledramaturgy production, since its emergence. It established specific and fixed schedules for telenovelas with a limited number of episodes. On a side note, we highlight that telenovelas use "chapters" as narrative units for they expand the plot to resolve the situations in the future, causing the product to have an exact time to finish (Vieira 2014). Nevertheless, in this chapter, we will refer to these units as episodes, as it is the preferred term in English.

Globo Network learned the trade from a more mature group of television professionals, as opposed to other broadcasters that came before. These Brazilian TV producers attended workshops at CBS and NBC in the United States and the broadcaster brought some professionals from the United States and Cuba to work at the TV station for a period. During this period, the construction of a Brazilian teledramaturgical language began. More than a medium, television started to configure modes of action. The television culture, aligned with the culture of consumption and the power of the image, was the main reference of appeal to audiences in Brazil (Caminha 2010).

Consolidated as the country's main broadcaster since the 1970s, Globo Network remains the leader in production and distribution of telenovelas in Brazil, and its main investment is focused on this kind of product. Its daily programming schedule has specific time slots for telenovelas and news programs that go on air interleaved. Currently, three telenovelas are being produced, occupying time slots from the afternoon until late in the evening. There is also a show in the early afternoon called *Vale a Pena Ver de Novo* [*Worth Seeing Again*] that replays successful telenovelas from the past. Globo Network's regular telenovelas air around 6pm, 7pm, and 9pm, and broadcast from Mondays to Saturdays, each one bringing a different kind of production, according to the specificities of the audience: (1) the 6pm telenovela concentrates most of Globo's historical productions; (2) the 7pm telenovela brings light and romantic stories, tempered with some humor; and (3) the 9pm telenovela (prime time) has the most investment, prestige, and social repercussion. They are telenovelas for the family in general, stories that focus on everyday life with relevant social issues. Eventually, there is also an 11pm telenovela, that brings topics not suitable for younger audiences, such as violence and sex. Telenovelas have an average of 200 episodes and each one is 50 minutes (Campedelli 1987; Hamburger 2005).

Brazilian telenovelas are respected and recognized worldwide, and Globo Network, although not a pioneer and not the only major broadcaster in the history of teledramaturgy, is the one with better know-how to make this type of entertainment product. The narrative form of telenovela is the strongest and most present dramaturgic reference in the life of Brazilian audiences. The most expensive and preferred TV programs in Brazil, they are audience leaders, making thousands of people watch the same story at the same time. Telenovelas launch fashion, induce behaviors, talk about controversies, provide services, and participate in the daily life of the country, playing an important role in the cultural, political, and behavioral life of Brazilian society (Sadek 2008).

Transmedia Content for Brazilian Telenovelas

Telenovelas are long serial narratives that have a great sentimental and popular appeal, and a number of production specificities. For instance, the fact that telenovelas are considered an "open work" for the influence of audience feedback on the author's creative process: "the time difference between an aired episode and a written episode is small, triggering a strong and rapid flow of feedback to the author" (Pallotini 2012, 13). The deep engagement Brazilians have with telenovelas represents a defining aspect that allows and instigates various possibilities of transmedia content development.

Within the context of media convergence and second screen phenomena, where the audience watches the show while tweeting and gives their opinion in online discussion forums, media companies seek to expand their reach and connect with other viewers through Internet. "It is the current economic and cultural context—large media conglomerates producing content in a coordinated and participatory manner with audience migratory behavior—that provides a fertile environment for the development of transmedia cultural products" (Lessa 2017, 90).

Therefore, combining new forms of communication and digital platforms with a product that naturally engages the public, it would be logical for telenovelas to seek new ways of interacting with the audience, increasing their engagement, narrative complexity, and the value of the product. With transmedia strategies—in its various nuances, for instance, from storytelling to branding—the audience has at its disposal a fictional world composed of a tangle of pieces spread across several media platforms (Lessa 2017).

However, the Brazilian television production of serial fiction is different in comparison to transmedia strategies of international productions, especially American TV series, whose products may consist of complex transmedia storytelling, involving web series, comics, short films, and consumer goods. In the case of Brazilian telenovelas, "the intense almost daily broadcast of episodes—except on Sundays—and the excess of plots, subplots and soap opera characters, leave little time and space for the creation of fictional transmedia extensions that are dedicated to narration of new stories" (Lessa 2017, 99–100). Thus, the most frequent transmedia extensions connected to telenovelas are fictional websites—of institutions that only exist within the plot—character's blogs and social media profiles. There is also the current use of a variety of game applications that promote the immersion at the fictional universe of the telenovelas, allowing the audience to explore details of the characters, situations, costumes, and fictional cities (Souza, Lessa, and Araújo 2013).

Regarding transmedia productions, Globo also stands out from other national broadcasters. That is why this chapter brings as a case study one of its telenovelas, *Cheias de Charme* [*Sparkling Girls*, 2012], considered a milestone in the transmedia content of the network. Since then, almost all Globo Network telenovelas have had an Internet team, not only to think about the transmedia content, but also for the paratextual content, which includes backstage videos, interviews, episode summaries, episode analysis by the characters, etc.

Globo's experience with transmedia narratives began with small actions. In *Viver a Vida* [*Seize the Day*, 2009–2010], a prime-time telenovela, one of the main plots was the story of Luciana, a young model who became quadriplegic after a serious accident. Globo then created Luciana's personal blog called *Sonhos de Luciana* [Luciana's Dreams] in which she talked about the changes in her life, her recovery, the difficulties and overruns. There was also a website called *Portal da Superação* [Overrun Portal] with testimonials and photos of real people who suffered physical accidents and managed to overcome the situation. Some testimonials were summarized and included in the telenovela.

Morde e Assopra [*Dinosaurs & Robots*, 2011], a 7pm production, marks the beginning of an Internet team for Globo Network telenovelas, as they were starting to use the online sphere as an ally of television shows. This team did not necessarily work within the transmedia aspects of all telenovelas, but expanded the action of Globo in other media, becoming an important asset for the broadcaster. The transmedia content of Dinosaurs & Robots involved *Caçadora de Dinossauros* [Dinosaur Hunter], a personal blog of a character called Júlia, a paleontologist, in which she shared a little about the adventure of finding fossils around the world. In addition, there was a game application called *O Segredo de Naomi* [Naomi's Secret], a mobile application to engage the audience around the secret of a character that would play an important role at the end of the story, reminding the public of facts that took place months earlier in the telenovela (Lessa 2017).

At the same year, *O Astro* [*The Illusionist*, 2011] engaged in a new experiment on transmedia content for Globo Network: a web series called *Sob o Signo de Ferragus* [*Under the Sign of Ferragus*], with seven short online episodes about the past of the male protagonist. In the same year, *Cordel Encantado* [*The Enchanted Tale*], a historic and fantasy telenovela, had a character directing a fictional documentary that was available online.

The transmedia of the 9pm telenovela *Avenida Brasil* [*Brazil Avenue*, 2012] brings a different aspect of this phenomenon. Besides some standard extensions such as the blog *Dicas da Monalisa* [Monalisa's Tips], with beauty tips related to the character Monalisa's beauty parlor, the broadcaster invested on transmedia content that could get the fastest repercussion online, such as GIFs and photomontages. The reception of this content engaged the fans to create their own, which ended up having as much impact as the official content. There were numerous fanpages and fake profiles of characters on Twitter and Facebook and several Tumblrs making GIFs and photomontages, bringing new situations to the plot.

After *Sparking Girls*, Globo Network developed more consistent transmedia strategies for some of the 7pm telenovelas. *Totalmente Demais* [*Total Dreamer*, 2015–2016], developed, beyond the fictional websites, a fashion magazine site and a character's blog, an unprecedented telenovela prequel and sequel. There was an episode zero with six-minute duration, a prologue where it was possible to know a little about the story and its characters, previously shown on the streaming service Globo Play, and a ten-episode web series called *Totalmente Sem Noção Demais* [*Total Clueless Dreamer*], launched on Globo Play the day after the end of *Total Dreamer*, also considered a spin-off.

Haja Coração [*Burning Hearts*, 2016], the successor 7pm telenovela, traced a similar transmedia path with a cartoon displaying Tito, a dog considered the telenovela's mascot. It also had an eight-episode web series about the character Teodora, showing her life on an island, while she had disappeared on the main plot. *Êta Mundo Bom* [*The Good Side of Life*, 2016], a 6pm telenovela, had as transmedia extensions a radionovela and a web documentary about the Brazilian cinema during the 1940s.

Other Brazilian networks are starting to modestly engage in diverse transmedia strategies for its telenovelas. For instance, Record Network created the web radio station *Ampola* as a transmedia extension for *Balacobaco* [*Tricky Business*, 2012–2013]. At radio *Ampola*, audiences could interact with the content of the telenovela 24/7 and listen to a selection of songs from the plot. The radio website also offered ways to share content and interact with social media networks. In addition, SBT Network stands out for its series and telenovelas targeting younger audiences, in which some transmedia extensions are implemented. In *Carrossel* [*Carousel*, 2012–2013], a Brazilian version of a Mexican telenovela, SBT created a transmedia franchise, containing diverse autonomous products that related textually and/or commercially: the telenovela; two movies that premiered in 2015; *Patrulha Salvadora* [*Salvage Patrol*, 2014–2015], a spin-off series; an animated cartoon aired in 2016, also available online; and a musical theater production that is currently active.

Case Study: *Cheias de Charme* [*Sparkling Girls*, 2012]

Sparkling Girls was a telenovela by Globo aired from April to September 2012. It was written by Izabel de Oliveira and Filipe Miguez, directed by Carlos Araújo, and had 143 episodes. The plot was focused on three housemaids—Maria do Rosário, Maria Aparecida, and Maria da Penha—who had different ages and lifestyles. They met in a police station because of different issues and from there on became good friends. Later in the story, they would become a musical group called *Empreguetes*, which is a term that plays with the Portuguese word for housemaid—*empregada*.

Sparkling Girls' relevance in transmedia terms relates to the fact that Globo took a chance with this product, creating many transmedia actions, using mainly the Internet to do so. They had character's blogs, websites enhancing the narrative, campaigns on Twitter, online music video launches, a book release, and, moreover, the story interacted with other products of the same broadcaster. It was a risky maneuver that paid off, as it contributed to the success of *Sparkling Girls*, which was the telenovela that had more audiences in four years than any other produced by the same broadcaster and during the same time slot (UOL 2012).

The Video Clips

Music was an important aspect within the narrative construction of *Sparkling Girls*. We argue that *Sparkling Girls* could be considered a musical telenovela, due to the presence of relevant characteristics of the musical genre, such as musical numbers and music/dance as important elements in the narrative (Grant 2012). Considering this relation, the video clip format was an important asset for the musical and transmedia purposes of this telenovela. The main characters were singers—the group *Empreguetes*, the villain Chayene, and the "prince of housemaids," Fabian. There were original songs composed specifically for the characters and two of them turned into video clips: *Vida de Empreguete* [Life of a Housemaid] and *Nosso Brilho* [Our Sparkle]. Both songs of *Empreguetes* became key to the construction of the musical group's trajectory, delineating the characters' anonymity first and successful career later.

The most important action within Globo's transmedia strategy within *Sparkling Girls* was the video clips creation: they were made public online before being aired on television, an unprecedented resource never seen in Brazilian telenovelas. The first one, *Vida de Empreguete*, had more impact. Within the plot, the protagonists were still anonymous housemaids, who had a reunion at Chayene's house—Rosário's employer at the time—in her absence, and decided to parody her and try on her clothes. Rosário, one of the *Empreguetes*, also played instruments, thereupon having the idea of a song called *Vida de Empreguete*, about the relationship between bosses and housemaids, their exhausting routine of household tasks, and the constant criticism they suffered. Since the singer's house had a recording studio, she suggested to record the song and make a music video to save the moment, with the help of their producer friend Kleiton. Since the video was made without the house owner's consent, and making fun of her as an employer, they decided to keep it to themselves. Here comes a plot twist: another character discovered the video clip's existence, stole the DVD that contained it, and showed it to Chayene, who decided to take legal action. In this process, the video was accidentally leaked online by the son of one of the singer's lawyers. Consequently, it went viral, transforming the three anonymous housemaids into music stars. This leaking process was shown in the telenovela in a mysterious way: some of the characters watched the video clip, but not the public. This strategy was central to the transmedia project: when Chayene and her team discovered that the video was online, at the final minutes of this episode (aired on a Saturday), Globo also "leaked" it to its website. The link was shown to the public for a few seconds through a close-up on the screen of a character's computer. During the final credits, the broadcaster took a more direct approach, writing the message: "Watch the video clip now at www.globo.com/empreguetes."

The video clip was only aired as part of the telenovela in the next episode, on Monday. It was a new strategy for telenovelas in Brazil. This video had over 12 million views, according to *Memória Globo* (n.d.), thus, the network gave the audience the possibility to access a new and essential content for the development of the narrative on the Internet prior to the TV launch. With the second video clip, *Nosso Brilho*, the transmedia strategy was similar. Globo also "leaked" this video online at its website few days before being aired on TV. Narratively, it was constructed in a similar matter, with suspense and with the link to the website being shown through a character's tablet, and later on reinforced at the final credits of the telenovela's episode. In *Sparkling Girls*, the transmedia narrative not just expanded the telenovela's universe to other platforms, but also helped to blur the boundaries between reality and fiction (Lima and Moreira 2012, 12). Therefore, the songs and video clips produced by the telenovela had fan-made versions distributed at video sharing websites such as YouTube, for example, and were played outside the fictional context, just as products made by artists in real life.

Websites: Tom's Stars and Domestic Worker

Nevertheless, Globo Network maintained its usual ways of implementing transmedia extensions, through the character Tom Bastos' website. He was the manager of all the main artists within the plot and posted information about them online. His website *Estrelas do Tom* [Tom's Stars] (Gshow n.d.a) presented each artist, their concert schedule, and shared information with the real fans of the telenovela. In addition, there was exclusive content, such as interviews made with singers and artists who participated in *Sparkling Girls*.

An important part of this website dealt with the contests promoted by Tom Bastos. They stimulated audience participation, and as the website would come up in conversations within the narrative, it opened a space for the contests' results to be shown at the telenovela. These kind of actions were promoted by Globo to try to control the fan made products, because "Globo acts judicially to restrain at social media networks the circulation of content arising from its programs, which is protected by copyright and intellectual property laws" (Castro 2012, 7).

When the manager signed the *Empreguetes* to his team of artists, he launched the contest *Empreguetes da Internet* [Internet Empreguetes], asking the fans to send versions and parodies of the video clip *Vida de Empreguete*. The best ones were posted at the website, with praise and tips from Tom Bastos, and shown to the *Empreguetes* by this character at the telenovela. Later, there was the *Concurso de Passinhos* [Dance Steps' Contest], a children's funk dance contest referring to one that happened in the narrative. It followed the previous one model with videos from the audience being uploaded to the website and aired on television.

Another website that enriched the telenovela's narrative and had a social agenda was *Trabalhador Doméstico* [Domestic Worker] (Gshow n.d.b). Since the telenovela's protagonists were housemaids, one of the main subjects discussed in the plot was the relationship between employers and their employees, abuses committed within this specific work relation, and the housemaid's legal rights. This website was part of a campaign that the character Penha, another *Empreguete*, made alongside a team of lawyers to inform this kind of workers of their rights. In the show, she was physically assaulted by Chayene, her boss, who threw a bowl of soup at her face. After Penha's legal complaint, Chayene was convicted, had to pay a fine and do community service for her crime.

Not often are houseworkers more than supporting roles in telenovelas. The authors of *Sparkling Girls* brought them to the center of the narrative, and even though there were problems with the portrayal of the class—for example, with mainly white women, not corresponding to the reality of a country with severe racial disparity such as Brazil, where poor black people are the majority in this kind of low-paying job—it was interesting nonetheless to see the less privileged social classes as protagonists. With this transmedia extension, the network opened a possibility to inform real houseworkers about their conditions of labor and the importance of working with a formal contract. The main voice of the page was the character Penha presented by means of videos, but there were also written articles.

Campaigns Free Empreguetes *and* Empreguetes Forever

Within the plot, in two occasions, the characters Kleiton, the *Empreguete*'s producer and friend, and Elano, Penha's brother, started campaigns that had audience interaction. That was the case of *Empreguetes Livres* [Free *Empreguetes*], which was launched at the time of the trio's prison due to the illegally made video clip. Their fan club fought for their release, protesting outside the prison and through the Internet, with the hashtag #Empregueteslivres. According to Castro (2012), it became a trending topic on Twitter at the time.

The other campaign that was promoted at the plot was called *Empreguetes para Sempre* [*Empreguetes* Forever] and was set off when the trio decided to split up because of work problems. The characters then created a website that received testimonials of artists and fans from Brazil. Even after the end of the telenovela some fans asked for the return of the group's activities in real life.

The participation of real Brazilian artists within the plot was an important factor to the public's immersion at the telenovela. Throughout *Sparkling Girls*, famous singers such as Ivete Sangalo, Alcione and Zezé di Camargo & Luciano, for example, interpreted themselves and interacted with the fictional singers as if they were their real colleagues. At the time of the *Empreguetes* Forever campaign, multiple artists' videos were posted at the official telenovela's website claiming for the return of the trio.

The Most Charming Housemaid in Brazil

A Empregada Mais Cheia de Charme do Brasil [The Most Charming Housemaid in Brazil] was a campaign that intertwined Sparkling Girls and another entertainment show from Globo, called *Fantástico*. This transmedia branding initiative invited the real housemaids in Brazil to send videos showing their artistic aptitudes. The only rules were that the video had to be recorded at their work space, and the participant had to have a formal contract as a housemaid. The prize was to participate in an episode of the telenovela, meeting the actors and being an *Empreguete* for a day. The videos were shown at *Fantástico*, and through popular vote, the winner was elected and then called to participate in an episode where she met the *Empreguetes* and danced with them the hit song *Vida de Empreguete*.

Book: Cida, the Empreguete, an Intimate Diary

The book *Cida: A Empreguete, um Diário Íntimo* [*Cida, the Empreguete, an Intimate Diary*] was an extension of the narrative released in 2012, written by Leusa Araújo, who worked as a text researcher in *Sparkling Girls*. Maria Aparecida (Cida), the youngest *Empreguete*, had a tragic life story: both her parents died when she was young, so she continued to live at her mother's employer's house in exchange for working as a housemaid since her childhood. Introspective and romantic, Cida had a diary, in which she wrote about her life and dreams, in a make-believe conversation with her mother.

With the goal of becoming a writer, at the end of the telenovela Cida succeeds and publishes her diary. Just after the end of *Sparkling Girls*, the book was actually printed, and it remains available for sale all over Brazil. In this book, there is a more thorough construction of Cida's character with details about her childhood, as well as her pathway to fame, her love life, and her life as a housemaid (Feitosa 2015, 87).

Conclusion

Regarding Brazilian telenovelas, broadcasters engaged in this type of content production since the 2000s with the goal to enrich their narratives, in order to try to match international productions. Globo Network, the country's largest producer, is to this day more dedicated to developing transmedia content for its fictional productions. In the last decade, it was remarked that the transmedia experiences in telenovelas increased in quantity of products, and became even more elaborate, including the participation of specialized consulting teams.

There are various possibilities of transmedia creation depending on the type of telenovela, for each one has particularities based on its themes, duration, and time of exhibition. It was observed that Globo Network has the tendency to propose most of its transmedia projects—also the more elaborate ones—as 7pm telenovelas, which have lighter plots and shorter narratives, and which more easily enable the development of transmedia extensions.

As for *Sparkling Girls*, the transmedia strategies were important for its success. According to Bieging (2013), this telenovela reached expressive levels of audience and public participation, standing out in the history of Brazilian teledramaturgy. It was a very successful product, that inspired Globo Network to continue investing in transmedia strategies within their telenovelas, which led, for example, to the transmedia strategies for *Total Dreamer*, that presented other unparalleled actions. Its successor telenovela, *Burning Hearts*, had a web series as well, showing that more elaborated content is in vogue regarding transmedia telenovelas in Brazil.

References

Balogh, Anna Maria. 2005. *Conjunções, Disjunções, Transmutações: Da Literatura ao Cinema e à TV* [Disjunctions, Conjunctions, Transmutations: From Literature to Cinema and TV]. São Paulo: Annablume.

Bieging, Patrícia. 2013. "Transmidiação como Ferramenta Estratégica: Cheias de Charme Explora Uma Nova Forma de Fazer Telenovela" [Transmediation as a Strategic Tool: Sparkling Girls Explores A New Way of Making Telenovela]. *Revista Novos Olhares* 2 (2): 60–71. Accessed August 17, 2017. www.revistas.usp.br/novosolhares/article/view/69828/72488.

Caminha, Marina. 2010. "A Teledramaturgia Juvenil Brasileira" [The Brazilian Youth Teledramaturgy]. In *História da Televisão no Brasil* [History of Television in Brazil], edited by Ana Paula G. Ribeiro, Igor Sacramento, and Marco Roxo, 197–215. São Paulo: Editora Contexto.

Campedelli, Sâmia Y. 1987. *A Telenovela* [The Telenovela]. São Paulo: Ática.

Castro, Gisela. 2012. "Cheia de Charme: A Classe Trabalhadora no Paraíso da Cibercultura" [Sparkling Girl: Working Class in Cyberculture]. Paper presented at the annual meeting of Intercom, Fortaleza, Ceará, September 3–7.

Feitosa, Klênnia. 2015. "Narrativa Transmídia e a Expansão do Universo Ficcional: Os Princípios e as Estratégias de Transmidiação da Telenovela Cheias de Charme" [Transmedia Narrative and Fictional Universe Expansion: Sparkling Girls Transmediation Principles and Strategies], Master thesis, Universidade Federal do Rio Grande do Norte.

Grant, Barry Keith. 2012. *The Hollywood Film Musical.* West Sussex: Wiley-Blackwell.

Gshow. n.d.a. "Estrelas do Tom" [Tom's Stars]. Accessed July 30, 2017. http://gshow.globo.com/novelas/cheias-de-charme/estrelas-do-tom/platb/.

Gshow. n.d.b. "Trabalhador Doméstico" [Domestic Worker]. Accessed July 2017, 2017. http://gshow.globo.com/novelas/cheias-de-charme/Trabalhador-Domestico/.

Hamburger, Esther. 2005. *O Brasil Antenado: A Sociedade da Novela* [Tuned Brazil: Telenovela Society]. Rio de Janeiro: Jorge Zahar.

Lessa, Rodrigo. 2017. "Explorações Conceituais Acerca de Narrativa Transmídia e Ficção Seriada Televisiva" [Conceptual Explorations About Transmedia Narrative and Television Fiction]. In *Ficção Televisiva Seriada no Espaço Lusófono* [Serial Television Fiction in the Lusophone Space], edited by Isabel F. Cunha, Fernanda Castilho, and Ana Paula Guedes, 87–105. Covilhã: Editora LabCom.IFP.

Lima, Cecília, and Diego Moreira. 2012. "Operações do Conceito de Hipertelevisão na Novela Cheias de Charme: A Criação de Universos Transmídias na TV Globo" [The Concept of Hipertelevision in Sparkling Girls: Creation of Transmedia Universes in Globo Network]. Paper presented at the annual meeting of Intercom, Fortaleza, Ceará, September 3–7.

Memória Globo. n.d. Acessed August 31, 2017. www.memoriaglobo.globo.com.

Pallotini, Renata. 2012. *Dramaturgia de Televisão* [Television Dramaturgy]. São Paulo: Perspectiva.

Sadek, José R. 2008. *Telenovela: Um Olhar do Cinema* [Telenovela: A Cinema Approach]. São Paulo: Summus.

Souza, Maria Carmem J., Rodrigo Lessa, and João Araújo. 2013. "Empresas Produtoras, Projetos Transmídia e Extensões Ficcionais: Notas Para um Panorama Brasileiro" [Networks, Transmedia Projects and Fictional Extensions: A Brazilian Panorama]. In *Estratégias de Transmidiação na Ficção Televisiva Brasileira* [Transmediation Strategies in Brazilian Fiction Television], edited by Maria Immacolata Vassalo Lopes, 303–344. Porto Alegre: Sulina.

UOL. 2012. "Capítulo Final de Cheias de Charme Alcança 32 Pontos no Ibope e Bate Três Antecessoras" [Final Chapter of Sparkling Girls Reaches 32 Ibope Points and Beats Three Predecessors]. *Uol TV e Famosos*, September 28. Accessed August 28, 2017. http://televisao.uol.com.br/noticias/redacao/2012/09/28/capitulo-final-de-cheias-de-charme-alcanca-32-pontos-no-ibope-e-bate-tres-antecessoras.htm

Vieira, Marcel. 2014. "Origem do Drama Seriado Contemporâneo" [The Origin of Contemporary Serial Drama]. Paper presented at the annual meeting of Compós, Belém, Pará, May 27–30.

5

TRANSMEDIA COMICS

Seriality, Sequentiality, and the Shifting Economies of Franchise Licensing

William Proctor

Historically, the comic book medium emerged out of the relationship between newspaper comic strips and the popular, much lambasted, pulp tradition of the 1920s, both of which introduced numerous trade characters to the popular imagination, including Tarzan, Popeye, Flash Gordon, and Buck Rogers. These were, however, preceded by Richard Felton Outcault's *The Yellow Kid* newspaper strips, often contentiously cited as the first comics in the United States, which developed an emergent transmedia presence in the late-nineteenth century with vaudeville plays, a short film, crossover appearances with Outcault's Buster Brown character, and a spate of merchandising products, such as gum, postcards, baby clothes, and household appliances. In the United Kingdom, the character Ally Sloper debuted in the pages of satirical magazine *Judy* in 1867 almost three decades before the first appearance of *The Yellow Kid*, and grew into "the first comics superstar" (Sabin 2003) and the first recurrent comic character—perhaps the earliest example of what would today be described as transmedia franchising, with a flotilla of texts and associated merchandising products, including music hall and street theater performances, village parades, advertising, film, and anthologies of previously published comic strips. Ally Sloper's popularity was so enormous and widespread that "it is no exaggeration to say that his visibility in [UK] popular culture would have been comparable to that of any Hollywood blockbuster creation" (Sabin 2003).

It was not until the early 1930s, however, that comic books began to serve an active role in the extension, elaboration, and enlargement of imaginary worlds via emergent licensing practices. As Avi Santos explains, it is through licensing that

> IP [intellectual property] owners are able to extend a property's reach into almost every area of consumer life without having to invest in manufacturing infrastructure or distribution networks. Licensing agreements typically involve a contract signed between a minimum of two parties, in which licensors give the licensee(s) permission to use the name and/ or image of their intellectual property for a specified purpose, for a limited amount of time, and within agreed-upon geographic and product market boundaries.
>
> *(2015, 7)*

From its inception, the comic book medium entered into an intensive dialogic relationship with other new media of the day, including radio, newspapers, television, and film, as well as the kinds of merchandising phenomena usually equated with media conglomeration and convergence in the contemporary moment. Over the decades since, comics of this kind have attracted a lion's share of critical scorn. As with other "tie in" products, transmedia comic "spin offs" are often pejoratively framed in purely commercial terms, as nothing but "a parasitic industry" leeching off the endeavors of legitimate creative agents and authors (Gaines, quoted in Santos 2015, 8). For much of the history of licensed comics, with few exceptions, these spin-offs "were crafted exclusively by the comic-book producers with only financial coordination with the intellectual property (IP) holders" (Clarke 2013, 27). As Kackman puts it, comics were certainly "not the product of smoothly engineered synergy," but "a profitable secondary market—a way to extract as much as possible from a popular media figure or text" (2008, 83). Essentially, comics were primarily seen as a valuable ancillary market through elongating the profit potential of popular character brands, such as the Lone Ranger (Santos 2015), the Shadow (Fast and Örnebring 2017), or Conan the Barbarian (Bertetti 2014), to name but a few. According to Henry Jenkins, the licensing system

> typically generates works that are redundant (allowing no new character background or plot development), watered down (asking the new media to slavishly duplicate experiences better achieved through the old), or riddled with sloppy contradictions (failing to respect the core consistency audiences expect within a franchise) … In reality, audiences want the new work to offer new insights and new experiences. If media companies reward that demand, viewers will feel greater mastery and investment; deny it, and they stomp off in disgust.
>
> *(2006, 105)*

To be sure, tie-in comics have certainly bore economic fruit, and it is true that narrative continuity between various iterations was almost non-existent. But here, Jenkins runs the risk of privileging a limited transmedia framework, "a monolithic view of 'old' licensing contrasted with 'new' coherent and integrated transmedia" (Hills 2012, 412). Jenkins' perspective is largely anchored to the notion that fan cultures desire and demand tightly orchestrated continuity systems, a phenomenon that did not yet exist in the early licensing era—and would not for several decades, arguably beginning with the development of the Superman mythology at the hands of DC editor, Mort Weisenger, in the 1950s, and the subsequent rise of Marvel in the 1960s. In so doing, Jenkins constructs a homogenously ideal and imaginary fan audience, who will only "stomp off in disgust" if they do not get what they want, disregarding the way in which the relationship between license parties has operated trans-historically. Specifically, it is mainly through licensing partnerships that character brands have been able to develop an accumulation of iterations by soaking up the fluids of transmedial exchange while sustaining its shelf life through multiplication.

That said, there have been a series of shifts and modifications in recent years undergirded by greater coordination, integration, and collaboration between license partners, with an emphasis on connecting primary and ancillary units according to principles of transmedia storytelling. Although this model has yet to become institutionally mandated as standard practice—indeed, there are still as many transmedia comics operating according to traditional licensing practices that have been ongoing for the better part of a century as there are experiments with transmedia storytelling, perhaps even more so—we are beginning to see significant maneuvers in this direction.

For the rest of this chapter, in fact, I will explore the shifting transmedia economies of franchised licensing as it relates to comic books, specifically the way in which film and television properties have been extended and franchised. To achieve this, the following sections include small case studies centred on licensed comics—mainly, Disney, *Star Trek*, and *Star Wars*—and should in no way be viewed as providing a complete account of the history of licensing across almost a century of such practices. Clearly, there is much work to be done by scholars in order to (more) fully historicise the licensing phenomenon which, as with transmedia storytelling (Freeman 2016), did not emerge *ex nihilo* from the wellspring of convergence and conglomeration in the contemporary moment.

Transmedia Disney

Beginning two years after his debut in the animated short *Plane Crazy* (1928), Disney's famous brand-mascot, Mickey Mouse, traveled across media to become the star of syndicated newspaper comic strips and played a key role in the economic fortunes of the burgeoning studio (Davis 2017, 56). Walt Disney himself had been keen on creating merchandise based on the anthropomorphic rodent as early as 1929 in order to stimulate "very good publicity" for further adventures on the silver screen—in turn, stimulating the cash nexus (Barrier 2007, 83). As money was especially tight for the studio during the period, Disney orchestrated several licensing deals for the purposes of merchandising products and novelties, but the first contract was borne out of his firm belief that "the character's regular presence in newspapers [would be] a key promotional strategy for his cinematic efforts" (Davis 2017, 56; see also Santos 2015, 37). So it was that on January 24, 1930, Disney entered into a licensing agreement with King Features Syndicate to begin producing newspaper strips and, soon enough, the zany mouse was featured in numerous daily publications as much as six times a week. Unlike later licensing partnerships, however, the relationship between Disney and King Features was based on production and distribution—the strips themselves were created "in house" and not farmed out to freelance writers and artists by the licensor. In essence, Disney remained responsible for creating its own narrative content.

The first story was a loose adaptation of animated short, *Plane Crazy*, but it was with the follow-up comic, titled *Mickey in Death Valley*, that the newspaper strips became more than just "cute animal antics and playful punch-lines" (Davis 2017, 56) and offered continuity strips that could last for several months. From January to March 1930, the strips were written by Disney himself and drawn by Ub Iwerks until April of the same year when he began sending plot ideas to animator Floyd Gottfredson, who took over completely the following month. Gottfredson was "given a broad license to freely adapt and expand concepts as he liked" (Booker 2014, 407), and would feature an irreverent brand of comedy, including one story that had Mickey repeatedly try to take his own life—and failing catastrophically each time—because he wrongly believed that he has seen Minnie kiss another mouse ("Without Minnie I might as well end it all!"). It is these kinds of zany escapades that saw Gottfredson's stories tampered with in later collected volumes (although this has since been redressed with Fantagraphics series of lavish hardback compendiums). Astonishingly, Gottfredson continued to work on Disney comic strips and books for 45 years until 1975, but as with legendary Donald Duck artist, Carl Barks, Gottfredson would not be recognised for his creative efforts during his own lifetime, as artists contributions tended to be uncredited during the period, with only Walt Disney's authorship serving as sole imprimatur. In recent years, however, Gottfredson has grown to be revered as "the young and unrecognised genius of graphic narrative" and "the definitive creative force" behind the evolution of Mickey Mouse, and indeed other Disney staple characters (Booker 2014, 407). Even in the early days of licensed comics, then, the question of authorship was freighted with paradox, despite Disney's signatory dominance. Both Gottfredson and Barks were able to invent new characters, some of which would become fully integrated members of the Disney catalogue (the most famous of which being Barks' Scrooge McDuck); and both Gottfredson and Barks are recognized contemporaneously as creating the "definitive" versions of the Mickey Mouse and Donald Duck characters, respectively.

The partnership between Kings Features and Disney would also produce popular adaptations of Disney's feature films, beginning with the syndication of *Snow White and the Seven Dwarfs* on December 12, 1937. Illustrated by Hank Porter and written by Merrill de Marris, the Snow White comic further tested traditional adaptation processes through the extension of key scenes—greatly expanding the role of Prince Charming, for example—and adding story elements not included in the source text (although these additive elements were not produced specifically for the comic strip, but were scenes left on the cutting room floor (see Booker 2014, 409)). It is worth noting that the strip preceded the theatrical release of the film by six weeks and by the time audiences queued up for tickets, they may have already become quite familiar with the story and characters.

King Features Syndicate had the option of licensing the Disney comic strips to other partners (which would effectively make them "sub-sub-licensees") and in doing so, "the acquisition of licenses with newspaper syndicates led to the production of comic books filled with reprinted material" (Duncan and Smith 2014, 198). In 1938, Dell Comics, an offshoot of Dell Publishing that had its genesis in pulp magazines, formed a partnership with Western Printing, who had obtained the license to publish Disney material. In this case, however, rather than reprinting daily newspaper strips in compilation formats, as was de rigueur at the time, Dell Comics started producing original Disney stories in the pages of anthology comic, *Four Colour* (1938–1968)—which incidentally remains the current record-holder for most "floppies" published in a comic series with 1,354 issues. With *Four Colour*, Dell would effectively use the publication as a litmus test for Disney characters, much in the same way that DC Comics would in the 1950s with the *Showcase* anthology. For example, the fourth issue featured Donald Duck, the success of which evinced that the character could be exploited more fully elsewhere—and so it was that eight months later, Dell published the first issue of *Walt Disney's Comics and Stories* (1940–1962), a title that eventually achieved the highest overall circulation of any comic book series in history (Benton 1993, 158). Disney's expanding catalogue of trade characters were not only transmedia figures in that sense, but also "transtextual" ones in that they often crossed over from one title to another within the same medium (Scolari, Bertetti, and Freeman 2014, 4).

Dell's success with the so-called "funny animal genre" led to other licensing coups with rival studios, such as Warner Bros. (Bugs Bunny, Porky Pig, etc.), and by the early 1950s, the publisher had steadily grown into "the largest publisher of comic books in the world" (Benton 1993, 109), well in advance of DC and Marvel's industrial domination. At its peak, *Walt Disney's Comics and Stories* sold in excess of four million copies per issue, a feat even Superman could not manage during his peak, and, as comic historian Mark Carlson emphasized, Dell's market-share may only have amounted to 15 percent of titles published, but it controlled almost a third of the total market and had "more million-plus sellers than any other company before or since" (2005). Clearly, then, the early Disney strips and comic books were essential elements of the studio's early transmedia presence—hardly redundant or solely economically driven, no matter what Walt Disney himself had originally envisioned. As Keith M. Booker explains:

> Studios were among the earliest to successfully exploit the creative and commercial potential of comic art, constructing wonderful worlds of comedy, fantasy, melodrama and adventure that would eventually produce a legacy of immense popularity, incredible financial success, and seemingly limitless imagination.
>
> *(2014, 406)*

As Fredric Wertham's crusade against comic books gathered steam in the early 1950s, leading to the establishment of institutional watchdog, the Comics Code Authority (CCA), as well as increasing competition from television, sales of comic books began to suffer accordingly. In order to face up to the challenge, comic publishers continued to accumulate licensed holdings to extend and augment film and television based properties. As the domestication of television accelerated in the 1950s and 1960s, the comic book industry responded to the changing cultural landscape, quite ironically, with a spate of licensed material based on popular television series and often associated with, but not limited to, the rising tide of "tele-fantasy" genres (Chapman 2011, 104). Perhaps the most interesting example during this period were comic strips and books based on the science fiction series, *Star Trek*, to which I now turn.

Television: The Final Frontier

In the United States, so-called "TV Comics"—basically, comics based on television properties—had multiplied exponentially by the 1960s. Prime-time series, some of which have since become seminal, such as *I Love Lucy*, *Sergeant Bilko*, and *Jackie Gleeson and the Honeymooners*, extended televisual worlds into comics, keeping the brand alive in the interstices between episodes and during the off-season. In 1962, the partnership between Dell Comics and Western Printing had come to an end, with the latter forming the Gold Key Comic imprint with which it continued to produce licensed material (although not as successfully as Dell). In 1967, one year after its debut on NBC, the *Star Trek* license was awarded to Gold Key and eventually became a largely faithful interpretation of its parent series. However, the first issues were marred by a lack of collaboration or dialogue between licensor and licensee(s), leading to the production of a comic curiosity.

Indeed, the first issues were drawn by Italy-based artist Alberto Giolotti without him actually "ever having seen a single frame of the on-air television program" (Clarke 2013, 27), using publicity shots as his chief guide. In the first issue, titled "The Planet of No Return," the *Enterprise* crew battle cannibal plants and giant trees, but this voracious vegetation's true threat is in the spores that they produce to infect hosts. "Giant trees are trying to germinate us!" exclaims Kirk dramatically. Now, while this may be rather silly, it is in the story's final moments that perhaps the most unfaithful interpretation of the program's political bent in the property's history occurs. To combat the threat, Spock recommends that the *Enterprise*'s weapons be used to eradicate all life on "that hideous little globe," an action far removed from the liberal humanistic agenda of Roddenberry's famous vision. As Darius elaborates on this point:

> And that's just what the *Enterprise* does. It burns *the entire planet from orbit*, flying around it and around it, killing and killing until it's exterminated all life. In the Captain's log that closes out the story, Kirk characterizes the mission as one of *"total destruction"*. And we're treated to the image of the *Enterprise*, firing phasers onto the burning surface, where the sentient trees are on fire, helplessly fleeing the eradication of their entire planet.
>
> *(Darius 2014, 31, italics in original)*

Over in the UK, comic books were usually published weekly, as opposed to the US monthly system, and were most often anthology titles wherein multiple stories featured a range of characters and storyworlds, a tradition which continues to this day in titles such as the long-running *The Beano* or influential science-fiction publication, *2000AD*. In the 1950s, the first television-themed anthology comics appeared, such as *TV Comic* (1951–1984) and *TV Fun* (1953–1959), but it was in the following decade that the genre boomed with the likes of *TV Express* (1960–1962), *TV Toyland* (1966–1968), *TV Tornado* (1967–1968) and, most pointedly, *TV Century 21* (1965–1971) (Chapman 2011, 104). The latter title, which was soon abbreviated to *TV21*, was created as a promotional tie-in vehicle for Gerry Anderson's various "Supermarionation" series (*Thunderbirds*, *Stingray*, *Captain Scarlet and the Mysterons*, and *Joe 90*, etc.). Esteemed comic book artist, Frank Bellamy, who had previously worked on Dan Dare for boys' comic, *The Eagle*, was recruited to draw the Thunderbirds and Lady Penelope strips before the series had even aired on British television (Chapman 2011, 105). Within its first year, *TV21* was selling 600,000 copies per week and also received two further spin offs in *Lady Penelope* (1966–1968) and *Joe 90 Top Secret* (1969), the latter of which was merged with *TV21* after only 34 issues. Eventually, *TV21* opened up its pages to other (Non-Andersen) US and UK television properties, including *The Saint*, *Land of the Giants*, and, perhaps the title's most historically fascinating attribute, the inclusion of a strip based on *Star Trek* which would have been British readers' first introduction to the crew of the Starship *Enterprise* given that the series did not air in the UK until 1969.

As with the Gold Key versions, the UK *Star Trek* comic strips are quite singular, thus again indicating the creative gulf between licensor and licensee during the period, yet the strip "proved to be one of the most vibrant and long-lived strips during a turbulent time for British comics titles" (Porter 2014, 36). The UK strips preceded the series' airing by six months and resulted in yet another oddity—as with Gold Key, writers and artists tried to adapt and extend *Star Trek* without even seeing the show. In the British context, Spock was a "central identification character" and was more representative of the British "stiff upper lip" stereotype than the dispassionate Vulcan of the television program. Scotty, meanwhile, was the strip's action hero, while the most egregious portrayal was Captain Kirk who was framed as inadequate, confused, and often a plain idiot, "the least effective command officer in Starfleet" (Porter 2014, 41).

Over the decades since, the *Star Trek* license has been passed from pillar-to-post: Gold Key (who folded in 1984 having lost their most valuable licenses), Marvel, DC, Malibu Comics, WildStorm, Tokyopop and, most recently, IDW, have all held the license to publish *Star Trek* comics across half a century and over 1,000 comic books (not to mention manga variations and graphic novels). Yet despite these comics far outweighing television episodes by a considerable margin in quantitative terms, none of these stories are considered to be (official) canonical extensions. In other words, the *Star Trek* primary text—the canon—is comprised of live-action material only, which, at the time of writing, consists of six television series and thirteen feature films. They are all of a piece. Hence, *Star Trek*'s armada of transmedia expressions, such as tie-in comics, novels, and even *The Animated Series*—which included the vocal performances of Shatner, Nimoy, and so forth, excluding Walter Koenig's Chekov, and featured the return of "classic" series alumni, D. C. Fontana and David Gerrod—are *non-canonical augmentations*, counter-factual narratives that do not represent the "true" continuity. The "real" Captain Kirk did not ever decimate an entire planet of cannibal plants.

This may not have mattered so much in the 1960s given that *Star Trek* was famously canceled by its third season, but the rise of Trek fandom in the early 1970s, leading to the series' resurrection as a blockbuster film franchise with *The Motion Picture* (1979), meant that continuity would become an important feature of imaginary worlds beyond the domain of superhero comics. Indeed, as many scholars have pointed out, serial continuity is one of the most pleasurable aspects associated with world-building for fan audiences (for example, see Reynolds 1992; Kaveney 2008), but, as with Jenkins' idealized fan community discussed earlier, this should in no way be embraced as a homogenous component of "geek" fan cultures. Put another way, whether or not *Star Trek* comics are viewed as "redundant" by fans or if "they storm off in disgust" at the disavowal of continuity conventions is another question entirely and beyond the scope of this chapter. It is worth noting, however, that it cannot be the case that all fans feel this way—for if the comics were not economically viable, it stands to reason that they would surely have been discontinued (which is eventually what would happen with comic strips). By the time *The Motion Picture* re-ignited *Star Trek*'s fortunes in 1979, another science-fiction property had already emerged that would have a tremendous impact on the cultural landscape, echoing across the decades since. That property was a little independent film written and produced by George Lucas.

Star Wars Comics: From Seriality to Sequentiality

In the summer of 1975, the release of Steven Spielberg's *Jaws* sent shockwaves through Hollywood, accumulating over 100 million dollars at the box office, and forging the template for the "event film" phenomenon. But it would be George Lucas' *Star Wars* that would become harbinger of the blockbuster franchise tradition, with all its attendant toys, action figures, spaceships, and merchandise, which have squeezed maximum profits out of the brand over the past 40 years—indeed, so much more than the films themselves. It would only be a matter of time before transmedia tie-in media would follow suit.

In 1977, Marvel obtained the license to publish *Star Wars* comics for a song, primarily because both Lucas and the head honchos at Twentieth Century Fox strongly believed that the film would tank at the box office, and if their predictions proved to be correct, then associated tie-in products would be the least of it all, especially considering that movie tie-in comics, especially those of a science-fiction bent, were no longer the sellers they once were. It was certainly a risk for Marvel—it had gone from industry savior in the 1960s to a struggling publication house defending against the forces of economic recession and stagnation in the early 1970s (as was its main competitor, DC Comics). Indeed, (then) editor Jim Shooter claims that the acquisition of the *Star Wars* license actually saved the publisher from bankruptcy (Proctor and Freeman 2016; Booker 2014, 447).

On April 12, 1977, the first issue of Marvel's *Star Wars* was published, and rapidly ascended the comic book charts almost six weeks prior to the film's release in theatres. This meant that readers would have already been introduced to a cast of characters, such as Darth Vader, R2-D2, C3P0, and Luke Skywalker (much in the same way that the Snow White daily strip would be the threshold into Disney's imaginary world for millions of readers, as discussed above). Written by Roy Thomas and drawn by Howard Chaykin, the first six issues of the series adapted Lucas' *Star Wars*, although reading it now may be rather discombobulating for those familiar with the film; like the *Star Trek* comics, Marvel did not have the luxury of seeing the finished product before starting work on the comic, and it certainly shows. The first issue of the comic sold over a million copies and, once the adaptation was completed, Marvel created new stories, new characters, and new mythological elements. Those working on the title, however, were creatively constrained by Lucas, who mandated that they "could use the main characters, but not infringe on the movies' developments" (Booker 2014, 447). At this point, of course, *Star Wars* had become perhaps the largest "sleeper hit" in Hollywood history and work had already started on what would become *The Empire Strikes Back* (1980).

The licensing arrangement between Lucasfilm and Marvel lasted for 107 issues, but following release of the final installment, *Return of the Jedi* (1983), and with no new film in the pipeline, the *Star Wars* brand entered its interregnum period—also known as "the dark age" in fan communities. In 1989, however, the force would be awakened once again as a new publisher, Dark Horse Comics, acquired the vacant license and the age of the *Star Wars* Expanded Universe (EU) began in earnest.

Founded in 1986, Dark Horse Comics would go on to become the third largest comic book publisher—behind Marvel and DC, of course, both of which have jockeyed for industrial dominance for the best part of six decades—mainly through the acquisition of blockbuster franchise licenses and associated characters, as well as television and video-game properties, which include the likes of: *Aliens*, *Transformers*, *The Terminator*, *Indiana Jones*, *Game of Thrones*, *Buffy the Vampire Slayer*, *The Legend of Zelda*, *Tomb Raider*, and, of course, *Star Wars*. In 1991, Dark Horse published the first issue in the *Star Wars: Dark Empire* series that, along with Timothy Zahn's sequence of sequel novels, catapulted the brand into popular consciousness once again. Between 1991 and 2014, Dark Horse published an armada of expanded universe comics, but, as with *Star Trek*'s tie-in media, these tales are not considered legitimate entries in the *Star Wars* canon, a sore point for many a fan reader, with some turning to the cyberspace chat-rooms of Web 1.0 to argue, debate, and deliberate—often quite vehemently—the finer points of canonical law (for more on the so-called "canon wars," see Brooker 2002, 101–114).

As we have seen throughout this chapter, the licensing system had not yet addressed the lack of continuity between various transmedia iterations and may not have mattered so much until later. As with many franchise licenses (if not all), it would be the film series—the primary text—that maintained its canonical hegemony and, should Lucas decide to create new *Star Wars* films that contradicted expanded universe elements, then they would take precedence and effectively erase them from inclusion altogether. To address such queries about canonicity, Lucasfilm created a hierarchical taxonomy with G-Canon (the "G" standing in for "George") being the apotheosis of official continuity (although the Ewok films, *Caravan of Courage* and *Battle for Endor*, were not granted this status). The *Star Wars* comics, be they from Marvel or Dark Horse, occupied not even the second tier, which only contained *The Clone Wars* animated television series (T-Canon), but the third rung (C-Canon).

In 2014, the new corporate owners of the *Star Wars* brand, Disney, committed hyperdiegetic genocide by declaring that the old licensing system was dead and buried, and that, from here on out, all transmedia *Star Wars* elements, including comics and novels, would be considered canonical, official components of a vast transmedia continuity system. As Matthew Freeman and I (Proctor and

Freeman 2016) have theorized elsewhere, this institutional decree represents a fundamental shift in the *transmedia economy* of *Star Wars*. Hence, the concept of seriality—most often used to detail the spreadability of imaginary worlds, whether or not such elements fit into a cohesive continuity system—gives way to "sequentiality": that is, a transmedia economy developed according to the principle of continuity between and across media.

From April 2014, *Star Wars*' new rulers mandated that the old system be replaced by a flattening of hierarchies and that new comic books (and novels, etc.) would henceforth be legitimately canonical. From a business perspective, this makes a lot of sense, especially considering that the comic license was taken from Dark Horse and awarded again to Marvel, a company that was already within the aegis of Disney. Some fans, however, were less than thrilled at being told that the comics (and novels etc.) that they had spent a long time reading and collecting were officially banished to a netherworld of falsehood and speculation. But it was not simply that these fans were concerned by the exclusion from official canon per se, but that the old EU would be re-branded beneath the "Legends" banner and, more importantly, that there were no plans to continue telling stories in what now effectively amounts to an alternative universe. In 2015, the so-called EU Movement raised funds via Kickstarter to advertise both their displeasure at Lucasfilm and to ask that the "Legends" series be continued. Contra Jenkins, the "old" licensing system matters more to these fans than a newly coordinated transmedia experience whereby everything matters and everything counts.

There have, however, been significant maneuvers and modifications in recent years toward a more closely integrated system of interlocking narrative regardless of media-specificity, a transmedia storytelling of the kind that Jenkins proposed in *Convergence Culture*. This in no way suggests that this shift has been adopted by all and sundry, nor that the old system has now been completely overhauled and replaced by the new. The shifts discussed here are continuing at the time of writing and it would perhaps be better to view the contemporary system as a conflict between old and new, not as a binary, but as a *spectrum of multiplicity* wherein alternative worlds, parallel universes, and counter-factual stories all co-exist with canonical continuity. That is to say that vast franchised narratives possess a range of continuities and several different canonical systems, largely dependent on the choices and positions of each individual reader, and fans are experts in navigating alternative worlds such as these. That said, it largely remains that the primary text, whatever that may be, maintains its power over the array of transmedia satellites orbiting the mothership.

In *Transmedia Television*, M. J. Clarke demonstrates these particular shifts by examining the production contexts relating to the television series *Heroes* (2006–2010) and *24* (2001–2010), and their respective comic book extensions. What is valuable here is the understanding that the contexts within which transmedia comics are produced might radically differ between license holders. The relationship between Fox and IDW regarding the *24* comics is quite different to that between NBC and Aspen's *Heroes'* extensions, but the principal shifts toward an inclusive dialogue between showrunners and subcontractors shares similarities:

> What these two ventures have in common is that they both operate via a combination of freelance creative labour and permanent supervision either closely or loosely affiliated with program producers. In the case of *Heroes*, scripts are drafted by either series writers, staff writing assistants, or freelance writers hired from outside the program staff. Yet, in all these cases, *it is the series writers who oversee and determine the content of each issue.*
>
> *(Clarke 2013, 32, my italics)*

The connective relationship between licensor and licensee—between showrunner and staff writers, on the one hand, and freelance creatives, on the other, serves an important authenticating role and a discursive production of cultural distinction, of "value." The involvement of production personnel, no matter how closely aligned or not, operates to ratify transmedia "micro-narratives" as legitimate installments of a piece with an overarching "macro-structure" (Ryan 1992, 373). This kind of thinking is illustrated by several transmedia comics that extend television series in canonical directions, usually branded as officially sanctioned continuations, such as *Buffy the Vampire Slayer* (subtitled Season Eight though Ten), all of which were either written, co-written, or overseen by Joss Whedon, the ultimate creative authority on all things Buffy-related. The attachment and transposition of television language into comics (the use of "seasons" is telling), as well as the attribution of Whedon's authorship is indicative of the shift in the transmedia economy of licensed comics. For all intents and purposes, the *Buffy the Vampire Slayer* "season" comics are "the definitive version of what happens after the television series ends" and are "absolutely canonical in the views of many fans" (Ford and Jenkins 2009, 305). The appending of authorship works to enhance the comic books' "aura" as authentic rather than illegitimate.

However, as we have seen earlier, fans may not simply and wholeheartedly embrace such extensions as legitimately canonical, especially given that the comic series introduces new generic elements into the mythos, such as Buffy being able to fly. Indeed, as Emma Beddows illustrates, fans may follow the adventures of the Scooby Gang across media, but it is less narrative continuity that concerns them than consistency of character, genre, and tone. Yet despite this flattening of hierarchies (theoretically) dissolving the borders between primary/secondary texts, and the evocation of authorship as a method of authentication and branding, the transmedia hierarchy is often re-ascribed and reaffirmed at the institutional level. In other words, the primary text, be it in film or television, retains its status as the imaginary "center," dislodging the poststructuralist decree that the center does not exist.

Consider IDW's *The X-Files: Season Ten* comic series, which was deemed as a canonical extension of the television series after over a decade in the cultural wilderness following the program's cancellation in 2002. The comics were branded as authentic continuations with showrunner Chris Carter's involvement appending an authorial aura to the project. But when Fox announced that the *X-Files* was to return to television screens in 2016 with what was described as an "event series," consisting of six episodes as opposed to a full-length season of 20-plus episodes, the comic books' status as the canonical "Season Ten" was canceled out and abolished. That is to say that what was once canon can just as easily be revised and re-positioned should the television text call for it, thus demonstrating that the textual, and media, hierarchy retains its primary status (see Proctor 2017).

Conclusion

These marked shifts in the transmedia economy of franchise licensing, from seriality to sequentiality, are not yet firmly embedded as institutional practice across the board. As an ancillary market, comic books continue to be valuable assets for extending not only narrative, but also the profit margins of franchise IP holders. For some fans, what counts as canonical— or not—matters a great deal, and it is by appealing to the fannish desire for continuity that the transmedia economy of licensed comic books has been undergirded by a series of modifications in the twenty-first century. As the relationship between producers and licensed subcontractors grows ever closer and more collaborative, the continuities between primary texts and secondary comics will also benefit from closely monitored orchestration in narrative terms (with the proviso that not all fan audiences care that deeply about canonicity, as demonstrated by the Star Wars EU movement, and the enormous success of Dark Horse's many franchised transmedia comics). Whether or not transmedia storytelling grows into an industry standard by directly appealing to continuity mavens as the pre-eminent factor governing consumption habits and voyages into imaginary worlds, however, remains to be seen.

References

Barrier, Michael. 2007. *The Animated Man: A Life of Walt Disney*. California: University of California Press.

Benton, Mike. 1993. *The Comic Book in America: An Illustrated History*. Dallas: Taylor Publishing.

Bertetti, Paolo. 2014. "Conan the Barbarian: Transmedia Adventures of a Pulp Hero." In *Transmedia Archaeology: Storytelling in the Borderlines of Science Fiction, Comics and Pulp Magazines*, edited by Carlos A. Scolari, Paolo Bertetti, and Matthew Freeman, 15–38. Basingstoke: Palgrave Pivot.

Booker, Keith M. (ed.). 2014. *Comics Through Time A History of Icons, Idols and Ideas*. Santa Barbara: Greenwood.
Brooker, Will. 2002. *Using the Force: Creativity, Community and Star Wars Fans*. London: Continuum.
Carlson, Mark. 2005. "Funny Business: A History of the Comics Business." *Nostalgia Zine* #1. Accessed May 28, 2017. http://archive.li/WyHb.
Chapman, James. 2011. *British Comics: A Cultural History*. London: Reaktion Books.
Clarke, M. J. 2013. *Transmedia Television: New Trends in Serial Network Production*. London: Bloomsbury.
Darius, Julian. 2014. "From Casual Galactic Genocide to Self-Referential Canon: Gold Key's Star Trek and the Evolution of a Franchise." In *New Life and New Civilizations: Exploring Star Trek Comics*, edited by Joseph F. Berenato, 23–36. Illinois: Sequart Organization.
Davis, Blair. 2017. *Movie Comics: Page to Screen/Screen to Page*. New Brunswick: Rutgers University Press.
Duncan, Randy, and Randy Smith (eds.). 2014. *Icons of the American Comic Book: From Captain America to Wonder Woman*. Vol. 1. Santa Barbara: Greenwood.
Fast, Karin, and Henrik Örnebring. 2017. "Transmedia World-building: The Shadow (1931-present) and Transformers (1984-present)." *International Journal of Cultural Studies* 20 (6): 636–652.
Freeman, Matthew. 2016. *Historicising Transmedia Storytelling: Early Twentieth-Century Transmedia Story Worlds*. London and New York: Routledge.
Hills, Matt. 2012. "Torchwood's Trans-Transmedia: Media Tie-ins and Brand 'Fanagement'." *Participations: Journal of Audience and Reception Studies* 9 (2): 409–428.
Jenkins, Henry. 2006. *Convergence Culture: Where Old and New Media Collide*. New York: New York University Press.
Jenkins, Henry, and Sam Ford. 2009. *Spreadable Media: Creating Value and Meaning in a Networked Culture*. New York: New York University Press.
Kackman, Michael. 2008. "Nothing on But Hoppy Badges: Hopalong Cassidy, William Boyd Enterprises and Emergent Media Globalization." *Cinema Journal* 47 (4): 76–101.
Kaveney, Roz. 2008. *Superheroes! Capes and Crusaders in Comics and Films*. London: I. B. Tauris.
Porter, Alan. 2014. "Flaming Nacelles and Giant Snails: The Unique Culture of the British Star Trek Comics, 1969–1973." In *New Life and New Civilizations: Exploring Star Trek Comics*, edited by Joseph F. Berenato, 37–51. Illinois: Sequart Organization.
Proctor, William. 2017. "Canonicity." In *The Routledge Companion to Imaginary Worlds*, edited by Mark J. P. Wolf, 236–245. London and New York: Routledge.
Proctor, William, and Matthew Freeman. 2016. "The First Step into a Smaller World: The Transmedia Economy of Star Wars." In *Revisiting Imaginary Worlds: A Subcreation Anthology*, edited by Mark J. P. Wolf, 221–243. London and New York: Routledge.
Reynolds, Richard. 1992. *Superheroes: A Modern Mythology*. Jackson: University Press of Mississippi.
Ryan, Marie-Laure. 1992. "The Modes of Narrativity and Their Visual Metaphors." *Style* 26 (3): 368–387.
Sabin, Roger. 2003. "Ally Sloper: The First Comics Superstar?" *Image and Narrative* 7. Accessed March 11, 2017. www.imageandnarrative.be/inarchive/graphicnovel/rogersabin.htm.
Santos, Avi. 2015. *Selling the Silver Bullet: The Lone Ranger and Transmedia Brand Licensing*. Austin: Texas University Press.
Scolari, Carlos A., Paolo Bertetti, and Matthew Freeman. 2014. *Transmedia Archaeology: Storytelling in the Borderlines of Science Fiction, Comics and Pulp Magazines*. Basingstoke: Palgrave Pivot.

6
TRANSMEDIA PUBLISHING
Three Complementary Cases

Alastair Horne

Much of the critical attention afforded to transmedia has focused upon its twenty-first century digital manifestations, following Henry Jenkins' (2003) initial association of the phenomenon with our present era of media convergence. Nevertheless, Jenkins (2011) himself has insisted that "historical antecedents for transmedia ... predate the rise of networked computing and interactive entertainment," and Matthew Freeman (2016) has argued for a consciously historicized understanding of the phenomenon in his analysis of the transmedia storyworlds of the early twentieth century.

Here, I explore the publishing industry's participation in transmedia storyworlds across two contrasting periods: our present digital phase, and the analogue era preceding it. I consider three complementary examples, the first two drawn from the past decade, the third an instance of transmedia storytelling spanning more than 50 years. The first, *Endgame*, was launched specifically as a transmedia storyworld; the second, *Harry Potter*, began in print before extending into film and digital media; the third, *Doctor Who*, started life on television before, like its hero, taking on new forms to survive catastrophe. My analysis situates all three within the context of the media practices of their times, exploring the business of transmedia storytelling from a publishing perspective. I derive my definition of "transmedia" principally from the ten points offered by Jenkins in his 2007 blogpost, "Transmedia Storytelling"; by "publishing industry," I mean that business defined, at least until recently, by the production and distribution of words on pages.

Transmedia Publishing in the Industry's "First Digital Decade"

Any understanding of how publishing has engaged in digital transmedia storytelling must be placed within the context of the industry's recent history. If we define the ten years since the launch in 2007 of the Kindle and iPhone—the first popular mass-market e-reading device and consumer smartphone, respectively—as the publishing industry's first "digital decade," then we see that the period has been characterized by a process of experimentation followed by retrenchment. Early attempts at developing digital content designed specifically to take advantage of the affordances offered by mobile electronic devices, such as interactivity and the ability to display multimedia, produced some notable critical and commercial successes, particularly from Faber & Faber. Their iPhone app *Malcolm Tucker: The Missing Phone*—a transmedia extension of the television satire *The Thick of It*—was the first app to be nominated for a BAFTA (British Academy of Film and Television Arts) award (Richmond 2011), while the digital edition of Eliot's poem *The Waste Land* recouped its costs within six weeks of publication (Dredge 2011).

More recently, however, mainstream publishers have increasingly turned their backs on digital experimentation to refocus their attention upon their core business of placing words on pages. The difficulties involved in recovering the high costs of developing digital content appear to have been the decisive factor in many such decisions. Developer Touchpress, which collaborated with publishers on several apps including *The Waste Land*, exited the market in 2015, its Chief Executive Officer asserting that "it's a challenge to explain to consumers why our apps are worth paying for when the majority of apps are free" (Page 2015). Few publishers now produce content specifically for digital devices, and even ebooks, essentially print works delivered digitally, have experienced declining sales (Tivnan 2016; Publishers Association 2017).

The publishing industry's recent forays into transmediality should therefore be understood against this background of unrewarded experimentation and consequently growing conservatism. Though advocates of transmedia publishing have argued in industry forums that the rising popularity of smartphones and tablets will lead to greater appetite for transmedia stories from readers (Celaya 2011), and that publishers should therefore innovate in such areas (Goerke 2015), they have also frequently acknowledged the challenges that publishers face in developing transmedia franchises—namely, that creating and promoting films and games is significantly more costly than for books, with little chance of success (Ramadge 2016; Celaya 2011). Attempts to resolve these problems have received little interest: production studio Kazap received only 11 pledges of support via crowdfunding site Indiegogo to extend its Transmedia Story Stream platform so that authors could build their own multimedia storyworlds (Snyder 2014a, 2014b).

Endgame

Thus, publishers' attempts at launching new transmedia franchises this past decade have tended to demonstrate the challenges more than the opportunities; the industry press for the period features more high-profile launches than longer-term successes. *Endgame*, launched in 2014 as an ambitious "innovative omni-platform endeavour" (Digital Book World 2014), by a coalition comprising author James Frey, co-writer Nils Johnson-Shelton, Frey's media and production company Full Fathom Five, publisher HarperCollins, developers Niantic Labs, and film studio 20th Century Fox, offers an instructive example. Up to a point, this alliance offers a textbook example of the "media consolidation" identified by Jenkins (2007) as a key factor in the economic logic of transmedia publishing. HarperCollins was formed when publishers Collins and Harper & Row were acquired by Rupert Murdoch's media conglomerate News Corporation in the late 1980s and then merged. 20th Century Fox is also owned by Murdoch, currently as part of the 21st Century Fox company spun off from News Corporation in 2013. Both author James Frey and his company Full Fathom Five have connections with Murdoch's properties, the former publishing a novel, *Bright Shiny Morning*, with HarperCollins, and the latter developing properties with both HarperCollins and Fox (Full Fathom Five n.d.). Only developer Niantic Labs stands entirely outside this example of horizontal integration; at the project's 2014 launch it was a start-up venture within Google, though it became a separate company the following year.

Endgame's planned components included a trilogy of young adult novels, each accompanied by an interactive puzzle and prize, 15 ebook novellas, an augmented reality app, several YouTube videos, and a series of film adaptations (Digital Book World 2014). Analysis of Goodreads (n.d.) and the HarperCollins website (n.d.) suggests that all three novels and the first nine novellas have been published, to diminishing returns, even given that later installments have been available for less time. *The Calling*, the first novel, published in 2014, has approximately 13,000 ratings on Goodreads and 2,000 reviews; the third, *Rules of the Game*, published in late 2016, has garnered only one tenth of that attention in a third of the time—a mere 1,370 ratings and 215 reviews; several of the ebook-only novellas have no reviews and fewer than ten ratings. Of the games, the first comprised a series of puzzles divided between the opening novel and the internet; a prize of $500,000 for anyone solving them all was claimed shortly before the second novel's launch (Heine 2015). The second game, the augmented reality app, *Endgame: Proving Ground*, using the platform Niantic had developed for their game *Ingress*, was apparently abandoned before the release of an invitation-only beta, though footage remains available on a YouTube channel set up to host video content from the original transmedia experience (Ancient Societies 2015). Speculation on the *Endgame* Reddit community (DoctorLemonPepper 2016) blames the abandonment variously on Niantic's separation from Google, and the company's decision to focus on developing the *Pokémon Go* game instead. Nothing seems to have been heard of the film in the three and a half years since it was announced.

One factor, it seems, in the failure of the *Endgame* franchise to match the success of its most obvious inspiration, Suzanne Collins's *Hunger Games* trilogy, may be its derivativeness. The most popular Goodreads review of *The Calling* was written several months before the book itself was published, accusing it of being "the victor of the 75th Huge ... Rip-off Games!" (Gillian 2014). Its attempt to create a genuinely transmedial experience within a *Hunger Games*-style storyworld by giving readers the opportunity to engage in similar challenges to those experienced by the novels' protagonists was principally hamstrung by Niantic's withdrawal from the project, though. The enormous success of the *Pokémon Go* game the studio released instead—10,000,000 downloads within its first week (Molina 2016)—demonstrated the huge potential audience for a location-based augmented-reality game allied with the right intellectual property. But without the game's support, the second novel failed to match the success of the first, whose prize competition helped build a significant media profile; the third received even less interest. The story underlines the importance of the synergies that media conglomerates bring to transmediality: since Niantic Labs was the one part of the coalition neither owned by nor closely connected to Rupert Murdoch's media empire, it therefore had far less incentive to stay with the project. Without that component, even a publisher the size of HarperCollins, part of the enormous Murdoch media empire, was unable to make a success of its transmedia-native franchise.

Pottermore

Attempts to launch franchises that are transmedial from conception are relatively new to publishing, which has tended historically to develop such properties either by licensing content from other channels, most commonly film or television (as in the case of *Doctor Who*), or by extending into other media a property originating in print. The most successful contemporary example of the latter type is *Harry Potter*. According to the Bloomsbury website, more than 450 million copies of the *Potter* novels have been sold worldwide (Bloomsbury n.d.); CNBC estimated that the entire franchise was worth $25bn, with book sales, cinema receipts, and toy sales each contributing between $7 and $8bn to the total (Wells and Fahey 2016). While these brand extensions were at first mere adaptations in Jenkins' (2011) terminology, over the past decade they have developed into genuinely transmedial components: the *Pottermore* website launched in 2012; the first in a series of film prequels, *Fantastic Beasts and Where to Find Them*, appeared in 2016, adapted from a non-narrative bestiary published in 2001; and the first stage play, *Harry Potter and the Cursed Child*, opened in July 2016, its script becoming the best-selling playscript since records began (O'Brien 2016).

Of these, perhaps the most interesting element is *Pottermore*, which is not just an instance of a franchise originating in print extending into digital media, but also an unusual example of an author, rather than a publisher, maintaining control over a franchise originating in print. Crucial to any understanding of the site is the fact that, exceptionally for an author, Rowling retained the rights to publish the digital editions of the *Harry Potter* novels herself, rather than licensing them to her print publishers (Solon 2011). This, combined with the enormous success of the franchise, and her consequent wealth, placed her in an exceptionally powerful position: she had a phenomenally desirable product, the opportunity to sell it directly to consumers, and the money to make it happen. *Pottermore* was therefore created as a means of doing so, serving simultaneously as an online retailer and a transmedia extension of the Potter storyworld, where users were able to read new stories by Rowling and experience a virtual Hogwarts; the transmedia content also acted as marketing materials for the ebooks. This may account for the fact that, as Brummitt (2016, 118–119) notes, the original iteration of the website focused almost exclusively upon Rowling's original writing, rather than its adaptations. Its adoption of a broader focus from September 2015, promoting content across several aspects of the franchise, follows a decline in ebook sales as initial demand was satisfied, a fall in site turnover partly attributable to the end of a licensing deal with Sony, and the decision to allow ebook retailers to sell the Potter novels directly (McLaughlin 2016). (Previously, retailers could list the ebooks in their stores but had to redirect buyers to *Pottermore* to complete their purchase; unprecedentedly, the likes of Amazon agreed to do so.)

Pottermore, then, is the product of an exceptional combination of circumstances. Publishers, prudently husbanding their resources between multiple brands, cannot justify spending £8m on a website promoting a single property; few even allow consumers to buy ebooks from their own sites, directing them instead to retailer sites to complete their purchases. None can bend Amazon to its will in the way that Rowling did. Though extending a franchise originating in print into other media remains a viable and popular mode of transmedia storytelling, publishers tend not to develop their properties themselves, relying instead upon partnerships with other media organizations. *Pottermore*'s change of focus post-2015 suggests that even a brand as powerful as *Harry Potter* struggles to go it alone.

Accidental Transmedia: Publishing and *Doctor Who*

As Freeman (2016) and Jenkins (2011) have both argued, versions or models of transmediality have existed before our present age of digital content and media conglomerates. In this section, I explore the changing role played by publishing within a franchise stretching from the analogue mid-twentieth century to our present digital moment. *Doctor Who* is often cited as an example of transmedia storytelling: Jenkins (2007), for instance, illustrates that brand extensions may serve various functions by referencing the radio dramas produced by the BBC (British Broadcasting Corporation) while the programme was off-air. However, much analysis of *Doctor Who* as transmedia focuses on digital content generated after the show returned to television in 2005: Perryman (2008), for instance, considers the transmedia strategies employed by the BBC in relaunching the programme; Hills (2016) how *Doctor Who* functions within the *Lego Dimensions* universe; and Freeman (2016, 198) how online-only episodes such as *The Night of the Doctor* reinforce hierarchies between old and new media.

My interest lies, however, in the long history of published *Doctor Who*: from the novelizations of on-screen stories beginning in the mid-1960s but mostly produced in the 1970s to 1980s by Target Books, via the original fiction published Virgin and BBC Books during the period between 1989 and 2005 when the programme was mostly off-air; and finally, the novels produced alongside the revived, post-2005 series. I shall explore how the relationship between the television series and published *Doctor Who* novels has developed from Jenkins' (2007) licensing model of transmediality, where extensions of a story into subsequent media are said to remain subordinate to the original "master text," into something more complex and organic.

The 1960s: Adaptation

The *Doctor Who* storyworld has encompassed an enormous range of platforms in its 54-year history, including comic strips, books, films, stage plays, audio dramas broadcast on the radio or for sale as CDs and downloads, games, animations, fan-made spin-off films, and several official spin-off television series. Launched in November 1963 as a television series, the program began extending into other media the following year. The first *Doctor Who* comic strip appeared in *TV Comic* on November 9, 1964; the first book, a retelling of the second episode of the television program, retitled *Doctor*

Who in an Exciting Adventure with the Daleks, was published five days later. The following August, the first *Doctor Who* film appeared, a second reworking of that same Dalek story, and in September the first *Doctor Who* annual was published, containing the first original prose fiction set within the larger universe.

These ventures enjoyed varying levels of success. *Doctor Who* comic strips have appeared almost uninterruptedly since 1965 in publications including *Countdown*, *TV21*, and, since 1979, the various incarnations of the official *Doctor Who Magazine*; currently, Titan Comics also produces regular comics devoted to the four most recent Doctors. Two further novels appeared in 1966, again reworkings of stories originally seen on screen, but it would be another eight years before a fourth book appeared. A second *Doctor Who* film, an adaptation of the next television story to feature the Daleks, was also released in 1966, but the option for a third was not taken up by producer Milton Subotsky (Haining 1988, 109). And *Doctor Who* annuals were published almost every year from 1965 to 1986, resuming publication when the television series returned in 2005.

The extent to which these initial forays into other media conform to accepted definitions of "transmedia" varies considerably. Though Christy Dena (Chapter 21, this volume) complicates the notion of a any kind if simple distinction between adaptation and transmedia storytelling in this very volume, on this instance the comic strips and the stories published in the annuals perhaps come closest to Jenkins' "unified and coordinated entertainment experience" (2007), extending the *Doctor Who* storyworld through original stories rather than reworkings of televisual adventures, and so diverging from the television version of the universe without explicitly contradicting it. Licensing agreements tend to restrict which characters from the television series feature in these new adventures, and though several television companions make appearances in the *TV Comic* strip between 1968 and 1971, and alien races the Zarbi, Voord, and Sensorites can all be found in the first *Doctor Who* annual, generally these media tend to generate their own characters, planets, and adventures. So, for his first three years in comic-strip form, the Doctor is accompanied by his grandchildren John and Gillian, rather than his television companions. Though fitting these stories into the universe established by the television series is not always easy, it is rarely logically impossible.

Contrastingly, the first *Doctor Who* novels and the feature films function rather as parallel or alternate storyworlds, sharing certain aspects with the television series but diverging markedly in others. The stories they tell are broadly the same as those told on television, casting them as adaptations in Jenkins' (2011) terminology, but there are certain significant differences between versions. For instance, the first novel, *Doctor Who in an Exciting Adventure with the Daleks*, retains the Doctor's televisual companions, teachers Ian and Barbara, and the Doctor's grand-daughter Susan, but brings them together in a way that contradicts their introduction on television, which had occurred in the story preceding this one; to fit this new introduction, it also changes Ian's occupation from teacher to scientist. Similarly, the first film changes the relationships between the Doctor's companions, while the second replaces them entirely. Each time, deliberate effort has been made to ensure that these stories in alternative media are not dependent upon a knowledge of the television series; by reintroducing the main series protagonists in ways that contradict the original series, they set themselves as removed from that series, attempting to create their own self-contained storyworlds, derivative from but not dependent upon the television series.

Their reasons for doing so are closely tied to the nature of media consumption in the 1960s, specifically the absence of video recording equipment from most homes. With very few exceptions—only eight episodes over seven years—episodes of *Doctor Who* broadcast in the 1960s were seen only once, on their date of transmission; even after repeats became more regular in the 1970s, they would involve only one or two stories per series: approximately eight episodes from a run of around 26. No television story would be made available on video until 1983, and so these episodes were not available and present to fans as they are now; the films and novelizations needed to be capable of standing alone.

The three *Doctor Who* novelizations were less successful than had been hoped, and no more were published until 1973, when a new children's imprint, Target Books, acquired the rights to publish paperback editions of the 1960s books, and then to produce their own series of adaptations of the television adventures. Seven new novelizations were published the following year, mostly featuring the then current Doctor, Jon Pertwee. Though the titles of several stories were changed to make them more attractive to readers, these were largely faithful renderings of what viewers might have seen on television, with few of the revisions that characterized *Doctor Who in an Exciting Adventure with the Daleks*. Essentially adaptations, according to Jenkins' (2011) terminology if not Dena's (Chapter 21, this volume), the novelizations nonetheless make occasional contributions to the television programme: notably, the device that (theoretically) enables the Doctor's TARDIS to change its appearance to match its surroundings was first identified as the "chameleon circuit" in Terrance Dicks' 1975 novelization *Terror of the Autons*, six years before the same name would be used on screen in the fourth Doctor's final adventure.

The Target Books series benefitted from a close relationship with the team behind the television program. Terrance Dicks, who wrote more than 60 of the novelizations and acted as an unofficial series editor (Howe 2007, 19), had served as script editor for the television series between 1968 and 1974. Many of the other books were written by the same writers who had created the original television scripts, who often took the opportunity to flesh out those scripts or to include elements that had not been achievable within the budget of the television show, exemplifying Jenkins' (2011) acknowledgment that adaptations can be transformative of their originals.

New Adventures in Time and Space

In 1989, the BBC cancelled the *Doctor Who* television series. With Target Books publishing a new novelization most months, and—from 1986—only four new stories being broadcast each year, fewer than 20 stories remained to be adapted by the time the final episode was broadcast. Most of these, moreover, were published within the next 12 months, with financial issues—primarily the low advances paid by Target Books—rendering the others unachievable (Howe 2007, 115). With no more television stories to adapt, then, Virgin Publishing, which had bought Target Books in 1989, obtained a license from the BBC to produce original fiction within the *Doctor Who* universe. Target Books had already published small amounts of new fiction within the *Doctor Who* storyworld in the second half of the 1980s, in the form of two novels featuring the Doctor's companions, *Turlough and the Earthlink Dilemma*, and *Harry Sullivan's War*, and adaptations of three unproduced television scripts were also published.

Thus, the moment at which the *Doctor Who* property became more definitively a transmedial storyworld, and the publishing component of that storyworld began to play a more active role within it, came as the result not of the synergistic economics of media consolidation that Jenkins (2007) identifies as a characteristic of contemporary transmedia storytelling. Rather, it resulted from the BBC's abdication of responsibility for its own intellectual property, and the commercial imperatives of an entirely separate and smaller-scale business, Virgin Books. With the television version of *Doctor Who* off-air, the published version took the lead. Virgin started publishing its *New Adventures* series in 1991, a collection ultimately comprising more than 60 novels over six years featuring Sylvester McCoy's seventh Doctor, the on-screen Doctor at the time of the program's cancellation, and initially also his companion, Ace. From 1994, the *New Adventures* were complemented by a second series of novels: 33 *Missing Adventures*, featuring previous incarnations of the Doctor and his on-screen companions, were published over three years.

With the series off-air, the novels were no longer subordinate to the program's "master text" (Jenkins 2007) and could therefore contribute their own original elements to the *Doctor Who* universe without fear of contradiction. In some respects, these novels mirror the more established comic strip's relationship with the *Doctor Who* universe, free to create its own companions and alien races.

Where they diverge from the model established by the comic strip, however, is in their development of the *Doctor Who* storyworld beyond what had been seen on television: the characters of both the Doctor and Ace are transformed through story arcs that take them far from their television originals. Ace leaves the Doctor to fight Daleks, and returns as a more cynical character, while the depictions of both the Doctor and his home planet Gallifrey developed them considerably beyond what was seen on television.

Notably, several of the authors of these new *Doctor Who* novels were writers connected to the television series: Marc Platt and Ben Aaronovitch had written stories for McCoy's Doctor, while Andrew Cartmel had been script editor when it was cancelled. Some of the *New Adventures* implement story arcs that Cartmel had planned for future series of the television show (Parkin 2007, 380), adding weight to the idea that this series of novels acts as the true continuation of the television programme. Other authors had written for the Target series of novelizations, like John Peel and Nigel Robinson. Terrance Dicks, author of the second of the *New Adventures* novels titled *Timewyrm: Exodus*, combined both these elements, having written for the television series, served as its script editor for six years, and having written more than 60 novelizations. Many of the writers, though, came from the fan community, as the participatory culture of *Doctor Who*—embodied in the fan fiction that had been published in magazines such as the Doctor Who Appreciation Society's *The Celestial Toyroom* since the 1970s—began to assume a more central and formally authoritative role within the *Doctor Who* storyworld, prompted at least partly by the difficulty in attracting more established writers due to the low advances offered by Virgin. *Timewyrm: Revelation*, for instance—Paul Cornell's first *New Adventures* novel—had begun life as a piece of fan fiction serialized in the fanzine *Queen Bat*; Marc Gatiss, Gareth Roberts, and Russell T. Davies were also fans who came to write for the Virgin series of novels.

In 1996, *Doctor Who* returned to television with the pilot episode for a proposed new series that signalled its continuity with the original television version by featuring the regeneration of McCoy's seventh Doctor into Paul McGann's eighth incarnation. Though no new television series resulted, the episode nonetheless marked a further shift in the balance of power within the *Doctor Who* storyworld, as the BBC withdrew Virgin's license to publish original *Doctor Who* fiction and began in 1997 to produce its own equivalents to the two Virgin series, using many of the same writers: the *Eighth Doctor Adventures* featured McGann's Doctor alongside a range of new companions, while the *Past Doctor Adventures*, like the *Missing Adventures*, featured previous incarnations of the Doctor, now including McCoy, mostly alongside their original television companions. Virgin Books, meanwhile, began to publish a new series of adventures based around Bernice Summerfield, a companion they had introduced in the *New Adventures* series, thus extending the *Doctor Who* storyworld beyond the BBC's control. Audio publisher Big Finish then began producing adaptations of the Summerfield stories before obtaining a license from the BBC in 1999 to create its own original audio adventures featuring the Doctor.

Participatory Culture Assumes Control

In 2005, *Doctor Who* returned again to television with a new series instigated by Russell T. Davies, a long-term fan who had written a novel for the Virgin *New Adventures* series; among his team of writers on the series were fellow *New Adventures* writers Mark Gatiss and Paul Cornell, Big Finish author and fan Rob Shearman, and Steven Moffat, another fan, who had written a parody episode of the programme, *The Curse of Fatal Death*, that had appeared as part of the BBC's programming supporting the Comic Relief charity. Harvey (2015, 24–25) has noted, with reservations, that such movement of writers from spin-off media to the original franchise is relatively uncommon, but this occurrence goes beyond movement between media. Though this was not quite the first time that fans had written for the television version of *Doctor Who*—18-year-old fan Andrew Smith contributed a story, *Full Circle*, to the show's eighteenth season—it did mark the beginning of a period in which the show was run by the very fans who had watched its original episodes and then written their own fictions within its storyworld: the participatory culture of fan fiction had assumed control over the franchise.

The new series' participation in digital forms of transmediality has been addressed elsewhere (Perryman 2008; Hills 2016; Freeman 2016), but its increasing willingness to engage with its print and audio versions in this period is also worthy of analysis. The new series borrows the idea of a "time war," whose aftermath provided the context for its first few new series before assuming center-stage in the fiftieth anniversary special, from the *Eighth Doctor Adventures* series, though Davies (2005) has stated that the two wars are not the same. And in an inversion of the practices of the 1960s to 1980s, several television stories from this period are adaptations of works from other media: Cornell's 2007 television story, *Human Nature*, featuring David Tennant's tenth Doctor, was an adaptation of his 1995 *New Adventures* novel of the same name, featuring McCoy's seventh; Gareth Roberts's episodes *The Shakespeare Code* and *The Lodger* both rework his comic strips for *Doctor Who Magazine*, while Rob Shearman's *Dalek* draws upon his audio adventure, *Jubilee*. These adaptations create further problems for anyone attempting to establish a consistent storyworld for *Doctor Who* across its various media, but they also demonstrate how the publishing component of the *Doctor Who* franchise has grown from being subordinate to the television program to become a significant creative contributor. For example, the *Eighth Doctor* and *Past Doctor* series of novels published by BBC Books were replaced shortly after the series' return to television by several new ranges of novels featuring the current Doctor at the time of publication and characterized by a close relationship with the television series. There was, however, no resumption of the novelization program, until the announcement in 2017 that adaptations of four stories from the post-2005 era would be published by BBC Books in 2018 (Fullerton 2017). The demand for such books, given that new televised stories are available on streaming services and released for sale on DVD and Blu-ray within months of their television airing, remains to be seen, but their existence lends this narrative a pleasing circularity.

Conclusion

These three examples of the publishing industry's participation in transmedia storyworlds offer contrasting perspectives on the challenges and opportunities for the industry. The decline of the *Endgame* franchise most pertinently demonstrates the difficulties faced by publishers in attempting to launch their own transmedia franchises: dependence on external partners can prove fatal to even a well-established publisher embedded within a massive media conglomerate. And while the success of *Harry Potter* shows that the industry is still capable of originating properties that can become massive transmedia franchises, the case of *Pottermore* reminds us that the development of those properties can prove a challenge beyond the capabilities of the publishers that helped originate them. The long history of *Doctor Who*, however, in all its contingency and unplanned nature, demonstrates how the contribution made by the publishing industry to a franchise originated elsewhere has proved capable not only of reinvigorating that franchise but even catalysing its development into a more genuinely transmedial storyworld, recalibrating the roles played by its different components and empowering its fans. As such, it may well offer a more positive model for the publishing industry's future efforts in this sphere.

References

Ancient Societies. 2015. "March 28th Live Stream from Pasadena." *YouTube*, March 28. Accessed April 14, 2017. www.youtube.com/watch?v=5GP-uPYVv1U&feature=youtu.be&t=45m22s.

Bloomsbury. n.d. "J.K. Rowling." *Bloomsbury Publishing*. Accessed April 4, 2017. www.bloomsbury.com/author/jk-rowling.

Brummitt, Cassie. 2016. "Pottermore: Transmedia Storytelling and Authorship in Harry Potter." *The Midwest Quarterly* 58 (1): 112–132.

Celaya, Javier. 2011. "Transmedia: A New World of Opportunity for Authors and Publishers." *Publishing Perspectives*, December 9. Accessed April 12, 2017. https://publishingperspectives.com/2011/12/transmedia-opportunity-for-authors-and-publishers/.

Davies, Russell T. 2005. "Production Notes: The Evasion of Time." *Doctor Who Magazine* 356: 66–67.

Digital Book World. 2014. "HarperCollins Partners with Digital Book Award-Winning Niantic Labs on New Transmedia Project." *Digital Book World*, January 15. Accessed April 14, 2017. www.digitalbookworld.com/2014/harpercollins-partners-with-digital-book-award-winning-niantic-labs-on-new-transmedia-project/.

DoctorLemonPepper. 2016. "What Happened?" Message posted to www.reddit.com/r/Endgame/. Accessed April 14, 2017.

Dredge, Stuart. 2011. "The Waste Land iPad App Earns Back its Costs in Six Weeks on the App Store." *Guardian*, August 8. Accessed April 12, 2017. www.theguardian.com/technology/appsblog/2011/aug/08/ipad-the-waste-land-app.

Freeman, Matthew. 2016. *Historicising Transmedia Storytelling: Early Twentieth-Century Transmedia Story Worlds*. London and New York: Routledge.

Full Fathom Five. n.d. "About Us." *Full Fathom Five*. Accessed April 1, 2017. http://fullfathomfive.com/about/.

Fullerton, Huw. 2017. "Steven Moffat and Russell T Davies are Writing Special Doctor Who Novels." *Radio Times*, November 15. Accessed April 11, 2017. www.radiotimes.com/news/tv/2017-11-15/doctor-who-books-steven-moffat-russell-t-davies/.

Gillian. 2014. "The Calling." *Goodreads*, January 14. Accessed April 13, 2017. www.goodreads.com/review/show/823646871.

Goerke, Jerome. 2015. "Publisher or Author? Whose Job Is it to Innovate Anyway?" *Publishing Perspectives*, August 13. Accessed April 6, 2017. https://publishingperspectives.com/2015/08/publisher-or-author-whose-job-is-it-to-innovate-anyway/.

Goodreads. n.d. "James Frey." *Goodreads*. Accessed April 14, 2017. www.goodreads.com/author/show/822.James_Frey.

Haining, Peter. 1988. *Doctor Who: 25 Glorious Years*. Chatham: W. H. Allen.

HarperCollins. n.d.. "Discover Author James Frey." *HarperCollins Publishers*. Accessed April 16, 2017. www.harpercollins.com/cr-103938/james-frey.

Harvey, Colin B. 2015. *Fantastic Transmedia: Narrative, Play and Memory across Science Fiction and Fantasy Storyworlds*. Basingstoke: Pan Macmillan.

Heine, Rachel. 2015. "Global Literary Puzzle Solved, Winner Takes Home 500K in Gold." *Nerdist*, October 1. Accessed April 1, 2017. https://nerdist.com/global-literary-puzzle-solved-winner-takes-home-500k-in-gold/.

Hills, Matt. 2016. "LEGO Dimensions Meets Doctor Who: Transbranding and New Dimensions of Transmedia Storytelling?" *Icono* 14: 8–29. doi: 10.7195/ri14.v14i1.942.

Howe, David J. 2007. *The Target Book*. Tolworth: Telos.

Jenkins, Henry. 2003. "Transmedia Storytelling." *MIT Technology Review*, January 15. Accessed April 3, 2017. www.technologyreview.com/s/401760/transmedia-storytelling/.

Jenkins, Henry. 2007. "Transmedia Storytelling 101." *Confessions of an Aca-Fan*, March 21. Accessed April 26, 2017. http://henryjenkins.org/blog/2007/03/transmedia_storytelling_101.html.

Jenkins, Henry. 2011. "Transmedia 202: Further Reflections." *Confessions of an Aca-Fan*, July 31. Accessed April 14, 2017. http://henryjenkins.org/blog/2011/08/defining_transmedia_further_re.html.

McLaughlin, Marty. 2016. "Harry Potter Digital Arm Loses £6m and Sheds Jobs." *The Scotsman*, January 17. Accessed April 12, 2017. www.scotsman.com/lifestyle/culture/books/harry-potter-digital-arm-loses-6m-and-sheds-jobs-1-4003232.

Molina, Brett. 2016. "'Pokémon Go' Fastest Mobile Game to 10M Downloads." *USA Today*, July 20. Accessed April 26, 2017. www.usatoday.com/story/tech/gaming/2016/07/20/pokemon-go-fastest-mobile-game-10m-downloads/87338366/.

O'Brien, Keira. 2016. "Cursed Child Racks Up Biggest Single-week Sales Since Deathly Hallows." *The Bookseller*, August 9. Accessed April 2, 2017. www.thebookseller.com/news/eighth-harry-potter-book-racks-biggest-single-week-sales-deathly-hallows-371826.

Page, Benedicte. 2015. "Touchpress Pivots Business, Selling Education Apps." *The Bookseller*, November 10. Accessed April 12, 2017. www.thebookseller.com/news/touchpress-pivots-business-selling-education-apps-316108.

Parkin, Lance. 2007. *A History: An Unauthorized History of the Doctor Who Universe*. 2nd ed. Iowa: Mad Norwegian Press.

Perryman, Neil. 2008. "Doctor Who and the Convergence of Media: A Case Study in Transmedia Storytelling." *Convergence* 14 (1): 21–39.

Publishers Association. 2017. "UK Publishing Has Record Year Up 7% to £4.8bn." April 26. Accessed April 14, 2017. www.publishers.org.uk/media-centre/news-releases/2017/uk-publishing-has-record-year-up-7-to-48bn/.

Ramadge, David. 2016. "Turning a Book into a Global Entertainment Franchise." *The Bookseller*, May 30. Accessed April 11, 2017. www.thebookseller.com/futurebook/how-turn-book-entertainment-franchise-330282.

Richmond, Shane. 2011. "Malcolm Tucker iPhone App Nominated for Bafta." *Daily Telegraph*, April 27. Accessed April 10, 2017. www.telegraph.co.uk/technology/mobile-phones/8475033/Malcolm-Tucker-iPhone-app-nominated-for-Bafta.html.

Snyder, Karen. 2014a. "Transmedia Story Stream: Don't Just Read a Book–Play It!" *Bleeding Cool*, January 18. Accessed April 9, 2017. www.bleedingcool.com/2014/01/18/transmedia-story-stream-dont-just-read-a-book-play-it/.

Snyder, Karen. 2014b. "Hello Transmedia Visionaries." *Indiegogo*, February 17. Accessed April 9, 2017. www.indiegogo.com/projects/transmedia-story-stream-don-t-read-a-book-play-it#/updates/all.

Solon, Olivia. 2011. "J. K. Rowling's Pottermore Details Revealed: Harry Potter e-books and More." *Wired*, June 23. Accessed April 14, 2017. www.wired.com/2011/06/pottermore-details/.

Tivnan, Tom. 2016. "E-book Sales Abate for Big Five." *The Bookseller*, January 29. Accessed April 12, 2017. www.thebookseller.com/blogs/e-book-sales-abate-big-five-321245.

Wells, Nick and Mark Fahey. 2016. "Harry Potter and the $25 Billion Franchise." *CNBC*, June 22. Accessed April 8, 2017. www.cnbc.com/2016/10/13/harry-potter-and-the-25-billion-franchise.html.

7
TRANSMEDIA GAMES
Aesthetics and Politics of Profitable Play

Helen W. Kennedy

The chapter presents a detailed examination of what I refer to as the "transmedia game/play continuum" drawing on various examples and case studies of transmedia games from the mainstream to the independent. I apply elements of "play" theory to the study of these transmedia games and revisit the much-debated narratology versus ludology trajectory which shaped the field of games studies as it emerged in 2001/2002. I then expand the notion of the "ludification of culture" and introduce the concept of "ludoaesthetics" as a lens through which transmedia play can be identified and more closely examined.

Transmedia Games

The transmedia game object in its most limited sense is a relatively straightforward media adaptation or expansion from film to game, television to game, or book to game, and are most frequently packaged as a platform-based or console-based format. The most dominant form of this instance of the transmedia game are those which are adapted from traditional Hollywood film. Examples of this "brand" of transmedia game would include the AAA franchise-based games of *Star Wars*, *Harry Potter*, and *Pirates of the Caribbean*—games which were all adapted from film-based intellectual property. There are of course those films which have themselves been adapted from computer games (such as *Tomb Raider*, *Silent Hill*, *Assassin's Creed*, and *Resident Evil*) where the playable experience is subject to an enhanced narrativization and cinematic rendering, but these are not examples considered within the definition being discussed here.

Transmedia games are characterized by an adaptation aesthetic, a process through which the images and characters from the film become playable and/or navigable. The shift in subjectivity from viewer to player is critical to understanding the distinctive nature of transmedia games. The definition and classification of the transmedia game in this sense thus focuses on the extent to which the text (or as we shall see—genre) is repurposed as a form of game or play which allows for a different perspective or level of engagement with the original text. This notion of a transmedia game does not necessarily accord with the conventional definition of transmedia storytelling (Jenkins 2006), since in these instances, these games do not necessarily offer their own distinct or unique elements of the storyworld to contribute to a unified whole.

In my use of the transmedia definition, I am shifting the focus from the text itself to the manifesting game form or the game mechanic that is being generated for the newly configured audience—the player. The text/story/images and to some extent the places of the film or the novel or the television

Table 7.1a Transmedia game genealogy

Board/table-top games/ *Dungeons & Dragons*, etc.	AAA console/ platform adaptations (most commercial)	Web games	Social media games	Live Action Role Playing Games	Street games	Alternate Reality Games	Experimental/ art games (least commercial)

Table 7.1b Transmedia game to transmedia play continuum

AAA console/ platform adaptations (most commercial) (game)	Web games	Social media games	Board/table-top games/ *Dungeons & Dragons*, etc.	Street games	Alternate Reality Games	Live Action Role Playing Games	Experimental/ art games (least commercial) (play)

show become "playable" and potentially navigable (or open to exploration), allowing for a different subjectivity and an alternate mode of engagement. These instances appeal to an expanded market and are clearly functioning within a structure of an expanded commodification of audience pleasures in extended engagement with particular well-loved texts or story worlds.

Table 7.1a shows the various instances of transmedia game "objects" that span a commercial/experiential continuum—with the exception of the board game and table-top games which have their own commercial axis such that on the one hand you have profit driven film/book overlays applied to board game classics such as a *Wizard of Oz* or *Star Wars* versions of the traditional Monopoly board game for instance. On the other hand, there are the influential subcultural non-profit *Dungeon & Dragons* table-top games.

Table 7.1a is also temporally and genealogically configured—with instances of transmedia games listed in the order (from left to right) in which they have manifested or gained prominence in the cultural domain. For instance, board games pre-date the introduction of computer platform/console games, web-based games pre-date social media games and so on. There are obviously some overlaps here as experimental or art games existed before the advent of computer games within movements such as Fluxus in the 1960s or the earlier dadaist absurdist practices. Here the genealogy is specific to how these forms relate to the production and instantiation of transmedia games or transmedia play.

Table 7.1b tracks the game to play continuum and is a useful means to begin to review the *game form* and player subjectivity that is being produced. The forms on the left are more bound and tightly constrained by the originary textual properties which determine and to some extent fix the character, behavior, and performance, whilst the forms on the right are more likely to engender freer play within a more loosely defined generic structure. Table 7.1b also tracks a shift from *textual* or *narrative specificity* which structures the player/participant behavior and interaction to *generic conventions* which are not attached to a specific text but that are sufficiently well-established as to be very easily evoked (such as the very familiar Zombie trope, the detective genre, the spy thriller, etc.).

Analysis of contemporary games has drawn heavily on the insightful typology developed by Roger Caillois which provides a structure within which to highlight the shifts between game-based structures and more playful formats. Caillois' table also affords an insight in to affective and dispositional elements of play and games (see Table 7.2).

Transmedia Play

Caillois' (1958, 36) typology helpfully establishes an axis that cuts across all kinds of games and play in the form of *Ludus* where we see the adherence to strictly enforced structure "calculation, contrivance

Table 7.2 Roger Caillois' typology of games and play

	Agon (*competition*)	*ALEA* (*chance*)	*MIMICRY* (*simulation*)	*ILINX* (*vertigo*)
Paidia	Racing	Counting-out	Children's initiations	Children's whirling
Tumult	Wrestling	Rhymes	Games of illusion	Horseback riding
Agitation	Athletics—not	Heads or tails	Tag, arms	Swinging
Immoderate Laughter	regulated		Masks, disguises	Waltzing
Kite-flying				
Solitaire	Boxing, billiards	Betting	Theater	Volador
Patience	Fencing, checkers,	Roulette	Spectacles in general	Traveling carnivals
Crossword Puzzles	football, chess,	Simple, complex		Skiing
Ludus	contests, sports in	and continuing		Mountain climbing
	general	lotteries*		Tightrope walking

NB In each vertical column games are classified in such an order that the paidia element is constantly decreasing while the ludus element is ever increasing.
*A simple lottery consists of the one basic drawing. In a complex lottery there are many possible combinations. A continuing lottery is one consisting of two or more stages, the winner of the first stage being granted the opportunity to participate in a second lottery.
Source: Caillois (1958, 36).

and subordination to rules," which for a transmedia game will relate to the extent to which player engagement is fully determined by the narrative world or intellectual property that is being opened up for engagement.

Paidia: is the "active, tumultuous, exuberant, spontaneous" (Caillois 1958, 53).

What the examples discussed within this chapter indicate is the interdependence that many experiences have on a complex interaction between both elements—the ludic and the paidiac.

In the argument for a clear distinction to be made between transmedia games and transmedia play, I am drawing on this attention to the experiential, this foregrounding of the playing subject relation. Clearly there is some cross-over in Table 7.1b—Live Action Role Playing (LARPs) games for instance are the most intensely rule bound of all the play forms to the right of the table in that the rules are very heavily proscribed at the outset. LARPs are an aesthetic form within which the participants engage in a series of activities in the real world but according to the rules established through an elaborate and carefully constructed fictional or historical narrative. Historical re-enactments with all their careful precision are one form within the LARPing continuum but LARPs take many guises from fantasy role playing to "dark play" LARPing engaging with sexual forms of role play (fetish, sadomasochism, etc.). Within the tight rules established by the LARP context there is a correspondingly wide arena of playful behavior afforded. However, Table 7.1b provides a good starting point to indicate the types and range of transmedia games that are under consideration here, whilst also understanding the application and nuances between the concepts of "game" and "play."

Another area of complexity is that in the most expanded and commercially successful types of transmedia storytelling, all types of games listed in the table could be included in the universe, for instance you may very well have a form of the *Harry Potter* or *Star Wars* or *Lord of the Rings* game in each of these categories.

As I have made clear, the notion of *transmedia play* relates to the subjective position and experience of the audience member (player). Where Table 7.1a shows the various different transmedia "objects," and I argue that each object affords a different type, mode, and depth of play, again which move along a spectrum—the most limiting and rule-bound mode of play being at the left-hand side of the spectrum, and the most freely exploratory and experimental modes being experienced by those at the right. At one end of the spectrum, it is the consuming subject that is most dominant and, at the other end, more of an engaged/player subject.

The tables also illustrate the nature of the theoretical discourse which has emerged to account for and articulate these instances, most notably a key point of disciplinary wrangling that emerged as the field of game studies was in formation in the period between 2000 and 2004. This came to be known as the "narratological versus ludological" debate. There have been a number of critical studies of this early moment of the evolution of a set of analytical, conceptual, and methodological approaches fit for the specificity of video games. Dovey and Kennedy (2006), among others, have summarized the instances of polarization within this period such that on the one hand there were theorists such as Espen Aarseth (2004) who were dismissive of the narrative element of games as irrelevant to their "gameness" and on the other hand there were theorists such as Janet Murray (2004), Barry Atkins (2003), and Diane Carr (2005) who engaged directly with the distinctive way in which stories were rendered and adapted within games and examined these as not wholly discontinuous with other media engagements and pleasures such as those related to comics, television, and film, for instance. They did so whilst also highlighting the distinctive nature of games as an interactive medium productive of new pleasures and aesthetics and requiring new analytical frameworks, methodologies, and conceptual vocabularies. Carr's (2005) work on the Matrix is particularly relevant to those studying transmedia games. (See Dovey and Kennedy 2006 for a full exploration of this debate.) Henry Jenkins' description of game design as "narrative architecture" offered a potentially useful path between these polarities by shifting a focus on games as spatialized narratives which the player explores, enacts, and creates through their gameplay:

> Game designers don't simply tell stories; they design worlds and sculpt spaces. It is no accident, for example, that game design documents have historically been more interested in issues of level design than on plotting or character motivation ... When you adapt a film into a game, the process typically involves translating events in the film into environments within the game.
>
> *(Jenkins 2004)*

Jenkins' contribution remains useful when considering the player experience of all the game forms described here. For an insightful discussion that explores the aesthetics of adaptation see Aubrey Anable's essay which is examines game designers RockStar's 2005 adaptation of the 1979 film *The Warriors* in which she examines how "seemingly similar representations of urban disorder take on different meanings in different historical moments as they move from film space to games space and from images to be viewed to images to be played" (2013, 87).

Within Table 7.3, the ordering of the game types is different to how they are presented in Tables 7.1a and 7.1b—this time to illustrate a media aesthetic to a ludo-aesthetic spectrum—an adaptation to experimentation continuum. From tight rules defined by or related to the originary text or narrative world to much more loosely applied genre conventions and greater freedom of engagement, interpretation and movement. The pleasures in engagement and participation also shift across the trajectory, with the pleasures of participation and engagement with specific well-loved texts, stories, or histories potentially dominating the left-hand side and genre pleasures dominating the right. This is where the notion of transmedia play is introduced since the type and format of game moves beyond a specific platform or console into the geographies and spaces of the material urban landscape. I would argue that is at the far right of the spectrum—within the domain of experimental/art games—that the most novel and innovative approaches to transmedia play have emerged. These approaches have then had a "ripple down" and diluted effect toward the left of the spectrum.

Table 7.3 From adaptation to experimentation: a table illustrating the characteristics across the transmedia games to transmedia play trajectory

Game type	AAA console adaptations	Social media games	Board games	Web games	Alternate Reality Games	Street games	Live Action Role Playing Games	Experimental/art games
Illustrative example	*Enter the Matrix*	*GoT – 19 Realms*	*LoTR*	*Inception*	*Batman*	*PacManHattan*	*D&D*	*A Machine To See With*
Subject position in game	First or third person character from the game world	You *are* the main character	Play as one of the Hobbits	Main character	"Yourself"—a broad character type such as "spy," "detective," "lover," etc.	Yourself in a displaced context	Player produced elaborate character	Range: prompts that the player/participant must inhabit, through to an elaborated character
Play form	Role play, mimicry	Competitive to collaborative	Alea (chance) Agon, (competition)	Agon	Community play – players deploy "collective intelligence"	Often features the topsy-turvy, fast-paced sense of ilinx	Combine a highly elaborate rule structure with capacity for open-ended play	May involve mimicry, role play, alea (chance-based elements) and also elements of agonistic pleasures. Often hybrid
Player motivation	Fan/consumer/explorer	Fan/consumer/explorer/social display	Fan—extending engagement	Social/fan	Exploration/experimentation/risk taking	Social/experimentation/risk taking	Comprehensive engagement with role/period/place	Experimentation/exploration
Location of engagement/consumption	Domestic space via console/PC Occasional arcade or social spaces	Across handheld, mobile, and computer devices	Domestic space, social gatherings.	Online/domestic/mobile	Wide variety of locations, disused warehouses, hotels, streets, bars, galleries, outdoors in parks, gardens, etc.	Urban streets	Variety of spaces and locations	Galleries, urban spaces, online/offline trajectories

Narrative spatialization	Navigable storyworld, levels and maps	Some engagement with the spaces/ topologies of the originary text	Space and navigation of specific story driven spaces key feature	Spatial navigation key element of interaction	Highly varied, space may be less relevant to the narrative	Space *as* narrative—embodied exploration of space generates story	Highly varied, space may be less or more relevant to the particular role play	Space *as* narrative—embodied exploration of space generates story
Producer motivation/ driver	Promotion and increased profit	Securing ongoing fan engagement with the franchise/ narrative world	Alternative ways of representing a storyworld, profit motivation	Securing ongoing or pre-release engagement w	Advance engagement with storyworld, advertisement, promotion, publicity	Experimentation with format, experience design experiments, occasionally promotional	Usually collective/ player produced. Not necessarily a commercial product	Artistic experimentation, political engagement, social awareness. Often funded/ supported by arts council or similar
Temporalities	Sell far beyond release of the original text, serve as promotional material to support further film, television or other media experiences within the narrative brand or "franchise"	Pre- or post-release of iteration within a franchise. Often time-limited	Durability and sustained engagement	Very short duration, timed to support promotional activities	Varied temporalities, can be longer duration if running in relation to a promotional campaign or they can be short-lived activities taking place over a specific location	Short duration—2–3 hours. Street game festivals / play festivals may feature a series of such activities to be played in short bursts over a couple of days	Highly varied—an entire weekend, others just a few hours	Dependent upon the motivation, player subjectivity and the funding protocols

For the purposes of classification and analysis I have proposed six key criteria for assessing where on the continuum the experience lies. These are:

- subject position in game
- play form
- player motivation/pleasures
- location of engagement or platform
- narrative spatialization
- producer motivation/driver
- temporalities.

These categories are intended to support a description and analysis of the aesthetics across the transmedia game to transmedia play continuum.

As Table 7.3 illustrates, transmedia play innovation has emerged from experimental, politically, socially, or artistically motivated designers such as Blast Theory, Splash and Ripple, Punchdrunk, Jane McGonigal, Ken Eklund, Conducttr, and Slingshot Effect. (With some exceptions, these designers have all created experiences which are not derived or anchored to existing brands but instead they provide experiences that extend audience and player engagements with familiar generic conventions, tastes, tropes, and images.)

The Alternate Reality Game (ARG) mechanic, although widely reported, highly publicized, and celebrated is a rarely used technique in transmedia storytelling instances, and limited to a few high-profile cases such as "The Beast" and "Why So Serious?" (see Chapter 1, this volume), and has been arguably deployed in a formulaic way. Rather, the more experimental, experiential, and playful transmedia game instances have been occurring in the lesser-known, fringe artistic practices of a handful of notable organizations and practitioners. But it has been in these spaces where techniques, practices, and principles have been evolved, refined, and defined (and then later adopted and co-opted by the mainstream).

Experimentation is often focused on innovative experience design and challenging or politically charged interactions from their audience/players/subjects (Adams, Ericsson, and Lantz 2009). The motivation for the design of the games or play form is rarely purely commercial and often these projects depend upon more uncertain, competitive, and insecure funding. These instances tend to be tightly temporally and spatially bound, with clear rules of engagement, but with freedom to explore the framework.

Within the Street Game genre—one of the most critically successful examples in the United Kingdom was Slingshot's zombie game *2.8hrs later* (see Kennedy 2018 for a full analysis) that very loosely applied fictional world/story or narrative. The player navigates through the terrain via a series of vignettes/set pieces with performances through which clues are given. These are not strictly speaking "interactive" as they do not depend upon or respond to the actions of the player as they unfold.

In *A Machine To See With* by Blast Theory, instructions are given to the player via a recording listened to through a mobile phone—again the interaction with the story is very much delivered as if a film direction provided to an actor by a director. It is within instances such as this that the "ludoaesthetic" tendency increases—the notion of ludoaesthetics refers to the extent to which the audience member/player is embodied as a playful subject. These hybrid experiences afford what Giddings and Kennedy (2008) describe as a "recombinatory aesthetic," an "amplification of affect" through this complex interplay of "tropes, literacies, energies and behaviours." See also Maria Chatzichristodoulou's essay on Blast Theory (2015) for an overview of their contribution to innovation within the theater sector.

Take Blast Theory's *A Machine To See With* (2010):

> It is a film where you play the lead. You sign up online and hand over your mobile phone number. On the day, you receive an automated call giving you the address you need to go

> to. Once you arrive on your allotted street corner your phone rings. From there a series of instructions lead you through the city. You are the lead in a heist movie; it's all about you. As you move from hiding money inside a public lavatory, to meeting up with a partner in crime and onwards to the bank, the tension rises. It's up to you to deal with the bank robbery and it's aftermath.

The experience lasts slightly less time than a feature-length film. The play/game takes place across your mobile device and across an urban space—for these experiences to work you need the infrastructure of a standard city scape for the unfolding story and actions to make sense. The instructions and directions received through your mobile phone function to transform the space around you in to a navigable and explorable game world. New rules are now in place (you are involved in a heist) and the environment is rendered unfamiliar (for instance, it is not clear who are fellow players or actors) and the affective register shifts dramatically in a number of critical ways. First, as a participant you are required to adopt what is described as the "lusory attitude" (Suits 2005 [1982]); this is the disposition of being *at play*: a critical subjective stance that enables the participant to enter in to the spirit of the game or experience. Second, the urban sprawl is now potentially fraught with threats and danger in a way relevant to the new reality triggered by the communication being received via the mobile phone.

> In this process, a participant's experience of the urban space and the awareness of strangers in that space is heightened, and the fissures between fiction and real life bring you into revealing encounters through a social drama which is … only fully realised once performed.
>
> *(Pereira Dias 2012)*

In the case of *A Machine To See With*, the familiar genre of the "heist" movie is loosely laid over a series of complex instructions and interactions to construct a playful experience that also brings together elements of theatrical performance with game and play forms such as the aleatory in the form of chance encounters, moments of good fortune, the agonistic in terms of moments of playful conflict or competition; as well as elements of mimicry when required to be immersed and act in "role" or loosely defined character; and illinx in the thrilling moments of confusion, vertiginous, and disturbing blurring of boundaries between fictional world and the situated environment. It is the complex combination of the different more rule-bound forms (instruction and direction) with the more playful forms of (role play, improvisation) that position these experiences at the edges of the continuum proposed within this chapter. In Blast Theory's more recent work, *Operation Black Antler*, the player participates in an experience that is traverses fictional generic conventions to include allusions to a relationship with timely factual reporting. Drawing on conventions of characterization from detective, noir, psychological thrillers, and conspiracy dramas:

> Operation Black Antler is an immersive theatre piece that invites you to enter the murky world of undercover surveillance and question the morality of state-sanctioned spying. In Operation Black Antler you are given a new identity as part of a small team; you are briefed and then sent into an undercover operation. From a first-hand perspective, you must make decisions and then reflect on the consequences of your decisions. What will you do when the power is in your hands?
>
> *(Blast Theory 2016)*

Player subjects are invited to perform a dual subjectivity of undercover agent and radical sympathizer, the briefing provides a very loose framework within which to improvise this performance. Again, there is no specific text that is drawn on or adapted, but a series of familiar genre conventions articulated within a framework that also immerses the participant in a murky world where they must take responsibility for moral, ethical, and legal decision-making. The experience sheds light on surveillance culture, the blurred lines between activism and terrorism, and empowers the player to shape the experience for others and to play a critical role in decision-making regarding the event outcomes.

Conclusion

This chapter has proposed a new way to configure and think about play in relation to transmedia games. The analysis of transmedia game and play undertaken here has situated the player behaviors and pleasures as dependent upon a ludic cultural imaginary, in which popular game tropes, affects, and energies underpin and determine the subject's inhabitation of the required subjectivities produced by these designs. Making stories playable is clearly a resurgent aesthetic form. It brings new aesthetic practices whilst also making new demands on the participant—the adoption of the lusory attitude is required to varying degrees, at the most subtle, audience members/players willingly submit to certain rules of behavior, and at the most extreme they will willingly engage in moments of imaginative performative improvisation.

Alongside their commercial and experiential function, transmedia games and transmedia play can be situated within a broader trend toward the ludification of cultural experience and are part of the burgeoning experience economy (Pine and Gilmore 2011)—a trend within which there is a greater demand for and celebration of interactive, playful experiences that extend our engagement with existing or novel cultural forms (see Kennedy's 2018 analysis of Secret Cinema, for instance). There is clear and widespread evidence for this wider ludification of cultural experience to be seen in the plethora of cultural experiences that are increasingly underpinned by principles of play where there is an underlying assumption of a play or games literate subject that underpins the interaction design and where this experimentation is clearly leading to new genres of experience design across and beyond the transmedia spectrum.

References

Aarseth, Espen. 2004. "Genre Trouble: Narrativism and the Art of Simulation." In *First Person: New Media as Story, Performance, and Game*, edited by Noah Wardrip-Fruin and Pat Harrigan, 45–55. Cambridge, MA: MIT Press.

Adams, Matt, Martin Ericsson, and Frank Lantz. 2009. "Art and Politics of Pervasive Games." In *Pervasive Games Theory and Design: Experiences on the Boundary between Life and Play*, edited by Markus Montola, Jaakko Stenros, and Annika Waern, 7–23. Burlington: Morgan Kaufmann.

Anable, Aubrey. 2013. "Playing (in) the City: The Warriors and Images of Urban Disorder." In *Game On, Hollywood! Essays on the Intersection of Video Games and Cinema*, edited by Gretchen Papazian and Joseph Michael Sommers. Jefferson: McFarland Press.

Atkins, Barry. 2003. *More Than a Game: The Computer Game as Fictional Form*. Manchester and New York: Manchester University Press.

Blast Theory. 2010. "A Machine to See With." Accessed December 20, 2017. www.blasttheory.co.uk/projects/a-machine-to-see-with/.

Blast Theory. 2016. "Operation Black Antler." Accessed December 20, 2017. www.blasttheory.co.uk/projects/%20operation-black-antler/.

Caillois, Roger. 1958. *Man, Play Games*. Paris: Librarie Gallimard.

Carr, Diane. 2005. "The Rules of the Game, the Burden of Narrative: *Enter the Matrix*." In *The Matrix Trilogy: Cyberpunk Reloaded*, edited by Stacy Gillis, 36–47. London: Wallflower Press.

Chatzichristodoulou, Maria. 2015. "Blast Theory." In *British Theatre Companies: 1995–2014*, edited by Liz Tomlin, 231–254. New York: Bloomsbury.

Dovey, Jon, and Helen W. Kennedy. 2006. *Game Culture: Computer Games as New Media*. Milton Keynes: Open University Press.

Giddings, Seth, and Helen W. Kennedy. 2008. "Little Jesuses and Fuck-off Robots: On Aesthetics, Cybernetics, and Not Being Very Good at Lego Star Wars." In *The Pleasures of Computer Gaming: Essays on Cultural History, Theory and Aesthetics*, edited by Melanie Swalwell and Jason Wilson, 13–32. Jefferson and London: McFarland.

Jenkins, Henry. 2004. "Game Design as Narrative Architecture." *Electronic Book Review*, July 7. Accessed August 20, 2017. www.electronicbookreview.com/thread/firstperson/lazzi-fair.

Jenkins, Henry. 2006. *Convergence Culture: When Old and New Media Collide.* New York: New York University Press.
Kennedy, Helen W. 2018. "Funfear Attractions: The Playful Affects of Carefully Managed Terror in Immersive 28 Days Later Live Experiences." In *Live Cinema: Cultures, Economies, Aesthetics*, edited by Sarah Atkinson and Helen W. Kennedy, 167–184. New York: Bloomsbury.
Murray, Janet. 2004. "From Game-story to Cyberdrama." In *First Person: New Media as Story, Performance, and Game*, edited by Noah Wardrip-Fruin and Pat Harrigan, 2–11. Cambridge, MA: MIT Press.
Pereira Dias, Marcos. 2012. "A Machine to See With (and Reflect Upon): Interview with Blast Theory Artists Matt Adams and Nick Tandavanitj." *Liminalities* 8 (1). Accessed July 12, 2017. http://liminalities.net/8-1/blast-theory.html.
Pine, B. Joseph, and James Gilmore. 2011. *The Experience Economy.* Boston: Harvard Business.
Suits, Bernard. 2005 [1982]. *The Grasshopper: Games, Life and Utopia.* Peterborough: Broadview Press.

8

TRANSMEDIA MUSIC

The Values of Music as a Transmedia Asset

Paola Brembilla

Faced with the technological, economic, and cultural evolutions of the entertainment world, the music industry has to engage with new business models and consumption habits—suffice it to consider the drop in album sales and the rise of subscription streaming services, such as Spotify. In the meantime, as this book demonstrates, industrial and cultural convergence has opened up new creative and commercial possibilities, allowing single industries to become part of larger networks of synergies that can create value from multiple businesses, outlets, and audiences by spreading one core idea, or concept, across different media. This is most evident in franchises based on audiovisual products, where the characters, along with the storytelling they generate, are the foundation of a transmedia storyworld. However, the music industry too has embraced not only cross-media distribution principles, but also world-building and transmedia storytelling models, mixing and remediating itself with other media languages in order to produce narratives and experiences that are centered on music. Thus, this chapter explores the ways the music industry can generate cultural and economic value through transmedia projects, and seek to assess the role of music in the diverse transmedia storyworlds that these projects generate.

Defining Transmedia Music

On April 23, 2016, Beyoncé released her studio album *Lemonade*. The release was accompanied by a one-hour film aired on US premium channel HBO—which, given its quality brand, therefore framed the project as "prestige" and placed it on a cultural Top Shelf. The film is a "visual album" (a visual representation of a concept album) divided into 11 chapters, one for each song, chronicling its concept: the singer's discovery of her husband's (rapper and producer Jay-Z) betrayal, and her road to healing. Between each song and visual segment, the album uses poetry and prose written by Somali poet Warsan Shire, who in turn gained instant popularity at the time of its release. In each chapter, Beyoncé wears dresses and accessories by high-end designers (Enrique Urbana, Roberto Cavalli, Yves Saint Laurent, among others), which also became instantly iconic best-sellers. In live performances on television and during the Formation World Tour, Beyoncé adopted the film's visuals and iconic aesthetics once again, repurposing outfits, atmospheres, and the poetry/prose interludes. Tapping into Beyconcé's fandom, the so-called Beyhive, the concept of the album triggered a public search to unveil the details of the husband's adultery, and particularly the name of the woman with whom he cheated on Beyoncé: a case that also oveflowed to the pages of magazines and tabloids, and ultimately boosted, once again, the popularity of the album. On June 30, 2017, Jay-Z released his fourteenth studio album, *4:44*, as an exclusive for the Sprint-Tidal partnership (Jay-Z's own streaming service). Soon after the release, many fans and critics noted that several songs are a response to Beyoncé's album, as it directly recalls some lines from *Lemonade*, stirring up interest in the couple's own narrative once again.

This cursory account of the *Lemonade* case, which is just a fragment of the whole Beyoncé/Jay-Z storyworld, exemplifies how transmedia strategies in the music industry can be diverse, combining old and new models, even within the same project. Here we have the narrativization of music (concept album), the visual representation of a musical concept (visual album), cross-marketing (fashion and literature, along with cross-promotion of two albums), branding (the iconicity and aesthetics of the "Beyoncé brand"), and industry synergies (the music and TV industries, with HBO). The connection of all these elements and strategies, as anticipated, ultimately constructed (and continues to construct) a storyworld based on Beyoncé and Jay-Z characters, functioning as a narrative that fosters fan engagement and social discourse.

Considering the bigger picture, transmedia in the music industry can take several, different shapes: a live concert can be experienced at the movies (The Flaming Lips, Justin Bieber, Kanye West); a live tour can turn into a transmedia treasure hunt (Nine Inch Nails); a band can exist in virtual reality only (The Gorillaz). Thus, there is no straightforward definition for "transmedia music." The whole concept is manifold and layered, and demands an analysis that takes several interweaving perspectives into consideration, connecting commercial and creative standpoints and bringing industrial and cultural relationships and dynamics to the fore.

A song, or an album, is not transmedia merely because it can be relocated and consumed via multiple platforms. According to Elizabeth Evans, the concept of transmediality is a key to understanding how new media technologies have inspired new business models, new forms of narrative content and audience engagement, through "the increasigly popular industrial practice of using multiple media technologies to present information concerning a single fictional world through a range of textual forms" (2011, 1). This directly connects to the well-known definition of transmedia storytelling by Henry Jenkins (2006). However, music does not necessarily equal storytelling, fictional worlds, characters, and so forth. Instead, one key requirement for music to be transmedia is that a music project must become a spreadable concept, the matrix of a narrative that unfolds through *content streaming* which is "innately liquid and multipurposable, one applicable across varied strategy, production and consumption contexts" (Murray 2005, 419). Before exploring how this can happen, let us return to the state of the industry, and pose a crucial question: why should music turn to transmedia production? As briefly mentioned in the introduction, the digital evolution of the media industries has caused market saturation, content abundance, the multiplication of outlets and players, anytime/anywhere modes of consumption, and the free circulation of files through illegal distribution platforms. And following the breakdown of the previous economic monopoly, held for decades by old producers and distributors, the music world needs to evolve its business models accordingly, aiming at competitive strategies that can ultimately bring an added value to their offer—and world-building and storytelling provide precisely this.

To understand the perks of these strategies, we turn to the notion of "total entertainment." According to Paul Grainge:

> Firstly, it can be seen as an industrial principle, describing the attempt by global media conglomerates to create an expansive entertainment and communication environment in which they have a disproportionate, near total, stake in terms of ownership and control. Secondly, it can be thought of as a particular form or horizon of cultural and textual practice, growing out of the permeable boundaries and newly "immersive" modalities of commercial entertainment media.
>
> *(2008, 54)*

Although this argument mainly refers to media franchises (in particular Disney), these two points help understand the logics of transmedia music as well. On the one hand, there is the need to maintain control over the circulation of content and share the risks of high-budget projects through corporate management, synergies, and partnerships. On the other, this fits into a broader trend where *content* becomes immersive and the *experience* become a product. On this point, as Frank Rose argues:

> a new type of narrative is emerging—one that's told through many media at once in a way that's nonlinear, that's participatory and often gamelike, and that's designed above all to be immersive. This is "deep media": stories that are not just entertaining, but immersive, taking you deeper than an hour long.
>
> *(2011, 3)*

In order for music to be more than a single or an album that is directly sold to consumers, it needs to appeal to the affective economies of cultural life, i.e., to understand the emotional underpinnings of consumer decision-making and to "shape brand reputations, not through an individual transaction, but through the sum total of interactions with the customer—an ongoing process that occurs across a range of different media 'touch points'" (Jenkins 2006, 62–63). To do this, music needs not only to shift into a concept and its storytelling, it also needs to reinforce the relationship with its audience through immersive, emotional, and long-lasting experiences. As we have seen with the case of *Lemonade*, the official narrative of the project is carefully crafted so as to generate more narrative through social discourses, tapping directly into participatory culture and those audience members that help generate interest in particular brands (Jenkins, Ford, and Green 2013, 7). Most importantly, the entire project is designed as a multiplatform narrative, made of several entry points that ultimately construct an immersive environment. However, as mentioned previously, there is no standard model for transmedia music. In the following sections, I identify a handful of cases that highlight this diversity, before providing final remarks on the role of music in transmedia projects.

One Direction: Superstar Brands, Transmedia Narrative, and Counter-Narrative

A narrative needs characters. And in a way, the music industry has always been in the business of producing "characters," also known as "superstars." Namely, stars are *collections of meaning*, as mediated identities, characters that embody images and personalities; *capital*, as labor for the industries; and *marketing tools*, in that their image can be capitalized in terms of a media text promotion (McDonald 2000, 6–13). So effectively, stars *are* characters around which industries can build a narrative and gain revenues. Although this is not a new phenomenon, with the evolutions and disruption of the media industry this "superstar design" has been further sharpened and enhanced. Peter Tschmuck notes that as major record labels replaced the push strategy with a pull strategy to make changes manageable, they also began to rely increasingly on music *concepts*, and to turn some performers into "superstar brands," with very specific music and interpretation styles (2012, 268). When talking about these "superstar brands," Roy Shuker stresses the importance of recognizing the "commerce mythology" around them, i.e., the constraints placed on the performers by the industry, as well as "the dynamics of the relationship between performers, their record companies and other aspects of the music industry (technology), and their fans" (2001, 116). One example is the Spice Girls, with each one of the girls becoming pervasive public signifiers through their iconic looks and nicknames, a brand image embodied by the "Girl Power" catchphrase, brand-extension and cross-marketing operations (books, TV specials, a film), and countless product endorsements (Shuker 2001, 129). In a way, this too was a kind of transmedia narrative. But what happens to industrially crafted strategies and narratives when, as suggested by Shuker, we take into consideration the fans too, and we add contemporary participatory culture to the mix? The boyband One Direction (1D) is one emblematic case.

Notoriously, 1D is a British boy band formed by Simon Cowell in the seventh series of *The X Factor* (UK), in 2010. After ranking third in the competition, the group signed with Cowell's label Syco Records and released, since then, five albums. Aside from this first synergy with the TV industry, the promotional activities for the band are not unlike those historically adopted for superstar brands: each member of the group embodies a personality and a set of values, recalled in every public appearance and music video. Around the band and their albums, we have a wide range of ancillary products: music videos, books, films, TV specials, products endorsements, etc. But what is really unique in this case, is how the fandom—namely, the directioners—played an active role in boosting the band's popularity, sales and in enriching their storyworld.

As Mark Duffett (2014) argues, with the Internet and new technologies, fans can now directly analyze, comment, transform, and share again pieces of their idols' universes, to such an extent that these communities have become a central element of record companies' marketing strategies. This has resonance with the example of 1D, who in 2015 released their fourth album, *Four*, containing the song "No Control." The song quickly became a fan favorite because of its overtly sexual lyrics, which were written by one member of the band—a first for 1D. Initiated by a Tumblr post by a London fan, a group of directioners started the *No Control Project*, a widespread and coordinated promo campaign to make "No Control" a single. The project involved the creation of posters, covers for the single and a music video; massive downloads of the track on iTunes to drive it up the sales chart; a Facebook-Twitter-Tumblr thunderclap (a syncronized post publication) of 55 million people, to get the song title trending worldwide; fans calling radio stations to play the song, which was ultimately played by 60 stations worldwide in one day, including BBC1. The project received great relevance in the media, was acknowledged by the boyband with public thanks and the song became an unofficial single—and was even included by James Corden in his "Carpool Karaoke" with the boys. The *No Control Project* is a case of what Patryck Galuszka (2015) calls the "new economy of fandom," where empowered fan communities use the potential of social media to cooperate with artists, ultimately working as sponsors, co-creators of value, stakeholders, and investors. Although the fans' primary intention was to shine a light on a "hidden" song that differed from the typical 1D brand, they ended up helping the industry generate revenues. Therefore, the cultural value assigned by the fans to the song actually turned into economic value for the industry.

A further explicit case of fans creating cultural value is through Larry Stylinson, and the creation of a counter-narrative. From the months of *The X Factor*, some attentive directioners noticed special affection between two members of the boyband, Louis Tomlinson and Harry Styles. Later on, these fans (that called themeselves the "Larries") began to collect evidences of a love affair between the two: they uploaded videos on YouTube with their most evident moments of sexual tension and loving looks during public appearances, and they named the couple Larry Stylinson (a mashup of the two full names). The theory peaked in 2011, with Louis Tomlinson's infamous tweet "Always in my heart @Harry_Styles"—which, in 2017, remains the third most retweeted tweet in history (see Luckerson 2017). Shortly afterwards, Harry and Louis began to be separated in interviews and to avoid one another on stage. Suddenly, the fans had no more material to make their videos and to support their theory. However, this sparked them further, feeding new theories that the couple was forced to stay separated and closeted by the management. Regardless of the truth behind these theories, it is interesting that a counter-narrative soon became canonical for a part of the fandom and that, in spreading across the Internet and even the trade press, it turned into a sort of urban legend. Further, the counter-narrative even contained social outcomes. Beyond fanart and fanfiction, the Larry Stylinson "division" of the fandom became a safe space for LGBTQ (lesbian, gay, bissexual, transgender, and queer) fans—mostly through Tumblr pages (like http://takemehomefromnarnia.tumblr.com) that act as communities offering moral support to directioners in need, and organizing meetings during the band tour dates. And as for the band, Harry Styles too started to show support for the LGBTQ community in interviews and during tour dates—he, in many cases, waived rainbow flags on stage during 1D's last tour, the On The Road Again Tour. The Larry Stylinson counter-narrative therefore became more than a byproduct of contemporary participatory culture; as Lisa A. Lewis suggests: "By participating in fandom, fans … enter a domain of cultural activity of their own making which is, potentially, a source of empowerment in struggles against oppressive ideologies and the unsatisfactory circumstances of everyday life" (1992, 3). While in immersive, branded environments, characters and plotlines can be extended and moved to other media to create a more robust world (Rose 2011, 19), the same holds true when these characters are moved to counter-narratives that expand and enrich the canonical storyworld with new, culturally/socially relevant meanings. As 1D are now on a hiatus and band members are pursuing solo carreers, the Larries still live on as a community engaged with support for LGBTQ youths—and producing 1D-themed merchandise for charity. Moreover, on the one hand, the directioners continue to provide an essential source of profit for each single member of the group, as they follow their solo careers. On the other, they keep the 1D storyworld alive through social discourses and active productions, turning the project into a long-term asset that, at some point in the future, could be picked up again by the industry.

Björk's *Biophilia*: A Multiplatform Project and Transmedia Experience

Donald Norman (2009) defines system-thinking as a key strategy in today's most succesful companies. System-thinking means selling not just a product, but a set of connected experiences revolving around a core business. In so doing, direct transactions give way to systemic experiences, which generate value through multiple touchpoints and by keeping users inside that system. The iPod is an ideal example: the business is not just about the product itself, but rather the Apple digital ecosystem surrounding it and the way users need to enter that ecosystem to make the product work. The same principle can be applied to the transmedia narratives, or immersive environments, we have seen so far: music is the core product, but the narratives built around it configures environments that keep audiences engaged and loyal.

Beyond superstar brands and narratives, there are more practical examples of how an artist can turn a single product, such as a concept album, into a multiplatform project and transmedia experience. One example is Björk's *Biophilia*. Released on October 5, 2011, *Biophilia* is the seventh studio album by the Icelandic artist. More than a concept album, *Biophilia* has been described as the first "app album," i.e., a multimedia project that explores the links between music, art, nature, science and technologies. It is in fact a transmedia experience, harmonically connecting and hybridizing heterogeneous languages and media.

From a musical perspective, each of the album's tracks ideally matches the intellectual purpose of the song and combines instruments and softwares. However, in an interview with *Wired*, Björk stated that she felt the songs could not stand on their own: "People are getting a lot of music for free by pirating it. But they are going to double [the number of] shows because they want a 3D, physical experience" (Burton 2011). The original idea to produce a three-dimensional film to complement the album was later set aside in favor of a mobile application launch. According to the same *Wired* article, Björk reasoned that, through the app, her audience would have interactive and educational experience at a premium, non-piratable value. For that reason, she turned to Apple: given the project's high costs, she ensured that the company would promote the app, and that it would find a spotlight within the iTunes Store. Apple therefore created a "super room" page, from which both audio and app versions of the album could be bought.

In terms of experience, the app is made of a "mother app" that functions as a constellation, containing ten separate apps, one for each song. They either work as games (e.g., in *Virus*, the user must protect cells from bacterias, as the song plays in the background) and/or music sequencers visually represented by instruments or natural elements, like piano strings or electrical lines (in *Moon*, the user can edit the original version of the songs by adjusting the notes). The app gameplay is presented by broadcaster and biologist David Attenborough, and the same introduction opens each date of the *Biophilia* Tour. The tour also incorporates educational lectures, nature footage and a publicity campaign featuring *National Geographic* and the scientific journal *Nature Medicine*. The project design and recording were later documented in the 2013 film *When Björk Met Attenborough*, and the tour became a concert film, *Biophilia Live*. In 2014, *Biophilia* became the first downloadable app in the New York MOMA permanent collection, signaling the album's official entry into the sphere of "high arts."

Clearly, this case recalls once again the importance of industrial synergies in a transmedia project—suffice it to recall the deal made with Apple. But what is most interesting here is the extension of the artist's brand and aesthetics to create a transmedia experience. Notably, Björk's brand is that of an eclectic, transmedia artist (singer, actress, writer, art performer) and it is extensively expressed through interest in nature and her concerns about the environment: she performed at concerts, wrote articles to raise awareness on environmental issues, she founded the Náttúra organization to support icelandic grassroots industries, and promoted it with a single of the same name. As in the cae of the iPod and its essential connection to the Apple ecosystem, the album is not the real product here. Through Björk's brand, the system-thinking design mentioned at the beginning of this section is implemented to create a whole ecosystem that the user can inhabit and experience, that is based on music, but actually harmonizes sounds and visuals, art and science, digital design and philosophy.

The Get Down: Music and TV Series

Throughout the chapter, we have observed transmedia narratives and experiences in which music remains the core product and the starting point of the project. However, music can also be employed as a shared asset with other entertainment industries, one that merges and adds value to other core products. On US television, for instance, musical shows have been very popular for quite sometime, in the form of talent/contest shows (e.g., *American Idol*), the return of live theater (e.g., FOX's *Grease Live*) and, more than anything else, in scripted series. In the latter case, music fits into ecosystemic architectures that often characterize contemporary TV shows, which are designed as ever-expanding storyworlds where single parts can be extracted and adapted to the needs of different media and audiences, each benefiting the other (Innocenti and Pescatore 2012). Such cases take advantage of industrial sinergies by placing music at the center of world-building processes in many ways. Music, both as a narrative and business asset, merges with TV language, to generate transmedia storytelling and transmedia revenues. A useful example is the Disney franchise *High School Musical*, where music was repurposed into live shows, international spin-offs, a reality contest, a show on ice, video games and so on. Another is ABC's *Glee*: the cover song licensed for the show was repurposed in live shows. In these cases, music drove not only the shows' narratives, but also the transmedia products' spread, becoming the core of their world-building process. Beginning with these considerations, I focus next on how music factors into the world-buiding process of a TV series by examining the case of Baz Luhrmann's *The Get Down* (Netflix, 2016), which allows us to explore the cultural and economic value of transmedia from yet another point of view. Set in the Bronx, New York City, in the late 1970s, the series follow the rise of disco music, hip-hop, and underground cultural trends through the eyes of a group of teenagers, also depicting the wars among gangs and the poverty faced by Bronx communities. In order to do so, *The Get Down* puts specific musical genres and their cultures at the center of its world-building process, illustrating an extremely accurate coordination of transmedia industries, assets, and creativity. From the industrial standpoint, the series relies on inter-industrial collaborations and brand-names in order to establish its reliable voice and then to reinforce, legitimate, and promote itself. *The Get Down*'s score was written by Elliott Wheeler, a composer and producer who specializes in unifying different genres and artists, as well as adapting and remixing pre-existing music to the needs of contemporary films and TV. The show features new music, cover versions and several songs from the 1960s and 1970s, and predominantly disco, rhythm and blues (R&B), funk, and soul. Importantly, Baz Luhrmann also taps directly into the real hip-hop community, having worked with technical advisers such as Nelson George and the pioneering hip-hop DJ

Grandmaster Flash—who is also portrayed in the series as a key character. Together with the DJ, rap legends Kurtis Blow and Nas hosted a Hip-Hop boot camp to educate the actors. The brand-name of Baz Luhrman plays an important role, too, as it helps to frame the show in the postmodern imaginery of his filmography, in which remixed contemporary music is essential in establishing the mood and aesthetics, and driving the narrative—this is manifested in his musical movies *Strictly Ballroom* (1992) and *Moulin Rouge!* (2001).

At the level of the text, in the mixture of true facts and heightened invention, music goes beyond its basic scoring fuction to become a storytelling driver. Beyond moving the narrative forward, on a macro-level, *The Get Down* operates the way original rap music did, by appropriating beats from earlier R&B and disco music. Similarly, on a micro-level, the two protagonists Mylene (a disco star wannabe) and Zeke (a poet and rapper) represent the evolution of two musical genres, as well as links to their respective communities. In fact, the show embraces the genres and cultures of disco and hip-hop, foregrounding their social and cultural value for emerging minorities (see, for instance, Krasnow 1993).

By representing the rise of disco and hip-hop as such, *The Get Down* actually "remediates" them. According to Jay Bolter and Richard Grusin (1998), *remediation* means reform. Simply put, remediation happens when one medium is represented within another and the second medium can improve or reform the former. This relates, therefore, to the notion of appropriation, insofar as it is based on the combination and integration of different media languages, each one enhancing the other. In this case, the language and imagery of two musical genres are appropriated into TV language, reforming it through aesthetic and cultural enrichment and ultimately increasing the overall value of the series.

Conclusion

Throughout this chapter, I have referred to core products and core concepts. However, the cases I analyzed demonstrate that although the idea of a transmedia project can be initiated by one industry, it becomes a sort of ecosystem that breaks down hierarchies, blurring the boundaries between the central text (e.g., a music album) and its transmedia extensions. In this sense, transmedia music operations are consistent with the ambivalence of transmedia storyworlds, based on the merging of texts and paratexts, products and by-products, cultural and commercial spaces (Gray 2010). Moreover, these projects also question the hierarchy subtending much scholarly work on music and audiovisual products in which sounds follow image, pointing instead to more systemic perspectives—Timothy Warner (2006), for instance, shows how the reverse holds true in pop videos that work toward an integrative function. As I argued at the beginning of this chapter, in fact, it is essential to examine transmedia music in the light of synergy networks, relationships, and dynamic processes, rather than single products that are simply connected with one another on a cross-media level. Since transmedia implies such concepts as convergence, mergences, and remediation, music should be seen as a transmedia asset on which cultural industries can capitalize in different ways: as a product to aid actual selling, but also/mostly as a means to unfold the plotlines of a storyworld, as the backbone of a transmedia project, as an aesthetic and a language that adds value to an audiovisual production. Combined with the systemic perspective, this view of music as a transmedia asset allows us to account for its versatility and ability to serve several purposes, and to understand better the creation of cultural and economic value in the contemporary mediascape.

References

Bolter, Jay D., and Richard Grusin. 1998. *Remediation: Understanding New Media*. Cambridge, MA: MIT Press.

Burton, Charlie. 2011. "In Depth: How Bjork's 'Biophilia' Album Fuses Music With iPad Apps." *Wired*, July 26. Accessed July 11, 2017. www.wired.co.uk/article/music-nature-science.

Duffett, Mark. 2014. "Introduction." In *Popular Music Fandom. Identities, Roles and Practices*, edited by Mark Duffet, 1–15. London and New York: Routledge.

Evans, Elizabeth. 2011. *Transmedia Television. Audiences, New Media, and Daily Life*. New York and London: Routledge.

Galuszka, Patryk. 2015. "New Economy of Fandom." *Popular Music and Society* 38: 25–43.

Grainge, Paul. 2008. *Brand Hollywood: Selling Entertainment in a Global Media Age*. New York and London: Routledge.

Gray, Jonathan. 2010. *Show Sold Separately: Promos, Spoilers and Other Media Paratexts*. New York: New York University Press.

Innocenti, Veronica, and Gulgluielmo Pescatore. 2012. "Information Architecture in Contemporary Television Series." *Journal of Information Architecture* 4: 1–2. Accessed July 11, 2017. http://journalofia.org/volume4/issue2/05-pescatore/.

Jenkins, Henry. 2006. *Convergence Culture: Where Old and New Media Collide*. New York: New York University Press.

Jenkins, Henry, Sam Ford, and Joshua Green. 2013. *Spreadable Media: Creating Value and Meaning in a Networked Culture*. New York: New York University Press.

Krasnow, Carolyn. 1993. "Fear and Lotahing in the '70s: Race, Sexuality and Disco." *Stanford Humanities Review* 3: 37–45.

Lewis, Lisa A. 1992. "Introduction." In *The Adoring Audience: Fan Culture and Popular Media*, 1–8. London and New York: Routledge.

Luckerson, Victor. 2017. "These Are the 10 Most Popular Tweets of All Time." *Time*, April 11. Accessed November 17, 2017. http://time.com/4263227/most-popular-tweets/.

McDonald, Paul. 2000. *The Star System: Hollywood's Production of Popular Identities*. London: Wallflowers.

Murray, Sarah. 2005. "Brand Loyalties: Rethinking Content Within Global Corporate Media." *Media, Culture & Society* 27: 415–435.

Norman, Donald A. 2009. "System Thinking: A Product is More Than the Product." *Interactions* 16: 52–54.

Rose, Frank. 2011. *The Art of Immersion: How the Digital Generation is Remaking Hollywood, Madison Avenue, and the Way We Tell Stories*. New York and London: W.W. Norton & Company.

Shuker, Roy. 2001. *Understanding Popular Music*. 2nd ed. London and New York: Routledge.

Tschmuck, Peter. 2012. *Creativity and Innovation in the Music Industry*. New York: Springer.

Warner, Timothy. 2006. "Narrating Sound: The Pop Video in the Age of the Sampler." In *Changing Tunes: The Use of Pre-Existing Music in Film*, edited by Phil Powrie and Robynn Stilwell, 167–179. Burlington: Ashgate.

9

TRANSMEDIA JOURNALISM

The Potentialities of Transmedia Dynamics in the News Coverage of Planned Events

Renira Rampazzo Gambarato

The transmedia storytelling concept was originally conceived in the fiction context (Jenkins 2003, 2006). However, the non-fiction realm embraced the transmedia dynamics as much and as profoundly as the entertainment sphere. Writers, novelists, and historians have long blurred the differentiation between fiction (the world of imagination) and non-fiction (the real world), and terms such as creative nonfiction have emerged (Clark 2001). News stories, however, traditionally are examples of non-fiction in the sense that "journalists should report the truth. Who would deny it? But such a statement does not get us far enough, for it fails to distinguish nonfiction from other forms of expression" (Clark 2001). Non-fiction can make false assertions, fabricate facts, and tell events from a certain perspective. Therefore, although the truth is not the main point in non-fiction, non-fictional stories, such as news, claim to describe reality to a certain extent (Gambarato and Tárcia 2017).

Kerrigan and Velikovsky (2016) argue that non-fiction transmedia forms have the same characteristics as fictional transmedia productions. "Non-fiction transmedia draws on the same definitions as fiction transmedia" (2016, 250), and "[n]on-fiction transmedia is an extant and ever-increasing phenomenon" (2016, 255). Gifreu-Castells, Misek, and Verbruggen (2016) remind us that "[a]udiovisual non-fiction is a vast field containing documentary, journalism, film essays, educational videos, museum exhibitions, scientific films, institutional, industrial or propaganda videos, etc." In this context, scholars such as Alzamora and Tárcia (2012), Canavilhas (2014), Gambarato and Alzamora (2018), Gambarato and Tárcia (2017), Moloney (2011), Pernisa Jr. (2010), Renó and Flores (2012), and Tellería (2016) have investigated to what extent transmedia storytelling characterizes contemporary journalism, which constructs a narrative that creates various entry points dispersed across multiple media platforms and involves different audience segments.

Similar to the transmedia dynamics observed in entertainment, transmedia journalism operates by expanding journalistic narratives on integrated platforms, in which the audience is involved in a committed way, adding and sharing content through digital environments, especially via online social networks.

Transmedia journalism, as Dominguez (2012) stated, is an elastic term with a wide variety of theoretical possibilities. For Tellería, transmedia journalism is "a field scarcely explored and with a wide range of possibilities to be implemented and tested" (2016, 71). Thus, this chapter first discusses the current contributions to the conceptualization of transmedia journalism and presents Gambarato and Tárcia's (2017) analytical and operational model that outlines the main features of transmedia strategies focused on the coverage of planned events in news media. Planned events are temporal occurrences that are usually well schematized and publicized in advance. For instance, the Olympic Games and other major sporting events, such as the FIFA World Cup, are recurring planned events. The relevance of this model is its contribution to overcoming the difficulties of transposing transmedia logic to the journalism realm. The difficulties comprise, for instance, the complexity of the journalism activity in general (Canavilhas 2014), the necessity of designing and planning distinct paths and content to be produced across multiple media (Renó 2014), and issues regarding journalism time-constrained brevity (Moloney 2011).

Transmedia Journalism

Scolari, Bertetti, and Freeman posit that "journalists have been producing transmedia storytelling for years, even before the arrival of the World Wide Web" (2014, 4). Since the advent of mass communication, news stories have been reported in diverse media, starting with radio, then television, and followed by next-day newspapers and weekly magazines. Engagement, at that time, occurred through telephone calls or letters to the newsroom. The arrival of the Internet, especially regarding online social networks, enhanced audience engagement through interaction and participation. Nevertheless, Gambarato and Tárcia emphasize that "although various media are present in journalism and journalists employ multiple practices to cover multifaceted media events, not every news production is necessarily transmediatic; thus far, the majority of the content spread across different media platforms is simply repurposed" (2017, 1385). Moreover, Tavares and Mascarenhas ponder that "it is not rare that a news story extends across multiple platforms with mass and post-mass functions. However, this is not enough to call an article as a transmedia narrative" (2013, 200). A crucial aspect to consider in this context is that transmedia storytelling is about expanding, not repeating, content.

One of the main approaches to transmedia journalism was proposed by Moloney (2011). He applied Jenkins' (2009a, 2009b) core principles of transmedia storytelling to journalism. Moloney (2011, 60–92) posits the following as characteristics of transmedia journalism: (1) spreadability (the spread of a story by users' sharing); (2) drillability (search for more details about the news; official content expansions, including social media networks); (3) continuity and seriality (maintain continuity and exploit the characteristics of each medium; keep the audience's attention for a longer period); (4) diversity (add other points of view, including those of the public); (5) immersion (generate alternative forms of storytelling for the public to delve deeper into the story); (6) extractability (apply the journalist's work in everyday life with the public commitment); (7) real world (show all shades of the news, without focusing on simplification); and (8) inspiration to action (pursue intervention by the public in real actions seeking solutions to problems).

Looney (2013) describes five ways to build transmedia news features: (1) keep the content unique (instead of repeating information on different media platforms, use different parts of the story to match the strength of each medium and maximize the user experience); (2) provide seamless points of entry (ensure that the media platforms involved in the transmedia news story offer the audience the possibility to engage in a simple manner); (3) partner up (transmedia news are often complex and require the involvement of other professionals and resources); (4) keep it cost-effective (although there are costly projects, this is not the only option because it is also possible to produce transmedia news cheaply, for instance, by introducing social media networks to expand the story); and (5) the story is number one (creative tools may do more harm than help; always put the story first).

Pernisa Jr. (2010) proposed an "opened monads model" for use in building transmedia news stories, "in which each medium would be taken as the smallest field in structure and would bind the other, forming a network of contextualized material for the user's query in various ways" (2010, 8). Furthermore, Canavilhas (2014, 60–64) describes the main characteristics of transmedia journalism as (1) interactivity, (2) hypertextuality, (3) integrated multimodality, and (4) contextualization. Gambarato and Tárcia conclude:

> In sum, we consider that transmedia journalism, as well as other applications of TS [transmedia storytelling] in fictional and nonfictional realms, is characterized by the

> involvement of (1) multiple media platforms, (2) content expansion, and (3) audience engagement. Transmedia journalism can take advantage of different media platforms such as television, radio, print media, and, above all, the internet and mobile media to tell deeper stories. Content expansion, as opposed to the repetition of the same message across multiple platforms, is the essence of TS [transmedia storytelling] and, therefore, should be the focal point of transmedia journalism as well. The enrichment of the narrative is facilitated by the extended content. Audience engagement involves mechanisms of interactivity, such as the selection of the elements to be explored, the option to read a text, watch a video, enlarge photographs, access maps, click on hyperlinks, and share information through social networks. Audience engagement deals with participation via, for instance, remixing content and creating original user-generated content.
>
> *(2017, 1386)*

The current news media patterns point to

> the continuous decline of print sales and the access to media by its homepage, mobile first acclaimed strategies, the ever-changing parameters of Social Media that directly affect the access and distribution of media content, the rising of podcasts, personalized news and content, video and add-blocks as well as the unstable ambient of apps.
>
> *(Tellería 2016, 68)*

This atmosphere is reflected in transmedia journalism, which is enriched by the current trends of contemporary contributions from immersive journalism (de la Peña et al. 2010), virtual reality (Matney 2017), podcasts (Barnathan 2014), slow journalism (Gambarato 2016), newsgames (Bogost, Ferrari, and Schweizer 2010), graphic journalism (Cute 2016), and comics journalism (Schlichting 2016), among others. In the midst of this reality, "[t]he purpose of a transmedia news story is to inform the readers in the best way possible" (Ford 2007). Transmedia journalism is relevant because of its refreshing approach to information architecture, visualization, and contextualization to foster the integrated transmedia flow of content across multiple media platforms with audience engagement (Tellería 2016).

Analytical and Operational Model of Transmedia News Coverage of Planned Events

Gambarato and Tárcia (2017) argue that transmedia storytelling's role in contemporary journalism relates in particular to the coverage of planned events. "Several studies of transmedia entertainment and transmedia narratives emphasize the planned, strategic aspects of their creation: media companies carefully structuring and portioning the narrative across different media platforms in order to maximize synergistic effects" (Fast and Örnebring 2015, 2). However, there is also the emergent/ad hoc nature of transmedia storytelling in which storyworlds can evolve over time and be created and co-created by professionals and amateurs alike. Therefore, Fast and Örnebring emphasize that transmedia storytelling has "(a) accrued characteristics that are more ad hoc/contingent than planned; and/or (b) contain disjunctions and contradictions that are actually the result of strategic planning decisions (i.e. strategic planning of transmedia worlds is not necessarily focused on creating a seamless, coherent world)" (2015). In this sense, transmedia journalism is the result of "carefully orchestrated company strategies" (2015) combined with ad-hoc aspects of transmedia storyworlds, especially related to the contributions provided via audience engagement (Gambarato, Alzamora, Tárcia, and Jurno 2017).

Planned events usually attract large domestic and international audiences, have the potential to integrate audiences in the news making (emergent/ad-hoc aspect), involve a substantial amount of human, technical, and financial resources (planned/strategic aspect), and provide abundant content, protagonists, and various stories. These characteristics constitute a fertile terrain for the development and growth of transmedia news production.

The proposed analytical and operational model of transmedia news coverage of planned events (Gambarato and Tárcia 2017) aims to contribute to a clearer understanding of transmedia news production and to foster and improve transmedia journalistic practices. The model addresses the specificities of such multiplatform news productions, clarifying how transmedia features are structured and implemented. The method draws on the transmedia project design model developed by Gambarato (2013) and establishes ten focal topics and subsequent practicable questions, regarding, for instance, news storytelling, media platforms, and audience engagement. The rationale behind the incorporation of characteristics commonly associated with fictional transmedia in transmedia journalism relies on the discussion presented earlier in the chapter about fictional and non-fictional transmedia productions sharing the same core principles and features (Kerrigan and Velikovsky 2016). A concise description of the model is provided in Table 9.1.

To briefly illustrate the application of the model for transmedia news coverage of planned events, the case of the Brazilian news coverage of the 2016 Rio Summer Olympics (Gambarato, Alzamora, and Tárcia 2018) is presented. The analysis examines ten features of the transmedia news coverage produced by the Brazilian official broadcaster of Rio Olympics—Globo Network (*Rede Globo*, in Portuguese): the premise and purpose, structure and context, news storytelling, world-building, characters, extensions, media platforms and genres, audience and market, engagement, and aesthetics.

The first feature is the premise and purpose of the coverage. Although the fundamental objects of the coverage were the competitions, the athletes, and their performances, the media also had to be aware of related issues, such as social and political demonstrations and controversies; an ongoing outbreak of the mosquito-borne Zika virus in Brazil; the pollution of Guanabara Bay, whose waters were used for sailing and windsurfing competitions; political instability and economic crises; and the Russian doping scandal and participation restrictions.

The second feature is the structure and context. More than 7,000 hours of video and audio coverage were produced and distributed to an audience of six billion people in 220 countries (Long 2016). Globo Network planned the transmedia coverage of the Games, including online streaming, social media networks, mobile applications, and websites.

The third feature is news storytelling. The primary stories broadcasted, especially on television, were directly related to the sports events, such as the Olympic competitions, the results, and the stories that portrayed the athletes as heroes. The secondary stories included, for example, the American swimmer Ryan Lochte fabricating a story of being robbed at gunpoint during the Games; two boxers arrested and accused of trying to rape housekeepers; and Patrick Hickey, a top Olympics official, arrested after illegally selling tickets to the Games. The secondary stories enriched the coverage and functioned as crucial elements of content expansion of the transmedia coverage.

The fourth feature is world-building. The Summer Olympic Games competitions occurred at venues throughout Rio de Janeiro and were big enough to support expansions across multiple platforms. Globo Group, comprising television, cable television, printed media, and radio, included several of the group's content platforms in this coverage, creating different concepts for each media platform, a core characteristic of transmedia journalism.

The fifth feature is the characters. The coverage focused on several characters, in addition to the array of international athletes, such as Brazilian President Michel Temer (who was loudly booed by Brazilians during the Opening Ceremony) and the Brazilian supermodel Gisele Bündchen (who walked the length of the Maracanã Stadium to the song "Girl from Ipanema" during the Opening Ceremony).

The sixth feature is extensions. The coverage had diverse extensions, including mobile applications, Internet live streaming, video-on-demand platforms, and profiles on social media networks. A highlight was the unprecedented partnership between Globo Network and Snapchat, which offered specific content for the feature "Live Stories" on Snapchat with photos and videos recorded behind the scenes of the event by journalists, commentators, and fans.

Table 9.1 Concise description of the analytical and operational model regarding transmedia news coverage of planned events (Gambarato and Tárcia 2017, 1389–1391)

Topic	*Practicable questions*
1. Premise and Purpose: The nature of the event, its magnitude, and comprehensiveness influence the journalistic coverage.	What is the planned event agenda? What is its core theme? What is the fundamental purpose of the event? What is the magnitude of the event? Is it a local, regional, or global event? Which areas are involved in the coverage (sports, culture, politics, economics, etc.)?
2. Structure and Context: The organization of the transmedia journalistic coverage, the professionals involved, and the infrastructure available depict how the operations were planned and executed.	Which media enterprise is covering the event? How big is it? What is the available coverage infrastructure offered by the event organizers? What is the media enterprise budget for the news coverage of the event? Is the journalistic coverage planned to be transmediatic? How does the coverage end? Do some extensions continue to be active after the event ends?
3. News Storytelling: The news coverage of the event involves primary and parallel stories.	Primary and Parallel News Stories: What elements of the news story (who, what, where, when, why, and how) of the event are involved in the coverage? What is the timeframe of the news story? Does the news coverage utilize gaming elements? Does it involve winning or losing? Is it possible to identify intermedial texts in the news stories?
4. World-building: The storyworld in which the news is placed should be robust enough to support multiplatform expansions.	Where is the event set? Does the storyworld involve any fictional characteristics? Are different time zones involved in the news coverage? If yes, what are the potential issues related to it and the alternative strategies for each platform? What are the regulations and policies related to the journalistic coverage? Is the event big enough to support expansions throughout multiplatform coverage?
5. Characters: The characters implicated in the coverage could be journalists, characters of the news stories per se, sources of information to be reported, and the audience as collaborators.	Who are the main characters presented by the coverage? How many? Are they aggregated to the coverage a posteriori? Who are the primary and secondary sources of information regarding the event? What is the approach of these sources? Are the sources official, nonofficial, or both? Can the audience be considered a character as well?
6. Extensions: News stories meant to spread throughout multiple media platforms should not simply transpose or repurpose the content from one medium to another but expand the news, taking advantage of the media platforms available.	How many extensions are included in the news coverage? Are the extensions mere reproductions of the same content or genuine expansions of the news stories across various media? Is there a plan to keep the content updated in each extension (for instance, on blogs and social media networks)? Do the extensions have the ability to spread the content and provide the possibility to explore the narrative in-depth? How long does the event last? If the event is overlong, how does the coverage proceed to maintain audience interest throughout the entire period?

Table 9.1 (Cont.)

Topic	*Practicable questions*
7. Media Platforms and Genres: In addition to telling news stories with more than one medium, transmedia news coverage can embrace several journalism styles, such as news articles, reports, and opinions; a number of journalism genres; and different technological devices.	What kind of media platforms (television, radio, print media, web, mobile) are involved in the news coverage? Which devices (computer, tablet, mobile phone, etc.) are required by the coverage? Is there a roll-out strategy for launching each coverage extension? If yes, what is the plan for releasing the platforms? Which journalism styles (news articles, reportages, opinions, etc.) are included in the coverage? Which journalistic genres (sports, celebrity, investigative journalism, etc.) are presented by the coverage?
8. Audience and Market: Scoping the audience is fundamental for a more appropriate delivery of the transmedia news coverage.	What is the target audience of the coverage? Who is the intended reader/user/viewer/listener? What kind of readers (methodical or scanner; intimate, or detached) does the project attract? Does other journalistic coverage like this exist? Do they succeed in achieving their purpose? What is the coverage business model? Does it involve open platforms, open television channels, cable TV, satellite, pay-per-view, monopoly, etc.? Is the event coverage successful revenue-wise?
9. Engagement: The relationship between the story and the people interested in it is an essential aspect of transmedia strategies.	What percentage of the public participates in the event *in loco*, and what percentage of the audience accesses the event via news coverage? What are the mechanisms of interaction in the transmedia strategy of coverage? Is participation involved in the coverage? If so, how can the *reader/user/viewer/listener* participate in the open system? Is there user-generated content (UGC) related to the event (parodies, recaps, mashups, fan communities, etc.)? Are there any policies restricting the disclosure of UGC? What activities are available to the audience within social media networks related to the event? Is there a system of rewards and penalties? For example, can the audience have its comments/photos published, can people get rewards for social media activities, and can they have comments blocked/removed?
10. Aesthetics: The visual and audio elements contribute to the news coverage and enhance the overall transmedia experience that unfolds across multiple media platforms.	What kinds of visuals (video, photo, infographics, news games, animation, holography, etc.) are used in the coverage? Is the overall look of the coverage realistic or a composed environment (usage of graphism, holography, immersive journalism, augmented reality)? Is it possible to identify specific design styles in the coverage? How does audio work in the coverage? Is there ambient sound, sound effects, music, and so forth?

The seventh feature is the media platforms and genres. The media platforms encompassed television, radio, the Internet, printed media, mobile media, and social media. It was noticeable the emphasis on second screen applications and geolocation, as well as the interaction on online social media networks, from the journalistic content offered by the broadcaster.

The eighth feature is the audience and market. Regarding the event's impact, televised broadcasting registered a 40 percent increase in the number of people reached compared to the 2012 London Olympics, and the Globo Network digital platforms registered 6.5 million users (Mermelstein 2016). These numbers show that the transmedia coverage strategy was successful: Globo Network beat audience records and led that segment. The broadcaster's robust and diversified editorial project reached a varied public.

The ninth feature is engagement. Globo Network attained the most engagement of any brand on online social media during the Summer Olympic Games (Soutelo 2016). Overall, the broadcaster's audience engagement strategy privileged interaction to the detriment of participation. This aspect should be improved in future transmedia coverages.

The tenth feature is aesthetics. The aesthetic perspective was characterized by technological innovations, for instance, incorporating virtual and augment reality, holographic projections, and the ultra-high-definition transmission 8K (a horizontal resolution of 7,680 pixels) format during the Opening and Closing Ceremonies.

This model was also applied to analyze the transmedia dynamics of similar milieus: Russian news coverage of the 2014 Sochi Winter Olympic Games (see Gambarato, Alzamora, and Tárcia 2016) and Brazilian news coverage of the 2014 FIFA World Cup (see Gambarato et al. 2017).

Conclusion

Transmedia journalism, as well as transmedia storytelling in general, essentially implies (1) multiple media platforms, (2) content expansion, and (3) audience engagement (Gambarato and Tárcia 2017). One of the fertile terrains for developing and implementing transmedia features in journalism is the coverage of planned events because this kind of media event presents emergent/ad hoc (large audiences and integration of the audiences in the news making) and planned/strategic (human, technical, and financial resources) aspects of transmedia productions. News coverage of planned events can vary, for instance, in terms of the size of the news company involved, the technological and financial resources available, the variety of professionals, and the strategies and goals to be achieved.

Noticeably, large broadcasters and media conglomerates (such as Globo Network) are as aware of the transmedia advancements in journalism as independent news professionals and companies (see Gambarato 2016). Transmedia journalism is already a reality that although likely more modest than comprehensive, is growing and improving. "Actually, our brain is transmedia" (Renó 2014, 8). "When we apply transmedia logic to journalistic online content, narratives and storytelling, this process would lead to exciting and engaging genres that are better able to inform citizens" (Tellería 2016, 74). In this scenario, the analytical and operational model of transmedia strategies for the news coverage of planned events presented in this chapter can contribute to this process of growth and improvement in the transmedia journalism realm. Transmedia journalism "covers an interactive content that grows and flows through the different media ecologies and technological environments, adaptable and flexible" (Tellería 2016, 77).

References

Alzamora, Geane, and Lorena Tárcia. 2012. "Convergence and Transmedia: Semantic Galaxies and Emergent Narratives in Journalism." *Brazilian Journalism Research* 8: 22–34.

Barnathan, Joyce. 2014. "Why Serial Is Important for Journalism: Transparency Is Key." *Columbia Journalism Review*, November 25. Accessed August 30, 2017. http://archives.cjr.org/the_kicker/serial_sarah_koenig_journalism.php.

Bogost, Ian, Simon Ferrari, and Bobby Schweizer. 2010. *Newsgames: Journalism at Play*. Cambridge, MA: MIT Press.

Canavilhas, João. 2014. "Jornalismo Transmídia: Um Desafio ao Velho Ecossistema Midiático" [Transmedia Journalism: A Challenge to the Old Media Ecosystem]. In *Narrativas Transmedia—entre Teorías y Prácticas*

[Transmedia Narratives—between Theories and Practices], edited by Carolina Campalans, Denis P. Renó, and Vicente Gosciola, 53–67. Barcelona/Barranquila: Editorial UOC/Universidad del Rosario.
Clark, Roy P. 2001. "The Line between Fact and Fiction." *Creative Nonfiction* 16. Accessed August 28, 2017. www.creativenonfiction.org/online-reading/line-between-fact-and-fiction.
Cute, Hillary. 2016. *Disaster Drawn: Visual Witness, Comics, and Documentary Form*. Cambridge, MA: Belknap Press.
De la Peña, Nonny, Peggy Weil, Joan Llobera, Elias Giannopoulos, Ausiàs Pomés, Bernhard Spanlang, Doron Friedman, Maria V. Sanchez-Vives, and Mel Slater. 2010. "Immersive Journalism: Immersive Virtual Reality for the First-Person Experience of News." *Presence* 19 (4): 291–301.
Dominguez, Eva. 2012. "Periodismo Transmedia, ¿Nuevo o Renovado?" [Transmedia Journalism, New or Renewed?]. *Revista de los Estudios de Ciencias de la Informacion e Comunicacion* 13. Accessed August 26, 2017. http://comein.uoc.edu/divulgacio/comein/es/numero13/articles/Article-Eva-Dominguez.html.
Fast, Karin, and Henrik Örnebring. 2015. "Transmedia World-building: The Shadow (1931-present) and Transformers (1984-present)." *International Journal of Cultural Studies* 20 (6): 636–652.
Ford, Sam. 2007. "Transmedia Journalism: A Story-based Approach to Convergence." *Futures of Entertainment*, April 5. Accessed August 30, 2017. www.convergenceculture.org/weblog/2007/04/ transmedia_journalism_a_storyb.php.
Gambarato, Renira R. 2013. "Transmedia Project Design: Theoretical and Analytical Considerations." *Baltic Screen Media Review* 1: 80–100.
Gambarato, Renira R. 2016. "The Sochi Project: Slow Journalism within Transmedia Space." *Digital Journalism* 4 (4): 445–461.
Gambarato, Renira R., and Geane Alzamora (eds.). 2018. *Exploring Transmedia Journalism in the Digital Age*. Hershey, PA: IGI Global.
Gambarato, Renira R., and Lorena Tárcia. 2017. "Transmedia Strategies in Journalism: An Analytical Model for the Coverage of Planned Events." *Journalism Studies* 18 (11): 1381–1399.
Gambarato, Renira R., Geane Alzamora, and Lorena Tárcia. 2016. "Russian News Coverage of 2014 Sochi Winter Olympic Games: A Transmedia Analysis." *International Journal of Communication* 10: 1446–1469.
Gambarato, Renira R., Geane Alzamora, and Lorena Tárcia. 2018. "2016 Rio Summer Olympics and the Transmedia Journalism of Planned Events." In *Exploring Transmedia Journalism in the Digital Age*, edited by Renira R. Gambarato and Geane Alzamora, 12–28. Hershey, PA: IGI Global.
Gambarato, Renira R., Geane Alzamora, Lorena Tárcia, and Amanda Jurno. 2017. "2014 FIFA World Cup on the Brazilian Globo Network: A Transmedia Dynamics?" *Global Media and Communication* 13 (3): 283–301.
Gifreu-Castells, Arnau, Richard Misek, and Erwin Verbruggen. 2016. "Transgressing the Non-fiction Transmedia Narrative." *View: Journal of European Television History and Culture* 5 (10). doi:10.18146/2213-0969.2016.JETHC108.
Jenkins, Henry. 2003. "Transmedia Storytelling: Moving Characters from Books to Films to Video-games Can Make Them Stronger and More Compelling." *Technology Review*. Accessed August 26, 2017. www.technologyreview.com/biotech/13052.
Jenkins, Henry. 2006. *Convergence Culture: Where Old and New Media Collide*. New York: New York University Press.
Jenkins, Henry. 2009a. "The Revenge of the Origami Unicorn: Seven Principles of Transmedia Storytelling (Well, Two Actually. Five More on Friday)." *Confessions of an Aca-Fan*, December 12. Accessed August 26, 2017. http://henryjenkins.org/2009/12/the_revenge_of_the_origami_uni.html.
Jenkins, Henry. 2009b. "Revenge of the Origami Unicorn: The Remaining Four Principles of Transmedia Storytelling." *Confessions of an Aca-Fan*, December 12. Accessed August 26, 2017. http://henryjenkins.org/2009/12/revenge_of_theorigami_unicorn.html.
Kerrigan, Susan, and J. T. Velikovsky. 2016. "Examining Documentary Transmedia Narratives through the Living History of Fort Scratchley Project." *Convergence: The International Journal of Research into New Media Technologies* 22 (3): 250–268.
Long, Michael. 2016. "Globo Gathering: How Brazil's Biggest Broadcaster Is Tackling Rio 2016." *SportsPro*, August 4. Accessed August 26, 2017. www.sportspromedia.com/magazine_features/globo_gathering_how_brazils_biggest_broadcaster_is_gearing_up_for_rio_2016.
Looney, Margaret. 2013. "5 Tips for Transmedia Storytelling." *Media Shift*, January 30. Accessed August 26, 2017. www.pbs.org/mediashift/2013/01/5-tips-for-transmedia-storytelling030/.
Matney, Lucas. 2017. "CNN Launches Dedicated Virtual Reality Journalism Unit." *TechCrunch*, March 7. Accessed August 26, 2017. https://techcrunch.com/2017/03/07/cnn-launches-dedicated-virtual-reality-journalism-unit/.
Mermelstein, André. 2016. "Olimpíada Impulsionou Globo em Diversas Plataformas" [Olympics Boosted Globo Network in Various Platforms]. *Converge Comunicações*, August 26. Accessed August 26, 2017. http://convergecom.com.br/telaviva/paytv/26/08/2016/olimpiadas-impulsionaram-globo-em-diversas-plataformas/.

Moloney, Kevin T. 2011. "Porting Transmedia Storytelling to Journalism." Master's thesis, University of Denver.

Pernisa Jr., Carlos. 2010. "Jornalismo Transmidiático ou Multimídia?" [Transmedia or Multimedia Journalism?]. *Revista Interim* 10 (2): 2–10.

Renó, Denis P. 2014. "Transmedia Journalism and the New Media Ecology: Possible Languages." In *Periodismo Transmedia: Miradas Múltiples* [Transmedia Journalism: Multiple Perspectives], edited by Denis P. Renó, Carolina Campalans, Sandra Ruiz, and Vicente Gosciola, 3–19. Barcelona: Editorial UOC.

Renó, Denis P., and Jesús Flores. 2012. *Periodismo Transmedia* [Transmedia Journalism]. Madrid: Fragua Editorial.

Schlichting, Laura. 2016. "Interactive Graphic Journalism." *View: Journal of European Television History and Culture* 5 (10). doi:10.18146/2213-0969.2016.JETHC110.

Scolari, Carlos A., Paolo Bertetti, and Matthew Freeman. 2014. *Transmedia Archaeology: Storytelling in the Borderlines of Science Fiction, Comics and Pulp Magazines*. Basingstoke: Palgrave Macmillan.

Soutelo, Nayla. 2016. "O Que as Redes Sociais Fizeram Pelas Olimpíadas e o Que Também Podem Fazer Pela Sua Marca" [What Have Social Media Networks Done for the Olympics and What They Can Also Do for Your Brand]. *SocialBrain*, August 26. Accessed August 30, 2017. www.socialbrain.com.br/inbound-marketing/o-que-as-redes-sociais-fizeram-pelas-olimpiadas-e-que-tambem-podem-fazer-pela-sua-marca/.

Tavares, Olga, and Alan Mascarenhas. 2013. "Jornalismo e Convergência: Possibilidades Transmidiáticas no Jornalismo Pós-massivo" [Journalism and Convergence: Transmedial Possibilities in Post Mass Journalism]. *Revista Famecos* 20 (1): 193–210.

Tellería, Ana S. 2016. "Transmedia Journalism: Exploring Genres and Interface Design." *Trípodos* 38: 67–85.

10

TRANSMEDIA SPORTS

The National Basketball Association, Emojis, and Personalized Participation

Ethan Tussey

Televised sports leagues, though different in many ways from other media content, are similar to most other popular media brands in that the meaning of the primary text (in this case the games) is influenced by peripheral or paratexts. Robert Brookey and Robert Westerfelhaus' (2002) work on DVD extras and Jonathan Gray's (2010) work on a variety of "paratexts" adapted the theories of Gèrard Genette in order to advocate for the consideration of peripheral content when assessing a media text. Not all paratexts are relevant to all studies of a media. Gray argues that for "different people and different communities, at different times, the hierarchy of value will be different" (Brookley and Gray 2017, 102). For transmedia strategies, those paratexts that contribute to the ongoing narratives of sports are the most relevant. Jason Mittel warns not to conflate paratexts with transmedia practices as he defines the former as those that introduce or promote a text and the latter as those "that function as ongoing sites of narrative expansion" (Mittel 2015, 293).

Sports leagues generate an enormous amount of paratexts but focusing on sports narratives reveals the transmedia texts. Sports leagues have narratives. These narratives are crafted by television producers, athletes, journalists, and, increasingly, by fans. The crafting of these narratives across multiple platforms is the topic of this chapter. The case study in this chapter centers on the development of the National Basketball Association (NBA) emojis (or NBAmoji) and their partnership with Snap Inc. as a social media tool that extends transmedia authorship to fans of the league. Transmedia sports strategies, like transmedia strategies generally, range from protectionist to inclusive. Suzanne Scott (2010) has written about the ways in which media companies constrain participation with transmedia texts via temporal control and unification of interpretative meaning. While most transmedia content does not give fans full creative freedom and opportunity for collaboration promised in the utopian predictions about transmedia storytelling, certain transmedia efforts, such as the Snap stories discussed in this chapter, offer more flexibility and creative license to fans that want to craft sports narratives.

Sports Narratives

Sports season are narratives in the same way that television seasons are narratives. Sports leagues even create new initiatives to extend their narratives into the off-season to keep their stories in the minds of their fans (Bontemps 2017). Journalists, radio sports talk hosts, and television producers each play a vital role in the maintenance and agenda-setting of sports narratives. Television producers, whether they be the league's own networks or their broadcasting partners, represent the most official narratives. The rise of digital platforms has been crucial to the extension and depth of these narratives. David Rowe and Brett Hutchins contend that digital media has transformed the "media sport content economy" "from the 'scarcity' that was typified by analogue television and radio, and print-based media, to the plentitude that is produced by convergent digital media forms that convert cultural data into material that can be reshaped and freely circulated in ways that are difficult to govern" (2013, 2). For example, social media platforms have offered athletes a larger say in sports narratives. Indeed, digital platforms have given voice to many constituencies that have made sports narratives more complex, shorter, and diverse.

Sports leagues use their transmedia strategies to attempt to engage with the digital audience and direct the conversation. A prominent transmedia strategy, especially in sports programming, is through narrative extensions or what Will Brooker (2004) has called "television overflow" in which the narratives on the primary text continue on digital platforms. For sports, overflow spills onto television post-game discussion, highlights, game recaps, analysis, and online commentary. Victoria E. Johnson (2009) argues that sports are particularly suited to digital overflow because games can be so easily dissembled into bite-sized pieces to create ancillary content across transmedia platforms. Leagues and television producers offer this ancillary content following the games and repeatedly throughout the week to channel fans from their television to their mobile devices and desktop computers and back to their televisions for the next game.

Research by Gambarato et al. (2017) on the transmedia storytelling efforts by Globo Network during the 2014 FIFA World Cup and by Gambarato, Alzamora, and Tárcia (2016) on the Russian news establishment during the 2014 Winter Olympics, detail the efforts of media networks, governments, advertisers, and sports administrations to unify the coverage of sports events around particular themes. For example, Gambarato et al. explain that the use of official hashtags during the 2014 World Cup was an effort by Globo Network to engage audiences around the narrative theme of "now we all are one" (2017, 287). The media company worked to promote this theme across media platforms to present a unified meaning about the event despite the public protests that surrounded the stadiums. Gambarato et al. (2017) point out that large planned events, like sporting events, are significant transmedia storytelling opportunities that require financial investment and planning by media companies. The goal of these media outlets is to effectively manage the perception of an event by engaging on a variety of platforms, encouraging and filtering audience participation to present a unified meaning of the event.

Transmedia storytelling is not limited to the event itself but extends across the days leading up to and following the event. A good example of this strategy was Fox Sports effort to create a digital programming segment dedicated to office workers called *Lunch with Benefits* (Fox Sports Digital 2009). The initiative was an effort to offer sports programming for sports fans during their lunch hour at their cubicle. Each weekday provided a different web series that provided strategic analysis of the sporting events, previews, humor, and commentary. I have argued that these workplace-focused transmedia efforts represent a "procrastination economy" designed to fill people's in-between moments with amusing diversions while promoting upcoming games (Tussey 2018). The procrastination economy is not limited to sports programming and has become a prominent strategy for transmedia efforts on mobile devices. Other examples of transmedia content designed for our in-between moments include podcasts, fantasy leagues, and social media content.

Sports leagues, broadcasters, and journalists have not only incorporated digital strategies but they have also adapted their narrative practices in response to changing technology. Emphasis has moved away from focusing on recapping the games and providing highlights as fans typically have access to this information instantly on their mobile devices. Instead, there is a premium on unique perspectives and "hot take" commentary that can foster discussion and analyze sports narratives from a variety of perspectives. Fox Sports went so far as to cut back on their editorial staff and "pivot to video" where they could feature their opinionated sports analysts repurposed from television debates on digital platforms (Shaw 2017). Brett Hutchins and Jimmy Sanderson (2017) point out that though there are more platforms for discussing the sports narratives, the stories still begin with the games. The interpretation of these games, and the creation of controversial perspectives have heighted the political heft of sports events. Where once official sports commentary was the domain of the journalist, ex-athlete, and strategy analysts, now pundits debate athlete motivations and the values of sports leagues across transmedia platforms. Through the growing volume of content and attention-seeking strategies, transmedia sports content has achieved some of Henry Jenkins' original hope for transmedia storytelling, that diffuse content on multiple media platforms could foster diverse meaning making practices (Jenkins 2006). If not diverse, there is at least much more time dedicated to discussing the meaning of sports narratives.

Protective vs Inclusive Transmedia Sports Social Media: The Case of NBAmoji

For all the transmedia contributions to sports narratives, the leagues and their television partners still maintain the majority of the control. The events of the game begin on television and the narratives they produce are debated and dissected across media platforms. On television, the sports leagues have talk shows and debate shows discussing the latest stories to come from the games. Written content from newspapers, magazines, and websites fuel the stories and expand their implications. Mobile video options provide highlights and updates about continuing stories in real time to keep consumers up to date. Social media is a place where leagues distribute highlights and promotions for their other content, but it is also the space where fans are most likely to enter into conversation and shape debate. The social media platforms Reddit and Twitter have become important clearinghouses for sports fan feedback and meaning making. The National Basketball Association (NBA) is especially attuned to the daily discussions on these platforms and has even revised rules and policy based on fan opinion (Wojnarowski 2017). Danny Chau of *The Ringer* argues that certain athletes, such as Lonzo Ball, are elevated in sports narratives because their skill sets and personalities are ideal for "the hyperspecific deep-dives that make the internet a wonderful and terrifying place" (2017). All leagues encourage fan engagement, but some are more willing to engage in the niche interests of their fans while other leagues are more focused on getting fans to join their preferred conversation.

For example, Major League Baseball (MLB) is much more protective of their content than other American sports leagues. Officials from the league office strictly monitor social media platforms like YouTube in order to remove any content that is uploaded without authorization (Brisbee 2012). By comparison, the NBA is much laxer in its copyright enforcement, allowing its fans to upload images and mixtapes of the leagues athletes and games. The relaxed policy has given birth to a cottage industry of fan made NBA YouTube channels that contribute to the leagues narratives (Winkie 2016). Ramon Lobato and Julian Thomas (2015) demonstrate how "informal media economies" like these YouTube mix tapes have transformed media industries practices and enabled media content to travel the global unencumbered by commercial efforts to divide the world into separate economic efforts. Informal media economies are often precarious, threatened by the legal actions of rights holders. The NBA's global ambitions may explain their decision to allow this more inclusive transmedia practice.

While sports leagues vary in their rights enforcement strategies, most leagues offer fans some form of digital invitation to engage with their branded content. Fantasy sports leagues are a lucrative platform in which league content, information, and narratives are coopted for the social purposes of its fans. In fantasy sports leagues, fans compete against each other by building rosters of players. The achievements of those players in the official games have dual meaning, as it affects the outcome of the actual games and the fantasy games arranged by groups of competing fans. It is possible that a fan's knowledge of a player is primarily informed by his/her experience playing fantasy sports rather than watching the actual games. Research on fantasy sports by Billings, Ruihley, and Yang (2016) has found that a fan's allegiance to their fantasy team can rival their rooting interest in their favorite real team. This fan practice makes viewers more interested in every game instead of just watching the games of their favorite team. More watching means higher television ratings and more subscribers to streaming services offered by sports leagues. Fantasy sports began as a part of the informal media economy but are now worth millions of dollars to sports leagues (Isidore 2015). MLB initially resisted fantasy leagues, eager to charge fans for access to the stats necessary for running a fantasy league (Greenhouse

2008). Since MLB's plan was rejected by the Supreme Court of the United States of America, sports leagues have embraced fantasy sports and routinely incorporate the narratives of "fantasy value" into their official coverage.

Photo-sharing social media platforms are the latest terrain for transmedia strategies. Social media companies like Snapchat and Instagram offer sports leagues an opportunity to reach a mobile audience with updates from live sporting events. Sports leagues try to engage these mobile users by offering the opportunity to upload their own contributions to the story of the sporting event. Snapchat calls this feature, appropriately enough, "stories," and each of the major American sports leagues have partnered with the social media app to produce stories (Constine 2016). For fans attending the games, a number of filters are connected via geo-tags to the sports venue. Acting as a collection of amateur sports reporters and commentators, sports fans can offer their own images and commentary on the action. The leagues receive the uploaded "snaps" and create a narrative about the game that attempts to capture the feeling of attending the event.

Sports stories on Snapchat are extremely formulaic, as anything controversial is typically filtered out of the officially branded feed. Suzanne Scott (2009) argues that rights holders offer these designated playgrounds with officially licensed content as an acknowledgment of the desire of fans to be a part of the meaning-making experience. According to Scott (2009), rights holders "gift" fans a few options for creativity and then filter fan content to select the most palatable options and "regift" this content to the mass audience as a representation of fan engagement. Jimmy Draper (2012) has argued that rights holders use these interactive opportunities as a way to confirm or deny narratives that strengthen their brand strategy. More specifically, Scott (2009) points out that the filtering process has a tendency to favor male fandom over female perspectives. These critiques are more than apt for the official sports league stories offered on Snapchat, but the official story is not the only place where transmedia content is distributed by fans.

Snapchat offers peer to peer sharing of "stories" as well as location based "stories." Additionally, "Snap Map" allows Snapchat users to travel the virtual globe looking for publicly posted images at different locations. A sports fan may not get their snap posted on the official NBA story but their snap would appear at the site of the game for any Snapchat user that wanted an unfiltered view. A similar unfiltered opportunity is available for snaps sent from one user to another or to their network of friends. In these instances, the filters and branding of the sports leagues are at the mercy of the fans and their own creative meaning-making practices. While leagues are invested in the creation of official Snapchat stories, the toolkit offered by Snapchat offers fans an opportunity to make diverse and creative narrative contributions.

The NBA, in particular, has led the way in developing more inclusive transmedia offerings through efforts like its licensed emojis. Semiotician Marcel Danesi claims that

> in the age of the "electronic global village" where people of different national languages and cultures are in frequent contact through online interactions, the emoji code might well be the universal language that can help solve problems of comprehension that international communications have always involved in the past.
>
> *(2016, vii)*

It is unlikely that the use of basketball emojis can accomplish that lofty goal, but Danesi's point about international comprehension is key to understanding why investment in emojis reflects an inclusive transmedia strategy. Emojis offer an open-ended, though typically limited to positive reactions, menu of expressive images (sometimes called stickers) and animations that can be used by fans in text-based messages. All of the major American sports leagues offer emoji keyboards but the NBA's is by far the most sophisticated and engaged with its fans.

Often with transmedia storytelling, the rights holders license their brand to third-party partners to produce these transmedia texts on particular platforms. Derek Johnson (2013) has written about the ways that media companies with long-standing brands rely on the creativity of licensors to provide guidance and expertise in unfamiliar markets. The autonomy, budget, and collaboration that goes into these partnerships reveals just how much effort and consideration sports leagues put into these transmedia texts. The National Football League (NFL), the most lucrative sports league in America, does not license to a third party but instead allows NFL Enterprises to design their emoji keyboard. The result is an emoji keyboard with very few options (only four unique player stickers per team) and very little investment. MLB similarly keeps its emoji keyboard in-house offering stickers and exclusive GIFs from broadcasts. The keyboard has many options and includes iconography that is meaningful to fans and their communications preferences. Still, the decision to keep the digital product in-house and leverage exclusive content reflects the league's larger digital philosophy. The National Hockey League (NHL) partnered with Molson Beer Company and Swyft Media Inc. to create their emoji keyboard which, according to iTunes App store, offers 64 "Molson Canadian and hockey themed emojis, including one for every NHL team." It is difficult to say if this emoji keyboard is more about drinking or hockey. The NBA emoji transmedia strategy is the most unlike its competitors and deserves to be considered as a harbinger of the league's unique approach to fan engagement.

The NBA partnered with YinzCam, a technology start-up out of Carnegie Mellon University, to create a series of emojis on different platforms. In its partnership with YinzCam, it updates its emojis regularly, offering animated stickers of players relevant to different NBA narratives. During the 2016 Olympics, the app featured the members of both the US men's and women's basketball teams. For the NBA Finals, emojis were created for each of the players on both teams with specific animations that reflected their characters within larger league narratives. For example, Lebron James was given a Crown and the title "King" while a flexing screaming Draymond Green reflected his enforcer persona (Beaumont 2016). These emojis provide NBA fans with illustrations for their conversations with friends. These transmedia tools contribute to the sports narratives constructed by a small circle of friends but they rarely extend to a wider public.

The more public facing version of NBA emojis are found in the league's partnership with Snapchat. The NBA licenses the logos of its teams to provide virtual clothing and poses that can be customized to Snapchat users' Bitmoji avatars. Bitmojis are personal emojis, animated versions of users that act as an avatar for a person in text chains and in Snapchat photographs. Bitmojis depict the users in dozens of regularly updated responses and actions that reflect common conversation on social media and text chains. Snap Inc purchased Bitmoji in 2016, bringing these animated creations to the photo-based social media platform. Bitmoji was the top downloaded app in five global markets in 2017, which reflects the international appeal of communicating through personalized emoji (Molla 2017). Through the NBA's Bitmoji transmedia texts, fans have the ability to represent themselves as part of the brand instead of simply integrating the logos and stickers into conversation. Beyond the ability to decorate pictures with licensed emojis, fans can situate their avatars into the game action using Snapchat's camera options. For users, checking for updates on NBA games via Snap Map, they will find personalized avatars encroaching on the court and contextualizing the game outside of the NBA's preferred narrative.

Elizabeth Evans (2011) identifies narrative, authorship, and temporality as the unified traits of a transmedia text. She argues that an audience is aware of the connections between text and its transmedia siblings because of the way transmedia texts directly engage with these three elements. Traditionally, these elements were guarded by the original storytellers, meaning that authorship and distribution are controlled by the rights holder. Indeed, the official NBA story is this kind of a transmedia text but the NBA's decision to offer branded bitmojis and site-specific geofilters allows fans to engage with these three elements by producing transmedia narratives alongside the official accounts using the same digital tools and markers of authenticity. Fans are then able to perpetuate or challenge narratives, that they author themselves, and distribute in the same temporal windows of the official account. Certainly, the options for using the branded iconography of the league are limited but a Snapchat user has control of the placement and timing of distribution in ways that are not as limited as the "regifted" official NBA Snap story. The possibilities for creativity and collaboration are far more inclusive than the more protectionist transmedia options offered to fans on other social media platforms.

For an example of how Snapchat allows NBA fans to present a diverse sports story consider the jersey retirement ceremony of Los Angeles Lakers guard Kobe Bryant. The official Snap story produced by the NBA for Snapchat features high-quality images, interviews with players, testimonials from celebrities, and archival footage of Bryant's career. The message of the short video is celebratory, congratulatory, and family oriented. This video is in stark contrast to the low visual quality of the Snap videos posted by people that attended the game. Some even used the NBA's branded emojis, filters, and lenses to critique Bryant's career including his treatment of teammates, his on-court demeanor, and the sexual assault allegations levied against him in 2003. Fans wanting to experience the event on Snapchat would have seen these user-generated videos and critiques of Bryant's villainy alongside celebrations of his career. The ambiguous story produced by the user-generated stories on Snapchat are much more nuanced than the simplistic celebration offered by the NBA's official Snap story and most coverage of the event by the mainstream sports press.

The audience's ability to co-opt the hashtag of an event to criticize the main narrative is not unusual. Indeed, Karin Fast and Henrik Örnebring (2015) have argued that emergent fan reactions that contradict or question the rights holder's narrative choices are a foundational reality of integrating audiences into collaborative storytelling. The NBA's collaboration of Snapchat seems to invite this emergent contradiction as they enable the audience to combine the cameras on their mobile devices with officially branded iconography of the league to present the events in diverse audiovisual ways. Thus, when a fan wants to see what is going on at a sporting event, they can go to Snapchat and watch as fans use the tools of the NBA to tell their own audiovisual stories.

Conclusion

The NBA's more inclusive transmedia strategy can be partly explained by the characteristics of their core audience and their desire to cater to a global fan base. All sports leagues have dedicated fans, but NBA fans are particularly engaged online. NBA Reddit, NBA Twitter, and the NBA blogosphere are among the most prolific fan communities of the major sports leagues. The NBA allows these online communities to remix their content on these platforms. Digital technology has also offered a glimpse at the players' lives in unprecedented ways. Each of the major American sports leagues has a social media policy for players but the NBA's is much less strict than its competitors, which may partially explain why Forbes ranked NBA players as four of the top ten sports Twitter accounts while leaving off any NFL players (DiMoro 2016). The power of the players in the league provides more flexibility and breadth in the stories about the league. The structure of the sport itself, with no helmets or equipment to obscure the players, and comparatively few athletes on the court at the same time, offers access to the players and their personalities in ways that other leagues cannot offer. The NBA encourages this through their creation of transmedia tools for its fans. These tools including emoji keyboards and Snapchat integration emphasize player personas and fan integration into the league's brands. These factors set them apart from other professional sports leagues.

Professional sports leagues have always been in the business of building narratives to bolster ticket sales, merchandising, and broadcast ratings. In the era of digital technology, building narratives requires a transmedia strategy. Some leagues choose to strictly constrain the narratives across their multiplatform efforts. While this may not restrict debate among fans, it certainly restricts the amount and type of branded and licensed content that fans can use to engage these narratives. The NBA, through its development of the NBAmoji app and its partnership with Snap Inc., charts a more inclusive course that provides some opportunities for creativity and collaboration. There are many factors that contribute to the decision to pursue a protective or inclusive transmedia strategy. League support by its most ardent fans may be one of those factors and the power and agency of the league's players may be another. Whatever the reason, the NBA's transmedia partnership with Snapchat reflects the democratizing potentials espoused by early theorists of transmedia storytelling practices.

References

Beaumont, Vanessa de. 2016. "NBA Debuts NBAmoji App in Time for Finals Reactions." *Bleacher Report.com*, June 10. Accessed May 12, 2017. http://bleacherreport.com/articles/2645620-nba-debuts-nbamoji-app-in-time-for-finals-reactions.

Billings, Andrew, Brody James Ruihley, and Yiyi Yang. 2016. "Fantasy Gaming on Steroids? Contrasting Fantasy Sport Participation by Daily Fantasy Sport Participation." *Communication & Sport*. doi: 10.1177/2167479516644445.

Bontemps, Tim. 2017. "How the NBA Made Summer League a Real Thing and Stole July in the Process." *The Washington Post*, June 20. Accessed May 11, 2017. www.washingtonpost.com/news/sports/wp/2017/07/20/how-the-nba-made-summer-league-a-real-thing-and-stole-july-in-the-process/?utm_term=.a3cb93cf0336.

Brisbee, Grant. 2012. "The Archaic YouTube Policy of Major League Baseball." *SBNation*, January 11. Accessed May 11, 2017. www.sbnation.com/2012/1/11/2699919/baseball-videos-youtube-removed-due-to-copyright-violation.

Brooker, Will. 2004. "Living on Dawson's Creek: Teen Viewers, Cultural Convergence, and Television Overflow." In *The Television Studies Reader*, edited by Annette Hill and Robert C. Allen, 569–580. London: Routledge.

Brookey, Robert, and Robert Westerfelhaus. 2002. "Hiding Homoeroticism in Plain View: The Fight Club DVD as Digital Closet." *Critical Studies in Media Communication* 19 (1): 21–43.

Brookley, Robert, and Jonathan Gray. 2017. "'Not Merely Para': Continuing Steps in Paratextual Research." *Critical Studies in Media Communication* 34 (2): 101–110.

Chau, Danny. 2017. "Lonzo Ball is a Superstar for the Reddit Generation." *The Ringer*, July 18. Accessed May 10, 2017. www.theringer.com/2017/7/18/16078122/lonzo-ball-is-a-superstar-for-the-reddit-generation-dd47b7db63c0.

Constine, Josh. 2016. "Snapchat Challenges Twitter and Facebook for Sports Talk." *TechCrunch*, January 29. Accessed May 3, 2017. https://techcrunch.com/2016/01/29/sportschat/.

Danesi, Marcel. 2016. *The Semiotics of Emoji: The Rise of Visual Language in the Age of the Internet*. London: Bloomsbury Publishing.

DiMoro, Anthony. 2016. "50 of the Best Athletes on Twitter in 2016." *Forbes*, December 19. Accessed May 22, 2017. www.forbes.com/sites/anthonydimoro/2016/12/19/50-of-the-best-athletes-on-twitter-in-2016/#48a6754656ae.

Draper, Jimmy. 2012. "Idol Speculation: Queer Identity and a Media-imposed Lens of Detection." *Popular Communication* 10 (3): 201–216.

Evans, Elizabeth. 2011. *Transmedia Television: Audiences, New Media, and Daily Life*. New York: Routledge.

Fast, Karin, and Örnebring, Henrik. 2015. "Transmedia World-building: The Shadow (1931–present) and Transformers (1984–present)." *International Journal of Cultural Studies* 20 (6): 636–652.

Fox Sports Digital. 2009. "Fox Sports Creates Digital Programming Unit." Press release, September 9. Accessed May 12, 2017. http://static.foxsports.com/content/fscom/binary/migrated/20027/10044456_37.

Gambarato, Renira R., Geane Alzamora, and Lorena Tárcia. 2016. "Russian News Coverage of 2014 Sochi Winter Olympic Games: A Transmedia Analysis." *International Journal of Communication* 10: 1446–1469.

Gambarato, Renira R., Geane Alzamora, Lorena Tárcia, and Amanda Jurno. 2017. "2014 FIFA World Cup on the Brazilian Globo Network: A Transmedia Dynamics?" *Global Media and Communication* 13 (3): 283–301.

Gray, Jonathan. 2010. *Show Sold Separately: Promos, Spoilers, and Other Media Paratexts*. New York: New York University Press.

Greenhouse, Linda. 2008. "No Ruling Means No Change for Fantasy Baseball Leagues." *New York Times*, June 3. Accessed May 11, 2017. www.nytimes.com/2008/06/03/sports/baseball/03fantasy.html.

Hutchins, Brett, and Jimmy Sanderson. 2017. "The Primacy of Sports Television: Olympic Media, Social Networking Services, and Multi-Screen Viewing during the Rio 2016 Games." *Media International Australia* 164 (1): 32–43.

Isidore, Chris. 2015. "How Fantasy Sports Changed the NFL." *CNN Money*, September 11. Accessed May 1, 2017. http://money.cnn.com/2015/09/11/news/companies/fantasy-football/index.html.

Jenkins, Henry. 2006. *Convergence Culture: Where Old and New Media Collide*. New York: New York University Press.

Johnson, Derek. 2013. *Media Franchising: Creative License and Collaboration in the Culture Industries*. New York: New York University Press.

Johnson, Victoria E. 2009. "Everything New is Old Again: Sport Television, Innovation and Tradition for a Multi-Platform Era." In *Beyond Primetime: Television Programming in the Post-Network Era*, edited by Amanda Lotz, 114–138. New York: Routledge.

Lobato, Ramon, and Julian Thomas. 2015. *The Informal Media Economy*. New York: John Wiley.

Mittel, Jason. 2015. *Complex TV: The Poetics of Contemporary Television Storytelling*. New York: New York University Press.

Molla, Rani. 2017. "Snapchat's Bitmoji is the No. 1 Most-downloaded App in Five Top Markets." *Recode*, April 7. Accessed May 24, 2017. www.recode.net/2017/4/7/15191464/snapchat-facebook-google-twitter-uber-bitmoji-app-downloads.

Rowe, David, and Brett Hutchins. 2013. "Introduction: Sport in the Network Society and Why it Matters." In *Digital Media Sport: Technology, Power and Culture in the Network Society*, edited by David Rowe and Brett Hutchins, 1–15. New York: Routledge.

Scott, Suzanne. 2009. "Repackaging Fan Culture: The Regifting Economy of Ancillary Content Models." *Transformative Works and Cultures* 3. Accessed May 23, 2017. http://journal.transformativeworks.org/index.php/twc/article/view/150/122%3B.

Scott, Suzanne. 2010. "The Trouble with Transmediation: Fandom's Negotiation of Transmedia Storytelling Systems." *Spectator* 30 (1): 30–34.

Shaw, Lucas. 2017. "Fox Sports Cuts Web Writing Staff to Invest More in Online Video." *Bloomberg*, June 26. Accessed May 2, 2017. www.bloomberg.com/news/articles/2017-06-26/fox-sports-cuts-web-writing-staff-to-invest-more-in-online-video.

Tussey, Ethan. 2018. *The Procrastination Economy: The Big Business of Downtime*. New York: New York University Press.

Winkie, Luke. 2016. "A Look into the Semi-Legal World of NBA Highlight Videos." *Sports Illustrated*, February 9. Accessed May 15, 2017. www.si.com/extra-mustard/2016/02/09/nba-highlight-videos-clips-youtube-legality.

Wojnarowski, Adrian. 2017. "NBA's Hope for Draft Lottery Reform Tied to Research Showing Fans Hate Tanking." *ESPN*, September 21. Accessed May 12, 2017. www.espn.com/nba/story/_/id/20774528/nba-%20draft-lottery-reform-rooted-fans-disinterest-watching-teams-tank.

11

TRANSMEDIA SOCIAL PLATFORMS

Livestreaming and Transmedia Sports

Portia Vann, Axel Bruns, and Stephen Harrington

Contemporary social media platforms provide clearly circumscribed media spaces in their own right; we speak of being "on Facebook" or "in the Twittersphere," for instance. At the same time, however, they are also densely interconnected with other parts of the broader media ecology, and enable rich transmedia experiences and engagement. Since the mass adoption of currently leading social media platforms such as Facebook, Twitter, and Instagram in the mid to late 2000s, media producers from many conventional media channels, as well as social media users themselves, have actively pursued the creation and further enhancement of such transmedia relationships. In doing so, they have sometimes worked in unison, and at other times clashed over their vision of what shape an engaging transmedia experience around a shared media text might take, and of which stakeholders might be in control of it.

Such transmedia practices have developed around a wide range of media texts, from the news through fiction content to live events. Of particular interest has been the role of social media in "connecting" audiences for live television, "amplifying" their collective voice, and harnessing their input in creative ways (Harrington, Highfield, and Bruns 2013, 405). In this chapter, however, we seek to move beyond a view of social media as a mere complement to the primacy of television, and examine on how transmediatization can produce distinct experiences and provide specific narrative contributions. Of particular note is how these forces collide in productive or challenging ways within the domain of live sports: a leading site of transmedia experiences (Hutchins and Rowe 2012).

The liveness of the central text positions the sporting event especially well for further engagement by sporting fans through social media: platforms like Twitter and (to a less extent) Facebook are themselves fast-moving, near-live media spaces, and are therefore well suited to sports fans seeking to follow and/or comment an unfolding sporting event in real time. Further, the lightweight, mobile nature of modern social media clients makes it possible for fans to engage with social media even as they are watching the event on television, via Webstream, or even in the stadium. But this has also encouraged the evolution of sport consumption from limited top-down models in the broadcast era to more open communication models in the digital era, now characterized by an abundance of information sources and distribution methods for sporting content (Hutchins and Rowe 2012; Rein, Kotler, and Shields 2006, 42), and this transformation has created a range of—as yet only partially resolved—tensions between the various stakeholders in any one sporting event. In this chapter, we show that niche sports, unencumbered by binding commercial and marketing arrangements, have at times been able to negotiate these tensions more proactively than their far better resourced mainstream counterparts.

The Trouble with Transmedia Sports

Frictions between sporting organizations' corporate interests and fans' expressions of engagement have become especially evident in the context of large-scale, mainstream sporting events such as the Olympic Games. As television networks incur substantial costs to secure exclusive broadcast rights to the Olympics and similar events, they (as well as the sporting organizations granting those rights) have aggressively employed copyright laws in order to protect these commercial agreements, and to ensure that the content they have licensed is not available through unauthorized distributors (Boyle and Haynes 2009, 38). However, this is exceptionally difficult in digital spaces, as fans and other media outlets now have the ability to share images and video from an event through a multitude of digital platforms. Thus, for sports organizations and television networks, attempting to control the dissemination of content that (potentially) infringes on broadcast rights is almost impossible without taking a heavy-handed approach (Hutchins and Rowe 2012).

For example, during the London 2012 Olympics, the International Olympic Committee (IOC) sought to prohibit ticket holders from sharing any images or videos taken at Olympic events on social media (Biggs 2012). Essentially, this meant that fans were not allowed to share personal photos or videos of the Games through their own social media accounts to document their Olympic experience—a restriction that many found understandably aggravating. In response to the backlash from sporting fans, the IOC relaxed these restrictions slightly, allowing event-goers to share *images* from the Games on social media; however, under current regulations, the IOC still prohibits ticket holders from sharing *video* content (Canton 2016). Given the multitude of fan content created at such major events, however, neither version of these rules appears particularly enforceable.

Similar tensions have also emerged in football, primarily at the Fédération Internationale de Football Association (FIFA) World Cup and in the English Premier League (EPL) (Hutchins and Rowe 2012; Statt 2014; Williams 2014). During the 2014 World Cup, many fans adopted the practice of sharing highlight-style video content on Vine after a goal was scored (Statt 2014). Owned by Twitter, Inc., Vine allowed users to easily create and upload six-second looping videos; many football fans created such content both by recording match events live in the stadium and by capturing replays at home from their television or computer screens. In response to this practice, organizing body FIFA and rights holders ESPN and Univision issued a number of takedown notices, while Vine suspended some of the offending accounts (Statt 2014). Following the Vine experience at the World Cup, the EPL subsequently similarly warned fans not to create and share short match videos online, explaining that the EPL was actively working to "curtail this kind of activity" (cited in Williams 2014).

The US-based National Football League (NFL) has extended this approach even further, policing the transmedia activities of ordinary fans as well as of teams and their players. In 2016, it introduced a strict video-sharing policy aimed at protecting the exclusive rights of its broadcast partners: the League prohibited teams from using platforms such as Facebook Live and Periscope to share live content during a game (from kick-off to one hour after the game finishes), creating and sharing any of their own highlights content, or turning any live video content into animated GIFs (Liptak 2016; Rovell 2016). Teams that violate this video-sharing policy face fines of $25,000 for the first offence, $50,000 for the second, and $100,000 for any further violations (Liptak 2016; Rovell 2016). The NFL created the video sharing policy to "maintain control of what is disseminated" by teams on social media (Rovell 2016) and "ensure viewers go through official NFL channels for video content" (Liptak 2016). In other words, in this case the organizing body is protecting rights holders from the unauthorized use of content even by the social media channels operated by teams legitimately participating in the League.

While IOC, FIFA, EPL, and NFL put these measures in place to protect their commercial interests and prevent fans (and teams) from distributing content from a venue or an official broadcast, it is hard to see how these organizations would be able to effectively restrict such widespread fan practices. Given the multitude of tools and platforms available to fans to create, share, or repurpose brand content, as well as the sheer number of fans who now engage in sharing content, successful enforcement would require substantial staffing and resources. Further, because of the liveness of sporting events, these crowdsourced alternatives to instant replay videos are likely to be watched mainly within hours or days of the event itself, namely by fans catching up with any sports developments they might have missed at the time. There is therefore also a need for organizations to respond to such possible infringements very rapidly. At the same time, however, such aggressive enforcement of commercial arrangements through blanket content restrictions and drastic retaliatory measures is also indicative of a "brand guardian" mindset (Christodoulides 2009, 141), which is very likely to alienate a fan community that considers itself of equal importance to the official sporting bodies in guarding the integrity of the sporting code (McCarthy et al. 2014, 183).

Transmedia Opportunities for Niche Sports

As we have noted, tensions around such spontaneous fan-driven activities, which implicitly seek to enhance the transmediality of live sporting events by accompanying the official, largely commercial live and broadcast media texts of the event with additional crowdsourced coverage, are especially prominent around major sporting events, tournaments, and leagues. This is unsurprising as many of the licensing arrangements governing the coverage and marketing of such events are long-standing and remain rooted in large part in a pre-digital broadcast mindset that positions the television coverage of the event as the undisputed core text, usually also creating distinct licenses for different geographical broadcast territories. This approach is no longer especially well-suited to audiencing practices that are increasingly incorporating alternative forms of live and time-shifted viewing (through broadband and mobile streaming) and assume comparable accessibility regardless of the geographic location of the viewer. Major sporting rights licensors including FIFA and the IOC continue to struggle to adjust their rights-granting approaches to such new audience engagement patterns; by contrast, our research has found that it is often smaller, niche sports that have been able to convert the disadvantage of their lack of established, lucrative broadcast contracts into a comparative advantage by exploring innovative new transmedia sports experience models.

Our observations here build especially on a major study of official social media communications activities around a comparatively niche sporting event in 2015: the Netball World Cup (Vann 2017), held in—and won by—Australia. Netball is played in fully professional annual leagues only in Australia and New Zealand, while teams elsewhere in the world are semi-professional or amateur. Even in its heartlands, the sport has traditionally received only limited mainstream media attention—in part this has resulted also from persistent media discrimination due to its status as a sport that is largely played by women only. However, this comparative lack of media interest also results in the potential for a relatively more flexible engagement with fans who seek to generate their own transmedia coverage of the sporting event, as well as a more proactive approach to enhancing the event experience with additional transmedia content produced by the event organizers themselves.

The Netball World Cup (NWC) had a limited domestic broadcast partnership with Australian pay-TV provider Foxtel, but no live broadcast arrangements in many other countries participating in the tournament. This arrangement not only freed it from the restrictions experienced by major events including the Olympic Games and FIFA World Cup, but indeed created an intrinsic need to explore approaches to providing at least some basic updates to fans in these otherwise unserved territories. Drawing on its partnership with Foxtel and the support of Twitter Australia, the NWC team therefore drew on the Twitter-owned SnappyTV tool to capture instant video replays from the Fox Sports broadcast stream and distribute them on Twitter and Facebook. Additionally, the NWC team also used Twitter, Inc.'s livestreaming tool Periscope to broadcast some post-match press conferences and other ancillary content from the tournament. Unlike the major sporting events discussed above, the NWC refrained from restricting netball fans from sharing their own coverage; rather, it sought to provide these fans with professionally produced, readily shareable content whose distribution would enhance the visibility and accessibility of the tournament even in the absence of substantial mainstream media coverage.

This approach shows an evolution beyond the blanket social media bans issued by FIFA and the IOC, and their media licensees. Mainstream media organizations pay significantly less for the broadcast rights to a niche event, but in turn also gain less power to restrict alternative, transmedia coverage. Ironically, niche sport events can therefore actually benefit from their lower ranking in the sports market: they retain at least some freedom to experiment with new forms of coverage, and to allow their fans to do the same. The NWC's broadcast partners did still restrict the social media team from creating its own livestreams (via platforms such as Periscope) of the live matches, but they allowed the team to share clips captured directly from the broadcast stream via SnappyTV. Further, the ban on streaming only applied to live games or immediate post-match content, and not to further livestreams designed to enhance fans' transmedia experience of the event. The central advantage of SnappyTV was its ability to offer accessibility to the event through broadcast-style content, compensate for the lack of easily accessible free-to-air coverage. The NWC team expected that a substantial number of fans would rely on its social media coverage to experience the event, as these fans could not access the event through mainstream media channels; its decision to use SnappyTV meant that fans could access content sourced from the pay-TV broadcast stream with only minimal time delays. Such short clips of the game could not replace a full broadcast, of course, but still offered fans a connection to the key moments in the Netball World Cup matches, and an opportunity to express their own fandom by sharing and commenting on the clips.

Niche Sports as Trailblazers of Innovation

The lack of funding and resources that niche sports commonly experience means that they are unable to fully utilize the technological opportunities available in a thoroughly converged transmedia environment—but it also compels them to find smart, agile, and innovative solutions that deliver benefits even amidst conditions of scarcity. Better established and resourced sporting codes, in turn, may be less agile and innovative due to their being locked into longer-term media partnership arrangements, but are also likely to gradually adapt the novel approaches that have been tried and tested successfully by minor sports. Indeed, in spite of the restrictive, broadcast-centric policies still embraced by some major global sporting events and leagues, there are signs that a shift in the balance between broadcast and digital media has begun to occur. Most centrally, a number of sporting codes are now experimenting much more openly with a variety of approaches to livestreaming their events to an undefined, global audience—a model that previously had been shunned as directly interfering with established territorial broadcast licensing arrangements.

As a result, many sports leagues, tournaments, and events have now created their own dedicated web platforms and/or smartphone apps to stream live content. Additionally, a range of livestreaming platforms, including Periscope, Facebook Live, YouTube Live, and embedded livestreaming directly on Twitter (an integration of Periscope into the platform itself), entered the market in 2015 and 2016. In fact, sports-specific livestreaming apps may therefore be a transitional phenomenon to widespread streaming on social media: with such apps, greater barriers of access continue to exist, whereas streaming on social media is more straightforward and often free. Consequently, livestreaming as an embedded transmedia experience on social media platforms is emerging as a legitimate option for sport organizations seeking to broadcast to an international audience. Fully developed, this model would reverse the conventional relationship between the sports broadcast as the central media text, and social media second-screening as an ancillary practice. Here, instead, the social media environment becomes the central platform of sports engagement, within which live and archived streams can be accessed on demand.

In its continuing efforts to broaden its appeal to more diverse audiences, Twitter, Inc. has been exploring these possibilities particularly aggressively. The NFL streamed ten games (out of 256) of its 2016 season on Twitter, in addition to live broadcasts on CBS, NBC, and the NFL Network (Stelter 2016). These games did not attract an especially large audience, however, with the initial games attracting some 243,000 viewers, compared to 15.4 million watching the game via simulcast on subscription-based services offered by CBS or the NFL Network (Wagner 2016b). Reportedly, sponsors regarded these viewing numbers as underwhelming: one advertising executive revealed that sponsors were "seeing a significant under-delivery from Twitter for our spots ... The problem is people aren't watching full games" (cited in Sloane 2016). This indicates two realities of the broader transmedia sport landscape. First, mainstream sporting organizations are beginning to use digital media platforms to broadcast live content, and fans are watching via these platforms; but second, consumption via television remains dominant for these mainstream sports, in spite of the increasing availability and accessibility of digital streaming options.

Nonetheless, Twitter, Inc. has struck similar agreements in the United States to livestream Major League Baseball (one game per week), the National Hockey League (one game per week), and the Professional Golfers' Association Tour (70 hours of coverage; Collins 2017; Wagner 2016c). Meanwhile, Twitter Australia partnered with the Victoria Racing Club to livestream the nation's major horse racing event, the Melbourne Cup, as a simulcast with free-to-air television broadcaster Seven and its own streaming platform, PLUS7 (Harley 2016; C-Scott 2016). While horse racing is not typically a mainstream sport, the Melbourne Cup carnival certainly enters a mainstream space. The event attracts mainstream media attention, a significant number of television viewers—1,986,000 in 2016 (Knox 2016)—and substantial commercial sponsorship; Twitter Australia's involvement is therefore a sign of a broader strategy to showcase sports livestreaming on the platform. Elsewhere, Twitter has also livestreamed content from around the grounds at Wimbledon, featuring interviews, analysis, match replays and highlights segments, but not live match coverage (Wagner 2016a). Similarly, the NBA has announced that it would "double the amount of digital content it creates for Twitter ... with more in-game highlights, behind-the-scenes shots, footage of player arrivals and livestreams of news conferences and interviews" (Koh 2016).

The continued dominance of broadcast television as the medium of choice for leading sports, and the easy accessibility of such broadcasts at least for domestic (as compared to international) audiences, however, means that although such major events might provide useful showcases for new approaches, their audience engagement patterns across different media platforms are unlikely to change dramatically within a short period of time. Rather, niche and minor sports will most probably reap more immediate benefits from these new opportunities: "unlike the major leagues, these sports generally don't have media rights that include TV broadcasts, nor the big budgets and sponsorship deals" (C-Scott 2016). Indeed, there is historical precedent for media transformation based on technological change creating greater opportunities especially for niche sports. In 1989, for instance, the European cable channel Eurosport was launched to accommodate the increasing amount of sport content acquired by the European Broadcasting Union's (EBU) member nations (Eurosport 2017b). On their own domestic channels, the national public broadcasters that constitute the EBU did not have the capacity to broadcast all of the sports content for which they had acquired the rights; thus, the pan-European dedicated sport channel of Eurosport was born.

The sports broadcast on Eurosport are typically not considered mainstream throughout many countries in Europe. Eurosport's current broadcasting program includes tennis (Australian Open, US Open), cycling, winter sports (for example alpine skiing, biathlon, and ski jumping), snooker, the FIFA Women's World Cup, the Union of European Football Associations (UEFA) European U-19 Championships, motorsports, weightlifting, and even the Australian Football League and US Major League Soccer (Eurosport 2017a). Eurosport's strategy was to obtain rights to a larger number of less mainstream sports throughout Europe so as to avoid the intense competition with local national broadcasters for rights of mainstream sports (Collins 1998). While the sports included in the channel's program may only have small audiences in individual countries, across the European market as a whole Eurosport's broadcast schedule was attractive to advertisers and sponsors. Therefore, its programming schedule provides access to a number of lower-cost, less mainstream sports, made possible due to technological changes that have fostered the globalized distribution of sport content.

Social media and digital streaming platforms may present similar opportunities for the distribution of niche sporting content. In a crowded sporting market, sports organizations—especially those representing niche sports—can now turn to new media technologies such as livestreaming on social media to bypass the traditional gatekeepers of sports programming. The confluence of niche sports needing media space, and the new media technologies providing that space, creates the chance for new sports communication models to emerge. While not all niche sport organizations will be willing or able to take advantage of the opportunities afforded by these digital channels, and may instead or in addition continue to pursue the (for now) greater accessibility and reach of mainstream television networks, the growth in livestreaming points to an evolution of transmedia sport experiences as innovative alternative models become more prominent.

Innovation Pressure from Fan Communities

Meanwhile, if such opportunities for a reconfiguration of sports coverage are not addressed by the sporting organizations themselves, it is likely that fans—who themselves have access to increasingly powerful devices and platforms for the (unauthorized) live coverage of sporting events—will continue to take matters into their own hands, and exert considerable pressure on sporting bodies to innovate. Unilateral action by fans is especially likely where fans feel that sporting organizations are not acting in the best interests of the sport, are failing to provide adequate coverage of their sporting events, or are making it unreasonably difficult or expensive for fans to access the sports they follow.

A well-publicized conflict between boxing fans and Australian pay-TV provider Foxtel provides a timely reminder of the amplified transmedia tensions caused by the increased televisual capacity of social media platforms. On February 4, 2017, Foxtel aired a fight between Australian boxing rivals Anthony Mundine and Danny Green on the pay-per-view channel Main Event. To view the fight, fans needed an existing subscription to Foxtel to access the channel as well as pay an additional A$59.95 to watch the bout (Harris 2017). Given the limited availability of Foxtel in regional and rural areas of Australia, Brisbane-based Foxtel subscriber Darren Sharpe decided to rebroadcast the Foxtel feed of the event via Facebook Live in order to enable his friends in regional areas to watch the fight with him (Brennan and Buttigieg 2017; Harris 2017). The feed went viral well beyond Sharpe's friends, however: eventually more than 150,000 viewers tuned in to Sharpe's Facebook Live stream rather than the official broadcast (Harris 2017), presumably in order to avoid the significant access fees charged by Foxtel.

Halfway through the broadcast, Foxtel contacted Sharpe and ordered him to terminate the stream, or face legal action. Reports have stated that other fans streaming the event on Facebook Live had their Foxtel services cut off all together (Brennan and Buttigieg 2017). The next day, Foxtel released a statement threatening legal action against all viewers who streamed the event on Facebook Live (Brennan and Buttigieg 2017; Harris 2017). Much as has happened with unauthorized download services for music and movies, such entirely punitive responses by rights holders are unlikely to stop circumventive actions by fans, however. Already, there is a broad range of long-established sites that—in analogy to torrent sites like The Pirate Bay—provide up-to-date lists of current and upcoming unauthorized fan broadcasts of live sports (Bruns 2008). The legal prosecution of individual fan re-broadcasters is likely only to drive others engaging in the same practice further underground, rather than to stamp out the practice itself. Much as the use of unauthorized music and movie download and streaming services has gradually declined as authorized services such as Spotify and Netflix became available, so will the fan-led re-broadcasting of sporting events be able to be addressed only as more sensible, official online sports streaming solutions enter the market. Twitter and Facebook Live may have a role to play in this context, but entirely new operators—the Netflix equivalent to Eurosport—might also emerge.

Indeed, it is possible that growth in the digital streaming space will not be driven by the usual players in sports broadcasting, but instead by side entrants into the market. In Australia, for example, Internet service providers have made a play for exclusive streaming rights in order to enhance the competitive standing of their services. In late 2015, broadband and mobile network provider Optus announced that it had secured rights to stream the EPL (commencing with the 2016/2017 EPL season) live through both its own app and digital television service, Fetch TV, for the following three years (Siracusa 2015). Until then, Foxtel had held the rights to the EPL in Australia (as the foundation rights holders for the competition), and broadcast the league as part of its pay-TV sports package (Siracusa 2015). Optus's move to secure exclusive coverage of the EPL in Australia is the first time we have seen an Australian Internet service provider break into traditional television's monopoly over sports coverage; it is now Optus that is sub-licensing one EPL match per round to free-to-air public broadcaster SBS (Special Broadcasting Service 2016), as well as delayed coverage of 12 matches to Foxtel's Fox Sports (Bradford 2016). This represents a previously inconceivable shift in the sports media landscape. In a similar situation in Germany, the exclusive domestic rights to the 2017 Men's Handball World Championship were acquired by a banking company, Deutsche Kreditbank (a long-term commercial partner of Germany's Handball Bundesliga), from international broadcaster beIN Sports. The bank then streamed the entire event on its website, drawing on YouTube as its technical partner (Krieger 2017).

Conclusion: Toward Live Sports' Netflix Moment?

As more such novel partnership arrangements become available and are trialed in individual events and tournaments, some sports organizations may eventually turn away from conventional television broadcasting altogether and toward what is known as stand-alone or "over-the-top" online broadcast. It remains unlikely, for now, that mainstream sports would take this route, since their lucrative broadcast contracts provide them with a strong incentive to protect the status quo. However, the proportionally more significant role of social media in the communication of niche sports, and the comparative freedom from long-term broadcast and sponsor partnership obligations, may lead to further changes especially in the distribution of niche sports content. While broadcast licensees still fervently guard the content generated from their big-ticket sports programming, they give niche sports more freedom to stream live or near-live content on a variety of social media platforms. In turn, many niche sports organizations cannot rely on mainstream television networks to provide access to their events and competitions, and so they have a pressing incentive to use digital and social media platforms to overcome such challenges. In netball, for instance, niche events such as the Netball Europe U21 Championship and the Australian Men's and Mixed National Association Championship have already been livestreamed on YouTube, the organizations' websites, Facebook Live, and the fan-run Website Netball Scoop.

Livestreaming of sports on social media and dedicated sites and apps remains in an experimental stage, and no consensus on best practice for digital streaming has yet emerged in the industry. However, with a growing number of attempts by emerging stakeholders, challenging the dominance of mainstream broadcasters, to reconfigure the top-down distribution of live sports content for the transmedia environment, combined with an increasing amount of bottom-up livestreaming by fans and niche sports organizations, the possibility exists for completely new social media-based transmedia sports models to emerge. New approaches to sports broadcasting may prove to be as disruptive as Spotify has been to music listening practices, or as Netflix has been to the consumption of television drama. However, sports come with their own sets of conventions, fan needs, organizational structures, and commercial considerations, and further longitudinal studies are needed to chart the ways in which these elements evolve and rearrange themselves as livestreaming on digital and social media platforms assumes greater importance, and to study the implications of such shifts for the sports media industry.

References

Biggs, John. 2012. "London Olympics to Visitors: Don't Share What You See." *Techcrunch*, April 24. Accessed September 7, 2016. https://techcrunch.com/2012/04/26/london-olympics-to-visitors-dont-share-what-you-see/.

Boyle, Raymond, and Richard Haynes. 2009. *Power Play: Sport, the Media and Popular Culture*. 2nd ed. Edinburgh: Edinburgh University Press.

Bradford, Kevin. 2016. "When Live Doesn't Mean Live: How Optus' Out-of-Sync EPL Coverage Is Letting Fans Down." *Mumbrella*, September 9. Accessed February 8, 2017. https://mumbrella.com.au/optus-out-of-sync-epl-coverage-letting-fans-down-394349.

Brennan, Ben, and Mel Buttigieg. 2017. "'This Is Stealing': Foxtel Threatens Legal Action after Men Stream Green-Mundine Showdown." *Yahoo!7 News*, February 4. Accessed February 8, 2017. https://au.news.yahoo.com/a/34338763/this-is-stealing-foxtel-threaten-legal-action-after-men-stream-green-mundine-showdown/#page1.

Bruns, Axel. 2008. "Reconfiguring Television for a Networked, Produsage Context." *Media International Australia* 126: 82–94.

Canton, David. 2016. "Rio Olympics Social Media Guidelines." *Slaw: Canada's Online Legal Magazine*, July 20. Accessed September 7, 2016. www.slaw.ca/2016/07/20/rio-olympics-social-media-guidelines/.

Christodoulides, George. 2009. "Branding in the Post-Internet Era." *Marketing Theory* 9 (1): 141–144. Accessed April 26, 2017. doi: 10.1177/1470593108100071.

Collins, Richard. 1998. "Supper with the Devil – A Case Study in Private/Public Collaboration in Broadcasting: The Genesis of Eurosport." *Media, Culture and Society* 20 (4): 653–663. Accessed February 7, 2017. doi: 10.1177/016344398020004008.

Collins, Terry. 2017. "Twitter Adds PGA Tour to Roster of Sports Live Streams." *CNET*, January 6. Accessed February 8, 2017. www.cnet.com/au/news/twitter-to-live-stream-early-rounds-of-the-pga-tour/.

C-Scott, Marc. 2016. "Twitter's Live Stream of the Melbourne Cup Could Change How We 'Broadcast' Sport." *The Conversation*, November 1. Accessed February 1, 2017. https://theconversation.com/twitters-live-stream-of-the-melbourne-cup-could-change-how-we-broadcast-sport-67291.

Eurosport. 2017a. "2017 Programme Guide." Accessed February 7, 2017. http://programmeguide.eurosport.com/en/home/.

Eurosport. 2017b. "Our History." Accessed February 3, 2017. http://corporate.eurosport.com/about-us/history/1989-2/.

Harley, Jonathan. 2016. "Announcing the #MelbourneCup Live on Twitter." *Twitter Blog*, October 12. Accessed February 1, 2017. https://blog.twitter.com/2016/announcing-the-melbournecup-live-on-twitter.

Harrington, Stephen, Tim Highfield, and Axel Bruns. 2013. "More than a Backchannel: Twitter and Television." *Participations* 10 (1): 405–409. Accessed August 2, 2017. =www.participations.org/Volume%2010/Issue%201/30%20Harrington%20et%20al%2010.1.pdf.

Harris, Lia. 2017. "Foxtel to Take Legal Action against Fan That Live Streamed Anthony Mundine v Danny Green on Facebook." *Daily Telegraph*, February 5. Accessed February 8, 2017. www.dailytelegraph.com.au/sport/boxing-mma/foxtel-to-take-legal-action-against-fan-that-live-streamed-anthony-mundine-v-danny-green-on-facebook/news-story/08f71587f21af7bf062ae9cfd9ec9310.

Hutchins, Brett, and David Rowe. 2012. *Sport beyond Television*. New York: Routledge.

Knox, David. 2016. "Tuesday 1 November 2016." *TV Tonight*, November 2. Accessed February 1, 2017. www.tvtonight.com.au/2016/11/tuesday-1-november-2016.html.

Koh, Yoree. 2016. "Twitter Signs NBA Deal to Stream New Shows." *Wall Street Journal*, July 19. Accessed February 8, 2017. www.wsj.com/articles/twitter-signs-nba-deal-to-stream-new-shows-1468933202.

Krieger, Jörn. 2017. "IHF Handball World Cup Moves to OTT in Germany." *Broadband TV News*, January 5. Accessed June 12, 2017. www.broadbandtvnews.com/2017/01/05/ihf-handball-world-cup-moves-ott-germany/.

Liptak, Andrew. 2016. "NFL Teams Could Face Huge Fines for Posting Game GIFs and Videos on Social Media." *The Verge*, October 9. Accessed January 18, 2017. www.theverge.com/2016/10/9/13217968/nfl-teams-social-media-fines-gifs-videos-football.

McCarthy, Jeff, Jennifer Rowley, Catherine Jane Ashworth, and Elke Pioch. 2014. "Managing Brand Presence through Social Media: The Case of UK Football Clubs." *Internet Research* 24 (2): 181–204. Accessed April 26, 2017. doi: 10.1108/IntR-08-2012-0154.

Rein, Irving, Philip Kotler, and Ben Shields. 2006. *The Elusive Fan: Reinventing Sports in a Crowded Marketplace*. New York: McGraw-Hill.

Rovell, Darren. 2016. "NFL Teams Can Be Fined for Posting Video under New Social Media Policy." *ESPN*, October 10. Accessed January 18, 2017. www.espn.com/nfl/story/_/id/17750196/nfl-starts-new-social-media-policy-teams-fined-posting-video.

Siracusa, Claire. 2015. "Optus Snatches English Premier League Rights from Fox Sports in Australia." *Sydney Morning Herald*, November 2. Accessed February 8, 2017. www.smh.com.au/business/media-and-marketing/optus-snatches-english-premier-league-rights-from-fox-sports-in-australia-20151101-gkoedn.html.

Sloane, Garrett. 2016. "Twitter's NFL Live-Stream Numbers 'Underwhelm' Some Sponsors." *Advertising Age*, October 7. Accessed February 1, 2017. http://adage.com/article/digital/twitter-s-nfl-live-stream-numbers-underwhelm-sponsors/306189/.

Special Broadcasting Service. 2016. "SBS Strike EPL Deal with Optus." *SBS*, March 17. Accessed February 8, 2017. http://theworldgame.sbs.com.au/article/2016/03/17/sbs-strike-epl-deal-optus.

Statt, Nick. 2014. "Vine in the Cross-Hairs as FIFA, ESPN Seek World Cup Takedowns." *CNET*, July 3. Accessed September 7, 2016. www.cnet.com/au/news/with-world-cup-takedowns-fifa-shows-social-media-paradox/.

Stelter, Brian. 2016. "Twitter to Live Stream Thursday Night Football Games." *CNN*, April 5. Accessed February 1, 2017. http://money.cnn.com/2016/04/05/media/twitter-nfl-thursday-night-football/.

Vann, Portia. 2017. "'Gateway to the Sideline': Organisational Communication on Social Media at Large-Scale Sporting Events." Ph.D. thesis, Queensland University of Technology, Australia.

Wagner, Kurt. 2016a. "Here's What Live Sports Looks Like on Twitter." *Recode*, July 6. Accessed February 8, 2017. www.recode.net/2016/7/6/12104722/twitter-live-stream-wimbledon-nfl.

Wagner, Kurt. 2016b. "More than Two Million People Watched Twitter's NFL Stream on Thursday Night." *Recode*, September 16. Accessed February 1, 2017. www.recode.net/2016/9/16/12943246/how-many-people-watched-nfl-twitter.

Wagner, Kurt. 2016c. "Twitter Will Livestream Some MLB and NHL Games Starting This Fall." *Recode*, July 25. Accessed February 8, 2017. www.recode.net/2016/7/25/12260820/twitter-livestream-mlb-nhl-games-mlbam.

Williams, Mike. 2014. "Premier League Warns about Posting Goal Videos Online." *BBC News*, August 15. Accessed September 7, 2016. www.bbc.co.uk/newsbeat/article/28796590/premier-league-warns-about-posting-goal-videos-online.

12
TRANSMEDIA CELEBRITY
The Kardashian Kosmos—Between Family Brand and Individual Storylines

Šárka Gmiterková

Successful and widely popular stars have always relied on transmedia functioning. Even in the early 1910s, a general precondition for stardom required that film incarnations matched with widely circulated notions of the off-screen personality of the performer (DeCordova 2001, 98–116). After a decade of disrupting the balance with various scandals concerning morality, sexuality, and substance abuse, revealing hidden facets of the then celebrated personalities (DeCordova 2001, 117–151), the 1930s introduced a carefully orchestrated and vertically integrated system. Central to such scheme was a narrative dispersed across movies, promotion campaigns, and publicity, and which also influenced a portfolio of carefully selected products endorsed by the star. For example, ethnical stars such as Dolores del Rio or Dorothy Lamour rose to fame due to the advent of Technicolor, favoring exotic and colorful locations, supported by the beauty industry promoting the tropical look of sophisticated ethnicity through these women (Bery 2004, 181–197). As a self-made star, Joan Crawford's image provided a happy ending to many of her films where their narratives did not; Crawford's frequent self-transformations were secured by her constantly changing wardrobe, thus making the promise of identity reinvention attainable and somewhat tangible (Allen and Gomery 1985, 172–186).

As Richard Dyer famously stated in *Stars* at the end of the 1970s, in order to fully grasp each individual star image, we first have to take into account four different types of material: promotion, films, publicity, and criticism and commentary (Dyer 1998, 60–63). While the first two are usually generated as a part of the deliberate manufacture of a particular image or image context for a particular star (Dyer 1998, 60), the last two escape the direct control of individual personalities. Although we may speculate to what extent gossip, scandals, and leaked stories concerning a star's private life are stirred up by the famous people themselves, it is only the combination of multiple narratives from various sources that forms the very complex intertext—the star image. As Dyer acknowledged himself, his analysis related to the classical Hollywood system rather than to later or non-American forms of renown, and indeed to cinema rather than television, sports, or fashion (Dyer 1998, 60–63). Subsequent conceptions of stardom did not completely redefine Dyer's insights, but offered more nuanced takes on post-studio and new media based fame.

Such elaboration started with broadening the existing paradigm. In contrast to classical, plastic modes of stardom, requiring a close fit between the typical part and off-screen persona, contemporary modes of stardom are best described as elastic. This adjective points to the continuous re-negotiating of public personality in order to match every new project and the star thus "has to have a wardrobe of identities connected to a product stream" (King 2003, 49). Since the break up of the studio system, stars now navigate their careers themselves, with the assistance of agents, managers, lawyers, and publicists. Therefore, the rather passive sounding term "image" made way for the "persona," signaling that stars now have to actively engage in the process of shaping their own brand (Shingler 2012, 121–126). No longer contractually tied to a single studio for a long period of time, stars are now choosing their own projects in tune with their preferences and needs and take on various creative positions as producers or directors.

Celebrity Culture

However, even the term "persona" did not fully accommodate emerging forms of renown. Instead, celebrity adequately conveyed a turn toward personality, epitomizing previously unprecedented coverage of the private lives of those in the spotlight. While traditional notions of stardom were associated with talent or some sort of performance skills, celebrity fame stems from personal qualities and information circulating in the media. The growth of mass media throughout the twentieth century facilitated the development of the Hollywood star system, but it was the 1990s trend toward technological and industrial convergence that gave celebrity culture an unprecedented boost. With the rise of media conglomeration starting roughly a decade earlier, tying film studios together with publishing, media, and entertainment corporations and other sectors of business, celebrity proved to be a useful way of connecting these cross-media processes (Turner 2004, 34–41).

Treating celebrity not necessarily as a strictly defined category, then, but rather as one of the facets of a successful star's career, Christine Geraghty's essay on re-examining stardom stands out as one of the examples when star theory tried to contain the phenomenon. Geraghty presented three categories—that of a performer, professional, and celebrity—based on the level of showcasing performance skills and the amount of publicity coverage devoted to private life. Depending on their career point, stars can "pick up" the appropriate celebrity narrative; for example, after a box-office flop they can steer media attention toward their private life and vice versa (Geraghty 2007, 98–110). In spite of this effort in academic circles to conceptually merge "star" with "celebrity," the boundaries separating these two concepts have persisted.

Instead of prioritizing talent, giving the performer a chance to transform again and again for the purpose of various roles, "celebrity" gains value from being an instantly recognizable personality at any time. While "star" seems to be better suited for the purposes of transmedia storytelling as it derives from a succession of different parts embodied in different narratives across media, it tends to highlight the transformative, actorly skills. On the other hand, the narrowly defined celebrity brand can migrate through media and social network landscape faster, easier, and have a greater impact. Surely, a strong undercurrent of cultural hierarchy flows under these star versus celebrity assumptions, with the star's persona built on agency, skills, and achievements, while the celebrity's persona is built on a passive lifestyle and accidental forms of fame.

As well as the gender dynamics that are associated with these particular discourses (active male star versus glamorous female celebrity), well-known people are usually associated with a particular media context. Back in 1991, Christine Gledhill stated that "the cinema still provides ultimate confirmation of stardom" (Gledhill 1991, xi), with other areas of popular culture accommodating celebrity based renown such as sports, music, or fashion. Commercial television stood out as one of the biggest celebrity manufacturers. In particular, reality television programs epitomized all of the troublesome aspects of this low-brow form of fame, namely its short-term nature and heavy dependence on a single show, whose different facets may lead to various spin-offs, but rarely offer the participants a chance to grow outside of the original entertainment format. As such, these personalities have only limited futures and are an easy target for exploitative and damaging management.

Now a decade-running reality television show, *Keeping up with the Kardashians* (KUWTK) departs from these limiting conceptions. First three and then five featured sisters led by a strong matriarch opened up a distinct territory of celebrity functioning. With *Big Brother* contestants and *The Real Housewives* participants they share several attributes, namely fame obsession, unabashed self-exposure, lack of any performing talent, exhibition of wealth, family bonding, and both enviable and pitiful lifestyles. What separates this clan from these short-lived celebrities are their marketing skills and their ability to turn the proverbial 15-minute reality television fame into a fully fledged Kardashian universe. An immensely profitable family brand that is currently estimated at US$450 million net worth is based on the long running show and its various spin-offs. However, only one-sixth (around US$80 million) of their yearly revenues comes from the actual series itself, and the bigger part of their fortune rests on their individual product endorsements, development of other content based on their personal storylines, and pursuing careers outside of the original program (Bruce 2017). Despite heading in various directions—Khloé capitalizing on her bodily transformation, Kylie extending her cosmetic empire, and Kendall staring in various fashion campaigns—the participants must keep in mind that the family is greater than the sum of its parts (Scheiner McClain 2014, 50). Preserving this equilibrium is crucial, otherwise the family business would dissolve, and it is precisely this balancing act that makes the Kardashian empire such a compelling case for transmedia analysis.

The World of the Family

Celebrity culture is dominated by the flow of stories and range of characters (Turner 2004, 3–4). Various personal narratives intertwine, and new personalities enter into already established storyworlds, thus spreading and modifying them. When Kris Jenner pitched the idea for a show based on the everyday reality of her own family back in 2007, she capitalized on her first husband's Robert Kardashian's proximity to O. J. Simpson during the infamous murder trial in the mid-1990s. As one of his lawyers, Kardashian's name gained pop-culture resonance; the attorney and businessman of Armenian descent passed away in 2003. In its first series, KUWTK thus employed commemorative narratives such as the sisters remembering their father on the anniversary of his death or the youngest, Khloé, struggling with his absence through partying and heavy drinking. In season six, Kris toyed with the idea of changing her surname back to Kardashian and two years ago Khloé and Kim visited Armenia in order to learn more about their origins. On numerous occasions KUWTK featured authentic home video footage of the siblings when they were little, accompanied by their loving father. Suturing the patriarch into the show's narrative does not simply benefit from Robert Kardashian's transient fame, but it also enhances the family values of the whole show.

Drawing on the notion of transmedial worlds having distinct mythos, topos, and ethos, the family background has been extremely important for the program from its start. Following Klastrup and Tosca's (2004) logic, mythos provides the Kardashian universe with a basic knowledge concerning a timeline, personalities, and their bonds; the topos grounds the storylines firmly in Los Angeles' Calabasas district, and the ethos encapsulates values, attitudes, and beliefs that the individual figures convey. Already the original credits present the family posing in front of the camera, with clearly designated roles and hierarchy protruding. For instance, Kris, as a matriarch of the group, bosses everybody around; Khloé, doubting the necessity of the wind machine, profiles herself as the funny one; and Kim, in tight-fitting red dress, usurping the spot at the center of the composition, is the vain star of the whole series. The Los Angeles-based show evenly covers all members of the clan, including Kris and Bruce's marriage, the growing pains of their youngest offspring, Kendall and Kylie, and the slow disappearance of Robert Jr. However, the most interesting relationships called for their own air time.

As Kris Jenner confessed, in 2009 she was losing faith in the show since it got stuck with a string of inconsequential conflicts. Then her daughters started to get married and have children and new storylines entered the frame. After Khloé's whirlwind marriage to Lamar Odom, E! cable network accommodated a show following the newlyweds for two seasons. The same thing happened after Robert Jr. conceived a child with model, entrepreneur, and former stripper Blac Chyna in 2016, when a series of eight episodes monitored the couple preparing for the birth of their first baby. Other spin-offs did not signal the central romantic narrative directly, but it is obvious that whenever KUWTK left the Los Angeles base in order to visit Miami, New York, or the Hamptons, Kourtney took the spotlight together with her then boyfriend Scott Disick. The oldest Kardashian sister revealed her first pregnancy during the first season finale of *Kourtney and Khloé Take Miami* and during the 2014 Hamptons summer spin-off announced her third baby. While the original series prominently features dramatic peaks together with mundane, corny, or even explicit details, the spin-offs give us a better understanding of the sisters' collapsing relationships. Following the season six special finale on Kim's fairytale wedding to Kris Humphries, resulting in the show's highest ratings up to date, consisting of 10.5 million viewers (Woodward and Hendin 2015), the fast dissolution of the marriage is addressed in the second season of *Kourtney and Kim Take New York*. Similarly, Scott and Kourtney's temporary separation in 2010 makes better sense when watching the second season of the Miami spin-off.

Although "the fam" stays at the heart of KUWTK, central relationships have largely dissolved. This is certainly the case for the marriage of Kris to Caitlyn (formerly Bruce) Jenner in 2014. Incidentally, multiplatform stories were generated by Bruce's transformation into Caitlyn, involving a book, docuseries, etc. For almost seven years the program emphasized their parenting skills, communication, and mid-life crises and clashes of their conflicting values and lifestyles—all of which helped to ground KUWTK in an ordinary and relatable reality, despite its extraordinary setting and display of wealth. Correspondingly, Kourtney's bond with Scott and their growing young family also gave the show a relatively stable point in an otherwise hectic universe. After Kim's very public split from Humphries she took a different turn with her next marriage to the rapper and fashion designer Kanye West. Apart from their spectacular engagement we saw very little of their life together. That is a striking difference from Kourtney's personal milestones, for example sharing the delivery of her first two children with the viewers. Since West did not wish to participate on the show, Kim diverted the attention to herself, detailing her struggles with getting pregnant, various business activities, and organizing family trips, talks, and celebrations.

The family narrative thus underwent a major development as the main relationships, securing the stability and coherence of the world and providing a specific set of conflicts, fell apart. Despite all the ruptures the notion of sisterhood persisted. KUWTK portrays not only a day-to-day experience of this bond, but also ties it up with numerous entrepreneurial activities. The Kardashian ethos dictates that all members should work together or at least support each other's various business endeavors. Such close, family-based cooperation differs from celebrity product endorsements since the K-labeled merchandise stems directly from the personal stories, physiques, and experiences of the three sisters. Kourtney, Kim, and Khloé collaborated on several products, which evolved around a set of attributes typically associated with them, namely body, beauty, and fashion. In 2011 the sisters launched a "Kardashian Kollection," a clothing line consisting of signature figure-hugging dresses, underwear, bags, shoes, and jewelry, later expanding into "Kardashian Kurves," introducing several pairs of jeans modeled upon the sisters' different body shapes. The trio also developed a make-up line called "Kardashian Khroma" and the sisters branched out into the publishing industry as well, when they co-wrote an autobiography *Kardashian Konfidential* in 2010, and a year later a fiction book called *Dollhouse*.

As well as these momentary enterprises, which were all depicted on and in the show, the sisters run a chain of DASH stores. The first boutique in Calabasas opened prior to the premiere of the show, subsequently expanding to Miami Beach, New York, and temporarily to the Hamptons. In the first years of airing KUWTK, the mere opening of a new DASH store warranted additional shows, as the first seasons of *Miami* and *New York* demonstrate. Moreover, in 2015 the E! network accommodated a program called DASH Dolls, chronicling lives of several Hollywood-based boutique employees. The stores are the only place that provide a tangible Kardashian experience because of their display of pricy fashion items together with KUWTK related souvenirs, thus attracting tourists and fans rather than solvent buyers. Despite the attention, the sisters struggled with generating profits for years and the situation escalated at the start of KUWTK season 13 with Khloé and Kim willing to sell the failing franchise. The rise and possible fall of DASH stores, the flagship of all the sister pact enterprises, may ultimately terminate the era when the trio willingly cooperated on various projects.

Such processes threaten to further undermine the family brand, one already jeopardized by divorces, break-ups, and restrained sharing of personal details.

The Individual Storylines

The season 11 promo aired in fall 2015, presenting individual women from the Kardashian–Jenner clan posing and strutting in front of a simple black or white background. The days of the raunchy but togetherness-themed opening credits were long gone and the Kardashian universe departed from accentuating the family brand in favor of building individual celebrities. In transmedia vocabulary this can be perceived as a turn away from world-building to prioritizing the singular characters. Characters over world is a valid strategy, but the question remains whether this is suited to a project such as the KUWTK. On one hand such, tactics naturally result from the development in the women's lives, supported by celebrity culture's broad inclination toward distinct personalities. On the other hand, such a strategy also contains risks and dangers. As Graeme Turner noted in relation to the music band Spice Girls, their message of "girl power" was more convincing and relevant when performed as a collective manifesto. Despite the members gradually profiling themselves as individuals rather than as parts of the branded ensemble, their solo careers ceased to have a larger impact (Turner 2004, 56–57). With the Kardashian–Jenner female clan similarly spreading in various directions, the central message about prioritizing your relatives as best friends, co-workers, and business partners might well get lost in the process of launching separate projects.

The sisters now develop and promote goods, apps, and additional content, which originates in their personal ups and downs as captured on KUWTK. For example, Khloé was for a long time labeled as the funny, yet least attractive, one of all the Kardashian sisters. After the visible toning of her body, Khloé capitalized on her physical transformation with a self-help book *Strong Looks Better Naked*, followed by a show *Revenge Body*, where she guided participants during their adoption of a healthier lifestyle. Kris Kardashian's younger daughters, Kendall and Kylie Jenner, who are in their early twenties, outgrew the closely cooperating sisterly phase quite rapidly. Their collaboration portfolio resembles their older siblings' ventures into fashion retail (collections for PacSun and Topshop brands) and the publishing industry (the dystopian fiction *Rebels: City of Indra*). Nevertheless, their conjoint activities seem much more strained as these sisters head in opposing directions. While Kendall continues to move away from reality television entertainment, associating herself with the world of high fashion, Kylie developed her own program for E! network, monitoring her life shared with a group of friends rather than with a family circle. With Kendall being praised by the modeling industry for her natural beauty and her lithe body, Kylie created a vastly different, "sexy siren" persona, supported by surgically enhanced looks and flashy, almost vulgar looks. Despite their differences, the older Kardashian sisters never had to face such contrasting positioning as all of them were labeled as distinctly lower class due to their association with sex tapes, a lack of sophistication and talent, and their frequently displayed sensuous figures.

Only time will tell if these individualistic tendencies pose a risk for the Kardashian ethos in terms of the transmedia brand. However, as the falling ratings demonstrate, no amount of drama can replace the once central tight family circle (Knibbs 2017). Since the women's lives are thoroughly documented and also shared through social media, hardly any plot featured on KUWTK comes as a surprising twist. As BuzzFeed demonstrated, Kourtney's first pregnancy also marked the first time when the family issues could translate into tabloid headlines (Woodward and Hendin 2015). Since then, the Kardashians have teased the media with snippets of information and saved the actual drama for the show. But as their lives became more and more scrutinized, keeping control over the course of events and their public circulation became almost impossible, with twists and turns revealed weeks or months prior to the show. For example, when Kourtney and Scott broke up during summer 2015, the series featured their separation in mid-November. Although KUWTK cannot compete with tabloids on the grounds of actuality, it does nevertheless offer a more intimate take on the exciting events, lending first-person voiceovers and giving primary access to the women's emotional and practical reactions to various events.

The Matriarch as an Author

Despite the sometimes-unfortunate evolution of the familial ties and a strong nod toward character-based content, one crucial aspect of KUWTK remained unchanged throughout the years. Since the Kardashian universe puts forward female protagonists, it has always attracted hyperfeminine aesthetics and behavior. Beauty and looks are key priorities to the Kardashian women as they spend a lot of time undergoing beautification procedures and make-up and hair-styling sessions in order to meet social expectations concerning an attractive female body. Sexuality is another essential feature, since the sisters have voluptuous, pneumatic bodies with famously rounded buttocks and parade their silhouettes in tight-fitting ensembles. Although Kim continues to be associated with a sex tape and the sisters frequently pose in nude photoshoots, the overall verbal discourse of the show emphasizes conservative values. Kardashian women vocalize traditional gender stereotypes concerning marriage, motherhood, and fidelity, yet they project an overtly sexualized image.

Such a close attention paid to female protagonists clearly values their perspective while excludes other storylines. Male characters usually fade in the background, are ignored, or disappear entirely—in short, they just cannot keep up with the Kardashian women. In its first seasons, KUWTK frequently showcased this gender role contravention. While Kris worked on establishing the family business, Bruce was left with traditionally female tasks of caretaking, completing errands, and running the household. The younger male KUWTK participants also clashed with the female-dominated system of governance on numerous occasions. Before Kim and Kris Humphries' wedding, Khloé reminded the cocky groom that men in the family have no say whatsoever; the program further depicted the athlete clearly not willing to spend his life outside of a conventional male role. Scott and Kourtney's longtime relationship was riddled with his substance abuse issues and although Kourtney tried to learn from her mother's mistakes with Bruce, she nevertheless remained firmly in charge of the household.

Kris embodies another paradox through her self-coined term "momager," pointing to her simultaneous position as a mother of the clan and its manager as well. This conflicting status is perfectly encapsulated in one of her statements at the show's start, commenting on Kim's leaked erotic video: "As her mother, I wanted to kill her. But as her manager I knew I had a job to do." Being frequently portrayed as an exploitative fame-monger by the press, Kris is also perceived as a master mind behind the whole show. While other Kardashian related narratives may have changed through time, that of Kris being in charge of the whole empire did not. As Jenner herself confessed in her 2011 memoir, around the fourth season, when her daughters started to settle down, she carefully orchestrated strategic moves in order to capitalize on the growing ratings numbers (Jenner 2011, 277). The operating business model that Kris came up with consisted of three cornerstones: there is the central successful reality television show, followed by social media nurturing even closer interaction with fans, and finally brand endorsements, which, as I already mentioned, generate most of the Kardashians' income.

At first, the Kardashian-branded products were easily recognized since they all bore double K initials as well as the sisters. Apart from the Kardashian Kollection, Khroma beauty line or the *Konfidential* biography, other products employed K-speak as well. For example, the Kardashian Kard, an ill-fated project of a prepaid card aimed at teens with rather substantial fees; Kardashian Khaos, a store with KUWTK souvenirs in Las Vegas, which closed in summer 2014, or the Kardazzle face palettes. This strategy is in line with Kris naming all of her daughters a name starting with the letter "K," thus lending the various goods and projects a more personal touch as well as securing the connection with the celebrity-filled family and embracing them into the already established brand. Currently sold Kardashian-related merchandise has decreased in numbers and also in the outlined

K strategy as the family brand made way for rather individualized labels such as KKW beauty, Kylie cosmetics, or Khloé's jeans and bodysuits company Good American.

Despite this evolution, Kris is still viewed as a unifying force behind the franchise. Not only on economic terms, but also through dealing with most of the negative publicity herself—in contrast to her vilified celebrity persona of a greedy, pushy momager, her daughter's images can appear more authentic and sincere (Leppert 2015, 133–150). It can therefore be argued that through Kris and her active role in creating, managing, and supervising her family's empire, it also gained a sense of authorship not readily associated with a female reality television participant. Undoubtedly, indeed, Kris' activities provide the KUWTK brand with qualities belonging to more traditional authorial works—authority, coherence, and legitimacy. Her case undermines existing paradigms on reality television fame with firmly installed hierarchies of the exploiting producers and exploited participants. The self-proclaimed momager complicates these conceptions as she occupies both sides of the spectrum. Jenner's visible, in-control position and management of her relatives produced not only a cultural narrative about women's prospects for successfully combining love and family with professional careers, but simultaneously created a unique reality television show, where the participants are not taken advantage of by a third party, but instead benefit from their own transmedial exploitation.

Conclusion

The Kardashian–Jenner women nowadays seem to be on top of their game as they conquered the Forbes top earning reality television star list in 2016 (Robehmed 2016). All of the sisters plus their mother further nurture their celebrity status through social media. As Scheiner McClain's analysis of Kim's Twitter account demonstrated, her various tweets fall under six different narratives—personal tidbits, lifestyle, promotional pieces, encouraging traffic to her website, and interaction with fans and other celebrities (Scheiner McClain 2014, 78–85). These categories closely resemble those presented in the television show, with one exception being the communication with followers. Social media permits an ostensible link between celebrities and audience members due to the continuous uploading of mundane facts, activities, opinions, or images by the famous personality herself. Thus, the Kardashians engage their fans in various conversations and decision-making processes, ranging from manicure trends, through finding adequate names for pets up to music suggestions. Although the Kardashian–Jenner sisters have massive followings on their social network accounts, with Kylie being the most popular person on Snapchat and Kim featuring in the top 20 for both Twitter and Instagram, it is also the case that without the reality television project on air their power would diminish rapidly. Such is the consensus among many commentators on the phenomenon, who mention Paris Hilton's step down from global fame after disappearing from reality television.

As I have demonstrated in the chapter, then, after a decade of steady growth the Kardashian world is currently struggling with finding the right direction in terms of maintaining a balance between keeping the family brand on track and allowing room for the individual aspirations to grow. The former CEO of E! networks Ted Harbert claimed that doing the show might be obnoxious for the family, but it is precisely the televisual presence that provides a basis for many other products (Bruce 2017). Whether the Kardashian clan will survive another decade of global fame depends on how they negotiate the wider family narrative with individual storylines, and how they handle the complexity of mixing television content that is gradually becoming increasingly personalized in its use of labels and social media features. However, the Kardashian women so far have arguably redefined celebrity television-based fame, particularly when examined through the lens of transmedia studies. What was only a decade ago perceived as a passive, ephemeral status, the Kardashians have stretched into an instantly recognizable, immensely profitable, highly participatory, and long-running branded celebrity storyworld.

References

Allen, Robert C., and Douglas Gomery. 1985. *Film History: Theory and Practice*. New York: McGraw-Hill Higher Education.

Berry, Sarah. 2004. "Hollywood Exoticism." In *Stars. The Film Reader*, edited by Lucy Fischer and Marcia Landy, 181–197. London and New York: Routledge.

Bruce, Leslie. 2017. "The Kardashian Decade: How a Sex Tape Led to a Billion-Dollar Brand." *The Hollywood Reporter*, August 16. Accessed February 13, 2017. www.hollywoodreporter.com/features/kardashian-decade-how-a-sex-tape-led-a-billion-dollar-brand-1029592.

DeCordova, Richard. 2001. *Picture Personalities: The Emergence of the Star System in America*. Champaign: University of Illinois Press.

Dyer, Richard. 1998. *Stars*. London: BFI.

Geraghty, Christine. 2007. "Re-examining Stardom: Questions of Texts, Bodies and Performance." In *Stardom and Celebrity. A Reader*, edited by Su Holmes and Sean Redmond, 98–110. London, New Delhi, Singapore, and Thousand Oaks: Sage.

Gledhill, Christine. 1991. *Stardom: Industry of Desire*. London and New York: Routledge.

Jenner, Kris. 2011. *Kris Jenner … And All Things Kardashian*. New York: Gallery Books.

King, Barry. 2003. "Embodying an Elastic Self: The Parametrics of Contemporary Stardom." In *Contemporary Hollywood Stardom*, edited by Thomas Austin and Martin Barker, 45–61. London: Arnold.

Klastrup, Lisbeth, and Susana Tosca. 2004. "Transmedial Worlds: Rethinking Cyberworld Design." Paper presented at the International Conference on Cyberwolds, November 18–20.

Knibbs, Kate. 2017. "The Dark Decline of Keeping Up With the Kardshians." *The Ringer*, May 4. Accessed March 11, 2017. www.theringer.com/2017/5/4/16044794/keeping-up-with-the-kardashians-ratings-season-13-3aa5b249eee7.

Leppert, Alice. 2015. "Mommager of the Brides: Kris Jenner's Management of Kardashian Romance." In *First Comes Love: Power Couples, Celebrity Kinships and Cultural Politics*, edited by Shelley Cobb and Neil Ewen, 133–150. London: Bloomsbury.

Robehmed, Natalie. 2016. "Top-Earning Reality Stars 2016: Kardashians, Jenners Combine for 122,5 Million." *Forbes*, November 16. Accessed February 12, 2017. www.forbes.com/sites/natalierobehmed/2016/11/16/top-earning-reality-stars-2016-kardashians-jenners-combine-for-122-5-million/#21b37eb7274d

Scheiner McClain, Amanda. 2014. *Keeping Up the Kardashian Brand: Celebrity, Materialism and Sexuality*. Lanham and Plymouth: Lexington Books.

Shingler, Martin. 2012. *Star Studies: A Critical Guide*. London: BFI and Palgrave Macmillan.

Turner, Graeme. 2004. *Understanding Celebrity*. London, New Delhi, and Thousand Oaks: Sage.

Woodward, Ellie, and Rebecca Hendin. 2015. "How the Kardshians Manipulated the Media to Become the Most Famous Family in the World." *BuzzFeed,* September 25. Accessed February 8, 2017. www.buzzfeed.com/elliewoodward/how-the-kardashians-manipulated-the-media-to-become-the-most?utm_term=.njZxjqxk1Z#.ko4zxKzAJo.

13

TRANSMEDIA ATTRACTIONS

The Case of *Warner Bros. Studio Tour—The Making of Harry Potter*

Matthew Freeman

The *Harry Potter* fandom is one of the biggest media fandoms around the world. The *Harry Potter* book series alone has sold more than 400 million copies worldwide and the first installment of the film series brought in over £600 million worldwide (*Calgary Herald* 2011). This fan base crosses borders, cultures, and socio-economic settings, providing a perfect opportunity for technology as well as additional platforms and spaces to be used to unite these fans. Not only are the entertainment industries interested in transmedia content, but marketing agencies alongside leisure and tourism developers are focusing their attention on using transmediality as a way of creating and expanding their brand universes. Transmediality provides a unique way of converging their products and brands with the stories and emotions of audiences (Jenkins 2006). It also affords a multiplication of revenue streams across multiple media platforms (Freeman 2016). What, however, is the relationship between transmediality and leisure and tourism-based attraction?

Using the *Warner Bros. Studio Tour London—The Making of Harry Potter* as a case study, this chapter aims to go beyond exploring the more apparent economic reasons for why leisure and tourism-based attractions are readily produced as extensions of media (Grainge 2007) and instead examines how and why this particular *Harry Potter* attraction takes the form that it does. Specifically, and by drawing on a mixed-methodology based on ethnography and surveys with both visitors and staff, I explore what impact this attraction has on how its promoters encourage engagement with the *Harry Potter* texts from which it derives, and also how its visitors characterize the functions of this space as a leisure-based extension of the *Harry Potter* storyworld. The chapter's claims revolve around theorizations of these perspectives, including the aesthetics and properties of the *Warner Bros. Studio Tour London—The Making of Harry Potter* as an extension of the storyworld. These comprise immersion, gamification, as well as the ways in which visitors are invited to go "through-the-looking-glass" (or "into-the-scenes") and "behind-the-curtain" (or "behind-the-scenes"). All direct quotes with visitors throughout this chapter are taken from an online survey I completed between the dates of January 31, 2017, and February 15, 2017. This survey aimed to capture a range of visitor responses and opinions about *Warner Bros. Studio Tour London—The Making of Harry Potter*. Respondents were happy to be quoted directly, but requested that they remain anonymous throughout this chapter.

Conceptualizing Leisure and Tourism Spaces as Transmedia Worlds

World-building, according to Henry Jenkins, concerns "the process of designing a fictional universe ... that is sufficiently detailed to enable many different stories to emerge but coherent enough so that each story feels like it fits with the others" (2006, 335). As has been established in the book's introduction as well as in other chapters, "to fully experience any fictional storyworld, consumers must assume the role of hunters and gatherers, chasing down bits of the story across media channels ... to come away with a richer entertainment experience" (Jenkins 2006, 21). In economic terms, then, transmedia world-building operates on the basis that audiences will gain a richer and fuller understanding of a fictional storyworld by consuming more and more media texts and products that each work to narrate adventures from that storyworld.

Building on Jenkins' emphasis on media texts, Matt Hills considers world-building from the perspective of leisure and tourism. Hills points out that "staging imaginary worlds in physical form has become a significant part of destination tourism, events entertainment and 'extended' film exhibition" (2016, 245). In fact, ever since what Michael Saler describes as "earlier resources of popular enchantment, such as street fairs, carnivals, circuses, panoramas, phantasmagorias, magic lantern shows, conjuring acts, and similar amusements" (2012, 45), spaces of leisure and tourism have long operated in close relationship with media production—and indeed as extensions of fictional storyworlds. Throughout both the twentieth and twenty-first centuries, theme parks, museums, conventions, galleries, and memorabilia fairs have all emerged as commercially viable extensions of media forms around the world.

Such extension spaces have been encapsulated by attractions such as Disneyland. Disney, in fact, was a pioneer of this domain. Paul Grainge has shown how in the 1950s Disney "linked film interests to the development of rides and to associated business concerns in real estate" (2007, 122). Grainge goes on to explain how

> the history of modern entertainment branding is inextricably linked with the Disney Company and its transition in the 1950s from a studio specializing in cartoon animation to a company whose activities would take place within, and in many ways herald, the postwar integration of leisure markets, connecting movie production to developments in television, tourism, theme parks and consumer merchandise.
>
> *(2007, 44)*

However, Disney may have created an all-encompassing consumer environment that Walt Disney himself described as "total merchandising" (Anderson 1994, 134), but how can we characterize the functions of such leisure spaces as a *transmedia storyworld*? In other words, what exemplifies or underpins the "experience economy" (Hills 2016, 244) of these kinds of leisure and tourism attractions in terms of how they operate as direct extensions of media-based worlds beyond the industrial-economic rationales and related commercial practices such as brand extension?

Approaching this question, then, one might posit that in terms of characterizing the type of world-building at work here—or rather the type of world-based experience that is presented to audiences—one concept, in particular, is important: *immersion*. Hills notes that theme-park-style attractions "involve participating in an immersive performance of a media property, or a repeatable walk-through 'adventure' ... that relies on audiences 'being there' and hence amassing embodied cultural distinction" (2016, 244). In the context of transmedia storyworlds, too, Jenkins defines immersion as "the consumer enter[ing] into the world of the story (e.g. theme parks)" (2009).

Later I will examine how this concept of immersion informs the creation and the expansion of the *Warner Bros. Studio Tour London—The Making of Harry Potter* as a place "grounded in ... the real, physical world [but] moved unmistakably not toward realism but toward a more convincing form of fantasy" (Kaufman 2011, 52). But for now it is important to acknowledge that immersion concerns the engagement of audiences *between* and *around* media texts—it is simultaneously a paratextual and an all-consuming practice. Media paratexts are of course defined as peripheral—though no less significant—items such as DVDs, promos, and online materials, which for Gray carry important meaning from films and television series beyond the actual texts and across into multiple media forms (2010, 34). Such materials may actively serve to build a given fictional storyworld and steer audiences across media and other artifacts and spaces—spaces that could well include an attraction. All of which is to say, as will be demonstrated shortly, that leisure-based media attractions can operate as both extra-diegetic consumer add-ons to media texts and as diegetic re-enactments for the fictional fabric of the storyworld itself. Building on his earlier definition, Jenkins also suggests that transmedia storyworlds are themselves based on a balance between immersion and extractability, claiming that "in immersion, the consumer enters into the world of the story (e.g. theme parks), while in extractability, the fan takes aspects of the story away with them as resources they deploy in the spaces of their everyday life (e.g. items from the gift shop)" (2009). Or to put it another way, transmedia world-building envisions a balance between fantasy and reality, between imaginary and real. But how does this balance between fantasy and reality manifest in the case of the *Warner Bros. Studio Tour London—The Making of Harry Potter*?

Behind-the-Scenes

The *Warner Bros. Studio Tour London—The Making of Harry Potter* attraction is an enormously successful walking tour attracting around 6,000 visitors per day, and which immerses guests into the world of *Harry Potter* and of the filmmaking that captured that world on screen. The attraction features authentic sets, costumes, and props that showcase the artistry, technology, skills, and talent that went into producing the world-famous and successful films, created at Warner Bros. Studios Leavesden. Opened in 2012 to the public, the Studio Tour frequently expands to offer new exhibitions and experiences—gradually the attraction has increased its sets by including the interiors of Privet Drive, the Muggle house where Harry Potter grew up, and most recently a new exhibition dedicated to Dobby the House Elf was added.

Perhaps most simply, the world-building provided by the *Warner Bros Studio Tour London—The Making of Harry Potter* can be best characterized as a combination of "behind-the-scenes" and "into-the-scenes"—a tension between offering visitors the chance to re-experience the fictional, media-based world of *Harry Potter* and a wish to learn more about said world extra-diegetically. Such a tension is emphasized in its newspaper promotion: one article, appearing in *The Express*, accentuated how visitors "enter the magical world of Harry Potter at *Warner Bros. Studio Tour London*, where you can see the original sets from the films and step aboard the Hogwarts Express train" (2016, 14). Thus there is the promotion of a balance between reality (film sets) and fiction (Hogwarts Express train). Similarly, in an advertisement titled "Experience The World of Harry Potter" published in the *Bristol Evening Post*, the attraction is described as "featuring Hogwarts in the snow, with sets decorated as they are for festive scenes and behind-the-scenes secrets will be revealed about how filmmakers created fire, snow and ice. Relive the magic through the eyes of the filmmakers who brought the Harry Potter film series to life" (2016, 46). There is indeed the sense of going both "behind-the-curtain" and "through-the-looking-glass" as visitors enter into the *Warner Bros. Studio Tour London—The Making of Harry Potter*, allowing guests to relish in an immersive imaginary world that abolishes binaries between real and imaginary and which goes right back to Disneyland.

However, to what extent has this attraction become an immersive blend of the real and the imaginary, as Jean Baudrillard famously proclaimed? And how so? In other words, how do visitors of the *Warner Bros. Studio Tour London—The Making of Harry Potter* characterize its function as an extension of the *Harry Potter* storyworld? In some respects, many visitors have tended to describe the so-called "experience economy" (Hills 2016, 244) of this attraction in terms of its paratextual or extra-diegetic relation to the *Harry Potter* world—in particular, as what one might call a kind of "media museum." And in characterizing it in this way, there is the sense that some visitors perceive the attraction as somehow less of an all-consuming experience for engaging with the fictional fabric of the *Harry Potter* world and instead as more of a paratextual experience that exists between the textual entry points of the storyworld.

For instance, many visitors commented on how the attraction has "increased [their] respect of the whole film industry." One fan remarked that "we learned so much about filmmaking and saw amazing props, costumes, sets, models, drawings and more." Another notes that "it was so cool to see the film sets and really appreciate the thousands of people and artists that make movie magic come to life. You don't have to be a Potterhead to learn how much work goes into each movie." This sense of learning how the "magic" of the films was achieved was articulated again and again:

> The tour itself is a must-do for any Harry Potter fan—you get a behind-the-scenes tour like if you were one of the actors on the movie, so exciting! Even if you are not a fan, it is quite impressive to see how they created the movies and all the hard work that goes into it.

It is somewhat telling, however, that while such film-industry-related education was seen as interesting by many of the visitors surveyed for this chapter, others used this emphasis as a basis for criticism: "Both my daughter and I are Harry Potter fans so seeing all the artifacts from the movies was great, but we did expect some more action. This was more like a museum rather than a theme park."

Distinguishing the museum from the theme park is an important description, since while the former is associated with history, with education, and with calm, the latter is associated with the present moment, with fun, and with a controlled sense of chaos. In other words, one might argue that there are more apparent parallels between the narrative world of the *Harry Potter* stories and that of a theme park, which both share characteristics of immediate excitement, adrenaline, drama, and spectacle. The aforementioned visitor responses hint that while the *Warner Bros. Studio Tour London—The Making of Harry Potter* gives fans the opportunity to "enter into the world of the story" (Jenkins 2009), its lack of what Hills described as a "repeatable walk-through 'adventure'" (2016, 244) is perhaps why its visitors signal the attraction as less of a "through-the-looking-glass-type" entrance into the *Harry Potter* imaginary world and more of a "behind-the-curtain-type" by-product of that imaginary world.

One should not assume that educational, museum-like experiences cannot occupy the status of fictional world-building. On the contrary, Jenny Kidd argues that we should think of the contemporary museum as a transmedia text, one that involves forms of interlinked storytelling extending across multiple platforms (2014, 68). Especially relevant to our discussions here is her suggestion that a more artifact-heavy approach to narrative at museums can result in visitors "encountering conflicting versions of events and not expecting them to be reconciled" (2014, 36). In the case of the *Warner Bros. Studio Tour London—The Making of Harry Potter*, for example, visitors are presented with unique access into the physical detail of the on-screen storyworld—such as the aesthetic function and design rationale of artefacts, backdrops and stylings. Some visitors remarked that the so-called "behind-the-scenes" insights offered here actually changed their understanding of particular scenes and story beats: "It's amazing and utterly impressive to see the detail that was put into things most people would consider 'background.' The richness of seeing Diagon Alley in the 'flesh' opened my eyes to just how magical this world really is!" In a very different sense, visitors also regularly commented on the fact that the attraction expanded their engagement with *Harry Potter*—not always with its imaginary world, perhaps, but certainly with its community of fans around the world: "It was gratifying meeting a young lady from Northern Europe who had stayed up all night to finish the last book so she would be fully prepared to take this tour. We will remain friends."

Into-the-Scenes

The experience economy of the type of world-building on display at the *Warner Bros. Studio Tour London—The Making of Harry Potter* is thus difficult to characterize. As was argued above, there is certainly the sense that the "behind-the-scenes" label is appropriate to characterizing this attraction, with many visitors indeed emphasizing the "up-close look at props, rooms sets and filmmaking techniques" as the reason to visit. But the industry-based insights into art and design do not tell the whole story. In other important ways, the *Warner Bros. Studio Tour London—The Making of Harry Potter* exemplifies the ways in which modern museums have come to operate as a "mashup, as a site of active consumption, micro-creation, co-creativity and remix" (Kidd 2014, 117). For Kidd, these changes have been brought about by new digital technologies and convergence culture, which—just as with transmedia—have all worked to empower audiences by giving them the "right to participate" (Jenkins 2006, 23). Scolari, Bertetti, and Freeman conceptualize transmedia production as "Media Industry (Canon) + Collaborative Culture (Fandom) = Transmedia Storytelling" (2014, 3). Scolari et al.'s formula echoes Kidd's aforementioned claims regarding the make-up of museums in the digital age, altogether suggesting that the role of leisure and tourism-based attractions in shaping and building transmedia storyworlds goes far beyond this chapter's earlier claims of merely going "behind-the-scenes."

What, then, does it mean for a leisure and tourism attraction to go "into-the-scenes" of imaginary worlds instead? For one thing, Kidd's proclamations of mashup, active consumption, micro-creation, and remix are certainly important. In this case, active consumption occurs via the visitor's extraction of purchasable physical artefacts from the storyworld, including the likes of Butterbeer, Gryffindor scarfs, Hogwarts mugs and wands which can be enjoyed "in the spaces of their everyday life" (Jenkins 2009). Micro-creation, too, occurs via photograph opportunities where visitors can dress-up in the full Hogwarts robes and pose against a green-screen that immerses visitors in the storyworld—quite literally, it seems, into the scenes of the story's fictional world.

But Kidd's idea that the modern museum is a space of mashup and remix is equally significant in terms of how we characterize the *Warner Bros. Studio Tour London—The Making of Harry Potter* as an expansion of the *Harry Potter* storyworld. Such ideas of mashup and remix can be argued to underpin the way in which the attraction was talked about in its promotion, which characteristically attempted to inform audiences how to re-engage with the *Harry Potter* texts, and doing so by promoting the *Warner Bros. Studio Tour London—The Making of Harry Potter* as a chance to head directly into the fantasy of the world, offering an augmented story experience.

And much of this promotion style stems from an attempt to gamify the storyworld. Gamification is the application of elements of game playing—such as point-scoring, competition with fellow players, rules of play, etc.—to other, non-game aspects of activity, most commonly as a way to encourage engagement with a product or service. More than encouraging engagement, the concept of gamification is used throughout the *Warner Bros. Studio Tour London—The Making of Harry Potter* as a means of both immersing visitors in the fiction of the storyworld and expanding the content of that storyworld. Consider, for example, the way in which the opening of the iconic Forbidden Forest sets was described in an advertisement for the *Daily Telegraph*:

> The Forbidden Forest, the mysterious woodland that borders Hogwarts, will be a new attraction for thousands of fans visiting Warner Bros Studio Tour London—The Making of Harry Potter at Leavesden in Herts. In the forest, visitors will come face-to-face with Aragog, the giant spider.
>
> *(2017, 32)*

The inclusion of Aragog, an imaginary creature, within the notably real sets of the Forbidden Forest was not only a way to engage visitors in a more spectacle-driven experience akin to the excitement of a theme park, but it was also a strategy that allowed the attraction to extend the narrative adventures of the storyworld across multiple platforms. For in an article published in *The Evening Standard*, *Harry Potter* fans were apparently "left mystified by #FollowTheSpiders teasers." The article read:

> These are challenging times for arachnophobic Harry Potter fans. Over the weekend, a flurry of posts encouraging fans to "Follow the Spiders" appeared on social media,

> prompting people to suspect that something new—and eight-legged—will be coming to the expanding Harry Potter franchise soon.
>
> (2017, 17)

These tweets mostly came from the marketing team based at the *Warner Bros Studio Tour London*, where on the attraction's Twitter account there was the mysterious message: "Looking for something? All you need to do is #FollowTheSpiders." Shortly after, a new Instagram account, called FollowTheSpiders, was set up, seemingly to post a series of images that matched up to make one giant spider face. Further details and images were staggered and timed to coincide with the Celebration of Harry Potter convention in Florida. Altogether, then, social media was used in close dialogue with the film sets constructed for the *Warner Bros Studio Tour London* as a tool for encouraging the "speed and movement that commonly accompanies digital practices" in the context of a leisure and tourism-based attraction (Kidd 2014, 34).

More than that, the integration of a leisure attraction with social media elevated the engagement of visitors into potential game players, enticing fans to participate in a series of tasks and clues that all pointed toward attendance at the *Warner Bros Studio Tour London*. As one fan remarked on Twitter, "#HPStudioTour is a ride through the Forbidden Forest maybe where Aragon lived." What's more, this gamification-led strategy also afforded a highly narrativized expansion of the fictional storyworld in ways that qualifies the attraction as a media platform in the larger transmedia story. Follow The Spiders is in fact a reference to the clue given by Hagrid in J. K. Rowling's second novel, *Harry Potter and the Chamber of Secrets*. In one scene, he tells Ron, Harry and Hermione to "follow the spiders" when the intrepid trio are looking for the monster that lurks in the tunnels underneath Hogwarts Castle. Here, they find Aragog, an enormous spider, or Acromantula, and learn that he is not involved in the Chamber of Secrets, and nor was a teenage Hagrid. All of which is to note that the *Warner Bros Studio Tour London* thus became part of the very fabric of *Harry Potter*'s transmedia storytelling, with the attraction contributing to the ongoing process the telling "stories that unfold across multiple platforms, with each medium making distinctive contributions to our understanding of the [story] world" (Jenkins 2006, 336). For as one fan also remarked on Twitter, "I love how this is an expansion of the tour!!"

Conclusion

So, what specific and emerging roles do leisure and tourism attractions continue to play in expanding fictional worlds across borders? This chapter has aimed to gain at least partial insight into this question, exploring the changing roles, forms, industrial attitudes, and audience responses to expanding storyworlds and their fictional spaces far beyond the borders of media itself and into the cultural spaces of leisure and tourism attractions. Amid a time of continued digitization as well as the increasing ease of availability of media worlds in the age of convergence culture and its interconnected screens, books, social media platforms, and other web content, the *Warner Bros. Studio Tour London—The Making of Harry Potter* ultimately paints a relatively complex picture concerning the ways in which we should go about defining the relationship between media production and leisure and tourism attractions.

Much like the Disneyland-inspired theme parks that came before it, it is certainly fair to understand the *Warner Bros. Studio Tour London—The Making of Harry Potter* in primarily economic terms—that is, as a form of "total merchandising" for a franchise now in its twilight years (Anderson 1994, 134). Even so, understanding this attraction as an economic success means characterizing its function as a more artful expansion of a fictional storyworld. And the type of world-building provided by the *Warner Bros. Studio Tour London* is indeed best described as a careful balance between the physical—i.e., film sets, industry insights, grounded in the real—and the imaginary, i.e., narrativized or gamified elaborations and re-enactments of story and character.

Different fans expressed interest and engagement in different ways at different times, sometimes characterizing the attraction as something that works best *between* the *Harry Potter* media texts or as one that exists *around* those texts as a component of the storyworld and its fiction. The "experience economy" (Hills 2016, 244) of the *Warner Bros. Studio Tour London* thus becomes akin to what Tony Bennett and Janet Woollacott defined as "inter-textuality"—that is, the diverse ways in which fictions may exist in the gaps in between their textual exploits, with those "in-between" pieces working to reshape how audiences read texts, adjusting their meaning (1987, 45). Such in-between pieces may refer to publicity, posters, fanzine articles, interviews with stars, promotional stunts, etc., which do more than merely "organise expectations in relation to a particular film," but instead work to "socially organise the relations between media texts" (Bennett and Woollacott 1987, 45). In their study of the James Bond phenomenon, Bennett and Woollacott explored "the respects in which, in adding to 'the texts of the Bond', [the films] contributed to a reorganization of the inter-textual relations to which both the films and the novels were read" (1987, 142).

Similarly, the *Warner Bros. Studio Tour London—The Making of Harry Potter* serves an intertextual function, reshaping and expanding how audiences engage with the *Harry Potter* world by inviting audiences to step outside of its fictionality and to consider its construction. And yet the most positive and enthusiastic responses to the attraction—at least those gained for this chapter—were in relation to its narrativized attempts to immerse visitors in the imaginary via the integration of sets and social media. As can also be described of the very best cases of transmedia world-building, the *Warner Bros. Studio Tour London* was found to be no more immersive than when it rendered itself somewhat invisible and attempted to take the *Harry Potter* story and make it bigger. Its contributions to the storyworld operate as a more physical and consumer-led form of intertextuality that simultaneously builds textual connections between stories while allowing those stories to escape textual borders and exist in between them in the form of a Studio Tour, folding paratext into text once and for all.

References

Anderson, Christopher. 1994. *Hollywood TV: The Studio System in the Fifties.* Texas: University of Texas Press.

Bennett, Tony, and Janet Woollacott. 1987. *Bond and Beyond: The Political Career of a Popular Hero.* London: Routledge.

Calgary Herald. 2011. "JK Rowling's Harry Potter Series of Books Has Sold More Than 400 Million Copies Worldwide." September 4. Accessed January 15, 2017. www.calgaryherald.com/entertainment/Rowling+Harry+Potter+series+books+sold+more+than+million+copies+worldwide/5077373/story.html.

Daily Telegraph. 2017. "Harry Potter and the Fantastic Flora." January 28.

The Evening Standard. 2017. "Harry Potter Fans Left Mystified by #FollowTheSpiders Teasers." January 23.

The Express. 2016. "Warner Bros. Studio Tour London—The Making of Harry Potter." September 11.

Freeman, Matthew. 2016. *Historicising Transmedia Storytelling: Early Twentieth-Century Transmedia Story Worlds.* London and New York: Routledge.

Grainge, Paul. 2007. *Brand Hollywood: Selling Entertainment in a Global Media Age.* London and New York: Routledge.

Gray, Jonathan. 2010. *Show Sold Separately: Promos, Spoilers, and Other Media Paratexts.* New York: New York University Press.

Hills, Matt. 2016. "The Enchantment of Visiting Imaginary Worlds and 'Being There': Brand Fandom and the Tertiary World of Media Tourism." In *Revisiting Imaginary Worlds: A Subcreation Studies Anthology*, edited by Mark J. P. Wolf, 244–263. London and New York: Routledge.

Jenkins, Henry. 2006. *Convergence Culture: Where Old and New Media Collide.* New York: New York University Press.

Jenkins, Henry. 2009. "The Revenge of the Origami Unicorn: Seven Principles of Transmedia Storytelling." *Confessions of an Aca-Fan: The Official Weblog of Henry Jenkins*, August 3. Accessed January 10, 2017. http://henryjenkins.org/2009/12/the_revenge_of_the_origami_uni.html.

Kaufman, J. B. 2011. "The Heir Apparent." In *Funny Pictures: Animation and Comedy in Studio-Era Hollywood*, edited by Daniel Goldmark and Charlie Keil, 51–68. California: University of California Press.

Kidd, Jenny. 2014. *Museums in the New Mediascape: Transmedia, Participation, Ethics.* Farnham: Ashgate.

Saler, Michael. 2012. *As If: Modern Enchantments and the Literary Prehistory of Virtual Reality.* Oxford: Oxford University Press.

Scolari, Carlos A., Paolo Bertetti, and Matthew Freeman. 2014. *Transmedia Archaeology: Storytelling in the Borderlines of Science Fiction, Comics and Pulp Magazines.* Basingstoke: Palgrave Macmillan.

PART II

Arts of Transmediality

14

TRANSMEDIA STORYTELLING

Character, Time, and World—The Case of *Battlestar Galactica*

Mélanie Bourdaa

It is claimed that the emergence of new digital technologies (Gillan 2010) alongside the increasing visibility of fan practices merged with traditional media production practices have shaped a culture of convergence (Jenkins 2006). In turn, this culture of convergence has led to new forms of serial media content. At the same time, strategies of transmedia storytelling have increased in television series, in particular, which offer stories that "go beyond the screen" (Peyron 2008, 337) in the form of "augmented storytelling" (Bourdaa 2012), with narrative universes scattered across several media platforms. While the term "transmedia" is not new (Kinder 1991), nor even the practice (Freeman 2016), the convergent media ecosystem of the present day favors the integration of production and narration strategies, thereby lending itself to the workings of transmedia storytelling.

This chapter analyzes how transmedia storytelling is used in contemporary television series, assessing how stories expand across different media platforms so to propose a typology of various transmedia storytelling characteristics. Using *Battlestar Galactica* (henceforth *BSG*) as a case study, I will identify three categories of transmedia storytelling that are used to structure this chapter: character, temporality, and storyworld, building on the conceptual characteristics previously identified by Evans (2011) and Freeman (2016). The case study's narrative extensions will be analyzed according to a specific methodology: each extension created around the series will be the object of a study according to a grid of criteria that I created. This grid is made up of three fundamental categories:

1. Importance of the media platform supporting the extension: what affordances?
2. Narration: characters, temporality, atmosphere, tone.
3. Canonical links: essential narrative additions and connection with the series.

This method will make it possible to highlight the contributions of each extension, the choices of the platform and more broadly the constitution of the extended universe of the series. A temporal thread can thus be drawn in order to understand at what moment each extension intervenes in the universe. Supporting this analysis, I also conducted interviews with writers from the series, as well as from the comic book issues and the web series in order to unravel the creative processes and understand how ancillary content fits into the *BSG* storyworld. Table 14.1 lists all of the transmedia extensions within the *BSG* universe that I categorized using the methodology explained above. For the sake of time, in this chapter I will only give details for specific extensions in order to make or illustrate a point.

Table 14.1 Transmedia extensions within the *Battlestar Galactica* universe

Platform	*Title*	*Seasons*	*Temporality*	*Narration*
TV	*Battlestar Galactica*	Mini-series	Fall of the Colonies	"Mothership"
	Battlestar Galactica	Season 1—Season 4	2nd Cylon War	"Mothership"
TV Film	*Razor*	Season 3	Pegasus Trilogy	Time related Transmedia
	The Plan	Season 4	Re-interpretation Season 1—Season 2	World-building
Webisodes	*BSG: Blood and Chrome*	Before the mini-series	1st Cylon War	Time related Transmedia
	The Resistance	Between Season 3 and Season 4	Occupation New Caprica	Time related Transmedia
	The Face of the Enemy	Between Season 4 (1) and Season 4 (2)	Occupation New Caprica and Season 4	Character-based Transmedia
Games	*Battlestar Galactica The board game*	Season 1—Season 4	Fall of the Colonies	World-building
	Battlestar Galactica Online	Season 4	After the destruction of the Cylon Resurrection Ship	World-building
Comic Books	*Cylon War*	Before the mini-series	1st Cylon War	Time related Transmedia
	BSG Season Zero	Before the mini-series	Before the fall of the Colonies	Time related Transmedia
	BSG Origins (Zarek, Baltar, Adama, Kara and Helo)	Before the mini-series	Between the 1st Cylon war and the fall of the Colonies	Character-based Transmedia
	The Final Five	Before the mini-series	Before the 1st Cylon War	Character-Based Transmedia
	Six	Before the mini-series	Before the fall of the Colonies	Character-Based Transmedia
	BSG Ghost	Mini-series	The Fall of the Colonies	World-building
	The Returners	Season 1	2ème guerre Cylon	World-building
	Pegasus	Season 3	Pegasus Trilogy	Character-based Transmedia

Character-based Transmedia: Giving Depth to the Cylons and the Humans

BSG tells the story of how Cylons, robots created and enslaved by men, rebelled and committed genocide, with the rest of humankind fleeing on-board Battlestars in search of a new Earth. Characters are central to the plot as viewers discover two factions fighting each other, humans and Cylons. Aaron Smith quoting John Fiske quite rightly says that

> viewers have strong emotional relationships with television characters because they can understand the characters' reaction to the problems they encounter. Fiske says that because television characters enter the lives of viewers every week, there is a sense of immediacy and of existence. The characters become familiar faces that return week after week, continuing to exist even when the TV is turned off.
>
> *(Smith 2009)*

In this traditional construction of the television series, based on a weekly broadcast, the recurring character becomes a fundamental element of the narration but also of identification for the audience. The character remains the force of fiction, the point of recognition of the universe. In a transmedia scenario, Paolo Bertetti asserts that "the transmedia character is a fictional hero whose adventures are told on several media platforms, each providing details about the character's life" (Bertetti 2014, 2344). It is interesting to note that even though Jenkins famously set up his definition of transmedia storytelling on the basis of narrative universes that lead to the development of franchises, his first proposal was centered on characters, with *Technological Review* article titled "Transmedia Storytelling: Moving Characters from Books to Films to Video Games Can Make Them Stronger and More Compelling" (Jenkins 2003). In this article, Jenkins underlines how the transmedia strategy is now deeply rooted in our entertainment culture since "younger consumers have become information hunters and gatherers, taking pleasure in tracking down character backgrounds and plot points and making connections between different texts within the same franchise" (2003). Jenkins' first attempt at defining transmedia storytelling was in line with the definition provided by Martha Kinder (1991) who, by developing the idea of transmedia superstructures, analyzed how fictional characters move from one medium to another, with audiences following them on their fictional journey.

For *BSG*, Ron D. Moore and David Eick, the showrunners, decided to develop the characters of the series even further and to give them depth and a history with added origins and motivations through the deployment of extensions on other media platforms—mainly comic books and webisodes. In collaboration with Dynamite Entertainment, an American comic book company, NBC Universal, which was broadcasting the series via its subsidiary Syfy, launched a series of comic book issues entitled *Origins*. Broadly, origins stories can be seen as a specific genre related to superhero franchises, mainly those of DC Comics and Marvel. In his analysis of transmedia extensions for television series, Clarke points out that

> the use of origins stories is influenced by the economic structure of the comic book industry, which continues to produce stories over years and decades ... By remaining faithful to the origins (which are frequently modified in their consistency), readers can discover a story without having to navigate in more than 400 numbers of comics.
>
> *(2013, 54)*

Similarly, in the case of *Origins*, the goal of these comic books is to create a "past" for the human characters that appeared in the series. The comic books thus examine five main characters across 11 issues, spread out over a year: William Adama, Zarek, Gaius Baltar, Kara "Starbuck" Thrace, and Karl "Helo" Agathon. These issues are then collected in an eponymous Omnibus. In terms of distribution, the series begins in *Media Res* and television viewers do not learn a lot of information during the show on who the characters were before the beginning of the show.

Importantly, the narrative elements displayed in the comic books reinforce a coherence with the television series and thus prove to be canonical additions to the storyworld, providing additional information to readers and viewers. To offer another example, the comic books focused on providing explanations for the origins of the Robot entities. During season 4 of the television series, some information and answers are given that fans have been waiting for a long time, but the showrunners did not explain these answers in details, leaving "narrative shadows" for the comic books to fill. Published by Dynamite Entertainment in 2009 in parallel to the broadcast of the last episode of the series, a comic book, *The Final Five*, offers "an interpretation of the history of the last five Cylons." The title of the comic book as well as the reprise of the Cylon's vignettes root the comic book in the narrative universe of the series that testifies its canonical belonging. Thus we are not dealing here with an interpretation of *BSG* history but instead *the* interpretation of its history. Nigel Barrucci, publishing director at Dynamite Entertainment, highlights the benefits of publishing this kind of transmedia narrative information in the form of comic books: "What comic books can do is amplify the aesthetic effects, lengthen and increase battles, expand universe like never before. Now that the series is over, we can continue to tell stories on this medium" (Brady 2008). Once again, the comic books served as a site for audiences to find Easter Eggs, but also to discover new and important narrative elements that enrich their understanding of the series and the characters.

Underpinning such narrative elaboration, the comic book platform itself offers a principle of seriality that echoes the serialized format of *BSG*. However, producing comic book extensions for the television series that were consistently coherent with said television series came with its own set of creative challenges. As Robert Napton (interview, 2014) explains in interview:

> We wanted to tell his life (Adama's life), fill in gaps concerning his past life, his career. We then based our scenario on what was known of him and from there we created new facts to fill in the narrative gaps. Our work was very complex because the comic books were produced in parallel with the broadcast of the series. I remember that I had to add scenes at the last minute because one of the episodes revealed that Adama was commander on another Battlestar, called the Walkyrie. It was new information about the character so I quickly incorporated it into the latest issue of the comic book series. It was an exciting challenge to stay consistent with the series under these conditions.

Napton's testimony about the collaborative process of creating narrative extensions provides us with insights into the high degree of coordination between the various writers of the comic books and the showrunners, but also the concern for narrative coherence within the diegesis.

In order to widen its storyworld and to broaden the narrative coherence of the diegesis, *BSG* also developed webisodes, short episodes specially created for the Internet, and launched online on the official streaming platform of the Syfy channel. Rather than simply offering trailers or promotional advertisements, easing the wait of the fans during the hiatuses between two seasons, the webisodes helped to advance the story. However, "in order not to alienate their traditional viewers, the showrunners ensured that the plot did not constitute an integral part of the series' narrative arcs, while advancing the intrigue and developing the complexity of the narrative universe of the series" (Door 2008). Following the broadcast of episode 11 of season 4 ("Sometimes a Great Nation"), which ends on a cliff-hanger before the final outcome of the series, a series of ten webisodes was put online, entitled *The Face of the Enemy*. Written by Jane Espenson, who scripted some of the television episodes, these webisodes focused on developing a character—in particular, Felix Gaeta—who became essential to the plot in the second half of season 4. Echoing the sense of mystery that surrounded the origins of many characters before they were fleshed out in the comic books, the character of Gaeta was similarly enigmatic on television, with viewers knowing very little about him on screen. Thus it was the role of the webisodes to "provide a story about [his] past, explore [his] motivations, and give clues to future plots, all in a short narrative" (Jenkins 2003). For Jane Espenson, the Internet's natural dissemination power offered new possibilities for extending the universe of *BSG*. As she explained in interview: "As the series, by definition, focuses on the main characters, we thought that webisodes would be a good opportunity to develop a supporting character" (personal correspondence, 2014). Two clear objectives stand out in Espenson's description: highlighting the story of a supporting character, and bringing a solution and a closure to a narrative arc that has not really been solved in the series, which in this case is the relationship between Gaeta and Baltar during the Cylon Occupation on New Caprica. From a production point of view, the webisodes were a challenge because of the limitations inherent in the project, the low budget and the few actors and teams involved. However, the webisodes gave the series an incredible online presence. Narratively, *The Face of the Enemy* resumes the plot nine days after the discovery of a devastated and irradiated Earth. In addition to the character of Gaeta, now missing his left leg, we find Colonel Tigh blind in one eye following the torture he endured on New Caprica, Sharon and Boomer the two Cylon number 8 models, and Lieutenant Hoschi. All of these characters are already prominent in the television series, which again serves to link these webisodes as part of the main *BSG* canonical narration, with new narrative elements that are only revealed in the webisodes which serve as a reward for the most invested fans. Such new narrative elements include details of the relationship that Gaeta has with Lieutenant Louis Hoschi. There is a sense that webisodes such as this one, produced on a lower budget and targeted toward the core fan base, can also work to re-engage and to please the desires of said fan base. For in the case of the *BSG* fan base, some fans had wanted Gaeta to be gay for a long time and regularly voiced their opinion on fan forums: "We, the fans of 'Gaeta is gay', did not have to suffer the outrage of a heterosexual character, and we finally have something to enjoy, and I am delighted that we have it at this advanced stage of history."

Time-based Transmedia: Telling the Past, Filling "Negative Spaces"

As has been shown, one of the objectives of all transmedia narrative strategies is to fill temporal voids, ellipses, between two episodes, between two seasons or to explain the genesis of the story. There is thus an important sense of temporality to transmedia storytelling (Evans 2011). A web-series called *The Resistance* was broadcast between season 2 and season 3 on the official website hosted by Syfy. This web-series had three goals: to alleviate the expectations of the fans during the hiatus between two seasons, to introduce the atmosphere and the universe of the series to the (new) users of the site, and to offer an explanation for the resistance movement that takes place on New Caprica during the Cylon Occupation, after they attack the planet in the finale of season 2. The ten webisodes, which last a total of 26 minutes, show how the Resistance is organised on New Caprica and what roles Tyrol, Callie, and Saul will play in it and what repercussions it will have in the television series. However, the scriptwriters took advantage of the online webisode format to develop two other secondary characters, whose actions have far-reaching consequences in the television series.

> We then chose to tell a complementary story in connection with some supporting characters: Jammer, a member of the technical team lead by Chief Tyrol, became a member of the police militia working with the Cylons on New Caprica. In the episodes of season 3, the reasons for this choice are never really revealed. And we worked on another season 3 character, Duck, who decides to commit suicide bombing at a police award ceremony in the series.
>
> *(Weedle, interview, 2014)*

Webisodes serve as a justification for the acts the characters perpetrate on television, notably Duck's suicide bombing, which had never really been explained except by the fact that he is an active member of the Resistance. These narrative extensions once again work to increase the diegesis of the series by exploring shaded areas and gaps in the narrative arcs. Bradley Thompson, who wrote the web series, confirms this function by detailing the production process of the webisodes he worked on: "Ron and David [Eick] were overseeing the scenarios, asking for changes, and then going to the studio and the channel to give their opinion. Ron then sent back the wish list of the channel with 'yes', 'no' or 'try to think about it'" (Napton 2014). David Weedle, another writer on the web series, compares the screenwriters to a jazz band, led by a maestro, here the showrunner Ron D. Moore. This confirms once again that the transmedia extensions were intended to fit perfectly into the universe of *BSG*, creating a complete and coherent narrative word linked by clear temporal bridges between platforms.

A second strategy linked to time extensions concerns the events that led to the situation described in the television series and the key narrative moments preceding the events that launch the television series. More than a focus on the characters, as was the case with the *Origin* comic books, these particular time-based narrative additions dwell on eventual explanations, to better explain how the narration came to be. For example, another comic book issue, still published by Dynamite

Entertainment and titled *The Cylon War*, explains in details the first war between the Cylons and the humans before the truce and the return of the Cylons that led to the genocide and the fall of the 12 colonies at the beginning of the series. This story recalls the events that took place before the beginning of the series and which are only evoked in a few episodes through dialogue, adding temporality and continuity to the television series. This important part of the story, effectively leading to the tragic events of *BSG*, is also told in the first scene of the mini-series via textual vignettes which sum up the situation. This four-part series of comic books functions as a prelude to the television series, and was published during the broadcast of season 4. The aim of this story is to establish a link between the original *Galactica* and the re-imagined version, showing the evolution of the metallic Cylon lineage and in particular how they were transformed into weapons attacking the humans of the colonies. Eric Nylund, the screenwriter, explains the origins of the project and the interweaving of this series of comic books into the universe of the series:

> With Joshua Ortega, we condensed a whole war in four issues, the rise and the disintegration of a one of the two civilizations, and the almost complete destruction of the human race ... You will see the evolution of the Centurions line described in these pages. It was something we really wanted to show. We show a bit of the second Cylon war, but the main story is why and how the robotic forces created by men become war machines, why and how they rebelled ... and how men survived.
>
> *(Ong, Kean, and Brady 2008)*

In this vein, Bandon Jerwa, a writer who has worked on several comic book extensions of *BSG* at Dynamite Entertainment, insists that *BSG* is a franchise with cross-platform extensions that must be seen as a rich, coherent, and expansive universe:

> The challenge is not only to write a prelogy but to enrol in a franchise mechanism and making sure that all the pieces of the puzzle match what was done before. All these things are placed in a continuity, and so there is a total synergy. The thing I particularly like about being a comic book screenwriter for *Battlestar Galactica* is working on describing a giant universe that fits into the even bigger universe of the series. Seriously, I promise you that in two years when a ton of *Battlestar Galactica* books will be in the hands of readers, they will be rewarded with richness and scriptwriting consistency.
>
> *(Jerwa 2009)*

The transmedia extensions Jerwa describes help to reinforce the mythology around the series by creating a prelogy and deploying augmented narrations between key episodes of the series. Of course, there is simultaneously a marketing agenda behind such so-called rewards of richness and consistency, as the channel tries to attract new audiences by multiplying the available platforms. In effect, each platform offers narrative additions complementary to the canon narrative, creating a larger audience by creating an extended universe for *BSG*.

World-based Transmedia: A Universe in Constant Expansion

The final transmedia narrative strategy employed by the makers of *BSG*, one which ultimately ties into the other two, was to develop an enriched storyworld that continued to grow even after the series had ended, bringing a narrative continuity to fans and viewers on other media platforms. For example, a series of comic books fits into this strategy of multiplication and starts from the well-known principle of "and if ... ," essentially offering the possibility of exploring a kind of alternate universe or at least a parallel one to the series. As Brandon Jerwa pondered for *BSG: Ghosts*: "And if a squadron of secret agents had also survived Cylon Attack?" For Jason Mittell, this type of extension, which he names the "What if?" extension, aims to "propell the extension beyond the main narrative into parallel universes, highlighting tone, atmosphere, characters, and style, rather than continuing with canonical plots" (Mittell 2012). In this category of transmedia extension, what tends to be privileged is therefore the atmosphere of the series, and not any particular narrative continuity. The aim is to offer readers alternate and parallel stories. *BSG: Ghosts* is composed of four issues released in 2008 between the first half of season 4 and the second half of season 4 which marks the end of the series. The action begins one day before the attack and the fall of the 12 colonies to continue in a parallel story with new characters. Even if the comic book issues focus on new characters, which will not be found in the television series, the Cylons, the Centurions as well as the humanoid ones, are still in the comic books. They appear from the very first pages of the first issue taking the unanimous decision to attack the colonies simultaneously. As in the television series, human-like Cylons may be sleeper agents infiltrated into the human fleet not aware of their state and then activated at a key moment in the narration. The human squadron consists of six Vipers, two armed Vipers, and one Raptor Ultra designed to carry 12 marines and a crew of two soldiers. They are under the command of Captain Alexander Chen and among their ranked officers are Lieutenant Dozil Pennit, a model Cylon number 2, an infiltrated sleeper agent. The story takes place in a hybrid present in the last pages of the second issue, and is one of survival, relying on the atmosphere of the beginning of the television series. The same themes are dealt with: the enemy is already among us since it is undercover among the crew, and the question of identity or rather how do our experiences transform us is also raised.

For author Brando Jerwa, the comic book format presents several interests in this case, especially regarding the extension of a canonical narration into other possible directions that exist on another media platform. He explains:

> *BSG* possesses a rich mythology that is more based on characters than on the epic side. It's certainly epic but I think it's more a human drama than a fantastic series, and that's what sets them apart. I am convinced that the end of the series will close the door to the development of the narration beyond this event, and in my opinion this is a good thing. The comic books will definitely have to evolve in order to survive because at some point we will end up exhausting the interest of the readers on the narrative continuity. Projects like *Ghosts* are definitely a good way to test public reaction to new ideas in a familiar environment.
>
> *(Brady 2008)*

In terms of carving a place for *Ghosts* within the narrative world of *BSG*, the intentions of NBC and Dynamite Entertainment were twofold: they wanted to revive the public's interest in the universe and atmosphere of the series, and to give it a whole new scenario, extending the franchise beyond the end of the television series by relying on parallel universes.

Conclusion

In her own analysis of the transmedia strategy for *BSG*, Suzanne Scott (2008, 210) claims that:

> *BSG* has aimed an unparalleled wealth of fan-oriented content at its audiences and, whether one chooses to view this as a dialogic departure from the producer/consumer binary or merely a tech-savvy marketing ploy, it is an integrated media model that is rapidly gaining popularity.

Transmedia storytelling for television series, and *Battlestar Galactica* is no exception, oscillates between marketing and economic imperatives for the channel, as indicated here by the creation of *BSG* webisodes between seasons or the producer's desire to continue the stories of the universe after the series had ended, and indeed the will to augment the storyworld by giving details, information, and depths to the main and supporting characters and hiding Easter Eggs that reward fans for their engagement in the narrative universe. *BSG* managed to expand its storyworld on numerous media platforms (web series, comic books, interactive games, board games, telefilms) focusing on three core aspects: characters, time, and world-building. Each of these three strategies exemplify approaches to transmedia storytelling at the present time, ensuring that the narration was coherently expanded and augmented and inviting fans to dig deeper into its storyworld, searching for clues. The mere existence of the Battlestar Wiki, created by fans, shows the importance of providing a detailed cartography and mapping of the universe of a given television series so that audiences can work to unravel all its secrets and understand all its mysteries and mythologies.

References

Bertetti, Paolo. 2014. "Toward a Typology of Transmedia Characters." *International Journal of Communication* 8: 2344–2361.

Bourdaa, Mélanie. 2012. "Transmedia Storytelling: Entre Narration Augmentée et Logiques Immersives." *InaGlobal*, June 13. Accessed August 27, 2017. www.inaglobal.fr/numerique/article/le-transmedia-entre-narration-augmentee-et-logiques-immersives.

Brady, Matt. 2008. "Jerwa: The Ghosts of Battlestar Galactica." *Newsarama*, July 15. Accessed August 13, 2017. www.newsarama.com/436-jerwa-the-ghosts-of-battlestar-galactica.html.

Clarke, M. J. 2013. *Transmedia Television: New Trends in Network Serial Production*. London: Continuum Publishing Corporation.

Door, Justin. 2008. "Weaving a Story through Webisodes." In *Finding Battlestar Galactica: An Unauthorized Guide*, edited by Lynette Porter, David Lavery, and Hillary Robson, 251–256. London: Sourcebooks.

Espenson, Jane. 2014. Personal correspondance.

Evans, Elizabeth. 2011. *Transmedia Television: Audiences, New Media and Daily Life*. London and New York: Routledge.

Freeman, Matthew. 2016. *Historicising Transmedia Storytelling: Early Twentieth-Century Transmedia Story Worlds*. London and New York: Routledge.

Gillan, Jennifer. 2010. *Television and New Media: Must-Click TV*. London and New York: Routledge.

Jenkins, Henry. 2003. "Transmedia Storytelling." *MIT Technology Review*, January 15. Accessed August 13, 2017. www.technologyreview.com/news/401760/ transmedia-storytelling/.

Jenkins, Henry. 2006. *Convergence Culture: Where Old and New Media Collide*. New York: New York University Press.

Jerwa, Brandon. 2009. *Battlestar Galactica Season Zero, Omnibus vol. 1*. London: Dynamite Entertainment.

Kinder, Marsha. 1991. *Playing with Power in Movies, Television, and Video Games: From Muppet Babies to Teenage Mutant Ninja Turtles*. Berkeley: University of California Press.

Mittell, Jason. 2012. "Narrative Complexity in Contemporary American Television." *MATRIZes* 5 (2): 29–51.

Napton, Robert. 2014. Personal interview.

Ong, Pang, Benjamin Kean, and Matt Brady. 2008. "Dynamite's 'BSG: Cylon war' Starts in Jan – First Look." Newsarama, October 16. Accessed August 11, 2017. www.newsarama.com/1299-dynamite-s-bsg-cylon-war-starts-in-jan-first-look.html.

Peyron, David. 2008. "Quand les œuvres deviennent des mondes. Une réflexion sur la culture de genre contemporaine à partir du concept de convergence culturelle." *Réseaux* 26: 335–368.

Scott, Susan. 2008. "Authorised Resistance: Is Fan Production Frakked?" In *Cylons in America: Critical Studies in Battlestar Galactica*, edited by Tiffany Potter and C.W. Marshall, 210–223. New York: Continuum.

Smith, Aaron. 2009. "Transmedia Storytelling in Television 2.0." Ph.D. diss., Middlebury University. http://sites.middlebury.edu/mediacp/2009/06/17/12-technological-convergence-content-in-a-multimedia-world/.

Weedle, David. 2014. Personal interview.

15

TRANSMEDIA WORLD-BUILDING

History, Conception, and Construction

Mark J. P. Wolf

Much of popular culture today involves world-based franchises, which include works in a wide variety of media and other merchandise. In this sense, imaginary worlds are being used as brands or sub-brands, with easily recognizable elements and designs, images and sounds, and continuing stories that entice consumers to keep returning to their favorite worlds, and spending their money on them—in whatever medium they may appear. An audience's engagement with a world is different from an engagement with characters or stories alone; a world is a place to be vicariously entered, an object of exploration and speculation, and refuge from the trials and troubles of the world in which we actually live. It is a virtual place to which fans keep returning, more than a story, and something they can often see, hear, and even interact with, depending on the medium used to encounter it. Author J. R. R. Tolkien referred to such an author-created world as a "secondary world," making it distinct from the "Primary World" in which we actually live, since all secondary worlds are, at least initially and in one way or another, patterned after the Primary World (Tolkien 1989). While an imaginary world can appear in a single medium, many of the largest ones being constructed today are transmedial worlds, with world materials, stories, and characters appearing across a range of different media.

The Road to Transmedial Worlds

Before the year 1900, imaginary worlds were mainly something encountered in books, and perhaps occasionally in paintings. Some books featured maps, woodcuts, and other kinds of illustrations to describe their worlds; these, in a sense, were the earliest transmedial worlds. By the late 1800s, we find not only a variety of maps, but also series of drawings, and more of them, depicting imaginary worlds. John Tenniel drew 92 drawings for Lewis Carroll's two Alice books, *Alice's Adventures in Wonderland* (1865) and *Through the Looking-Glass and What Alice Found There* (1871). French author and illustrator Albert Robida drew his own imagery for his futuristic trilogy of books, *Le Vingtième Siècle* (1883), *La Guerre au Vingtième Siècle* (1887), and *Le Vingtième Siècle. La vie Électrique* (1890), which showed wild cityscapes, flying cars and other vehicles, and fantastic electrical technologies; today, his books are remembered mainly for his images. And L. Frank Baum worked closely with W. W. Denslow to design the pages of *The Wonderful Wizard of Oz* (1900), in which text and image are closely intertwined and designed with each other in mind.

The Land of Oz, indeed, which grew substantially as it came to include other imaginary lands from Baum's books, which he attached to Oz geographically, hoping they would be more likely to find the same popularity the Oz books had found, would also become the first great transmedial world. Baum adapted his Oz characters and stories for films, comics, stage plays, games, and merchandise, not merely adapting his stories, but telling new ones in each medium, resulting in a world spanning multiple media in which the works in each medium contributed something new to the world. During the first two decades of the twentieth century, Baum's success provided an example for other intellectual property owners, who came to understand how popular characters and worlds thriving in one medium could make the jump to others. Newspaper comics could make the jump to animated shorts, and animated characters could likewise appear in comic strips. Novels were being adapted into live-action films, just as they had been adapted into stage plays. For example, Alex Raymond's *Flash Gordon* started as a comic strip in 1934 and was adapted to three serial films, the 13-chapter *Flash Gordon* (1936), 15-chapter *Flash Gordon's Trip to Mars* (1938), and 12-chapter *Flash Gordon Conquers the Universe* (1940); a 26-episode weekly radio serial, *The Amazing Interplanetary Adventures of Flash Gordon* (1935); a novel, *Flash Gordon in the Caverns of Mongo* (1936); and even a ride at the 1939 World's Fair (see Santo 2015).

For some time, transmedial growth followed a pattern; a world or set of characters would find success in its medium of origin, and then get adapted into other media, once success seemed certain. During the 1970s, blockbuster films began to change Hollywood in the 1970s, and *Star Wars* (1977) demonstrated that an interesting world could find wild success even cast largely with unknowns, and even make more money from merchandising than from box office ticket sales (and this despite *Star Wars* having record box ticket sales to begin with). The 1970s also saw the rise of the popularity of sequels, perhaps starting with *The Godfather* (1972) and *The Godfather: Part II* (1974), the first time a film and its sequel both won a Best Picture Academy Award, although technically, both were from the same novel. As cross-media ownership increased in the late 1970s and into the 1980s, the idea of planning releases in multiple media simultaneously became easier and more practical, and new venues for films, like cable television and videotape, kept films in the public eye longer and gave studios more ways to earn money from their films, as well as merchandising deals. Cable and videotape also encourage sequels, namely by giving audiences more chances to see the first film in a series, in case they had missed it in theaters.

By the late 1990s and early 2000s, the organization of multiple-media release plans were becoming more common, as entertainment franchises were coordinated for simultaneous releases in various media. For example, the film *The Matrix Reloaded* (2003) came out the day after the video game *Enter the Matrix* (2003), and both works had concurrent stories, and used the same actors and actresses (Jenkins 2006). Simultaneous releases became possible due to the success of the original work (or works) that started a successful franchise, as well as advertising plans designed to increase audience anticipation (which did not always succeed). And, instead of being treated as separate stories, the works involved in simultaneous releases taken together usually suggest a single world, one in which multiple stories occurred, and which could be viewed through multiple media windows.

Multiple Windows on the Same World

By the second half of the twentieth century, and particularly after the spread of television, many Americans knew most of what they knew about the real world or "Primary World" not through first-hand knowledge, but through mass media. Certainly, the amount and scope of first-hand experience had also increased; car culture and the new interstate system of highways made travel easier, commercial airline travel made international travel affordable, and a series of wars in which Americans were involved, from World War II, to the Korean War, to the Vietnam War, gave many Americans a wide range of first-hand experiences of the world. But films, television, newspapers, and magazines could bring you the world as well, more cheaply and without all the inconveniences of actual travel. Better still, they could take you to places that actual travel could not; not just the moon landing, limited to an elite few, but worlds of the imagination that were beyond anything you could experience in the real world. The ubiquity of second-hand experience of the world gained through media normalized the experience, making the imaginary worlds experienced through multiple media more like the way the real world was experienced through multiple media. For example, anyone who has never been to Mongolia only knows it through second-hand or mediated experiences, which is also the only way one can experience Oz or Middle-earth.

And over time, the windows on the worlds, both real and imaginary, grew in size and transparency. Black-and-white film and television gave way to color imagery, and the tiny screens of early television began to grow in size from the 1960s onward. And from the 1970s onward, video games began to offer increasingly complex worlds in which users could interact with objects, and both video games and television shows were produced in high-definition imagery during the 2000s, and some in 4K imagery after that. Interaction and high-resolution imagery create the need for the inclusion of more detail, since audiences can explore a world's visual elements better than they could previously; video games like *Grand Theft Auto V* (2013) let players walk around their worlds and examine them closely.

As media franchise releases become more coordinated toward simultaneous releases in multiple media venues, then, the worlds they depict are seen through a variety of media windows, having two particular impacts on the worlds observed through them. The first impact is that the multiple perspectives on these worlds together create the illusion of an actual world which is being described and reported in multiple ways; an experience that can become very immersive the more material is produced about a particular world. The more unified and consistently the world is presented, the more it seems like an object viewed through transparent windows which has an existence of its own, rather than something the existence of which is only a virtual one located in the windows themselves. The world of the *Star Wars* franchise has shown us enough different aspects of its worlds—urban, rural, military, domestic, public, and private—that we feel as though we have a good idea what it would be like to live there.

The second impact on the world is the potential for a clash of styles and representations, which will at best be a variety of different interpretations that results from appearing in multiple media venues. Scenes of the world may appear in the form of live-action, hand-drawn illustrations or animations, computer graphics, or even as toy play sets or other merchandise. Even within each of these methods of representation, the stylization of the imagery can vary considerably. For example, in the *Star Wars* franchise, we have the live-action films, live-action television specials, computer-animated television series (and one feature film), comics books in various drawing styles, video games with varying levels of realism, and other adaptations such as LEGO Star Wars, and several decades' worth of *Star Wars* toys and merchandise. Throughout these manifestations and iterations, we find a wide range of stylizations, from the dramatic to the comedic, photorealistic to the abstract, and characters being used in non-canonical material such as television ads, marketing campaigns, and cross-overs with television shows and trans-franchise appearances, such as Luke Skywalker, Chewbacca, C3PO, and R2D2 appearing on *The Muppet Show* (1976–1981) in 1980, or more recent appearances with Disney characters after Disney purchased the *Star Wars* franchise in 2012 (by which time Disney had already owned the Muppets). This clash of difference styles for the same characters and world has the effect of making the windows through which the world is seen potentially less transparent, as attention is called to the differences in representation; or, at the very least, audiences would compare the different depictions found in different windows. But, instead of altogether destroying the consistency of the world, this also has the effect of presenting the world as something that can be interpreted by different artists and other craftspeople; and as an object of multiple interpretations, it becomes similar to the Primary World, but in a different way. Characters like Batman and the Joker, for example, have demonstrated just how much variation is possible in a character's depiction, while still keeping enough central elements unchanged. The presence of both similarities and differences, then, is one of the great strengths of transmedial worlds.

More of the Same, But Different

Imaginary worlds in general, and transmedial imaginary worlds in particular, are very good at balancing the two basic needs of audiences, that of novelty and familiarity. First, there is the balance mentioned above, that every imaginary world has enough content within it similar to our world to make it relatable, while at the same time changing enough world defaults in such a way as to produce novelty, and often a great deal of it. When popular characters or worlds are encountered by an audience, they will often want more, and this is where the balancing act between novelty and familiarity comes in. Early on, sequels were one of the first ways to provide more, wherein characters, and often the world they lived in, would be reused in a new story. This was also an economical move from a world-building perspective, because much of what was needed for the story would already be created. The idea of a series went even further, giving the audience multiple stories with the same characters (and often, world).

Transmedial worlds, however, not only give an audience more stories with the same characters and world, but also different media experiences. For example, while an audience may have first encountered a world and its characters by reading about them in a novel, they could next see what they looked like in a film or comic book, and hear what they sounded like in a film or radio play. Each medium provided a distinctly different way of encountering the same material, which means that it was technically not quite the same, as it included different aspects of that material (such as image, sound, and motion). Combined, all of these experiences added to the illusion of a complete world, as described previously. At the same time, however, the requirements of image, sound, and motion meant that, unlike worlds found in books, multiple people would be needed to make the world come about in visible and audible form, requiring collaboration and coordination to keep the world and its design consistent. Thus, transmedial imaginary worlds, unlike those found in books and comic books, almost always require multiple authors or a hierarchy of authors overseen by the originator of the world, without whom the world could not be constructed.

The Construction of Transmedial Worlds

Transmedial worlds often begin in a particular medium of origin, in which they are introduced to audiences, and if they find popularity, then they will be given the funding and attention needed for expansion in other media. It is perhaps no surprise, then, that many worlds begin in novels, short stories, or comic books, all of which could be made by a single author, and made without a large financial outlay. Some worlds are introduced to audiences in films, television shows, or video games, but even these will likely also begin as written or drawn descriptions in the form of scripts and storyboards, until they are able to attract the interest of someone to finance them (so in this sense, they, too, are first introduced in written form to a very particular audience, one which has the power to make them in audiovisual form). When mass popularity seems assured, along with the necessary fiscal success for a world's continuation, then finally can the release into multiple media be planned.

Still, depicting a world in multiple media often means shifting assets from one medium to another. Elsewhere, I have suggested the different processes that occur when such shifts are made: *description, visualization, auralization, interactivation*, and *deinteractivation* (Wolf 2012, 250–264). *Description* occurs when something must be described in words; novelizations of films or video games, or even descriptions of sound effects, mean having to articulate enough of the features of the thing being described to evoke it in the reader's imagination. There are advantages to using description over actually showing something, since one can suggest in the description the emotions experienced by an onlooker; for example, when J. R. R. Tolkien describes Saruman's voice as:

> low and melodious, its very sound an enchantment. Those who listened unwarily to that voice could seldom report the words that they heard; and if they did, they wondered, for little power remained in them. Mostly they remembered only that it was a delight to hear the voice speaking, all that it said seemed wise and reasonable.
>
> *(Tolkien 1994, 452)*

Instead of merely attempting to reproduce a sense of its actual sound, a large part of Tolkien's description of Saruman's voice is the effect that his voice has on those who hear it; thus a description can accomplish what an actor speaking can only attempt to do, which naturally is still only one possible interpretation of what Tolkien intended.

Visualization and *auralization*, meanwhile, both involve taking a written description and turning it into actual, concrete imagery and sound, respectively. As the example of Saruman's voice demonstrates, this can sometimes be a difficult thing to do. Such a transmedial move involves interpretation, since there are almost always multiple ways that something can be visualized or auralized. The shift from the conceptual (verbal description) to the perceptual (images and sound) forces specificity; it is inevitable that additional detail of some kind will need to be supplied which was not in the original description. On the other hand, providing an audience with imagery and sound gives an author more control over how a world looks and sounds, since it is not left to the imaginative abilities of the individual audience members. This also means that an audience member's experience of a world will vary less from one person to another, since the visual and sonic material can simply be seen and heard, requiring no imagination on the part of the audience (this can also be seen as a critique, of course, since such authorial control reduces the need for an audience's imaginative ability).

There are also multiple forms of visualization, each of which have different demands when assets are adapted to fit them. Visualizations can be two-dimensional or three-dimensional, with the latter requiring designs to be extended into three dimensions, which involves working out a consistent geometry for the objects in question. Considering time as the fourth dimension, we could even add four-dimensional visualizations which add movement and require animators to decide how something moves and what limitations it may have. Finally, there is the question of representational style, which can range from abstract to photorealistic, and includes caricatures and stylized versions of world assets. For example, due to their transmedial nature, the characters from the aforementioned *Star Wars* franchise have appeared as live-action characters played by human beings, radio voices, hand-drawn comic book characters, photorealistic computer-generated characters, simplified, stylized computer-generated characters (as in *The Clone Wars* (2008–2014) television series or in many video games), and even as LEGO minifigures, which appear as both physical objects in LEGO building sets and as digital models in LEGO films and video games. While having multiple visual incarnations of the same characters (or objects) may seem bad for consistency and canonicity, comparisons of these incarnations reveal the distinct and unique combination of elements which identify each character, in much the same way that caricatures can be identified more quickly than photographs of the people on which they are based (Mauro and Kubovy 1992). Transmedial adaptations, then, can result in assets which are more iconic and easier to evoke in the minds of audience members due to the heightened specificity of their designs.

Finally, *interactivation* is the addition of interactivity to a set of assets (for example, when film locations, objects, and characters become used in a video game), whereas *deinteractivation* is the opposite process, in which interactivity is removed from a set of assets (for example, when a video game is adapted into a movie or novelization). Interactivity is itself something that can appear in multiple media, such as a choose-your-own-adventure book, a film which allows the viewer to choose between narrative pathways, or a video game or website. When something is adapted into an interactive medium, how much it will need to be changed will depend on the kind of interactivity that is present. Interactivity which includes controlling an avatar in the diegetic world and altering the direction of the story is quite different from interactivity which allows a viewer to read various informational screens, bring up details, and explore locations visually without doing so in the context of a game. Likewise, the removal of interactivity usually means that a particular story becomes a fixed series of events, as opposed to the ones chosen by the player. The notion of interactivity also brings up another issue pertaining to the experiencing of transmedial worlds, that of audience participation.

Forms of Audience Participation

The audience's participation in any transmedial world will naturally include integrating materials experienced in different media together into a unified experience, which should not be difficult if consistency is a priority of the world-builders. More direct participation in the world will depend on the role given to the audience, which may range from observer (in worlds in which stories have already been constructed for audiences) to participant (as in single-player video game worlds) to virtual inhabitant (as in massively multiplayer online role-playing games (MMORPGs)), where hundreds of thousands of player-characters interact and determine the direction of events occurring in the world, as opposed to merely experiencing events which are determined ahead of time by an author.

The nature of this participation and the kind of experience it creates also varies considerably, ranging from the physical to the virtual, and from the abstract to the representational. Participation could involve the playing of a table-top board game like *Forbidden Island* (2011), in which the player's avatar and events are highly abstracted during gameplay. Video games represent everything virtually, in designs from the very abstract (like 1970s arcade games), to highly photorealistic three-dimensional environments. Finally, audiences can physically enter theme park settings, like "Pandora—The World of Avatar" which opened in Disney's Animal Kingdom in 2017, or "Star Wars: Galaxy's Edge" set to open in Disney Hollywood Studios in 2019, which are both designed to represent the experiencing of an imaginary world through a form of pretend that is similar to that found on the theater stage.

In most cases, transmedial worlds are made up of both interactive and non-interactive experiences or works, which are understood to refer to the same world. Typically, the division is clearly understood by audiences, whose expectations largely depend on the medium in which a work appears. For example, an audience's expectations involving photorealistic imagery are reduced when a video game is encountered, as opposed to the level of realism that an audience expects to find in a live-action film. But does the variety of media increase or decrease the overall immersive intensity for a world? Certainly, a novel like *The Lord of the Rings* (1954–1955), can be immersive by itself, whereas even a physical walk-through experience can fail to be immersive if it is badly designed. Having a world present, in some way, in multiple media can enhance the immersive experience by providing a variety of windows through which a world is seen, but it may also underscore its artificiality by depicting its elements in many different styles and by using its imagery on such a wide array of merchandise that the original meanings and emotions attached to the characters are diluted into merely a commercial brand. Villains like Darth Vader may lose some of their menace when encountered on children's pajamas or as LEGO minifigures. Here, too, however, the actual content of the franchise will partly determine what sort of balance can be had. A franchise like David Lynch's Twin Peaks, created for an adult audience (and perhaps a particular one at that), has appeared as two live-action television series (or one with a 26-year hiatus), a live-action feature film, a few books, and a trading card set; but one does not expect to see action figures or cartoons made from it. Whereas a franchise like Pixar's Toy Story, whose characters are already children's toys (which even appear themselves as merchandise diegetically within their own world), is more likely to be transformed into cartoons, video games, and a wide array of children's toys and merchandise, but is not a franchise we are likely to see made into a live-action feature film.

Conclusion

Transmedial franchises, especially world-based ones, are the direction that much big-budget entertainment has taken. As has been shown, the building of a transmedial world goes beyond an individual story, medium, or author, and provides the basis, and in many cases even a brand, that audiences can return to multiple times as well as anticipate future releases set in the same worlds. The transmedial nature of these franchises works well with the convergence of media technology, and new audiences' interest in a franchise can often mean more sales of older works set in the same world. Basing entertainment on an imaginary world is also a sound financial strategy, since works set in a given world can continue to be produced almost indefinitely, beyond what is possible for a single author, actor or director; the world extends beyond all of these. Both the advantages and popularity of this model of entertainment production suggests that transmedial world-building will continue to be the dominant one for some time to come.

References

Jenkins, Henry. 2006. *Convergence Culture: Where Old and New Media Collide.* New York: New York University Press.
Mauro, Robert, and Michael Kubovy. 1992. "Caricature and Face Recognition." *Memory & Cognition* 20 (4): 433–440.
Santo, Avi. 2015. *Selling the Silver Bullet: The Lone Ranger and Transmedia Brand Licensing.* Austin: University of Texas Press.
Tolkien, J. R. R. 1989. "On Fairy-Stories." *Tree and Leaf.* Boston: Houghton Mifflin Company.
Tolkien, J. R. R. 1994. *The Lord of the Rings* (one-volume edition). Boston: Houghton Mifflin Company.
Wolf, Mark J. P. 2012. *Building Imaginary Worlds: The Theory and History of Subcreation.* London and New York: Routledge.

16

TRANSMEDIA CHARACTERS

Additionality and Cohesion in Transfictional Heroes

Roberta Pearson

Vast and expansive fictional storyworlds built upon an accumulation of multiple texts have existed for millennia; the Greek gods, the Christian God, Robin Hood, and King Arthur are but a few instantiations of humanity's propensity for the narrative form. For example, Jesus Christ had his textual origins in the four gospels of the New Testament but the character almost immediately spans out across successive periods' available media, from painting to sculptures to illuminated manuscripts to stained glass windows and eventually to analogue and digital screens. In the late nineteenth and early twentieth centuries, interconnected cultural industries such as publishing, newspapers, advertising, and the cinema gave rise to industrially produced storyworlds such as the Wizard of Oz and Tarzan (Freeman 2016). Beginning in the 1980s, the media and industrial convergence of the cultural industries established expansive storyworlds as a dominant narrative form—from *Harry Potter* to *Star Wars* to the Marvel Cinematic Universe and beyond.

Narratologists and media studies scholars have addressed the narrative and industrial relationships in some expanded storyworlds (see Bertetti 2014; Denson 2011; Doležel 1999; Jenkins 2006; Ryan 2013; Scolari 2009; Thon 2015). However, little consideration has been given to the specific narrative and industrial factors that determine particular producers' strategies for creating character-based additions which consumers are likely to accept as part of the previously established transmedia storyworld (but see Harvey 2014). This chapter offers some tentative hypotheses concerning producers' strategies for additionality and cohesion for transfictional characters in different types of fictional storyworlds, using as its case studies Sherlock Holmes, Batman, and *Star Trek*. But prior to discussing these particular instantiations of the broader narrative form, some definitions first, followed by a taxonomy.

Rather than continuing to use the term "storyworld," the chapter draws upon Marie-Laure Ryan's concept of transfictionality, defined as "the migration of fictional entities across different texts" (2013, 383). Transmedia transfictions are a subset of transfictionality, crossing over two or more media. This chapter concerns factors that may influence the additionality and cohesion of single-medium transfictions as well as multiple-media transfictions, although most contemporary high profile transfictions are transmedial, as are my three case studies.

Two reasons motivate my use of the term additionality rather than expansion. First, the term addition does not necessarily imply a narratively meaningful expansion—that is, one that enlarges or reworks a transfiction's previously established settings, events, and characters. Jan-Noël Thon says that "two single works" within a transfiction can be defined, first, by a relation of redundancy, when one is aiming to represent the same elements of a storyworld that the other represents; second, by a relation of expansion, when one is aiming to represent the same storyworld—and the characters within in—that the other represents but adds previously unrepresented elements; and, third, by a relation of modification, when one is aiming to represent elements of the storyworld represented by the other but adds previously unrepresented elements that make it impossible to comprehend what is represented as part of a single, non-contradictory storyworld (Thon 2015, 33).

Two adaptations of the same Holmes story add to the transfiction but do not expand it in terms of new events, new settings, or new character details—this is a relationship of redundancy. But additions to the Holmes transfiction may (and frequently do) include new events, settings or character details, as for example in the Granada television series of the 1980s and 1990s starring Jeremy Brett—this is a relationship of expansion. Additions may also (and frequently do) rework previously established events, settings and character details, as for example in *Sherlock* (BBC, 2010–)—this is a relationship of modification. Additionality refers to all these cases, with expansion a subset of a broader industrial practice.

Second, expansion seems implicitly to imply cohesion whereas additionality does not; an addition can have fairly minimal points of contact with the previously established transfiction. Referring to newly added texts as additions rather than by another label such as installments avoids implications of narrative continuity. Speaking of iconic characters such as Holmes and Frankenstein, Shane Denson hypothesizes that these figures "exist as the concatenation of instantiations that evolves, not within a homogeneous diegetic space, but *between* or *across* such spaces of narration" (2011, 536). Denson continues: "These characters … carry traces of their previous incarnations into their new worlds, where the strata of their previous lives accrue in a non-linear, non-diegetic manner" (2011, 537).

Denson's traces of previous incarnations are my points of contact—the overlaps with previous texts that identify an addition as part of an established transfiction. Maximum points of contact lead to strong cohesion, while minimal points of contact lead to weak cohesion; the degree of overlap establishes a spectrum between strongly and weakly cohesive transfictions. Ryan's distinction between logical and imaginative storyworlds speaks to the opposite poles of this spectrum. If a text rewrites an existing narrative, modifying the plot and ascribing different features or destinies to the characters, it creates a new storyworld that overlaps to some extent with the old one. While a given storyworld can be presented through several different texts, these texts must respect the facts of the original text if they are to share its logical storyworld. In an imaginative conception, by contrast, a storyworld consists of named existents and perhaps of an invariant setting (though the setting can be expanded), but the properties of these existents and their destinies may vary from text to text (Ryan and Thon 2014, 5). This chapter discusses those factors that may result in the construction of logical versus imaginative storyworlds and their characters or in my terms, the points of contact between an addition and cumulative previous additions.

I propose three different structuring factors, two narrative and one industrial; the first two concern the differences between types of transfictions and the third transfictions' intellectual property status. The consensus amongst narratologists is that a storyworld consists of settings, events and characters. In some transfictions, strong or weak cohesion results from the points of contact between additions and previously established settings and events (timeline). These are time/place transfictions, such as the world of *Star Trek*. In other transfictions, strong or weak cohesion arises primarily from points of contact with a previously established character. These are character transfictions such as Batman and Sherlock Holmes.

The definition of events and settings is fairly non-contentious, but narratologists have argued for decades about the definition of character (see Chatman 1978; Rimmon-Kenan 1983; Bal 1997, 2011). Let us approach the problem using the Holmes character. Most would agree that character name (Sherlock Holmes or some variant thereof as in *Sherlock Hound*, the 1980s Japanese animated series) and narrative function (detection) are fundamental elements of a fictional character. In my previous work, I have argued that television characters are composed of a character template composed of six elements; this applies equally well to the transmedial Holmes. The Holmes character's psychological traits/habitual behaviors would include: intelligent; non-emotional; plays violin; smokes pipe; takes drugs; relentless curiosity; easily bored by lack of action, etc. His physical traits/appearance would be: tall; thin; aquiline nose; deerstalker, etc. His speech patterns and dialogue: "Elementary, my dear Watson" and other characteristic phrases. His interactions with other characters: Watson; landlady Mrs. Hudson; police; Moriarty; brother Mycroft, etc. His environment: Baker Street/London in original texts but setting can vary in additions. And his biography: relatively little in original texts but additions often fill in the backstory.

Some of these elements originated in the Conan Doyle stories while others originated in additions that have achieved canonical status, becoming widely accepted as defining elements of the character. For example, Conan Doyle's Holmes never wears a deerstalker; Sidney Paget, the illustrator for the *Strand Magazine*, added the iconic headgear. Similarly, Conan Doyle's Holmes never utters the famous phrase "Elementary, my dear Watson"; various reports attribute these words to actors William Gillette or Basil Rathbone. To use Denson's formulation, the deerstalker and the phrase are the traces of previous incarnations that the Holmes character frequently carries into his new worlds. But the character template is not coterminous with the character; a Holmes character in a single addition cannot manifest all the potential character elements established in the myriad additions to the transfiction. The particular combination of elements manifested by the Holmes character in a new addition results from the producers selecting suitable elements for the intended audience and omitting unsuitable elements. For example, *Sherlock Hound*, intended for a child audience, does not reference the detective's cocaine habit, while *Elementary* (CBS, 2011–), intended for an adult audience, makes drug addiction a central attribute of the character's biography.

The second narrative factor addresses the ontological status of the transfiction; is it realist or fantastic? Rather than relying on genre theory, I turn to Ryan's concept of possible worlds. Ryan argues that all fictions entail the creation of possible worlds that are linked to the actual world by an "accessibility relation"—various similarities/dissimilarities of logical principles, physical laws, material causality, geography or history, populations of natural species, stages of technological development, human inventory, and the like (Ryan 2005, 446). The Holmes transfiction is a realist one, strongly linked to the actual world, while Batman and *Star Trek* are fantastic transfictions, diverging from the actual world in many respects including physical laws, geography, and history.

The third factor is an industrial one, revolving around the transfictions' intellectual property status. The ownership of copyright and trademark enables the proprietors, either individuals or corporations, to augment transfictions with legally authorized additions, to license other individuals or corporations to produce legally authorized additions and to prohibit individuals or corporations from producing non-authorized additions. These are proprietorial transfictions, such as *Star Trek* and Batman. Other transfictions have no one central holder of the intellectual property (IP); these are non-proprietorial or public domain transfictions such as Sherlock Holmes. As of 2015, Arthur Conan Doyle's texts have almost all entered into the public domain (with the exception of a few of the later stories in the United States under copyright until 2023), although Doyle's descendants made no concerted attempt to police the nature of additions even when they owned the IP to all the author's Holmes texts (see Pearson 2015). Producers of additions such as the BBC's *Sherlock* own the copyright and trademark for the new text and, like any other holders of intellectual property, can themselves produce new additions to and both authorize and prohibit new additions by other parties.

And now for the promised taxonomy. The two narrative factors, time/place versus character and realist versus fantastic, together with the industrial factor, proprietorial versus public domain, produce eight types of transfictions set out in Table 16.1.

From now on, this chapter uses the three case studies of *Star Trek*, Batman, and Sherlock Holmes to consider producers' strategies for additionality and cohesion in different types of transfictions. First it compares a time/place transfiction (*Star Trek*) to character transfictions (Holmes and Batman). It then compares a realist character transfiction (Holmes) to a fantastic character transfiction (Batman). It

Table 16.1 Taxonomy of transfictional characters and worlds

Transfiction type	*Example*
Proprietary realist time/place	*The Sopranos, Downtown Abbey*, etc.
Public domain realist time/place	Austen, Dickens
Proprietary fantastic time/place	*Star Trek*
Public domain fantastic time/place	Greek mythology
Proprietary realist character	James Bond
Public domain realist character	Sherlock Holmes
Proprietary fantastic character	Batman
Public domain fantastic character	Frankenstein

concludes by comparing proprietorial transfictions (*Star Trek*, Batman) to a public domain transfiction (Holmes), although given the character's vexed and complicated copyright status, it is more accurate to call the Holmes' transfiction a semi-proprietorial transfiction that has for several decades operated like a public domain transfiction. Some tentative hypotheses arise from these comparisons that may be more broadly applicable to similar types of transfictions.

Time/Place Transfictions versus Character Transfictions

Producers of time/place transfictions create additions through two primary strategies: extending the timeline and establishing new settings. One example is *Star Trek: Discovery*, a CBS production distributed via its on-demand service, CBS All Access, in 2017. *Star Trek: Discovery* takes place in the twenty-third century, sitting in the *Star Trek* transfiction's timeline after the events of the original series and before the events of *The Next Generation*, sometime around the events of the sixth *Star Trek* feature film *The Undiscovered Country* (Nicholas Meyer, 1991). Previous *Star Trek* producers have traditionally augmented the transfiction with new settings. The three shows occurring in the twenty-fourth century, *The Next Generation*, *Deep Space Nine*, and *Voyager*, were set respectively on a starship in the Alpha Quadrant of the Milky Way galaxy, on a space station in the far reaches of the Alpha Quadrant near a wormhole to the Gamma Quadrant, and on a starship in the Delta Quadrant.

Producers of character transfictions also extend timelines and create new settings. Sherlock Holmes and Batman can be teenagers, as in *Young Sherlock Holmes* (Barry Levinson, 1985) or *Gotham* (Fox Broadcasting, 2014–), or old, as in *Mr. Holmes* (Bill Condon, 2015) and Frank Miller's graphic novel *The Dark Knight Returns* (DC Comics, 1986). Producers can also move the characters to new settings: *Sherlock* updates the character to twenty-first-century London and *Elementary* to twenty-first-century New York City. DC relocated Batman to nineteenth-century Gotham, in *Gotham by Gaslight* (1989), the first of its Elsewhere series. But producers of character transfictions can also employ a strategy not so readily available to producers of time/place transfictions—new embodiments of a character. In the rebooted *Star Trek* film series, Chris Pine plays James T. Kirk, but the actor channels William Shatner, the first Captain Kirk, both in appearance and acting style and will most likely continue to play the role throughout the run of the series. As I was researching this chapter, the sad death of Anton Yelchin, Pavel Chekov in the new *Star Trek* films, was announced. Should the film series continue, the producers will face the difficult decision of either providing a narrative explanation for the character's absence or of embodying the character in a new actor. Character transfictions generally do not establish such a strong equivalence between actor and character as to cause such dilemmas. Wikipedia's undoubtedly incomplete list of the actors who have played Sherlock Holmes in film, television, radio, and on the stage runs to over 90 entries, while five actors have portrayed the Dark Knight in the Warner Bros. films alone (http://en.wikipedia.org/wiki/List_of_actors_who_have_played_Sherlock_Holmes). While new Holmes or Batman embodiments may have based elements of their performance on previous embodiments of the Great Detective or the Dark Knight, they do not have to strongly resemble previous embodiments in appearance and acting style. On a side note, proposed and actual reimbodiments have recently generated controversy in fan communities and the media. Will audiences accept a black James Bond, Captain America, or Hermione Granger as having sufficient minimal points of contact with previous embodiments? This raises issues well beyond the scope of this chapter.

Hypotheses:

1. Cohesion in time/place transfictions arises primarily from points of contact between additions and the transfiction's previous events (timeline) and settings. *Star Trek* additions take place in the timeline established by the television series and feature films and must have a family resemblance to previous settings in terms of physical laws, institutions, aliens, history, technology, and design. But cohesion does not arise primarily from the characters—an addition does not require the presence of Captain Kirk, Captain Picard, Mr. Spock, or any other character as a point of contact with the transfiction.
2. Cohesion in character transfictions arises primarily from points of contact between the character in the addition and the character name, function, and template established by the transfiction.

Realist Character Transfictions versus Fantastic Character Transfictions

Are there differences between realist and fantastic character transfictions' strategies for additionality and cohesion? As noted above, both the Holmes and Batman transfictions move the character to new settings, but Holmes additions seem to require fewer points of contact with the established character template than Batman additions. It is telling that texts that expand the timeline or create new settings for the Dark Knight, such as the television series *Gotham* and the comic *Gotham by Gaslight*, point to the established transfiction through the fictional city's name—Batman and Gotham are coterminous. As the villainous Riddler put it, "When is a man a city? When it's Batman or when it's Gotham. I'd take either answer. Batman is this city" (Gaiman 1989; see also Uricchio 2010). Holmes has strong associations with London but is not coterminous with it; a Holmes addition requires neither Baker Street nor London as a point of contact with the character. In *Elementary*, Holmes relocates from London to New York City but continues to perform his narrative function of detection. By contrast, the Dark Knight could not perform his narrative function of crime fighting in *Elementary*'s realist metropolis where only the deluded pursue careers as masked vigilantes (see "You've Got Me, Who's Got You?" season 4, episode 17). Holmes, however, can perform his narrative function not only outside of London but even in a fantastic world, as he does in the graphic novel series *Victorian Undead* or in the animated television series *Sherlock Holmes in the 22nd Century* (Scottish Television, 1999–2001).

Hypotheses:

1. Specific environments may constitute a more essential element of the character template for fantastic characters than for realist characters.
2. Realist characters can function in fantastic worlds but fantastic characters cannot easily function in realist worlds.
3. As a result, realist character transfictions have more strategies available for additionality and cohesion than fantastic character transfictions.

Proprietorial Transfictions versus Public Domain Transfictions

Do additionality and cohesion strategies differ between proprietorial and public domain transfictions? In both cases, producers of additions can use paratexts to signal alignment with the transfiction, one strategy being the paratextual invocation of the author function. However, it should be noted that the creator/author is not necessarily the holder of the transfiction's intellectual property. During his lifetime, Conan Doyle held undisputed rights to the Holmes' character, which then descended to his heirs upon his death. By contrast, National Allied Comics, the precursor of DC Comics, held the IP for Batman, not the original writer Bob Kane nor the original illustrator, Bill Finger, while Gene Roddenberry sold the rights to *Star Trek* at a difficult moment in his career. In keeping with cultural propensities to valorise individual authorship, both proprietorial and public domain transfictions invoke revered original author/creators, not the current producers or intellectual property holders, to achieve cohesion.

Additions to the *Star Trek* and Batman transfictions respectively include the credit lines created by Gene Roddenberry and created by Bob Kane and, as of 2015, the latter also credit Kane's co-creator Finger (McMillan 2015). Promotional paratexts such as interviews with producers, directors, and stars also invoke the author function. Leora Hadas has demonstrated that Paramount's promotion for the first Abrams *Star Trek* film (2009) aimed paratexts mentioning Gene Roddenberry at *Star Trek* fans, presumably more alert to indices of cohesion than the broader audience (Mac 2014). The Roddenberry name holds such value that Gene's son, Rod, serves as executive producer on the new CBS series; the motivation for adding another Roddenberry to the credits probably stems from a desire to appeal to the core fan base of Trekkers. Screen additions to the Holmes transfiction frequently include the credit "based upon characters created by Arthur Conan Doyle," although promotional paratexts do not consistently mention Sir Arthur. For example, Steven Moffat, *Sherlock*'s showrunner, constantly refers to Conan Doyle's importance and influence, whereas Guy Ritchie, director of the Warner Bros. Holmes films, and Robert Doherty, *Elementary*'s showrunner, speak of the author far less frequently. Proprietary transfictions may be more inclined to invoke the author function than are public domain transfictions. Proprietorial transfictions such as *Star Trek* and Batman have a direct line of descent through intellectual property from the original author(s) (whether or not they originally owned the IP) to the current producers; producers may invoke the author function to implicitly signal the construction of a logical storyworld with multiple points of contact with the established transfiction, thus appealing to the established fan base. By contrast, producers of public domain transfictions may elide the author function to implicitly signal the construction of an imaginary storyworld with relatively few points of contact with the established transfiction, thus appealing to new audiences.

Proprietorial transfictions also employ promotional paratexts to define the precise narrative relationships between new additions to a transfiction originating from different producers simultaneously exploiting the same intellectual property. Fox Broadcasting has specified that *Gotham*, which it produces under license from Time-Warner, will not impinge upon the narrative continuity of the Warner Bros. Batman films. Kevin Reilly, former chairman of Entertainment for the Fox Broadcasting Company, said that, "Warner Brothers manages the entire franchise and its one of their top global franchises of all. So there will be an awareness of both and we'll have to coordinate when we're in the market place, but the productions are not piggy-backing off one another" (Cannata-Bowman 2016). Similarly, CBS, which owns the rights to all *Star Trek* television, must coordinate with Paramount, which owns the rights to any *Star Trek* films. CBS President Les Moonves announced that the new *Star Trek: Discovery* television series would not appear until six months after the release of the Paramount film *Star Trek Beyond* (2016). Said Moonves: "Our deal with [Paramount] is that we had to wait six months after their film is launched so there wouldn't be a confusion in the marketplace" (Hadas 2017, 53). Producers of public domain transfictions, or even the semi-proprietary-but-acts-like-a-public-domain Holmes transfiction, do not need to engage in such paratextual coordination. Both the *Elementary* and *Sherlock* producers received permission from the Conan Doyle Estate to make their programs, but CBS has no need to specify its program's relationship to the BBC's program, despite the two companies' simultaneous exploitation of the same intellectual property. The Holmes transfiction as a whole lacks the industrial convergence and mutual licensing agreements that underpin most contemporary transmedia transfictions—hence its greater resemblance to a public domain transfiction than a proprietorial one.

Finally, what of textual strategies for additionality and cohesion? Producers of proprietorial time/place transfictions who own the rights to all previous texts can weave additions together through dense intertextual webs linking back to the transfiction's established history. The *Star Trek* transfiction has, for example, used character cross-overs to launch new additions (the original series Doctor McCoy appeared in the first episode of *The Next Generation*) or to celebrate anniversaries (original series characters appeared in episodes of both *Deep Space Nine* and *Voyager* in 1996 to mark the franchise's thirtieth year). But such flashy narrative machinations are usually reserved for special occasions, the intertextual family resemblance between events and settings otherwise sufficing.

By contrast, producers of additions to the Holmes transfiction must rely upon the character as the primary connective tissue. The thousands of additions to the Holmes transfiction that have accumulated since the 1890s have all incorporated character name, function, and minimal elements from the character template to persuade consumers that their story was about Sherlock Holmes. As stated above, each producer selects those elements from the template most appropriate to the intended audience. The Sherlock Holmes character thus manifests extreme divergence: he can be young or old, white or black (as in New Paradigm's comic *Watson & Holmes*), live in London, New York, the nineteenth, twenty-first, or twenty-second centuries, and even be transformed from human to canine, as in *Sherlock Hound*. But the addition must quickly establish connections to the established transfiction through name, narrative function, habitual behaviors, appearance, speech, and sometimes, interactions with other characters and environment.

Hypotheses:

1. Proprietary transfictions avail themselves of paratextual strategies for additionality and cohesion to a greater extent than public domain transfictions.
2. Public domain character transfictions exhibit greater reliance upon textual manifestations of cohesion.

Conclusion

This chapter has concerned additionality and cohesion in transfictions, arguing that cohesion depends upon points of contact between the addition and the transfiction. However, it has also argued that there is a spectrum between strongly and weakly cohesive transfictions. To conclude, I want to make clear that strong cohesion (maximum points of contact) is not necessarily "better" than weak cohesion (minimal points of contact)—both strategies can attract audiences in an interplay of familiarity and differentiation. Maximum cohesiveness has its uses and pleasures but so does minimal cohesiveness; producers must assess their audience to determine the most successful strategy. For example, the Basil Rathbone and Benedict Cumberbatch Holmes embodiments have a complicated relationship with the canonical deerstalker that frequently serves as a primary signifier of the character. Rathbone reaches for it in *The Voice of Terror* (John Rawlins, 1941) only to have Watson (Nigel Bruce) remind him of his promise not to wear it. Cumberbatch rejects the proffered headgear in "The Reichenbach Fall" (season 2, episode 3). In both cases, the absent deerstalker indicates the relocation of the character to new settings in the 1940s and the 2010s, the producers quickly differentiating their texts from the original texts and the many additions to the transfiction. Both *Elementary* and *Sherlock* establish a relationship of modification to the original texts and the many additions to the transfiction by updating the character to the present. Audience pleasure derives from seeing the character function in this new setting, although *Sherlock* includes many modified points of contact with the original character template, for example, by having him contemplate a three (nicotine) patch problem rather than a three pipe problem. *Elementary* audiences derive pleasure from the gender modification that turns John Watson into Joan Watson. Long-lasting and ubiquitous transfictions may benefit as much from weak cohesion as from strong cohesion, purposively omitting or modifying events, settings, and characters as they seek to retain old audiences and attract new ones.

And finally, a call for further research. This chapter has used the three case studies of Sherlock Holmes, Batman, and *Star Trek* to offer some tentative hypotheses concerning additionality and cohesion in character-based transfictions. The case studies were selected partly to represent three types of transfictions from my eight-type taxonomy, but were primarily selected because they are the transfictions I know best both as fan and as scholar. Decades old, or even centuries old, transfictions comprised of hundreds or even thousands of accumulated additions pose a methodological challenge; analysis requires detailed familiarity which can only be individually acquired for a limited number of transfictions. Hence the tentative nature of the hypotheses, which must be confirmed or disproved by collaborative research among scholars well-versed in multiple and different transfictions.

Acknowledgements

The core content of this chapter was previously published as Pearson, Roberta. 2017. "Additionality and Cohesion in Transfictional Worlds." *The Velvet Light Trap* 79: 113–119.

References

Bal, Mieke. 1997. *Narratology: Introduction to the Theory of Narrative.* Toronto: University of Toronto Press.

Bertetti, Paolo. 2014. "Towards a Typology of Transmedia Characters." *International Journal of Communication* 8: 1–20.

Cannata-Bowman, Nick. 2016. "Star Trek: Everything We Know About the New TV Series." *TV Cheat Sheet,* May 30. Accessed April 21, 2017. www.cheatsheet.com/entertainment/star-trek-everything-we-know-about-the-new-tv-series.html/4/.

Chatman, Seymour. 1978. *Story and Discourse: Narrative Structure in Fiction and Film.* New York: Cornell University Press.

Denson, Shane. 2011. "Marvel Comics' Frankenstein: A Case Study in the Media of Serial Figures." *Amerikastudien/ American Studies* 56 (4): 531–553.

Doležel, Lubomir. 1999. *Heterocosmica: Fiction and Possible Worlds.* Baltimore: Johns Hopkins University Press.

Freeman, Matthew. 2016. *Historicising Transmedia Storytelling: Early Twentieth-Century Transmedia Story Worlds.* London and New York: Routledge.

Gaiman, Neil. 1989. "When Is a Door: The Secret Origin of the Riddler." *Secret Origins Special #1.*

Hadas, Leora. 2017. "A New Vision: J.J. Abrams, Star Trek and Promotional Authorship," *Cinema Journal* 56 (2): 44–66.

Harvey, Colin B. 2014. "A Taxonomy of Transmedia Storytelling." In *Storyworlds Across Media: Toward a Media-Conscious Narratology,* edited by Marie-Laure Ryan and Jan-Noël Thon, 219–228. Lincoln: University of Nebraska Press.

Jenkins, Henry. 2006. *Convergence Culture: Where Old and New Media Collide.* New York: New York University Press.

Mac, Danny. 2014. "Fox's Gotham is NOT Connected to the Warner Brothers DC Cinematic Universe." *Comic Book Movie,* January 13. Accessed April 21, 2017. www.comicbookmovie.com/fansites/DannyMac/news/?a=92884.

McMillan, Graeme. 2015. "DC Entertainment to Give Classic Batman Writer Credit in 'Gotham' and 'Batman v Superman.'" *Hollywood Reporter,* September 18. Accessed April 21, 2017. www.hollywoodreporter.com/heat-vision/dc-entertainment-give-classic-batman-824572.

Pearson, Roberta. 2015. "Sherlock Holmes, a De Facto Franchise?" In *Popular Media Cultures: Fans, Audiences and Paratexts,* edited by Lincoln Geraghty, 186–205. London: Palgrave Macmillan.

Rimmon-Kenan, Shlomith. 1983. *Narrative Fiction: Contemporary Poetics.* London: Methuen.

Ryan, Marie-Laure. 2005. "Possible-Worlds Theory." In *Routledge Encyclopedia of Narrative Theory,* edited by David Herman, Manfred Jahn, and Marie-Laure Ryan, 446–452. London: Routledge.

Ryan, Marie-Laure. 2013. "Transmedial Storytelling and Transfictionality." *Poetics Today* 34 (3): 361–388.

Ryan, Marie-Laure, and Jan-Noël Thon (eds.). 2014. *Storyworlds Across Media: Toward a Media-Conscious Narratology.* Nebraska: Nebraska University Press.

Scolari, Carlos A. 2009. "Transmedia Storytelling: Implicit Consumers, Narrative Worlds, and Branding in Contemporary Media Production." *International Journal of Communication* 3: 586–606.

Thon, Jan-Noel. 2015. "Converging Worlds: From Transmedial Storyworlds to Transmedial Universes." *Storyworlds: A Journal of Narrative Studies* 7 (2): 21–53.

Uricchio, William. 2010. "The Batman's Gotham City™: Story, Ideology, Performance." In *Comics and the City: Urban Space in Print, Picture and Sequence*, edited by Jörn Ahrens and Arno Meteling, 119–132. London: Bloomsbury.

17
TRANSMEDIA GENRES
Form, Content, and the Centrality of Memory

Colin B. Harvey

In this chapter, I explore genre in relation to transmedia storytelling. I look at existing definitions of genre specific to different forms of media, and examine how these definitions can inform an understanding of genre in relation to the operation of storyworlds that spread across multiple media forms. In particular, I look at the dominance of science fiction and fantasy genres within the transmedia sphere, examining potential explanations for why this might be the case. As I will show, recurring tensions around form and content which characterize existing genre theory in a variety of media are rendered more complex still by the inherent nonlinear properties of transmedia networks and the technological components involved in audiences' manipulation of instances of digital transmedia.

A consistent theme to emerge in my analysis concerns memory, both in terms of the generic traits which are remembered from artifact to artifact across different media, but also in terms of the remembered expectations with which audiences encounter and engage with generic artifacts. As I will show, what this suggests, consistent with the views of many genre theorists, is that a purely textual appreciation of genre is not enough, and that genre needs to be understood as part of a much larger network of relations. I will further suggest that this relationship needs to be understood in affective, materialist, and energetic terms, and that associated form and content issues should be framed in this way.

Finally, I will conclude by moving on to explore emergent transmedia genres, drawing on Janet Murray's seminal work on cyberspace narrative to look at the ways in which existing, established genres might be translated into transmedia networks, but also the potential for both subverting genre conventions and creating new genre formations capable of specifically exploiting the unique characteristics of transmedia storyworlds.

Genre Defined

Drawing upon Heather Dubrow's work, scholar of medieval literature Simon Gaunt identifies Aristotle and Plato as the first theorists of genre in the Western culture tradition, advancing broad categories for literature that include "comedy, tragedy and epic" (Dubrow 1982, 45–52; Gaunt 2005, 5). In his wide-ranging overview of genre, Daniel Chandler suggests that the term is now widely used in the fields of "rhetoric, literary theory, media theory and more recently linguistics" (1997, 1). Chandler discusses the "family resemblance" approach by which categories are identified as such by their shared traits, but highlights the inherent subjectivity involved in such a process of categorization (1997).

A significant point relates to the positioning of a particular genre in relation to other genres, but also in relation to other textual formations and what we might term the wider "network of relations."

John Frow suggests that genre cannot be understood as obtaining merely at the textual level: beyond the text itself is the importance of thinking about the text's function but also differing audience structures and patterns of reading (2015, 1). He goes on to argue that genre can "almost" be defined as the "*relationship* between textual structures and the *situations* that occasion them," with the caveat that the situation in question has to be very carefully delineated (2015, 14).

This network of relations can assume manifold guises. For instance, a further important point Gaunt highlights in relation to Frederic Jameson's writing on the subject is the ideological basis for genre, and the corresponding way in which specific genres might historically emerge to engage with societal concerns and tensions (2005, 7–8). Indeed, Will Wright in *Six Guns and Society*, his pivotal structuralist study of the Western, notes the ability of the Western film to adapt itself to changes in American society (1977, 210–211) (an ability this particular genre now seems to have forsaken, despite its preceding longevity).

Form and Content

In his account, Gaunt goes on to suggest that the study of genre is dominated by two approaches, one founded around form and the other centered around content. He maintains that Jameson has been most active in exploring the relationship between the two, although Mikhail Bakhtin and Tzvetan Todorov have also discussed the interplay. As Gaunt notes, Jameson characterizes Northrop Frye as primarily articulating a "semantic approach" to the study of genre, while Vladimir Propp opts for a "syntactic approach." Jameson himself, however, argues that genre is constituted by both syntactic and semantic elements, so that genre can be analyzed using either one of these approaches (Gaunt 2005, 6).

The complex relationship between form and content is also articulated in more industry-led approaches to describing transmedia storytelling. In their book *Storytelling Across Worlds*, Tom Dowd et al. specify 22 categories ranging from "Action" and "Adventure" through "Film noir," "Romance" and "War," "Western," and "YA or young adult" (2013, 57–59). Dowd et al. stress the importance of understanding the ways in which the different platforms involved in transmedia storytelling utilize the "Key Story Elements" of "Story/Theme," "Plot," "Characters," "Setting," and "Style/Tone" (2013, 48–57). They also, however, note the fluidity of genre categorization among websites like IMDB and Wikipedia, and among other individuals and parties with an interest in placing works into genre groupings, such as film critics and marketing teams (2013, 58).

This ambiguity around genre is something Frow also alludes to (2015, 13) and which Brigid Cherry discusses in her analysis of fan definitions of horror films (2007, 214). Indeed, Chandler observes that some approaches to genre theory prefer an approach which utilizes the "psycholinguistic concept of prototypicality," by which some texts are seen as being more indicative of a genre than others because of the number and type of traits demonstrated by the text in question (1997, 2–3). This approach means that genre categories should be viewed as "fuzzy." Brian Attebery utilizes the idea of the "fuzzy set" coined by George Lakoff and Mark Johnson to define the fantasy genre, seeing prototypical examples as sitting in the center of a genre definition, with those with ambiguous features sitting at the boundaries of the category (1992, 12). In her analysis of fantasy types, literary critic Farah Mendlesohn refines Attebery's approach to argue that fantasy should be viewed as multiple fuzzy sets (2008, xvii).

The specificity of newer kinds of digital interaction, in which a participant might be asked to manipulate complex, multi-modal interfaces, complicates the form and content discussion of genre still further. Writing in 2001, Mark J. P. Wolf proposed a taxonomy of 42 video games differentiated by kinds of interactivity, but which crucially were intended to be read in conjunction with existing "thematically based genres (like those of film)" (2001, 116–134). In this context, then, most of the *Halo* games can be viewed as science fiction first-person shooters (FPS), although the *Halo Wars* spin-off series consists of science fiction real-time strategy games (RTS).

In my own work, the importance of identifying media specificity and the role it plays in participant engagement with transmedia storytelling led me to identify "transmedia configuration"—that is to say the processes of manipulation and negotiation by which participants engage with transmedia networks (Harvey 2015, 121–123). This might extend to deciding which element of a transmedia network to engage with—for instance, whether to watch the animated television series *Star Wars Rebels* or play the console game *Star Wars Battlefront*—but also the configurative strategies required to engage with a particular element of the transmedia network in question. Clearly the particularity of the narrative experience involved in engaging with a comic or novel is going to be quite different to the kind of engagement involved in manipulating a DVD or video game, in terms of material, energetic and affective relationships.

Those transmedia networks which include digital media platforms such as video games explicitly recast Jameson's dialectic between the semantic and the syntactic in materialist and energetic terms. Yet this is also true of those elements within a transmedia network which are articulated in ways in which the configurative engagement asked of the participant—be they viewer, listener, reader—is altogether less complex. As Emma Beddows notes in her phenomenological study of *Buffy the Vampire Slayer*, fandom, audience expectation of the particularities of engagement with one medium, such as a television series, may not translate to their engagement with transmedia material articulated in a different medium, such as a comic. The individuals Beddows studied brought their own expectations from the television series and indicated dissatisfaction when other media disrupted their experience by not obeying the same conventions (2012, 149–150).

Genre and Memory

Significantly, Gaunt points to Todorov's observation that genre is "not only an intersection of social and formal properties, but also a fragment of collective memory" (Gaunt 2005, 7). This extends to the expectations an audience possesses prior to engaging with a genre artifact, expectations which are learnt—*remembered*—from previous engagements. As Stanley Fish observes, "a description of genre … can and should be seen as a prediction of the shape of response" (1980, 95).

While some genres have the audience's affective response factored in—"horror" being a good example, to build upon a point made by John Clute and John Grant (1997, 337)—all genres ought to be understood as experiential, framed by emotional responses such as anticipation and suspense, and more broadly informed by our memories and expectations of preceding generic experiences. There is, of course, a strong element of subjectivity involved, influenced by the extent of the individual's experience; literary critic Gerard Genette talks about the "narrative *competence*" of the reader, "arising from practice" which allows he or she to decipher the code of the genre in question (1983, 76–77).

The experiential nature of genre is something narratologist Monika Fludernik has discussed in her wider examination of narratology. Fludernik argues that our exposure to the generic conventions associated with written texts is now so profound that they have moved from being consciously negotiated and instead become "cognitive schemata" (2005, 45). While this particular position seems dangerously close to a reassertion of the Cartesian split, the experiential nature of *all* narrative engagement is difficult to refute without also returning to a dualist position of mind and body.

In a transmedial context, however, articulations are both remembered from outside the network in question and *within* the network. In the case of *Star Wars*, for instance, the character of C3PO is remembered from the films into other media such as the animated television series *The Clone Wars* and *Rebels*, novels, comics and video games. Though C3PO undergoes a transformation as part of this process, becoming stylized in the case of the animated series or represented via words in the case of the novels, he is still fundamentally the same character, which Beddows identifies as a "functional motif" able to "reconcile tonal and aesthetic differences" between different media platforms (Beddows 2012, 149–150).

At the same time, and beyond the formal expression of the character of C3PO in the medium in question, however, are the co-existence of the character's story signification as the specific character of C3PO remembered within a medium but also across other media, and a connected level of generic signification in which C3PO is a robot. The latter signification clearly identifies the text in question as belonging to the genre of science fiction or science fantasy, connecting the text to a wider network of relations and invoking certain expectations while rejecting others.

Dominant Genres

At the time of writing, the arena of popular culture is dominated by a large number of high-profile science fiction and fantasy transmedial franchises, or at least franchises demonstrating prototypical characteristics most readily associated with science fiction and fantasy genres (Chandler 1997, 2–3). The films *Guardians of the Galaxy* (2017), *Spiderman: Homecoming* (2017), and *Black Panther* (2017), all set within the Marvel Cinematic Universe and supported by varieties of transmedia expansion, have either been released or are about to be released. The long-running British science fantasy series *Doctor Who* (1963–present) continues on television, the *ur-text* of a complex web of transmediality including spin-off television shows, novels, audio plays, comics, web material, and video games. The Assassin's Creed franchise, originally founded on a series of video games (inspired in turn by the novel *Alamut* by Vladimir Bartol), has spread to comics, toys, a feature film, and a forthcoming animated television series. The X-Men series owned by Fox and informed by Marvel's long-running comic series has generated feature films and television series such as *Legion* and *The Gifted*, set within the same fictional milieu. The *Star Trek* franchise, too, which saw its fiftieth anniversary in 2016, continues as both a high profile film series and television series, as well as in multiple other transmedial and licensed guises.

In discussing *The Matrix* film sequels and its supporting transmedia infrastructure, Jenkins famously observes that in order to get the most from these kinds of fictional environments, consumers must become "hunters and gatherers," seeking out material across media channels, comparing notes and collaborating with fellow fans to maximize the quality of their experience (2008, 20–21). Elsewhere, Jenkins notes that *Star Wars* fans were amongst the first to engage with new technologies, using merchandising as resources for their web-based fan movies (2006, 144). The fan creativity that has long characterized science fiction and fantasy fandoms and which Jenkins documents in *Textual Poachers* (1992) and much of his subsequent writing fits naturally—*logically*, indeed—with the active and creative nature of transmedia storytelling, and which intellectual property (IP) holders increasingly seek to engage with and exploit in terms of community engagement and competitions.

Yet a significant other reason for the domination of science fiction and fantasy genres in the arena of transmedia storytelling might lie with the diegetic traits that often characterize these two genres. Such traits might include parallel universes, time travel, or magic, and can provide a means by which inconsistencies and contradictions arising from the nature of the licensing arrangement can be explained away, by either IP holders, fandom, commentators, or a combination of these agents. These traits are particularly the case in the past when transmedia networks were altogether less well-integrated and were tended to utilize much looser licensing arrangements than is currently the case.

A key example is *Doctor Who*, which has now existed for over 50 years and accrued a highly complex transmedia network. A key aspect of *Doctor Who*'s diegesis is time travel, and parallel universes also frequently feature, not to mention the sheer scale of the universe the main character is able to explore. Inconsistencies occurring across the wider transmedia storyworld—a key character dying who then reappears in other media without explanation, a similar fate happening to a fictional planet—can then be explained away by the contradictions inherent in the idea of time travel, parallel universes and the scale of time and space, and so remain diegetically consistent.

Often these inconsistencies are navigated exclusively by the fan base, but in *Doctor Who*'s case the television show itself has made increasing efforts to assist in this process. For instance, the recurring villains the Cybermen have been given multiple origins across *Doctor Who*'s transmedia network, including the television programme itself, comics, and audio. In the 2017 conclusion to the tenth series of the revived, post-2005 version of the program, the main character of the Doctor offers the explanation of "parallel evolution" having occurred to account for these multiple contradicting origin stories.

In a comparable vein, the ABC television show Marvel's *Agents of SHIELD* (2013–present), set within the same Marvel Cinematic Universe as films including the *Iron Man*, *Captain America*, and *Guardians of the Galaxy* film series and a range of tie-in comics, approaches the possibility of such a diegetic contradiction head-on. *Agents of SHIELD* is centered around the character of Agent Phil Coulson (played by Clark Gregg), despite the character having perished in *The Avengers* (2012), which happens chronologically prior to *Agents of SHIELD*. Coulson's resurrection becomes an ongoing plot point within the first season of *Agents of SHIELD*, and is revealed as part of a covert plan by Samuel L. Jackson's character Nick Fury to use alien DNA to bring Coulson back to life.

Where transmedia production occurs in-house, it is possible to ensure synchronicity in terms of the storyworld (this was my own experience working as a Narrative Designer on video games and related transmedia material for the British company Rebellion Developments, as opposed to my prior experience undertaking transmedia media work as the writer of licensed fiction for properties such as *Doctor Who* and *Highlander*). In other, generally larger contexts, where material is licensed to external transmedia producers, it is much more challenging to ensure consistency between the various parts of a transmedia network. In the contemporary era, where integrated transmedia networks are considered much more desirable than in the past, IP holders have investigated various methods of ensuring franchises are sufficiently consistent. The Star Wars Story Group, founded after the purchase of Lucasfilm by Disney, is a prominent example of one approach to this problem.

Though they may not be overtly articulated in the same way as the *Doctor Who* and Marvel franchises, fans of other franchises in which time travel, parallel universes and magic routinely feature are able to use similar strategies to explain away inconsistences. Notable examples include *Buffy the Vampire Slayer*, *Ghostbusters*, *Highlander*, and the *Star Trek* franchise. Interestingly, such approaches tend not to be evident with regards to the *Star Wars* saga, which does not feature parallel universes or time travel and which portrays its magic system—the Force—in carefully delimited ways. On deciding to "de-canonize" the vast bulk of the existing Expanded Universe of non-film media following the purchase of Lucasfilm by Disney, the decision was made to re-brand this material under the banner "Legends" (Star Wars.com 2014). Narratively speaking, this approach offers similar advantages to that of parallel universes, arguably awarding the Legends range a mythical status within the broader *Star Wars* storyworld, akin to the way Greek or Norse myths are positioned in the real world.

A further reason for the dominance of science fiction and fantasy in the transmedia sphere might obtain in the world-building characteristics of fantastic genres. Wolf talks extensively about the world-building techniques employed by J. R. R. Tolkien in adumbrating Middle-earth, in the process making the distinction between "Storytelling and World-building" (Wolf 2012, 29–33). Significantly, the novel series *The Lord of the Rings* and its prequel *The Hobbit* have lent themselves to multiple adaptations in different media, and most notably in terms of the films directed by Peter Jackson, also provided impetus for manifold kinds of transmedia expansion. However, Dowd et al, while accepting transmedia storytelling is "*world-driven*," suggest the importance of understanding the way in which different platforms engage with story elements, also suggesting that "Genres are an example of plot-driven stories" (2013, 57).

Of course, the world-building techniques employed by novels are not exclusive to fantasy and science fiction genres. Yet the fact that other genres like romance, thrillers, and war stories operate in recognizable, mimetic settings rather than completely invented ones arguably means there is less necessity for dense world-building. In the context of other media, like the comics published by DC and Marvel, the sheer longevity of the storyworlds in question might explain why these storyworlds have spread transmedially, as well as provided the basis for shared cinematic and televisual storyworlds with their own varying degrees of transmedial engagement. In other words, the intricately described storyworlds which often characterize commercially and critically successful fantasy and science fiction storytelling in one medium might explain why such storyworlds make such good fodder for adaptation *and* subsequent transmedial expansion.

With regard to other media, a related point is that we have arguably reached a point thanks to computer generated imagery whereby believable fantastic worlds can be convincingly built in a variety of audiovisual contexts including film, television, and video games, finally catching up with the evocative and immersive capabilities of the novel. These factors, and the related importance of the ways in which fantastic genres lend themselves to merchandising as well as "spreadable media" (Jenkins, Ford, and Green 2013), might explain the particular contemporary dominance of the fantastic in the transmedia sphere. It is possible to speculate, however, on which existing largely mono-media genres might make the translation into cross-media forms, and how this translation might occur.

Genre on the Holodeck

The inherent nonlinear quality of transmedia storytelling arguably makes it a form of interactive fiction. Generally speaking, participants can elect to engage with transmedia networks in whichever way they choose, deciding whether to watch a film or play the video game or read the comic in the order they decide. This is not to deny that franchises will often push participants toward preferred readings, or at least preferred *sequences* of reading. Marvel, for instance, often publishes preview comics tied to forthcoming film releases which both enlarge the storyworld of the MCU but also act as cross-media promotion for the film in question. However, there is nothing to stop an MCU fan either engaging or re-engaging with the preview comic having seen the film, illustrating the inherent nonlinear quality of a transmedia network.

Writing in *Hamlet on the Holodeck* (1999), a seminal text exploring the potential of computer-based interactive narrative, Janet Murray suggests ways in which digital storytelling might operate in relation to a range of genres. As well as discussing science fiction and fantasy forms, Murray also identifies more realist forms of drama including soap opera as ripe for enhancement and expansion by digital techniques. In many ways the techniques Murray discusses might be seen as anticipating some of the increasingly ubiquitous approaches adopted by contemporary transmedia producers.

For instance, Murray talks about ways in which a show like *ER*, a hugely popular American television medical drama that ran from 1994 to 2009, might be expanded by digitality. She suggests that specific locations frequently seen in the series could be presented virtually for participants to explore, expanding existing storylines or previewing forthcoming storylines, or providing more background on specific characters (Murray 1999, 255–256).

The commercial success of tie-in games related to existing crime and thriller franchises, from *CSI* to Sherlock Holmes, suggests the potential for detective fiction and its related sub-genres to become increasingly dominant formations within the transmedia sphere. Indeed, the BBC's *Sherlock* (2010–present) has been transmedially articulated in multiple ways, from the websites written by fictional characters (including Dr Watson's blog, which the character is seen writing on screen) to the accompanying app produced by The Project Factory. In the Hollywood context, a television series set in the same storyworld as the John Wick films is also currently in development (Flook 2017).

Anticipating Fludernik, Murray suggests that she is versed in how to engage with interactive stories within certain genres by her prior exposure to such genres (1999, 192). Murray observes that she knows she must question each of the suspects in a murder mystery CD-Rom adventure, that she will be expected to "shoot at the bad guys in a Western" and that she knows to "enter the haunted house in a horror story" (1999). Of course, such expectations can clearly afford a short cut for creators and audiences, of particular use in interactive narratives where the participant is also expected to learn *how* to interact. At the same time, the subversion of the tropes and techniques we expect to find in genre-based stories can clearly enhance the experience, as journalist Steve Rose explores in relation to "post-horror movies" (Rose 2017).

In this vein, arguably one of the key strengths of transmedia storytelling lies in its ability to utilize the nonlinear aspects inherent in the form to surprise participants. This is most obviously the case with regard to plotting. For instance, a revelation about a particular character's motivations might occur in one medium in a transmedia storyworld, only to be set in context by another part of the storyworld in a different medium. The order in which the two elements are engaged with will clearly affect the participant's understanding. Similarly, the ability to provide unexpected juxtapositions might also afford opportunities to subvert generic expectations, using the formal, nonlinear potential of the transmedia network to exploit audience expectations around content.

Conclusion

Memory emerges as a central aspect of genre within the transmedia sphere. When we engage with a cultural artifact positioned in a particular genre, our expectations are conditioned by previous experience of that genre, but also of experience of other genres—in other words a genre is determined as much by why it is not as by what it is. The form and content discussions characteristic of genre theory become more complex in a transmedia context, in which the participant is asked to "configure" their engagement with the transmedia network, and by the specific material and energetic conditions of the network, some or all of which might be articulated through digitality.

The dominance of science fiction and fantasy genres in the realm of transmedia storytelling can be attributed to a number of reasons. While arguably these genres attract the kinds of audiences interested in seeking out new material, these genres come with "in-built" solutions to the fuzziness that can arise when storyworlds spread across multiple media forms. Such solutions are not necessarily open to more mimetic, realist genres. The supremacy of fantastical genres might also be attributable to the merchandising power of these kinds of storyworlds (and their appeal to children), but also to the kinds of world-building involved in constructing such storyworlds, and the concomitant pleasures such world-building offers fan bases.

However, if audience appetite and commercial imperatives continue to point toward increasingly integrated transmedia storytelling, it may become the case that science fiction and fantasy storyworlds become less prevalent, as the need for the diegetic "sticking plasters" these genres can provide becomes less important. Equally, increasing experimentation with less explored genres amongst independent transmedia producers may result in more mainstream transmedia franchises also seeking to explore such genres.

The operation of genre in those transmedia networks that incorporate newer digital media forms is therefore intersected by a technological imperative. In other transmedia contexts, however, the material and energetic expression of the work in question is also vitally important to how the story element is understood, whether it is a comic book, a novel, or audio play. Jameson's point that genre is constituted by the ongoing dialectic between the semantic and syntactic remains as fundamental to an understanding of how genre operates in transmedia networks as it does to mono-media examples. Equally, genre needs to be understood in affective and experiential terms, as a particular form of narrative framing embedded in subjective and communal remembering.

References

Attebery, Brian. 1992. *Strategies of Fantasy*. Indiana: Indiana University Press.

Beddows, Emma. 2012. "Buffy the Transmedia Hero." *Colloquy: Text, Theory, Critique* 24: 149–150.

Chandler, Daniel. 1997. "An Introduction to Genre Theory." *Visual Memory*. August 11. Accessed January 5, 2017. http://visual-memory.co.uk/daniel/Documents/intgenre/integenre.pdf.

Cherry, Brigid. 2007. "Subcultural Tastes, Genre Boundaries and Fan Canons." In *The Shifting Definitions of* Genre, edited by Lincoln Geraghty and Mark Jancovich, 201–215. London: McFarland.

Clute, John, and John Grant. 1997. *The Encyclopedia of Fantasy*. London: Orbit.
Dowd, Tom, Michael Fry, Michael Niederman, and John Steiff. 2013. *Storytelling Across Worlds: Transmedia for Creatives and Producers*. Abingdon: Focal Press.
Dubrow, Heather. 1982. *Genre: The Critical Idiom*. London and New York: Methuen.
Fish, Stanley. 1980. *Is There a Text in This Class? The Authority of Interpretive Communities*. London: Harvard University Press.
Flook, Ray. 2017. "John Wick Prequel Series 'The Continental' Offers Big Stories, Less Wick." *Bleeding Cool*, June 14. Accessed May 12, 2017. www.bleedingcool.com/2017/06/14/john-wick-prequel-series-continental-offers-big-stories-less-wick/.
Fludernik, Monika. 2005. *Towards a Natural Narratology*. Abingdon: Routledge.
Frow, John. 2015. *Genre*. Abingdon: Routledge.
Gaunt, Simon. 2005. *Gender and Genre in Medieval French Literature*. Cambridge: Cambridge University Press.
Genette, Gerard. 1983. *Narrative Discourse: An Essay in Method*. New York: Cornell University Press.
Harvey, Colin B. 2015. *Fantastic Transmedia: Narrative, Play and Memory Across Science Fiction and Fantasy Storyworlds*. London: Palgrave Macmillan.
Jenkins, Henry. 1992. *Textual Poachers: Television Fans & Participatory Culture*. London: Routledge.
Jenkins, Henry. 2006. *Fans, Bloggers, and Gamers: Exploring Participatory Culture*. New York: New York University Press.
Jenkins, Henry. 2008. *Convergence Culture: Where Old and New Media Collide*. New York: New York University Press.
Jenkins, Henry, Sam Ford, and Joshua Green. 2013. *Spreadable Media*. New York and London: New York University.
Mendlesohn, Farah. 2008. *Rhetorics of Fantasy*. Connecticut: Wesleyan University Press.
Murray, Janet H. 1999. *Hamlet on the Holodeck: The Future of Narrative in Cyberspace*. Massachusetts: Massachusetts Institute of Technology.
Rose, Steve. 2017. "How Post-Horror Movies Are Taking Over Cinema." *Guardian*, July 6. Accessed July 28, 2017. www.theguardian.com/film/2017/jul/06/post-horror-films-scary-movies-ghost-story-it-comes-at-night.
Star Wars.com. 2014. "Biggest Star Wars Moments of 2014." December 23. Accessed January 7, 2017. www.starwars.com/news/biggest-star-wars-moments-of-2014.
Wolf, Mark J. P. 2001. "Genre and the Video Game." In *The Medium of the Video Game*, edited by Mark J. P. Wolf, 116–134. Austin: University of Texas Press.
Wolf, Mark J. P. 2012. *Building Imaginary Worlds: The Theory and History of Subcreation*. London: Routledge.
Wright, Will. 1977. *Six Guns and Society: A Structural Study of the Western*. London: University of California Press.

18

TRANSMEDIA WRITING

Storyworlds and Participation at the Intersection of Practice and Theory

Donna Hancox

As transmedia projects continue to flourish and experiment with different styles and structures so too does our knowledge about the ways in which writing and storytelling function within transmediality. What has become clear from the emerging theory and practice of transmedia storytelling is that existing methodologies—literary theory, narratology, semiotics, film theory, media studies, and so on—all contribute important perspectives to the scholarship and the practice of transmedia storytelling, but that none is sufficient on its own. As such, transmedia storytelling necessarily encompasses a broad range of theoretical, philosophical, and creative approaches and continues to develop in ways that expand our understanding of story and writing.

This chapter explores the theory and practice of writing in transmedia storytelling, and the potential it holds as an innovative creative practice to challenge and shift existing views of writing for multiplatform projects. It is not only *how* the act of writing is transformed for these new kinds of stories that is of interest, but also considerations of authorship and readership for collaborative, participatory narratives. Transmedia writing is more than a practice or a process centered around digital technology; it represents an original way of thinking about and doing the practice of writing that encompasses traditional literary understandings of writing, design thinking, user experience, and collaboration. It can also be understood as a framework or a philosophy in which to consider modes of storytelling and methods of dissemination, and brings together the theory of writing and the practice of writing with considerations of audiences and the embedded qualities of media platforms.

In 2004, Hayles stated that "Literary criticism and theory are shot through with unrecognized assumptions specific to print. Only now, as the new medium of electronic textuality vibrantly asserts its presence, are these assumptions clearly coming into view" (67). This awareness of the specific attributes of individual media (film, audio, text, blogs, photographs, email, games, hypertext, for example) and how they work with various forms of storytelling (documentary, fiction, webisodes, advertising, television series) continues to grow and become more sophisticated. Elwell (2013) argues that the distance between the discernible differences between embodied existence and representations of that existence is being diminished through digital mediation and the creation of new kinds of narratives can exploit the possibilities. To truly interrogate the complexity of this media ecology, we must think beyond particular devices and traditional understandings of platforms as distinct and separate entities, and instead embrace a more dynamic and porous description that disrupts the most common definitions of platforms or mediums. Ryan (2016) suggests that a medium is best understood as an "inherently polyvalent term whose meaning involves technological, semiotic and cultural dimensions" (5). This is a useful lens through which to view the form of transmedia storytelling explored in this chapter.

In short, this chapter will focus on elements of writing that are necessary for transmedia writing: participation and interactivity, design thinking, and world-building. These features and considerations sit on top of the traditional story elements that all writers work with, and this combination may be seen as a new type of ecology of storytelling in which writing and storytelling tools can be leveraged in different ways for individual projects. The incorporation of design thinking and user experience into the writing process for transmedia storytelling (beyond the previous role of design such as font and typesetting) presents avenues to address these new challenges and offers a contemporary collaborative model of writing. The experimental transmedia project *Welcome to Pine Point* offers a compelling case study through which to explore these ideas and to consider what might be in store for future writers of transmedia stories.

As acknowledged in this book's Introduction, Henry Jenkins brought the term "transmedia storytelling" into mainstream media studies as a way to describe, rather than define, an approach to storytelling that exploited the growing ubiquity of information and communication technologies and responded to a new generation of audiences. Ever since, scholars and practitioners have grappled with ways of analyzing and critiquing the process, and the products that have resulted. Some of the most influential writing about narrative design in the past five years has been focused on transmedia storytelling, and has addressed the innovations in books, film, and television narratives, with scholars such as Bolter (2014), Linnell (2014), and Punday (2011) offering overviews of the ways in which technology has affected narrative theory and the culture of remediation across platforms. Similarly, Hayles (2004) and Ryan (2005) exemplify of the most challenging and original writing about narrative theory in recent years; they do, however, diverge in interesting ways in their consideration of the inherent uniqueness of different media and how this affects narrative design across them.

What is sometimes lost in translation with these conflicting perspectives is that each medium is embedded with nuanced and distinct implications for the nature of the environments they are trying to explain. None of this is necessarily new, the practice of developing narratives across media and forms is "as old as media themselves; think of paintings dramatizing biblical scenes or the iconic nineteenth-century characters such as Frankenstein or Sherlock Holmes whose narrative scope transcends any single medium" (Mittell 2014, 253). This may well be true; however all *digital* media arrives with new considerations and more importantly, new audiences.

Participation and Interactivity

Transmedia writing demands an attention to audience in a way that previous forms of writing have not, and at the foundation of Jenkins' ongoing theorization of transmediality is that understanding audiences is a major consideration for successful transmedia writing. This makes transmedia storytelling a unique field with fluid borders, and subsequently it has created entirely new forms that may incorporate diverse, hybrid narrative modes and a distinctive role of fans and audiences.

By privileging audiences in the writing process, transmedia stories necessarily hold many voices within each project. In *Hybrid Stories*, Tom Abba states that "new media favour a multiplicity of voices (a digital version of an iteration of Mikhail Bakhtin's notion of multivocality)" (2009, 61), and Jenkins (2009) discusses the concept stating "multiplicity allows fans to take pleasure in alternate retellings." Jenkins adds that "the concept of multiplicity paves the way for us to think about fan fiction and other forms of participatory expression as part of the same transmedia logic" (2013, 170) and acknowledges that transmedia writing requires dynamic innovation rather than conformity and adherence to tradition. Transmedia storytelling and writing represents a unique challenge to combine form and content that re-imagines the accepted elements of writing—character, plot, setting, etc. Anyone who has taught a creative writing class knows that we routinely espouse well-worn, and often useful, statements about rules for writing and the craft of writing. Andrea Phillips claims in *A Creator's Guide to Transmedia Storytelling* that "accepted writing rules include: everything should reveal character, advance plot or support theme. For transmedia, I'd add one more item: adding color to your world" (2012, 77). Phillips goes on to elaborate about what "adding color to your world" might entail, but what becomes evident is that transmedia storytelling is an exercise in open-ended storytelling and writing, in being "boundless where a traditional single medium story is finite" (2012, 79).

This boundlessness encourages transmedia stories to strive to be dialogic spaces where, similar to Abba's assertion of multi-vocality, many voices are present and co-exist without needing to privilege one over the other. If the nature of transmedia stories favors multiple voices, then interactivity and participation enable them to be each fully realized and important. However, interactivity and participation are not the same activity or phenomenon consequently differing in their intentions, and subsequently their usefulness for certain projects. Interactivity can mean audiences engaging with the text in multiples ways: navigating the narrative via a range of choices, choosing the order in which to experience the story, making decisions for characters from a series of options are some of the most common forms of interaction. Conversely, participation generally entails audiences contributing to the text by attending real world events, uploading their own stories, re-mixing the text. The two activities can exist in one project but have very different functions. One function, surely, is to keep audiences reading, playing, engaging for the entirety of the story, while the other function is to keep the text alive outside of the boundaries of the project and allowing for diffuse and personal experiences of a text.

Carpentier stresses the distance between participation and interaction in relation to digital media projects, whereby participation requires input into decision-making and interaction is associated with sociocommunicative relations (2013, 275). He also presents two modes of participation: in and through the media. It would seem that these are not oppositional avenues of participation, but his definition offers further insight into the multitude of ways in which individuals are able to participate. While participation in the media can be clearly understood as the ability to make decisions about the media products while participation through media opens up another field of the participatory process—namely, in other areas of decision-making, which have more to do with how people can enter public spaces and use media to enter into societal debates, dialogue, and deliberations (2013, 274). As will be discussed later in relation to *Welcome to Pine Point*, interaction and participation can be very subtle and nuanced, and ideally are a reflection of the aesthetic, philosophy, and world of the narrative.

World-building

Detailed and expansive storyworlds are acknowledged as being a crucial element to transmedia story telling and are evident in the early, commercial examples of transmedia projects. Transmedia writer and theorist Christy Dena says:

> If you're playing transmedia bingo, "worldbuilding" scores 10 points and one of those little jelly desserts from the kitchen ... There is a reason for this: it is a helpful metaphor for understanding and communicating that a transmedia project involves many stories and media and there is a whole ecology operating.
>
> *(2012)*

Transmedia writing is heterogeneous and has a multiplicity of features, however it can be argued that all transmedia writing must involve a narrative universe, or storyworld. Saldre and Torop state that discussions of transmedia storytelling are dominated by a cognitive spatial lexicon, evidenced by metaphors such as environment, landscape, and maps. The confluence of these metaphors necessitates a theoretical distinction between the shape of the narrative being told and the shape of the narrative sites that enable the telling to occur. Dena highlights this idea as a distinction between story structure and story creation, and posits that the unity of the two is a "design problem concerned with the distribution of time and place across different sites that take into account the real world as a vantage point from which the storyworld is viewed" (2009, 263–268). Being able to create an environment that can encompass the variety of experiences and points of view that exist in the real world through different forms of media creates the possibility for the general public to engage and interact with stories in new ways. "A storyworld is not just the spatial setting where a story takes place; it is a complex spatio-temporal totality that undergoes global changes" (Ryan 2016, 13). In her earlier work on storyworlds across media, Marie-Laure Ryan with Thon (2014) claims:

> the replacement of narrative with storyworld acknowledges the emergence of the concept of "world" not only in narratology but also on the broader cultural scene. Nowadays we have not only multi-modal representations of storyworlds that combine various types of signs and virtual online worlds that wait to be filled with stories by their player citizens but also serial storyworlds that span multiple instalments and transmedial storyworlds that are deployed simultaneously across multiple media platforms, resulting in a media landscape in with creators and fans alike constantly expand, revise and even parody.
>
> *(1)*

Building on Ryan's theoretical claims from a producer's point of view, Robert Pratten (2011, 70) believes that future storyworlds will have the following characteristics:

Pervasive: the story will be built around the audience—connecting with them across devices.
Persistent: the story evolves over time, reacting to audience engagement.
Participatory: the audience interacts with characters and other audience members.
Personalized: the story remembers decisions and conversations and becomes tailored to each.

Meanwhile, Mark J. P. Wolf says in his book *Building Imaginary Worlds: The Theory and History of Subcreation* that "worlds, unlike stories, need not rely on narrative structures, though stories are always dependent on the worlds in which they take place. Worlds extend beyond the stories that occur in them, inviting speculation and exploration through imaginative means" (2012, 32). Importantly, these imaginative means of world-building rest on the shoulders of the audiences, authors, and designers.

Design and Collaboration

Consistent technological developments have allowed for more sophisticated location-based storytelling, enhanced animation, real-time engagement, novel and intuitive predictive narrative generation, and a collapsing of boundaries between books, games, and film, and have resulted in an environment more receptive to experiments with writing, storytelling, and ways of considering interactivity. The early experiments in electronic literature have revealed that storytelling utilizing digital technology and multiple platforms is no longer the product of a singular author working within a model of practice, which pre-dates the computer. Instead what is brought to the fore is the way in which digital artifacts are "designed" and the specific nature of this design process in respect to writing. It is increasingly clear that the futures of stories are collaborative and interactive and require teams of people with divergent skills to move the forms and the practices forward. There is also a focus on the ways in which writing and design intersect to allow for genuinely new modes of creative communication. It prefigures a future where readers and writers are fluent in this new language, just as we are fluent in the current forms of written expression. It is in this context that we can study new audiences, new practices, and the interaction between writers and designers, and in doing so aim to expand upon the limited knowledge in respect to the ways in which this move toward collaborative writing models will create new experiences of reading, innovations in the practice of writing. Hypertext as a form of literature has a long history that pre-dates digital technology but arguably found a new prominence within the World Wide Web. The term hypertext is most clearly defined by Nelson as "non-sequential writing" (1981) and his vision of hypertext is that of a system that "branches and allows choices to the reader" (2), so that the user can move within a hypertext system according to their rationale. Montford (2003) provides a summary of the history of early electronic literature and draws connections to philosophers such as Eco (the open text) and Barthes (pleasure of writing and writerly texts). However, Montford takes a step in his work that sets the trajectory for the following decade, where he theorizes that interactivity as that which needs rules and directs audiences.

As writing met and continues to meet the digital age, two very different histories of practice collide, that of literature and that of digital design, or design of the Human Computer Interface. While these may appear on the surface to be unrelated fields of endeavor, they both share a focus on the role of language in the creation of experience, and in the exchange and sharing of knowledge. In the context of transmedia writing, this is most evident in the prevalence of cinematic/televisual experiences over the written word. However, the design of Human Computer Interfaces stands on language and human ability to describe the "new" through forms of rhetoric. In this way new creative products and conceptual frameworks emerge when writers and user experience designers collaborate.

Transmedia writing brings with it a range of challenges for the writers, designers, developers, publishers, readers, and the wider industry that will review, describe, recommend, and disseminate these projects. Transmedia stories come into existence as a result of the combined efforts of a team of designers and developers working with the writer or writers and produce. This type of co-creation can be foreign to many writers, and so a number of university writing programs are taking steps to address this by creating formal connections between writing and design. "Aligning writing studies with art and design rather than literature challenges the entrenched perceptions of the field—at least perceptions commonly held by those outside of the field and academia" (Purdy 2014, 613). Marback argues that appealing to the concept of design is a way to solve "wicked problems" in writing studies, particularly for those "teaching writing in digital media," and design thinking understands and allows for the reality that wicked problems are not just solved once by finding new information, they must be solved over and over again (2009, 399). Arola (2010) explains:

> Today our students still choose photographs, words, sounds, and hyperlinks (clearly all rhetorical choices) but they choose colours, fonts, and shapes less and less. Instead, the platform, or more specifically the design template, is chosen for them. Those of us engaged with digital rhetoric continue to acknowledge the need to allow students to, in Rea and White's terms, "experiment with new forms of writing"
>
> (421)

Kjartan Müller, further, argues that "in a digital platform an underlying layer defines the design space for the layer above. This design space is defined negatively by the constraints set by the underlying layer and positively by the possible space it creates for design. Design, in this case, is a neutral term that covers hardware architecture, system design and text composition" (2011, 186). Both design and transmedia require a clear intentionality about the purpose of the work and a bringing together of skills for enhanced user experience.

Case Study: *Welcome to Pine Point*

Through video footage, photographs, audio, and text, the creators of *Welcome to Pine Point*, Paul Shoebridge and Michael Simons, present the story of a town that no longer exists. Pine Point was a community planned around an open-cut mine, and when the mine closed down in 1988, the single-industry town also closed. Shoebridge and Simons manage to capture not just the geographical details of the town so that the audience understands its location and unique characteristics; they also capture the social and cultural details of the time and place that turn out to be universal and instantly recognizable even for audiences outside of the United States.

The project is designed to resemble a photo album from the 1980s, and using this nostalgic aesthetic they are able to convey a mood and a spirit that is necessary to truly understand the cast of characters who share their experiences of growing up in Pine Point and their feelings about its eventual demolition. The details of the town are the key to the universality of this story and to the connection created with audiences. This is a story not only about place and memory but is also a larger story about the macroeconomics that influence lives and how individual fates are tied to corporate decisions. When a town closes down or loses an influential industry, the everyday stories about the lives affected and the struggle to continue are often lost.

The storyworld created in *Welcome to Pine Point* shows that storyworlds are larger than what is directly shown in the text, larger than the narrative "here" and "now" (Ryan 2016, 4). *Welcome to Pine Point* reveals what existed before the town died, meaning that audiences can connect with the real lives and dreams that lived and died in that town.

> I think we just told the story how we thought we could tell it. We think it's more part memoir for people growing up at that time and feeling things about what memory was to us, what tangible objects meant to us, and how memory gets flaky but interesting and romantic. Sometimes concrete and sometimes evocative.
>
> *(Macaulay 2012)*

The first image in *Welcome to Pine Point* is a crudely drawn video cassette over a black background, which for a certain generation—the generation this story is about—is an immediately recognizable object that brings with it memories. The video cassette represented a new age of home movies, user-generated content, and more accessible home entertainment that exploded in the 1980s, and this one image signals to the audience that this project is firmly rooted in that amateur, homemade tradition (despite the makers being media professionals). What follows is an intriguing mix of sound and image that further locates the audience in the era and emotional landscape of the story in complex ways that utilize a visual and audio language that communicates the small individual stories and the details of life to explore and extrapolate much larger themes of home, belonging, memory, and so on. In these very particular ways, *Welcome to Pine Point* exemplifies the strengths of transmedia storytelling by allowing each medium to do what it does best, and for the storyworld to be communicated through the relationship between the authors and the audiences. It can be easy to overlook the connections and thoughtfulness embedded in this project that is often either admired or misunderstood.

One of the most challenging and original aspects of *Welcome to Pine Point* is the way that it disrupts widely accepted views about transmedia platforms. Rather than existing across distinct media platforms that audiences need to migrate across, platforms are instead re-defined as modes of storytelling (visual, text, audio, etc.) and forms (documentary, oral history, memoir, photo album) for audiences to then consider the vastly different meanings and traditions in each form, despite all of them existing in one digital site.

A common criticism, however, of *Welcome to Pine Point* is that the narrative is essentially linear, which for some, at least, suggests that the creators have not taken full advantage of the possibilities of transmedia storytelling.

> When people think of digital interactive media, one of the first things they say is: "It's going to have multiple entry points, and you can go wherever you want to." And sure, you can deliver certain kinds of information like that, but it's not super-great for stories, at least in our experience.
>
> *(Pitzer 2011)*

Phillips (2012), further, adds to this argument that nonlinear narratives are not suitable for every transmedia story and also the idea that audiences always experience a narrative in a linear way regardless of where and when they enter the story. Phillips explains:

> So regardless of how we encounter the pieces of the story our experience is always linear. This has caused Janet Murray to coin the term "multi-sequential" meaning that there are multiple linear sequences the audience can experience. I like this term a lot and often use it but "open storyworld" is a more familiar term which is why I've chosen it here.
>
> *(2012, 30–31)*

More fundamentally, in his seminal 1964 work *Understanding Media*, McLuhan deploys the framing metaphor of the human central nervous system, the "electric network that coordinates the various media of our senses" (2012 [1964], 47) to describe the relationship between the message and the medium. McLuhan argues that media technologies are metaphorical extensions of an embodied understanding of stories that allows for multiple ways to convey a story, linked to the capacity of whichever media is being utilized. As Shoebridge states: "We kept a lot of the old handmade book-like things, in keeping with that medium-is-the-message concept. We tried to emphasize what each medium does well" (Pitzer 2011).

Transmedia narratives therefore reflect distinct media traditions (film, literature, games, social media, for instance), but also exist at the intersection of these media, particularly in the way "information gets dispersed" across all of them (Jenkins 2011, 955). The label of "interactive documentary" for *Welcome to Pine Point* is thus quite misleading and displays an unwillingness to embrace broader ideas of platforms, storyworld, interactivity, and linearity that are themselves integral parts of future writing for transmedia storytelling.

Conclusion

In fact, *Welcome to Pine Point* encapsulates so many opportunities and obstacles associated with transmedia storytelling by working deeply with each medium and allowing a strong visual aesthetic to support the writing style and to do some of the heavy lifting of metaphor, lyricism, cadence, etc. that is associated with text dominant stories. It also uses an innovative approach to the role of character and characterization, and the ways in which they are created and shown to audiences. *Welcome to Pine Point* offers a unique narrative that is both linear and interactive and does not require audiences to solve a mystery or to unlock keys along the way. Instead, it immerses audiences in a human story filled with the emotions, surprises, and banalities of the everyday. More so than many other contemporary transmedia projects, indeed, *Welcome to Pine Point* illustrates a new understanding of platforms and of writing in a transmedia environment that re-imagines the intersection of media, genre and form to present an entirely new approach to writing that differed dramatically from dominant understandings of transmedia writing. "While stories are transmitted by discourse, which means by text, they remain inscribed in our mind long after the signifiers have vanished from memory. This means that a story is a cognitive rather than a linguistic construct" (Ryan and Thon 2014, 87). Looking forward, then, the understanding that transmedia stories are at least as cognitive and emotional as they are linguistic may well be the indicator of success in capturing, entertaining, and maintaining audiences who are used to existing in a crowded and noisy media ecology.

References

Abba, Tom. 2009. "Hybrid Stories: Examining the Future of Transmedia Narrative." *Science Fiction Film and Television* 2 (1): 59–75. doi: 10.3828/sfftv.2.1.4.

Arola, Kristin L. 2010. "The Design of Web 2.0: The Rise of the Template, the Fall of Design." *Computers and Composition* 27 (1): 4–14.

Dena, Christy. 2009. "Transmedia Practice: Theorising the Practice of Expressing a Fictional World Across Distinct Media and Environments." Ph.D. diss., University of Sydney.

Dena, Christy. 2012. "Some Things I've Learned from Transmedia Worldbuilding." *Christy's Corner of the Universe*. Accessed August 12, 2017. www.christydena.com/publications/transmedia-worldbuilding.

Elwell, J. Sage. 2013. "The Transmediated Self: Life Between the Digital and the Analogue." *Convergence: The International Journal of Research into New Media Technologies* 20 (2): 233–249. doi: 10.1177/1354856513501423.

Hayles, Katherine. 2004. "Print is Flat, Code is Deep: The Importance of Media-specific Analysis." *Poetics Today* 25 (1): 67–90. doi: 10.1215/03335372-25-1-67.

Jenkins, Henry. 2009. "The Revenge of the Origami Unicorn: Seven Principles of Transmedia Storytelling." *Confessions of an Aca-Fan: The Official Weblog of Henry Jenkins*. December 12. Accessed February 20, 2017. http://henryjenkins.org/2009/12/the_revenge_of_the_origami_uni.html.

Jenkins, Henry. 2011. "Transmedia 202: Further Reflections." *Confessions of an Aca-Fan: The Official Weblog of Henry Jenkins*. August 11. Accessed November 2, 2017. http://henryjenkins.org/2011/08/de ning_transmedia_further_re.html.

Jenkins, Henry. 2013. "Theorizing Participatory Intensities: A Conversation about Participation and Politics." *Convergence: The International Journal of Research into New Media Technologies* 19 (3): 165–186.

Linnell, Sheridan. 2014. *Art Psychotherapy and Narrative Therapy: An Account of Practitioner Research*. Sharjah: Bentham Science Publishers.

Macaulay, Scott. 2012. "The Beauty of 'Welcome to Pine Point'." *Filmmaking*, July 15. Accessed December 11, 2016. http://filmmakermagazine.com/48259-the-beauty-of-welcome-to-pine-point/#.WfNNkGiCxPZ.

McLuhan, Marshall. 2012 [1964]. *Understanding Media*. London: Routledge.

Marback, Richard. 2009. "Embracing Wicked Problems: The Turn to Design in Composition Studies." *College Composition and Communication* 61 (2): 397–419.

Mittell, Jason. 2014. *Complex TV: The Poetics of Contemporary Television Storytelling*. New York: New York University Press.

Montfort, Nick. 2013. "Toward a Theory of Interactive Fiction." December 19. Accessed March 4, 2017. http://nickm.com/if/toward.html.

Müller, Kjartan. 2011. "Genre in the Design Space." *Computers and Composition* 28 (3): 186–194.

Murray, Simone. 2012. *The Adaptation Industry: The Cultural Economy of Contemporary Literary Adaptation*. London and New York: Routledge.

Nelson, Tim. 1981. *Literary Machines*. Swarthmore: Self Published.

Phillips, Andrea. 2012. *A Creator's Guide to Transmedia Storytelling: How to Captivate and Engage Audiences Across Multiple Platforms*. New York: McGraw-Hill Company.

Pitzer, Andrea. 2011. "The Goggles on 'Welcome to Pine Point': Digital Narrative Chases Memory and Loss." *Nieman Storyboard*, February 4. Accessed July 3, 2017. http://niemanstoryboard.org/stories/the-goggles-on-welcome-to-pine-point-digital-narrative-chases-memory-and-loss/.

Pratten, Robert. 2011. *Getting Started in Transmedia Storytelling: A Practical Guide for Beginners*. London: CreateSpace Independent Publishing Platform.

Punday, Daniel. 2011. *Narrative Bodies: Toward a Corporeal Narratology*. Basingstoke: Palgrave Macmillan.

Purdy, James. 2014. "What Can Design Thinking Offer Writing Studies?" *College Composition and Communication* 65 (4): 612–641.

Ryan, Marie-Laure. 2005. "On the Theoretical Foundations of Transmedial Narratology." In *Narratology Beyond Literary Criticism: Mediality and Disciplinality*, edited by Jan Christopher Meister, 1–24. Berlin: Walter de Gruyter.

Ryan, Marie-Laure. 2016. *Narrative Space, Spatializing Narrative: Where Narrative Theory and Geography Meet*. Columbus: Ohio State University Press.

Ryan, Marie-Laure, and Jan-Noël Thon. 2014. "Storyworlds Across Media: An Introduction." In *Storyworlds Across Media: Towards a Media-conscious Narratology*, edited by Marie-Laure Ryan and Jan-Noël Thon, 1–24. Lincoln: University of Nebraska Press.

Wolf, Mark J. P. 2012. *Building Imaginary Worlds: The Theory and History of Subcreation*. New York: Routledge.

19

TRANSMEDIA PHOTOGRAPHY

Implicit Narrative from a Discrete Moment

Kevin Moloney

On February 1, 1968, a bullet crashed through the skull of Nguyễn Văn Lém, a Viet Cong operative, in front of Associated Press photographer Eddie Adams and NBC television cameraman Vo Su. Adams' still photograph, of Vietnamese chief of national police Nguyễn Ngọc Loan executing a handcuffed Lém, burns itself into the visual memory of most who have seen it. Its horror is inescapable. The morning after the image was made, it appeared in newspapers around the world, fueling public exhaustion with the apparent quagmire of the Vietnam War. "Saigon, 1968" has become a global icon of a life-wasting conflict (Zelizer 2010, 225–229; Morris 1998, 240–241; Sontag 2003, 59–60).

Vo Su's 16mm movie camera rolled on the same scene and the footage was broadcast to an estimated 20 million people a day after Adams' photo appeared in print (Bailey and Lichty 1972). Su's film shows the speed at which the entire situation unfolded. Only nine seconds pass between Loan pushing his way through the arresting soldiers to Lém laying on the street with blood gushing from his head. This is a different horror. Where the still image seems to emphasize premeditated vengeance and anger, the motion footage startles the viewer as a sudden gory outburst. It is Adams' still image that is remembered by nearly anyone who sees it, however. The 16mm motion footage is rarely remembered. It is that ability to stare in wonderment and horror at the moment of death that makes Adams' image so memorable to all who see it. By contrast, the speed of action in Su's film limits our consideration of context and consequence in the moment. It gives only shock and gore.

Though Adams' photograph and Su's film are visual journalism of the exact same event, the stories they tell unfold in different ways and with different results. "Photographs may be more memorable than moving images because they are a neat slice of time, not a flow" wrote Susan Sontag. "Television is a stream of underselected images, each of which cancels its predecessor. Each still photograph is a privileged moment, turned into a slim object that one can keep and look at again" (1978, 17–18). Both photographs and moving images demonstrate the divergent strengths of each media form in a transmedia story. If transmedia storytelling is a narrative composed of smaller discrete narratives, or an argument composed of smaller discrete arguments, then visual stories are an important piece in the development of a transmedia project. Unlike language-based texts, images communicate quickly and without translation, and they evoke emotional responses of which words often prove incapable (Smith-Rodden and Ash 2017). This chapter examines still photographs as discrete stories, capable of implying a complex narrative of events though they are only frozen moments sliced from the otherwise unstoppable flow of time.

The Photograph as a Narrative

In narrative, a story is traditionally delivered by a narrator relating a series of events that present a lesson to be learned, as in the case of Æsop's famous fables (1848), or as fantasy and entertainment as in the case of Cervantes' *Don Quixote* (2005). A narrator is not required, though, as most modern novels, films, or even the plays of Shakespeare show. How then can a still photograph, an image captured of a single arrested moment, act as a narrative? Narratologist Marie-Laure Ryan (2004) defines a narrative:

1. A narrative text must create a world and populate it with characters and objects. Logically speaking, this condition means that the narrative text is based on propositions asserting the existence of individuals and on propositions ascribing properties to these existents.
2. The world referred to by the text must undergo changes of state that are caused by nonhabitual physical events: either accidents ("happenings") or deliberate human actions. These changes create a temporal dimension and place the narrative world in the flux of history.
3. The text must allow the reconstruction of an interpretive network of goals, plans, causal relations and psychological motivations around the narrated events. This implicit network gives coherence and intelligibility to the physical events and turns them into a plot (2004, 8–9).

In simple terms, a narrative story is built on characters who come into conflict within a setting, and in which is developed a contextual understanding of motives and the effect of the conflict. Though traditionally a story is the work of written or oral language and with a linear timeline, that understanding is too limiting.

Ryan also differentiates between media forms that *are narratives* and those that *have narrativity*. Texts (or images in this case of this chapter) that *are narratives* are designed to evoke a script in the mind of the audience—they explicitly direct the reader to observe the characters, setting an order of events through which a story emerges (2004, 9). Of the visual media forms, motion pictures—cinema, video, animation—are the champions of these explicit narratives. Events are usually laid out in order and the viewer follows the narrative arc as designed by the producer. In cases where that order of events is not chronological—the films *Pulp Fiction* or *Memento* are excellent examples—viewers still prove quite adept at cognitively reconstructing order from the nonlinear. Vo Su's film footage of the 1968 Saigon execution, as short and sparse as it is, still walks the viewer through a series of events where two characters come into conflict with a new world order as a result. Still photographs presented in series are equally capable of delivering a linear narrative as events progress from one image to the next. This is seen fully developed in the great photographic essays presented in the picture magazines of the mid-twentieth century. The photo essay is to motion pictures as panel comics are to the animated cartoon.

Having narrativity, Ryan describes, is the ability to evoke a narrative script where one may not be intended. Here she includes pictures, music, dance, and life itself as examples of media forms that have narrativity without being narratives (2004). They are implicit rather than explicit. Adams' still photograph is a frozen pattern on paper or screen, and from that pattern the reader internally reconstructs characters, setting, conflict, and the actions leading up to and down from that climactic moment. Ryan describes "the fullest form of narrativity" as both directing the reader to the structure of the intended story and evoking a deeper experience of story that is not in the text (2004, 9–10). It tells both explicit and implicit stories.

Art critic John Berger posited in response to Sontag's *On Photography* that one contextualizes a photograph from his or her own memories. The memory, he argued, is a nonlinear system that draws from a complex network. "A radial system has to be constructed around the photograph so that it may be seen in terms that are simultaneously personal, political, economic, dramatic, everyday and historic" (1980, 60–63). On their position in time, he wrote:

> Photographs are relics of the past, traces of what has happened. If the living take that past upon themselves, if the past becomes an integral part of the process of people making their own history, then all photographs would re-acquire a living context, they would continue to live in time, instead of being arrested moments. It is just possible that photography is the prophecy of a human memory yet to be socially and politically achieved. Such a memory would encompass any image of the past, however tragic, however guilty, within its own continuity. The distinction between the private and the public uses of photography would be transcended.
>
> *(1980, 57)*

He added, "In general the better the photograph, the fuller the context that can be created. Such a context replaces the photograph in time—not its own original time for that is impossible—but in narrated time" (1980, 61).

A relevant example of this phenomenon is Dorothea Lange's famous 1936 image of a migrant worker mother and three of her young children made in during the refugee crisis caused by the combined economic collapse of the Great Depression and environmental disaster of the Dust Bowl (Figure 19.1). In the image, Florence Owens Thompson looks off camera with a hand held to her cheek. She appears to be between 30 and 40 years old. An infant lays in her arms and two young children turn away from the camera and behind her shoulders. Her clothes are worn and soiled. The facts are limited in this simple portrait. When one sees this image more than 80 years later, it is contextualized with all one was taught about the Depression and Dust Bowl era, it is compared

Figure 19.1 "Migrant Mother," 1936, by Dorothea Lange.

to similar global circumstances or experiences from lives lived and the rest of the story is assumed. The image is most often read as a portrait of despair. The hand to the cheek symbolizes worry about her economic circumstance and tenuous future. The hidden faces of the older girls imply that they are crying, hungry, and deprived. All of this is a construction, however. Examination of Lange's few outtakes shows the two girls smiling and giggling just moments before, revealing that the girls are likely hiding their faces in giggly shyness in the famous image. The concern on the mother's face may be due to her worry about her partner's late return for dinner, or that a stranger is aiming a large and intimidating camera at her and her children. It is unlikely that she is ruminating on the state of the economy or of her plight in it as is most often assumed from decades distant. It is an icon of an era because complex contextualization of the image makes it so. Right or wrong, its narrative is a construction. Sontag (1978, 23) wrote:

> The ultimate wisdom of the photographic image is to say: "There is the surface. Now think—or rather feel, intuit—what is beyond it, what reality must be like if it looks this way." Photographs, which cannot themselves explain anything, are inexhaustible invitations to deduction, speculation and fantasy.

Though the photograph is not in and of itself a narrative, it does, as Ryan suggests, have narrativity. Through reading of the isolated and frozen moment one contextualizes, emotes, and intuits a fully fleshed narrative from the sparse hints contained therein.

The Photograph on Freytag's Pyramid

A narrative or story is a reporting of connected events commonly understood to have a beginning, a middle, and an end. Of the dozens of defined forms stories can take, the drama is among the most compelling. A good drama rivets the viewer to the pending conflicts, the collision of opposing characters, and then the aftermath of their actions. Dramas are lessons, vicarious experiences of the lives of others that amuse, horrify, and teach people about their own lives. Consciousness researcher Owen Flanagan describes that people cast their lives as a story to better understand its series of connected events, from beginning to end. "Evidence strongly suggests that humans in all cultures come to cast their own identity in some sort of narrative form," he states. "We are inveterate storytellers" (1992, 198). One will find a classic drama in nearly any series of events, intuitively hunting for the narrative arc in its telling.

In the nineteenth century, Gustav Freytag (1872; Herman, Jahn, and Ryan 2010) distilled the structure of drama to a triangle organized along the structure of the classical five-act tragedy. *Freytag's*

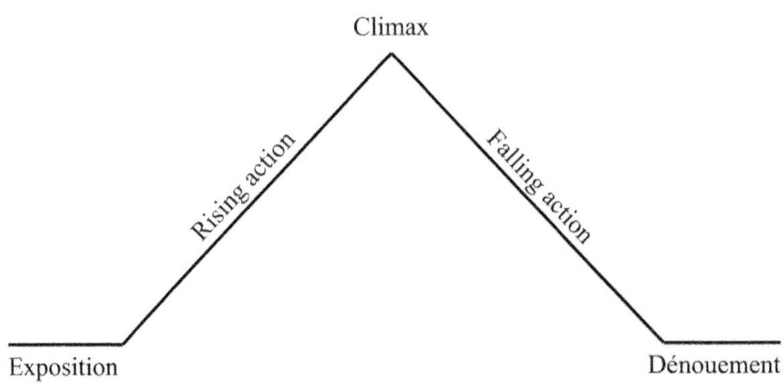

Figure 19.2 Freytag's pyramid.

pyramid (Figure 19.2) begins at the left base of the triangle with the *introduction* or *exposition* of the story. One gets to know the characters and who and where they are. In the *rising action*, tensions between the protagonist and antagonist rise and the reader anticipates the pending conflict. At the top of the triangle the protagonist achieves the pinnacle of success, *conflict* erupts, and the characters face turning points and the beginning of declining fortunes. As one reads or watches the story he or she moves on to the *falling action* where one learns of the consequences of the conflict, of who will suffer and who will rise. The five-act drama then ends with *catastrophe* or *dénouement*. Here the story settles into the new normal as viewers understand how the storyworld has irrevocably changed from the actions of the characters and the conflicts they faced. The human species has loved this structure since the dawn of human culture. It attracts attention and holds it through the storytelling, whether on screen, in print, or in games, regardless of genre or purpose.

Unlike motion pictures, however, a photograph cannot in and of itself be a fully fleshed narrative. As an isolated moment, the still photograph can only be one point on Freytag's pyramid or any other narrative structure. In the case of "Saigon, 1968" described above, the reader sees only the absolute climax of the story, the moment at which the two characters come into deadly conflict. In the case of "Migrant Mother, 1936" also described above, the reader sees only the exposition. There, one is introduced to the characters and a tiny bit of their environment. From these isolated points on the triangle, the reader or viewer implies the rest of many possible stories, or from where the characters have come and what brought them into the conflict—as in "Saigon, 1968"—or to where they might be going and what conflicts they will face—as in "Migrant Mother, 1936." As this author and others (Ranta 2013; Steiner 2004; Kafalenos 2001, 1996; Lessing 1853) argue, the rest of the stories are inventions, created through real-world experiences, what has been learned or known already about the situations they represent, or what may simply and baselessly be presumed. The photograph is an implicit, emergent story.

At first blush, this idea of an implicit narrative may seem as a condition only of nonverbal media, such as painting, sculpture, photography, and others that—as isolated discrete pieces rather than a series as in a comic book or photographic essay—do not convey a chain of events over time. These media forms, however, might better be understood as one end of a spectrum in which all storytelling forms contain gaps. Even in the real-time experience of life as a narrative, some elements not directly experienced are implied or assumed as the reader creates an understanding of experience. If real-time experience is one end of a spectrum spread between information-rich stories and sparse stories, then film arguably lies in the middle as a form understood as *being a narrative*. With a few rare exceptions such as *Logistics*, a 37-day-long real-time experimental film (Magnusson and Andersson 2012), traditional narrative films construct a chain of events in compressed time, the archetype of which is the montage editing pioneered by Sergei Eisenstein in the 1920s. To compress the time represented in film, a novel, a game, or comic, gaps in the story must be filled by our imaginations or experiential contextualization. As David Bordwell argued, "narration is better understood as the organization of a set of cues for the construction of a story" (1985, 62).

Discussing the distinct lack of time passing in a still photograph, Marshall McLuhan (1964, 188) wrote, "It is one of the peculiar characteristics of the photograph that it isolates single moments in time." However, for professional photographers such as this author, who spent 30 years as an international photojournalist, this literal interpretation of photographic communication is simplistic. As a still photograph implies a point on a narrative arc such as Freytag's pyramid, it also implies the passage of time through several important photographic techniques: frozen action, blurred motion, and layered content.

Action

When a reader opens the sports section of any news website, the action of the recent or currently unfolding game is documented with dramatic images of players frozen midair, faces grimacing with

Figure 19.3 Young football players compete for the catch at a high school football game in the United States. Photograph © Kevin Moloney, all rights reserved.

athletic strain, sweat beads breaking free from furrowed brows. The classic sports action photograph (Figure 19.3) demonstrates the power of frozen action to communicate drama. Yet footballers do not levitate. People understand their relationship to earthly physics and though the players appear to float in midair, the reader knows this is a slice of a continuum. As such, the actions that lead up to the captured moment and away from it are imagined on a timeline. A brief narrative is reconstructed.

Blurred Motion

Humans have a remarkable capacity to reconstruct reality from sparse clues. In the early days of photography, long exposure times resulted in photographs of blurred action as the subject moved too quickly to be captured with acute fidelity. This compromise soon became a tool for photographers to demonstrate speed and motion in their images. A person's remarkable ability to recognize the human form from unclear details allows these photographs to be read as a demonstration of speed and motion (Figure 19.4). As with frozen action, blurred motion becomes a clue for the reader that time is unfolding in the scene if not in the actual photograph. Again, a brief narrative is reconstructed.

Layered Content

The artist's effort to escape the temporal bonds of the canvas long predate the photographer's efforts at the same. Medieval and early Renaissance painters often battled the constraints of the

Figure 19.4 A tuba player tunes backstage as a violinist passes before a performance with Dutch conductor and violinist Andre Rieu.
Photograph © Kevin Moloney, all rights reserved.

still image by incorporating multiple moments from an unfolding story on one canvas, as Wendy Steiner (2004, 161–162) points out in analysis of such works as Benozzo Gozzoli's *The Dance of Salome and the Beheading of St. John the Baptist.* A subject may appear multiple times at different points in the work, each appearance representing a different moment in the narrative. Though this is certainly possible to accomplish in photography, its execution requires extensive work either in a controlled scene or in digital post-production. That added labor makes this a rare technique in photography. Photographers are more often drawn to the instantaneous nature of the medium, preferring images captured in a single snap as pioneered by the Eastman Kodak company in the late nineteenth century with its famous slogan "You push the button and we do the rest" (Jaeger 2007, 6).

The snapshot ethos of photography never impeded many photographers from the pursuit of a narrative within a single frame, however. Using a technique often described as layering, a street photographer or photojournalist such as this author will simultaneously watch the foreground, middle ground, and background of a scene for individual moments to occur. When captured this layering of moments that each could be good individual photographs provide greater complexity for the reader of the photo. As the reader's gaze moves from one moment in the photograph to another, those simultaneous moments seem to unfold in series as if they were different points in a process or a narrative arc. In the author's image above (Figure 19.5), the multiple individual characters are captured at different stages of the process of collecting water from a community pump in Burkina Faso. A man arrives by bicycle for a turn at the pump (left), a woman waits for her turn (right) as another jumps into the air while pumping (center). At rear, another walks away with a fresh load of water. The complete process is represented in a single frame through multiple characters; the time spent by readers examining the image lengthens; and the illusion of time passing is achieved.

Figure 19.5 Local women push their weight into a pump to collect water in their northern Burkina Faso village.

Photographs in the Transmedia Narrative

Historically, photographs have most often been used to illustrate a point raised in another media form. This chapter is no exception to that example as the images used here simply repeat visually the points made in the text. To see photographs simply as illustrations is simplistic, however. The photograph is capable of complex and compelling narratives, delivered almost instantaneously to readers. Though the images that accompany this chapter do illustrate points made in the text, they also present their own self-contained narratives that diverge from the core message of the text. They are their own stories.

In entertainment transmedia storytelling, the still photograph is a rarely used media form. The core stories of *Star Wars* or any of a number of Marvel Comics franchises are predominantly told through cinema, video, comics, and games. Still photographs are used more sparingly and may only appear on a child's lunch box or in promotional materials like movie posters, advertisements, or media reviews of a film. However, these limited uses can still be powerful storytelling tools, pulling fans into the storyworld through strong character development, a hint at a pending turn of plot, or simply a dramatic moment. One notable example is *Vanity Fair* photographer Annie Leibowitz' 1999 group portrait of the cast of the HBO drama *The Sopranos* positioned in an ekphrastic reflection of DaVinci's *The Last Supper* (Wolcott and Leibovitz 1999). In the image, character Tony Soprano sits at center in the position of DaVinci's Jesus. From his right, his genetic family looks on and from his left his crime family. The image is only one of uncountable pop culture references to the famous Renaissance painting (Plumb 2012), all of which use the familiar image and its far-reaching cultural implications to connect a current story with it, as drama or satire. In this case, Tony Soprano's character conflicts and complexities are cast in a disturbing light and motivations of other characters are echoed. The greatest storytelling strength of the image, however, is its shocking irreverence, preparing the reader of the photograph for the drama, brutality, and intrigue of the television series. It is not a simple entry point to the storyworld, but a story in its own right.

While still photographs may be relatively rare in transmedia entertainment, they are much more influential in its journalistic or documentary equivalents. In these genres of storytelling, the still photograph performs an important role of visually documenting fleeting moments in circumstances where the reporters must be fleet of foot, where subtlety and a not-intimidating presence may be necessary, or for publications designed for a speedy read, such as newspapers, many magazines. In journalism and documentary storytelling, still photographs also provide the reader with a chance to stare at a moment that may be critical in the cultural, political or societal order, as was Eddie Adams' image discussed at the beginning of the chapter.

One transmedia organization notable for its use of still photographs is the National Geographic Society. Though it started as a scholarly journal, the *National Geographic* magazine revolutionized the publication of photography early in the twentieth century. By the 1930s, the photographs had supplanted text as the central storytelling form. Photographers working for the society are considered to be among the world's best, and their images rich with storytelling even when separated from the rest of the publication's content. In the 1960s, the National Geographic Society began its transformation into a transmedia empire that now publishes content on more than 40 different digital, analogue, and brick-and-mortar media channels (Moloney 2015, 80). Through nearly all of those channels, the still photograph continues to be a central media form, holding central stage in the printed magazines and mobile apps, standing alone in books and gallery displays, and fueling conversation around Society-wide reporting topics such as 2016's national parks series and 2014's "The Future of Food" Project. The "Future of Food" published 3,555 still photographs between April and December, 2014, and included both images made by Society professionals and amateurs who contributed their work when invited. Despite its long-running prowess with the finely crafted images that grace its magazine pages, the National Geographic Society also effectively feeds the Instagram social media service with images like those found in the magazine as well as behind-the-scenes looks at the production of the content. Photographers on assignment are now given access to the @NatGeo account to post images from the field as they report their stories. Readers could follow along as if they were in the field, and their excitement for the final story was fueled early. Social network analysis of the "Future of Food" project revealed that the National Geographic Society's 131 Instagram posts attracted more social interaction and influence than the polished pages of the magazine (Moloney 2015, 89). By comparison, when commercial film producers use Instagram they add to the storyworld with remix and reinterpretation of scenes in the movie, and only rarely with material that is additive to the storyworld. They are platforms for marketing rather than continued storytelling.

Conclusion

It is simple to understand that photographs are one of many media forms a producer might use in a transmedia project. One produces transmedia to reach new audiences in multiple ways, reaching people through varying interests, tastes, contexts, learning styles, and allowing for different engagement depths. It is less simple to understand a photograph as more than an illustration of a point made in another form, such as text, lecture, museum or gallery display. The photograph is a self-contained story. It works independently of its companion media forms as much as it complements them. For producers and critics of transmedia storytelling in any genre, the critical thinking about photographs must not only be how they interact with other media forms used in a project, but how they are also autonomous stories, capable of rich, immersive narrative, fine detail, and visual fact presentation.

In this chapter the author examined how a still photograph can imply the passage of time even though it may be frozen by a fast shutter or let blur with a slow one. Though it may be only a thin slice of the flow of time, a photograph can occupy any of the points on a narrative structure like Freytag's pyramid. When its composition becomes skillfully complex, that single photograph may hold many points on the pyramid within its frame. The frozen moment of peak action also informs the reader in ways that other visual media cannot. Eddie Adams' "Saigon, 1968" forces itself into memory and emotion not only through its socio-political context but because we can stare at its horrifying moment for as long as we can stomach it. Once seen, that image will never let go of the viewer's memory. It is the climax of a drama that the reader reconstructs with every seeing of the image. To paraphrase Marshall McLuhan: It is one of the peculiar characteristics of the photograph that it allows one to see the invisible, stare at a decisive moment and reconstruct its narrative through both real and imagined experiences.

References

Æsop. 1848. *Æsop's Fables*. New York: Grosset & Dunlap.

Bailey, George A., and Lawrence W. Lichty. 1972. "Rough Justice on a Saigon Street: A Gatekeeper Study of NBC's Tet Execution Film." *Journalism Quarterly*: 221–229.

Berger, John. 1980. *About Looking*. New York: Pantheon Books.

Bordwell, David. 1985. *Narration in Fiction Film*. Madison: University of Wisconsin Press.

Cervantes, Miguel. 2005. *Don Quixote*. Translated by Edith Grossman. New York: Harper Perennial.

Flanagan, Owen J. 1992. *Consciousness Reconsidered*. Cambridge, MA: MIT Press.

Freytag, Gustav. 1872. *Die Technik des Dramas* [Technique of the Drama]. Leipzig: Hirzel.

Herman, David, Manfred Jahn, and Marie-Laure Ryan (eds.). 2010. "Freytag's Triangle." In *Routledge Encyclopedia of Narrative Theory*, 189–190. London and New York: Routledge.

Jaeger, Anne-Celine. 2007. *Image Makers, Image Takers: Interviews with Today's Leading Curators, Editors and Photographers*. New York: Thames & Hudson.

Kafalenos, Emma. 1996. "Implications of Narrative in Painting and Photography." *New Novel Review* 3 (2): 53–66.

Kafalenos, Emma. 2001. "Reading Visual Art, Making, and Forgetting, Fabulas." *Narrative* 9 (2): 138–45. doi:10.2307/20107239.

Lessing, Gotthold Ephraim. 1853. *Laocoon: An Essay on the Limits of Painting and Poetry*. Translated by E. C. Beasley. London: Longman, Brown, Green and Longman.

McLuhan, Marshall. 1964. *Understanding Media: The Extensions of Man*. New York: McGraw-Hill.

Magnusson, Erika, and Daniel Andersson. 2012. *Logistics*. Documentary. Accessed April 15, 2017. https://logisticsartproject.com/.

Moloney, Kevin. 2015. "Future of Story: Transmedia Journalism and National Geographic's Future of Food Project." Ph.D. diss., University of Colorado.

Morris, John G. 1998. *Get the Picture: A Personal History of Photojournalism*. New York: Random House.

Plumb, Alastair. 2012. "12 Examples of The Last Supper in Pop Culture." *Empire*, August 7. Accessed May 2, 2017. www.empireonline.com/movies/features/last-supper-pop-culture/.

Ranta, Michael. 2013. "(Re-)Creating Order: Narrativity and Implied World Views in Pictures." *Storyworlds: A Journal of Narrative Studies* 5 (1): 1–30.

Ryan, Marie-Laure (ed.). 2004. *Narrative Across Media: The Languages of Storytelling*. Lincoln: University of Nebraska Press.

Smith-Rodden, Martin, and Ivan K. Ash. 2017. "The Effect of Emotionally Arousing Negative Images on Judgements About News Stories." *Visual Communication Quarterly* 24 (January–March): 15–31.

Sontag, Susan. 1978. *On Photography*. New York: Picador USA.

Sontag, Susan. 2003. *Regarding the Pain of Others*. New York: Picador.

Steiner, Wendy. 2004. "Pictorial Narrativity." In *Narrative Across Media*, edited by Marie-Laure Ryan, 145–177. Lincoln: University of Nebraska Press.

Wolcott, James, and Annie Leibovitz. 1999. "The 1999 Hall of Fame." *Vanity Fair*, December.

Zelizer, Barbie. 2010. *About to Die: How News Images Move the Public*. Oxford: Oxford University Press.

20
TRANSMEDIA INDIE
Creativity Outside Hollywood

Erica Negri

On January 25, 1999, a tiny independent horror film produced and directed by two unknown filmmakers from the University of Central Florida School of Film was presented at Sundance Film Festival. The film, actually a feature, was made to look like a low-budget documentary about three film students who on October 1994 got lost in the woods in Maryland while shooting a documentary about a local legend, intentionally blurring the border between fiction and reality. During the festival, flyers were handed out asking viewers to come forward with any information about the "missing" students who were the protagonists of the film. Produced on a mere US$60,000 budget, the film grossed more than US$140,000,000 in domestic sales, over US$248,639,000 worldwide. That film was *The Blair Witch Project*, which today is regarded as one of the of first and most groundbreaking transmedia storytelling projects developed outside the mainstream arena.

The film managed to create a fictionalized reality, presenting itself as the actual recorded evidence of a ghostly urban legend, exposing the disappearance of three people and then screening it in theaters nationwide. It spread like wildfire by word of mouth and the style of the film was so realistic and visually intriguing that the audience all around the world started to debate over whether the film was a real-life documentary or a work of fiction. To reach this goal, all the textualities in and around the film had to work together to convey the illusion of reality: the locations, the directing, the acting—mostly based on improvisation, the casting, the found footage packaging as well as the tie-ins.

It was, of course, the early days of the Internet and Daniel Myrick and Eduardo Sánchez, the creators of the film, sensed that somewhere in the Internet was the key to reaching viewers in an unprecedented and engaging way. As such, during the summer of 1999 the www.blairwitch.com website was launched, which included fake police reports, "newsreel-style" interviews of the parents, the biography of each of the missing students, a complete mythology of the appearances of the Blair Witch and pictures of the objects found in the woods, such as the camera, the tapes, and the diary of one of the protagonists. Everything looked extremely real and, above all, it was all directly accessible by the audience, who could explore a variety of different textualities that went deeper into and expanded the mythology and the storyworld through the web.

But what made this particular strategy peculiar in comparison to other marketing approaches of the era was that, since the beginning, the website was not intended as a paratext or a merely promotion-based platform, but rather as a text with a diegetic function, a place where the audience could find content that would carry on the story beyond the end credits of the film. The website was conceived as part of the story itself; it expanded the story of the Blair Witch in astonishing ways considering we were in a pre-social networks and pre-YouTube era. As Ernest Mathijs and

Jamie Sexton point out, the website for the film set a mythical yet highly realistic storyworld that created a backdrop against which the film was just one specific artifact. Taken together, the film and the website were constructed to interrelate with each other so that each segment added to the experience of discovering more about this particular mythical storyworld, allowing the audience to become immersed within an active search for information. Adhering to Jenkins' later theorization of transmedia storytelling (2006), then, both segments added to the experience, but audiences did not have to experience all of the mediated Blair Witch modes in order to appreciate the story (Mathijs and Sexton 2011).

What's more, in the September of the same year, another diegetic expansion was published: the book *The Blair Witch Project: A Dossier*, which looked like a real police report, with unedited pictures, interviews, and newspaper articles regarding the missing students, treating the premise of the film as a real fact. Several other books and graphic novels followed, artifacts that similarly connected to the backstory of the film and the mythology of the Blair Witch, all of which led to the release of the sequel *Blair Witch 2* (2000), a film that was not as successful, and which was most recently followed by the *Blair Witch* reboot (2016).

In effect, even though the Blair Witch intellectual property (IP) gradually became increasingly confined to a franchise logic—not to mention influencing those franchise logics— the creators of the original film clearly had been driven by an innovative idea: that since they were not financially supported by a film studio, a transmedia approach which integrated promotional communication and narrative could be the key to its success and to finding an audience. They opted for an expanded narrative strategy, distributing different textualities on different platforms with the aim of enriching the storyworld, and, importantly, increasing the credibility of the narrative. And that was before "transmedia storytelling" was even theorized or studied. For the purposes of this chapter, it is important to note that such a winning strategy required a very low budget but a high level of control from an authorial perspective; being independent from a major studio was itself key to the filmmakers being able to achieve the goal of creating a story that was far bigger than the film itself and yet consistent in all its components.

Contextualizing the Independent Approach to Transmedia Storytelling

Since the 1990s, the process of digitalization and media convergence has had a strong impact on modes of production, distribution, and reception for narrative content, both at a mainstream and indie level. Phenomena such us the application of transmedia storytelling practices, the hybridization of forms of discourse, the integration of interactive elements within traditionally linear narratives, and the growing importance of world-building as part of the creative process all started to impact different levels of media production and the media industry's approach to generating new textual forms, namely:

- Alternate Reality Games (e.g., *Perplex City*);
- multiplatform IPs with a high-level of interactivity (e.g., *Lost*);
- transmedia film franchises (e.g., *The Matrix*);
- video games mixing game and narration (e.g., *Assassin's Creed*);
- independent films using transmediality both as a narrative asset and as a marketing tool (e.g., *The Blair Witch Project*);
- pioneering interactive properties (e.g., *Collapsus*);
- transmedia narratives spanning across several media forms (e.g., Collider).

In the light of the intrinsic semiotic variety that defines the post-digital media environment, Jason Mittell points out the necessity to distinguish between an ideal form of *balanced transmedia* in which no medium or text serves a primary role over the others and where texts form a single narrative, and the most commonly applied model of *unbalanced transmedia*, characterized by a clearly identifiable core text and a number of peripheral transmedia extensions that might be more or less integrated into the narrative whole, and that expand or further explore the storyworld. Arguably most examples of contemporary transmedia projects fall somewhere on a spectrum between balance and unbalance (Mittell 2015).

Echoing Mittell's sentiments, Christy Dena distinguishes between two main typologies of transmedia objects: the first, called *intracompositional transmedia*, refers to a collection of mono-media stories that interlace synergistically to build a whole storyworld; this is the case with many mainstream transmedia IPs like *Heroes* or *Lost*. The second typology, defined *intercompositional transmedia*, refers to a single story told through a collection of media and texts. While the first typology generally involves a self-contained story for each medium, meaning that it is not necessary for the audience to go through all of the textualities in order to understand the story, the second typology implies that the story can be fully understood only by consuming all of the texts (Dena 2011). This is what Geoffrey Long calls *transfiction*:

> In its purest form, a transmedia franchise engages in transfiction, wherein the first chapter is told in one media type, then leads straight into a second chapter in a second media type, which then cliffhangers straight into a third chapter in a third media type.
>
> *(Long 2007)*

Hence, a transfiction, or *transmedia narrative*, differs from a transmedia franchise owing to three key characteristics. A transfiction is:

- *retro-active*, i.e., a transfiction is thought to be transmedia since the beginning and it is designed accordingly (Davidson 2010);
- *centripetal*, i.e., it is made up of several textualities that are not necessarily expansions of a central main text, such as a film or television show (Mittell 2015);
- *transcendent*, i.e., it is aimed at creating a new unity that transcends the singularity of each text purposely distributed on different platforms (Jenkins 2009).

The Industry Impact on Indie and Mainstream Transmedia Production

The emergence of transmedia storytelling as a cultural phenomenon with a variety of objectifications and discourses—spanning transmedia franchises, transfiction, ARG, and so forth—cannot be analyzed without framing such forms within the specific industry contexts that generated them.

With that in mind, it is important to note that, since the franchise has become the central paradigm of the contemporary Hollywood film industry, the business logic that has guided the creation and distribution of all media content (significantly called "properties" or IPs) entails huge limitations, mainly because the possibility to experiment and hybridize narrative forms is so closely contingent on an ability to generate multiple sources of income. Paradoxically, the chance to experiment has proven to be greater inside the independent film industry, where less money has come to mean fewer restrictions.

In fact, most of today's Hollywood-produced digital and cross-media outputs appear to be almost exclusively based either on the duplication of the same content on different media assets—through, for example, practices of intertextual adaptation, licensing and the creation of paratextual extensions—and/or the extensive exploitation of the story universe of a particular IP through the production of narrative extensions, spin-offs and reboots, where the ultimate "goal" is always essentially commercial. This model is arguably the case with 90 percent of Hollywood franchises, including Marvel and DC superhero movies and successful action-based IPs such as *Fast & Furious*, *Terminator*, *Pirates of the Caribbean*, *Madagascar*, and so forth. It is clear that, even within projects that include transmedia storytelling practices, what prevails in Hollywood is the commonplace model of the aforementioned unbalanced transmedia, with a clearly identifiable core text and a number of peripheral transmedia extensions that might be more or less integrated into the narrative whole (Mittell 2015). And thus if it is true that practices of transmedia storytelling are applied mostly within Hollywood franchises as a commerce-driven model, then a claim can certainly be made that those practices are rarely aimed at the creation of proper "transmedia narratives," but instead occupy the status of forms of "narrative transmediality," with an arguably "purer" transmedia approach more commonly being applied to marketing and communication (Negri 2015).

To explain, this distinction is due to the fact that most production companies and film studios are part of media conglomerates horizontally and vertically integrated, whose objective is the commercial exploitation of the IPs on the largest number of markets. For the IP owners, then, the goal is to increase the profit through diversity, extending the commercial profitability of the franchise and creating strong feelings of identity and ownership in its consumers (Lemke 2004). Hence, the predilection is for practices of cross-media extension and exploitation of the IP brands though licensing deals aimed at the creation of content and products for different media devices and different markets, but without pursuing an integration at a narrative level. Moreover, even when a transmedia storytelling approach is fully implemented within the mainstream, commercial media environment, the studios tend to produce a consolidated canon of "official" texts that in fact tend to discourage or discredit unauthorized expansion or speculation by fans (see Scott 2013). Within such a structured industrial model, transmediality will always presuppose a central text (usually a film or television series), aimed at defining the canon and launching the property, while the other texts will serve the purpose of expanding that nucleus and capitalizing on other markets.

The independent film sector, on the other hand, even though it appears to be economically penalized since it does not benefit from the financial security and the communicative force given by a strong distribution partner such as a Hollywood studio, has always appeared to be a more fertile ground for the production of pioneering transmedia projects. This is due to the fact that, as stated before, being outside the studios' sphere of influence quite often translates into having fewer limitations and constraints, both from an economic (funding, business models, distribution channels) and from a creative point of view. For example, it is within this indie context that in recent years the most innovative transfiction projects have emerged, such as *Collapsus*, *Alpha 0.7*, *Pandemic 1.0*, and *Final Punishment*, to mention a few.

Collapsus is a transmedia project directed by Tommy Pallotta and produced by Submarine Channel, with the Dutch broadcaster VPRO, which combines animation, interactive fiction and documentary. This story follows how the impending energy crisis affects ten young people, while international powers battle with political dissension and a fearful population during transition from Fossil fuel to alternative fuels. *Collapsus*' pioneering approach lays in the will to blend different forms of discourse—video blogs, interactive maps, fictional newscasts, live action footage, and animation—with strong elements of interactivity in order to immerse the player in the narrative. The project requires the player to access additional information and interact with the story, making decisions about the storyworld's energy production at both a national and global scale.

Alpha 0.7, meanwhile, is a transmedia narrative that takes shape through various media platforms, composing a multi-tiered structure. Developed as a transmedia property from the outset, *Alpha 0.7* describes a transparent society where fear of terror attacks has led to personal freedom being eroded away in the name of homeland security. It asks the question: "What if all that data we share so readily, ends up in the wrong hands? What if it is used against us?" At the heart of the project there was a six-episode fantasy mini-series aired in 2010 on the German regional channel SWR. At the same time, the story was enriched by an Alternate Reality Game, a radio show called "SWR 2 Jungle," narrative podcasts and various other forms of content available on the Internet.

Pandemic 1.0, by comparison, is a transmedia experience created by Lance Weiler and presented at the Sundance Film Festival in 2011. The goal was to provide a more engaging experience of the storyworld, to test new business models and study the social interactions among the participants involved. Medic Mobile and FreedomLab collaborated on the project by modeling various types of game interactions in the hope of gaining insights into how narrative elements develop in social environments. Pandemic's immersive experience took off with a short film, which was part of the official selection of the Festival and told the story of two brothers whose mother had been infected by a dangerous virus. The experience continued out of the screening room through a comic, a scavenger hunt happening through various devices, online and offline contents, a website, and a series of tweets.

Final Punishment, finally, is an interactive transmedia thriller produced by beActive for OI, the main Brazilian telecom, telling the story of eight detained women imprisoned in a high-security penitentiary in Brazil; when the security system breaks down, they start to be the object of a series of horrible and mysterious murders. Nominated at the International Digital Emmy Awards 2010, Final Punishment consists of an online television series and an Alternate Reality Game that stimulated the participation of the viewers to free the prisoners through several mobile and online content.

What is notable about all of these independent productions, besides the higher level of interactivity, is the fact that they are balanced and centripetal transmedia formats, characterized by high integration and consistency among all of the texts, with the aim to create a single narrative.

To mark these two distinct approaches, Andrea Phillips talks about the idea of "West Coast transmedia" versus "East Coast transmedia." West Coast-style transmedia, more commonly known as Hollywood or franchise-based transmedia, consists of multiple big pieces of media: feature films, video games, etc., all of which are grounded in big-business commercial storytelling practices, with the stories in these projects interwoven, but lightly. In this model, each piece of media can be consumed on its own and audiences will still come away with the idea that they were given a complete story. On the other end of the spectrum, East Coast transmedia (and European transmedia) tends to be more interactive and much more web-centric. It thus overlaps heavily with the traditions of independent film, theater, and interactive art. These East Coast projects make heavy use of social media, and often run over a limited period of time rather than persisting forever. The plot in some cases is so tightly woven between different media that you might not fully understand what is going on if you do not actively seek out multiple pieces of the story (Phillips 2012). Another peculiarity of the East Coast approach to transmedia is the search for modes of audience participation, which takes place in three modalities:

- through the active exploration of the different texts and content that compose the project, skipping through several media platforms (participation as *exploration*);
- through the interaction with other participants (participation as *social interaction*);
- through the interaction with the narrative itself (participation as *world-building*).

Beyond Franchise: *Collider*, an Indie European Transfiction

A particularly interesting case study of an independent transfiction that emerged within a European context is *Collider*, a transmedia property developed and produced by Nuno Bernardo's Irish-Portuguese company beActive. *Collider* constitutes an example of a virtuous accommodation of narrative and economic bids within a non-Hollywood multiplatform audiovisual project. As a validation of this fact, it is significant to report that *Collider* was nominated for several prizes, including a Digital Emmy Award in March 2014.

Marketed as "*Terminator* meets *Lost*," *Collider* is a science-fiction multiplatform project that combines television and web series, mobile games, and online graphic novels available on the App Store and Google Play, which offer exclusive access to both pre-series storylines and the feature film.

The narrative world of *Collider* revolves around six characters, coming from different years and places, who find themselves mysteriously transported to a post-apocalyptic future of which they know nothing and in which apparently they are the only humans left alive. Their aim is to find out how they got there, discover what happened to the Earth, and go back in time to save the world; all of this without getting themselves killed by the mysterious creatures that roam outside the hotel in which they woke up.

The first point of entry into the *Collider* storyworld are six online comic books, one devoted to each character within the story, which each cover their backstories before they jump to the future, revealing their past life and their unsolved issues. The subsequent web-series, featuring Peter Ansay, the European Organization for Nuclear Research (CERN) scientist, covers the inciting incident of the story, describing his decision to sabotage the Large Hadron Collider after he learns about the dangers of the experiment. This sabotage attempt will result in the creation of the wormhole that will transport the protagonists in time and space. Moreover, Peter's blog is available on Wattpad.com, an online platform for self-publishing. Next, the mobile game (*Collider Code Breaker*) is set in the future and allows the player to act like one of the characters, who are trying to reach the Hadron Collider, unlocking doors and finding the right path beneath the city of Geneva.

BeActive's key concern was to make all of these various story elements self-contained and, at the same time, to make them complementary to each other, like different pieces of the same puzzle, in order to make the full experience much richer. Since the beginning, *Collider* was conceived as a *transmedia narrative* combining different texts to create a distributed diegesis that meets the audience where the audience is (on smartphones, tablets, computers, theater) and exploiting the opportunities of immersion and interactivity offered by digital assets in order to generate a loyal fan base deeply involved with the property.

The peculiarity of *Collider*, however, is that it looks like a hybrid between the multiplatform franchise model typical of commercial Hollywood properties and the independent production model, which, as already stated, implies the absence of economic strength and a major promotional or distribution backing. In this case, the implementation of transmedia storytelling practices aimed to create a loyal fan base that progressively engaged with the narrative world and which was encouraged to proceed from one medium to another in search of content, but within a framework that did not allow for direct monetization (the content was available for free online).

At the same time, the project aimed to expand its potential audience as much as possible, creating content for different market niches: a comic book reader is not necessarily an app-gamer or a viewer of web series, but those who discover a property within their own niche are more likely to move to a different media platform (one that they might not naturally approach) in order to find more material from the same fictional universe. But unlike most transmedia Hollywood franchises, the narrative structure of *Collider* was *distributed* and *balanced*—that is to say, it lacked a central predominant text defining the canon of the entire narrative world, but that canon was the result of the sum of several texts that worked synergistically to create a single narrative track (or storyline), retraced in its entirety only by moving from one platform to another. In these terms, *Collider* configured itself as a bona fide example of *transfiction*.

Central to this notion of transfiction is the way in which a given transfictional text is intended to reinforce multiple sets of content by placing them within a larger narrative. This point is confirmed by BeActive's decision to create unique and original content for each platform utilized, planting specific pieces of information in each on these platforms so that the viewer is naturally inclined to want to explore the other textualities in order to know more about the narrative world.

Ironically, however, it should also be observed that, although the film was the text within this narrative world that required the largest production effort and investment, the wider narrative and distributive structure of *Collider* remained essentially horizontal. That is to say that the various media and content that the project supported were in fact deliberately placed on the same "level." In fact, Nuno Bernardo's goal with the project was to create a story universe and then divide that universe in different, but self-contained, pieces and distribute them to each media form, namely in order to connect with the audience and fans on all of the different media that an audience would make use of during an average day. In this way,

> the story can follow the audience and be there with the audience, whatever they are doing, whether browsing on the Internet, using a mobile phone or just reading a book … So each of the elements of *Collider* work as stand-alone pieces, but if the audience watch, read and play all of them they will have a better understanding of the world we created and connect at a more engaging level with the characters and the story.
>
> *(Author interview, 2013)*

Another specificity of *Collider* concerns its strong internal *consistency* on different levels: narrative, authorship, and temporality, which were all arguably much stronger than with many Hollywood-based cross-media IPs. From an authorship point of view, *Collider*'s internal consistency resulted from the presence of a single creator responsible not only for setting the canon in the central content, but also for developing, producing and distributing all of the extensions. No licensing was involved in the creation of the fictional world. Comics, video games, web series, movies and all forms of content and paratextual materials were developed and produced by the beActive team, under the supervision of Nuno Bernardo, who maintained direct control over each step of the creative process. As Nuno Bernardo puts it,

> the advantage to develop, produce and distribute a transmedia project independently is that we don't have to respond to a major studio or financer that, for the fact of having invested capital on the project, has the right to interfere with the strategy planned by the story architect.
>
> *(Author interview, 2013)*

In essence, the spirit of transmediality is about collaboration, or at least multiple pieces of content being farmed out across different divisions within a media corporation, with multiple sets of (often competing) authorships at play. Bernando's point is thus a simple one—that the creative freedom of independent production is arguably better suited to the needs of transmedia storytelling.

This degree of internal consistency derived also from the control of the producer over the roll-out plan, which was designed strategically to enhance each specific piece of content within the project at the same time. As mentioned previously, the decision to release first the digital content, which was more interactive and less economically demanding (comics, video games, and web series) and only after the traditional ones (film/television), was highly strategic as it contributed to the gradual creation of a fan base, in anticipation of the theatrical release.

All in all, then, *Collider* stands out from other, more commercial cases of transmedia storytelling on account of a few specificities—namely, a balanced and horizontal structure; an intercompositional configuration; a high consistency on the levels of narrative, authorship, and temporality; the absence of redundant content; the positioning of all texts on the same level (no extensions); and the absence of licensing deals.

These particular specificities thus work to differentiate *Collider* from the traditional Hollywood transmedia IPs, characterizing it as a genuine *transmedia narrative*, as it does not exploit different media as mere means of distribution for purely promotional content (like in most Hollywood licensing-based franchises), nor does it not conceive of the different textualities that make up the storyworld as extensions of a core text, but instead builds a balanced, non-centrifugal, and intertextual set of media content in which the principles originally postulated by Henry Jenkins (2006) are fully realized.

Conclusion

Digitalization and the proliferation of new portable technology has profoundly altered the media landscape, altogether working to produce a context that is characterized by cultural and economic convergences that are reshaping the experiences and textualities of media. This digital shift has also accelerated the challenges faced with researchers working within the fields of traditional semiotic and narratological research, and has done so by calling into question, on the one hand, the cultural and cognitive practices of interaction with the media and reception of media content, and, on the other hand, the value and the semiotic identity of that media content, and the relative narrative structures around them. As Martin Rieser and Andrea Zapp point out, we find ourselves in the middle of an era of "narrative chaos," in which traditional narrative structures prove inadequate to the emergent media forms, while the relationship between these forms and the audience keeps changing. It is in this situation of narrative chaos that experimentations and new approaches to the art of storytelling have started to emerge, in the attempt to conciliate narrative forms and digital technologies (Rieser and Zapp 2002).

Transmedia storytelling is one of these attempts. Ideally, all texts involved in a transmedia storytelling process should be put on the same level, nullifying the distinction between texts and paratexts, as well as between central textualities and extensions. However, throughout this chapter I have suggested that the economic and industrial conditions in which Hollywood-based commercial transmedia projects operate unavoidably affects the discursive structure as well as the intertextual nature of those projects—arguably for the worse. In particular, the logic of profit that drives the film and television business, especially in Hollywood, influences the role and use of narrative extensions, leading to them often being treated as part of branding or franchise strategies. This commercialism implies the prevalence of unbalanced transmedia structures characterized by a main item of central content and a number of "lesser" media extensions, which may be more or less integrated into the main narrative. Consequently, the commercial logic guiding the production of such audiovisual content in Hollywood seems to compromise the possibility of creating proper *transmedia narratives*, as the textualities will never be at the same level, with one medium always being dominant, central, defining the canon and launching the property, with the role of the other items of media content being to merely "expand" that core for primarily commercial reasons.

Conversely, the independent film sector arguably presents far fewer restrains on transmedia storytelling practices from a commercial, financial, narrative, and aesthetic point of view, proving a less vitiated environment to study the peculiarities of transmedia narrative forms. It is within this independent context, in fact, that in recent years several pioneering organic transmedia projects have emerged—projects whose structures have resulted in more balanced manifestations of transmedia storytelling, and whose different texts are more diegetically integrated be being placed on the same level as each other. However, these projects, such as those examples explored throughout this chapter—though while more courageous from the point of view of experimentation than many Hollywood projects—are typically weaker in terms of reach and cultural value, which often leads them to fail in their attempts to engage their audience. This "failure" is especially evident if the engagement strategy of such projects implies high interactivity, as was certainly the case for the aforementioned *Collider*.

References

Davidson, Drew. 2010. *Cross-media Communications: An Introduction to the Art of Creating Integrated Media Experiences*. Pittsburgh: ETC Press.

Dena, Christy. 2011. "Do You Have a Big Stick?" In *Hand Made High Tech: Essays on the Future of Books and Reading*, edited by Simon Groth, 47–50. Brisbane: Institute for the Future of the Book in Australia.

Jenkins, Henry. 2006. *Convergence Culture: Where Old and New Media Collide*. New York: New York University Press.

Jenkins, Henry. 2009. "The Aesthetics of Transmedia: In Response to David Bordwell (Part one)." *Confessions of an Aca Fan*, September 10. Accessed June 22, 2017. http://henryjenkins.org/blog/2009/09/the_aesthetics_of_transmedia_i.html

Lemke, Jay. 2004. "Critical Analysis Across Media: Games, Franchises, and the New Cultural Order." Paper presented at the First International Conference on CDA, Valencia, May 5–8.

Long, Geoffrey. 2007. "Transmedia Storytelling: Business, Aesthetics and Production at the Jim Henson Company." Master's thesis, Massachusetts Institute of Technology.

Mathijs, Ernest, and Jamie Sexton. 2011. *Cult Cinema*. Oxford: Wiley-Blackwell.

Mittell, Jason. 2015. *Complex Television: The Poetics of Contemporary Television Storytelling*. New York: New York University Press.

Negri, Erica. 2015. *La Rivoluzione Transmediale* [The Transmedia Revolution]. Turin: Lindau.

Phillips, Andrea. 2012. *A Creator's Guide to Transmedia Storytelling: How to Captivate and Engage Audiences Across Multiple Platforms*. New York: McGraw-Hill.

Rieser, Martin, and Andrea Zapp. 2002. *New Screen Media: Cinema / Art / Narrative*. London: BFI Publishing.

Scott, Suzanne. 2013. "Who's Steering the Mothership? The Role of the Fanboy Auteur in Transmedia Storytelling." In *The Participatory Cultures Handbook*, edited by Aaron Delwiche and Jennifer Jacobs Henderson, 43–52. New York: Routledge.

PART III

Practices of Transmediality

21
TRANSMEDIA ADAPTATION
Revisiting the No-Adaptation Rule

Christy Dena

A long-held view by transmedia theorists and practitioners alike is that adaptation is not part of the phenomenon. While there are those that have argued against this exclusion (Dena 2009; Ruppel 2012), and there has been a softening of the arguments (Jenkins 2011, 2017; O'Flynn 2013; Harvey 2015), the no-adaptation approach has helped many identify and understand the area. Given the consensus regarding excluding adaptation, this chapter investigates the underlying assumptions informing the exclusion and looks at the costs involved. What do we miss when we exclude adaptation from transmedia projects, and is there another way we can distinguish transmedia phenomena? We begin this investigation by revisiting the adaptation exclusion arguments.

The Adaptation Exclusion Arguments

In 2003, off the back of attending an industry event with Hollywood and games producers, Jenkins shared his manifesto of transmedia storytelling. This beginning piece, not coincidentally, establishes the exclusion of adaptation:

> We need a new model for co-creation-rather than adaptation-of content that crosses media. The current licensing system typically generates works that are redundant (allowing no new character background or plot development), watered down (asking the new media to slavishly duplicate experiences better achieved through the old), or riddled with sloppy contradictions (failing to respect the core consistency audiences expect within a franchise). These failures account for why sequels and franchises have a bad reputation. Nobody wants to consume a steady diet of second-rate novelizations!
>
> *(Jenkins 2003)*

Jenkins continued with these sentiments in 2006 with the release of his popular book, explaining how transmedia storytelling is unlike the current licensing system, which "typically generates works that are redundant" (Jenkins 2006, 105). Any composition that does not make a distinctive and valuable contribution does not offer a "new level of insight and experience" (2006, 105). Indeed, anything that allows "no new character background or plot development" is therefore "redundant" (2006, 105). Redundancy, Jenkins summarized, "burns up fan interest and causes franchises to fail" (2006, 96).

After the book, Jenkins continued to argue against adaptation while noting that some considered adaptation transmedia: "for many of us, a simple adaptation may be 'transmedia' but it is not 'transmedia storytelling' because it is simply re-presenting an existing story rather than expanding and annotating the fictional world" (Jenkins 2009). Jenkins' MIT student at the time, Geoffrey Long, continued the delineation: "Retelling a story in a different media type is *adaptation*, while using multiple media types to craft a single story is *transmediation*" (Long 2007, 22, original emphasis); and Aaron Smith likewise felt the "distinction must be made between transmedia extensions and adaptations" (Smith 2009, 24).

In her 2011 book on the topic, Elizabeth Evans likewise corroborates the idea that "[t]ransmedia elements do not involve the telling of the same events on different platforms; they involve the telling of *new* events from the *same* storyworld" (Evans 2011, 27, original emphasis). Evans argues there are three key characteristics that, while appear to varying degrees, distinguish transmedia: narrative, authorship, temporality. They represent the "key ways in which texts become transmedia, rather than function as marketing, spin-offs, or adaptations" (2011, 28).

Indeed, we see this sentiment with the majority of academic works published in the years that followed. For example, Carlos Alberto Scolari argues that transmedia storytelling "is not just an adaptation from one medium to another" (Scolari 2009, 587). Mélanie Bourdaa notes that

> [w]hen it comes to narration, transmedia storytelling is richer than crossmedia adaptations since it develops a whole universe instead of only adapting the same storylines to different platforms. Each platform represents a new entry into the universe and not just the same story.
>
> *(Bourdaa 2013, 205)*

The view is not for academics only either, it pervades industry texts on the subject too. Jeff Gomez has discouraged the use of direct adaptations with clients, and on his company website describes transmedia as distinguished from:

> A "cross-platform" or "cross-media" approach to communications was established in the 1960s and '70s, when televised content was repurposed for cable or satellite broadcast, or advertisements were repurposed for print, radio and television ... Transmedia is a subset of cross-media in that the story itself is distributed across a variety of media. Each piece of the story feels at least somewhat complete and adds to the audience's concept of the characters and story world.
>
> *(Starlight Runner 2017)*

Tom Dowd, in both his books (Dowd et al. 2013; Dowd 2015), says that "in general we do not think of adaptation as a transmedia process because it implies we're simply taking a story and presenting it in a different medium" (Dowd 2015, 22). Dowd continues, labelling adaptation a "translation process," that produces a "derivative work" (2015, 22).

Andrea Phillips does not go into detail about what transmedia storytelling is. But after citing Jenkins, does briefly state there are three criteria for transmedia storytelling: multiple media, a single unified story or experience, and avoidance of redundancy between media" (Phillips 2012, 15). Howard Houston writes that unlike transmedia, "a multimedia approach would tell a story in a film and then retell the story in a book novelization and then retell the story yet again in a comic book" (Howard 2013). More recently on the influential *Film Courage* YouTube channel, Houston explains that it is not just a definition, but "the marketplace has been going for new stuff rather than repurposed content" (Howard 2017a). Alternately, Anne Zeiser quotes Jenkins saying "[d]on't be too hung up on definitions," and offers the following salve: "If you set aside semantics, then transmedia can be what you want, as long as it tells a broader story across multiple platforms and engages your project's audiences in that story" (Zeiser 2015, 20). So while Zeiser does not call out adaptation, your work would only qualify if it tells a broader story.

An exception includes Nuno Bernardo, who instead goes straight to the different strategic types of transmedia he uses with no qualification of what constitutes this practice or not (Bernardo 2014). Then there is myself, a practitioner-academic, who argues for adaptation in both my academic (Dena 2009) and industry work, which Marc Ruppel also does with his analysis of repetition effects (Ruppel 2012). Practitioner-academic Colin Harvey cites my argument and further interrogates adaptation and transmedia (Harvey 2015), and Siobhan O'Flynn cites my argument in her epilogue to Linda Hutcheon's book on adaptation (O'Flynn 2013). More on Harvey and O'Flynn's thoughts shortly.

While the influence of Jenkins as an academic and industry advocate for transmedia storytelling is clear, not all of those who reject adaptation are doing so because they are following suit. The majority of academics and practitioners alike (including those who are both, such as Geoffrey Long), feel that adaptation should not be situated as part of the transmedia phenomenon, and not included in practice. Why?

The Functions of Adaptation Exclusion

Before interrogating the stated reasoning put forward for academic exclusion (much of which you have been shown in the previous section), we need to address the function of excluding adaptation from transmedia theory and practice. This is because the greater issue informing the need for adaptation exclusion is an uncontested space: we all agree on there being an issue with understanding. There are genuine concerns about understanding and comprehension of what transmedia is, as a literacy, and a theoretically defined phenomenon. This section looks at the pedagogical and differentiation functions of the no-adaptation argument.

You may recall from the earlier quotes, that a common refrain of the no-adaptation argument is to distinguish it from marketing, the usual licensing approach, spin-offs, novelizations, etc. This attempt at differentiation is an important one, as it helps people understand why a new theory has been proposed. Transmedia is a particularly difficult phenomenon to highlight, as people have been employing multiple media platforms for the same storyworld for as long as memory permits. Industry has been doing it for a long time, and so too have independents. Further to this, there are existing, well developed theories that may already address the phenomenon with aplomb. It is the task, therefore, of those proposing a new term, a new area of inquiry, to justify its introduction.

Jenkins' approach is to distinguish transmedia from the "'business as usual' projects which are not exploring the expanded potential of transmedia, but are simply slapping a transmedia label on the same old franchising practices we've seen for decades" (Jenkins 2011). This is a concern for media studies scholars whose object of study is primarily the "intercompositional," "West Coast," or "portmanteau" transmedia phenomenon (Dena 2009; Clark 2011; Pratten 2015 respectively), which means looking at the relationships between mono-medium texts. The no-adaptation rule cuts straight to what is unique. Adaptation has always been happening, but continuations across media less so. Sequels and prequels happen within the same media platform, but not as often cross them. These discussions about the nature of what is different about transmedia help progress the field, especially in the context of those producing theories, conferences, and research groups whose titles could be substituted with "new media" without any perceptible difference.

Then there is the pedagogical role of the no-adaptation rule. While I include adaptation as part of transmedia practice, I find describing transmedia as "a story starting in one platform and continuing in another" the most efficient way to explain the concept to laypersons. Likewise, in 2016, Jenkins leveraged the known form of a TV series to explain the concept in an online course for the University of New South Wales (UNSW): "So if we think of transmedia as a system where the story is dispersed not across episodes of a television series, but across multiple media platforms, the story is the total information we assemble by going to those various media platforms" (Jenkins 2016). These descriptions help the student, the newcomer, get the basic concept quickly.

So we have three important tasks being performed by the no-adaptation rule: highlighting how there is a different way to approach multiplatform practice, distinguishing it from pervasive industry practices; providing media studies a narratological rationale for introducing a new theory to already advanced conversations on related phenomena, specifically differentiating the area from the established field of adaptation studies; and providing an effective shorthand for newcomers to grasp. What I wish to challenge here, perhaps controversially, is the idea of exclusion being so necessary to achieving these goals that the consequences are considered collateral damage, or worse not considered at all.

Let us first address the pedagogical or discursive effectiveness of situating transmedia as an extension rather than an adaptation process. It is effective in what it does. But I consider such an approach to be a gateway rhetoric—where we strategically use language and concepts that help people understand it. It is when scholars and practitioners consider gateway rhetoric the *conditio sine qua non* of transmedia, we have short-term gains with long-term costs. Gateway rhetoric is meant to leverage progressive disclosure, by "separating information into multiple layers, and only presenting layers that are necessary or relevant" (Lidwell, Holden, and Butler 2003, 154). It will not help, for instance, to explain a film to someone and list all the genres and all the different media forms it can take right at the beginning. But when these handy descriptions are taken as representing the limits of truth about a phenomenon, about a practice, then complexity and the reality of use is lost.

Take for example the Producer's Guild of America's 2010 guidelines updated to recognize a Transmedia Producer (Producers Guild of America 2010). To be accepted in the Guild, one must have worked on a "Transmedia Narrative Project franchise" that consists of "three (or more) narrative storylines existing within the same fictional universe" (Producers Guild of America 2010). With these simple words, producers of two-screen experiences, for example, are excluded. I voiced these concerns and others in a blog post, and posited that I suspect the three-platform was for pedagogical reasons. Gomez, who was key in developing this much-needed credit, wrote back in my blog comments:

> To respond to your concerns, the minimum of three in terms of both platforms and storylines was indeed designed as a kind of educational tool to distinguish this form from the standard movie/website mentality that had been driving Hollywood marketing for well over a decade. At the same time, I think it's an easy enough goal to hit for transmedia producers, since very nearly anything can be counted as a third leg on that stool.
>
> *(Gomez 2010)*

While the requirement of three is softened with the realization that perhaps any element can be included, the attempt to invoke strategic understanding with an industrial barrier to entry is concerning. Indeed, practitioner Houston Howard instructs in his book: "Remember, to be industry-compliant you need *at least three.* I have no idea why they ended up deciding on three, but three is the magic PGA [Producers Guild of America] number for whatever reason" (Howard 2017b, 38, original emphasis). Professional membership, funding, partnerships, and education are all affected when pithy rhetoric becomes codified. As Steven Maras, in his discussion about the historical effects of screenwriting norms "can marginalise other understandings or working with film and produce a narrow conception of the possibilities of the medium" (Maras 2009, 4). Jenkins responds in 2011, noting that the amount of platforms is not an issue for him. Instead it is more how scholars are thinking about media platforms and not the relations between them:

> I have been troubled by writers who want to reduce transmedia to the idea of multiple media platforms without digging more deeply into the logical relations between those media extensions. So, if you are a guild, it matters deeply that you have a definition which determines how many media are deployed, but for me, as a scholar, that is not the key issue

> that concerns me. As we think about defining transmedia, then, we need to come back to the relations between media and not simply count the number of the media platforms.
>
> *(Jenkins 2011)*

Indeed, within the discourse of the transmedia extension arguments is a concern about comprehension and competency. Jenkins implies it when he states that scholars who do not put a media relations provision in, are not understanding what transmedia is. Harvey states that

> [w]ith at least one notable exception, most transmedia commentators agree … that retelling a story in a different medium involves a distinct set of creative and consumptive processes that are different from telling a new story set in a consistent storyworld but utilising a different medium.
>
> *(Harvey 2015, 64)*

The "notable exception" may be the vocal scholars and practitioners like myself, who argue for adaptation inclusion. But this is a misrepresentation of the pro-adaptation argument and reflects the common belief that people who argue for adaptation do not have a transmedia literacy. It is impossible, the thinking goes, to think adaptation is part of the transmedia phenomenon and understand what transmedia is.

Dowd softens the view and acknowledges:

> there may be times when adapting material may involve just one story within a larger intellectual property universe, so we cannot say that adaptation is never a part of transmedia, just that it cannot be the only approach. If it is, then it does not meet the accepted definition of transmedia storytelling.
>
> *(Dowd 2015, 22–23)*

It is true that adaptation has been the default mode of multiplatform use (besides repurposing) in film, TV, theater, literature, and beyond. And so if a practitioner employs adaptation, it is likely they do not realize all platforms can be used for different means; and if a transmedia scholar looks just at adaptations over the centuries, they may be inaccurately ascribing a technique. This returns us to the primary function of the no-adaptation rhetoric: to differentiate the phenomenon.

Like the rhetorical tactic, the differentiation approach can have a short-term gain for a long-term cost. The cost comes at differentiating a phenomenon through exclusion. Indeed, the easiest way to describe how something is different is to position it as not being the same. This is quite simply a means-end strategy, where a problem state is compared to the goal state, and then the quickest actions are taken to reach the end (Chi, Glaser, and Rees 1982; Larkin et al. 1980). These arguments also often come with an air of supercession: not only are they different, they are better, replacing what has come before. TV is different to radio, it is better because it has images; games are different to TV, they are better because they are interactive; transmedia is different to franchises, it is better because it extends stories.

Such instincts are not limited to media industries. Semiotician Yuri Lotman observed such processes: "Every culture begins by dividing the world into 'its own' internal space and 'their' external space. How this binary division is interpreted depends on the typology of the culture" (Lotman 1990, 131). The focus on differentiation via exclusion happens quite naturally. It is an efficient way to distinguish something and help someone recognize it. It is an approach, however, that may be facilitated when you seek to differentiate via the traits of an object.

A textual or object traits approach is first- and second-order design thinking, as distinct from third- and fourth-order thinking, which is exemplified with thinking of activity and systems (Buchanan 1998, 2001). What does it mean when we view a phenomenon as an object with traits? The connection between understanding things (second-order design) and differentiation via exclusion is made clear through a discussion about how this has had detrimental effects in games (Dena 2017a). Game theorist and journalist Dan Golding, for instance, reviewed the history of video game scholarship and found a textual definition of games became the primary approach to understanding what games are for scholars and practitioners alike (Golding 2013). To Golding, this "notion of configurability has had long-standing repercussions across a significant range of videogame scholarship" (2013, 33). Golding argues that the repercussion is that "it carries with it assumptions and preconceptions that emphasize formal and textual processes to the detriment of experiential factors and the act of play" (Golding 2013, 37). Likewise, Brian Upton, ex-Senior Designer at Playstation, has explained that

> One of the drawbacks of associating games so closely with interactivity is that it biases design away from stillness. It encourages the construction of games that are action-packed, with lots of short-term business for the player to attend to ... And it's hard to think about the deeper meaning of a play experience if your entire attention is required merely to sustain it.
>
> *(Upton 2015a, 78)*

Likewise in transmedia, a whole generation of transmedia practitioners, scholars, and students do not understand why adaptation would be employed. They think that not only is it something you should not do if you identify as studying or working in transmedia, but more so that it is something no one should be doing. This is because throughout the no-adaptation arguments are two repeated claims: that adaptations are redundant and simple retellings.

Debunking the Redundancy and Simple Retelling Arguments

Jenkins and many others state there is a reason why adaptations should not be used: they are redundant. This has a strategic impact in that "[r]edundancy burns up fan interest and causes franchises to fail" (Jenkins 2006, 96). Unfortunately, despite there being no proof offered that audiences find adaptations redundant, this argument is echoed by many. To refute this claim, consider first the popularity of adaptations. If we look only at the ticket sales of feature films, as they have high cost of production compared to a book or graphic novel, and even many TV shows, this would give us a good idea as to whether an adaptation is of economic value. From 1995 to 2016, the top-grossing movie for each year is from an existing media property (Nash Information Service 2017). Juxtapose this with the stat that between 2005 and 2014, 39 percent of the top movies released were truly original (Follows 2015). So from this we can see that yes, adaptations do make money (people are buying tickets), *and* they are not just popular because there are no other options for consumers. Indeed, over one-third of feature films are truly original.

Interestingly, we have actually seen an increase of "intracompositional," "East Coast," "complex portmanteau transmedia" (Dean 1997; Clark 2011; Pratten 2015 respectively) utilizing existing media properties, as exemplified by the 2012 Emmy-award-winning webisode and social media adaptation of *Pride and Prejudice: The Lizzie Bennet Diaries* (*LBD*). Since *LBD*, the variously labeled "transmedia literary adaptations" (Lockwood 2017, 5) and "social media fictions and distributed adaptations" (Berryman 2017) have seen over 80 productions adapted from public domain novels and plays (Lockwood 2017, 10).

But there is another claim that transmedia adaptations are for new audiences, and so do not qualify as adaptations (there is no redundancy if you are experiencing the work for the first time) (Harvey 2015, 91; Lockwood 2017). The strategy behind this approach would be one of audience acquisition, not retention. So why invoke an existing property if you don't want fans of that property? Internally it saves ideation time, but in terms of audience strategy it is a way to facilitate the decision-making process for consumers who are engaging in "limited problem solving" (LPS) (Howard 1977, 306). A LPS scenario is when the consumer doesn't want to expend too much effort in their decision-making, and so the known brand of an existing property assists the process. In other words, the familiar nature of the literary work being adapted provides consumer appeal. But this still doesn't explain the exclusion of fans.

One way we can test whether existing fans are targeted, whether retention is part of the strategy, is through an analysis of fidelity. If these transmedia literary adaptations were for new audiences, then fidelity would be less of an issue. No one would care if they were the same because no one would know. In his analysis of the adaptation changes in *LBD*, Cameron Cliff notes the reduction of five sisters to three, and what was needed to make the world interactive, but concluded that "the strategy for LBD is undeniably to produce a finite and faithful, if modern, adaptation" (Cliff 2017, 136). Indeed, it would not be prudent to exclude a market, and so instead producers ideally should aim to appeal to existing fans *and* newcomers, audience retention and acquisition. What of the next reason for not including adaptation in transmedia practices: they are simple retellings?

Theorists and practitioners alike have invoked the notion of a "simple retelling" alongside "redundancy" as a way to justify the exclusion of adaptation. As explained previously (Dena 2009), adaptation rarely involves a simple retelling. Adaptation theorist Linda Hutcheon explains how adaptation always involves "interpreting and creating something new," as creators always need to add and subtract (Hutcheon 2006, 20). Indeed, adaptation theorists are scratching their heads wondering why such an argument was put forward in the first place: "the definition of adaptation as 'retelling' seems unnecessarily limiting," especially as "adaptation and television constantly seek to balance out these two elements [familiarity and novelty]" (Wells-Lassagne 2017, 89–90). Further to this, it is exactly what we don't understand about adaptation, "the heresies committed by the practice and criticism of adaptation that we learn most about literature and film in relation to each other" (Elliott 2004, 239). The realization that adaptation is rarely retelling has thankfully been recognized by some transmedia theorists (Jenkins 2011; O'Flynn 2013; Harvey 2015), but still pervades texts and discussions.

Indeed, a key theme of the adaptation-exclusion reasoning is the unfounded claims, inaccurate framing of what adaptation is, and the denigration of adaptation practices. This is so common, adaptation theorist Eckart Voigts calls this the "evaluative" and "not just" rhetoric of transmedia scholars (Voigts 2017). Transmedia storytelling is *not just* adaptation, it is something better. "In the vein of Jenkins, transmedia scholars view traditional adaptations as marked by the problematic processes of cross-platform compromise and the clumsy semiotic or textual rearrangements of texts, whereas transmedia story-worlds create a sustained and intensified experience of fictional worlds" (Voigts 2017). This performance of cultural disdain of adaptation, it should be noted, is in the face of countless economically and critically successful adaptations in all media.

Finally, the framing of adaptation as an unappealing and non-strategic approach is also contrary to what we know about aesthetic preference. Multiple studies have shown us the pleasures and design benefits of experiencing a work again, such as the "mere exposure effect" and how it engenders a familiarity preference (Zajonc 1968, 2001), how "rereading" is an attempt to capture the experience of the first reading (Călinescu 1993), how with "prototypicality" we have a preference for things that are representative of what we usually engage with (Martindale and Moore 1988), the "dependable pleasure" of reruns (Weispfenning 2003), and the appeal of the "cognitive fluency" familiar works afford us (Reber, Schwarz, and Winkielman 2004).

The Effects of Exclusion

The effects of this differentiation via exclusion rhetoric, which often includes derision of what is being excluded, is manifold. The detrimental effects we have seen in games are the culture wars where people assign hierarchical value and identity to different kinds of games, and enact boundary policing as a consequence; that in the face of object or textual traits, the experiential nature of the form is devalued; and practitioners end up biased against techniques that are at times essential for producing good work. Design is contextual. So while it is important to highlight what is unique about a phenomenon, the manner in which one does impacts the cultures being included and excluded. Perhaps it is not necessary to identify a phenomenon through a unique trait? As Upton explains, "interactivity is a thing that games can DO. It is not what games ARE" (Upton 2015b, original emphasis). The expansion of stories across platforms is what transmedia can do, not what it is.

Indeed, as I mentioned earlier, I am not alone in my contention that these approaches do more harm than good. Game theorist Espen Aarseth, who wrote a pivotal book differentiating interactivity at a textual level (Aarseth 1997), has now 20 years later released a special statement in *Game Studies* arguing against this differentiation via exclusion practice:

> In fact, "game" and "play" are not scientific terms, but vernacular words whose meaning changes over time, and is given by ordinary people through their use of language. To try to fix them is a good intellectual exercise, but not something a researcher will ever be able to do. Nor is it vital for the health of the field, but more likely quite counter-productive, if it could be done. Other fields, from literature and media to planetology and even biology, cannot sufficiently define their central objects either, and they are none the worse for it … The move from computer games to games in general is about inclusion. The affected research communities already overlap, to the extent that it makes no sense to formally exclude them with the "computer" or "digital" label, which never made much historical, scientific, or intellectual sense at all.
>
> *(Aarseth 2017)*

It is worth noting that recently too, Jenkins himself recognized the consequences of his differentiation approach:

> This focus on extension and additive comprehension was originally a way for transmedia theorists to criticize the redundancy that marred so many mindless novelizations of existing media franchises in favor of works that explore new creative possibilities. But the result has been to keep adaptation studies and transmedia studies at arm's length for more than a decade.
>
> *(Jenkins 2017)*

More than studies, the exclusionary qualification has kept transmedia practitioners at arm's length from adaptation for over a decade. If Douglas Adams or J. K. Rowling tapped on the door of a transmedia consultant, hoping to find out how to best strategize a move into the multiplatform space, we never would have had the pleasure of watching Arthur Dent deal with the bulldozers; or seeing Potter run at the wall at platform 9 and ¾ for the first time. Adaptation would never have been allowed. This is because a transmedia design strategy, according to accepted wisdom, is only effective if it is different to what creatives have previously done. But this boundary policing and constructed knowledge scarcity is the low-hanging fruit of consultant economics, and the antithesis of good design.

Possible Futures of Differentiation, Theory, and Practice

So what can we do? We do need a way to discuss transmedia as a unique prospect in both scholarly and practitioner settings. This is what drives the adaptation-exclusion argument, but as we have seen that approach is both unfounded and potentially harmful. Aarseth proposes a "perspective approach" where games are not viewed as an object or an activity, but instead looking at whether a phenomenon is "interesting as ludic" (Aarseth 2017). Perhaps there could be a transmedia perspective?

One could still explore traits in this context, but interrogate how they have changed over time and in different contexts. Consider this with the traits Evans has put forward with narrative, authorship, temporality focus (Evans 2011); and Matt Freeman's diachronic analysis of character-building, world-building, and authorship (Freeman 2017). At present, they exclude adaptation though, but as Jenkins now concedes, "those of us who study transmedia (and fan fiction) and those who study adaptation are asking a related set of questions" (Jenkins 2017). Indeed, as adaptation theorist Wells-Lassagne muses, adaptation can assist with transmedia's future:

> [t]ransmedia ensures the coherency of the narrative, which continues outside the source, filling in the gaps of the source text. Ironically, then, for a form whose very nature as adaptation is in question, the transmedia text, in particular, is plagued by the *běte noire* of adaptation, fidelity.
>
> *(Wells-Lassagne 2017, 92)*

An interesting work by Marta Boni puts aside allegiances to field definitions, and instead looks at many theories to find which best explains different aspects of the same project (Boni 2013). Harvey does an admirable job of attending to different adaptation theories, to "help further illuminate the question of where transmedia storytelling ends and adaptation begins" (Harvey 2015). But, while a fruitful investigation, this is a less interesting and less helpful task compared to the potentially illuminating task of establishing what degrees of familiarity versus novelty operate best for audiences in what contexts.

Industry practitioner Robert Pratten has also championed the experiential aspect of transmedia as being critical to practice: "The problem with the traditional definition is … it describes the production and not the consumption" (Pratten 2015, 2). Pratten continues, "transmedia storytelling is a design philosophy" (2015, 2). The most promising area I have seen that does away with boundary work is that of looking at transmedia through the lens of social science and aesthetic preference, bridging design and audience. Susana Tosca calls it "transmedial desire," and describes it as "a pulsion common to enthusiastic audiences that are driven to seek more contact with the fictions they love" (Tosca 2016). These include the desire for "experiencing more" (which correlates with transmedia extensions and new insights), "inhabiting our favourite worlds" (which correlates with immersion), and "transforming the transmedial worlds ourselves" (which correlates with participatory practices) (Tosca 2016).

I too have made the switch to thinking about design through audience behavior, and have named them the "revealing," "inhabiting," "molding" and of course "reliving" drives (Dena 2017b, 2017c). As we have noted, audiences do find adaptations appealing, and part of that attraction is being able to relive the experience. If transmedia creators do not understand that there are many reasons why audiences cross media, then how will they truly understand why extensions work? Indeed, for the past two years I have been teaching these drives to students and industry alike. We discuss when to use what strategy, and I highlight how the revealing drive (extensions) has been underutilized. Here is just one example of an approach that helps with our shared goal of distinguishing the phenomenon while at the same time acknowledging the realities of practice.

Conclusion

I began this chapter referring to the not coincidental time Jenkins put forward his no-adaptation rule after coming from a frustrating Hollywood and game industry event. The desire to have the phenomenon properly recognized, especially when surrounded by those who refuse to see or understand it, is something we all feel. It makes sense to double-down and create a simple criterion of exclusion, so people understand and recognize what you see.

But we do so at the cost of understanding the complete picture. Irrespective of the function of the exclusion, the stated reasons—redundancy and simple retelling—are shown to be false not just because of the technical inaccuracy, or even due to the popularity of adaptation, but also due to the pleasures aesthetic preferences studies have shown us.

There are costs to differentiating transmedia via the exclusion of adaptation, as witnessed in games and in transmedia: assigning hierarchical value and identity to artforms and the resultant industrial impact of this; thwarting understanding of the experiential nature of the phenomenon; and being oblivious to the full range of strategic and design decisions available to creatives. Focusing on a particular aspect of practice, in this case extensions, is a worthy and needed field of inquiry. It may be that transmedia can continue to be the name of that particular focus. But to claim it encapsulates what creatives do, and is the only valid design choice for multiplatform-thinking, is a misrepresentation of best practice.

References

Aarseth, Espen. 1997. *Cybertext: Perspectives on Ergodic Literature.* Baltimore: Johns Hopkins University Press.

Aarseth, Espen. 2017. "Just Games." *Game Studies* 17 (1). Accessed July 31, 2017. http://gamestudies.org/1701/articles/justgames.

Bernardo, Nuno. 2014. *Transmedia 2.0: How to Create an Entertainment Brand Using a Transmedial Approach to Storytelling.* Lisbon: Beactive Books.

Berryman, Rachel. 2017. "Fictionalising Re(a)lationality: The Social Media Storyworld of *Nothing Much to Do*." Master's thesis, University of Auckland.

Boni, Marta. 2013. "Romanzo Criminale: Transmedia and Beyond." *Innesti/Crossroads 1.* Venice: Ca' Foscari.

Bourdaa, Mélanie. 2013. "Following the Pattern: The Creation of an Encyclopaedic Universe with Transmedia Storytelling." *Adaptation* 6 (2): 202–214.

Buchanan, Richard. 1998. "Branzi's Dilemma: Design in Contemporary Culture." *Design Issues* 14 (1): 3–20.

Buchanan, Richard. 2001. "Design Research and the New Learning." *Design Issues* 17 (4): 3–23.

Călinescu, Matei. 1993. *Rereading.* New Haven: Yale University Press.

Chi, Michelene, Robert Glaser, and Ernest Rees. 1982. "Expertise in Problem Solving." In *Advances in the Psychology of Human Intelligence*, edited by Robert Stemberg, 7–75. Hillsdale: Erlbaum.

Clark, Brian. 2011. "Reclaiming Transmedia Storyteller." *Facebook*, May 2. Accessed February 10, 2017. www.facebook.com/notes/brian-clark/reclaiming-transmedia-storyteller/10150246236508993/.

Cliff, Cameron. 2017. "Transmedia Storytelling Strategy: How and Why Producers Use Transmedia Storytelling for Competitive Advantage." Ph.D. diss., Queensland University of Technology.

Dean, Russell T. 1997. *Improvisation, Hypermedia, and the Arts since 1945.* Amsterdam: Harwood Academic Publishers.

Dena, Christy. 2009. "Transmedia Practice: Theorising the Practice of Expressing a Fictional World across Distinct Media and Environments." Ph.D. diss., University of Sydney.

Dena, Christy. 2017a. "Finding a Way: Techniques to Avoid Schema Tension in Narrative Design." *Transactions of the Digital Games Research Association (ToDiGRA)* 3 (1): 27–61. doi: 10.26503/todigra.v3i1.63.

Dena, Christy. 2017b. "A Narrative Designer's Experience of Amazon's 'The Man in the High Castle' Resistance Radio Campaign." *Medium*, March 10 (last modified March 11). Accessed February 12, 2017. https://medium.com/@christydena/a-narrative-designers-experience-of-amazon-s-the-man-in-the-high-castle-resistance-radio-72100489552d.

Dena, Christy. 2017c. "Comparing World Recentering Practices in the USA and Australia." Paper presented at the Australian and New Zealand Communication Association Conference (ANZCA), University of Sydney, July 4–7.

Dowd, Tom. 2015. *Transmedia: One Story, Many Media.* Boca Raton: CRC Press.

Dowd, Tom, Michael Niederman, Michael Fry, and Josef Steiff. 2013. *Storytelling Across Worlds: Transmedia for Creatives and Producers.* Burlington: Focal Press.

Elliott, Kamilla. 2004. "Literary Film Adaptation and the Form/Content Dilemma." In *Narrative Across Media: The Languages of Storytelling*, edited by Marie-Laure Ryan, 220–243. Lincoln: University of Nebraska Press.

Evans, Elizabeth. 2011. *Transmedia Television: Audiences, New Media, and Daily Life.* New York: Routledge.

Follows, Stephen. 2015. "How Original are Hollywood Movies?" *Stephen Follows*, June 8. Accessed July 19, 2017. https://stephenfollows.com/how-original-are-hollywood-movies/.

Freeman, Matthew. 2017. *Historicising Transmedia Storytelling: Early Twentieth-Century Transmedia Story Worlds.* London and New York: Routledge.

Golding, Daniel. 2013. "Moving Through Time and Space: A Genealogy of Videogame Space." Ph.D. diss., University of Melbourne.

Gomez, Jeff. 2010. "PGA's Transmedia Producer!" *Christy's Corner of the Universe,* April 6. Accessed January 21, 2017. www.christydena.com/2010/04/pgas-transmedia-producer/#comment-1813.

Harvey, Colin. 2015. *Fantastic Transmedia: Narrative, Play and Memory Across Science Fiction and Fantasy Storyworlds.* New York: Palgrave Macmillan.

Howard, Houston. 2013. *Make Your Story Really Stinkin' Big: How to Go From Concept to Franchise and Make Your Story Last for Generations.* Studio City: Michael Wiese Productions.

Howard, Houston. 2017a. "Biggest Challenge Storytellers Face Today" *YouTube,* July 11. Accessed February 22, 2017. www.youtube.com/watch?v=JWigfZVCqoU.

Howard, Houston. 2017b. *You're Gonna Need a Bigger Story: The 21st Century Survival Guide to Not Just Writing Stories but Building Super Stories.* Middletown: One 3 Creative.

Howard, John. 1977. *Consumer Behavior: Application of Theory.* New York: McGraw-Hill.

Hutcheon, Linda. 2006. *A Theory of Adaptation.* New York: Routledge.

Jenkins, Henry. 2003. "Transmedia Storytelling". *MIT Technology Review,* January 15. Accessed June 4, 2017. www.technologyreview.com/s/401760/transmedia-storytelling/.

Jenkins, Henry. 2006. *Convergence Culture: Where Old and New Media Collide.* New York: New York University Press.

Jenkins, Henry. 2009. "The Revenge of the Origami Unicorn: Seven Principles of Transmedia Storytelling." *Confessions of an Aca-Fan,* December 12. Accessed February 20, 2017. http://henryjenkins.org/blog/2009/12/the_revenge_of_the_origami_uni.html.

Jenkins, Henry. 2011. "Transmedia: Further Reflections." *Confessions of an Aca-Fan,* July 31. Accessed February 11, 2017. http://henryjenkins.org/blog/2011/08/defining_transmedia_further_re.html.

Jenkins,Henry.2016."HowaTransmediaStrategyEnrichesStory."*Coursera*AccessedJune10,2017.www.coursera.org/learn/transmedia-storytelling/lecture/oJHbb/henry-jenkins-how-a-transmedia-strategy-enriches-a-story.

Jenkins, Henry. 2017. "Adaptation, Extension, Transmedia." *Literature/Film Quarterly* 45: 02. Accessed July 31, 2017. www.salisbury.edu/lfq/_issues/first/adaptation_extension_transmedia.html.

Larkin, Jill, John McDermott, Dorothea Simon, and Herbert Simon. 1980. "Models of Competence in Solving Physics Problems." *Cognitive Science* 4: 317–348.

Lidwell, William, Kritina Holden, and Jill Butler. 2003. *Universal Principles of Design.* Gloucester: Rockport Publishers.

Lockwood, Elise. 2017. "Peter Pan has a Blog and Jane Eyre has Twitter: Applying Adaptation Theory to Transmedia Adaptations of Classic Literature." Master's thesis, Ball State University.

Long, Geoffrey. 2007. "Transmedia Storytelling: Business, Aesthetics and Production at the Jim Henson Company." Master's thesis, Massachusetts Institute of Technology.

Lotman, Yuri 1990. *Universe of the Mind: A Semiotic Theory of Culture.* Bloomington: Indiana University Press.

Maras, Steven. 2009. *Screenwriting: History, Theory and Practice.* London: Wallflower Press.

Martindale, Colin, and Kathleen Moore. 1988. "Priming, Prototypicality, and Preference." *Journal of Experimental Psychology: Human Perception and Performance* 14 (4): 661–670.

Nash Information Service. 2017. "Domestic Movie Theatrical Market Summary 1995 to 2017." *The Numbers.* Accessed July 4, 2017. www.the-numbers.com/market/.

O'Flynn, Siobhan. 2013. "Epilogue." In *A Theory of Adaptation,* 2nd ed., edited by Linda Hutcheon, 179–206. Oxon: Routledge.

Phillips, Andrea. 2012. *A Creator's Guide to Transmedia Storytelling: How to Captivate and Engage Audiences Across Multiple Platforms.* New York: McGraw-Hill.

Pratten, Robert. 2015. *Getting Started with Transmedia Storytelling: A Practical Guide for Beginners, Second Edition.* London: CreateSpace.

Producers Guild of America. 2010. "Code of Credits: New Media." *Producers Guild of America.* Accessed February 5, 2017. www.producersguild.org/?page=coc_nm#transmedia.

Reber, Rolf, Norbert Schwarz, and Piotr Winkielman. 2004. "Processing Fluency and Aesthetic Pleasure: Is Beauty in the Perceiver's Processing Experience?" *Personality and Social Psychology Review* 8 (4): 364–382.

Ruppel, Marc. 2012. "Visualizing Transmedia Networks: Links, Paths and Peripheries." Ph.D. diss., University of Maryland.

Scolari, Carlos Alberto. 2009. "Transmedia Storytelling: Implicit Consumers, Narrative Worlds and Branding in Contemporary Media Production." *International Journal of Communication* 3: 586–606.

Smith, Aaron. 2009. "Transmedia Storytelling in Television 2.0: Strategies for Developing Television Narratives Across Media Platforms." Master's thesis, Middlebury College.

Starlight Runner. 2017. "Transmedia Service." www.starlightrunner.com/transmedia.

Tosca, Susana. 2015. "We Have Always Wanted More." *International Journal of Transmedia Literacy* 1 (1): 35–43. doi: 10.7358/ijtl-2015-001-tosc.

Upton, Brian. 2015a. *The Aesthetic of Play*. Cambridge, MA: MIT Press.

Upton, Brian. 2015b. "The Play of Stillness: Designing Experiential Play Spaces." Paper presented at *Game Developers Conference*, San Francisco, March 2–6.

Voigts, Eckart. 2017. "Memes and Recombinant Appropriation: Remix, Mashup, Parody." In *The Oxford Handbook of Adaptation Studies*, edited by Thomas Leitch, 285–304. New York: Oxford University Press. doi:10.1093/oxfordhb/9780199331000.013.16.

Weispfenning, John. 2003. "Cultural Functions of Reruns: Time, Memory, and Television." *Journal of Communication* 53 (1): 165–177.

Wells-Lassagne, Shannon. 2017. *Television and Serial Adaptation*. New York: Routledge.

Zajonc, Robert Boleslaw. 1968. "Attitudinal Effects of Mere Exposure." *Journal Personality and Social Psychology Monograph Supplement* 9: 1–27.

Zajonc, Robert Boleslaw. 2001. "Mere Exposure: A Gateway to the Subliminal." *Current Directions in Psychological Science* 10: 224–228.

Zeiser, Anne. 2015. *Transmedia Marketing: From Film and TV to Games and Digital Media*. Burlington: Focal Press.

22

TRANSMEDIA DEVELOPER

Success at Multiplatform Narrative Requires a Journey to the Heart of Story

Jeff Gomez

There is the initial impression by outsiders, novices, and many of potential corporate clients that the job of a transmedia storyteller is to somehow break apart a sprawling, epic story into smaller pieces, and assign those pieces to various media platforms. The most important storyline? The movie. The big battles, where most of the bad guys can be mowed down with big guns? The video game. The melodramatic prequel? The comic book, of course.

But we contend that this level of understanding is the equivalent of how a writer gets the differences between novels, short stories, and, well, comic books. It is one thing to know the broad strokes of how story operates on different media, but it is quite another to be able to craft multiple streams of narrative so that they each remain true expressions of the core creation (the story world as envisioned by its creator), and yet act in concert, weaving a tapestry of story that surrounds, immerses, and interacts with the audience. Such is the job of the transmedia developer.

This chapter examines the New York production company Starlight Runner Entertainment's evolution from experimental transmedia storytellers to professional transmedia developers, focusing on the critical case study of their experience with the Walt Disney Company's *Pirates of the Caribbean*. It will reveal aspects of the company's proprietary process that are rarely described, but have proven to be highly effective over the years. They are the tools that any developer must comprehend and employ during the earliest possible phases of the transmedia production.

The Challenge

Disney's marketing division was faced with an unusual predicament: success. The feature film *Pirates of the Caribbean* (2003), based on the theme park attraction of the same name, was an unexpected blockbuster hit. But with no division of Disney prepared for the box office windfall, there had been no products developed in-house and no lucrative third-party licensing deals to capitalize on the property.

In 2005, with two more films now in pre-production, Disney's new Chairman and CEO Bob Iger expressed a strong desire for the company to avoid making this mistake a second time. The responsibility fell on Disney Chief Creative Officer Oren Aviv, and Christine Cadena, Senior Vice President Marketing Synergy & Franchise Development to assign and oversee development of *Pirates of the Caribbean* content across nearly every division of the company.

This content, which would include an array of licensed and merchandised products, would become available in tandem with the release of *Pirates of the Caribbean: Dead Man's Chest* (2006) and *Pirates of the Caribbean: At World's End* (2007).

On assessment, however, key Disney executives realized that, although there were now a number of *Pirates*-related projects in early development, the level of quality, consistency, and content of the products in development were problematic. Challenges included the following:

- Ambiguity over the films' darker and more horrific aspects, and how these gelled with the family-friendly Disney brand.
- Existing toys and souvenirs that did not reflect the level of quality, depth, and detail of their counterparts in the films.
- Story content that did not convey the characters or storyworld accurately, or placed them wildly out of context—they did not *feel* like *Pirates of the Caribbean*.
- The addition of myriad fantastical elements that diminished the property with a "kitchen sink fantasy" (anything goes) approach.
- Jack Sparrow depicted in a wide variety of ways, at times having him exhibit overtly villainous behavior.
- Key art and imagery that reflected the old theme park ride and not the characters and situations depicted in the films.

At the time, Senior Vice President Marketing & Business Development Gordon Ho was visiting with Disney Publishing in search of solutions on behalf of Aviv. There he learned about the work that Starlight Runner Entertainment had done for Mattel and the Hot Wheels brand. In 2002, Mattel had commissioned CEO Jeff Gomez and his company Starlight Runner to develop a storyworld based on the die-cast metal cars.

Instead of a simple style guide, Gomez had created a "trans-media [*sic*] bible" fully describing every element of a new Hot Wheels "mythology," covering everything from driver character, to super-powered cars, to the various science fictional aspects of a race through multiple dimensions. This single Mythology document would then be used by every division of Mattel to create packaging, comic books, video games, promotional tie-ins, an elaborate website, and a computer-animated television series.

With Cadena, Ho approached Starlight Runner with the idea of creating a *Pirates Mythology*, a guidebook that could do the same for Disney's *Pirates of the Caribbean*.

Linking the Story to the Storyteller

After securing the assignment, Starlight Runner quickly understood the challenge being faced by Disney. By its nature, *Pirates of the Caribbean* was a dense, peculiar, convoluted storyworld, with an almost dreamlike sense of logic. On close analysis of the original film and the upcoming scripts, Gomez's team understood that there was a strange rhyme and reason to the world, but could not immediately grasp how those mechanisms worked, nor how to communicate effective storytelling to others based on these mechanics in order for them to generate transmedia content. For the first time, Starlight Runner would have to request access to some of a franchise's key creatives to clarify its unique perspective.

Here, the team learned their first key lesson, which is applicable to any transmedia developer: if the goal is to create quality content outside of feature films (the franchise's driving platform), it would be risky or even reckless to guess at what made an established storyworld tick.

Recognizing the complexity of the challenge, Aviv and Cadena arranged a series of interviews with the filmmakers and Disney stakeholders—an unusual level of access at the time. But their perspectives were instantly helpful.

One recurring stumbling block encountered by the team was the unusual tone of the films. Although a Disney-branded film, *Pirates of the Caribbean* was scary. The films were full of frightening supernatural beings, edgier violence, and an atmosphere of menace. Some in the company called the films "Disney Dark," meaning that the tone fell on the higher end of the age-range for the studio's family fare. Others wondered about whether the films danced over that edge, making it difficult to gauge what kind of ancillary content and licensed product ought to be derived from the films in order to grow the franchise.

It became important for Disney Marketing and Starlight Runner to understand what linked the films to the Disney brand; why *Pirates* was distinctly Disney. Sifting through dozens of pages of interview notes, Gomez stumbled upon a quote attributed to Walt Disney himself: "Doing the impossible is kind of fun."

The quote was meant by the interviewee as a descriptor of the essence of all great stories to come out of the studio, but Gomez saw it as the essence of the Disney company itself. His team also saw it as a descriptor of protagonist Captain Jack Sparrow, and the *Pirates* films themselves. The term pegged the fact that the films behaved like thrilling theme park dark rides—scary enough to be unsettling at times, but also full of humor, action, and near-miraculous outcomes.

The missing link between story and storyteller had been found; its source, the entire company's founding storyteller. This was instantly approved by all involved. It was a small but vital foundation upon which to build the project and surmount the challenge.

Deriving Storyworld Essence

With ubiquitous broadband connectivity still years away, it was decided that Starlight Runner would essentially tour the world in order to examine all of the marketing, in-house content and consumer products, theme park elements, and licensed products, wherever they were in development or production. It became clear that *Pirates of the Caribbean* was subject to a vast range of interpretations, much of which widely differed from the films.

Some saw Jack Sparrow as an historically accurate pirate: a vicious, violent brigand who might just as soon let an innocent die if it will spare his own skin. Others set the characters to flights of fancy, allowing them to travel through time, ride winged horses, and fire laser guns. Still others avoided the characters and events of the films altogether, choosing to focus on the broader and more cartoony characters and events featured in the Disney theme park ride.

The team intuited that fans yearned for narrative and visual continuity. They wanted characters to speak and behave consistently. Differing depictions of a property like *Pirates* could create confusion, making the characters harder to recognize, ultimately splintering or fracturing a brand that was still in the process of establishing itself.

Here, the team learned a second key lesson that is applicable to any transmedia developer: rich transmedia story worlds require a core set of narrative tenets, a system to which various creative stakeholders can adhere in order to generate consistent and persistent content.

The challenge here is that Starlight Runner could not be so bold as to assert an opinion on the executives of the Disney company and the producers of the films. Instead, the *Pirates* narrative, particularly as depicted in the film trilogy, would have to be deconstructed. Essential narrative components would have to be accurately determined and extracted. Starlight Runner calls this the "narrative essence" or "brand essence" of the storyworld.

In recent years, various marketing, advertising, and transmedia agencies each have their own definition of the essential factors that make a brand distinct. Gomez chose to adhere to purely narratological elements, in keeping with the assignment, parsing the films for their archetypes, thematic messaging, mythic and cultural resonances (how the films reflect contemporary society at large), and through lines of wish fulfillment (what Starlight Runner calls "aspirational drivers").

Focus of attention would obviously be the character of Jack Sparrow, who was so frequently misunderstood by stakeholders. The Starlight Runner team determined his primary archetype to be the Trickster, backing their findings with research that included Joseph Campbell's (1949, 1990; Campbell and Moyers 1991) seminal writings on this mythic persona.

But while the Trickster provided clues that could help creatives better determine what Jack might and might not do, the team could not yet pin down how Jack Sparrow resonated with the entire universe of *Pirates of the Caribbean*, nor could they connect how the various *Pirates* narratives (particularly those in ancillary content, which may not feature the character) could be linked thematically with Captain Jack.

In short, a "unified theory" of the *Pirates* universe was necessary in order for narrative content to genuinely satisfy the audience by "feeling like" a *Pirates* story. It was not enough for Starlight Runner as transmedia developers to tell the client "that content does not come across as authentically *Pirates*." What was necessary was to provide the client with a single central message that was indeed quintessentially *Pirates*, and then a set of guidelines to help realize that message in narrative form.

The Primal Message

Although the Starlight Runner team had successfully untangled the myriad plot threads, backstories, and mythos of the three *Pirates of the Caribbean* films (the latter two based on shooting scripts and early footage alone), and they had discerned large portions of the story world essence, they had yet to fully "crack the franchise," as they put it.

In his transmedia master class, Jeff Gomez frequently discusses the fact that the most successful storyworlds—fictional universes that have stood the test of time and incarnated themselves across multiple media platforms—most often contain a singular and unique aspirational message. This is "a piece of advice," he says, "that would improve any individual's life, and if adopted by all of us, would make the world a better place in which to live." Gomez calls this the central theme of the storyworld. Its primal message.

But what unique advice does cynical, conflicted, romantic *Pirates of the Caribbean* have to offer? To answer that at first, the Starlight Runner team researched the actual Pirate Code, often referred to in the films as the Code of the Pirate Brethren. Although this research paid off, allowing Starlight Runner to furnish Disney with an actual historically based code, its transactional nature, cynicism, and simplicity was of little use in terms of the primal message.

Eventually, the team found themselves back at the booted feet of the Jack Sparrow character. If Jack's true motivation could be discerned—the events that shaped him, the motivations that drive him, the underpinnings of the humanity that had so charmed audiences, despite his detached and selfish behavior—perhaps an answer could be found.

But neither the films, stakeholder interviews, nor any supplementary materials were revealing. Gomez finally asked to speak with Sparrow actor Johnny Depp himself, a request that initially met with understandable resistance. After all, Depp was in the middle of a massive film shoot.

After days of wrangling, permission was finally granted, with the caveat that Gomez would have no more than 15 minutes with the actor. As a rule, transmedia developers who must work with visionaries and highly placed stakeholders whose time is precious, need to carefully prepare in order to maximize the effectiveness of brief interviews.

Once with Depp, Gomez explained that the path to the heart of all of *Pirates of the Caribbean* led to a deep and hidden conflict within Jack Sparrow, one that seemed to manifest itself in the unusual choices he made as an actor. "From a character perspective, can you give me any insight into the nature of those choices?"

Depp's response was surprisingly personal. After the birth of his daughter, Depp was determined not to pass down to his own child the anger and negativity that had haunted him since he was a boy. Learning yoga, meditation, and other techniques, he purged himself, reaching a state of serenity. But then, during his tense first few days on the set of *Pirates*, he had difficulty "finding" the character.

One evening, in crisis, Depp gazed into a mirror and asked himself, "How can I balance what is noble in myself with the wildness I need to bring my characters to life?" Depp held up two hands, balancing them like scales, spreading them wide, as if drunkenly walking a tightrope. This was all the Starlight Runner team needed: *To achieve a balance within ourselves between nobility and savagery is to reach a state of grace.*

This primal message worked for Jack, but remarkably it also worked for the entire franchise. Born a poet in the regimented United Kingdom, all Jack Sparrow yearned for was freedom. He found it in the Seven Seas, and survived his journeys in the guise of a ferocious pirate. He would tread a line between the higher aspirations of civilized humanity and the feral lawlessness of the untamed Caribbean, thumbing his nose at both sides, Trickster that he is.

At a more cosmic level, this is also the story of the three films: the disciplined, regimented forces of Western Europe sought conquest over the chaotic wilderness of the Caribbean. In their clash, supernatural forces are unleashed, threatening the world. And there is Captain Jack, sailing between them, mocking them, the results of his actions somehow enabling our heroes (Elizabeth and Will Turner) to set things right.

Here, the team learned a third key lesson that is applicable to any transmedia developer: the answers to the most trenchant mysteries of a multiplatform story world may not be easily recognized in the content, but can almost certainly be discerned in the storyteller.

While great storytellers are under no obligation to spell out the messages and meanings of their narratives to audiences, leaving them the pleasure of making their own interpretations, transmedia developers cannot afford such luxuries. They have to be right, because the correct meaning—primal messages, brand essence—must be infused into all extensions of the story world if the franchise is to feel authentic and continue to achieve success. Fail to do this, and it won't be long before the franchise loses its way, withers, and dies.

The Guidebook

Surveying the wealth of materials assembled by Starlight Runner, Christine Cadena approved the final assembly of the Pirates Mythology, a franchise story bible designed to exhaustively survey the universe of *Pirates of the Caribbean*. The document would act as a resource for any and all stakeholders, to be parsed out in smaller sections to some, or used in its entirety by others.

The Mythology differed from standard Disney style guides in that its focus was on the details of the story world, as opposed to the image and design aesthetics of the property. It differed from the standard Hollywood show bible in terms of how deeply it explored every facet of the world, the entirety of the known history of the *Pirates* universe, and the story world essence.

Further, the Mythology leveraged Starlight Runner's unique capability to deconstruct and explain the cosmology of fantasy worlds: how myth, magic, and morality applied to *Pirates of the Caribbean*; the nature of the greater world beyond the series' Caribbean setting; the origins of various iconic characters and situations many of which were only obliquely alluded to in the films.

Initially, however, Starlight Runner received some pushback by a few stakeholders over the document's inclusion of "abstract," "metaphysical," or "academic" passages. They were perceived as dense or difficult to understand, and perhaps unnecessary. Some were concerned that this would paint future screenwriters and filmmakers into a corner, limiting possibilities or curtailing imagination in future installments.

Here, the team learned a fourth key lesson that is applicable to any transmedia developer: any story world worthy of standing the test of time, no matter how fantastical, deserves to be taken seriously as a work of depth and integrity.

Gomez explained to Disney that there are many stakeholders who need to understand how a universe as seemingly inscrutable as *Pirates* actually works. If a line of novels, for example, were to be true to the films, guidelines not just about what Jack Sparrow might and might not do, but also about how the franchise universe works, were required. A high-end video game series could generate dozens of hours of content, much of which would be reliant on this kind of information. Cadena ultimately supported the argument, and the sections remained intact.

Cadena also favored the term *distant mountains*, which Gomez borrowed from J. R. R. Tolkien, author of *The Lord of the Rings* (1954). Starlight Runner had requested to include in the Mythology a lengthy list of interesting but unexplored story points suggested by the films. The team cited Tolkien's tendency to seed his Middle-earth stories with bits of lore that both made his creation feel more real, and inspired curiosity and excitement from the reader. Distant Mountains sections would become a staple in Starlight Runner franchise story bibles.

Establishing the Clearinghouse

Transmedia development becomes transmedia production at the point at which stakeholders from across divisions and across media channels are provided storyworld guidelines and green-lighted to generate content. In an organization as vast as Disney, there was no previously established process by which to do this easily. A solution was urgently necessary.

Historically, like most major Hollywood studios, the Walt Disney Company leveraged its narrative properties in either linear serial fashion (sequels), or by replicating the narrative across different media. The story of *The Lion King* (1994) was established in a feature film, and would be adapted as story books, comics, video games, theme park attractions, and a Broadway play with each of its major narrative beats replicated intact. The success of the film spawned a series of direct to video sequels. However, there had been no concerted effort to expand and leverage the story world of *Lion King* with different stories across different media. It would be different for *Pirates*.

With *Dead Man's Chest* and *At World's End* in full production in the fall of 2005, Cadena was facing the challenge of communicating and coordinating logistics across the siloed divisions of the company. Starlight Runner suggested that the conferences could also function as a kind of franchise clearinghouse for story, used to encourage executives to take inspiration from the Pirates Mythology and push for unique, "additive" content that, over time, could eventually shift emphasis from the films' movie star actors (transient) to the films' sprawling mythos (timeless).

Cadena's Pirates Task Force took flight, functioning as the heart of the franchise, pumping the Pirates Mythology lifeblood across the company and out to licensees. The clearinghouse also facilitated a streamlined system of approvals, expediting a complex process that involved attorneys, agents, actors, and various levels of executives. This would address some of the hesitation expressed by content creators to include character likenesses or direct references to the films in their storylines.

Here, the team learned the fifth and final key lesson that is applicable to any transmedia developer: buy-in is cultivated and reinforced through the validation of all stakeholders and their participation in the development process.

While Starlight Runner did not necessarily involve video game licensees in the transmedia development process of the Pirates Mythology, the Mythology did address concerns raised by the interviewed producers of video game and interactive content. They saw their concerns reflected in the document—their language incorporated in Starlight Runner's presentations—and became convinced that the transmedia approach would facilitate and enhance their work.

Starlight Runner found that many stakeholders, particularly marketers and creatives, appreciated the novel fact that they were "helping to tell the story"—in other words, contributing to the official canon of the *Pirates* universe with content that "mattered." The *Young Jack Sparrow* novels from Scholastic Books did not simply convey whimsy about these characters. Instead, they told stories that contributed to the formative years of the character, and did so in a way that was accessible to an audience that might not even be old enough to see the movies.

Conclusion

The transmedia development work done for *Pirates of the Caribbean* resulted in an immediate elevation of quality, consistency, and quantity of franchise extensions. Adherence to the tenets of the Pirates Mythology allowed for Disney to juvenilize the property (children's comics, the aforementioned *Young Jack Sparrow* chapter books) without alienating older fans. Interactive projects course-corrected and hewed closer to the look and feel of the storyworld. Toys and action figures were accompanied by bits of lore, giving them a more authentic feel. The theme park ride itself was updated based on inspiration from this work, integrating characters and events from the films and distant mountains to bolster its narrative.

Disney was satisfied with these results, retaining Starlight Runner to do similar work for such projects as *Disney Fairies* and *Tron Legacy*. Under Bob Iger, the transmedia development approach became infused into the company's DNA, generating homegrown properties out of such unexpected platforms as the Disney Channel (*High School Musical*) and ABC television (*Lost*).

Perhaps more importantly, these techniques laid the groundwork for Disney's handling of the Marvel Cinematic Universe and *Star Wars*, two of the most successful transmedia franchises in history. While visionaries such as Kathleen Kennedy (Lucasfilm) and Kevin Feige (Marvel) bought into the transmedia story world concept early on, their efforts would not have been nearly as successful if these processes were unfamiliar to Disney's corporate hub. The narrative-based, licensing and merchandising extensions of these properties are handled by Disney marketing.

In summary, these are the five key lessons Starlight Runner took from its seminal work with Disney as transmedia developers:

1. Remove as much guesswork from your assessment of the source narrative, and instead use narratological analysis to understand the storyworld without personal bias.
2. Transmedia developers must elicit a core set of narrative tenets from the storyworld, a system to which various creative stakeholders can adhere in order to generate consistent and persistent content.
3. The answers to the most trenchant mysteries of a multiplatform storyworld may not be easily recognized in the content, but can almost certainly be discerned in the storyteller.
4. Any storyworld worthy of standing the test of time, no matter how fantastical, deserves to be taken seriously as a work of depth and integrity.
5. Buy-in is cultivated and reinforced through the validation of all stakeholders and their participation in the development process.

These lessons have been applied to Starlight Runner's work on some of the world's most popular entertainment properties (such as *Transformers*, *Men in Black*), but they also have been used on the team's corporate and brand work (Coca-Cola, Pepperidge Farm), on its work in documentary (*In My Lifetime: The Nuclear World*) and philanthropy (World Vision), and on its geopolitical and crisis intervention projects.

References

Campbell, Joseph. 1949. *The Hero with a Thousand Faces*. New York: Bollingen Foundation.
Campbell, Joseph. 1990. *The Hero's Journey: Joseph Campbell on his Life and Work*. Novato: New World Library.
Campbell, Joseph, and Bill Moyers. 1991. *The Power of Myth*. New York: Bantam Doubleday Dell Publishing Group.
Tolkien, John Ronald Reuel. 1954. *The Lord of the Rings*. London: Allen & Unwin.

23
TRANSMEDIA PRODUCTION
Embracing Change

Robert Pratten

It is an uncommonly sunny day in London and my clients are still sleeping in California. The time difference works to our advantage because we have the whole day to work toward the weekly 5pm meeting. My team and I are working on many projects for various clients but, in these pages, I am going to compare and contrast two active projects: a personal finance experience for VISA and a location-based game for Kodansha, one of the world's largest manga publishers.

About the Team

There are seven people in our company—three work in the "back office," developing the core transmedia engine (Conducttr), and four of us work in the "front office" with clients. My role is typically listed in the proposal as Quality Assurance, but I take an active involvement in the project acting as collaborator, sounding board, connector, risk assessor, and commercial coach. Everybody in our team has multiple roles and multiple skills that combine arts and technology. Everybody uses the Conducttr (n.d.) software and that has been a key recruitment requirement since we have been in operation.

On the projects, the core roles are:

- Creative technologist: responsible for front-end web development and apps.
- Author: an all-encompassing title that means writing the story and developing the game. We do not distinguish between experience development and story writing—they are two sides of the same coin.
- Project manager: the person watching the deadlines and the deliverables.
- Quality assurance: the person that acts as eyes, ears, heart, and soul for both client and the audience.

Video production and graphics work is outsourced to other companies, including on occasion to the client. Our company mission is make everyone's life an adventure. We do this through our cloud-based transmedia story-game engine, Conducttr, and that is where we focus our resources.

About the VISA Project

The goal of the VISA project is to help 13–18-year-olds become more financially astute. VISA does a lot of work with communities through its Practical Money Skills initiative and we were commissioned to develop a scenario-based training experience playable online and in classrooms. This is an international project that will play around the world directly from VISA's website and also re-badged by VISA clients (potentially playable from the client's website).

The project also needs to play in locations without Internet access—a problem for web-based technology. Our solution for this is to allow the project to run on our product Conducttr Local—a high-powered laptop with its own private Internet router. Being able to run on the public Internet and on a private network creates its own problems because URLs (Uniform Resource Locator) to images, videos, audio and such like need to be easily switched depending on where they are hosted.

The narrative of the experience is that you play as a wannabe video blogger and must decide which of two media contracts to sign. To make an informed decision, players need to get up to speed with practical financial matters like budgeting, saving, investing, and insuring.

About the Kodansha Project

The goal of the project is to connect with fans of the popular manga and anime series Attack on Titan (AoT) and give away a digital code to download a free comic on Comixology. The narrative of the experience is that the convention center for Anime Expo, LA is "inside the wall" and Little Tokyo is "outside the wall." Consistent with the AoT storyworld, inside the wall is positioned as a training exercise requiring players to find five missing soldiers while outside the wall is a full-on mission to defeat increasing stronger Titans.

The outside quest uses a Web application built on our TeamXp application framework (with back-end powered by the Conducttr real-world gaming engine) and requires the player to be in LA and to enable location services on their browser. The inside quest is a more familiar SMS-based (short message service) scavenger hunt. Part of the reason for the switch of channels between quests is expectations of poor WiFi and Internet reception that is typical of most conference venues. While the inside quest requires players to follow a pre-defined path (as discovering each checkpoint reveals a clue to finding the next), the outside quest allows a more free-flowing route with players able to visit Titan hotspots in the order of their choosing.

Project Organization

Internally, we use Slack (a cloud-based set of collaboration tools) for team instant messaging, a private Google Drive for shared project files, and software called Teamworks to share project plans and tasks with clients and external partners. We also use Jira (an issue tracking software) for fault reporting and improvement requests and customer service software from the Teamworks guys called Helpdocs for customer care.

Asset Numbering

Project assets are rich media content like PDFs, videos, images, and audio files. These are usually made and re-made several times during a project—first in draft and then steadily as improved production versions. There can also be lots of them: a project for our Brazilian partner, Lifelike, has more than 120 video files with each one selected based on the audience' personal journey.

This makes it vitally important to track which assets are which and to be able to easily update and replace them.

When working with production partners, it is common that we will use their asset numbering because for us it does not matter too much so long as there is a numbering scheme. On the VISA project however we need an international numbering scheme that will allow us to easily switch between different regional media. This has resulted in a spreadsheet to map between the production company asset numbering and our global numbering. It is not ideal, and it can lead to errors, but it is workable.

Languages

During the project definition stage, we identify how many languages the project will need to work with. While both projects are American English only, the VISA project must be capable of localization. We have two mechanisms to allow a project to run in multiple languages—the first uses groups and the second uses arrays. With groups, the player is placed into a language group and whenever content is published, Conducttr takes the right content for that group.

In more complicated projects where we are already using a lot of groups to track decisions and other personalizations, we will use arrays and append a language code (from an audience attribute) to a content identification to create a unique index which hence publishes the correct content in the correct language. This again makes the use of asset numbers important.

Design and Other Documents

The first design document is the proposal. The proposal contains the creative concept and the commercial constraints. This is what has been agreed with the client and it is what we must deliver.

After receiving the client's go-ahead, the project lead—who could be the Author or Project Manager—sets about creating ideas for how we might deliver what we have promised. A lot of the technical aspects have already been discussed and agreed with the team before the proposal was submitted but there could be some promised functionality that requires new development.

New developments of our technology are addressed in our normal line of business and clients do not pay for this, but improvements do need to be scheduled as part of our product development process and migrated from the development system to the production system. This takes coordination across the business.

A key to success is simplicity. Complexity creates cost and confusion, and both are to be avoided. Complex projects can also create resistance to change, which is what we want to avoid. There are usually several great ideas that get discarded either because they are too complicated to pitch to the client or to the player or might be longer to fault fix should there be issues after the project goes live.

Right-sizing the Documentation

It is very easy to produce too much documentation and pretend that it is productive work. It is important to remember that documentation is a means to an end, it is not an end in itself. The key purpose of documentation is to make sure that everyone who needs to know (or remember) is aware of how a project works or should work and can easily pick up the reigns to take over a project should the original author fall ill or be on vacation.

For a project like Kodansha's, it will only run for several weeks so the amount of documentation needs to be sized for that period. The project manager can easily make sure that there is sufficient cover for the project for that short period and can talk others through the design so that everyone is fully aware.

For a project like VISA, however, which will run all year, and which has new feature developments, the documentation is quite significant, and a good attempt is made to keep up with documentation while the project is being implemented. Note that neither project has a "transmedia bible" with character descriptions, factions and such like because it is not deemed necessary: the world of AoT is already well-known and documented elsewhere in comics and fan wikis and the VISA project.

A key role for Quality Assurance is to make sure all documentation is up to date and sufficient for business continuity. This also includes comments and labeling inside Conducttr and later, as the project moves into formal testing, change control. Change control is extremely important. This tracks what changes are made, to what, by whom and when.

All projects are documented on an internal HelpDocs portal and this provides key information such as Conducttr project number, author, client contact details, traffic profile and launch and end dates. It also describes the mechanics for player registration, activation, and operation.

Key Design Features—Attack on Titan

Obey the Lore

When working with an existing intellectual property, it is important that any narrative does not violate the canon and can only impact the canon with express client permission. For AoT, we had no freedom to add to the canon, so several creative ideas were rejected because of this and we decided to focus more on the game aspects and less on the narrative.

For a game, it is important that the mechanics of the game are consistent or analogous to what is known about the storyworld. This is evident here in the different activities given for the two game spaces inside the wall (inside the conference center) and outside the wall (outside the conference center) and in the way that Titans are killed.

Defeating Titans

Those familiar with AoT will know that to kill a Titan requires the soldier to slice the back of the neck. In our game, we test the player's sword skills with increasingly complex swipe patterns against a vanishing fight period. This simple mechanic combined with the minimally frustrating requirement to maintain a sufficient gas supply (the swords are gas-powered) delivers an extremely fun experience. The swiping motion and the gas supply constraint both obey the lore of AoT.

Character Phone Calls

A highlight of the SMS experience is several phone calls from the popular AoT character, Mikasa. The calls were recorded by the same actress that is the voice-over artist for the anime which screens on Funimation. This means the voice is familiar to the show's millions of viewers and when heard directly spoken to YOU via your phone creates an incredible intimacy that is actually quite mind-blowing. From a production perspective, both features are straightforward to implement. We can use a draft audio recording up to the last minute and upload the final production version as soon as it becomes available.

Key Design Features—VISA Finance Game

The VISA experience should be classed as a "serious game" or a "learning game" but it is also an "interactive narrative" rather than a financial simulation. Our approach is to immerse the player in a first-person experience in which they play the role of someone making decision as if it were real life rather than a game.

Maintaining a first-person perspective throughout an experience is difficult for any game but particularly tricky for games where we must demonstrate learning: there is always a balance to be struck between the immersion and instructional design.

Several key design choices in this game that were made to meet the challenges of engagement and teaching the curriculum are:

- the use of cut scenes;
- the use of an automated personal assistant (i.e., like Alexa or Siri);
- the use of embedded mini-quizzes.

Cut Scenes

A cut-scene is a video that progresses the story without player interaction. They are usually considered problematic in an interactive experience because they prevent the player interacting. However, they are helpful in the VISA project in advancing the story by taking the player from a subjective point of view to an objective one. That is, the cut-scenes position the player as an observer, a fly on the wall, watching a development in the narrative. They also give us the advantage of expanding beyond the confines of an interactive narrative to deliver a lot of story in an entertaining, minimal effort (on the part of the player) and immediate way.

In this experience, they are mainly used for story advancement rather than "teaching." In fact, we have tried to avoid instruction as much as possible, allowing the player to make their own mistakes and learn by experience.

Personal Assistance

In most of our games we will have a "mentor" character that helps the player when times are tough or if we think they might need guidance. In the VISA game, we created MAI—mobile artificial intelligence—as a personal assistant who finds useful information ahead of when you need it. We felt this was essential in a game like this and allows us to implement an approach based on the 4C/ID (Four-Component Instructional Design) model (Van Merriënboer 1997; Van Merriënboer, Kirschner, and Kester 2003; Van Merriënboer and Sweller 2005; Van Merriënboer and Kirschner 2007) for complex learning which aims to reduce the cognitive load on players.

The 4C/ID model comprises four core components:

- learning tasks
- supporting information
- procedural information
- part-task practice.

Learning tasks are the core experiences that someone should learn from. In our case, the goal was money management, so tasks included financial planning, investing, buying insurance, buying products, choosing a university course (because this impacts future earnings).

Supporting information is provided before the task and is intended to help with decision-making and making sense of the situation. For example, before choosing a university course the players are given information about how likely a certain career is to be replaced by artificial intelligence and how to assess the information universities present on tuition and accommodation fees and grant repayments. In the case of the university decision, information is given in the form of a mini-quiz by a careers chatbot (of which more below).

Procedural information was not given in this project for the purpose of money management but sometimes how-to instructions were given on how to use the app.

Part-task practice is practice of routine activity that over time will become second nature. This took the form of different but repeated examples of buying products from a fictional store and having to make choices about whether to spend money on discretionary items or not. Going to the simulated bank account was often a requirement to give players familiarity with how debits, credits, and current balance are presented.

Embedded Quizzes

MAI is not the only assistance in the game and we make frequent use of in-world chatbot-type personas to guide players through certain financial choices such as choosing a career, buying a product or investing. A chatbot is a program that talks to the player in a conversational style. At the time of writing these are quite commonplace in customer service websites and it was this we sought to mimic.

The chatbots allow us to easily present a one-question quiz in amongst their other chatter.

Although, in this case, not used directly for assessment, in classroom environments the player choices are shared in real-time with the teacher both as an individual player choice and in aggregate for that class as a pie chart. This can then be used as a class discussion point and hence allows the tutor to expand further or dive deeper into a topic, asking students to reflect further on their decision and its consequences.

Production Workflow

Our end-to-end workflow follows a process called the Active Story System (Pratten 2015). Rather than go into its details here, I will cover some top-tips that allow us to be fast and agile.

Premise and Purpose

A key requirement for any project is to have a very clearly defined purpose and an equally clearly defined premise:

- Why does the client want us to create this experience?
- How will this experience add value to the client and to the client's clients?
- What message does the experience need to convey and to whom does it need to convey that message?

Everyone on the team must understand this and even though it is the job of the proposal-writer to understand this upfront, it is always worth confirming again at the start of the project.

The shorter and simpler that the premise and purpose can be stated, the better. Without locking this down, the project has no direction and no guiding principal by which to determine which ideas are good ones and which should be discarded.

Early Demo

Projects with many stakeholders require tollgates and sign-off stages to ensure that we do not go too far in a direction that all stakeholders cannot agree on. It is also true that the closer to finished implementation a project is, the costlier it is to make changes.

Our approach to avoiding surprises is to allow stakeholders to see an early demo or prototype and check that everyone is on the same page. While this sounds like a good idea, accomplishing it is harder than one might imagine, and it can produce its own problems. Some ideas, for example, do not sell themselves well in draft form and the team needs to decide how and when to take the gamble that the unfinished, unpolished version is good enough to have the desired effect of communicating our intentions.

In our experience, it is best to run the demo with stakeholders watching so that we can answer any questions or misunderstandings immediately rather than just email a link to the demo. Note that this is not supposed to be a play test, it is a progress review.

Prioritization

When I first started making films, an experienced producer told me to organize a shoot around the most expensive resource. Sometimes this might be a location, sometimes equipment, sometimes a cast member.

I have carried this same thinking into transmedia production where "most expensive" usually translates to "most difficult to change." In our projects, the costliest changes are often changes to the user interface (UI). Seemingly small requests to "move the logo three pixels to the left" can cause more pain for us than changing a storyline. This is because each change needs to be tested on a range of devices and device modes. This makes it important that the UI is agreed and signed-off early before we get too deep into the implementation of the interactive experience.

Because client sign-off on the UI can take a while, we typically use a common or temporary development UI for the team so that we can continue with the project foundations while waiting for the final front-end to be agreed and implemented.

Testing

Testing creates its own needs for flexibility and agile changes. With the Attack on Titan project, for example, we initially placed Titans around our office in London and then in San Francisco for client testing and then finally in LA. Thinking about how the project would be tested and tweaked was built into the design so that the experience would be easy to change. Hence, we added the ability easily upload Titan types, swipe patterns, delays, and Global Positioning System (GPS) coordinates from a spreadsheet. This made changes very easy to accomplish which would not have been the case if the details were "hard coded" into the project.

We also had an "online mode" allowing us to play the game from a comfy seat rather than physically go to a Titan location but licensing restrictions ultimately prevented us from offering this mode to the public.

Where we often encounter issues is when the client tests multiple times with different accounts because Conducttr attempts to understand who the person is behind the accounts and will try to merge the audience record—that is, it creates a single representation of that person and attributes them with multiple contact details such as two email addresses, two mobile numbers, and so on. Because Conducttr is trying to orchestrate a single user journey, activity on one email address such as the player choosing "left" instead of "right," is now known by the second email address because after all it is the same person. Therefore, the client's attempt to be multiple people for testing is thwarted unless that use contact details are truly exclusive.

Another testing issue is the use of caching. We use a content distribution network (CDN) to improve performance around the world and it means that local versions of files are stored locally (known as "caching") in the network. The player's browser will also cache local files. During testing, this can be problematic because we can update UI or certain files like images, but testers report no change or no improvement. To overcome this, we can "flush the cache" so that recent files are used and ask the client to use incognito mode on their browser as this mode does not cache files locally.

Resetting and Replaying

The ability to replay a computer game is common practice but it is not very common at all in transmedia projects. I am referring here to Alternate Reality Game (ARG)-type transmedia projects where content might be published on social media and, having been made public, it is unusual to expect to play through the experience again.

We have always had to address this issue because even if the project is not live with a real audience, it does need to be tested and hence reset back to the beginning. For resetting during testing, we use several approaches but the most common is to use multiple private social media test accounts and then, after testing, switch the accounts to the live production accounts. Technically, this is straightforward, but it can be troublesome if the client owns the social media accounts because this requires it to authenticate Conducttr with its account. However, the client representative involved in the project may not be responsible for social media.

Another issue is erasing all the personalization history which builds up during a play-through of the experience. Usually during testing it is easiest to generate a new audience record and play "fresh" with no history. But if the experience is replayable, then we need to make it possible at the end of the experience to hit a single kill switch to erase the personalization. Resetting an attribute sounds like it ought to be easy and it is except when that attribute is holding a simulated message feed, references to other team mates and so on. This makes the sequencing of which attributes to reset important too.

For the VISA experience, players can choose to play as Alex or Jess and then at the end of the game opt to replay as the other. In true transmedia fashion, each protagonist reveals a different perspective on the world and in this case also on the training scenario, so it is quite beneficial to play through again. Here the issue was how much of the past life should be forgotten because part of the learning might be to compare each subsequent run-through with the previous run-throughs … but if we did this, would it kill the immersion? The solution was to re-publish the final summary email as though it were addressed to the other character, preserving the data and walking the line on in-world versus out-of-world.

Review and Approval

A final area to address is that of sharing the final content with the client for approval and sometimes for sign-off by the lawyers. Although the story and experience have been signed off at earlier tollgates, when the time comes to share the final content it is usually easier to ask the client to play through the experience than send them a document to read.

Central to the problem of sharing the final content is the nature of interactive multichannel narratives. They are not easily presented in simple Word documents and therefore the challenge remains to find a presentation format that works for the layperson versus the production team.

The Conducttr platform currently offers two formats—a "script report" which looks like a movie script and a "table report" which presents the content in a Word-like table. To the untrained eye neither looks very friendly to read and when clients want to make changes or raise questions, while it is helpful to do it in a shared Google doc, transferring the amendments back into the production system (Conducttr) is more time-consuming than I would like.

At the End of the Day

The video conference calls over, everyone in Europe packs away their laptop and heads for home while our clients and partners on the West Coast tackle their to-do lists. Tomorrow morning, I will wake at 5am to check my email for links to new cuts of the videos, new image assets, and any questions that came up during the day Pacific Standard Time (PST). There is an hour or two now to chat in real time about any tasks we need to complete before we are on our own again for another day.

I set to work on the privacy statement that will accompany the experiences and the opt-in statement needed before player registration. Both will be tweaked by the clients' commercial teams but this will give them a good start in understanding the information we are collecting and why we need it.

The new European General Data Protection Regulation (GDPR) (see Information Commissioner's Office n.d.) has become law as of May 2018, and we have already made changes to the UI to allow for the improved opt-in requirements.

Conclusion

For the most part, transmedia production does not look too dissimilar to other single-channel production processes and that should be reassuring to newcomers because any skills they have in media production or project management are transferrable.

Where transmedia usually stands apart is in the additional design documents required for the multi-channel or multi-part integration and particularly the infamous "story bible." While the bible is often required, I have attempted to explain that they are a means to an end and their size and complexity should be matched to the needs of the project.

Transmedia projects lend themselves well to learning environments because they more closely imitate real life than typical eLearning projects and the 4C/ID model which is intended for complex learning tasks is a handy model to use for transmedia learning because it is based on experiential learning and interactive projects should be about experiences.

The type of transmedia projects that my team and I create are living, breathing worlds that span countries, languages, platforms and time. We view them as dynamic systems in implementation as well as in operation. This understanding—that change is the only constant—demands that we design from the outset for modifications, tweaks, updates, revisions, and improvements.

Transmedia production workflows should be ductile and bend with the winds of change.

References

Conducttr. n.d. Accessed March 15, 2017. www.conducttr.com/.

Information Commissioner's Office. n.d. "Overview of the General Data Protection Regulation (GDPR)." Accessed March 11, 2017. https://ico.org.uk/for-organisations/data-protection-reform/overview-of-the-gdpr/.

Pratten, Robert. 2015. *Getting Started in Transmedia Storytelling: A Practical Guide for Beginners*. 2nd ed. London: CreateSpace.

Van Merriënboer, Jeroen J. G. 1997. *Training Complex Cognitive Skills: A Four-Component Instructional Design Model for Technical Training*. Englewood Cliffs: Educational Technology Publications.

Van Merriënboer, Jeroen J. G., and Paul A. Kirschner. 2007. *Ten Steps to Complex Learning: A Systematic Approach to Four-component Instructional Design*. Mahwah: Erlbaum.

Van Merriënboer, Jeroen J. G., and John Sweller. 2005. "Cognitive Load Theory and Complex Learning: Recent Developments and Future Directions." *Educational Psychology Review* 17 (2): 147–177.

Van Merriënboer, Jeroen J. G., Paul A. Kirschner, and Liesbeth Kester. 2003. "Taking the Load off a Learner's Mind: Instructional Design for Complex Learning." *Educational Psychologist* 38 (1): 5–13.

24

TRANSMEDIA COMMODIFICATION

Disneyfication, Magical Objects, and *Beauty and the Beast*

Anna Kérchy

In spring 2017, coinciding with the release of Disney's live-action cinematic remake of *Beauty and the Beast*, a limited-edition budget item offered by Primark stores provoked an authentic fan frenzy. Chip purse, a coin holder in the shape of a cute anthropomorphic tea cup from the movie (according to the plotline, a little boy turned into tableware under a magic spell), sold out within minutes, fueled a bidding war on online auction sites selling for several times its original price, and generated heated discussions on social media platforms. The discussions revolved around proud owners posting photos of the much-sought-after product and others, who could not get hold of it, venting their fury about their dissatisfied consumer demands and the madness dubbed #ChipGate. I wish to argue here that the Chip purse is much more than just a lovable collectible: ironically, as an object that stores money to buy further products with—Primark have expanded its *Beauty and Beast* line to include homewares, bedding, tea cups and pots, as well as clothing, beauty accessories, and décor knick-knacks as "part of the magical range" (Jones 2017)—it can be easily interpreted as an emblem of the fetishist commodification of fantasy that takes place via a transmedia storytelling experience paradigmatic of post-industrialist consumer societies of spectacle.

Beauty and the Beast fans queued in front of stores before opening hours to make sure they could take home their tangible memorabilia of magic, but their obsession was not stimulated by any of the original literary source texts penned by French novelist Gabrielle-Suzanne Barbot de Villeneuve in 1740, rewritten and abridged by Jeanne-Marie Leprince de Beaumont in 1756, and published in English in Andrew Lang's *Blue Fairy Book* in 1889. Customers were likely just as much unconcerned about the written versions of this animal bridegroom fairy tale, initially designed to prepare aristocratic young girls in eighteenth-century France for arranged marriages by celebrating the transformative powers of love, as they were forgetful about its classic adaptations: Jean Cocteau's surrealist poetic fantasy film *La Belle et la Bête* (1946) or Philip Glass's opera (1994) inspired by it. As attested by the avalanche of Twitter, Instagram, and Facebook hashtags accompanying pictorial evidence of the collectible's acquisition, including #instadisney, #disnegram, #disnerd, and #disneylifestylers, the adoring audience emphatically positioned themselves within the Disney fandom, attributing their enchantment to Disney's latest film, 2017 computer-generated imagery (CGI) enhanced live action musical romantic fantasy, a remake of Disney's own 1991 animation, yet another original reiteration of a "tale as old as time." A film that has grossed over $1.2 billion worldwide, and surpassing the original film, made it into the highest-grossing films of 2017, the tenth highest-grossing film of all time, and the fastest selling family film in the company's history in the United States (Disney.wikia.com).

Disneyfied Magic Merchandise: Simulating "Capitalist Realist" Reveries for Infantilized Consumers

Consumer choices are always manipulated by complex power struggles concerning the authority over narrative meanings and the appropriation of the prestigious label of originality, which hold the stakes of financial profitability. According to postmillennial reception studies (Stephens and McCallum 2000, 160), children nowadays are likely to first encounter literary classics, deprived of the original authors' names, mediated by popular film industry in the interpretation of media moguls like the Walt Disney Company that claims exclusionary proprietary ownership over stories customized to meet the uniform house-style marked by its brand name. Disney's formative role of contemporary cultural imaginaries has often been criticized for colonizing individual fantasies homogenized by shallow scopophiliac pleasures of ready-made clichés governed by the ideological and commercial interests of the company. "Disneyfication" has become a household term to denote the company's notorious adaptation strategy whereby stories are rendered "safe" for juvenile audiences through the removal of undesirable plot elements (themes of sexuality, death, or moral ambiguity), the emphasis on the clear-cut dividing line between good and evil, additions of light entertainment features (Broadway-style music hits, talking animal sidekicks), and reworkings necessary for happily-ever-after endings. Many have problematized the "saccharine finales" for imperceptibly transmitting capitalist, patriarchal, colonialist, and a fundamentally normative "petit bourgeois" ideology, with the aim to maintain the political status quo, discourage social transformation, and most importantly raise a generation of obedient consumers (see Bell, Haas, and Sells 1995; Budd and Kirsch 2005; Giroux and Pollock 2010; Zipes 2011).

Certainly, the 2017 Disney adaptation designed for family fun ignored the most troubling underlying central themes which make *Beauty and the Beast* such a special story: interspecies romance, anxieties about bestial sexuality, female rite of passage initiated by an encounter with terrifying otherness (Warner 1995, 276), patriarchal oppression of woman reduced to a sacrificial object of exchange, or social marginalization of disability remained taboos left off the screen. It was much more important for the company to associate straight away in the very first opening shots the Beast's enchanted fortress with the Disney logo, the trademark emblem of the Disneyland Magic Kingdom Castle and, hence, extend the movie's entertainment quality with a "commercial intertext" (Maltby 1998, 27), turning the cinematic fictional reality into an advertising site for an actual touristic spot families can visit to gain an enhanced first-hand interactive experience of *living* the fairy tale.

According to Maltby, the placement of consumer products in high-budget Hollywood movies function "as budgetary instruments but also as a form of *capitalist realism*" (1998, 27), simulating the veracity of the filmic fantasy by evoking in spectators the real-life referent of the fictional fortress. By connecting on- and off-screen realities, the make-believe realm of Disneyland theme park—postmodern philosopher Baudrillard (1994 [1981]) called a par excellence example of simulacra's referentless reality abundant in illusory needs propagated by commercial images of a desirable lifestyle—also becomes fantastificated as an authentic locus of magic where all dreams can come true with a little imagination and money to spare. Disneyland offers you an unforgettably vivid experience of the tale: you can meet in person and take selfies with your beloved characters, dine sumptuously at the Red Rose Tavern, or enjoy a comic performance of the love story at the Royal Theatre, even a theme park ride on spellbound tableware awaits you from 2020; and surely you can buy Belle's ball gown, cute stuffed toy Beasts, or jewelry in the shape of the rose of true love, along with a plethora of other Disney-themed merchandise (with more than 250 products matching the search *Beauty and the Beast* at disneystore.com from bed sheets to iPhone covers).

The Disney Empire strategically indoctrinates children to become insatiable consumers of commodified fantasies, sharing adult appetites for newer and newer releases of collectibles. Youngsters constitute the most lucrative target-market for many businesses today because they integrate three markets in one: (1) *current* mini-consumers with own funds to spare on items of their choice encouraged by their elders to master consumer skills, money awareness, and economic responsibility at their earliest convenience; (2) *future* customers cultivated now to build a brand awareness for tomorrow; and (3) *influencers* who cause many billions of dollars of purchases among their parents (McNeal 1991). Disney's child-focused advertising is systematically pitched at very young audiences in Saturday morning television commercials and extra features available on kids' websites, but the company also exploits the generational nostalgia of parents lured back to the delight they took as kids in Disney's animation that provided the source material for the modernized live action facsimile. Mature viewers prove to be complicit in playing their role in the socio-economic fabric of consumers while embracing the identity of the eternal child (the archetype of the *puer eternus*), as many grown-up fans' online self-denominations attest in infantile nicknames like #foreverabigkid or #minniemousemom.

The "crossover appeal" (Beckett 2009) to multiple age groups increases commercial success, media attention, critical legitimacy, as well as intergenerational bonding that functions as a stable sedimentation of fandom. However, there is also a false promise of democracy lurking in marketing rhetoric, which address a dual audience with similar slogans: Primark's women's line urges adult females "to rival Sleeping Beauty with PJs fit for a queen, or be the belle of the ball in serious fan-girl T-shirts" (Jones 2017) and promotes its kids' collection by teasing the "moms' mini-me-s" to "make their dreams come true as they get ready to be the Belle of the Ball" or "open the pages of the beauty world with an enchanting mix of magical make-up products, fit for little princesses" (Shah 2017). The right bargains hold the promise of opening the gates of Wonderland for style-conscious parents and their offspring who are both the most vulnerable to commercial abuses and, as digital natives, the most complicit in participatory cultural activities.

Transmedia Extension and Market Expansion

Disney adopts various strategies to keep its story brands alive: since the 1960s it has been releasing in 25-year intervals the classic animated features either theatrically or on home entertainment platforms, and as old films entered the public domain, devoted attention to the release of live-action remakes, like *Maleficent* in 2014, *Cinderella* in 2015, and *Beauty and the Beast* in 2017. Disney's illusory status as the authentic teller of the original story has been repeatedly confirmed by the multiple adaptations, interconnecting story versions tailored to different media, as well as remakes of its own products (like 1991 animation Chip, 2017 CGI Chip, and Primark's Chip merchandise as well as innumerable tie-in products) coupled with commercial intertextual allusions. The critically minded ponder if these remakes are "meant as homages, updates, 'brand deposit' reminders of existing franchises, or just high-profile cash grabs?" (Robinson 2017). However, most importantly, these dialogic intertexts also belong to that particular type of postmillennial fantasy world-building Henry Jenkins, one of today's most influential critics of popular culture, has called "transmedia storytelling" (2006, 2007). Throughout the "transmedia expansion" of fictional realities, integral elements of a make-believe universe "get dispersed systematically across multiple delivery channels for the purpose of creating a unified and coordinated entertainment experience [where] each medium makes it own unique contribution to the unfolding of the story" (Jenkins 2007).

"Transmedia extensions" facilitate "additive comprehension," they make the story more immersive by adding new pieces of information which invite us to revise our understanding of the fictional reality. They might offer insight into characters, motifs, unelaborated plotlines, bridge events, fill in gaps or resolve excesses in the unfolding of the story, flesh out unknown aspects of the imaginary world, add a greater sense of realism and augment fantastic effects, too. Disney's 2017 film adaptation invested a lot of energy in extending the background stories of characters. The pre-title sequence centers on the original hubris of the proud prince turned Beast. It is noticeable that, in the 1991 animation, his tragic flaw is summed up in a stained-glass-window-style tableau, one can actually visit in Disneyland, as a prominent example of the commercial transmedia extension of the narrative.

Beauty and the Beast subplots shed light on unresolved traumas, including the mysterious loss of Belle's mother; her father never ceases to grieve and refuses to explain to his daughter troubled by his silence; macho villain Gaston's war veteran past, and his gay sidekick's hopelessly unrequited love for him; or the enchanted objects/servants tolerance in face of their master's animalistic temper explained by their remorse felt over not intervening in the moral corruption of the young prince by his tyrant father. These extensions create different points of entry for different audience segments to expand the artwork's potential market. Prehistories of vulnerability support the psychic involvement of spectators, while the political correctness implied in the Beast's multiracial staff, the loveable goofy gay figure, and the bookworm feminist warrior Belle targets the liberal-minded millennial generation. The addition of the ultimately cute figure of the little boy magically metamorphosed into a chipped tea-cup holds universal appeal for young and old because of its associations with loveable naughtiness, cosy homeliness, and even British nationalistic pride—all easily commodifiable qualities as attested by the commercial taglines of the related merchandise: you can "fill this little guy [Chip cup] with yum hot chocolate [or a five o'clock tea!] for the ultimate quiet night in" (Jones 2017) or more miraculously "switch to Mrs Potts [Chip's mom] mode and become mamma of the year! [by purchasing all the *Beauty and the Beast* goodies]" (Shah 2017).

The extensions' interconnections with one other, the tie-in products they generate, and with the original mastertext(s) stimulate audiences to interact with the story world larger than the single story. The collective transmedial (de)construction of communal knowledge within a networked society that can visualize, digitalize, and commercialize the initially orally circulated fairy-tale cultures is exemplified by subtle gestures like Disney naming Belle's picturesque village Villeneuve as an homage to the writer of the literary source text. Yet Gabrielle-Suzanne Barbot de Villeneuve's authorial name likely remains imperceptible, unreadable for most of the viewers, overshadowed by spectacular transmedia crossover allusions to the brand name labelling the adaptation: during Belle and Beast's waltz the letters WD can be seen in the coat of arms design on the floor with reference to Walt Disney, and the lock of the carriage in which Belle is imprisoned is decorated by a hidden Mickey Mouse motif.

Transmedia storytelling can segment and disseminate an idea into multiple media installments within one single work: internal monologues encapsulating characters' background stories are often told in song-and-dance numbers which complement the narrative filmic diegesis with an acoustic affective charge, and turned into new media applications like mobile phone ringtones can be integrated into fans' daily lives. The proliferation of tie-in products also functions by means of transmedia extensions, frequently allowing for a conjoining of old and new media through a process Bolter and Grusin (2000) call remediation: one can purchase actual print-and-paper specimen from the Beast's fantastic three-dimensional (3D) CGI-simulated library that Belle falls in love with in the movie. Fictitious titles brought for real include *Belle's Library* with a foreword by scriptwriter Linda Woolverton, a collection of inspiring quotes from Belle's favorite books, her own notes and colorful drawings, as well as an enchanted book called *Nevermore* featuring in *Beauty and the Beast: Lost in a Book* that can take readers on time-travel adventures just like the Beast's magic book does in the 2017 movie (transporting Belle back to a plague-ridden Paris where she can find out about her mother's past). The recurring festishization of the analogue book format invested with magical qualities—equally articulated in Disney's 1997 *Beauty and the Beast's Enchanted Christmas* animation sequel, in *Disney's Beauty and the Beast: Magical Ballroom* computer game and girl's educational software (2000), and Disney's 2017 live-action 3D CGI *Beauty and the Beast* movie enhanced by new media technology—neatly demonstrates the transmedial spreading of the same story about the wonders of storytelling, too.

Prosumer Networks: Audience Engagement and/or Ideological Containment?

The popularity of a fictive world's extension into multiple media and "advanced moving image" formats can be explained by the (post)modern narrative condition Matt Hanson calls "screen bleed," arguing that today's digital media consumers' immersion in 3D worlds reflects the need for all-encompassing mythologies multimedially involving contemporary audiences in "interactive online worlds, where each strand of narrative offers a new dimensional layer" (2003, 47). Post-industrialist consumer societies' hybrid new subjectivities elicited by new media's interactive potentials have been recently referred to as "prosumers" (Toffler 1980; Ritzer and Jurgenson 2010) or "produsers" (Marshall 2004): compound words made up of the fusion of producer and consumer, to denote the activity of browsing through media contents to make sensible choices that can eventually prove to be transformative of the meanings to be generated. *Beauty and the Beast*'s new (trans)medial extensions like Snapchat filter applications, Facebook quizzes, or YouTube fan videos allow interpreters to maximalize their engagement with the storyworld through individual choices, additions, and creative retellings. These activities tie in with the sociological concept of McDonaldization (Ritzer 1993), a cost-efficient process for the brand-owner corporations since prosumers work actively and free of charge in producing the services and goods they buy and consume. Prosumers also take part in viral marketing practices like the reproduction of memes which keep the hype alive via digital word-of-mouth tactics, spreading messages consumers cannot resist sharing with friends, who in turn share with more friends, fueling the cycle of exponential growth and reinforcing personal bonding of a community of the initiates. The dissemination of user-generated content on social media network websites or the blogosphere, like sharing *Beauty and the Beast* related do it yourself (DIY) project hints on Pinterest, uploading data onto Disney Wikipedia, or contributing to the online fan-fiction community's collective corpus or the organization of live-broadcast flashmobs could be examples for the unpaid labor of enthusiasts. Although prosumers may gain a relative empowerment like Twitter-using conference audiences, whose tweets posted online reach a global group of recipients who witness and respond to messages, which may spiral out of the control of the initial producer of meaning, but with free promotion they still serve its economic interests. Moreover, prosumer networks' subversiveness is mostly ideologically contained by transmedia commodification practices like online shopping sites' suggestions to let your friends know about your consumer activities driven by the aim to make the appetite for consumption reach epidemic proportions.

The (neo)Marxist critique of capitalist consumer culture is called commodities compensatory fetish objects, which stimulate a constant desiring that can never reach satisfaction. Commodity fetishism masks a primordial lack caused by the disruption of direct interpersonal relationships and an alienating division of labor, whereby "the appearance of goods hides the story of who made them and how they made them, making invisible the economical exploitation and social injustice" (Lury 1996, 41). Zygmunt Bauman (1987) differentiated between two social groups: (1) "the seduced" who are incorporated into consumer culture to make illusory decisions in the market area, and (2) "the repressed" who, devoid of economic and cultural resources, are excluded from the market as subjects objectified to the bureaucratic control organizing the state provision of services. Celia Lury's (1996) focus on the first group points out that acts of consumption rarely satisfy actual basic needs and are rather meant to express social status, cultural style, or being in the know, and hence, belonging to a group of like-minded consumers for whom products are worthy more for their symbolic value than their market or exchange value. Throughout a commodity aesthetics permeating packaging, promotion, and advertising, merchandise is associated with illusory cultural meanings: the promise of a lifestyle is being sold as "images of romance, exotica, fulfilment, or high life" are associated with mundane consumer goods as soap, washing machines, housewares, cars, or alcoholic beverages (Lury 1996, 42). Just how much commodities hold an "identity value" (Featherstone 1991)—paradoxically selling the promise of the expression of a uniquely individual self with mass-produced, one-size-fits-all consumer goods—is reflected by the compulsive online sharing of news about novel acquisitions: selfies of proud owners with cult products, hashtags expressing consumer identities (#disneyfan), or the publicization of purchases on social media like Facebook profile updates.

From Princess Industrial Complex to Feminism Appropriated for Marketing Purposes

Sugary fantasies about upward social mobility and wealth conjoined with the myth of romantic true love constitute the number one bestselling commodity of the "Disney Princess Industry." As Jack Zipes' (1995) socio-historical overview of the changing function of folk and fairy tales argued, the oral wonder tale storytelling's democratic quest for a communal harmony was gradually lost with the medial shift to written and printed tales that the rising Victorian bourgeoisie exploited for the sake of solidifying its social class status through marginalizing the illiterate from the privileged elite of readers to reach a "control over imagination and desire within the symbolic order of Western culture" (19). The contemporary Disney Princess Industry's "monologue of self-praise of the ruling classes" (24) masked by the illusory promise to "eliminate social and class conflicts forever" (26) with rags-to-riches fantasies seems a logical continuation of this self-establishing gesture of the middle class, constituting the main social corpus of the seduced consumers.

According to Helen Pilinovsky (2011), the control of powerless juvenile audience's, mainly little girls', consumer tastes is predicated upon the Princess Industrial Complex system, whereby the ideological basis of the persona of the heroine of the commodified fairy tale—deprived of the political criticism and proto-feminist concerns of the original *contes de fées*—is used to reinforce the values of heteronormative, patriarchal, capitalist society and "sell products associated with them, beginning with the tales themselves, and continuing through their trappings and accessories" (19), peaking in the Wedding Industrial Complex with the indoctrinated spectators/consumers' coming of age. The classist commodification of the Happily-Ever-After scenario is perfectly illustrated by the lure of high life voiced in advertisement of Primark's *Beauty and the Beast* budget items. "Ladies, upgrade your little ones from princess to queen as they enter the Primark palaces and grab these must-haves. They'll be ready to rule their kingdom in complete, beautiful style!" (Shah 2017). The intended audience of little girls can become princesses at the price of becoming playthings in the consumer mothers' hands.

Since Belle is one of the original members of the Disney Princess line launched in the early 2000s, the spell of her feminization is hard to break, and the 2017 Disney adaptation's attempts at feminist revisions remain largely ineffectual. Emma Watson cast as Belle is an ideal representative of postmillennial girl power. Renowned as brave, smart, tomboyish Hermione Granger in the *Harry Potter* series, a United Nations Women's Goodwill Ambassador, and an activist for the *He for She* campaign, her public persona matches the rebellious, creative, feminist aspect of Disney's Belle, who invents a washing machine to have more time for books, teaches a young village girl to read, refuses to marry macho Gaston, wears boots instead of glass slippers, and mocks the Wardrobe's attempts at dressing her up as a princess. Yet the emphasis on Belle's persona as "a soul ahead of her time" might just as well function as a marketing strategy to attract wider audiences by means of the commodification of feminism. Ironically, the moral of the original *Beauty and the Beast* story—never judge by outward appearances, true love looks beyond deceptive surfaces—gets lost in Disney's classist, lookist version where the romantic central couple meets conventional beauty standards. The leonine Beast is beautifully sublime and never hideous, and in the end he loses all his bestiality to become a clean-shaven human prince charming, while Belle is endlessly radiant and always prepared for a dance—just like the brand related enchanted products, which can come in the most bizarre forms like the "socks that look like stained glass in which you can practice the royal waltz around your bedroom," or princess lip balms which can grant "chapped smackers a 'happily ever after'" (Shah 2017).

Neurotic Hedonists

The commodification of fantasies of enduring beauty and eternal love are particularly interesting when contrasted with consumer culture's *carpe diem* philosophy dictating an accelerated rhythm of purchases driven by an insatiable hunger for novelty. Although consumption is stimulated by a glamorization of a hedonistic lifestyle allegedly made available with new acquisitions, buyers experience a constant dissatisfaction due to the ever-expanding range of goods on sale. Sociologist-philosopher Renata Salecl (2010) located the communal neurosis of advanced post-industrialist capitalist consumer societies in the psychological burden caused by the freedom of choice: the endless series of open options provokes feelings of inadequacy, guilt, and anxiety about possibly making the wrong individual choices, while forgetting about more important collective decisions and our power as politically responsible social thinkers. For Bauman (1987), the simultaneous compulsion and inability to choose results in melancholy, a sense of infinite connectivity while being hooked up to nothing. In more paranoid readings of the pitfalls of the mythicized choice, material obsession emerges as another form of child abuse, comparable to fast food and porn industries (Salecl 2013). The temporal confusion caused by the accelerated consumption of timeless values in the privileged here-and-now of a one-of-a-kind bargain occasion is a side effect of the neurotic experience of the compulsion to choose and buy, abundantly present in *Beauty and the Beast* commercial campaign with lines like: "Quick! Time is running out, Cogsworth's hands are ticking, and these beauties won't be in store forever …" (Jones 2017), "These Beauty and the Beast collectibles are everything. Until the last petal falls, purchase literally everything on this list" (Baardsen 2017). While in the original, true love must blossom before the last petal falls from the enchanted rose, throughout the transmedia commodification of the fairy tale, all goods must be purchased before the stock runs out. The rat race with time is for materialistic rather than idealistic purposes.

The Secret Lives of Things: Enchanted Object or Replaceable Merchandise?

The most popular musical hit of the 1991 Disney animation and the 2017 live-action remake was "Be Our Guest," a song-and-dance number performed by enchanted live tableware and household utensils who invite spectators like Belle, the only human character in the CGI enhanced scene, to embark on interactive participation, to accept the invitation, and fully submerge in the delights of the dinner and the story. The song—written by Howard Elliott Ashman and Alan Menk, and visualized as a grand scale Moulin Rouge cabaret revue vaudeville performance—voices a veritable hymn of consumer culture, addressing customers from the point of view of products, services, and sellers who keep the market going in a confidence trickster fashion with promises of a happily-ever-after, and illusory subservience masking omnipresent omnipotence. This interpretation can be perceived by the following lines of the lyrics: "Be our guest/ Be our guest/ Our command is your request/ …/And we're obsessed/ With your meal/ With your ease/ Yes, indeed, we aim to please/ While the candlelight's still glowing/ Let us help you/ We'll keep going."

Magically animated objects in fairy tales have a potential to free things from the functional or fetishist roles assigned to them by consumerism, and act in accordance with the tenets of "thing studies," or postmillennial object-oriented ontology that attributes to objects a proactive existence, a psychic reality of their own, independent of meanings projected on them by human cognition (Bennett 2010). In Madame de Beaumont's original version of the *Beauty and the Beast* tale, magic objects versus commodities play a significant role as narrative engines of the text. A merchant, the father of six siblings, goes bankrupt, so that the family has to move to the country; when he travels back to the city to take care of business affairs, his offsprings ask for gifts; his sons want weapons and horses, his older daughters jewels and the finest dresses; it is only the youngest girl, his dearest child Beauty, who asks him for a single rose, that is not precious because of its market value but its symbolical, aesthetic worth, and erotic-amorous implications. The rose, asked with a touch of Oedipal desire from the father but eventually received from a monster turned prince charming, is a magic object inducing a family drama with fatal transformative consequences. As the emblem of the hope for true love which can break the curse, the rose is the hexed Beast's most precious object—in Disney, too, an enchantress disguised as a beggar offers the flower in exchange for shelter and when the prince refuses it, transforms him into a beast—so when the father plucks it from his garden as a souvenir for his daughter, it is only fair that the monster asks for the merchant's most precious belonging in return, hence turning the intended receiver of the gift into a immeasurably valuable gift herself. In Beaumont, the Beast's magic objects are invested with a romantic emotional charge and a potential of connectivity: the rose represents yearning, a mirror helps Beauty to see what is happening in the Beast's castle, and a ring allows her to return there in an instant, whereas the magnificent gowns Beast gave Beauty turn into rags at her sisters' touch and restore their splendor in her hands, reflecting thing studies' philosophical assumption of objects choosing their owner, relativizing human agency. The castle's household utensils, the Beast's enchanted servants, keep their anthropomorphic qualities: they can talk, sing, dance, love, and display a wide range of human emotions from flirtatiousness and humor to anxieties about being trapped in object status, the uncanny state of animate inanimation, and losing their humanity forever (unless the Beast learns to love another and earns her love in return before the last petal falls).

In Disney's 2017 film adaptation, few objects do preserve their enchanting qualities like the clockwork machinery Belle's father cannot cease craftworking upon since the object (containing a miniature replica of himself painting his wife and daughter) functions as a storehouse of both traumatic and blissful memories, reminding him of his lost spouse via a materialized flashback of a happy past—providing an exciting counterpart to the daughter's upcoming happily-ever-after. Yet most of the wondrous possibilities promised by the object-oriented ontological view are neutralized due to objects' predominant use as spectacular devices of stage magic, like in the above-mentioned "Be Our Guest," a tagline of the 2017 film adaptation. The exotic flavor of the French culinary expressions (which sound like fantastic brand names), the rhetorical questions, emphatic repetitions, and direct invocations in the song are poetic devices which clearly resonate with advertisements' discourse of commodification. "Soup du jour/ Hot hors d'oeuvres/ Why, we only live to serve/ Try the grey stuff, it's delicious/ Don't believe me? Ask the dishes"—cry out ecstatically the household utensils and tableware ... and fans rush off to Primark to purchase their Chip cup, taste their Grey stuff ice-cream at Disneyland theme park, or compile their Belle menus with the help of online computer games. The "politics of wonder" rooted in traditional fairy tales' quest for communal harmony and psychologically efficient treatment of collective trauma are transformed into a hegemonically calculated, "commercialized poetics of magic" (Bacchilega 2013, 5), creativity contained via an "idiotic pleasure of consumption" (Žižek 1989) as self-serving prosumers willingly gift themselves with an ever-expanding, beastly selection of beauties.

Conclusion

As Karin Beeler and Stan Beeler (2014) argued, postmillennial popular cultural studies, and more specifically screen studies' scrutinization of participatory culture must pay close attention to how new technological devices radically altered viewing platforms; however, the other most influential factor necessitating new ways of discussing spectatorship proves to be "the youthful segment of the audience" (2014, 1), who have become a significant economic factor in successful productions, a consumer group increasingly independent of parental supervision. Instead of waiting for a family outing to the local cinema, children can watch and rewatch a film on their mini DVD players or play with tablet applications or e-book adaptations of the same film on handheld devices in the comfort of their own rooms (Beeler and Beeler 2014). We can only wonder how this generation of children—digital natives and natural born prosumers with an unprecedented new media literacy—will contribute to the visual narratives of the future following a "digimodernism" Kirby (2009) calls a new cultural paradigm of the twenty-first century. It is to be subject of further investigations how they will reform the art of storytelling they have been trained to understand in terms of a multimodal process involving "dispersed media content" (Jenkins 2006, 3) they can make connections with while seeking out new information about the artwork that can be bypassed throughout creative interactions with special features (like added bonus contents at Disney Movies Online) they might reinvent as they like, once they have learnt to negotiate media transition as "mix of tradition and innovation" (Jenkins and Thorburn 2003).

Göran Bolin (2007) pointed out the liberating potential of transmediation technologies: just like the arrival of mechanical movable type printing freed textual components from their authors and introduced the Gutenberg Galaxy era of mass communication, the advent and quick proliferation of digital production, distribution, and reception procedures liberated information from the dependence on any given medium, as the computer has melted several older media into its technology (Jensen in Bolin 2007, 238). According to Bolin, the transmedia commodification of fictional characters can be regarded as a further "liberation of textual components" (2007, 243): throughout market convergence and cultural synergy, media and non-media enterprises coordinate their initiatives for mutual economic benefit and allow cultural items to freely move among different media platforms. As my initial example of ChipGate (and any Primark merchandise using Disney logos or characters) has shown, this cross-platform circulation opens up possibilities of multiple readings which simultaneously hold the potential for creative innovations and comprise a cunning marketing strategy to encourage brand loyalty and addictive consumer behavior by offering an abundance of media content that can be decoded only by purchasing more media technologies. Amazement results from the maze-like experience of ever-expanding, self-deconsructing transmedia storyscapes which invite audiences to get *all* the magic yet clearly prevent ultimate satiety by promising newer and newer adventures to collect.

References

Baardsen, Dana. 2017. "These Beauty and the Beast Collectibles are Everything." *Bestproducts*, March 8. Accessed July 29, 2017. www.bestproducts.com/parenting/kids/g2387/disney-beauty-and-the-beast-products/.

Bacchilega, Christina. 2013. *Fairy Tales Transformed? Twenty-First Century Adaptations and the Politics of Wonder*. Detroit: Wayne State University Press.

Baudrillard, Jean. 1994 [1981]. *Simulacra and Simulation*. Translated by Sheila Glaser. Ann Arbor: University of Michigan Press.

Bauman, Zygmunt. 1987. *Legislators and Interpreters: On Modernity, Postmodernity, and Intellectuals*. Cambridge: Polity.

Beckett, Sandra L. 2009. *Crossover Fiction: Global and Historical Perspectives*. New York: Routledge.

Beeler, Karin, and Stan Beeler. 2014. *Children's Film in the Digital Age: Essays on Audience, Adaptation and Consumer Culture*. Jefferson: McFarland.

Bell, Elizabeth, Lynda Haas, and Laura Sells (eds.). 1995. *From Mouse to Mermaid: The Politics of Film, Gender, and Culture*. Bloomington: Indiana University Press.

Bennett, Jane. 2010. *Vibrant Matter: A Political Ecology of Things*. Durham, NC: Duke University Press.

Bolin, Göran. 2007. "Media Technologies, Transmedia Storytelling and Commodification." In *Ambivalence Towards Convergence: Digitalization and Media Change*, edited by Tanja Storsul and Dagny Stuedahl, 237–248. Gothenburg: Nordicom.

Bolter, Jay D., and Richard Grusin. 2000. *Remediation: Understanding New Media*. London: MIT Press.

Budd, Mike, and Max H. Kirsch. 2005. *Rethinking Disney: Private Control, Public Dimensions*. Middletown: Wesleyan University Press.

Cocteau, Jean, dir. *La Belle et la Bête*. DisCina, France. 1946.

Condon, Bill, dir. *Beauty and the Beast*. Walt Disney Pictures, Mandeville Films, US/UK. 2017.

Disney's Beauty and the Beast: Magical Ballroom. Action Strategy PC game for Windows. Disney Interactive Studios, 2000.

Disney, Walt, dir. *Beauty and the Beast's Enchanted Christmas*. Walt Disney Pictures. 1997.

Featherstone, Mike. 1991. *Consumer Culture and Postmodernism*. London: Sage.

Giroux, Henry A., and Grace Pollock. 2010. *The Mouse that Roared: Disney and the End of Innocence*. Plymouth: Rowman & Littlefield.

Glass, Philip. *La Belle et la Bête*. Opera, 1994.
Hanson, Matt. 2003. *The End of Celluloid: Film Futures in the Digital Age*. Brighton: Rotovision.
Jenkins, Henry. 2006. *Convergence Culture: Where Old and New Media Collide*. New York: New York University Press.
Jenkins, Henry. 2007. "Transmedia Storytelling." *Confessions of an Aca-Fan. The Official Weblog of Henry Jenkins*, March 22. Accessed July 9, 2017. http://henryjenkins.org/2007/03/transmedia_storytelling_101.html
Jenkins, Henry, and David Thorburn (eds.). 2003. *Rethinking Media Change: The Aesthetics of Transition*. Cambridge, MA: MIT Press.
Jones, Jody. 2017. "A Tale as Old Time." *Primark*, March 13. Accessed July 3, 2017. www.primark.com/en/features/women/2017/march/beauty-and-the-beast.
Kirby, Alan. 2009. *Digimodernism: How New Technologies Dismantle the Postmoderm and Reconfigure Our Culture*. New York: Continuum.
Lury, Celia. 1996. *Consumer Culture*. Cambridge: Polity Press.
McNeal, James U. 1990–1991. "Children as Consumers: Insights and Implications." *Media & Values*, Fall-Winter: 52–53. Accessed June 20, 2017. www.medialit.org/reading-room/savers-spenders-how-children-became-consumer-market.
Maltby, Richard. 1998. "Nobody Knows Everything. Post-Classical Historiographies and Consolidated Entertainment." *In Contemporary Hollywood Cinema*, edited by Steve Neale and Murray Smith, 21–46. New York: Routledge.
Marshall, David P. 2004. *New Media Cultures*. London: Hodder Arnold.
Pilinovsky, Helen. 2011. "*Salon des Fées*. Cyber Salon: Re-Coding the Commodified Fairy Tale." In *Postmodern Reinterpretations of Fairy Tales*, edited by Anna Kérchy, 17–33. Lewiston, Lampeter: The Edwin Mellen Press.
Ritzer, George. 1993. *The Macdonaldisation of Society*. Newbury Park: Pine Forge Press.
Ritzer, George, and Nathan Jurgenson. 2010. "Production, Consumption, Prosumption: The Nature of Capitalism in the Age of the Digital 'Prosumer.'" *Journal of Consumer Culture* 10 (1): 13–36.
Robinson, Tasha. 2017. "The Beauty and the Beast Remake is a Long Series of Wasted Opportunities." *The Verge*, March 17. Accessed July 14, 2017. www.theverge.com/2017/3/17/14962212/beauty-and-the-beast-review-remake-gay-lefou-bill-condon-controversy.
Salecl, Renata. 2010. *Choice*. London: Profile.
Salecl, Renata. 2013. "The Myth of Choice for Children and Parents: Why We Deny the Harm Being Caused to our Children." In *Exploiting Childhood: How Fast Food, Material Obsession and Porn Culture Are Creating New Forms of Child Abuse*, edited by Jim Wild, 98–109. London: Jessica Kingsley.
Shah, Freya. 2017. "Belle's Beauty." *Primark*, June 28. Accessed June 13, 2017. www.primark.com/en/features/beauty/2017/june/beauty-and-the-beast.
Stephens, John, and Robyn McCallum. 2000. "Film and Fairy Tales." In *The Oxford Companion to Fairy Tales: The Western Fairy Tale Tradition from Medieval to Modern*, edited by Jack Zipes, 160–164. Oxford: Oxford University Press.
Toffler, Alvin. 1980. *The Third Wave*. London: Pan Books.
Warner, Marina. 1995. *From the Beast to the Blonde: On Fairy Tales and their Tellers*. London: Vintage.
Wise, Kirk, and Trousdale Gray, dir. *Beauty and the Beast*. Walt Disney Pictures, US/UK. 1991.
Zipes, Jack. 1995. "Breaking the Disney Spell." In *From Mouse to Mermaid: The Politics of Film, Gender, and Culture*, edited by Elizabeth Bell, Lynda Haas, and Laura Sells, 21–43. Bloomington: Indiana University Press.
Zipes, Jack. 2011. *The Enchanted Screen: The Unknown History of Fairy-Tale Films*. New York: Routledge.
Žižek, Slavoj. 1989. *The Sublime Object of Ideology*. London: Verso.

25
TRANSMEDIA FRANCHISING
Driving Factors, Storyworld Development, and Creative Process

Peter von Stackelberg

Transmedia franchising has grown rapidly in number and sophistication since 2010, when the transmedia revolution began to gain momentum. The pattern of "reboots, remakes, prequels, sequels, adaptations, and shared cinematic universes rule the day" (Burt 2016) as the film industry seeks new creative and business approaches in a rapidly evolving media environment. This changing media environment is driving the development of transmedia franchises, which are likely to absorb more traditional types of media franchises as that evolution continues.

Various types of media franchises have been in place since at least the late nineteenth century and early twentieth century. The emergence of transmedia storytelling in the late nineteenth and early twentieth century was a result of technological, economic, and social changes during the Second Industrial Revolution. The mass production and promotion of cultural products and new forms of advertising in the early twentieth century supported the development of early transmedia franchises. The emergence of mass audiences during this period went hand in hand "with the alignment of fictions told across platforms" (Freeman 2014, 2263).

> There are certainly precursors to the mass culture media-franchise phenomenon. In the past, popular children's films often inspired comic books, or vice versa. Some of these early franchises extended to an adolescent market (e.g. Superman, Spiderman, Batman & Robin).
>
> *(Lemke 2004, 6)*

Many early literary, film, and game franchises were focused on a main character. When Sir Arthur Canon Doyle wrote *A Study in Scarlet* in 1886 and published it the following year, he introduced Sherlock Holmes and created an enduring literary franchise that included multiple books and, later, films and television programs. Other literary franchises based on fictional characters—for example, Miss Marple and Hercule Poirot created by Agatha Christie or Nancy Drew written by Carolyn Keene (a pseudonym used by multiple authors who wrote books for the franchise)—emerged during the early to mid-twentieth century. Contemporary writers like James Lee Burke (author of the Dave Robicheaux mystery series), Kathy Reich (creator of a series of mystery novels featuring the character Temperance Brennan), and Daniel Silva (author of a series of action-spy novels) are among those using fictional characters around which they build literary franchises that typically involve multiple novels. Many such literary franchises have been expanded to include films and/or television series.

Film franchises like Tarzan, the Lone Ranger, and the Ma and Pa Kettle emerged during Hollywood's film studio era. Character-based film franchises continue to thrive, with some like James Bond and Jason Bourne, eclipsing the books from which they originated. The emergence of video games in the 1980s led to the development of franchises such as Super Mario and Kong as game producers sought to replicate early successes with follow-up games.

An early and notable exception to this character-based structure for media franchises was the world of Oz created by L. Frank Baum, author of *The Wonderful Wizard of Oz*. This storyworld was the basis for an early transmedia franchise that included four novels, a theater production, a comic strip, and a mock newspaper. Between 1900 and 1907, Baum created an "expansive, unfolding transmedial storyworld" that stemmed at least in part from Baum's "apparent aptitude to develop his fictional creations according to the emerging commercial systems of modern advertising" (Freeman 2014, 2366).

The 1970s saw the beginning of a shift away from character-based franchises. *Star Trek* and later *Star Wars* began to place a greater emphasis on the storyworld. While both of these franchises initially featured a cast of characters that remained largely the same (although, in the case of some *Star Trek* films, with the same cast of characters replaced with different actors), new characters joined the storyworld in later productions. The shift from character-based to storyworld-based transmedia, at least in the case of *Star Wars*, was not so much a planned extension of the franchise as an ad hoc process resulting from contingency planning over a period of years. Freeman argues that uncertainty about whether *A New Hope*, the first of the *Star Wars* movies, would be a commercial success resulted in the development of *Splinter in the Mind's Eye* as a world-building contingency (Freeman 2017, 66–67) that could be used as the basis for a low-budget sequel if the initial *Star Wars* film was commercially unsuccessful.

The importance of storyworld is apparent with the many film franchises that have been created over the first two decades of the twenty-first century. Fourteen films in Marvel's Cinematic Universe franchise grossed more than $4 billion between 2008 and 2016; nine films in the *Star Wars* franchise grossed $3.7 billion; and six films in *The Lord of the Rings* franchise grossed more than $1.8 billion (Dirks n.d.). This shift from character-based to storyworld-based franchises also occurred in the video game industry, with characters like Mario and Kong giving way to the increasingly sophisticated storyworlds needed for first-person games like *Grand Theft Auto*, *Halo*, and *Call of Duty*.

While the shift to storyworld-based transmedia franchises, the role of strong characters in these franchises continues to be important. C-3PO and R2-D2, the two droids who were the first characters introduced in *Star Wars: A New Hope*, are important figures in representing and maintaining narrative cohesion and stability across various elements of the *Star Wars* transmedia franchise (Lomax 2017, 41–42). Other characters in the *Star Wars* storyworld have been used in novels, comics, and animated series in addition to the various films (Geraghty 2017, 117–118).

Transmedia Franchises versus Traditional Media Franchises

The emergence of transmedia franchises is blurring the line between traditional literary, film, and game franchises and franchises that encompass a broad range of media. Media franchises have been defined as a "systemic structure, or network of texts that work together across a variety of platforms under a single unifying brand name, image, or concept to create an imaginary world" (Johnson 2009).

Johnson's definition of "media franchise" is very close to the generally accepted definition of "transmedia." One definition of transmedia storytelling is that it is a process in which episodes of a story are spread across multiple platforms (Gambarato and Nani 2016). This focus on the narrative design of transmedia storytelling is central to many definitions of "transmedia." Slota's definition is similar: "Transmedia storytelling is the distribution of a particular narrative (i.e., the holistic story universe, not to be confused with an individual plot or theme) across multiple delivery channels and technologies" (Slota 2017).

A focus on the nature of the commercial aspects of transmedia narratives has also been used to define what they are.

> Transmedia storytelling is characterized by an often complex network of interrelationships involving license holders and licensees, producers, consumers, and prosumers. Such interactions are further complicated by processes of production, distribution and consumption, and by the relationships across specific media platforms utilized by the franchise or project in question.
>
> *(Harvey 2013, 115)*

Robert Pratten notes that many traditional definitions of transmedia storytelling focus on the methods of production, instead of taking an audience-oriented perspective that focuses on the consumption of a narrative. He states that "transmedia storytelling is a design philosophy" that attempts to create synergy between the content of the narrative and an emotional, participatory experience for the audience (Pratten 2015, 2–3).

It is important to note, however, that not all transmedia projects are franchises and not all media franchises are transmedia. Transmedia storytelling spans a broad range of storytelling projects, from marketing and advertising campaigns to public service projects that intend to educate, engage, and inspire to cross-platform entertainment projects. *The Blair Witch Project*, *Year Zero*, *The Lizzie Bennet Diaries*, and many other projects "could all be reasonably called transmedia even though they're wildly different in structure, content and scope" (Burns 2016).

Many early transmedia narratives—Robert Pratten's *Lowlifes*; the alternate reality game (ARG) *Year Zero* based on Nine Inch Nails album of the same name; and *Welcome to Pine Point* by Paul Shoebridge and Michael Simmons—were single stories delivered across multiple media (i.e., books, graphic novels, film, video, live events, video games, etc.) or, increasingly, on a single platform like a smartphone, tablet, or smart television that presents a single narrative using multiple modes of communication (i.e., text, still images, video, audio, and animation).

The term "small-scale transmedia storytelling" has been applied to projects that integrate a short story with a soundtrack and comic to define various facets and perspectives with a larger narrative universe (Slota 2017). This definition of small-scale transmedia storytelling should be broadened to include any project that uses transmedia techniques to tell a single story, not just short stories. Many early transmedia narratives were "small-scale" and current independent transmedia projects also tend to be small-scale.

Transmedia franchises, on the other hand, typically consist of multiple fully developed narratives. The term "large-scale transmedia storytelling" has been used for complex, multi-author narrative universes that span films, books, comics, games and—at the highest level—concerts, theme parks, and themed vacations (Slota 2017).

Factors Driving Development of Transmedia Franchises

Media and transmedia franchises have gone from being a minor factor in Hollywood to near domination (Nicholson 2016). The emergence of these modern transmedia franchises is the result of several factors, including:

- technological change;
- proliferation of media channels but declining audiences;
- commercial considerations; and
- fan experiences.

Technological Change

Technological change has been a key factor driving the development of transmedia franchises for more than a century. The technological changes of the Second Industrial Revolution—a period rapid industrialization in most Western societies that occurred from the 1820s through the 1940s (von Stackelberg 2014)—drove the emergence of mass communication on a scale never seen before. Technologies like photoengraving, commercial lithography, mass-produced newspapers (known as the "penny press"), paperback books, the rotary letterpress, hot metal typesetting machines, and many other print technologies enabled mass production and distribution of shared information on a nationwide scale (American Printing History Association n.d.). Likewise, technological change has helped the emergence of transmedia storytelling and transmedia franchises as we now know them during the Information Age—a period of more than a century between the 1920s and the 2040s, during which radio, television, computers, the Internet, and a myriad of other communications and computing technologies emerged (von Stackelberg 2014).

Since the 1960s, wave after wave of change has swept over the publishing, film, and television industries as new information technologies were introduced, disrupting and destroying not just individual companies but entire segments of those industries. The convergence of various modes of communication to single platforms like smartphones, tablets, and smart televisions now makes it technically feasible and commercially viable to deliver high quality text, still images, video, audio, and animation on a single device, eliminating the need for users to hop from one media platform to another.

Proliferation of Content Choices; Declining Audiences

In addition to changing the nature of media platforms, the technologies of the Information Age dramatically increased the choices audience have. The number of television channels increased a hundred-fold or more from the limited number of broadcast networks of the 1950s and 1960s to hundreds of cable TV channels in the early twenty-first century. The average American household had access to 206 different television channels in 2016, an increase of more than 50 percent from 2008, but the number of channels actually viewed was stagnant (MarketingCharts.com 2016).

The number of books published each year in the United States in 2016 exceeded 1 million, with more than 750,000 of them being self-published. This is a dramatic increase from the roughly 400,000 books published in 2007 (Piersanti 2016). While the number of books available increased dramatically, book industry sales have fallen or been flat since 2007 (Piersanti 2016).

The film industry is also experiencing declining audiences. The number of tickets sold to films shown in theaters in the United States and Canada has declined from more than 1.5 billion in 2002 to 1.2 billion in 2017 (TheNumbers.com 2017). The number of movies given wide release by the six major studios—Warner Brothers, Walt Disney, 20th Century Fox, Paramount Pictures, Sony Pictures, and Universal—has dropped from 110 in 1995 to 79 in 2017, while the total movie releases for other studios only increased from 28 in 1995 to 47 in 2017 (an overall decline of 12 films per year) (TheNumbers.com 2017). These trends are driving the television, book, and film industries to look for new ways to attract and hold audiences, with transmedia franchises becoming a primary focus.

Commercial Considerations

Commercial considerations are a key driver of transmedia franchise development. The use of modern advertising, licensing, and cross-sector industry partnerships has characterized transmedia practices over time (Freeman 2017, 62). Many funding bodies require transmedia strategies to expand a production across multiple platforms (Burns 2016). The globalization of entertainment is driving the creation of large entertainment franchises. Lynda Obst, author of a book on the film industry's "sequel mania" is quoted in the *Financial Times* as saying:

> You can't make movies the same way internationally as you do in the U.S. You can't pay for television advertising in every city in the world, so you become dependent on pre-awareness of the movie. The more the international audience is familiar with a title, the more they look forward to seeing it again. In Hollywood, familiarity breeds success, not contempt.
>
> *(Garrahan 2014)*

Accompanying the emergence of transmedia franchises has been the inclusion of many different stakeholders who are dependent on the profitability of the franchises they work within (Freeman 2017, 64). The development of long-term transmedia franchises is seen as one way to bring a measure of stability to a film industry which has become dependent on periodic big returns from a small number of blockbuster films. The Walt Disney Company has recognized the importance of transmedia franchises by embedding Transmedia Producers in all their franchise teams—building on a model of franchise development that has expanded to incredible profit in recent years (Burns 2016).

Transmedia franchising has extended the commercial lifespan of many entertainment products. "DVD release, TV broadcast, internet streaming, and transmedia franchisation of [intellectual property] have extended the afterlife of films into infinity" (Nicholson 2016).

Fan Experiences

As noted earlier, the proliferation of media channels and content has resulted in a crowded market for information and entertainment products. "The growing diverse competition is a big reason that sequels and franchises dominate the box office; it's enough that it's a struggle drawing audiences to theaters without the trouble of persuading them that original content is worth their while" (Nicholson 2016). The same challenges of drawing audiences face book authors and publishers, creators of graphic novels, and practitioners working with other forms of media.

> Until recently, the goal of developing [media] franchises was usually monetization through repurposing, whereas the concept of transmedia storytelling focuses primarily on the (more costly) expansion of the story itself. The question that then remains is—will consumers be satisfied with repurposed content (franchise development), or will they demand more (transmedia)?
>
> *(Weitbrecht 2011)*

From a business perspective, consumers' familiarity with a creator, her cast of characters, and a known storyworld can be a significant benefit of a media and transmedia franchise when competing in a crowded marketplace. The proliferation of content and the number of channels available for content distribution has made discoverability a significant challenge for all media products, whether they are films, television programs, books, or some other form of media. The development of a loyal base of fans for a transmedia franchise can help mitigate the challenges of discoverability by creating both a recurring source of revenue and opportunities to use fans' personal and social media connections to promote the franchise. "By having multiple connections on a personal level with audiences/consumers as part of the transmedia franchise, a company can begin to structure itself around those relationships" (*Evo News* 2017).

The audience experience created by transmedia franchises that span multiple media and extend over long periods of time far exceed encounters with unfranchised print and broadcast media (Lemke 2004). In a competitive media environment, this high-quality audience experience is a significant advantage for a transmedia franchise by creating a "strong sense of ownership and identification with them and with their points of view" (Lemke 2004, 4).

Transmedia Franchises and Storyworlds

The development of strong contemporary transmedia franchises requires the creation of sophisticated storyworlds; without expansive storyworlds, transmedia franchises would be much more limited in scope. The creation of sophisticated imaginary worlds has been central to some of the greatest epic fantasy and science fiction stories ever written (von Stackelberg and McDowell 2015). While the process of world-building—the creation of imaginary worlds with coherent geographic, social, cultural, and other features—has a long history, it is reaching new levels of sophistication in twenty-first-century science fiction. A storyworld is a cognitive construct that consists of recurring elements, among them, characters, significant objects, settings (Ryan and Thon 2014, 129). Rich storyworlds—the "universes" within which stories are set—provide detailed contextual rules-sets that develop a larger reality that extends beyond a single story, while potentially providing a deeper understanding of the underlying systems that drive these worlds (von Stackelberg and McDowell 2015).

As noted previously, early media franchises tended to be character-based, with a single character—usually the protagonist—typically carrying the franchise. Storyworld-based franchises were less common although some, like J. R. R. Tolkien's Middle Earth and Frank Herbert's Arrakis, developed as the desert planet in *Dune*, have existed since at least the early to mid-twentieth century. When *Star Trek* debuted in 1966, the television series was more character-based than it was storyworld-based. However, as the Trekkers cult phenomenon around *Star Trek* grew, an extensive storyworld was developed as much by fans as by the franchise owners and licensees. While *Star Trek*'s initial development as a storyworld-based transmedia franchise may have been largely unplanned (given that the original series was unceremoniously cancelled after its third season only to return stronger than ever because of fan pressure), storyworlds are now vitally important aspect of franchise development.

> Learning from success like *Pirates of the Caribbean*, *Disney Fairies*, *Tron Legacy*, and others, Disney put storyworlds first in its decision making. It's heavily invested in purchasing or expanding rights in existing storyworlds, Pixar's intellectual properties, *Star Wars*, the Marvel Cinematic Universe.
>
> *(Burns 2016)*

Developing, maintaining, and growing a healthy storyworld is essential to keeping a narrative franchise commercially viable. For example, what made *Star Wars* so influential was the existence of an entire world that characters—and the fans who followed them—could have adventures in (Rose 2011).

Gomez (2017) states the development of effective large-scale transmedia franchises requires extensive preparation of the storyworld and its underlying mythos. This can require a substantial investment in what is in essence a massive blueprint, not just a stack of manuscripts or screenplays. A well-developed storyworld should help the creators of individual stories understand how things work in that particular universe. It needs to have its internal logic and that logic must be applied consistently across all of the products in the transmedia franchise. "Great fictional worlds provide not just a source of characters, settings, and events that writers can draw from; they provide powerful messages in ways that help audiences realize those messages in their individual lives" (Gomez 2017). Careful development of the storyworld must be central to the creative efforts of a transmedia franchise, Gomez (2017) argues.

Transmedia Franchises and the Creative Process

The emergence of transmedia franchises has made possible "widespread creativity through the collaborative creation and collective consumption of narrative worlds" (Ciancia 2015, 131). This perspective that the development of transmedia franchises fosters collaborative and collective creativity is a broad generalization that speaks to the possibilities but to which there are many notable exceptions. For example, the relationship between *Star Wars* creator George Lucas, co-creators, and fans was at times contentious as Lucas fought to maintain creative control over many aspects of the franchise (Lomax 2017).

The role of the lone author, creative teams, and even fans in the creation of content for transmedia franchises involves a delicate balancing act. Solitary can provide a strong vision and a unified voice at critical points in the early development of a franchise. As a transmedia franchise grows, however, the creative processes will be significantly different from the processes used to create an individual book, graphic novel, television program, film, or video game.

Even a small-scale transmedia project can easily require far more skills than any one author, illustrator, or screenwriter can muster. The rapid evolution of digital media technology makes transmedia storytelling a viable option for many small projects. Professional and prosumer hardware and software provides sophisticated, low-cost tools for the production of transmedia stories (von Stackelberg and Eira Jones 2014, 59). While the cost of sophisticated media technologies has dropped rapidly, the development of the many skills needed to use these tools in the service of effective transmedia storytelling and franchise development is still a significant barrier. Finding good storytellers that can meet the needs of transmedia storytelling, particularly with large-scale franchises, is essential for the development of transmedia franchises, posits transmedia expert Jeff Gomez in interview to Andersen (2010). Gomez states unequivocally that good storytellers are relatively rare and we should cherish them, but stewards—people who fundamentally understand the vision behind the narrative and the characters and can make sure that vision is adhered to—are also essential as a transmedia franchise grows (Andersen 2010).

Talented writers, illustrators, cinematographers, directors, and other creative personal are in relatively short supply. Individuals who are skilled in multiple areas—for example, writing, visual design, game development, and ebook publishing—are even more scarce. As Gomez notes, "it takes a different creative skill set to devise a massive fictional universe than it does to write a screenplay" (Gomez 2017).

> What you need are [transmedia] writers who know how to create sprawling, massive storyworlds from whose strengths conventional screenwriters can mine great individual (smaller) stories. Shared universe writers' rooms require experts in long-form storytelling, storyworld and universe design, game design, and mythology.
>
> *(Gomez 2017)*

In addition, Gomez (2017) mentions that a major skill required for creators of a transmedia franchise is a working knowledge of epic storytelling.

> Classic epic storytelling is about how an entire people must grapple with their morals, values, and afflictions, in order to gain a distinct and extraordinary identity. They are about the birth of entire cultures, and how they aspire for shared and common fulfillment.
>
> *(Gomez 2017)*

The many stories that emerge from a shared storyworld do not need to answer all of the questions at the core of that universe, but they do need to present mysteries in a grand and passionate way "that your audience will drive themselves crazy trying to answer" (Gomez 2017).

The interactive nature of transmedia storytelling will require creators working within a franchise to accommodate the needs and interests of their audience in ways that different from more traditional media franchises. Transmedia designers will have to manage their relationship with fans and be prepared to modify or alter stories in response to audience interests and focus (*Evo News* 2017).

The participation of fans in the creation of content—i.e., fan fiction—has occurred with numerous transmedia franchises. A number of transmedia franchises, *Iron Sky* and *The Cosmonaut* among them, have taken fan participation to a new level through the use of crowdfunding of pre-production and production costs (Ciancia 2015, 138). The extent of the control transmedia franchise creators or owners have over their storyworlds is a complex exercise of balancing the need to engage and build a fan base by allowing those fans to use the storyworld for the creation of their own content while continuing to maintain control over both the canon of the storyworld and legal ownership of the franchise's intellectual property (Lindsay 2014, 61–62).

The growth of transmedia franchises has prompted numerous concerns that they will kill creativity and originality in storytelling. "(W)hen George Lucas introduced us to the Star Wars universe, it also opened the doors to sequels and franchises, blotting out drama and original content" (Nicholson 2016). While creativity and originality can suffer as a result of massive transmedia franchises, that has more to do with the individuals working on a franchise than nature of media and transmedia franchises themselves.

Transmedia design, whether an individual project or an entire transmedia franchise, requires the development of a business model, narrative context, and media structure (Ciancia 2015, 140–141). A more comprehensive framework for a transmedia project identifies ten points that work equally well for small-scale transmedia projects and sprawling large-scale transmedia franchises. The ten points creators must address are premise and purpose of the transmedia franchise; the narrative the franchise will tell; the storyworld the franchise will encompass; the characters who will inhabit the franchise's storyworld; extensions to the franchise's overall narrative; media platforms and genres the franchise will use; audience and market for the franchise; how the franchise will engage its fans; the overall structure of the franchise; and the aesthetics that contribute to the overall cohesiveness of the franchise (Gambarato 2013, 90–95).

Von Stackelberg proposes an approach to transmedia narrative design that can be applied to either individual transmedia projects or to a series of projects across a transmedia franchise. He identifies three key design phases (von Stackelberg 2011, 115–116):

- engagement design, which focus on the users' cognitive engagement with and participation in the narrative;
- narrative design, which focuses on the story elements of the transmedia narrative; and
- interaction design, which focuses on how users interact with the narrative's interface and navigate through the narrative.

Four levels of design tasks are identified—project, storyworld, story, and scene/sequence (von Stackelberg 2011, 161–281). Two of these levels—project and storyworld—are particularly relevant to the development of transmedia franchises. Within these levels, the tasks identified correspond closely with the ten-point framework proposed by Gambarato (2013).

Conclusion

As technological change and consumers tastes change, traditional approaches to storytelling will continue to rapidly evolve. The emergence of the transmedia franchise is a key element of that evolution. While media franchises are not new, the cross-media nature of transmedia franchises means they will differ substantially from the literary, film, video game, and other media franchises that preceded them. Commercial pressures will drive the adoption of transmedia franchises across the various media sectors, resulting in the integration of a variety of media platforms to an extent not seen in the past. This shift to transmedia franchises will in turn, drive the need for new creative approaches that mean authors, illustrators, filmmakers, and others will need to become skilled in a number of areas and be able to work effectively in cross-disciplinary teams.

As with most technology-driven change, the shift to transmedia franchises will present significant challenges to those working in various forms of media. It means the traditional divisions between the film, television, book publishing, and video game industries will continue to be erased. It will also mean that some organizations will not survive as they fail to make the transition to a world in which transmedia franchises are dominant. For those who can adapt, however, the growth of transmedia franchises will open new commercial and creative opportunities.

References

American Printing History Association. n.d. *History of Printing Timeline*. Accessed December 15, 2017. https://printinghistory.org/timeline/.

Andersen, Michael. 2010. "Jeff Gomez Reveals Secrets to Transmedia Franchise Development at Cinekid." *Wired*, November 18. Accessed June 11, 2017. www.wired.com/2010/11/jeff-gomez-reveals-secrets-to-transmedia-franchise-development-at-cinekid/.

Burns, Caitlin. 2016. "Transmedia: Art Forms Created in Real Time." *Immerse*, November 15. Accessed July 12, 2017. https://immerse.news/transmedia-art-forms-created-in-real-time-4943648389a4.

Burt, Kayti. 2016. "Why the Franchise Is Hollywood's New Genre System." *Den of Geek!* July 22. Accessed August 15, 2017. www.denofgeek.com/us/tv/movies/256966/why-the-franchise-is-hollywoods-new-genre-system.

Ciancia, Mariana. 2015. "Transmedia Design Framework: Design-Oriented Approach to Transmedia Research." *International Journal of Transmedia Literacy* 1 (1): 131–145. https://mafiadoc.com/transmedia-design-framework-design-oriented-led-on-line_59d6d19d1723dd.

Dirks, Tim. n.d. "All-Time Top Film Franchises." *AMC Filmsite*. Accessed July 12, 2017. www.filmsite.org/series-boxoffice.html.

Evo News. 2017. "Transmedia Storytelling to the Rescue." June 25. Accessed August 5, 2017. https://evonews.com/entertainment/2017/jun/25/transmedia-storytelling-to-the-rescue/.

Freeman, Matthew. 2014. "Advertising the Yellow Brick Road: Historicizing the Industrial Emergence of Transmedia Storytelling." *International Journal of Communication* 8: 2362–2381.

Freeman, Matthew. 2017. "From Sequel to Quasi-Novelization." In *Star Wars and the History of Transmedia Storytelling*, edited by Sean Guynes and Dan Hassler-Forest, 62–72. Amsterdam: Amsterdam University Press.

Gambarato, Renira R. 2013. "Transmedia Project Design: Theoretical and Analytical Considerations." *Baltic Screen Media Review* 1: 80–100. doi:10.1515/BSMR-2015-0006.

Gambarato, Renira R., and Alessandro Nani. 2016. "Blurring Boundaries, Transmedia Storytelling and the Ethics of C.S. Peirce." In *Ethics in Screenwriting: New Perspectives*, edited by Steve Maras, 147–175. Melbourne: Palgrave Macmillan.

Garrahan, Matthew. 2014. "The Rise and Rise of the Hollywood Film Franchise." *Financial Times*, December 12. Accessed August 15, 2017. www.ft.com/content/192f583e-7fa7-11e4-adff-00144feabdc0.

Geraghty, Lincoln. 2017. "Transmedia Character Building: Tracking Crossovers in the Star Wars Universe." In *Star Wars and the History of Transmedia Storytelling*, edited by Sean Guynes and Dan Hassler-Forest, 117–128. Amsterdam: Amsterdam University Press.

Gomez, Jeff. 2017. "10 Keys to Building Successful Shared Universe Movie Franchises." *Medium*, June 12. Accessed June 14, 2017. https://medium.com/@Jeff_Gomez/10-keys-to-building-successful-shared-universe-movie-franchises-a9d983884ad3.

Harvey, Colin B. 2013. "Transmedia Storytelling and Audience: Memory and Market." In *Digital World: Connectivity, Creativity and Rights*, edited by Gillian Youngs, 115–128. London: Routledge.

Johnson, Derek. 2009. "Franchising Media Worlds: Content Networks and the Collaborative Production of Culture." Ph.D. diss., University of Wisconsin-Madison.

Lemke, Jay. 2004. "Critical Analysis across Media: Games, Franchises, and the New Cultural Order." *MafiaDoc.com*. Accessed July 13, 2017. https://mafiadoc.com/critical-analysis-across-media-games-franchises-and-the-new-_59c22f9f1723ddc052bf1dd3.html.

Lindsay, David. 2014. "Franchises, Imaginary Worlds, Authorship, and Fandom." In *Law and Creativity in the Age of the Entertainment Franchise*, edited by Kathy Bowrey and Michael Handler, 52–74. Cambridge: Cambridge University Press.

Lomax, Tara. 2017. "Thank the Maker!" In *Star Wars and the History of Transmedia Storytelling*, edited by Sean Guynes and Dan Hassler-Forest, 35–48. Amsterdam: Amsterdam University Press.

MarketingCharts.com. 2016. "Who's Watching How Many TV Channels?" *Marketing Charts*, October 3. Accessed December 12, 2017. www.marketingcharts.com/television-71258.

Nicholson, Emma. 2016. "New Hollywood: The Franchise Factory." *Digital Killed the Video Star*, November 29. Accessed July 12, 2017. https://digitalkilledthevideostar.com/2016/11/29/new-hollywood-the-franchise-factory/.

Piersanti, Steven. 2016. "The 10 Awful Truths About Book Publishing." *Berrett-Koehler Publishers*, September 26. Accessed December 15, 2017. www.bkconnection.com/the-10-awful-truths-about-book-publishing.

Pratten, Robert. 2015. *Getting Started with Transmedia Storytelling: A Practical Guide for Beginners (2nd Edition)*. London: CreateSpace Independent Publishing Platform.

Rose, Frank. 2011. "The Art of Immersion: The Star Wars Generation." *Wired*, March 9. Accessed November 12, 2017. www.wired.com/2011/03/star-wars-generation/.

Ryan, Marie-Laure, and Jan-Noel Thon. 2014. *Storyworlds Across Media: Toward a Media-Conscious Narratogy*. Lincoln: University of Nebraska Press.

Slota, Stephen. 2017. "When You Wish Upon a Star (Wars): Transmedia Storytelling and Design." *In Sync Training*, May 11. Accessed June 14, 2017. http://blog.insynctraining.com/when-you-wish-upon-a-star-wars-transmedia-storytelling-design.

TheNumbers.com. 2017. "Domestic Movie Theatrical Market Summary 1995 to 2017." Accessed December 28, 2017. www.the-numbers.com/market/.

von Stackelberg, Peter. 2011. "Creating Transmedia Narratives: The Structure and Design of Stories Told Across Multiple Media." Master's thesis, State University of New York Institute of Technology. Accessed September 21, 2017. www.transmediadigest.com/wp-content/uploads/2012/02/Creating-Transmedia-Narratives-Thesis.pdf.

von Stackelberg, Peter. 2014. *Technology & the Future: Managing Change and Innovation in the 21st Century*. Kindle Edition. Alfred, NY: Jericho Hill Interactive.

von Stackelberg, Peter, and Ruth Eira Jones. 2014. "Tales of Our Tomorrows: Transmedia Storytelling and Communicating About the Future." *Journal of Futures Studies* 18 (3): 57–76.

von Stackelberg, Peter, and Alex McDowell. 2015. "What in the World? Storyworlds, Science Fiction, and Futures Studies." *Journal of Futures Studies* 20 (2): 25–46.

Weitbrecht, Christine. 2011. "Transmedia vs. Franchise Development." *Christine Weitbrecht Blog*, October 16. Accessed May 11, 2017. http://christineweitbrecht.com/2011/10/transmedia-vs-franchise-development/.

26

TRANSMEDIA DISTRIBUTION

From Vertical Integration to Digital Natives

Elizabeth Evans

Distribution has historically been an often-overlooked part of media scholarship, with a greater amount of attention being paid to processes of production, reception, and occasionally exhibition. However, there is a growing recognition, especially within media industries research, that distribution is not only a vital component of media operations, but also a rich site for the exploration of how industrial and political structures shape our experiences of media content. Denise Mann, for instance, argues that "[d]istribution, the ability to provide audiences with access to content, is at the very heart of the entertainment industries" (Mann 2014a, 18). Alisa Perren demonstrates the centrality of distribution through her description of the work that media distributors do:

> Distribution companies have been labeled as "middlemen" and their employees as "intermediaries" responsible for ensuring that media find an audience. Yet distributors handle a large variety of different tasks. Overviews of the media industries typically identify the following as the primary emphases of distributors: assembling financing, procuring and/or licensing rights for projects for various platforms (e.g., iTunes, Netflix) or markets (e.g., Japanese theatrical, Latin American satellite television), managing the in flow and out flow of income from various corporate partners, designing release schedules and marketing strategies to establish and sustain audience awareness, and building and managing libraries.
>
> *(Perren 2013, 166)*

Distribution therefore focuses on when, how, and where audiences are provided with access to content within a global and increasingly technologically dispersed marketplace. Perren goes on to argue that the changes wrought by digital technologies (those same technologies that have facilitated a burst of transmedia logics) have placed a spotlight on the key area of concern when thinking about distribution: "the ways that content moves through space (flows) and time (windowing)" (Perren 2013, 167). In taking Perren's definition of distribution, the connections to notions of transmediality become clear. Transmedia logics are also about the ways in which content moves through space and time. The various ways in which transmediality manifests that are explored in this volume are fundamentally tied to practices of distribution. Transmedia storytelling or marketing, for instance, rely on distribution strategies that carefully spread content across different media platforms and spaces. The temporality of transmedia content (whether deliberately strategized or emerging more organically) is equally key to creating transmedia experiences. Transmediality is inherently *about* distribution.

Transmedia distribution can, to a certain extent, be defined as the placing of media or content on multiple different distribution platforms. A "television episode," for instance, may appear via broadcast, DVD/Blu-ray, streamed via a VOD service, or downloaded via a digital retail store; it may be experienced on a television set, laptop or computer, tablet or mobile phone. At the same time, though, critical thinking around transmediality has also sidelined certain ideas of distribution, especially the idea that the same content may be available on multiple platforms. Henry Jenkins, for instance, labels the same content on different platforms as "cross-platform," as there is "no opportunity here for additive comprehension, no real reason for the same consumer to visit these various hubs to learn anything new" (Jenkins 2016). Placing the same content on different platforms is clearly not the same as placing different pieces of a storyworld, brand, performance, or similar on different platforms. Importantly, transmedia distribution does not equal transmedia *storytelling*. However, in terms of how cultural content is increasingly experienced, the fact that content is not just available via one distribution outlet still has important consequences for understanding processes of transmediality. From the perspective of the audience, the connection between platform proliferation and transmedia experiences becomes clear and any understanding of transmediality as a lived experience by audiences and as a strategic approach by the media or cultural industries must embrace the full consequences of transmedia distribution. This includes the idea that experiencing a transmedia televisual experience like *Doctor Who* may involve never turning the television set on because its episodes are available via video on demand (VOD) services on a laptop or tablet. Similarly, experiencing the full transmedia version of the Marvel Cinematic Universe (comprising film, web videos, video games, and comics) may never involve going to the cinema, watching the television set or even reading a comic or graphic novel made out of paper. What is crucial here is that we are not talking about adaptation (a possible source for the uneasiness of considering such phenomenon as transmedia). This is not content being adapted from one form to another. This is a piece of content being explicitly placed on multiple platforms either consecutively or simultaneously, opening up ever more ways in which both content and audiences can move transmedially.

This chapter will consider the proliferation of (primarily digital) platforms that have reshaped the nature of media distribution in the twenty-first century and in turn acted as the foundations of contemporary transmedia strategies. These platforms, and the way they are being utilized by content creators and owners, are contributing to media culture becoming increasingly and inherently transmedial. As audiences, we no longer experience media within rigid platform-defined ways; the choices of distribution methods available to us means that we can move across and between different media platforms in ways of our choosing, turning even single media texts into *trans*media experiences. Michael Curtin, Jennifer Holt, and Kevin Sanson identify these shifts as a "distribution revolution," one that acts as "the latest iteration of an on-going tension between the diverse desires of audiences for cheap and easy access and the twentieth-century business models that sought to manage media flows and audience consumption" (Curtin, Holt, and Sanson 2014, 6–7). In order to streamline the discussion, the focus will mainly be on the United States (US) and United Kingdom (UK) services relating to film and television content, however consideration will also be made of the gaming and publishing sectors where appropriate. Despite this focus, many of the issues raised by transmedia distribution and discussed below function as useful starting points in their application to other media locales and sectors.

Mapping Transmedia Distribution

In *Transmedia Television,* I posited the notion of distribution as being a key part of transmediality by exploring how digital technologies were positioned as alternatives to the television set (Evans 2011, 40). At that point, in the early 2010s, the television industry was beginning to develop transmedia distribution strategies along a set of key categories. The Internet had created opportunities for audiences to become distributors of copyrighted content more easily, something that caused much consternation within the media industries (see also Lobato 2012, 96). On the one hand, the industry responded through legal attempts to shut down such sites that provided access to copyrighted content and public statements on how much piracy cost the creative economy (see, for example, Lodderhose 2014). At the same time, however, they began to develop their own distribution strategies in order to colonise the spaces of online media piracy. A key categorization for these early services, therefore, was the relationships between different stakeholders and the comparative legality or legitimacy of each platform. Such categorizations are particularly important given what Ramon Lobato calls the "'grey' services that sit somewhere in between legal online movie rental/download-to-own services ... and the more proudly piratical peer-to-peer networks" (2012, 96). Nascent official platforms such as broadcaster catch-up services sat alongside "guerilla networks" such as peer-to-peer downloading services and collaborations both within the industry and with external third parties.

The rapid expansion of digital platforms requires this model to be reconsidered in order to map the distribution networks that sit underneath contemporary transmedia strategies and the consequences of these networks for how we can understand transmedia culture. It is possible to categorise the increasingly transmedial nature of the media industries in multiple ways. The legality of services remains a key factor in distinguishing the myriad of ways in which audiences can access content (see Crisp 2015), however other factors have also emerged. Business models form a key point of differentiation with subscription-funded video on demand (SVOD) and advertising-funded video on demand (AVOD) acting as the dominant two categories. Similarly, the type of access offered varies, with some services offering permanent access ("ownership") and others only temporary access (rental or catch-up services). However, as distribution is ultimately about the pathways in which content creators provide audiences with access to their content, I will focus here on industrial ownership. Much has changed over the past decade and three new trends are emerging that shape transmedia distribution within the media industries: new forms of vertical integration and conglomeration, collaborations, and the emergence of digitally native providers.

New Forms of Vertical Integration and Conglomeration

The first trend speaks to the re-emergence and strengthening of organisational structures that characterized Hollywood in the twentieth century: vertical integration and conglomeration. Vertical integration describes how film studios in the 1930s and 1940s owned the entire production, distribution and exhibition pipeline. In 1948, a US Supreme Court ruling against Paramount Studios required the film studios to sell the cinemas that they owned, ending the period of vertical integration. The period after the Paramount Decree not only saw the breaking up of the studios' monopolies over film production, distribution and exhibition but eventually the emergence of newer media and distribution options in television, home entertainment (VHS, DVD and then Blu-ray), and the Internet. During the 1980s the US-based media industries shifted to a different, horizontal, form of integration as film studios and television broadcasters were increasingly bought by large multi-national corporations through processes of conglomeration. This process was not confined to film and television, as these growing media empires also sought to expand into music, gaming, publishing, and leisure spaces. The result has been six major, global corporations (Comcast, Disney, Vivendi, Time Warner, 21st Century Fox, and CBS) that maintain control over the vast majority of US media production and that operate across individual media sectors. Although mainly examined in terms of the US media industries, the global nature of these conglomerates impacts on sectors outside of the United States, for instance in Fox's involvement in satellite television across Europe and Asia via Sky Plc.

To return to transmedia distribution, both vertical integration and conglomeration are concerned with securing access over multiple different connection points between audiences and content. In essence, they are both about consolidating power over distribution networks, be those networks medium specific or sitting across multiple media sectors. It is within this context that recent transmedia distribution strategies continue the traditions of vertical integration and conglomeration. The historically well-established film studios and television broadcasters, for instance, have increasingly developed or bought VOD services in order to control online access points to their content. Film studio-owned VOD services such as Flixster (co-owned by Comcast/NBCUniversal and Time Warner) and Disney Life both keep ownership of online access channels in house. In the UK, the television-oriented VOD landscape remains dominated by the major broadcasters, most notably the BBC with its iPlayer service, but also the ITV Hub, Channel 4's All 4, and Sky's Sky Go and Now TV. These services expand the broadcasters' distribution networks away from television broadcasting and involve the creation of large libraries of content that can be accessed when audiences choose (see Evans and McDonald 2014). Within the video game sector, both Sony and Microsoft have developed online stores (PlayStation Store and Xbox One Games) that not only keep control of content distribution but also tie that distribution to their proprietary consoles. In comics, Disney-owned Marvel have created Marvel Unlimited Comics, a subscription app where customers can download digital copies of Marvel comics. Just as vertical integration offered the film studios of classical Hollywood consolidated power over their entire product chain (from production, through distribution to exhibition), contemporary transmedia distribution strategies do the same. Established media industries organizations have used transmedia distribution strategies to maintain control over how, where, and when their audiences access content.

Intra-Industry Collaborations

The second category of transmedia distribution is closely related to notions of vertical integration and conglomeration but is defined through deliberate collaboration between content creators and owners. Such services have primarily emerged within the film and TV sectors. The most publically visible are US streaming service Hulu (co-owned by Disney, Fox, NBCUniversal, Warner Bros) and Ultraviolet, a collaboration between 20th Century Fox, Warner Bros, Paramount, Sony Picture, Lionsgate, and the BBC, which allows consumers to access digital versions of the DVDs they purchase. Other forms of collaboration also emerge, however, for instance in ownership of streaming service Flixster (NBCUniversal/Comcast and Time Warner) and Britbox (a collaboration between UK broadcasters the BBC and ITV to distribute their content in the United States). Such collaborations are most common within the film and television industries, but a 2017 collaboration between video game streaming site Twitch and games developer Blizzard (Bryant 2017) suggests that such collaborations are not exclusively film or television oriented. These services demonstrate attempts by the established film and television industry to build greater capacity through collaborative distribution and investment in new platforms. By pooling content together, it becomes possible to both share development and marketing costs for a new service and offer a larger, and more attractive, library of content to audiences. In many ways, this acts as an extension of the strategies discussed above in terms of vertical integration and conglomeration, but with media corporations entering into partnerships (rather than purchasing deals) to acquire distribution outlets. This then gives them both an outlet for their own content and a means to generating financial return on the distribution of content produced by other studios or broadcasters.

Digitally Native Global Players

If the previous two trends have been about the established media industries consolidating their power during a period of technological change, it is also necessary to recognise how the scope of the industry itself has been shifting. As a number of scholarly collections have explored, the media industries have undergone an intense period of challenge and change in response to the development of digital distribution platforms, especially in the United States (see, for example, Mann 2014a; Holt and Sanson 2014; Lotz 2017). Most notably, digitally native companies have emerged and begun operating as distributors of content and, as a result, offer a challenge to the historically dominant studios, broadcasters, and publishers. While the emergence of digital platforms and technologies capable of displaying media content has offered audiences more choice about how, where and when they access content, they also have greater choice over who they access that content from (and who they pay—or do not pay) for it. Not only have distribution strategies changed, but who the distributors are has also changed. There has been a long history of illegitimate media distribution, stretching back to the pirating of 35mm films and continuing through (see, for example, Lobato 2012; McDonald 2007; Crisp 2015). More recently the boundaries of legitimate distribution have dramatically changed as digitally native companies, some whose primary business is decidedly un-media related have begun operating as media distributors.

The most well-known of these services are Netflix, Amazon Prime Video, YouTube, and Apple's iTunes and Apple TV, but we can also consider other services such as Steam (video games), Comixology (comic books), and Google Books within the same category. All emerge from companies with a primary focus on the digital economy (online shopping, Internet searches, computing) and so can, to a certain extent, be considered digitally native. They therefore offer new distribution channels for the established media industry players, but also mean significant competition for those players. This sense of competition is also increasingly shifting to production. The major global VOD services, Netflix, Amazon, and Apple TV, have all evolved from being acquisition only spaces to commissioning and producing their own content (Ullin 2014, 369–371). On a side note, it is relevant to consider that this pattern is not exclusive to VOD spaces, with digital television channels in the UK and US also beginning to move from repeats of bought-in programs to original production. In an alternative (though currently unique) model, the BBC has sought to re-position its youth-oriented channel BBC Three as a digitally native space by ceasing broadcast transmission and producing original content for initial release via the iPlayer. In addition, Denise Mann, for example, has described the emergence of YouTube-based channels, or multi-channel networks (MCNs) as allowing the development of "grassroots creativity" (Mann 2014b). By initially focusing on distribution as a strategy, these companies have been able to establish themselves as key media institutions and develop an audience base before the heavy investment required for producing original content. Ultimately, the emergence of digitally native players has not only shaken up the parameters of transmedia distribution but also demonstrated how distribution can be understood as a key precursor strategy for content production.

Breaking and Reforming the Bottleneck: Understanding Transmedia Distribution

The proliferation of distribution avenues evident in the above categories demonstrates how audiences have multiple options for accessing content and how increasingly our experiences of media culture inherently involve moving across different platforms and technologies. Media culture is ever more defined by that movement and, consequently, is becoming ever more defined by notions of transmediality. This has had two interconnected consequences for our understanding of the media industries and how the underlying structures of the industry shape the experiences of audiences.

The first consequence is a loosening of the oligopoly over how audiences access content that had previously been defined by the major film and broadcast companies that were established in the first half of the twentieth century. In discussing the US television industry of the late 1990s, Timothy Todreas identified the "distribution bottleneck," which he described as "the stranglehold that stations and cable operators have had in the television business. Ownership of the bottleneck allowed the distributors of video to capture most of the profits available to the industry" (1999, 9; see also Evans 2011, 41–42). For Todreas, the web offered the potential for that bottleneck to be broken: "if creators merely want a presence on the Web, all they have to do is hire designers and programmers in order to build their own sites" (1999, 100). Todreas was writing in the closing years of the twentieth century, just as the infrastructure companies behind the Internet were beginning to develop its capacity to delivery high-bandwidth content such as audiovisual material. Since then, the expansion of the Internet's capabilities has been exponential and whilst Todreas's prediction has come true to a certain extent, the distribution bottleneck has not completely broken. Whilst it is now easier to launch content without the backing of a major studio, broadcaster, or publisher, a small number of companies remain the dominant providers of media content. That number may have expanded to include new players such as Netflix and Amazon but any loosening of the distribution bottleneck has occurred only within certain parameters. As Denise Mann notes, the development of transmedia storytelling through formats such as alternate reality games has reinforced established hierarchies, arguing that "[l]ess well-funded and well-known digital creators find themselves shut out by Hollywood's institutionalized gatekeepers because their work lacks clear-cut paths to monetization or ignores copyright restrictions" (Mann 2014a, 4–5). As Jean Burgess and Joshua Green have argued, even YouTube (the bastion of user-generated content) has a "double function as both a 'top-down' platform for distribution of popular culture and a 'bottom-up' platform for vernacular creativity" (2009, 6). Although the options for accessing content may continue to multiply, the dominance of established players and certain digitally native newcomers remains strong.

Instead, the distribution bottleneck has broken in a slightly different way, one that relates to the increasingly globalized strategies prioritised by the new, digitally native content distributers. This is the second consequence of transmedia distribution. Whereas non-digital media have traditionally had nationally specific boundaries, digital media have merged with a more fundamentally global outlook. Such a merger is complex, however. On the one hand, the adherence to national boundaries has persisted, with geo-locking technologies allowing web-based services to limit access based on where a user is located. On the other hand, the global reach and outlook of VOD services, especially digitally native services, have opened up the scale at which content can be rolled out globally. Television is a good example here as broadcasters are often bound by national borders. Although BBC broadcast signals do reach across the border to the Republic of Ireland or across the English Channel to northern Europe, they are primarily bound to the British Isles in terms of target market, customer base, and government policy. This territorial protectionism extends to the BBC's own online spaces, with the iPlayer only available in the UK. BBC content in non-BBC owned online spaces, however, has a much more global reach. BBC content can be accessed via Netflix globally, with short clips also available via the Corporation's YouTube channels. Sam Ward, for example, has argued that Netflix in the UK opened up a space for US content that had been unable to find a broadcast home such as *Breaking Bad* and "has presented itself explicitly as an importer of content" (Ward 2016, 310). A similar situation has also occurred in the opposite direction, with Netflix in the US becoming a home for content from around the world, most notably non-English language television series that have not traditionally been found on US broadcast television such as *3%* from Brazil and *Nobel* from Norway. Todreas's vision of any producer being able to use the Internet to get their content to audiences has therefore, to a certain extent, happened, with user-generated content platforms such as YouTube. However, at the same time the distribution bottleneck has been broken in a slightly different way, instead around the kind (and origin) of professionally produced content audiences can now access.

Transmedia as Distribution Logic: A Case Study of "Carpool Karaoke"

Many of the issues at stake in transmedia distribution can be seen through the example of *The Late Late Show with James Corden* (CBS, 2015–), and its distribution strategy for the "Carpool Karaoke" segment. US late night chat shows offer a ready-made example of content that can be re-packaged for transmedia and "spreadability" purposes. The division of such programs into short sections of semi-discrete content (monologue, short sketches, and interviews with individual guests) immediately creates YouTube-ready short videos. As Myles McNutt has argued in reference to *The Late Late Show*'s "Carpool Karaoke" section:

> Rather than functioning as a "late night snack" from the show, then, "Carpool Karaoke" registers as its own standalone series, with each "episode" featuring a clear beginning (Corden picking up, or beginning his conversation with, the artist), middle (the interview and performance elements), and end (typically, either Corden thanking his guest or, in some cases, a sketch element arriving at Corden's studio).
>
> *(McNutt 2017, 582)*

Treating each segment as a coherent, self-contained unit of content allows for flexibility in distribution strategies that can exploit transmedia opportunities. By releasing segments online in advance of the program's broadcast (McNutt 2017, 579), such videos naturally provide marketing opportunities. This becomes more important for content that airs in the post-11pm "shoulder" period of the US television broadcast schedule, when viewers must be encouraged to stay up and watch rather than going to sleep (Ellis 2002, 132). By releasing further segments post-broadcast, the program's content almost automatically turns into additional marketing material for audiences who either miss the broadcasted airing, or choose not to watch it.

In the case of "Carpool Karaoke," the success of such a transmedia distribution strategy also opened up both the originating program and the segment to a global audience. McNutt argues that the VOD framing of "Carpool Karaoke" in particular has encouraged a break from its broadcast origins. He argues that

> Notably absent, however, is any reference to the show and its linear broadcast, which are not included in the end tag—"next time," in this case, is understood to be on YouTube, shifting away from YouTube as a promotional space to YouTube as the primary space of engagement for the "Carpool Karaoke" audience.
>
> *(McNutt 2017, 583)*

The separating out of this particular segment indicates the transmedia and global transformation of "Carpool Karaoke," but at the same time that transformation has remained bound up with a similar transformation for its *Late Late Show* origins as well. The global success of "Carpool Karaoke" on YouTube has led to new distribution deals for *The Late Late Show* in non-US territories, including those without the traditions of daily late-night comedy talk shows. In Germany, commercial free-to-air broadcaster RTL bought the broadcast rights and began airing un-dubbed episodes one day after their US airing. In the UK, Sky (who were already invested in Corden's star persona through his hosting of its comedy-sports quiz *A League of Their Own*, 2010–) acquired rights to the program in mid-2016. Ironically, those rights were limited to its VOD service Sky Go, with only a *Carpool Karaoke Special* on its broadcast Sky One channel (tx. 19/07/2016, plus three episodes filmed in London 9/6/2017-11/06/2017). What was broadcast content in the US was transformed into VOD-only content in the UK. In perhaps a clearer example of the inextricable link between content production and distribution within the transmedia turn, the transmedially distributed segment then became its own piece of full-length digitally-native content, when rights to a "Carpool Karaoke" series were bought not by a television broadcaster but by on-demand provider Apple TV.

Conclusion

The international and transmedia journey of "Carpool Karaoke" thus demonstrates many of the characteristics of transmedia distribution and the consequences of transmedia distribution strategies on the way we can understand the media industries and their practice. It involves a well-established media company (CBS) branching out into new online spaces in order to promote traditional television content and capture online audiences. This then created a greater global reach for the content, including in countries where CBS has no established broadcast base, further widening the content's audience through distribution channels that are already global. This in turn consolidated CBS's position by allowing them to create very traditional transnational sales deals with local broadcasters. The segment's (so far) final transformation into a standalone program highlights both the collaborations that are beginning to form between the old and new distribution channels (CBS and Apple) and the strategy for platforms that had previously focused solely on acquisition of content produced elsewhere into original production. As "Carpool Karaoke" makes clear, the emergence of transmedia logics is reshaping the media landscape, and distribution sits at the very heart of that process.

References

Bryant, Jacob. 2017. "Twitch and Blizzard Announce Two-Year Worldwide Collaboration." *Variety*, June 20. Accessed August 15, 2017. http://variety.com/2017/digital/news/twitch-blizzard-streaming-deal-overwatch-hearthstone-1202471111/.

Burgess, Jean, and Joshua Green. 2009. *YouTube: Online Video and Participatory Culture*. Cambridge: Polity Press.

Crisp, Virginia. 2015. *Pirates and Professionals: Film Distribution in the Digital Age*. London: Palgrave.

Curtin, Michael, Jennifer Holt, and Kevin Sanson. 2014. *Distribution Revolution: Conversations about the Digital Future of Film and Television*. Berkeley: University of California Press.

Ellis, John. 2002. *Seeing Things: Television in the Age of Uncertainty*. London: I. B. Tauris.

Evans, Elizabeth. 2011. *Transmedia Television: Audiences, New Media and Daily Life*. London and New York: Routledge.

Evans, Elizabeth, and Paul McDonald. 2014. "Online Distribution of Film and Television in the UK: Behavior, Taste and Value." In *Connected Viewing: Selling, Streaming and Sharing Media in the Digital Age*, edited by Jennifer Holt and Kevin Sanson, 158–179. London and New York: Routledge.

Holt, Jennifer, and Kevin Sanson. 2014. *Connected Viewing: Selling, Streaming & Sharing Media in the Digital Age*. London and New York: Routledge.

Jenkins, Henry. 2016. "Multichannel Networks and the New Screen Ecology." *Confessions of an Aca-Fan: The Official Weblog of Henry Jenkins*. April 21. Accessed February 24, 2018. http://henryjenkins.org/blog/2016/04/multichannel-networks-and-the-new-screen-ecology-an-interview-with-stuart-cunningham-and-david-craig-part-one.html.

Lobato, Ramon. 2012. *Shadow Economies of Cinema: Mapping Informal Distribution*. London: BFI.

Lodderhose, Diana. 2014. "Movie Piracy: Threat to the Future of Films Intensifies." *Guardian*, July 17. Accessed August 10, 2017. www.theguardian.com/film/2014/jul/17/digital-piracy-film-online-counterfeit-dvds.

Lotz, Amanda D. 2017. *Portals: A Treatise on Internet-Distributed Television*. Ann Arbor: Maize Books.

McDonald, Paul. 2007. *Video and DVD Industries*. London: BFI.

McNutt, Myles. 2017. "Classroom Instruments and Carpool Karaoke: Ritual and Collaboration in Late Night's YouTube Era." *Television & New Media* 18 (7): 569–588.

Mann, Denise. 2014a. *Wired TV: Laboring Over an Interactive Future*. New Brunswock: Rutgers University Press.

Mann, Denise. 2014b. "Reinventing TV for the Digital Age: Multichannel Networks." *Transforming Hollywood*, September 10. Accessed August 15, 2017. www.transforminghollywood.tft.ucla.edu/author/dmann/

Perren, Alisa. 2013. "Rethinking Distribution for the Future of Media Industry Studies." *Cinema Journal* 52 (3): 165–171.

Todreas, Timothy M. 1999. *Value Creation and Branding in Television's Digital Age*. Santa Barbara: Quorum Books.

Ullin, Jeff. 2014. *The Business of Media Distribution: Monetizing Film, TV and Video Content in an Online World Second Edition*. Burlington, MA: Focal Press.

Ward, Sam. 2016. "Streaming Transatlantic: Importation and Integration in the Promotion of Video on Demand in the UK." In *The Netflix Effect: Technology and Entertainment in the 21st Century*, edited by Kevin MacDonald and Daniel Smith-Rowsey, 219–234. London: Bloomsbury Academic.

27

TRANSMEDIA BRANDING AND MARKETING

Concepts and Practices

Max Giovagnoli

Not all stories can be translated into transmedia storytelling projects, and nor can all brands. First and foremost, this is because both transmedia projects and brands share the need for complex storyworlds: not all the stories that a brand can potentially tell can correctly be developed and spread on different media in an articulated way. Moreover, since transmedia properties are expansive narrative engines built to deliver almost unlimited stories across multiple media platforms, such richness and complexity can be a too wide objective for a company's communication strategy. Transmedia brands endure through the maintenance of a one-to-one relationship between the brand and the individual consumer, and the "one consumer" needs participation, synergism, and mid-term conversation. The one consumer represents millions of consumers who do not speak the same language but can meet for some time in different media simultaneously thanks to transmedia universes and experiences. This brings us back to all the observations in terms of design and editing, technique and management, and budget. These conditions contribute to the creation of the true, tangible difference between the marketing mix used to tell a brand/product, and transmedia used to embody and "become" the brand/product itself (see Giovagnoli 2017).

Moving from the role of authors and brands to that of the public of transmedia and looking for the most useful definitions referring to today's "participative consumers," it is relevant to mention Robert V. Kozinetz's *E-Tribalized Marketing*, which divides them—according to their active involvement and to their proactiveness—into the following categories: (1) *tourists*, (2) *minglers*, (3) *insiders*, and (4) *devotees* (from the least involved users to those most involved in the communication and in the brand content) (1999, 252–264). Consumers who cooperate within the range of complex storyworlds, so that—if we compare those processes with traditional advertising and customer relationship management—the success of the promotional action shall be evaluated not only in terms of customer loyalty and approval rating, but also according to proactiveness and active engagement.

In the context of transmediality, the relationship that has traditionally existed between a brand and its consumer's *emotional repertoire* is transformed by developing the brand into a whole made of more narrative storyworlds, and by developing the consumer into an experimenter, a tutor, a player, a supporter, and much more. Moreover, the constant and increasingly frequent shift from "me" to "you" and vice versa we are witnessing in today's interactive fruition of stories and brands is leading to the strengthening of the power of the narrative, if compared to that of the desire to possess a given object or good. As Barry Stamos, CEO and founder of Videoo has it:

> Consumers are now asking publishers to bring them not just the facts but the social story. Show me what others, like me, have to say, how they feel and what they're doing about an

> issue. This is social video transcending "my" story and "my" view and transitioning to "our" story and "our" view.
>
> *(Stamos 2016)*

When applying this concept to a brand, it works to grow its importance enormously as the very idea of ownership amidst contemporary audiences has changed, and, as Simon Staffans poses:

> We're looking at generations of people NOT looking to own a new car or a fancy apartment … but instead looking to have experiences to enrich their lives. As storytellers, that's exactly what we should be providing with – access to experiences on different scales, with different demands, different possibilities to dive in and engage, different communities, different niches.
>
> *(2015)*

An example? In 2012, New York-based agency Barbarian Group created *GE Show* for American Electricity giant General Electrics, eager to develop a more friendly, ordinary public image closer to its brand. *GE Show* is an online multimedia container which has—over time—collected and hosted documentaries, games, and apps which aimed to show the public all the activities of the business group, from motor to services to companies and hospitals, from aeronautics to renewable energy. The result was 300 million contacts in less than one year.

The *GE Show* project was intended to create and diffuse a different public idea about the corporation by also showing a parallel universe to those established by official media channels, broadly targeted in terms of the age of the audience. *GE Show* was meant to quickly go beyond its objective and give the company a great opportunity for brand activation to its potential future public: young families, influencers, and young people into new media. It was thanks to a supportive system that *GE Show* has exploited a crucial feature typical of transmedia: the creation of real experiences where, as it is with a mosaic, all the different parts build on each other and, in turn, offer a total overview that is greater than the sum of its components.

Deodorizing and Merging

Still, not all brands or institutions possess the necessary narrative and technological characteristics to effectively use transmediality. Furthermore, because of its high visibility, transmediality may even appear to be somewhat cumbersome, out of context, or invasive to the product or the service to be promoted or re-imagined. As I have stated above, transmediality is not a tool for simply promoting, but rather to become part of a product. Eventually, the brand, form, and content for a product's transmedia communication system all need to shape their content according to the various market contexts where it operates. Such a process is called *deodorizing*; it was applied in the above-mentioned example and is aimed at avoiding issues of cultural compatibility in specific community actions carried out in territories or times different from the original ones. Transmedia deodorizing of a brand is thus a revision process which may be applied with different degrees of intensity, through three essential operations:

- *camouflaging* or *censoring* of elements which might hurt other, "local" markets' sensitivity (religious, political, cultural, etc.);
- *combination* of different cultural traditions or the *temporary transfer* of the brand to alternative universes which get integrated in terms of plot, characters and location, describing the brand through a merging of different languages;
- a *new reading* or the *explicit violation* of the brand, aimed at producing a "surprise effect" in its positioning on the market.

Among the different activities, the last two are without a doubt the most used by transmedia marketers in terms of their application to institutions and brand communication and promotion. Let us now explore two examples to better highlight these operations.

First, the English–German brand Lynx. It produces deodorants and has a very young target market. Lynx created a curious transmedia project for its debut on the Chinese market in 2011. After having identified with accuracy the primary target of the communication (male, 20-something, eager to please and seduce, college education, digital media fan spending more than 30 hours per week on these platforms), the project was structured in three steps:

1. The creation of expectations through viral videos showing kinky situations positively finalized thanks to the use of the deodorant and presenting a live commentary by a fake focus group; it was an explicit *merging* of Western television imagery—for example by quoting English television series *Skins* (2007–2013) and movie imagery (with guests such as the US actress Angelina Jolie).
2. The announcement of the launch, made with alluring posters and videos placed on skyscrapers evoking the seductive power of the product.
3. The education of the new public, seduced by the messages of the campaign, focused on national pride (success vs. failure) and on the individual dimension of consumption (sexy consumer vs. loser).

Now, the second example. Between 2001 and 2007 German car brand BMW created a transmedia project called *The Hire*, whose "rabbit hole" was a series of eight "provocative" short movies sent online in two different seasons on BMW's official channel (bmwfilm.com). The objective of the project was to refresh in a sporty and fashion style the brand's image on the international market. At the same time, the short movies aimed at a process of cinematic merging, with the use of contaminations between crime and action genres and the presence of various showbiz stars, and with an extravagant and ironic transfiguration of the brand's traditional image based on reliability and comfort. The short movie *Star*, for example, was directed by Guy Ritchie and starred popstar Madonna and actor Clive Owen. It told the story of the temporary kidnapping of the popstar by a fearless driver who would take her on a joyride in an aggressive BMW M5 and then he would literally kick her out of the car in front of the paparazzi waiting for her in front of the main entrance of her concert venue.

The transmedial universe of *The Hire* was completed by four subplot films where BMW imagery was in part traced back to the brand's traditional image and by an alternate reality game where, through links across different websites such as Apple, Starbucks, and others, a call to action was launched. By using clues from the short movies this action would bring the users to find phone numbers and solve an enigma in order to win the BMW Z4, object of the campaign. The 250 finalists of the game were then "mysteriously" contacted via mobile phone where a voice message would invite them to meet in Las Vegas at a VIP party where the prize would be given to the only winner of the competition with a very exclusive ceremony. This final action of the project would grant the brand a sporty and elite image at once, one elegant and mundane, reliable and unexpected. Eventually, in 2004, a comic book series as well as a series of audiobooks to be listened to in the car while driving were added to the products described above. The result was a hundred million views for the videos, one million DVDs sold, and 17 percent global sales growth of the two models involved in the project.

In terms of the language of the different transmedia communities involved in editorial projects such as Lynx and *The Hire*, this is an action which must be carried out mostly by its addressees so that it shall become, from the very beginning, an important amalgam and the best pidgin for sharing the narrative. Therefore, although swimming in the waters of advertising, synthesis, shortness, and simplicity are neither the only nor the best solution possible in transmedia project applied to brands and institutions. Also in this case, imaginative universes, storyworlds, and design all play a crucial role: one inalienable, one which prefers complexity over simplicity, engagement to passive and general consumption.

Brand Stories

But let us go back for a moment to consider the role of experiences on transmedia storytelling. In marketing and advertising campaigns, in fact, the main features of a transmedia brand can be turned into narrative matter for participative authors who, on their own initiative, decide to manipulate that content in order to make them personal, or create new narratives independent from the original. As Anne Zeiser reports in her book *Transmedia Marketing*:

> When a brand touches audiences, it activates the senses. How it looks, how it sounds, how it feels, even how it smells are part of its identity ... The visual identity of a brand is created through consistent use of visual elements such as fonts, colors, and graphics that are specific to a brand.
>
> *(2015, 124–125)*

Once again, then, the combination of design and storytelling are key. Transmedia for brands needs to be memorable, timeless, versatile, and appropriate. And in telling the story of a brand, an institution, a product, or service with transmedia communicative systems, brand stories mainly work to enhance the value of:

- a product's name, its brand image and its brand identity, that is the universe of reference created by the company;
- a brand's overall image and knowledge, including ideas, attractiveness, and its consumers' "historical" expectations;
- brand value, that is the brand's value and reputation in its own market segment.

More specifically, the added value given to transmedia projects using brand story can be measured in terms of:

- *brand experience*, either in terms of engagement or of length, quality and satisfaction through consumption developed onto all the platforms of the communicative system;
- *brand activation*, since brand stories are often aimed at encouraging new customers to change their consumer behavior (within creative spaces identified by the company), through the transmedia system;
- *brand franchise*, through the audience's response to the new narratives and to the brand stories.

In order to take an active part in those different aspects of the brand in and across all of the media platforms involved in the project, the "participative consumers" and the brand stories mainly use:

- structure of the story with different narrative layers corresponding to different communicative registers and experiential opportunities for the public;
- the presence of *early adopters* (see Jenkins 2006), *devoted fans*, and *influencers* potentially able to become the brand's sounding board in order to have the most appropriate interpretation of the communicative action, even the most original and provocative;
- the use of few characters or maybe just one character (*testimonial*) in order to favor the self-identification of the consumer and leading him or her into the story.

To showcase these systems, allow me to present two examples. In 2008, Ileana Douglas created the web series *Easy to Assemble*, which tells the adventures of an IKEA clerk and her bizarre colleagues, all with insuppressible artistic ambitions. Its value was enhanced by the presence of famous actors and directors, and the series uses a brand story which makes a very smart merging of different television genres within each episode. Its main objective was related to the brand's vicinity to families, even though the series is supported by transmedia strategies only in particular cases, such as, in one example, the possibility for the public to rent a space in their houses where the IKEA catalogue could be displayed in plain sight, in the living room, in the bedroom, and so on.

Meanwhile, the example of the marketing agency Campfire in 2012 is a good example of brand exploitation achieved via a transmedia narrative. Harley-Davidson wanted a new and younger audience to be introduced to their timeless American brand. They wanted to create a destination for a new generation of fans to come together and interact. Campfire concocted both *The Ridebook* and *The Rideline* as digital environments to bring the brand to life in a new way. *The Ridebook*, for instance, was conceived as the riding manual for the ones who cherish the experience of having the wind in their faces. As these fan communities grew, the creators aimed to show how Harley-Davidson owners' lives were intertwined with the iconic motorcycle company's past, present, and future. Strong characters, old-styled environments, and traditional storytelling bridging from the web to television and digital platforms were the points of strength for the project and for the brand exploitation operation.

It is well known that the practice of transmedia loves celebrations and is eager to find sudden and valuable ways to express its potential at its best. This eagerness is even more true when it comes to the fashion world, a universe spinning around rituals such as runway shows, seasons, collections, anniversaries, and prizes, and yet has very weak fictional foundations, usually limited to a given stylist, a fashion house, or a product's brand stories. Nevertheless, fashion brands similarly need to create lifelong relationships on multiple production lines and collections.

An example? In 2013, Burberry wanted to celebrate 150 years of being in business and decided to create "Burberry World," a concept that has become a sort of "permanent philosophy" for communicating and promoting the brand since then. It all started with the ebook *Open Space*, which contains animations and old photographs of the brand, in order to make the whole project a way of promoting Burberry's founding value: trust. This was applied not only to consumers but to associate business partners and investors as well (in different countries and continents at the same time). In cooperation with Google and Grow, Burberry launched the campaign "Burberry Kisses," an app that allowed the user to send their pictures and interact with the official advertising of the campaign by simply wearing a Burberry item of clothing, and kissing the screen of the mobile phone. Then it was music's turn to become the primary asset in the communicative system and the engine of the brand's Britishness and behavior. On the company's website, for example, a section called "Burberry Acoustic" was created, hosting young British music talents performing on video. Those clips were then brought into the real world thanks to outdoor concerts, worldwide in-store performances, and LED wall screenings. This was all essentially done to lower the average age of Burberry's target consumers and then lead all of their audience toward the creation of an emotional bond that embodies the brand and its history.

Different to Burberry, by contrast, the luxury French brand Chanel used a character-based form of transmedia storytelling in its 2012–2014 campaigns, but it was more digital than real-world oriented. Linear formats spread the character Coco Chanel across documentaries and social media activities, and utilized touch points either in pre-purchase and post-purchase experiences. The campaign "Inside Chanel" re-interpreted Coco Chanel's role and biography with a multi-strand series. Each episode began with the words "Once upon a time" and used archive footage. As Stine Johansen reports:

> Chapter 12 featured a tour guide through the streets where Chanel was originally founded. Anchoring the story at a specific place allows users to move from the digital platform to their own reality, binding those two worlds and sending an enforced message.
>
> *(2017, 83)*

Facebook, YouTube, and other social media platforms, none of which possessed the aesthetic characteristics of a luxury brand, still used the black/white, tweed, and pearl color branding in the art direction.

Meanwhile, a different kind of transmedia example is represented by another successful fashion brand: that of the leader in sportswear for athletes and prosumers, sponsor of the Danish national football club, Hummel. At the end of 2016, a small group of young creatives linked to VIA University—Fashion & Transmedia of Aarhus—created and developed for Hummel the project *Change the World through Sport*. The project had two goals: to create a new story angle for the brand perception and to celebrate the effective social impact of the "Hummel Universe."

The project consisted of: a 3D Video Mapping experience to be performed in larger Danish cities, a teaser film on Facebook, pics-video on Instagram (#Makehistory), a Portrait Film of Hummel Ambassadors, the video "Discover your own story" and user-generated "Win a Sponsorship" videos. A final contest for small sports clubs would then launch a call to action for sports clubs promising an annual Hummel sponsorship to fulfill all their needs, which was awarded to the best five teams. The asset "Hummel Ambassadors," meanwhile, would use five athletes to identify and share positive messages with the audience. Finally, all the content produced would be saved as part of the "Hummel Universe" and celebrated in an annual event for testimonials and fans—the first "transmedia party" in Hummel's history.

Gamification

Going further, translating a company's brand identity into gaming-oriented actions involves the integration of play dynamics into communication and promotion of products or services. A significant part of the public today gets fond of a given brand in a more long-lasting and effective way than before, and it prefers to "play" the brand rather than just listen to its slogans. The aim of this kind of strategy is quite clear: to get the public closer to more complex messages in an easier and friendlier way; to restructure subjects which were felt as "too ordinary"—or conversely not popular enough—and, at the same time, enhance and refresh the brand's intimacy with the public through games. Specifically, gamification offers transmedia promotion campaigns:

- informal, spontaneous and positive memorization (*mark up*) of the brand, granted by the positive interaction given by the experience of gaming;
- a deeper experience of the brand between consumer and product, thanks to the game;
- development of a different brand awareness by the consumer/player about the brand's universe (for example with experience and educational games);
- the creation of a database of consumers which would be impossible to extend out in any other way, but very present in the advergame microcosm.

In terms of transmedia practice, a brand's gamification, or rather the gamification of a narrative experience, can range from the proposition of a simple interaction with a given story to the creation of advergames or free creative experiences such as *open worlds* and *sandboxes* (narrative spaces where an audience's engagement is encouraged without limitations of creativity and self-expression). Within transmedia communicative systems, then, the elements of a game can be exercised through two basic activities:

- the creation of a brand gamification project whose task is to reinterpret the activity, the role, and the image of the company through games that are used to illustrate, describe, or directly experience its actions;
- the enrichment of promotional and advertising offers with forms of brand experience such as events, team games, urban quests, and so on.

An example of the first point is the gamified activity initiated by LEGO in 2012 to celebrate the fiftieth anniversary of its Australian market launch. LEGO created a "Festival of Play," which moved around the country, launched by an online trailer and immediately transferred into the real world thanks to the creation of urban forests and pop-up play-pits inside and outside city centers. At the same time, the public would create posters dedicated to the idea of "play" and "Play Days" were organized in schools with the aim of celebrating the milestones of Australian history, all translated into LEGO style. The students' creations were published on the project's official website, while an m-site (mobile-optimized website) would update the public on the events all over Australia in real time. In the digital world, meanwhile, an app would transform every piece of LEGO into a game. The project eventually ended with an interactive Christmas card, "The Lost Brick." The result? $15 million profits in terms of communication; 400,000 visitors to the events; a sale rise on the local market of 18 percent in a year.

Now, to give an example of brand experience created for the "inner" audience of a brand we must take a step back in time and deal with the case of one of the tech world's giants. In 2010, after the great success of its previous *The Threshold*, Juxt Interactive asked No Mimes Media to partner with them in the creation of a transmedia experience designed to entertain and inform Cisco's Global Sales Force. *The Hunt* put employees at the center of a thriller where characters sent and responded to their emails, left phone messages, communicated through Facebook and Twitter, even asking them to retrieve items from a dead drop and to send them photographs and information. And while helping fictional characters Isabel and Keith escape an ancient secret organization, the sales force also learned about new Cisco technologies coming to the market. Cisco had new demands for the 2010 experience. A geographically and culturally dispersed sales force raises challenges when it comes to introducing dozens of new products and technologies each year. Cisco wanted *The Hunt* to have global reach, to educate, to build collaboration, and to be fun. This demanded new ways of storytelling and new ways of thinking. *The Hunt* was quick and intense, unfolding in real time in just two weeks. *The Hunt* involved audience members from countries around the world, including China, India, the Netherlands, Germany, Norway, Pakistan, Japan, the United Kingdom, and the United States. It highlighted new Cisco technologies like Pulse and Mediator, painlessly engaging the audience in what those technologies do and how they work. How? Players collaborated across silos, creating networks of cross-disciplinary experts. *The Hunt* pushed the boundaries of storytelling with events unfolding on Twitter and Facebook, and in the real world where the audience had to use social engineering to find and secure a package with vital information.

As I have noted, then, the most important actions necessary to revisit and customize a gamified product or an image of a company are not hugely different from the traditional ones, although their application to multiplatform narrative and technological strategies does make precise strategic and editorial interventions essential for two reasons. First, in terms of the study of a brand's strengths and inadequate parameters to the new market, and, second, in terms of the creation of cross-cultural universes and characters which can have a positive impact on the imagery of different types of users in terms of consumption and media habits. These actions require a close and accurate study of the technological platforms, the networks, and the consumption attitudes of the potential target for the experience.

Here's an example. In 2008, GMP, a recruitment and human resources giant, launched a transmedia campaign for the Asian market, and it did so with almost no costs at all. It had two main objectives: to increase the brand's diffusion and to win the loyalty of a new public—namely, people in search of a job. The campaign was based on a deeply heart-felt subject matter intended for the Asian population—that being the relationship between quality of life and the incessant rhythms of work. The project's "rabbit hole" was a video uploaded onto YouTube where two Singapore clerks who were working long hours in their office are visited by a ghost. But the two clerks are so tired they do not even notice the ghost's presence. The company's video cameras—on the other hand—*capture* everything on video, which becomes viral and was viewed 500,000 times in ten days. After that, and on the association's official blog, which was investigating the apparition, three ghostbusters are hired and a call to action on the web and on social networks was launched. The immediate result was the spontaneous, operative, and communicative contribution of associations specialized in paranormal phenomena, including newspapers and television shows all over the world, as well as users who believed that they have lived similar experiences, with workers' unions, bloggers, and even students studying special effects all trying to reveal the "technical secrets" of the video. Finally, on May 1 that year, it was revealed that they were beyond such an action, which triggered an amused media grapevine echoing all over the world, which was based on the key message of the story: no one should work late hours. The result was a 30 percent increase in sign-ups on the company's website and the equivalent of US$500,000 in terms of media coverage—all stemming from a campaign that was completely free.

Conversely, if a company wants to conquer newer portions of the same market it operates on, then a good example to demonstrate this approach would be "The Pink Squad" experience, a transmedia project realized in 2009 in Slovakia by insurance company Union Insurance. "The Pink Squad" aimed to denounce dangerous driving behaviors responsible for the rising cost of insurance policies, and to finally give those consumers burdened by those few irresponsible drivers a new, audible voice. This is a clear example of a two-goal structure. The project was successful thanks to the intervention of a task force of activists who, wearing quirky pink masks, went around for numerous weeks theatrically "punishing" those responsible for such dangerous behaviors. The campaign's "rabbit hole" was a viral video presenting the Pink Squad manifesto, broadcast in a seemingly illegal way via the country's main broadcasters. The following steps came from the squad's website along with several videos showing the nastiest punishments carried out by the squad. It was only after two weeks of posting videos online and broadcasting on the main national channels—followed by thousands of support messages posted by consumers on the main social networks—that Union Insurance openly acknowledged that they were behind such an action. And it was only then that official television spots and traditional communication modes started.

The results? A million webpage views, 80,000 friends on Facebook and, above all, a free media campaign estimated at 500,000 euro. And yet the story of Pink Squad was not over, as the fictional group became the promoters of a "National Weekend Without Road Accidents," which was followed by thousands of Slovak drivers who were invited to display something pink on their cars. This action resulted in a diffusion and greater brand awareness and, most importantly, in a significant 80 percent reduction of car accidents over that year's All Saints weekend. Finally, mechanics dressed in pink for that occasion would also become the company's testimonials in all the car repair garages throughout the country. Together, these elements are a reflection of another important aspect of transmediality when applied to brand and corporate storytelling: as Simon Staffans puts it, "in order for the audience to fully immerse themselves in what the company, brand, product or service is about, we first need interconnected stories that support each other and build over the long run" (2015).

Conclusion

In a nutshell, the examples I have described throughout this chapter all clearly demonstrate how all the social and cultural differences a brand or a transmedia campaign face in a moment of transition from one market to another, may nowadays be a resource and a basket full of opportunities, rather than an operative or editorial limit for transmedia authors and its audiences.

Acknowledgments

The core content of this chapter was previously published (with full permissions) in Giovagnoli, Max. 2017. *The Transmedia Way: A Storyteller's, Communicator's and Designer's Guide to the Galaxy*. Halifax, Canada: ETC Press.

References

Giovagnoli, Max. 2017. *The Transmedia Way: A Storyteller's, Communicator's and Designer's Guide to the Galaxy*. Halifax, Canada: ETC Press.

Jenkins, Henry. 2006. *Convergence Culture: Where Old and New Media Collide*. New York: New York University Press.

Johansen, Stine. 2017. "Transmedia and Fashion: Case Studies and Potentials." In *Fashion, Film and Transmedia: An Anthology of Knowledge and Practice*, edited by SPOTT Trends & Business, 83–96. Aarhus: VIA Film & Transmedia Research & Development Centre. Accessed March 6, 2017. www.northsearegion.eu/media/3115/createconverge-fashion-transmedia-anthology-2017.pdf.

Kozinetz, Robert V. 1999. "E-Tribalized Marketing? The Strategic Implications of Virtual Communities of Consumption." *European Management Journal* 17 (3): 252–264.

Staffans, Simon. 2015. "One Year in Now Media." *Simonstaffans*, May 3. Accessed March 8, 2017. https://simonstaffans.com/2016/01/04/one-year-in-now-media-2015/.

Stamos, Barry. 2016. "5 Rules for Adapting Your Company to the Age of Group-storytelling." *Venturebeat*, March 5. Accessed March 7, 2017. https://venturebeat.com/2016/03/05/5-rules-for-adapting-your-company-to-the-age-of-group-storytelling/.

Zeiser, Anne. 2015. *Transmedia Marketing: From Film and TV to Games and Digital Media*. London and New York: Routledge.

PART IV

Cultures of Transmediality

28
TRANSMEDIA ARCHAEOLOGY
Narrative Expansions across Media Before the Age of Convergence

Paolo Bertetti

The idea of transmedia archaeology arises from the observation that transmedia storytelling practices similar to those first described by Henry Jenkins (2003) in relation to the current panorama of industrial and technological convergence had actually been present for a long time in relation to more traditional media. Scolari, Bertetti, and Freeman (2014, 6), who first opened this debate, state that

> if we consider transmedia storytelling as an experience characterized by the expansion of the narrative through different media and, in many cases, by the participation of the users in that expansion, then we could say that this is not a new phenomenon.

Indeed, its origins can actually be traced back almost to the beginning of the modern cultural industry at the turn of the twentieth century.

Although most scientific publications about transmedia storytelling focus on contemporary productions (or at best date back to narrative universes created in the 1960s, such as *Star Trek* or *Doctor Who*), we can find many popular narratives dating back as early as the 1920s and 1930s, such as Batman and Mickey Mouse, which spread across different media (comics, pulp magazines, radio, etc.) and were eagerly consumed by the fans that sometimes contributed significantly to the expansion of these imaginary worlds (see Bertetti 2011). From this standpoint, transmedia storytelling is "a trans-historical practice of media production" (Scolari, Bertetti, and Freeman 2014, 8), both of a narrative and social nature that emerges "as a fundamental component of popular culture in the twentieth century and even before" (Scolari, Bertetti, and Freeman 2014, 76).

Oz is a good example of an archeological transmedia universe. Conceived long before the concept even existed, this make-believe place has been studied exhaustively by Matthew Freeman (2014a, 2014b, 2016) and Frank Kelleter (2012), and defined "the first great transmedial word" (Wolf 2012, 117). This fictional world created by Frank L. Baum makes its first appearance in 1900 in the novel *The Wonderful Wizard of Oz*—the first in a series of 14 novels—and already began its narrative expansion across media as early as 1902, when a musical based on the book was staged in Chicago and New York. In 1904, long before modern transmedia franchises came into being, the second novel, *The Marvelous Land of Oz*, was published. On this occasion Baum developed a genuine cross-promotion mechanism that expanded his world across a variety of different media: comics (*Queer Visitors from the Marvelous Land of Oz*, 1904), illustrated children's books (*The Woggle-Bug Book*, 1905, published along with a musical play of the same title), a movie, theatrical plays, various kinds of merchandising (postcards, buttons, card games, etc.) and even a multimedia travelling show (*The Fairylogue and*

Radio-plays, 1908) including an orchestra, over 20 actors, magic lantern slides, and film clips. Such media extensions of Baum's original creation were not simply transpositions of the literary works, but deliberate narrative expansions. As Wolf observes:

> Oz did not simply originate in Baum's books and then get adapted to other media: new Oz stories could begin as books, musicals, comic strips, or plays and then be adapted across media, and those adaptations would often add new material, events, and characters as well, making Oz a truly transmedial world.
>
> *(Wolf 2012, 118–19)*

Oz is certainly a particularly complex case, similar in some ways to modern practices of transmedia expansion (as Wolf 2012 notes, it is a rare case of world-based narrative expansion in the first half of the twentieth century). But it is not alone. The circulation of narratives, characters, and fictional worlds through different media is a practice that has a long history, one that still to be fully fleshed out.

Roberta Pearson (2009) suggests that biblical stories, often experienced by different audiences through the written word, drama, visual arts, etc., can be understood in terms of transmedia storytelling. Similarly, Elizabeth Evans (2011) quotes myths such as King Arthur or Robin Hood. Carlos Scolari (2013) has also contributed to the argument by analyzing how the adventures of Don Quixote de la Mancha spread among illiterate people. Colin Harvey (2015, 57–58) also contributes to this debate, observing how the advent of Penny Dreadfuls in the 1840s saw the emergence of several notorious characters such as Sweeney Todd, who nimbly shifted from one medium to another. Despite the obvious migration of plots and characters across different media, it remains to be seen in these cases to what extent we are witnessing simple and more traditional forms of intersemiotic translation and adaptation rather than genuine narrative expansions.

A more cautious analysis by Matthew Freeman (2014a, 2016) draws our attention to the turn of the twentieth century and the development of the modern cultural industry as well as a series of major transformations, especially in the United States. It is, in fact, the rise of specific industrial-cultural configurations in those years as well as the alignment of different media that boosted the evolution of transmedia storytelling (Freeman 2016). For Freeman (2016), indeed, three decisive factors spurred this development: (1) industrialization (especially the development of new industrial production technologies creeping into the cultural environment), (2) the growing consumer culture (increasingly large numbers of people who were consuming a broader range of new cultural products), and (3) the development of new laws and policies of media regulation.

A New Perspective on Transmedia Storytelling

One of the newest dimensions of contemporary transmedia entertainment is our recognition of it as such (Johnson 2017). As with every new theoretical concept or analytical category, the concept of transmedia storytelling allows us to re-read some aspects of our cultural history in a new light. Its emphasis on the distribution of content across media affords new perspectives on the history of the media industry. At the same time, an historical view of transmedia storytelling allows us to more fully understand transmediality itself. As Freeman observes:

> only by looking to the past can we fully see the contingencies of the present, and by searching for historical precedents it can force us to be far more nuanced in describing what is truly specific to our present media moment.
>
> *(2016, 3)*

In particular, increasing our knowledge of older and different forms of transmedia expansion allows us to recognize that the current model is only one of many possible models, thus enabling us to better identify fundamental and recurring characteristics. Here, Freeman (2016) conceptualizes transmedia storytelling as a system that serves to build variation on sameness, one that seems to have the following three characteristics: (1) character-building, (2) world-building, and (3) authorship (2016, 9).

Transmedia archaeology therefore compels us to reconsider our idea of transmedia storytelling. First, we need to develop a more flexible conception than the one originally described by Jenkins (2003, 2006), "which refers to the systematic unfolding of elements of a story world across multiple media platforms, with each platform making a unique and original contribution to the experience as a whole" (Jenkins 2011). In other words, transmedia storytelling involves a unique and coherent narration developing across different media text, each one offering a distinct, significant, and non-redundant content. Definitions like this emphasize the *planned, strategic* aspects of transmedia creation (Fast and Örnebring 2015, 2). This implies starting from a narrative matrix (a plot, or more often a storyworld) and elaborating a complex multiplatform project in order to create specific products for every type of user (Innocenti and Pescatore 2008, 71), as with Jenkins' studies of *The Matrix* (2006). Early twentieth-century transmedia storytelling instead "accrued characteristics that [were] more *ad hoc/contingent* than planned" (Fast and Örnebring 2015, 2).

Jenkins himself (2006) admits that his ideal model of transmedia storytelling corresponds only partially to concrete production practices, where contradictions and overlaps seem inevitable. Years later, in a well-known synthesis of the principles of transmedia storytelling (Jenkins 2009), the author speaks rather of an opposition: *continuity vs. multiplicity*. Jenkins uses the term "multiplicity" primarily to refer to mashups and user-generated content or to programmatic alternative storylines such as those of the DC Comics' Elsewhere series, but also to less planned productive practices that allowed developers to avoid strict adherence to details (Jenkins 2009). Notably, Fast and Örnebring (2015) observe that the term multiplicity involves a shift in focus towards the *emergent* (as opposed to the planned) narrative aspects of transmediality.

We must not forget that the "strategic planning of transmedia worlds is not necessarily focused on creating a seamless, coherent world" (Fast and Örnebring 2015, 2). If this is true for contemporary productions, as Fast and Örnebring discuss in their analysis of the *Transformers* franchise, it seems particularly evident in older transmedia expansions where the exigencies of coherence and consistency of the storyworld or the continuity requirements in the storyline are often considered less important compared with the affordances of different media or "writer/company goals, which may include, for example, the broadening of an audience segment" (Fast and Örnebring 2015, 13). In other words, a study of transmedia storytelling should emphasize not so much the construction of coherent universes and non-redundant content as the media expansion itself and, in particular, the creation of an audience experience based on a recurring fictional world and/or narrative characters across different media. From this perspective, absolute consistency is a negligible characteristic of narrative extensions, but *recognizability* is essential: an audience must be able to recognize the storyworld or the characters on which this expansion is based. As long as expansion does not overstep certain limits, audiences are elastic and tend to accept development patterns and variations (Bertetti 2014; Marrone 2003).

Be that as it may, there are actually different models of transmedia storytelling: Jenkins' original model sought to describe and conceptualize some strategies of narrative expansion that seemed strictly related to contemporary digital transformations and developments in media industries. But we may consider different models that offer a recursive (even if not necessarily coordinated) transmedial entertainment experience and different strategies of developing media franchises, as I will now demonstrate.

Three Approaches to Historicizing Transmediality

Research into the historicization of transmediality has really only just begun, but in recent years several scholars have started to delineate a fairly clear framework. Even Jenkins (2006, 2009, 2011, 2014), who originally suggested that the innovative features of transmedia storytelling are essentially a new means of organizing converging media content, recognizes the legacy of older and more traditional forms of media transcendence. Other scholars (Evans 2011; Pearson 2009) have also highlighted the importance of studying the historical backgrounds of modern forms of transmediality. Derek Johnson (2017), in particular, urges the importance of historicizing transmedia storytelling, pointing out its relationship with long-established models of media franchising and traditional processes of adapting and translating content.

Broadly speaking we can identify three different interwoven approaches to studying transmedia archaeology. The first is a textual, narratologic (and sometimes semiotic) approach, focusing on the ways narratives are organized as transmedia expansions. This involves relating recent transmedia practices to the forms of serial narration that have developed in popular culture since the 1800s, predominantly in a single medium, given that transmedia narratives are none other than a particular case of transtextuality. In particular, some scholars (Scolari, Bertetti, and Freeman 2014; Bertetti 2016) have highlighted how certain modes of narrative organization (such as expansion and disimplication, see Scolari 2009), which form the basis of today's transmediality, are rooted in serial forms developed on the pages of pulp magazines in the first half of the twentieth century (as in the case of serialized characters such as Conan, Tarzan, and John Carter). It is precisely the type of narrative expansion facilitated by serialization that made it possible to create more extensive and detailed storyworlds on the pages of pulp magazines, creating storyworlds that would go on to become the settings of novels—for example, Edgar Rice Burroughs' Barsoom, Robert E. Howard's Hyborian Age or Philip Francis Nowlan's twenty-fifth-century Earth, the home of Buck Rogers (Wolf 2012, 120). Successively, models of seriality and narrative formulas as well as characters and storyworlds once associated with the worlds of pulp magazines surfaced in other media, too, including radio programs, films (and later television series), and comics (Scolari, Bertetti, and Freeman 2014, 74). Regarding the latter, a number of key critical questions relating to transmedia storytelling emerged for the first time in relation to comics—namely, questions to do with narrative continuity across parallel series and the overall consistency of shared diegetic universes (Scolari, Bertetti, and Freeman 2014, 76).

Beyond a focus on narrative continuity and world-building, many studies of transmedia archaeology tend to include an analysis of fictional characters, exploring the ways in which they are built and how they spread across different media. Following Bertetti, who defines a transmedia character as "a fictional hero whose adventures are told across different media platforms, each one giving more details on the life of that character" (Bertetti 2014, 3344), if contemporary transmedia storytelling is "the art of world-building" (Jenkins 2006, 166), then older forms of transmedia franchises can be based on a more simple logic centered on character sharing. Jason Scott (2009) introduces a similar notion of character-oriented franchising after tracing the origins of transmedia productions back to the age of silent cinema, where he discovered economic and promotional strategies common to contemporary media franchises. Similarly, Jenkins (2009) connects today's transmedial practices to other older ones, distinguishing between modern transmedia figures who are carrying the timeline and the world depicted on the "mothership" text with them and older characters such as Felix The Cat that are not bound to any specific narrative context (Jenkins 2009). With a focus on transmedia characters, too, Fast and Örnebring (2015) have also analyzed the practice of moving storyworlds across media. Their study of The Shadow character traces the development of this super-powered detective from his origins as a narrating voice in a radio drama to a full-fledged character in his own right in a series of pulp novels and later appearances in comics, television series, and films. As noted above, Freeman (2016) has also recognized that character-building is a fundamental component of all transmedia storytelling, but his analysis of transmedia characters such as Tarzan or Superman focuses on industrial contingencies and practices that allowed particular strategies for actually building those characters across media in the past and their differences across time.

In doing so, indeed, Freeman's book provides us with an excellent example of the second approach to transmedia archaeology. This approach connects the different forms of transmedia storytelling to the wider industrial and cultural context. At the same time, it investigates the socio-historical conditions and the economic, legal, and productive mechanisms that fostered its development. When taking this second approach, many scholars (Jenkins 2014; Johnson 2013; Santo 2006, 2015) have underlined the key role played by licensing. In fact, "through licensing, intellectual property owners were able to extend a property's reach into almost every area of consumer life without having to invest in manufacturing infrastructure or distribution networks" (Santo 2015, 7). Licensing led to a rise in transmedia franchises before the rise of conglomeration or technological convergence. In his study, Johnson focuses on the post-World War II period, when franchising became a key corporate strategy (2013, 41). But the licensing model started long before, with significant examples in the 1920s and 1930s. Kristen Thompson (2007), for instance, has studied Walt Disney's model to produce spin-offs and merchandising and Avi Santo (2006, 2015) investigates the development of *Green Hornet* and *The Lone Ranger* franchises, underscoring the role played by independent licensors such as George W. Trendle (or, as we will see, John F. Dille) in extending a brand across media. Santo highlights the fact that the economic and legal aspects of franchising and licensing were inseparable from creative ones; licensing contracts were indeed "far from being *only* economic agreements ... [they] were also clearly sites of cultural and creative deliberation" (Santo 2015, 12).

Taking a broader perspective, Freeman (2016) distinguishes three different macro-phases of transmediality that ran from the beginning of the twentieth century to the 1950s. Besides above-mentioned management practices based on licensing, which Freeman characterizes as the second phase of transmedia storytelling and which was associated with the period of the late 1910s to the late 1930s, he also identifies an earlier model centered on *advertising* practices of world-building. Underscoring "the importance of colour, spectacle comics strip characters and also posters and reviews as key promotional mechanism for building story worlds and pointing audience across media" (Freeman 2016, 100), the author shows how Frank L. Baum made Oz a transmedial world through the adoption of cross-promotion strategies, as we have seen before. Finally, Freeman's third phase, which he claims lasted from the early 1940s until the 1950s, was driven by a model based on industrial partnerships—as in the case of DC comics' Superman—stemming from different factors, including the governmental intervention in the production of war-propaganda messages during World War II and, after the war, the presence of marginal productions like independent B-movies and telefilm series.

Besides text-centered and production-centered approaches, however, a third approach to studying transmedia archaeology is to focus on the role of audiences, their experiences across media, and their contribution to transmedia constructions. To date, few studies have dealt with the different historical forms of participatory culture; an exception is Scolari, Bertetti, and Freeman (2014), who analyzed the political re-appropriation of *El Eternauta* by young radical movements in 1950s Argentina and the contribution fans played in establishing the canon for Conan the Barbarian in the 1930s. Conversely, current studies have tended to focus on understanding the textual features of historical transmedia stories and the cultural and industrial infrastructures that made them possible (Freeman 2016, 11).

Textual Expansions of a Transmedia Hero: Buck Rogers in the 1930s

By way of example, I will now move on to take a close look at one of the most popular fictional American icons of the 1930s so to demonstrate how a popular character was subjected to transmedial expansion processes during the first half of the twentieth century. Buck Rogers is said icon, a character that was a science fiction comic strip hero and the mainstay of a character-oriented franchising operations.

Buck Rogers in the 25th Century A.D. made its debut in newspapers on January 7, 1929, the first science fiction comic strip in history. But the origins of Buck Rogers go back much further to popular literature. In fact, Anthony Rogers (not yet "Buck") was the main character in two novellas published in Amazing Stories: *Armageddon 2419 A.D.* (August 1928) and its sequel *The Airlords of Han*

(March 1929). In the first novella he is a veteran World War I pilot who falls into a state of suspended animation after breathing radioactive gas while exploring an abandoned coalmine in Pennsylvania and awakens 492 years later in the year 2419. Shorty after coming round, he meets the young Wilma Deering—his future girlfriend—who explains that America has been conquered by the Mongolian Han Lords. She invites him to join the resistance, hoping that Rogers' experience of warfare will serve their struggle. In the sequel, six months have passed, and our hero is now leading a group of resistance fighters to the final victory over Asian invaders.

Realizing the commercial potential of Nowlan's creation, John F. Dille, then director of National Newspaper Syndicate (later John F. Dille Company), came up with the idea for an adventure comic strip. Dille convinced Nowlan to write the texts of the strip and hired staff-artist Richard Calkins to do the drawings. The resulting comic strip is neither a transmedial expansion of the novellas nor an example of intersemiotic translation, but rather a loose adaptation that preserves the two main characters (Antony Rogers, with the new name of Buck, and Wilma Deering), the original beginning and some narrative episodes. The main differences occur in the storyline. While the initial situation is also the same (a struggle between Americans and Hans), in the extension it soon starts to diverge dramatically as does the nature of the diegetic universe. In the comic strips the war with the Han is not fought only in the northeast United States but includes other nations such as Chile and Canada. Then, after the cycle of wars against the Han, Buck Rogers' narrative universe expands across the entire solar system and beyond; moreover, many aspects of the storyworld are very different from the original one. A third difference concerns the system of characters: in the comics Buck and Wilma are joined by various recurring side characters not present in the novellas—the ultra-villain Killer Kane and his sidekick Ardala Valmar, Buddy Deering (Wilma's brother), and Dr. Huer, an ingenious inventor of astonishing technological gadgets.

As is clear, a transmedia analysis of Buck Rogers should therefore start with the comics, since this is where the expansion of the storyworld and the franchise began. Even from a commercial and legal perspective, the novellas and the comic strip were two different entities: intellectual property rights for the former were the property of Nowlan and his family, while Dille owned the rights to the latter, although Nowlan continued to write the Buck Rogers comic strip texts until the end of the 1930s. By licensing these rights, the John F. Dille Company managed to create the franchise and to control—albeit to a limited extent—its expansion. As with Burroughs' Tarzan (see Freeman 2016, 108–144), Buck Rogers perfectly illustrates the ability of an industrial configuration based on licensing to afford transmedia storytelling and for audiences to be steered across media. But it also reveals—as we shall now see—its limits in terms of controlling extensions on different media.

While Buck Rogers is a typical example of a character-oriented franchise, it is clear that Dille was keen to maintain a coherent narrative universe. This is evident in the first transtextual expansion, published as a full-page weekly comic strip in March 1930, which focused on the world rather than the character. Initially the series featured Bud Deering and another supporting character, Princess Alura from Marsin, in leading roles; only later did Buck Rogers become the main character. Giving prominence to the diegetic coherence of the fictional world was necessary to avoid having to juggle with different character-based storylines. In fact, the full-page Sunday comic strip and the shorter daily comic strip were published in different newspapers, making it impossible to develop a single storyline based on the same figure. The need to maintain consistency in the storyworld determined similar expansions on different media in years to come. The transmedia expansion of Buck Rogers started in 1932 when he made his first appearance in the radio drama series, *The World in 2432*.

Written by Jack Johnstone and directed by Carlo De Angelo, the show was aired intermittently until 1947, first by CBS (from 1932 to 1936) and then by Mutual (from 1939). The storyline echoed those of the comic strip and featured the same characters. Its popularity can be gauged by the huge number of listeners who wrote in asking for a premium giveaway offer. In 1933, for example, Cocomalt, which was then sponsoring the series, offered a free "solar map" of the planet to any listener writing in to the show and requesting a copy. The sponsor was deluged with more than 125,000 requests, a staggering affirmation of Buck's popularity (Lucanio and Coville 2002, 37). Later, premium items—such as a cardboard rocket pistol and a space helmet—generated more than 140,000 proofs of purchase. Moreover, weapons and other super scientific gadgets were central in the episodes of the serial (DeForest 2008, 181) and their presence, as we shall see, was crucial to promoting merchandise products.

Among the sponsor's gifts, the Big Little Book offered in 1933 by Kellogg's is particularly significant in terms of the expansion of the narrative universe. In this small autobiographical book, containing Nowlan's texts and Calkins drawings, Buck provides readers with a detailed description of his world and the technological discoveries of the twenty-fifth century. As well as systematizing the narrative universe until that date, the book also evidences the authors' interest in constructing a coherent reference storyworld to be used across media. Even the numerous aforementioned merchandising products—in particular, toys, produced under license of J.F. Dille Company—are also parts of this universe. As Santo (2006, 41) observes, Dille recognized

> that its greatest profits lay in merchandising the many science fiction gadgets the hero used in his adventures. Buck Rogers ray guns were amongst the most successful toys sold throughout the 1930s, as were Buck Rogers space suits and toy spaceships.

Dille was also well aware of the importance of cross-promoting merchandise through media. A ten-minute film, *Buck Rogers in the 25th Century: An Interplanetary Battle with the Tiger Men of Mars*, shown at the 1934 Chicago World's Fair, generated dozens of new licenses (Santo 2006, 41) and was later used in department stores to further promote merchandise. Directed by Harlan Tarbell and written by Nowlan, this sort of homemade production features John Dille Jr. (the son of John F. Dille) in the role of Buck Rogers alongside the strip's main characters (Wilma, Dr. Huer, Killer Kane and Ardala). This short film depicts the future world and shows the Tiger Men of Mars attacking Earth before they are eventually defeated by Buck in a star battle.

In 1939, further, Universal Pictures produced a 12-part Buck Rogers movie serial. This time, however, the result broke with the comic strip, introducing numerous changes to the storyline and characters. In the serial, Buck Rogers and Buddy Wade (based on the comic character Buddy Deering) awake after 500 years spent in suspended animation in a world ruled by the tyrannical dictator Killer Kane. Soon they join the resistance movement lead by the scientist Dr. Huer in the "Hidden City" and later travel to Saturn with Wilma Deering to establish an alliance. This deviation from the original storyworld is not surprising. 1930s and 1940s movie serials like *Buck Rogers* were low budget B-unit affairs whose producers, Blair Davis notes, "were not as concerned with issues like narrative fidelity and successfully capturing the 'essence' of source material as much as they were with delivering a fast-paced film on time and within budget" (2017, 40). Furthermore, Buck Rogers found himself on the movie screen only after the wide success of the serial *Flash Gordon* (1936) and the sequel *Flash Gordon's Trip to Mars* (1938); although the more adventurous Flash Gordon was clearly an imitation of Buck Rogers—albeit with less technological gadgetry—the follow-up Buck Rogers serial inevitably "had no choice but to follow the Flash Gordon construct, and as such it became hardly distinguishable from the *Flash Gordon* serial" (Lucanio and Coville 2002, 41). Incidentally, casting Bustin Crubble as Buck Rogers after he had played Flash Gordon, coupled with the fact that the producers re-used scene materials from *Flash Gordon's Trip to Mars* to save money, doubtless contributed enormously to their uncanny likeness.

More generally, as Freeman (2016) observes, in the first half of the twentieth century a coordinated system of transmedia storytelling came up against numerous obstacles when approaching cinema. In particular, in the 1930s

> the system of vertical integration that came to characterize the major Hollywood studios meant that these studios occupied a producer-distributor-exhibitor model and had

> therefore grown accustomed to working internally ... it was much more difficult for creative personnel to author storyworlds that crossed in and out of the cinema.
>
> (Freeman 2016, 193)

Despite such industrial obstacles, Buck Rogers adventures would continue long after World War II: the comic strip ran until 1967 and was also adapted for a television series in 1950–1951. Even in recent years the character has been rebooted, linked to the television series of the late 1980s and to a new comic book series that started in 2009.

Conclusion

Although limited to 1930s productions, my analysis has evidenced some strengths and weaknesses of the licensing model that allows for the circulation of worlds and characters through media, albeit at the expense of preserving authorial control and the integrity of intellectual property when comes up against the productive constraints of different media industries.

References

Bertetti, Paolo. 2011. *Conan il Mito. Identità e Trasformazioni di un Eroe Seriale tra Letteratura, Fumetti, Cinema e Television* [Conan the Myth: Identities and Transformations of a Serial Hero among Literature, Comics, Film and Television]. Pisa: ETS.

Bertetti, Paolo. 2014. "Toward a Typology of Transmedia Characters." *International Journal of Communication* 8: 2344–2361. Accessed August 30, 2017. doi:1932/8036/20140005.

Bertetti, Paolo. 2016. "Personaggi seriali e mondi transmedali: I pulp, Tarzan e le origini del Transmedia Storytelling." *Mediasacapes Journal* 6: 155–167.

Davis, Blair. 2017. *Movie Comics: Page to Screen/Screen to Page*. New Brunswick, NJ: Rutgers University Press.

DeForest, Tim. 2008. *Radio by the Book: Adaptations of Literature and Fiction on the Airwaves*. Jefferson and London: McFarland.

Evans, Elizabeth. 2011. *Transmedia Television: Audience, New Media and Daily Life*. New York: Routledge.

Fast, Karin, and Henrik Örnebring. 2015. "Transmedia World-building: The Shadow (1931–present) and Transformers (1984–present)." *International Journal of Cultural Studies* 20 (6): 636–652.

Freeman, Matthew. 2014a. "Advertising the Yellow Brick Road: Historicizing the Industrial Emergence of Transmedia Storytelling." *International Journal of Communication* 8: 2362–2381. Accessed July 26, 2017. doi:1932/8036/20140005.

Freeman, Matthew. 2014b. "The Wonderful Game of Oz and Tarzan Jigsaws: Commodifying Transmedia in Early Twentieth-Century Consumer Culture." *Intensities: The Journal of Cult Media* 7: 44–54.

Freeman, Matthew. 2016. *Historicising Transmedia Storytelling*. New York and London: Routledge.

Harvey, Colin. 2015. *Fantastic Transmedia: Narrative, Play and Memory Across Science Fiction and Fantasy Storyworlds*. Basingstoke: Palgrave Macmillan.

Innocenti, Veronica, and Guglielmo Pescatore. 2008. *Le Nuove Forme della Serialità Televisiva. Storia, Linguaggio, Temi* [New Forms of Seriality. History, Language, Themes]. Bologna: Archetipolibri.

Jenkins, Henry. 2003. "Transmedia Storytelling: Moving Characters from Books to Films to Video Games Can Make them Stronger and More Compelling." *Technology Review*, January 15. Accessed August 30, 2017. =www.technologyreview.com/s/401760/transmedia-storytelling/

Jenkins, Henry. 2006. *Convergence Culture: Where Old and New Media Collide*. New York: New York University Press.

Jenkins, Henry. 2009. "The Revenge of the Origami Unicorn: Seven Principles of Transmedia storytelling (Well, Two Actually. Five More on Friday)." *Confessions of an Aca-Fan: The Official Weblog of Henry Jenkins*, December 12. Accessed August 30, 2017. http://henryjenkins.org/2009/12/the_revenge_of_the_origami_uni.html.

Jenkins, Henry. 2011. "Transmedia 202: Further Reflections." *Confessions of an Aca-Fan: The Official Weblog of Henry Jenkins*, July 31. Accessed August 30, 2017. http://henryjenkins.org/2011/08/defining_transmedia_further_re.html.

Jenkins, Henry. 2014. "The Reign of the 'Mothership': Transmedia's Past, Present, and Possible Futures." In *Wired TV: Laboring over an Interactive Future*, edited by Denise Mann, 244–268. New Brunswick: Rutgers University Press.

Johnson, Derek. 2013. *Media Franchising: Creative License and Collaboration in the Culture Industries*. New York: New York University Press.

Johnson, Derek. 2017. "A History of Transmedia Entertainment." *Spreadable Media*. Accessed July 26, 2017. http://spreadablemedia.org/essays/johnson/#.WV-l0YXZc-8.

Kelleter, Frank. 2012. "'Toto, I Think We're in Oz Again' (and Again and Again): Remakes and Popular Seriality." In *Film Remakes, Adaptations and Fan Productions*, edited by Katleen Loock and Constantine Vervis, 19–44. London: Palgrave Macmillan.

Lucanio, Patrick, and Gary Coville. 2002. *Smokin' Rockets: The Romance of Technology in American Film, Radio and Television, 1945–1962*. Jefferson and London: McFarland.

Marrone, Gianfranco. 2003. *Montalbano. Affermazioni e Trasformazioni di un Eroe Mediatico* [Affirmations and Transformations of a Media Hero]. Rome: Rai-Eri.

Pearson, Roberta. 2009. "Transmedia Storytelling in Historical and Theoretical Perspective." Presentation at The Ends of Television conference, University of Amsterdam, June 29–July 1.

Santo, Avi. 2006. "Transmedia Brand Licensing Prior to Conglomeration: George Trendle and the Lone Ranger and Green Hornet Brands, 1933–1966." Ph.D. diss., University of Texas.

Santo, Avi. 2015. *Selling the Silver Bullet: The Lone Ranger and Transmedia Brand Licensing*. Austin: University of Texas Press.

Scolari, Carlos. 2009. "Transmedia Storytelling: Implicit Consumers, Narrative Worlds, and Branding in Contemporary Media Production." *International Journal of Communication* 3: 586–606. Accessed August 30, 2017. doi:1932/8036/20090586.

Scolari, Carlos. 2013. "Don Quixote of La Mancha: Transmedia Storytelling in the Grey Zone." *International Journal of Communication* 8: 2382–2405. Accessed August 30, 2017. doi: 1932–8036/20140005.

Scolari, Carlos, Paolo Bertetti, and Matthew Freeman. 2014. *Transmedia Archaeology: Storytelling in the Borderlines of Science Fiction, Comics and Pulp Magazines*. Basingstoke: Palgrave Macmillan.

Scott, Jason. 2009. "The Character-Oriented Franchise: Promotion and Exploitation of Pre-sold Characters in American Film, 1913–1950." *Scope: An Online Journal of Film and Television Studies* 15: 34–55. Accessed August 30, 2017. www.nottingham.ac.uk/scope/issues/2009/october-issue-15.aspx.

Thompson, Kristiane. 2007. *The Frodo Franchise: The Lord of the Rings and Modern Hollywood*. Berkeley: University of California Press.

Wolf, Mark J. P. 2012. *Building Imaginary Worlds: The Theory and History of Subcreation*. New York and London: Routledge.

29
TRANSMEDIA HERITAGE
Museums and Historic Sites as Present-Day Storytellers

Jenny Kidd

> Visitors ... fit their experience of [an] exhibition into their own experiences of everyday life, and in so doing construct, as bricoleurs, their own fragmentary, but meaningful, rhetorics and narratives from the materials which confront them.
>
> *(Silverstone 1988, 235)*

This chapter considers the extent to which the transmedia storytelling concept (Jenkins 2006, 2011; Pratten 2011; Rose 2011; Clarke 2013) might be applied to heritage practice. The above quote from Roger Silverstone in 1988 should suggest that transmedia storytelling is not such a radical proposition within these contexts. By way of clarification, the chapter takes institutionalized and formalized heritage practices as its focus, namely the work of museums and heritage sites, rather than (for example) public or community heritage initiatives. In what ways, it asks, might the approaches utilized by museum professionals echo those introduced in other contributions in this volume? Why is it useful to consider the experiences historic sites facilitate as transmedia? And what might be the limitations of that approach? This chapter will use two cases to illustrate the potentials of the transmedia concept within contexts often typified as traditional in both their outlook and the methodologies they employ. In so doing I expose those stereotypical perceptions of heritage work as (often now) outmoded, and suggest instead that many museums are pursuing a much more radical agenda.

I have written elsewhere about museum makers as transmedia storytellers (in Kidd 2014), and since that time their continued experimentation with form and content would seem to further bolster that approach. Globally, those working at museums and heritage sites have been engaged in searching analyses of their roles and impacts including—but not exclusively—as educators, and this has often led to more considered reflection on their potentials as storytellers (Rowe et al. 2002; Henning 2006; Kelly 2010; Chan 2012; Parry 2013). Many of them have been looking to re-orient what they do away from interpreting the stories of monarchs and aristocrats (for example), toward giving more of a lead to those traditionally understood as supporting and/or minor characters from the past, a move that quite profoundly mirrors the ambitions of transmedia storytelling. In addition, the outputs of museums and heritage sites can be considered "nuclear texts" and/or "intertexts" (Zorrilla Abascal 2016, 2632) in relation to other (hi)stories that might be being constructed elsewhere. As such, the dynamism of the narratives they work with is quite remarkable.

Heritage institutions are now often seen (not uncritically) as part of what has been termed the "experience economy" (Pine and Gilmore 1998). As they have sought to position themselves alongside and in relation to other kinds of experience, they have been investigating the potentials of a variety of digital, analogue, and mixed media. It has been noted that the transmedia approach *can* help institutions (and perhaps heritage institutions also) to distinguish themselves within a "highly fragmented contemporary marketscape" (Ilhan, Kozinets, and Otnes 2013, 529) for cultural encounters, and that different kinds of value might circulate around that practice. Certainly, this is of interest to heritage institutions, as is building and maintaining a strong brand identity and community, something that again it is understood focused transmedia approaches might help to facilitate (Simons 2014, 2221). Following Bourdaa (2013), transmedia activity might help extend the "mythology" of these (now often) globally recognizable brands and institutions, but it can perhaps also frustrate it (see Hills 2012 on "fanagement").

In heritage contexts, engagement has typically been defined in terms of visitation, but this is increasingly considered a problematic concept. It is now common for the boundaries around a visit to be incredibly blurred so that we might ask when does it begin and when does it end? A visit might start and stop online, and may or may not include a trip to a physical venue somewhere in the middle. A visit can meander from a site's What's On pages, to TripAdvisor, to Facebook, to a physical museum at which point the visitor might start sharing photos via Twitter or Instagram. They might opt to use an audio guide, a mobile application, to buy a catalogue or site map. They may consult Wikipedia as they stroll around the site, or email themselves a photo of a web link they have found on an interpretation panel. They might leave a lengthy comment in a visitors' book. Afterwards they may find themselves back on the TripAdvisor site, or on the Facebook page recommending (or indeed criticizing) the site, the café, or the public conveniences. A historic site or museum visit has become rather a complex thing to grapple with, and is often far less linear than the one I have presented here. Echoing Clarke's definition of transmedia storytelling, a museum visit is *already* best conceived as "a diversity of media complicating and complementing one another" (2013, 209). It is "dispersed" (Jenkins 2010), "distributed" (Walker 2004), and "networked" (Zapp 2004), so that "the story we construct depends on which media extensions we draw upon" (Jenkins 2011). Increasingly the above terms are being applied within heritage contexts where talk about "connection" (Drotner and Schrøder, 2013), "participation" (Simons 2014), and "multimodality" (Kidd 2017) are becoming normative.

The above attests to the sheer number and complexity of media one might encounter in the navigation of a heritage experience, and does little to convey the enthusiasm and playfulness found in acts of creativity that surface around heritage content whether online or offline. The stories these institutions construct have become multifaceted, and often inseparable from the stories that are constructed *about* them by their users especially within the digital domain. Here we are reminded of Donna Hancox's (2017) assertion that what is especially interesting about transmedia is the decentralization of the concept of authorship, something many museums have been keen to explore. To extend Hancox's theory here to museum practice, it is something of a present-day aspiration to foreground storytelling "that does not privilege one voice, one part of the story or one platform over another" (2017, 50). Srividya Ramasubramanian (2017) has also noted these potentials indicating that transmedia storytelling can challenge traditional representations and provide space for alternative voices in ways that (to apply within the museums' context) traditional modes of interpretation often struggle to. In doing so transmedia projects can create "alter-spaces"; "alternative community-oriented spaces for affirming cultural identities" offering more "thoughtful, complex, and multifaceted representations of [in Ramasubramanian's analysis, but not limited to] ethnic/racial minorities" (Ramasubramanian 2017, 340). Significantly, within a museum's context and remit, there might be even *more* scope for the kind of politically and culturally radical re-scripting of accepted narratives about people, place, and our pasts that Ramasubramanian (2017) is calling for.

There is emerging scholarship about how transmedia might be utilized within educational environments, which is worth reviewing in light of museums' and heritage sites' core learning remits. Scholars note that using multiple media platforms within the classroom has become the norm (Rodriguez-Illera and Castells 2014, 4), as it is now in museum education where learning experiences (formal and informal) now often include such things as hands-on sessions, forms of making, using interactive screens, participation via social media, and playing digital games. They recognize that most visitors and audiences are seeking learning experiences that in some way mirror their everyday experiences within our "complex media world" (Pence 2012, 134). This fits within a broader re-appraisal of learning (and museum learning) as constructivist (Hein 1995; Hooper-Greenhill 2007). Dudacek notes that "Forcing pupils and students to know facts without understanding why they need to know it is contra productive and unnecessary. Much more effective is to encourage their curiosity" (Dudacek 2015, 695). Such approaches do however come with a warning from Shalom M. Fisch who notes that whilst "Data from several empirical research studies have indicated that cross-platform approaches have the potential to promote significantly greater learning than use of any one media component in isolation" it is important not to "simply flood children with a greater number of media products; 'more' is not always better" (2013, 226). Henry Jenkins (2006) and others (Sweller 2005; Zorrilla Abascal 2016) have noted that redundant materials can provide significant interference in learning contexts. This is something that museums and heritage sites will need to consider as they embrace multiple channels and media in interpretation.

Having reviewed a number of potentials for the transmedia storytelling concept within these contexts, I come then to an assessment of its limitations, noting first that current definitions of transmedia do not map neatly onto heritage practice. According to Colin Harvey (2015, 1), some scholars have proposed fairly "strict criteria" on what constitutes transmedia storytelling. To Ilhan, Kozinets, and Otnes (2013, 529), the transmedia approach indicates three central tenets; media coordination, world-building, and what they call "negative capability," that is the capacity for users to explore gaps and mysteries within a narrative. Heritage practice has the latter of these in spades. At its very core, it is all about gappage, and many institutions now seek to foreground ambiguity in the way they talk about processes of interpretation and history "making." A museum or site visit has always been about constructing a narrative between different elements or fragments of experience, or "patching" as Ilhan, Kozinets, and Otnes call it; "uniting, joining and affixing story elements together to produce narratives whose elements are drawn from different media platforms" (2013, 529). But the extent to which this might be considered a premediated, coordinated or built assemblage of bits, or indeed a story "world," is perhaps less easy to determine. Unlike a slick campaign to accompany a new TV show, game or film, museums and heritage sites have legacy content, buildings, branding, and exhibitions which could be decades, if not centuries, old. When looking at high-profile examples such as Doctor Who or Batman "the boundaries of the storyworld can be difficult to discern" (Harvey 2015, 186), but within heritage contexts parameters can be nigh on impossible to locate. For example, what are the boundaries around Hampton Court's cross-media world of the Tudors, and how does the world they construct intersect with other representations of the lives of those monarchs? Could the Science Museum's high-profile Robots installation (2017) be understood as a transmedia exhibition and if so, how does it intersect with the larger storyworlds containing the robots it re-presents? What is the relationship between fact and fiction in both of the above, and within current definitions of transmedia? Does all of this bend and flex Harvey's concept of "detached transmedia storytelling" (2015, 188) or Pence's notion of "framework (or open) transmedia" (Pence 2012, 135) beyond breaking point? Producing a "transmedia topology" of museum narratives (to use the terminology of Hook, Barrios-O'Neill, and Dyer 2016) is tricky indeed, involving both lateral and diachronic exploration. In addition, the concentration of transmedia scholarship on franchises and storyworlds that are copyrighted legal entities, and where licenses and tie-ins are identifiable (Harvey 2015), is also complicated within heritage contexts where authorship and ownership are often less clear, and where materiality takes quite a different form.

These are challenges, and (as the old adage goes) also opportunities.

To explore these themes further, I introduce two cases in the remainder of this chapter. The first is a site which can be readily understood as itself a storyworld; St Fagans National History Museum, part of Amgueddfa Cymru—National Museum Wales. Second, I explore the activity of literary museums and houses such as the Charles Dickens Museum and Jane Austen's House Museum; sites which themselves might be conceived of as "intertexts" of the larger storyworlds constructed by and around literary figures. For me, what emerges as interesting in this latter analysis is not only how sites can intersect powerfully with storyworlds from the past, but also how they can do so with our unprecedented contemporary socio-political landscape too.

St Fagans National History Museum

St Fagans National History Museum is part of the Amgueddfa Cymru—National Museum Wales consortium and the most popular visitor attraction in Wales. It is a co-location of more than 40 re-erected buildings from all over Wales in the grounds of an Elizabethan manor house known as the Castle. These buildings include iron age roundhouses, nineteenth-century mills, a row of cottages (each one furnished according to a different era from 1800 to 1985), a series of workshops, a general store and bakehouse, a pigsty, an air raid shelter, a "House of the Future," and a pub from the Adamsdown area of Cardiff, which is currently being relocated to the site. As an assemblage of parts, it is a curious fiction, one that can be experienced physically on site, but also virtually in its various digital manifestations; the website, the online collection, oral testimonies, the social media channels, and the blog for example. To borrow from Henry Jenkins, we might note that the story of St Fagans is "so large" that it "cannot be covered in a single medium" (2006, 95) or indeed in one language, it is bilingual through and through.

St Fagans is, I would contend, a "storyworld," but one that has been designed and scripted through periods of significant re-appraisal over time. The heritage being presented here is not one in stasis, it is evolving all of the time, including at the current moment through a multi-million-pound redevelopment project. It is a storyworld produced through both coordination and serendipity. St Fagans National History Museum was formerly the Museum of Welsh Life and before that, the Welsh Folk Museum (Mason 2005), and one constant has been the ambition to construct an overarching narrative about Wales and Welsh identity as a site that continues to celebrate "Welsh traditions and lifestyles" (Amgueddfa Cymru 2017).

Rhiannon Mason notes in her discussion of St Fagans that "there is a danger of reading museums as too internally coherent, too unitary in their meanings" (2005, 19). There have been concerns in the past that the presentation at St Fagans is too static, unified, and simplified (see Mason 2005 and Dicks 2000 for an overview of these criticisms) but Mason contends instead that "the text of this museum is far more organic, open-ended and internally contradictory" (2005, 22), an assertion which makes room for more playful and complex readings of that narrative of Welshness. Indeed, Mason asserts, St Fagans "represents a meeting point for competing ideas about national identities" in Wales (2005, 29). At St Fagans, the narratives on offer under this rubric are many, and they intersect in ways that are impossible to pre-empt. Stories about Welsh rural life, "ordinary" industrial working life, Welsh textiles, horticulture, the Earl of Plymouth and his family, and even war (the Castle became a convalescent hospital during World War I). But it is also the focal point for unlimited individual stories and memories that visitors bring with them; for frequent visits to St Fagans are a feature of schooling and family life for many residents of the Cardiff city region. School groups participate in re-enactments, families attend weddings on site, groups come for ghost walks, festivals, and music events, and children count lambs being born on the online LambCam every spring. Few people do all of these things—that is not the ambition of the site—and as a result the narratives constructed about it are many and diverse.

This ambiguity is picked up in the latest extension of narrative at St Fagans. In 2017, in partnership with creative marketing company yello brick and Cardiff University, St Fagans launched *Traces* (*Olion* in the Welsh language), a site-specific storytelling subtle mob which constitutes an expressly performative encounter within the space of St Fagans. The mobile application plays with boundaries between fact and fiction, past and present, and connections between being and feeling in place. It pushes visitors to engage with the site's stories in new ways, and to explore their own assumptions about what using or visiting a museum is. This is something we hope people will respond positively to, but this outcome is of course no certainty. Nele Simons (2014, 2231) notes of her own investigations into audiences for transmedia storytelling that there remain deep waters between what is possible and what is desirable for most people when it comes to their uses of technology. People tend to be rather conservative in their expectations, in our case here of what it means to use or visit a museum or heritage site.

St Fagans is then an assemblage of fragments, a storyworld that can be accessed in various conditions of completion, one containing conflicting truths that will not be reconciled, and one which has the capacity to surprise at every corner. Framed in this way, it begins to sound very like the examples recounted elsewhere in this volume.

Literary Houses and Museums

The second case under scrutiny here is that of a broader genre of museums, those dedicated to the heritage and mythology of noted literary figures from the past. There are a great number of houses and museums which intersect with literary heritage around the globe, and perhaps in the United Kingdom especially this is something of a phenomenon. It might be obvious how such sites intersect with (incredibly well known) storyworlds from literature, but I contend that how such storyworlds are activated in the present is worthy of attention, especially for those interested in the civic value of museums and heritage sites.

The Charles Dickens Museum for example, situated at 48 Doughty Street (Dickens' London home where he wrote *Oliver Twist*, *The Pickwick Papers*, and *Nicholas Nickleby*) will be a part of the tapestry of experiences of those texts which are patched together by visitors in their making of meaning about those stories. This patchwork *might* include reading the novels and enjoying film and television adaptations, adaptations which themselves often re-package and hybridize the novels in interesting ways (as with the *Dickensian* series on BBC in 2016). As is now normative, a visit to this museum or its website might be complemented by different media extensions (the mobile application, YouTube videos, the blog, Instagram, Facebook, or Twitter). It might include participation in a tour, or watching a performance. The varying storyworlds which Dickens created, and which have re-surfaced and been re-created multiply since that time, are opened up to users and visitors in their negotiation of meaning.

Similarly, at Jane Austen's House Museum in Hampshire, England, visitors will interact with complex mixed media storyworlds whether online, offline or somewhere in between. The online collection, learning resources, Instagram, Facebook, and Twitter posts all make large the connections between Austen's contemporary environment and the storyworlds she created. Visitors will bring with them their perceptions about Austen formed in large part by the books themselves, staged, film, and televisual adaptations, but also perhaps memes and YouTube parodies.

But what is interesting to note in relation to these examples is how museum staff can be seen attempting to activate the storyworlds of Dickens and Austen (for example) to connect with real-world concerns and agendas in the present. As noted previously, transmedia storytelling can be part of a move toward a more socially and representationally just vision for our present (Ramasubramanian 2017), and museums are embracing that potential. At Jane Austen's House Museum, there has been an active campaign to promote women's rights. Some of this activity has circulated around recognition for Jane Austen herself, including a campaign to get Austen celebrated via a portrait on new British £10 notes (a campaign that was not uncontentious). At the Charles Dickens Museum, in the aftermath of the Grenfell Tower fire in London June 2017, a simple quote from *Martin Chuzzlewit* shared on Twitter contributed to heated debates that were taking place about equality, fairness and responsibility; "Charity begins at home, and justice begins next door." Here, spokespeople for the museum connect powerfully the infamous storyworld created by Dickens to that of our present.

In these moments, transmedia heritage can be seen at its most subtle, ambiguous and powerful. Connecting across communities (variously defined) and timeframes, museum communications become embedded within (arguably) more consequential narratives. Working across media extensions and intersecting with stories unfolding in real time in the news, these museums re-activate storyworlds from the past with renewed social purpose.

Conclusion

The above examples echo and evidence Antero Garcia's assertion that "transmedia is about connecting new modes of storytelling" (2017, 715). In doing so, museums and heritage sites are having to let go of (some of) their control over the narratives produced in and around them, a shift that many in the sector would argue is long overdue. They are also, as has been seen, seeking to connect the storyworlds that are at their nexus, with those that circulate around them within contemporary society; including debates about identity, rights, and justice. Whether and how these developments in turn challenge and disrupt the very idea of "the museum" remains to be seen.

Research into museums as storytellers will continue to be fertile territory, and so this chapter closes with propositions for further debate and scrutiny. One promise of transmedia is that it can, and indeed "*has to* reward consumers' efforts" (Bourdaa 2013, 206, my italics), but what "reward" looks like within museum and heritage contexts is more ambiguous than within other environments where "consumer" motivations are perhaps easier to define. What then, we should be asking, do visitors/users/audiences for heritage encounters take away from these media-rich experiences? And in what ways are they meaningful to them (or not)? Second, we might ask if there is a value in identifying a canon of work as transmedia heritage, or whether the joy of studying these contexts is precisely freedom from that very notion, a liberation perhaps not afforded to those who research storyworlds within other contexts. Third, and most excitingly, it would seem pertinent to ask what the possibilities are for the storyworlds circulating around and within heritage contexts. Given their cultural and representational significance, and given their role in identity and nation building, museums are ripe sites for exploration of the liminal spaces between known and unknown, past and present, fact and fiction. Those kinds of storyworlds might emerge as active, rewarding, and consequential indeed.

References

Amgueddfa Cymru. 2017. "St Fagans National Museum of History Highlights." Accessed July 4, 2017. https://museum.wales/stfagans/about/.

Bourdaa, Melanie. 2013. "'Following the Pattern': The Creation of an Encyclopaedic Universe with Transmedia Storytelling." *Adaptation* 6 (2): 202–214.

Chan, Seb. 2012. "On Storyworlds, Immersive Media, Narrative and Museums – an Interview with Mike Jones." Accessed September 19, 2013. www.freshandnew.org/2012/10/storyworlds-immersive-media-narrative-interview-mike-jones/.

Clarke, M. J. 2013. *Transmedia Television: New Trends in Network Serial Production*. London and New York. Bloomsbury.

Dicks, Bella. 2000. *Heritage, Place and Community*. Cardiff: University of Wales Press.

Drotner, Kirsten, and Kim Christian Schrøder (eds.). 2013. *Museum Communication and Social Media: The Connected Museum*. New York and London: Routledge.

Dudacek, Oto. 2015. "Transmedia Storytelling in Education." *Procedia – Social and Behavioral Sciences* 197: 694–696.

Fisch, Shalom M. 2016. "Introduction to the Special Section: Transmedia in the Service of Education." *Journal of Children and Media* 10 (2): 225–228.

Garcia, Antero. 2017. "Transmedia: Redefining Where and How Stories Are Told." *Journal of Adolescent and Adult Literacy* 60 (6): 715–717.

Hancox, Donna. 2017. "From Subject to Collaborator: Transmedia Storytelling and Social Research." *Convergence* 23 (1): 49–60.

Harvey, Colin B. 2015. *Fantastic Transmedia*. New York and Hampshire: Palgrave.

Hein, George. 1995. "The Constructivist Museum." *Journal of Education in Museums* 16: 15–17.

Henning, Michelle. 2006. *Museums, Media and Cultural Theory*. New York: McGraw-Hill Education.

Hills, Matt. 2012. "*Torchwood's* Trans-Transmedia: Media Tie-Ins and Brand Fanagement." *Participations* 9 (2): 409–428.

Hook, Alan, Danielle Barrios-O'Neill, and Jolene Mairs Dyer. 2016. "A Transmedia Topology of Making a Murderer." *VIEW* 5 (10).

Hooper-Greenhill, Eilean. 2007. *Museums and Education: Purpose, Pedagogy, Performance*. Oxon and New York: Routledge.

Ilhan, Behice Ece, Robert V. Kozinets, and Cele C. Otnes. 2013. "Transmedia Consumption Experiences (TCE): Patching as a Narrative Consumption Practice." *Advances in Consumer Research* 41: 529–531.

Jenkins, Henry. 2006. *Convergence Culture: When Old and New Media Collide*. New York: New York University Press.

Jenkins, Henry. 2010. "Transmedia Education: The 7 Principles Revisited." Accessed July 26, 2013. http://henryjenkins.org/2010/06/transmedia_education_the_7_pri.html.

Jenkins, Henry. 2011. "Transmedia 202: Further Reflections." Accessed July 13, 2013. http://henryjenkins.org/2011/08/defining_transmedia_further_re.html.

Kelly, Lynda. 2010. "How Web 2.0 is Changing the Nature of Museum Work." *Curator* 53 (4): 405–510.

Kidd, Jenny. 2014. *Museums in the New Mediascape*. New York and London: Routledge.

Kidd, Jenny. 2017. "*With New Eyes I See*: Embodiment, Empathy and Silence in Digital Heritage Interpretation." *International Journal of Heritage Studies*.

Mason, Rhiannon. 2005. "Nation Building at the Museum of Welsh Life." *Museum and Society* 3 (1): 18–34.

Parry, Ross. 2013. "The Trusted Artifice." In *Museum Communication and Social Media: The Connected Museum*, edited by Kirsten Drotner and Kim Christian Schrøder, 17–32. New York and London: Routledge.

Pence, Harry E. 2012. "Teaching with Transmedia." *Journal of Educational Technology Systems* 40 (2): 131–140.

Pine, Joseph, and James Gilmore. 1998. "Welcome to the Experience Economy." *Harvard Business Review*. Accessed June 26, 2017. https://hbr.org/1998/07/welcome-to-the-experience-economy.

Pratten, Robert. 2011. *Getting Started in Transmedia Storytelling*. London: CreateSpace.

Ramasubramanian, Srividya. 2017. "Racial/Ethnic Identity, Community-Oriented Media Initiatives, and Transmedia Storytelling." *The Information Society* 32 (5): 333–342.

Rodriguez-Illera, José Luis, and Núria Molas Castells. 2014. "Educational Uses of Transmedia Storytelling: The Ancestral Letter." *Journal of Educational Multimedia and Hypermedia* 23 (4): 1–22.

Rose, Frank. 2011. *The Art of Immersion: How the Digital Generation is Remaking Hollywood, Madison Avenue, and the Way We Tell Stories*. New York and London: Norton & Company Ltd.

Rowe, Shawn M., James V. Wertsch, and Tatyana Y. Kosyaeva. 2002. "Linking Little Narratives to Big Ones: Narrative and Public Memory in History Museums." *Culture and Psychology* 8 (1): 96–112.

Silverstone, Roger. 1988. "Museums and the Media: Theoretical and Methodological Exploration." *The International Journal of Museum Management and Curatorship* 7 (3): 231–241.

Simons, Nele. 2014. "Audience Reception of Cross- and Transmedia TV Drama in the Age of Convergence." *International Journal of Communication* 8: 2220–2239.

Sweller, Joanne. 2005. "Cognitive Load Theory and Instructional Design: Recent Developments." *Educational Psychologist* 38 (1): 1–4.

Walker, Jill. 2004. "Distributed Narrative: Telling Stories Across Networks." In *Internet Research Annual*, edited by Mia Consalvo and Kate O'Riordan, 91–103. Brighton: Peter Lang.

Zapp, Andrea. 2004. *Networked Narrative Environments: As Imaginary Spaces of Being*. Manchester: Cornerhouse Publications.

Zorrilla Abascal, Maria Luisa. 2016. "Transmedia Intertextualities in Educational Media Resources: The Case of BBC Schools in the United Kingdom." *New Media and Society* 18 (11): 2629–2648.

30

TRANSMEDIA FANDOM AND PARTICIPATION

The Nuances and Contours of Fannish Participation

Paul Booth

The relationship between transmedia textual networks and participatory fandom is more complex than it may first appear. Common perceptions of transmedia texts indicate the crucial participation of fans for the success of a narrative, although other scholarship maintains the industrial focus of transmedia, outside of audience interpretation. This chapter interrogates this relationship by asking: to what extent are fans part of a transmedia experience, and to what extent are they separated from it? In posing this dichotomy, this chapter asks how precisely fans are discursively constructed in the digital age: as viewers, as consumers, as active participants, as all or even as none of the above? At the same time, by constructing fandom in relation to transmedia, we also have to construct transmedia in relation to fandom. This chapter thus also interrogates the notion of transmedia itself by examining the crucial role that audiences play in its determination.

More specifically, this chapter will examine previous scholarship on transmedia fandom through two different lenses. As Göran Bolin (2011) notes:

> Transmedia storytelling ... has the dual quality of being both market and non-market motivated, or, to put it the other way around, it is driven by both artistic and non-artistic motivation. And transmedia stories can also result from the engagement of both the media industry (for economic reasons) and non-market motivated fans.
>
> *(98, quoted in Hills 2012)*

The first lens comes at the relationship from an industrial perspective to view fans as the ideal audience for transmedia narratives. As a marketing strategy and brand development practice, transmedia reaches its audience—an audience increasingly made up of fans—via multiple media outlets. The second lens highlights the fan as instrumental to—and productive within—a transmedia framework. In this sense, creative fan work becomes part of a transmedia franchise, extending and developing the larger narrative world. Dipping back into the fan-as-folk-culture argument developed by Henry Jenkins (1992), this view articulates fans as crucial co-contributors to "transtext" franchises (Derhy Kurtz and Bourdaa 2017).

Yet, in presenting such a bifurcated argument, this chapter falls into a trap that I hope, in the end, to counter: that viewing any simple two-dimensional model of transmedia and fandom (at either end of a spectrum) ignores the radical multi-dimensionality of the participants (see Stein 2017). Indeed, as Harvey (2015) notes, transmedia storytelling can more often be seen as "*relational*, emphasizing the relationship between a particular transmedia articulation such as a comic book or website with the wider storyworld in question, and by extension the wider culture" (2). There is a slippery slope here—if we see fan work as situated within an industrial transmedia strategy, then how "strategic" can we find the industry? In other words, in an era where fan theories are routinely ignored or contradicted by media producers (Scott 2013), does it make sense to say that fan work can be—or even should be—part of an industrial strategy of linked narratives and thematic content? Academic and popular writing on transmedia tends to discursively frame this relationship between these two polarities (see Derhy Kurtz and Bourdaa 2017, 5–6), and yet, at the same time, "attempts at understanding audience engagement are rendered more complex still by the multiple kinds of participatory activity afforded by crossmedia projects" (Harvey 2015, 4). I want to counter this discursive construction through developing an examination of the nuances and contours of fandom within transmedia narratives and extensions (see Richards 2017).

Specifically, then, I will be reframing the relationship between transmedia narratives and fan work through both a review of contemporary literature on transmedia/fandom and an analysis of texts that do not fall neatly into this continuum of industrial text to fannish work. One of the more commonly discussed examples of a transmediated text is the BBC's *Doctor Who*, because of its longevity (having been a cultural presence since 1963), its plethora of transmediated elements (television shows, audio adventures, books, DVDs, comics, etc. (see Perryman 2008)), and its unique status as a text without an "authorized" canon (Cornell 2007; Harvey 2010; *contra* Derhy Kurtz 2017). Using both *The Five-ish Doctors Reboot* (2013) and two *Doctor Who* documentaries as my case studies, I hope to show that revising the relationship between fandom and transmedia reveals a more complex interaction, where fans are constituent of transmedia franchises just as much as transmedia stories are dependent on fannishness.

Transmedia and Fandom: An Industrial Perspective

Emphasizing the "industry" aspect of transmedia highlights the marketing potential of a transmedia text for a particular fandom. Henry Jenkins (2014) argues that the branding logics of the Hollywood system "always return to the idea that, ultimately, transmedia is a mode of promotion designed to intensify audience engagement" (260). Indeed, although fandom has always been discussed in relation to transmedia, it is not always as part of transmedia franchises itself. It is useful to return to Jenkins' influential early scholarship on transmedia, if only to reflect on the ways that fandom has been conceived of as an audience for, not a partner in, the transmediated text. In *Convergence Culture* (2006), Jenkins writes that

> A transmedia story unfolds across multiple media platforms, with each new text making a distinctive and valuable contribution to the whole … Reading across the media sustains a depth of experience that motivates more consumption. Redundancy burns up fan interest and causes franchises to fail.
>
> *(95–96)*

Here the discussion of transmedia references the construction of a vast narrative (Harrigan and Wardrip-Fruin 2009), but fandom itself is a form of consumption that influences the success or failure of the franchise—fandom is not integrated into transmedia texts, which remain the purview of professional creators. Indeed, as Jenkins (2017) later notes: "By itself, the word *transmedia* tells us little about the media involved (transmedia does not necessarily involve digital), [or] about the relationship between producers and consumers (transmedia is not necessarily interactive or participatory)" (220). In other words, *transmedia* as a term is merely a descriptor, with little explanatory power (especially regarding the relationship between its users and its creators).

Looked at through the lens of the industry, then, fandom becomes merely a marketing strategy, a way to garner a greater, more passionate audience. Fans are a crucial part of the transmedia franchise—gone are the days when fandom per se was seen in a negative or pathological light. That being said, as Busse (2013) points out, not all fandoms are treated equally, and many non-mainstream fan audiences are still derided. For Karen Hellekson and Kristina Busse (2014), too, "fans have moved from being ignored or merely tolerated by producers to being important and sought after" (135). For example, Aaron Delwiche (2017) describes how "educators, media practitioners and marketing professionals have been quick to embrace the term 'transmedia storytelling.' Strategic communicators understand that corporate identity and product brands are 'stories' constructed across a range of platforms" (34). *Doctor Who* presents a noteworthy example, as the BBC has been quick to emphasize the multi-media modality of this flagship enterprise; but as Delwiche (2017) also notes, this type of "soft" transmedia features "a shared fictional world [that] unfolds across media channels but there are relatively few narratives links between the channels" (37). Hoping to garner fan interest, the BBC creates additional content for the show, but the show itself remains primary. As Perryman (2008) analyzes:

> The franchise had been successful in a variety of different media platforms (books, CD, radio, comics and the web), while the writers and producers of the new series had written for all of them. As Russell T. Davies [the New series' first showrunner] said prior to the show's return: "This is a show now owned by its fans", and the fans were already used to regarding *Doctor Who* as a transmedia franchise that could be linked together to form a coherent and satisfying whole.
>
> *(25–26)*

For Perryman (2008), the activity of *Doctor Who* fandom demonstrated to the BBC that a passionate fan audience existed, especially since "grassroots amateur fan-fiction and semi-professional, low-budget video productions helped to plug the gap during the early years of *Doctor Who*'s Interregnum" (23). However, once the show came back on the air, the BBC fandom returned to being an audience:

> the BBC has successfully created a transmedia world that … allows passive audiences to simply sit back and enjoy the parent show in blissful isolation, while at the same time it gives active, migratory and participatory audiences opportunities to engage in a rich, and extended multimedia experience.
>
> *(37)*

Participation is filtered through the non-primary texts.

At the same time, Suzanne Scott (2013) has shown that fandom can be pigeonholed into consumptive realms, rather than being seen as a productive partner in transmedia creation:

> Transmedia stories … produce a consolidated canon of "official" texts that frequently discourage or discredit unauthorized expansion or speculation by fans. The danger here is that, despite transmedia stories' collaborative narrative design, the media industry frequently equates fans' "participation" with their continuous consumption of texts that narratively and financially supplant a franchise.
>
> *(43)*

Similarly, Matt Hills (2012) describes how the transmedia function of the *Doctor Who* spin-off *Torchwood* was less to do with a sense of power for fans, and more to do with disciplining, what he calls *fangagement*, a way of "responding to, and anticipating, fan criticisms, as well as catering for specific fractions of fandom who might otherwise be at odds with the unfolding brand, and attempting to draw a line under fan resistance to diegetic and production changes" (410). In this case, transmedia becomes a policing mechanism by which fans are directed toward particular texts and outlets "in order to protect the 'hub' of a TV series and its brand value" (425). Aaron Taylor (2014) takes this idea of fangagement one step further to note how the "Marvel 'Cinematic Universe' represents a canny attempt to co-opt and exploit the fannish expertise and subcultural influence of cultic (i.e., knowledgeable, fan-based) communities" (181). In other words, the Marvel Cinematic Universe (MCU) franchise *relies* on fans to work as ambassadors with the knowledge required to understand the complexity of the franchise.

This type of brand ambassadorship is not that different from brands using the language of fandom to co-opt audience labor (Chin 2014), and as Charles Davis (2013) puts it, "it has become very common to involve audience members in the co-creation of value, typically by employing audience labour to create content for commercial purposes" (181). Viewed through the lens of the industry, fandom works within transmedia franchises both because engaged, participatory audiences seek out more content, but also because they function as a type of advertising for the franchise, speaking out about it or promoting it through social media. The television show *Glee* has been highly influential in this form of marketing (Stork 2014), for example. Megan Wood and Linda Baughman (2012) describe fan-created Twitter accounts where individual fans roleplay as characters from the show. Although fan-created, these accounts also expand "consumer/viewer investments through new marketing strategies, [and] networks and advertisers are able to reap the benefits of 'collective intelligence' while allowing audiences to shape the nature of what it is they are consuming/viewing" (329). Even within this marketing strategy, the work of fans plays an important role within transmedia franchises, helping not only to establish the larger context of the text, but also the content that gets to be termed "transmedia."

Fandom and Transmedia: A Fannish Perspective

While transmedia storytelling can be seen from a marketing perspective, aimed at a participatory and engaged audience, another perspective sees the mighty influence of fans within a transmedia franchise. This influence can take many forms, including both the drive and engagement of fans to seek out additional components of a transmedia story, and also the inclusion of fan creative work within a transmedia franchise. Indeed, in their book *The Rise of Transtexts*, Benjamin W. L. Derhy Kurtz and Mélanie Bourdaa (2017) describe a new interpretation of transmedia textuality, the *transtext,* that "could—and should—[include] … fan-produced transmedia texts … [The] transtext account[s] for both the institutional (or industrial) contributions and user-made ones in relation to the cross-media expansion of a story universe" (1, 5). Stein (2017) argues that "if we focus on this … fantext not as peripheral but as a primary site of transmedia authorship, we discover a very different picture from the traditional notion of transmedia storytelling as a targeted and uniform, intentional form" (71). That is, focusing on the fannish side of transmedia storytelling reveals a more intricate, grassroots, and folk antecedent for storytelling.

The study of fans and fandom thus indicates the constructive power of audiences; fans are not just passive audiences, but engaged and active users of media. In *Textual Poachers*, Jenkins describes how fans create communities, write original texts, and engage in traditional folk practices. Indeed, fandom is adjacent to folk culture, "since motifs and themes from the mass media are often attached to tunes scavenged from popular or folk music, frequently with a keen awareness of the meanings that arise from their careful juxtaposition" (Jenkins 1992, 257–258). As a type of audience, fans are ideal to explore transmedia narratives, as their engagement and passion encourages more time spent with the text(s) and more participation with cross-media content. "The oral tradition of folk tale," as Richards (2017) articulates, "similarly relies on close interaction between storyteller, text, and audience" (17). In *Digital Fandom*, I (2010) also noted the way fans can augment stories through transmedia mechanisms, especially using social media like blogs, wikis and social network sites (see also Hills 2015, 153). Fans' original work—fan fiction, fan edited videos, cosplay, fan art, etc.—has largely been seen as ancillary to the original text, an augmentation but not a continuation (see Scott 2010). However, as Stein (2017) has shown, especially in connection with some newer texts like *Supernatural* or *Welcome to Nightvale*, "the multiplicity of fan works coalesce into a transtext that is immense, flexible, and powerful, if contradictory and contentious … it is an ongoing, dynamic, creative process of collective authorship that spans commercial, independent, and fan production" (71–72). Suzanne Scott's (2010) work on transmedia fandom also demonstrates that fan-created work helps to "decentralize authorship and promote collaboration, both between creators in different mediums and creators and fans" (30). In other words, at a discursive, rather than a textual, level, fandom *does* become constitutive of transmedia work.

Different media therefore invite different levels of fan participation and engagement (Delwiche 2017). Hellekson and Busse (2014) describe how social media has helped to facilitate greater fan engagement with transmedia franchises: "Fans engage in transmedial role-playing on various social media platforms … where they perform various character identifies and use textual play to collectively create narratives" (197; see Booth 2010). Kohnen (2012) also recognizes social media as a means for "official and fan-produced transmedia [to] increasingly share the same media spaces. Both fans and those who address fans through marketing use these spaces because they make sharing media easy" (3). Yet, social media is not the only outlet for fans to engage in transmedia practices; as Stein and Busse (2012) describe of the historicity of the character of Sherlock Holmes, he is "an evolving transmedia figure, at the center of myriad cultural intersections and diverse representational and fan traditions" (10, see also Evans 2012). They describe how fans have developed narratives that have grown the Sherlock Holmes character over the past century, and then go on to note that "*Sherlock* fans have congregated in a host of other online interfaces to engage with the series and with each other, building their own transmedia web of text and image" (13).

Can fans' work be seen as part of a transmedia franchise, even if it is not necessarily planned or invited? Louisa Stein (2017)'s discussion of the television series *Supernatural* indicates so. She argues that "Many participants in digital fan spaces … may only encounter the fan-authored transtexts for a particular piece of media, and never see the 'original' commercial text" (73). In the case of *Supernatural*, "fan-authored transmedia texts can transform the concerns and politics of the source" (73). One example of this is the "Wayward Daughters" spin-off, a fan-posited idea about taking three women characters from the show and creating a new show for them. Producers of the show supported the idea and in July 2017 "confirmed the rumor of an upcoming backdoor pilot called *Wayward Sisters*" (Stein 2018, 412; Kennedy 2017). Stein (2017) also mentions *Welcome to Nightvale*, a podcast that has incorporated fan art into the show. Another example is the Alternate Reality Game (ARG), a type of transmedia game involving both mediation and real-world play where fan engagement is often relied upon to build the game (Booth 2010). As Jeff Watson (2017) describes, "by designing the core (or 'official') game system around procedures, rather than curated content, designers of ARGs can lay the groundwork for players to further iterate and repurpose the game according to their own desires" (206). Relying not on the narrative itself, but on the discursive mechanisms by which that narrative gets understood, helps to reframe the way players—fans—are invoked and involved in the storytelling process (much like the oral traditions of fandom itself). And other franchises confuse the matter in different ways: as I will discuss in the next section, the BBC's *Doctor Who* has never had an established "canon" so what is "official" and what is "unofficial" is not dictated. Rather, it emerges from the co-interest of both the BBC and *Doctor Who* fandom.

Doctor Who and Issues of Canonicity

The BBC show *Doctor Who* has been noted for its transmedia qualities before (Perryman 2008; Hills 2010; Derhy Kurtz 2017), and so in some respects this is a well-trodden case study. But within that, there are multiple points of intersection between fandom and transmedia storytelling that *Doctor Who* illustrates in productive and generative ways. Perhaps the most relevant for this chapter is the notion of canon, the established, authoritative understanding of what constitutes the text. Colin Harvey (2015) argues that "For transmedia fandoms, issues of memory are often expressed through arguments over 'canon': in other words, which elements of a particular storyworld are 'genuine' or 'authentic' and which are non-canon" (3). For *Doctor Who* in particular, the BBC has never established a clear canon, although "fandom has used canon debates to enact value judgements over what 'is' and 'is not' proper *Doctor Who*" (Hills 2010, 148). Derhy Kurtz (2017) argues that these fandom debates focus on three elements: consistency, authenticity markers, and authority figures.

Part of the reason *Doctor Who* functions well as a case study is because of its long history as both a television show and with additional products. The show started in 1963 and ran until 1989; during that time multiple ancillary texts were released by the BBC and also by other companies that augmented the storytelling of *Doctor Who*, including annual books, novelizations of the episodes (some of which differed tremendously from the episode's narrative), comics, games, magazines, and films. As Perryman (2008) argues, this is not what we might term transmedia today, as "little or no collaboration existed between the BBC and the books' publishers, and more often than not the spin-offs were riddled with contradictions and surreal interpretations of the show's protagonist" (23). Yet, they built up a fanbase for *Doctor Who* that, importantly, demonstrated a willingness and a desire to participate with alternate versions of the story.

During 1989–2005, in what is known as the hiatus period, the show was off the air (barring one BBC/Fox co-production television movie starting Paul McGann), and an array of novels picked up where the show left off and developed the narrative in new directions. When the show was announced to return in 2005,

> One of the most complex forms of transmedia storytelling occurred via a series of meta-textual websites and blogs that were produced by the BBC to accompany the return of *Doctor Who*. These websites were a perfect example of what Brooker (2003: 323) describes as "television overflow": "the tendency for media producers to construct a lifestyle experience around a core text, using the Internet to extend audience engagement and encourage a two-way interaction."
>
> *(Perryman 2008, 29)*

Since the return of the series, there have been more traditional types of transmedia texts—licensed webisodes, comics, books, and video games. But, as Perryman (2008) notes:

> There are limitations to how far the BBC, as a public service broadcaster, can take the concept of transmedia storytelling ... they can never be integrated into the over-arching narrative to such an extent that it impacts directly upon it. In short, these platforms must stand alone.
>
> *(34)*

Of particular interest in terms of fandom, however, is the rise of what Hills (2010) calls the "Doctor Who Mafia," or a dispersed group of fans of the original show who have now become creators of the new series. Many of the ancillary products that were not-quite transmedia in the hiatus period were written by fans like Paul Cornell, Justin Richards, Russell T. Davies, Mark Gatiss, Steven Moffat, Robert Shearman, and Gareth Roberts—all fans who have transitioned to work on the show now. Additionally, some fans (Nicholas Briggs, David Tennant, Peter Capaldi) have acted in the show. Briggs helped found Big Finish, a company that has a license with the BBC to make *Doctor Who* audio adventures staring original cast members (as well as new ones) and takes place within the same universe (Hills 2007), but moves the story in radical new directions. Yet, some of Big Finish's original characters have been named in BBC-produced texts (e.g., *The Night of the Doctor* webisode, see Stoeber 2013), confusing the canonicity further.

So all this means that *Doctor Who*'s texts exist in a nebulous realm between canon and non-canon; it may be "soft" transmedia, but there are enough links between the BBC-produced television text and non-BBC, fannish texts that create nuances with the fan/transmedia relationship. Two texts demonstrate this nuance: the twin documentaries *Perfect Scenario: Lost Frontier* and *Perfect Scenario: The End of Dreams*, from the DVDs of *Frontier in Space* and *Planet of the Daleks* respectively; and *The Five-ish Doctors Reboot*, a 2013 special film created by Fifth Doctor Peter Davison. Both these texts complicate the fan/producer relationship in multidimensional ways.

The two-part documentary *Perfect Scenario* appears on two different *Doctor Who* DVDs, and tells one story:

> In the far future, the remaining population of an oxygen-depleted planet Earth lies in enforced stasis in The Field of Dreams, their minds kept active through the work of scenariosmiths. Looking for ideas to help him re-connect to his captive audience, Zed, a young scenariosmith, turns to the world of Doctor Who for inspiration ... What he finds will have a profound effect on the lives of all of the remaining sleepers in The Field of Dreams.
>
> (*whospecialfeatures.wikia.com 2017a; whospecialfeatures.wikia.com 2017b*)

In effect, the documentary portrays the narratives of *Doctor Who* as fodder to keep a future society entertained and sated. "Zed" learns about the popularity of the program and its cultural relevance through interviews with the stars and behind-the-scenes personnel, as well as with actors pretending to be experts from the future (like Aloysius Kahn, Emeritus Professor of Tele-Sci-Fi from the year 2234). The traditional documentary format common to *Doctor Who* DVDs is upended, as special effects and live-action acting complement the storytelling and non-fiction elements. The documentary offers not just a glimpse of *Doctor Who* production, but an original transmedia story element that adds *Who*-related content—*Doctor Who* becomes a text that, in the future, can be analyzed for its larger storytelling discourse. In addition, because it takes the corpus of *Doctor Who* itself as a focus, it establishes *Who* as a central narrative around which other ancillary content organizes. It forces us to read *Doctor Who* as a whole transmedia text, not just constructed by the different ancillary texts, but also by different time frames within which *Doctor Who* is situated.

A second complication within transmedia fandom of *Doctor Who* is the fiftieth anniversary special *The Five-ish Doctors Reboot*, a semi-parody starring Classic Doctors that were not involved in the lavish BBC-produced anniversary special *The Day of the Doctors*. A mockumentary that spoofed the pageantry surrounding *The Day of the Doctor*, as well as giving the actors a chance to celebrate the anniversary in their own special, *The Five-ish Doctors Reboot* "blurs real-world production/industry personae with humorously constructed portrayals" (Hills 2014b, 108). Although created by the stars of *Doctor Who*, the mockumentary "draws on long-term fan sentiments" (Hills 2014b, 108) and "spoke most keenly to fan knowledge" (Hills 2014a, 169). In this way, although it is not a fan-film (in that it was not made by fans per se), it relies on a *discourse* of fandom for its effect. In addition, *The Five-ish Doctors Reboot* "reveals" that, despite the fact the audience didn't see the Classic series Doctors (Davison, Colin Baker, and Sylvester McCoy) in *The Day of the Doctor*, they were in fact included—shrouded under sheets in a key scene. The mockumentary thus forces a fannish re-read of the actual anniversary special, establishing a transmedia-like relationship between the parody and the original.

Both the documentaries and the parody are thus situated in nebulous regions between fannish production and transmedia content. They are not made by fans (at least, they are made by professionals who are also fans), but they are also not content considered part of a canon (even if that canon itself doesn't exist except in fannish discourse). Far from illustrating a clear-cut relationship between fandom and transmedia, these two texts demonstrate a more nuanced reading.

Conclusion

Reading the relationship between transmedia and fandom can lead in different directions. On the one hand, transmedia texts seemingly require a fan audience, or at least an audience primed to participate in fannish ways. On the other hand, fanwork itself may be considered part of transmedia, as a specific "transtext" that transgresses traditional boundaries of consumption and production. As a conclusion, I want to assert that perhaps the contrast between both views of the transmedia/fandom relationship stems not from either side, but from an antiquated discourse about media creation itself. As Marwick, Gray, and Ananny (2014) note:

> Transmedia studies often think first and foremost about texts' origins and how they transcend venues, screenings, or broadcasts. But this approach is no longer sustainable. Scholars interested in flows of media across sites of engagement must imagine audiences as mobile meaning-makers embedded in shifting contexts that are not defined by a single or static set of circumstances.
>
> *(631)*

We no longer live in a "one-way" broadcast flow of media information. Stein and Busse (2012) argue that a "looser" definition of transmedia could integrate fannish work with professional creators. But what if instead of loosening the definition of transmedia, we instead refocus on the media environment, as Hills (2015) suggests? The issue is not necessarily that fandom and transmedia creation are opposing, but rather "our very sense of what narrative 'is' might need to be reconstructed" (Hills 2015, 151). We live in a "changing mediascape, in which the meaning of authorship appears to be shifting to a more visibly collaborative and ongoing process" (Stein 2017, 72).

What does this reconstruction look like? In some ways, we are already seeing it—television series ask audiences to tweet fan theories, or integrate fannish work into the series itself. Films require a level of content knowledge that necessitates fannish work to uncover the larger meaning. But we may be in for more multidimensional transmedia/fannish content, as fans are becoming more mainstream and more engaged with their media. Co-productions between fans and creators, or collaborations between the different parties, might engender types of overlap we have not even considered yet. In some ways, the contemporary trend of second screen viewing and online discourse means that we are all already participating in transmedia production. In this, transmedia storytelling might be the future of all content.

References

Bolin, Göran. 2011. *Value and the Media: Cultural Production and Consumption in Digital Markets*. Farnham: Ashgate.

Booth, Paul. 2010. *Digital Fandom: New Media Studies*. New York: Peter Lang.

Brooker, Will. 2003. "Overflow and Audience." In *The Audience Studies Reader*, edited by Will Brooker and Deborah Jermyn, 322–335. London: Routledge.

Busse, Kristina. 2013. "Geek Hierarchies, Boundary Policing, and the Gendering of the Good Fan." *Participations* 10 (1): 73–91.

Chin, Bertha. 2014. "Sherlockology and Galactica.tv: Fan Sites as Gifts or Exploited Labor?" In "Fandom and/as Labor," edited by Mel Stanfill and Megan Condis, special issue, *Transformative Works and Culture* 15. doi: 10.3983/twc.2014.0513.

Cornell, Paul. 2007. "Canonicty in Doctor Who." *Paul Cornell*, February 10. Accessed February 13, 2017. www.paulcornell.com/2007/02/canonicity-in-doctor-who/.

Davis, Charles H. 2013. "Audience Value and Transmedia Products." In *Media Innovations*, edited by Tanja Storsul and Arne H. Krumsvik, 175–190. Gothenburg: Nordicom.

Delwiche, Aaron. 2017. "Still Searching for the Unicorn: Transmedia Storytelling and the Audience Question." In *The Rise of Transtexts: Challenges and Opportunities*, edited by Benjamin W. L. Derhy Kurtz and Mélanie Bourdaa, 33–48. London: Routledge.

Derhy Kurtz, Benjamin W. L. 2017. "Set in Stone: Issues of Canonicity of Transtexts." In *The Rise of Transtexts: Challenges and Opportunities*, edited by Benjamin W. L. Derhy Kurtz and Mélanie Bourdaa, 104–118. London: Routledge.

Derhy Kurtz, Benjamin W.L., and Mélanie Bourdaa. 2017. "The World is Changing ... and Transtexts are Rising." In *The Rise of Transtexts: Challenges and Opportunities*, edited by Benjamin W. L. Derhy Kurtz and Mélanie Bourdaa, 1–11. London: Routledge.

Evans, Elizabeth Jane. 2012. "Shaping Sherlocks: Institutional Practice and the Adaptation of Character." In *Sherlock and Transmedia Fandom*, edited by Louisa Ellen Stein and Kristina Busse, 102–117. Jefferson: McFarland.

Harrigan, Pat, and Noah Wardrip-Fruin, eds. 2009. *Third Person: Authoring and Exploring Vast Narratives*. Cambridge, MA: MIT Press.

Harvey, Colin B. 2010. "Canon, Myth, and Memory in Doctor Who." In *The Mythological Dimensions of Doctor Who*, edited by Anthony Burdge, Jessica Burke, and Kristine Larsen, 22–36. Crawfordville: Kitsune Books.

Harvey, Colin. 2015. *Fantastic Transmedia: Narrative, Play and Memory across Science Fiction and Fantasy Storyworlds*. Houndmills: Palgrave.

Hellekson, Karen, and Kristina Busse. 2014. "Fan Communities and Affect." In *The Fan Fiction Studies Reader*, edited by Karen Hellekson and Kristina Busse, 131–137. Iowa City: University of Iowa Press.

Hills, Matt. 2007. "Televisuality without Television? The Big Finish Audios and Discourses of 'Tele-centric' *Doctor Who*." In *Time and Relative Dissertations in Space*, edited by David Butler, 280–295. Manchester: Manchester University Press.

Hills, Matt. 2010. *Triumph of a Time Lord: Regenerating Doctor Who in the Twenty-First Century*. London: I. B. Tauris.

Hills, Matt. 2012. "*Torchwood*'s Trans-transmedia: Media Tie-ins and Brand 'Fanagement.'" *Participations* 9 (2): 409–428.

Hills, Matt. 2014a. "The Year of the Doctor: Celebrating the 50th, Regenerating Public Value?" *Science Fiction Film & Television* 7 (2): 159–178.

Hills, Matt. 2014b. "When *Doctor Who* Enters Its Own Timeline: The Database Aesthetics and Hyperdiegesis of Multi-Doctor Stories." *Critical Studies in Television* 9 (1): 95–113.

Hills, Matt. 2015. "Storyselling and Storykilling: Affirmational/Transformational Discourses of Television Narrative." In *Storytelling in the Media Convergence Age*, edited by Roberta Pearson and Anthony Smith, 151–173. Houndmills: Palgrave.

Jenkins, Henry. 1992. *Textual Poachers: Television Fans and Participatory Culture*. New York: Routledge.

Jenkins, Henry. 2006. *Convergence Culture: Where Old and New Media Meet*. New York: New York University Press.

Jenkins, Henry. 2014. "The Reign of the 'Mothership': Transmedia's Past, Present, and Possible Futures." In *Wired TV: Laboring Over an Interactive Future*, edited by Denise Mann, 244–268. New Brunswick: Rutgers University Press.

Jenkins, Henry. 2017. "Transmedia Logics and Locations." In *The Rise of Transtexts: Challenges and Opportunities*, edited by Benjamin W. L. Derhy Kurtz and Mélanie Bourdaa, 220–240. London: Routledge.

Kennedy, Michael. 2017. "*Supernatural* Spinoff Wayward Sisters 'Evolves Organically.'" *Screen Rant*, July 25. Accessed February 13, 2017. http://screenrant.com/supernatural-wayward-sisters-spinoff-evolves/.

Kohnen, Melanie E.S. 2012. "Creating a Spark: Official and Fan Produced Transmedia for *The Hunger Games*." *Antenna*, May 11. Accessed February 13, 2017. http://blog.commarts.wisc.edu/2012/05/11/creating-a-spark-official-and-fan-produced-transmedia-for-the-hunger-games/.

Marwick, Alice, Mary L. Gray, and Mike Ananny. 2014. "'Dolphins Are Just Gay Sharks': *Glee* and the Queer Case of Transmedia as Text and Object." *Television & New Media* 15 (7): 627–647.

Perryman, Neil. 2008. "*Doctor Who* and the Convergence of Media: A Case Study in 'Transmedia Storytelling.'" *Convergence: The International Journal of Research into New Media Technologies* 14 (1): 21–39.

Richards, Denzell. 2017. "Historicizing Transtexts and Transmedia." In *The Rise of Transtexts: Challenges and Opportunities*, edited by Benjamin W. L. Derhy Kurtz and Mélanie Bourdaa, 15–32. London: Routledge.

Scott, Suzanne. 2010. "The Trouble with Transmediation: Fandom's Negotiation of Transmedia Storytelling Systems." *Spectator-The University of Southern California Journal of Film and Television* 30 (1): 30–34.

Scott, Suzanne. 2013. "Who's Steering the Mothership? The Role of the Fanboy Auteur in Transmedia Storytelling." In *The Participatory Cultures Handbook*, edited by Aaron Delwiche and Jennifer Jacobs Henderson, 43–52. New York: Routledge.

Stein, Louisa Ellen. 2017. "Fandom and the Transtext." In *The Rise of Transtexts: Challenges and Opportunities*, edited by Benjamin W. L. Derhy Kurtz and Mélanie Bourdaa, 71–89. London: Routledge.

Stein, Louisa Ellen. 2018. "Of Spinoffs and Spinning Off." In *The Wiley Companion to Media Fandom and Fan Studies*, edited by Paul Booth, 401–403. Oxford: Wiley.

Stein, Louisa Ellen, and Kristina Busse. 2012. "Introduction: The Literary, Televisual and Digital Adventures of the Beloved Detective." In *Sherlock and Transmedia Fandom*, edited by Louisa Ellen Stein and Kristina Busse, 9–24. Jefferson: McFarland.

Stoeber, Jenna. 2013. "The Cultural Lives of *Doctor Who:* 'The Night of the Doctor.'" *Antenna*, November 21. Accessed February 13, 2017. http://blog.commarts.wisc.edu/2013/11/21/the-cultural-lives-of-doctor-who-the-night-of-the-doctor/.

Stork, Matthias. 2014. "The Cultural Economics of Performance Space: Negotiating Fan, Labor, and Marketing Practice in *Glee's* Transmedia Geography." *Transformative Works and Cultures* 15. doi: 10.3983/twc.2014.0490.

Taylor, Aaron. 2014. "Avengers Dissemble! Transmedia Superhero Franchises and Cultic Management." *Journal of Adaptation in Film & Performance* 7 (2): 181–194.

Watson, Jeff. 2017. "Games Beyond the ARG." In *Alternate Reality Games and the Cusp of Digital Gameplay*, edited by Antero Garcia and Greg Niemeyer, 187–210. New York: Bloomsbury.

Whospecialfeatures.wikia.com. 2017a. Frontier in Space. Accessed February 13, 2017. http://whospecialfeatures.wikia.com/wiki/Frontier_in_Space.

Whospecialfeatures.wikia.com. 2017b. Planet of the Daleks. Accessed February 13, 2017. http://whospecialfeatures.wikia.com/wiki/Planet_of_the_Daleks.

Wood, Megan M., and Linda Baughman. 2012. "*Glee* Fandom and Twitter: Something New, or More of the Same Old Things?" *Communication Studies* 63 (3): 328–344.

31
TRANSMEDIA PARATEXTS
Informational, Commercial, Diegetic, and Auratic Circulation

Matt Hills

In an influential discussion of transmedia storytelling, Henry Jenkins suggests that such stories unfold

> across multiple media platforms, with each new text making a distinctive and valuable contribution to the whole. In the ideal form ... each medium does what it does best—so that a story might be introduced in a film, expanded through television, novels, and comics; its world might be explored through game play or experienced as an amusement park attraction.
>
> *(2006, 95–96)*

The implication is that all these different media-traversing components will come together as a textual whole, each having made its own valuable narrative contribution. In such an "ideal form" transmedial textuality can thus be pieced together without any media hierarchy; different media can offer up the various parts of an overarching narrative. *Blade Runner's* "origami unicorn" is cited as a precursor for this textual philosophy; the inclusion of a moment where Rick Deckard discovers the origami figure in *The Director's Cut* aids "additive comprehension," i.e., inviting viewers to understand the textual whole differently by making it more likely that Deckard is actually a replicant. Jenkins quotes Neil Young of video games company Electronic Arts: "That changes your whole perception of the film, your perception of the ending ... The challenge for us ... is how do we deliver the origami unicorn, how do we deliver that one piece of information that makes you look at ... [a] film ... differently"? (2006, 123).

In contrast to this ideal of a single text scattered across media, and pieced together via audiences' "additive comprehension," transmedia storytelling has often fallen back into a media hierarchy, with one originating medium—e.g., film/TV—representing a textual/canonical hub, whilst other media do not add to an overarching textual whole so much as supplement or support this core media textuality. As Jason Mittell has observed, this center–periphery scenario represents a

> creative challenge that plagues the entire transmedia enterprise: how do you create narrative extensions from an ongoing core franchise that reward fans seeking out canon but do not become essential consumption for single-media fans ...? In other words, the constraints of the television industry and norms of television consumption insist that transmedia extensions from a serial franchise must reward those who partake in them but cannot punish those who do not.
>
> *(Mittell 2015, 303)*

Rather than dispersing a coherent narrative across media, this "mothership" model instead treats one medium as a narrative baseline. Extra material can circulate across media but, as Mittell notes, it cannot be essential to narrative comprehension. Such a scenario is not only grounded in film and TV industries per se, however. It has also been enforced in relation to norms of public service television in the United Kingdom (UK), meaning that charging people to purchase narrative "extensions" that would be essential to understanding a publicly funded TV narrative has been prohibited by the BBC.

Such a "mothership" model of transmedia world-building, where additional content outside the originating medium is positioned as an optional extra rather than as a vital instance of "additive comprehension" means that transmediality has frequently become *paratextual* in character instead of acting as a form of distributed textuality. This aspect of transmedia study fits alongside the fact that the "primacy of the filmic text has been tentatively softened in film scholarship in recent years" (Zahlten 2017, 6), as textual analysis has given way to more extensive paratextual exploration, not only of posters or trailers, but also merchandising, comics, games, tie-in novels, and so on.

Paratexts were initially addressed by Gerard Genette in his (1997) literary-theoretical *Paratexts: Thresholds of Interpretation*. Here, Genette considers how meaning-making is never restricted only to "the text," but also occurs through a "heterogeneous group of practices and discourses" which constitute a

> zone between text and off-text, a zone not only of transition but also of *transaction*: a privileged place ... of an influence ... that ... is at the service of a better reception for the text and a more pertinent reading of it (... in the eyes of the author and his allies).
>
> *(Genette 1997, 2)*

This literary theory has been developed in relation to media by the likes of Jonathan Gray (2010), who considers an array of paratexts such as action figures, DVD/Blu-ray extras, and marketing hype. Gray argues that paratexts "contribute to the text and are often vital parts of it," meaning that they are "not just marketing 'add-ons' and 'ancillary products,' as the media industries and academia alike have often regarded them" (2010, 208). Against this explicit assertion, Alexander Zahlten reads Gray's work as "retain[ing] the division and hierarchy between text and paratext" (2017, 6). This suggests that paratexts can call into question the notion of "primary" versus "secondary" textualities, blurring these in a series of ways yet without entirely collapsing them together, and hence preserving a liminal "zone between text and off-text," as Genette puts it.

Media franchises and their associated brands can involve "world-sharing among creative workers" (Johnson 2013, 109) as production communities working in different media contribute to transmedia world-building, and as "paratextual industries" (Consalvo 2007, 183) supplement textual meanings. Indeed, it has even been argued that specific paratexts can also be interpreted as vital to the development of transmedia storytelling and branding—dubbed a "media mix" ecology in the Japanese context—with certain magazine titles having provided "nodes" of transmedia-related information addressed to fan subcultures (Steinberg 2017, 151). In these kinds of examples, paratexts do not simply offer up non-essential narrative supplements which fans can choose to consume (or not); they also catalogue, promote, and identify transmedia narrative extensions, as can be seen in the previews/reviews pages of official magazines such as *Star Wars Insider* or *Doctor Who Magazine*. Such cases therefore act, somewhat recursively, as informational "para-paratexts" (Hills 2015, 16); that is, they frame a range of other transmedia paratexts such as toys, games, and tie-in novels, representing a mode of fan-cultural or subcultural "transmediaphilia" (Collins 2017).

Alongside this informational/cataloguing type of transmedia paratext, some such paratexts are primarily diegetic—raising the issue of how, and to what extent, they fit into the established canon of a franchise (Derhy Kurtz 2017)—and some are primarily auratic (Gray 2010, 83), seeking to lend cultural value to the franchise concerned. Still others are concerned first and foremost with a franchise's brand identity, and so carry little to no narrative/auratic content, instead being firmly tied to commerciality.

Transmedia paratexts usually exploit niche economies (of established fans) and cost-effective forms that are not dependent on sizeable production teams and/or notably high production costs. Comic books have been a popular way of "continuing" cult film/TV series for these very reasons—*Buffy the Vampire Slayer, The X-Files*, and J. J. Abrams' *Star Trek* have all been developed in this manner, for instance. And audio adventures, sometimes featuring the original voice cast, have also worked effectively for niche economies of cult fandom; the producer Big Finish has worked on *Blake's Seven, Doctor Who, The Prisoner, Sapphire and Steel*, and *Torchwood* audios, among others. Maura Edmond argues that radio "is experimenting with ever more complex cross-media practices. These types of activities have been analysed at length with regard to commercial film, television and gaming, but much less is understood about radio-based approaches to transmedia engagement" (2014, 2), although this downplays scholarship on audio drama paratexts which would not necessarily be positioned as "radio." Another medium that does not carry prohibitive overheads and production costs, and hence lends itself well to acting as a transmedial paratext, is the novel, sometimes termed a media "tie-in" (Clarke 2013).

Tie-in novels, like commercial fan magazines, are typically aimed at established fan bases and hence subcultural readers. In rare cases, the size and scope of a fandom can be enough to propel such books on to bestseller lists, as has happened for *Star Trek* and *Star Wars* novels in the United States. Yet licensed franchise novels tend to find themselves in something of a diegetic paradox, as M. J. Clarke has outlined. They are called upon both to supplement their canonical film/TV series, and at the same time, not to significantly diverge from this "parent" text. As Clarke summarizes this tension: "tie-in writers … must not replicate the on-air series while simultaneously not being able to add anything different to the series" (2013, 82). Negotiating these restrictions means that tie-in writers are often fans themselves with a detailed knowledge of a franchise's diegetic world; they draw on this in-depth fan knowledge but must not engage in fan speculation or transformative possibilities, having to remain strictly "true" to canonical stipulations (Clarke 2013, 76).

It is worth noting, however, that a tie-in novel's temporality can play a role in loosening some of these restrictions (unless it is knowingly a "what if?" transmedia narrative extension, playing with an alternate reality view of canon; see Mittell 2015, 315). If a transmedia paratext is coterminous with a contemporary franchise installment—perhaps released in the build-up to a blockbuster film—then it will tend to be highly integrated into diegetic canon. If, on the other hand, a transmedia paratext follows years after a film/TV series, when the franchise is perhaps largely culturally dormant, then there can be far greater scope for diegetic transformations or contradictions. While *Doctor Who* was off-air as a TV series in the 1990s, the Virgin New Adventures, licensed by the BBC, adhered to specific guidelines but were nevertheless able to promote themselves as telling stories "too broad and too deep for the small screen." And some 13 years after *Blade Runner*'s 1982 box office flop, K. W. Jeter began a series of sequel *Blade Runner* novels, authorized by the Philip K. Dick estate. These novels playfully combined elements of Dick's original novel, *Do Androids Dream of Electric Sheep?*, and the film's vision, whilst also self-reflexively engaging with the notion of creating a filmic version of events within the diegetic world (Gray 2005, 149–152). Rather than displaying the "canonical verifiers" (Derhy Kurtz 2017, 116) that are typical of coterminous franchise novels, i.e., markers of authenticity regarding character and diegetic world, Jeter's novels toyed instead with canonical non-identifiers, weaving together different source texts (Philip K. Dick/Ridley Scott) into a new hybrid rather than an "authentic" imitation of either the original novel or the film. Arguably, *Blade Runner*'s lack of commercial presence and franchise authority by 1995 meant that Jeter had far greater freedom than he would have done had he been writing in the shadow of *Blade Runner 2049*, say, or in 1981–1982. In any case, Colin B. Harvey's excellent work on fantastic transmediality has demonstrated how fans and producers can strategically "non-remember" transmedia paratexts (and even official textual reboots) through the creation of new, official, and canonical narratives (Harvey 2015, 2). Canonical verifiers and non-identifiers alike can be lost in time, reconfigured, reclaimed, or rejected by subsequent franchise "production communities" (Johnson 2013, 123).

By purely targeting fan cultures rather than the coalition/mass audiences that are characteristic of franchise texts, transmedia paratexts can display forms of "fanagement" (Hills 2012). That is, they can engage in feedback loops with vocal sections of fandom, giving these fans what (it is assumed) they want, e.g., re-uniting characters whose relationship had been championed by fans, such as Captain Jack Harkness and Ianto Jones of the BBC TV series *Torchwood*. With Ianto having died in the canonical TV series, a radio play/audio adventure *The House of the Dead* brought Captain Jack and Ianto back together one last time via supernatural means, enabling a greater sense of narrative and emotional closure than had previously been evident in *Torchwood: Children of Earth* (Hills 2013, 75). And Torchwood's Cardiff-based setting—which had been a key part of the show's brand identity in series one and two—is commemorated in the tie-in novel *Long Time Dead* (Hills 2013, 79), allowing canonical developments, and the loss of Cardiff as a crucial setting, to be symbolically reversed. However, these extended exercises in fan service are only present at the level of transmedia paratexts rather than in the canonical TV text itself. Sections of fandom are thus catered for or responded to, but they are not given what they actually want, i.e., changes to the television canon of *Torchwood*. Instead, fanagement via transmedia paratexts seems to be more concerned with heading off or containing fan criticisms of the official text and its changing format.

Being subculturally or fan-culturally targeted means that transmedia paratexts can sometimes be more challenging and progressive than "mothership" franchise texts that are aimed at a mass audience. Like fanagement, this too is a double-edged sword: including progressive content can be a positive step, but restricting it to transmedia paratexts means potentially remaining locked into an industrial system where "mainstream" audiences are imagined as incapable of appreciating progressive developments, whether in representations of gender, race or sexuality. For instance, Carolyn Cocca argues that *Star Wars*' comics and "expanded universe" novels (1977–2014, when Disney definitively de-canonized the EU and rebranded it as "Legends") depict Princess Leia as not "objectified in dress, framing, posing, or description. This is the case even as … [she has] been written and drawn mostly by men" (2016, 96). Cocca attributes this partly to the character's depiction in filmic canon, but also to the fact that *Star Wars*' popularity as media science fiction meant that comics creators were not compelled to cater to the "narrow fan base of 1990s–2000s superhero comics" (2016), where female objectification would have been rather more rife. In Cocca's argument, transmedia paratexts continue to be concerned with imagining and targeting specific fan readers/audiences, but this becomes a case of media science fiction fandom versus superhero comics fandom, rather than targeted fandom versus "mainstream" audiences per se.

If diegetically based transmedia paratexts tend to be targeted at fandoms, then extra-diegetic paratexts crossing media platforms can sometimes be aimed not only at subcultural legitimation, but also at enhancing a franchise's cultural status. Jonathan Gray has written about "behind-the-scenes" paratexts that "actively create artistic aura for their associated text … insist[ing] on its uniqueness, value, and authenticity in an otherwise standardized media environment" (2010, 82). *Star Wars* has been particularly successful at this, with its entry into museum spaces being premised, in part, on its paratextual positioning as a "mythic" series of films (Hills 2003). Indeed, George Lucas is currently engaged in developing the Lucas Museum of Narrative Art (Bartolomé Herrera and Keidl 2017, 155), whilst the Museum of Popular Culture in Seattle (formerly the Experience Music Project) hosts traveling/exclusive media franchise exhibits and holds a permanent collection featuring props and costumes from the likes of *Star Trek* and *Blade Runner*.

Museums of all sorts have increasingly become spaces of transmedia extension, multiplicity, and paratextuality, but it remains fruitful to consider not just the "transmedia museum" in general (Kidd 2016, 23), but also how pre-existent transmedia storytelling, branding, and paratexts—linked to specific franchises/texts and fandoms—have intersected with contemporary museums' curatorial practices. As Beatriz Bartolomé Herrera and Philipp Dominik Keidl have observed, specifically in relation to *Star Wars*, this can mean redirecting "debates of cultural and educational value to a discussion of exhibitions' role in Star Wars's transmedia economy ... Star Wars-themed exhibitions function as a space for managing and refocusing transmedia expansion" (2017, 156). Such exhibitions can emphasize George Lucas's authorial role, for instance, managing transmedia paratexts via a powerfully corralling "meta-paratext" (Hills 2015, 14) which works by subordinating an array of other paratexts to its collective (re)framing.

Furthermore, the auratic function of transmedia paratexts in a museum setting draws on "museums' roles as 'memory institutions', [where] transmedia memory is therefore key to examining how ... exhibitions produce, commodify, and circulate a distinct remembrance" (Bartolomé Herrera and Keidl 2017, 162–163; Harvey 2015, 39). Fans' memories of merchandizing and other material paratexts (Hills and Garde-Hansen 2017) can thus be drawn on and re-activated by museum displays:

> Unlike immersive museums that use roleplaying as a pedagogical tool for exploring real-world events, those that focus on popular culture need to find a way to bring a sense of reality to the fantastical that provides a balance between play (i.e. the visitor's usual state when interacting with the fictional text) and learning.
>
> *(Peters 2015, 97)*

And this "sense of reality" typically involves focusing on the profilmic fashioning and crafting of iconic media texts; even the *Star Wars: Identities* exhibit which has toured internationally, and which seeks to use storyworld planets and species to spark visitors' reflections on the contexts underpinning cultural "identity" in its broadest anthropological and psychological senses (Peters 2015; Bartolomé Herrera and Keidl 2017) nevertheless includes production design sketches, costumes, spacecraft models, and so on. What has been termed "tactile transmediality" (Gilligan 2012, 25) hence becomes especially important to the embodied practice of "being there" at museum exhibits involving these props and costumes, as well as featuring physical replicas and reconstructions that are typically accompanied by rich narrative and production information (Hills 2017, 247; Kidd 2016, 28). Transmedia paratexts take on an auratic status here through museums' twinning of materiality and memory. Visitors can engage with "original" production materials (even when there may have been multiples made for filming), as well as merchandise recontextualized in relation to popular-cultural history and (implicitly) biographical self-narrative. Diegetic meanings may well be evoked in relation to such paratexts, but when physical "reality" becomes a crucial marker of authenticity then tactile transmediality becomes less about diegesis and more about auratic cultural value. The museumification of screen-used/production-used props, models, and costumes seems to offer a privileged moment of non-mediation rather than transmediation, allowing visitors to step outside textual consumption and into an encounter with the raw materials of popular culture. Yet the aura and performative cultural value of museum-displayed paratexts depends on their connections to mediation, rendering this physical presence (usually behind glass) as just another kind of transmedial platform-crossing where profilmic elements can be "extracted" from the textual instead of merely preceding it (Freeman 2016, 30).

Merchandise Beyond Rebranding and Reviews Beyond Textual Framing: Transmedia Paratexts as Sites of Audience Struggle

Transmedia paratexts, in the form of merchandise, have often been dismissed as pure commerciality and branded hype, viewed as a matter of crude economics and franchise (as well as fan) exploitation. Even work that seeks to revalue and take seriously such blatantly commercial paratexts occasionally gives way to their dismissal, as in Jonathan Gray's analysis of "unincorporated" paratexts—his example is a Domino's "Gotham City pizza" linked to the release of *The Dark Knight*—which add nothing to the storyworld and text concerned (2010, 208–210). Keith Johnston contests this dismissal, however, arguing that, from a reception studies' viewpoint, such a "commercial relationship would be seen as an important part of the 'consumable identity' of that film, one of the 'multiple avenues of access' viewers might have" into the text (2011, 421). The merchandising of movie blockbusters and cult texts frequently generates "sizeable posses" of paratexts (Gray 2010, 114), meaning that any such "consumable identity" can range from the edible—an unusual form of transmediality, perhaps—through to low-end, mass-produced toys, games, and action figures, high-end collectors' limited editions of all sorts, and even souvenirs specifically linked to theme park attractions or museum exhibits.

Viewed primarily in relation to commerciality rather than through diegetic/auratic lenses, transmedia paratexts can work collectively to reinforce brand value and omnipresence. As Hélène Laurichesse observes, "a media franchise that launches a licensing policy in several economic sectors is termed a 'brand'" (2017, 193). And "transmedia brand licensing" (Santo 2015) requires that cultural intermediaries such as brand managers will solicit, approve, and police an array of licenses in order to safeguard the brand consistency and diegetic/auratic functions of paratexts. However, assessing transmedia paratexts only as a way of framing texts, contributing to their storyworlds, or offering multiple forms of access into a franchise, neglects the extent to which brands can become more powerful than texts, operating through the (re)configuration of paratexts and texts in order to establish new, updated, and bounded meanings: "In branding there is no longer an immutable text ornamented or regulated by its paratexts" (Aronczyk 2017, 113). Instead, franchises can rewrite mutable texts and paratextual surrounds in order to establish their branded "now-ness."

Star Wars offers a good example of this, with *The Force Awakens* having been textually (re)constituted in relation to earlier installments in order to far more prominently feature a powerful female lead via the character of Rey (Daisy Ridley). *The Last Jedi* then took this even further, detaching Rey from the patriarchal logic of the previous films as a kind of "Skywalker saga" which had been focused on the lineage of male figures Anakin (Hayden Christensen) and Luke Skywalker (Mark Hamill). But Melissa Aronczyk views branding logics of textual and paratextual reconfiguration as more coherent and potent than they have proven to be, just as Jonathan Gray has perhaps overly emphasized paratexts as textual framings. Contrasting with each approach, the #wheresrey controversy in 2015 brought into focus how transmedia paratexts aimed at child consumers—in this case, action figures, toys, and games—could become the domain of influential audience criticism and push-back (Scott 2017). The hashtag #wheresrey denoted audience critique of the fact that Rey's character, despite her prominence in film *Star Wars: The Force Awakens*, had been markedly marginalized in commercial paratexts, especially action figures and games. Ostensibly this was due to well-established industrial lore that boys would be the primary consumers of the franchise's toys, and that, in turn, they would not purchase female action figures. The result, however, was a reconfigured text lacking in equally reconfigured, gendered transmedia paratexts (i.e., material merchandise). *The Force Awakens*, as identified by the #wheresrey social media campaign, amounted to an incoherent rebranding, and one marred by sexist transmedia paratexts (Brown 2017, 7). Despite their different theoretical assumptions, both Aronczyk (2017) and Gray (2010) allocate insufficient agency to audiences' engagements with transmedia paratexts. By utilizing Twitter and other social media, #wheresrey was amplified as an audience response, also then being taken up as a narrative by "old" media such as press and broadcasters. Paratexts are no longer merely powerful tools of branding or textual (pre-)interpretation; instead they have been repositioned as a new terrain for audience struggles over feminist, progressive meaning-making.

Moreover, such struggles across and around transmedia paratexts have not been limited to material merchandise. The online aggregator of critics' and audiences' reviews, Rotten Tomatoes, has also become a significant site of paratextual tensions and audience conflict. When the remake of *Ghostbusters* starring women in what had previously been male lead roles scored a remarkably low rating on Rotten Tomatoes' audience scoring, it was suggested that male fans opposed to the reboot (see Proctor 2017) and may have tried to game/skew the site in order to harm the film's paratextual reputation, and validate their own opposition. Similar accusations arose around *The Last Jedi*, where again Rotten Tomatoes' critic and audience scores were wildly at odds. A Facebook group entitled "Down With Disney's Treatment of Franchises and its Fanboys" claimed to have sabotaged the film's audience score via the use of bots, review-bombing the site with negative ratings (Davies 2017). Rotten Tomatoes issued a public statement refuting this (Davies 2017), and took further action when the same group set up a Facebook event targeting the audience ratings for Marvel's black superhero film *Black Panther* (Virtue 2018). This time, Rotten Tomatoes' statement read:

> We … are proud to have become a platform for passionate fans to debate and discuss entertainment and we take that responsibility seriously. While we respect our fans' diverse opinions, we do not condone hate speech. Our team of security, network and social experts continue to closely monitor our platforms and any users who engage in such activities will be blocked from our site and their comments removed as quickly as possible.
>
> *(McClintock and Jarvey 2018)*

With Rotten Tomatoes being used in Hollywood films' own marketing campaigns when its audience/critic scores are unusually strong, fans wishing to register their opposition to specific film titles have evidently felt that gaming the site's scores—or even merely threatening to, and hence generating uncertainty over their own cultural power or the site's validity—can act as a badge of status. In effect, aggregated reviewer paratexts such as Rotten Tomatoes have, like #wheresrey, become another flashpoint for paratextual struggle in the online culture wars surrounding progressive/reactionary viewpoints.

Far from merely framing preferred readings of a text (Gray 2010), enforcing rebranded meanings (Aronczyk 2017), or providing multiple access points to a film's reception (Johnston 2011), transmedia paratexts have now become sites of cultural-political struggle in their own right. This is how cultural studies analyzed media texts in the 1980s and 1990s, suggesting that textual struggles over meaning have been reflexively extended to paratextuality. Alongside the ongoing informational, commercial, diegetic, and auratic roles of transmedia paratexts—through which subcultural audiences of fans can be targeted, and cultural value can be performed—we also need to analyze transmedia paratexts as official materials that audiences can actively seek to influence.

References

Aronczyk, Melissa. 2017. "Portal or Police? The Limits of Promotional Paratexts." *Critical Studies in Media Communication* 34 (2): 111–119.

Bartolomé Herrera, Beatriz, and Philipp Dominik Keidl. 2017. "How Star Wars Became Museological: Transmedia Storytelling in the Exhibition Space." In *Star Wars and the History of Transmedia Storytelling*, edited by Sean Guynes and Dan Hassler-Forest, 155–168. Amsterdam: Amsterdam University Press.

Brown, Jeffrey A. 2017. "#wheresRey: Feminism, Protest, and Merchandising Sexism in *Star Wars: The Force Awakens*." *Feminist Media Studies*. doi: https://doi.org/10.1080/14680777.2017.1313291.

Clarke, M. J. 2013. *Transmedia Television: New Trends in Network Serial Production*. New York and London: Bloomsbury Academic.

Cocca, Carolyn. 2016. *Superwomen: Gender, Power and Representation*. New York and London: Bloomsbury Academic.

Collins, Jim. 2017. "Transmediaphilia, World Building, and the Pleasures of the Personal Digital Archive." In *World Building: Transmedia, Fans, Industries*, edited by Marta Boni, 362–376. Amsterdam: Amsterdam University Press.

Consalvo, Mia. 2007. *Cheating: Gaining Advantage in Videogames*. Cambridge, MA and London: MIT Press.

Davies, Megan. 2017. "Rotten Tomatoes Denies Claims that The Last Jedi Scores Were Rigged by Bots." *DigitalSpy*, December 22. Accessed February 6, 2018. www.digitalspy.com/movies/star-wars/news/a846065/rotten-tomatoes-denies-claims-last-jedi-scores-rigged-by-bots/.

Derhy Kurtz, Benjamin W. L. 2017. "Set in Stone: Issues of Canonicity of Transtexts." In *The Rise of Transtexts: Challenges and Opportunities*, edited by Benjamin W.L. Derhy Kurtz and Melanie Bourdaa, 104–118. London and New York: Routledge.

Edmond, Maura. 2014. "All Platforms Considered: Contemporary Radio and Transmedia Engagement." *New Media Society* 17 (9): 1566–1582.

Freeman, Matthew. 2016. *Historicising Transmedia Storytelling: Early Twentieth Century Transmedia Story Worlds*. New York and London: Routledge.

Genette, Gerard. 1997. *Paratexts: Thresholds of Interpretation*. Cambridge: Cambridge University Press.

Gilligan, Sarah. 2012. "Heaving Cleavages and Fantastic Frock Coats: Gender Fluidity, Celebrity and Tactile Transmediality in Contemporary Costume Cinema." *Film, Fashion & Consumption* 1 (1): 7–38.

Gray, Christy. 2005. "Originals and Copies: The Fans of Philip K. Dick, *Blade Runner* and K. W. Jeter." In *The Blade Runner Experience: The Legacy of a Science Fiction Classic*, edited by Will Brooker, 142–156. London and New York: Wallflower.

Gray, Jonathan. 2010. *Show Sold Separately: Promos, Spoilers, and Other Media Paratexts*. New York: New York University Press.

Harvey, Colin B. 2015. *Fantastic Transmedia: Narrative, Play and Memory Across Science Fiction and Fantasy Storyworlds*. Basingstoke: Palgrave Macmillan.

Hills, Matt. 2003. "*Star Wars* in Fandom, Film Theory, and the Museum: The Cultural Status of the Cult Blockbuster." In *Movie Blockbusters*, edited by Julian Stringer, 178–189. London and New York: Routledge.

Hills, Matt. 2012. "*Torchwood*'s Trans-transmedia: Media Tie-ins and Brand 'Fanagement'." *Participations* 9 (2): 409–428.

Hills, Matt. 2013. "Transmedia *Torchwood*: Investigating a Television Spin-off's Tie-in Novels and Audio Adventures." In *Torchwood Declassified: Investigating Mainstream Cult Television*, edited by Rebecca Williams, 65–83. London and New York: I. B. Tauris.

Hills, Matt. 2015. *Doctor Who: The Unfolding Event—Marketing, Merchandising and Mediatizing a Brand Anniversary*. Basingstoke: Palgrave Macmillan.

Hills, Matt. 2017. "The Enchantment of Visiting Imaginary Worlds and 'Being There': Brand Fandom and The Tertiary World Of Media Tourism." In *Revisiting Imaginary Worlds: A Subcreation Studies Anthology*, edited by Mark J. P. Wolf, 244–263. New York and London: Routledge.

Hills, Matt, and Joanne Garde-Hansen. 2017. "Fandom's Paratextual Memory: Remembering, Reconstructing, and Repatriating 'Lost' *Doctor Who*." *Critical Studies in Media Communication* 34 (2): 158–167.

Jenkins, Henry. 2006. *Convergence Culture: Where Old and New Media Collide*. New York: New York University Press.

Johnson, Derek. 2013. *Media Franchising: Creative License and Collaboration in the Culture Industries*. New York: New York University Press.

Johnston, Keith. 2011. "Review: Jonathan Gray, *Show Sold Separately: Promos, Spoilers and Other Media Paratexts*, 2010." *Screen* 52 (3): 419–422.

Kidd, Jenny. 2016. *Museums in the New Mediascape: Transmedia, Participation, Ethics*. London and New York: Routledge.

Laurichesse, Hélène. 2017. "Considering Transtexts as Brands." In *The Rise of Transtexts: Challenges and Opportunities*, edited by Benjamin W. L. Derhy Kurtz and Melanie Bourdaa, 187–203. London and New York: Routledge.

McClintock, Pamela, and Natalie Jarvey. 2018. "Facebook Deactivates Anti-Black Panther Group." *The Hollywood Reporter*, February 1. Accessed February 6, 2018. www.hollywoodreporter.com/heat-vision/black-panther-rotten-tomatoes-denounces-group-taking-aim-at-movie-1081081.

Mittell, Jason. 2015. *Complex TV: The Poetics of Contemporary Television Storytelling*. New York: New York University Press.

Peters, Ian. 2015. "Hello Shoppers? Themed Spaces, Immersive Popular Culture Exhibition, and Museum Pedagogy." Ph.D. diss., Georgia State University. Accessed May 19, 2017. http://scholarworks.gsu.edu/communication_diss/60.

Proctor, William. 2017. "'Bitches Ain't Gonna Hunt No Ghosts': Totemic Nostalgia, Toxic Fandom and the Ghostbusters Platonic." *Palabra Clave* 20 (4): 1105–1141.

Santo, Avi. 2015. *Selling the Silver Bullet: The Lone Ranger and Transmedia Brand Licensing*. Texas: University of Texas Press.

Scott, Suzanne. 2017. "#Wheresrey? Toys, Spoilers, and the Gender Politics of Franchise Paratexts." *Critical Studies in Media Communication* 34 (2): 138–147.

Steinberg, Marc. 2017. "Platform Producer Meets Game Master: On the Conditions for the Media Mix." In *World Building: Transmedia, Fans, Industries*, edited by Marta Boni, 143–163. Amsterdam: Amsterdam University Press.

Virtue, Graeme. 2018. "Claws Out: How Black Panther Fought Off a Toxic Ghostbusters-style Online Campaign." *Guardian*, February 5. Accessed February 6, 2018. www.theguardian.com/film/filmblog/2018/feb/05/black-panther-fought-off-a-toxic-ghostbusters-online-campaign-rotten-tomatoes.

Zahlten, Alexander. 2017. *The End of Japanese Cinema: Industrial Genres, National Times and Media Ecologies*. Durham, NC and London: Duke University Press.

32

TRANSMEDIA POLITICS

Star Wars and the Ideological Battlegrounds of Popular Franchises

Dan Hassler-Forest

"We are the resistance." During the historic Women's March on Washington of January 21, 2017, many protestors held aloft photos of Carrie Fisher in her career-defining role as Princess Leia in the *Star Wars* franchise. Created as a popular expression of political dissent after the election of a US president who had notoriously bragged about sexual assault, these homemade tributes resonated at multiple levels at once. Perhaps most obviously, they emphasized the enduring strength of a female character whose defining characteristic within the *Star Wars* trilogy was her rebellious sensibility. But at the same time, the posters paid homage to the actress who played Leia, as Carrie Fisher's unexpected death had refocused public attention on her off-screen life as an outspoken feminist, activist, and mental disability advocate. And third, the placards perfectly fit the ideological direction the Disney-era *Star Wars* franchise had taken, foregrounding female characters in positions of leadership and narrative agency, while their Rebellion has taken aim at villainous political organizations with strikingly obvious white supremacist overtones.

This particular use of an iconic character in one of the most recognizable transmedia storyworlds points toward a few key aspects in the complex intersection between politics and transmediality as a form of cultural logic. It shows how transmediality is more than just a textual practice, in which a single narrative—or, more appropriately, *storyworld*—is strategically disseminated across multiple media (see Jenkins 2006). It is also a dialogic form that is constantly being reconfigured by competing forms of audience appropriation, technological transformations, and changing industrial practices. In this example, the image becomes meaningful to large groups of people at this specific historical moment due to a constantly shifting and expanding constellation of politics, economics, social practices, and entertainment.

In this chapter, I will untangle a few of the most crucial strands in this complex web that unites the term transmedia with politics. Since *Star Wars* is not only one of the most popular, recognizable, and enduring transmedia franchises, I will draw on this famously expansive space opera as my main example. It will provide examples that illustrate three distinct approaches to the politics of transmediality: first, the tradition of Marxist ideology critique, which in the case of *Star Wars* yields fruitful but also ambivalent or even contradictory conclusions; second, the framework of political economy, which approaches transmedia franchises based on questions of ownership, distribution, and intellectual property (IP); and third, a consideration of the political uses to which the franchise is put in the context of media convergence and participatory culture—ultimately bringing us full circle to ideological readings of the core text, and how its structural ambivalence can be understood within twenty-first century brand culture. But first, I offer a brief discussion of this chapter's basic terms, beginning with *transmedia* and *politics*.

Transmedia as Cultural Form and Industrial Practice

As the many other chapters in this collection demonstrate so vividly, there exists a multitude of approaches to the term *transmedia*. But at its most basic level, we can understand it as a term meant to indicate the dominant textual practices within a larger convergence culture, in which entertainment properties so often spiral outward across multiple media. Since this chapter deals specifically with transmedia politics, it is most important for our purposes here to distinguish between two broad approaches to this basic cultural logic. First, we must clearly acknowledge transmediality as a *cultural form*, emerging organically and spontaneously from audiences' interaction with commercially produced entertainment properties. And second, transmediality also has a long and increasingly visible history as an *industrial practice*, as authors and media companies have produced and/or licensed expansions of their original IP across a variety of media.

The key difference between these two conceptions of transmediality can be articulated as the distinction between *bottom-up* and *top-down* approaches, each of which has obvious political implications. The former has been associated—most famously by Henry Jenkins—with forms of *textual poaching*, as audiences take command of their culture and become active producers of fan fiction, mashups, supercuts, cosplay, and other forms of transformative appropriation. Jenkins' main case study in his hugely influential book *Textual Poachers* (1992) is the *Star Trek* franchise and its active fan cultures. In his 2002 study *Using the Force*, Will Brooker took a similar approach to *Star Wars* fandom, foregrounding the many ways in which elements drawn from the commercial franchise were made meaningful through acts of creative appropriation.

The more top-down understanding of transmediality as an industrial practice is often referred to as *media franchising*, taking a more political-economic approach to the industrial production of commercial entertainment and its endless varieties of merchandising, spin-offs, reboots, sequels, prequels, and re-imaginings. While Derek Johnson's book *Media Franchising* (2013) emphasizes how this industrial process is much more unpredictable and dialogic than its top-down perception would suggests, he nevertheless concludes that "the collaborative networks constituting media franchising have been persistently structured by hegemonies of cultural power" (2013, 238). And *Star Wars* is of course commonly seen as one of the most obvious examples of this very kind of industrially produced entertainment, as Lucasfilm has produced, licensed, and developed transmedia expansions ranging from novels and television shows to amusement park rides and tabletop role-playing games (see Guynes and Hassler-Forest 2018).

Ideological Criticism: *Star Wars* as an Expression of Politics

While these two separate (though often co-existing) transmedia frameworks are themselves associated with rather different political models, the term "politics" itself must also be specified further in order to unpack its most prominent intersections with transmediality. The first distinction to make is again twofold, expressed in the difference between *ideology criticism* and *political economy*. Scholars working within the tradition of Marxist literary criticism and cultural theory have long emphasized that politics should be defined in its most basic sense as the organization of power within any given human society. Therefore, while we may be tempted to see politics as a separate societal sphere that may or may not be specified within a transmedia storyworld, there is in fact no way of telling a story or even of representing a world that is not ideologically determined, and therefore fundamentally political. Recognizing and decoding the ideological structures that make up any given text's politics has been described by Marxist theorist Fredric Jameson's as identifying its *political unconscious*—the ways in which the organization of power is communicated implicitly via the text's content and form. This kind of analysis therefore must begin "with the recognition that there is nothing that is not social and historical—indeed, that everything is 'in the last analysis' political" (Jameson 1981, 5).

To clarify this distinction with an example from the *Star Wars* franchise: while the original trilogy has been widely perceived as a mythological depiction of a timeless battle between good and evil, the prequels were criticized for introducing a complicated political system into this galaxy far, far away. Plot elements like trade negotiations, senate votes, and Palpatine's Machiavellian machinations were perceived as explicitly "political" and therefore less appropriate for escapist and child-friendly entertainment. As a 2016 article in *Forbes* expressed this, "it's fun to get lost in an epic lightsaber duel without worrying too much about intergalactic trade routes" (Di Placido 2016). In other words, the representation of politics as an explicit structural component within a fantasy world is often seen as less "fun"—something that intrudes upon a supposedly apolitical depiction of a self-contained fantastic storyworld.

But whether or not the political organization of a transmedia storyworld is depicted, explained, or even referenced, the cumulative choices that make up the act of world-building are themselves grounded in an ideologically determined understanding of human social relations. The function of ideological criticism is therefore to interpret, decode, and make visible the political unconscious that underlies any given text. Since this decoding is obviously an interpretive act, there are no truly definitive readings, nor can we reach any kind of ultimate agreement on what kind of politics a given text represents. The original *Star Wars* from 1977, for instance, has often been identified as an early expression of 1980s Cold War conservatism, as "both George Lucas and Reagan understood the value of packaging a utopian vision as lapsarian rather than as progressive, and did so by resuscitating the West through the redemption of the Western" (Nadel 2012, 199–200). But others have also seen elements from the original trilogy as critical or even radically progressive reflections on American military intervention, most specifically the war in Vietnam (Booker 2006, 116).

One of the problems when applying this kind of ideology criticism to a transmedia multitext is clearly that of scale: as challenging as it is to decode the political unconscious of a single film trilogy, the problem clearly expands exponentially when faced with a transmedia storyworld that is not only made up of hundreds of different texts, but which are also mediated via many different platforms. And since mediation itself is ideological and political in the sense that representational fiction both simplifies and reifies existing sets of social relations, the ways in which these ideas are *transcoded* into a particular medium contribute enormously to their political implications (Jameson 1981, 25). In other words, watching the battle on the ice planet Hoth in *The Empire Strikes Back* can express a very different kind of politics than playing a video game in which the same locations, characters, and story events are recreated. So even if a transmedia storyworld remains entirely consistent in terms of its politics throughout its history (which is rarely—if ever—the case), the use of multiple media platforms inevitably complicates such an analysis, which will therefore always be selective and limited in its conclusions about the franchise's politics (Pearson 2013, 214).

In the case of *Star Wars*, the multitude of contradictory political readings clearly also results from the ways in which the storyworld's abstract distinction between good and evil keeps breaking down, as the franchise somehow always finds "a way to have your authoritarian cake and eat it too" (Rubey 1978, 12). Both in its ideological investment in heroic individualism and its "machine aesthetic" that is always intended to be "as overwhelming as possible" (Rubey 1978, 9), the net effect of this conflict bereft of meaningful politics has been described as a form of "friendly fascism":

> In the end, STAR WARS embraces by implication all the things it pretends to oppose. The Nuremberg rally scene is a fitting conclusion coherent with the film's fascination with speed, size, and violence, and with the mysticism that cloaks the film's patriarchal power structures. The romance plot incorporates sexism and racism and supports a hierarchical social system that glamorizes those at the top and literally turns those at the bottom into machines.
>
> *(Rubey 1978, 14)*

Rubey's profoundly insightful reading of the film, written before *Star Wars* had ballooned into the indescribably vast transmedia storyworld it has since become, most accurately pinpoints the very contradiction that has made any single reading of its politics so elusive. Little wonder, then, that it has been interpreted, appropriated, and mobilized in such contradictory ways, and for such wildly contradictory political agendas.

Political Economy: *Star Wars* as a Capitalist Commodity

This then brings us to the second form of political analysis, which approaches these transmedia franchises from a political-economic perspective. Within the field of transmedia studies, political economy focuses primarily on material questions of ownership and production, most frequently turning to "the study of how media conglomerates like Disney and Time Warner establish monopolies over cultural and communicative resources and exert corporate control over their use" (Johnson 2013, 9). While one could justifiably make the point that such approaches often ignore or skim over both the contents and the reception of cultural texts, political economy does have the advantage of clarifying very precisely how expansive transmedia storyworlds most commonly exist as commodities that circulate within the larger dynamic of global capitalism. And from this perspective, their primary political function is to maintain the existing power relations within a capitalist society (Murdock and Golding 1973, 232).

Thus, in the case of *Star Wars*, such an analysis would be much less interested in the internal political organization of the storyworld, and more in the ways in which the franchise's success helped spawn George Lucas' empire of media production and IP licensing. It might map out how the production of *Star Wars* led to the establishment of key players in the entertainment industry like Lucasfilm, Industrial Light and Magic, Skywalker Sound, LucasArts, and Pixar. A political economy of *Star Wars* would therefore in the first place attempt to map out its franchise's "universe" as a valuable form of IP that became central to a complex constellation of producers, licensees, rights holders, distributors, managers, and consumers (Wasko 2001, 28–29). The extent to which the ideological organization of the narrative is significant for this kind of analysis can vary substantially in studies like this. And this point has become especially important since the media landscape has shifted in the twenty-first century from a relatively homogeneous mainstream culture that expresses a "dominant ideology" (Althusser 2014, 137) to the much more fragmented context of "on-demand culture" (Tryon 2013, 5–9).

A good example of both these issues has been the Disney-era incarnation of the *Star Wars* franchise. The Walt Disney Company further solidified its position as the world's most powerful media conglomerate with its $4.05 billion purchase of Lucasfilm in October 2012, thereby adding the company's IP to its growing portfolio of popular transmedia properties. Having collaborated previously on successful ventures like the *Star Tours* amusement park ride, Disney saw in *Star Wars* an opportunity to replicate its own success developing its stable of Marvel superheroes into another highly profitable transmedia universe—with annual blockbuster films, new television series, comic books, video games, and, of course, elaborate amusement park rides, and high-profile theme park expansions.

Many critics, scholars, and fans have responded enthusiastically to this most recent revival of the now 40-year-old media property. In particular, the new films' more progressive approach to gender balance and ethnic diversity has been hailed as a long-overdue improvement, while the ideological position the new films have taken toward the rising tide of global fascism has similarly been celebrated. In my own review of *Rogue One* (2016) for the *Los Angeles Review of Books*, I praised the film's politics for offering up "a story in which political resistance isn't a matter of fate, but of political choices made at tremendous personal cost" (Hassler-Forest 2016). Therefore, in terms of ideological criticism, the series has not only changed over time, but also appears to have become more politically progressive—something that would seem to be all the more surprising for the fact that the franchise's ownership transitioned from a (nominally) independent media company to the world's most powerful media conglomerate.

This apparent contradiction is easier to understand from the "macro-perspective" of political economy. Going beyond a textual analysis of the transmedia storyworld, this framework takes into account the strategic ways in which media producers develop IP within a specific set of (historical, cultural, social, industrial) circumstances in an attempt to reach a particular audience. More than merely expressing a certain cultural "Zeitgeist," it means constructing a brand identity that maximizes the relevant IP's value in terms of its realizable cultural *and* economic capital. For the Disney-era *Star Wars* franchise, this brand identity is intentionally ambivalent, positioning itself as simultaneously artificial (in its material nature as a mass-produced commercial commodity) and authentic (in its appeal to fans' memories of the original trilogy). This structural ambivalence results in an internal contradiction that defines what Sarah Banet-Weiser has described as *brand culture*:

> This affective sentiment, the feeling of authenticity, often does the cultural work of an inducement, attracting and retaining consumers as loyal members of a brand culture. Individual consumers trust the affective knowledge offered by brand cultures, even as they are aware that brand marketers carefully cultivate this trust, even when this knowledge is recognized by consumers as "irrational" or "emotional."
>
> *(Banet-Weiser 2012, 219)*

When viewed as a franchise that is not only an elaborate transmedia storyworld but also a global transmedia brand, the politics of Disney-era *Star Wars* therefore clearly take on rather different dimensions. As one of the world's most profitable brands, *Star Wars* consists primarily of a range of commodities that are produced and distributed within global capitalism's normative brand culture, which "more often than not reinscribes people back within neoliberal capitalist discourse rather than empower them to challenge or disrupt capitalism" (Banet-Weiser 2012, 221). Thus, even if ideological criticism of *Star Wars*'s Disney-era could lead us to celebrate its "subversive" ideological values, the political economy within which it exists simultaneously clarifies and explains the brand's structural ambivalence.

Meme Wars: Negotiating the Political Meanings of *Star Wars*

When the first trailer for *Star Wars Episode VII: The Force Awakens* (2015) appeared, a "boycott Star Wars" campaign was launched by fan groups on the far right. One year later, a similar controversy surrounded the release of *Rogue One* (2016), which was targeted by Trump supporters and the neo-Nazi "Alt-Right movement" with the same kind of online campaign. A tweet that (falsely) claimed that parts of the movie were rewritten and reshot "to add in anti-Trump scenes" circulated widely, resulting in a flurry of online outrage and debate, especially within far-right-leaning online forums like Reddit and 4chan. This time, screenwriters Chris Weitz and Gary Whitta involved themselves in the debate, writing tweets stating that "the Empire is a white supremacist (human) organization," and that it is "opposed by a multicultural group led by brave women" (Siegel 2016). While these explicitly political tweets were quickly deleted and Disney CEO Bob Iger insisted publicly that "there are no political statements in it, at all" (Loughrey 2016), the ensuing debate demonstrates how strongly transmedia franchises like *Star Wars* are understood, used, and referenced in political terms.

As I have explained in this chapter's previous sections, the ideological organization of a transmedia storyworld can be decoded via textual analysis, while the larger political economy within which these texts circulate over-determines its political orientation in a different (and sometimes contradictory) way. In this sense, Iger's seemingly disingenuous claim about *Rogue One* is clearly also accurate: like so many other prominent pop-cultural texts, *Star Wars* has cultivated a thoroughly ambivalent brand identity in which audiences recognize and respond to a variety of different meanings that coexist within the storyworld's larger multitext. In this chapter's final section, I will therefore examine how the franchise also operates as a public battleground, as fans from hugely divergent ideological beliefs waged war over the cultural meaning of *Star Wars* within a shifting political context. My discussion will show how the appropriation of characters, plot elements, and the iconography of the franchise in the context of political activism contributes to the flexible and constantly contested cultural meanings of this popular entertainment franchise.

Within the growing interdisciplinary field of fan studies, the figure of the media fan tends to be approached as politically progressive and culturally productive almost by default. Henry Jenkins' enduring influence has helped privilege the academic perception of the fan—certainly within social studies and the humanities—as that near-utopian figure of the "produser" within convergence culture: the cultural magpie and textual poacher whose active and creative engagement with fictional texts provides an inspiring model for democratic and civic involvement. And in the early days of the Internet, when online fandom was a diverse and lively but relatively "niche" phenomenon, one can easily see how this association between certain kinds of fandom and politically progressive attitudes could be credibly maintained.

But as we entered the age of ubiquitous computing, corporate-owned social media, and the "mainstreaming" of fan culture (Scott 2013), the political diversity of online fandom has also become much more apparent. A generation of digital natives has emerged, demonstrating cultural and political sensibilities that run sharply counter to Jenkins' own descriptions of radically inclusive and progressive fandoms surrounding franchises like *Star Trek* and *Harry Potter*. The expressions of fan culture that were fostered on platforms like Reddit and 4chan were dominated by a sarcastic and often aggressive attitude toward cultural and political debates. Sharply distinct from the general climate of sincerity and inclusiveness that dominates the often female-driven forums so often privileged by fan studies scholars, these environments were fed instead by the tech-savvy and straight white male-dominated sensibilities of gamers, coders, and software engineers.

The culture that developed on 4chan (and later 8chan) would prove to be especially influential for the growing visibility of this kind of fandom. The now-ubiquitous cultural form of the *meme* (a recognizable and often-repurposed image accompanied by a comical inscription) best typifies this environment's cultural and political attitude. Where fan-cultural traditions like the writing of fanfiction and elaborate cosplay foreground deep immersion and a strong, highly personal commitment to a particular transmedia storyworld, meme culture can be described as its polar opposite. Rather than displaying the individual's deeply felt investment in a particular property, the meme circulates as an anonymous and widely appealing shard of ephemeral "spreadable media" (see Jenkins, Ford, and Green 2013).

In terms of its politics, meme culture appears on the surface to be an empty vessel: a way of using the vocabulary of popular culture and the infrastructure of social media to comment sarcastically on anything, from every possible political perspective. But as Fredric Jameson tirelessly reminds us, form is as much a form of politics as content, and the meme as a cultural form is therefore just as strongly determined in its political bearings as, say, a modernist novel. As what has quickly become one of the defining cultural forms of digital transmedia culture, the meme has a form of politics ingrained into its very structure. For while it performs many obvious functions, including the demonstration of cultural capital, the successful meme is made up of two basic components: a hyper-recognizable pop-cultural image (sometimes consisting of two or three separate images in juxtaposition) alongside a prominently placed caption that provides a humorous variation on a familiar theme.

To illustrate this description with a well-known example: one of the classic forms of the Internet meme is "Condescending Wonka" (sometimes also referred to as "Creepy Wonka"). The meme consists of an image of Gene Wilder in his iconic role as Willy Wonka, accompanied by a two-part caption indicating the condescending expression of a deeply insincere interest. The first part of the caption is a rhetorical question—for example: "So you hate how George Lucas keeps changing the Star Wars films?" And the second part offers a condescending answer, usually beginning with the phrase "Please tell me more"—in this example, "Tell me again how you helped finance them."

The image and its history, both as a meme and as a reference to a classic children's movie, provides a low-threshold form of accessible and easily shared critical commentary with the potential to reach an audience of millions within a matter of hours.

But as this example shows, the typical attitude fostered within these memes is indeed one of sarcastic condescension. While clearly steeped in fan culture's participatory tradition, its highly developed media literacy, and its foregrounding of various forms of pop-cultural capital, meme culture simultaneously distances itself from the emotional involvement that underlies fandom's older forms of transformative appropriation. The politics of meme culture are therefore best understood in terms of the *affect* produced by its most typical combination of form and content. This affect is best described as one of emotional distance, insincere appreciation, and condescension—both toward cultural objects and toward other users. This formal structure of condescension, which emerged from 4chan and Reddit to thrive amidst the winner-takes-all model of platform capitalism and algorithmic media culture (see Srnicek 2016), therefore feeds and sustains a *politics of superiority*.

Unlike the gift economies and potluck cultures associated with fandom (Turk 2014), meme culture's politics of superiority aligns comfortably with the position of white supremacists, neo-Nazis, men's rights activism (MRA) groups, and other far-right factions associated with the twenty-first-century "alt-right" movement. Like the movement's icon—the meme-like figure of Pepe the Frog—this loose affiliation of groups consistently demonstrates cultural attitudes and political preferences that are united in their opposition to progressive political values. While the commonly used term "for the lulz" (i.e., "just for laughs") automatically dismisses any deeply held beliefs or politics behind meme culture, 4chan users have in practice launched organized harassment campaigns, most expressed under the moniker of free speech fundamentalism:

> The accepted standard was a sort of libertarian "free speech" banner, in which isolated man-boys asserted their right to do or say anything no matter someone else's feelings. This meant generally posting pornography, swastikas, racial slurs, and content that reveled in harm to other people.
>
> *(Beran 2016)*

This toxic mix of aggression, condescension, and far-right politics became more publicly visible with Gamergate's online bullying of women involved in the video game industry, with 4chan users' organized hate campaign directed toward the gender-swapped *Ghostbusters* remake, and in the aforementioned *Star Wars* boycott.

In support of the boycott of *The Force Awakens*, many virulently racist and anti-Semitic memes were circulated on social media, many of which were accompanied by the hashtags #boycottStarWarsVII and #whitegenocide. One widely circulated tweet, posted by a Twitter user going by the name "End Cultural Marxism," simply stated "#BoycottStarWarsVII because it is anti-white propaganda promoting #whitegenocide." Following the same logic, images were produced that used the familiar *Star Wars* font to display the words "Race Wars," with fascist phrases like "Europeans Awaken" in the place where the actual films' episode title would be located.

While this campaign was obviously waged by some of the most extremist groups within the far-right, they point toward the troubling ways in which political and cultural wars are now waged in the digital arena of popular culture. Media-industrial practices and storytelling choices are connected directly by these "unruly" fans to political and ideological debates, resulting in a public tug-of-war over the franchise's multiple possible meanings. Many of these self-made images demonstrate how the franchise's visual depiction of the totalitarian Empire—including the design of the stormtrooper suits, the Imperial officers' uniforms, and the iconic figure of Darth Vader—resonate within what we might term "fascist fandom": a form of fan culture for which the explicitly fascist aesthetics of the storyworld's evil Empire is clearly more appealing than the Rebellion's more "multicultural" earth tones and rounded, organic structures.

On the other hand, if we look at a hugely popular meme that circulated on the other side of the political spectrum in the aftermath of Donald Trump's election, we find similarly problematic undercurrents of fascist thought. In this meme, the original trilogy's narrative structure (as well as its color scheme and iconography) was projected onto the American electoral cycle: Barack Obama (2008) represented "A New Hope," Donald Trump (2016) stood for "The Empire Strikes Back," and Bernie Sanders (2020) would bring redemption as the "Return of the Jedi." Similar in tone, audience, and political orientation was the "Star Wars safety pin" avatar, introduced by screenwriter Chris Weitz in an attempt to wed the franchise's spirit of rebellion to anti-racist and anti-fascist activism. Following the social and political upset of the Brexit referendum in the UK, the safety pin had been introduced as a way to express solidarity with and support for immigrants, refugees, or other minorities who felt unsafe. Countering 4chan users' attempts to claim *Star Wars* as a recently-tarnished icon of white supremacy, this widely circulated avatar positioned the films' Rebellion as a symbolic extension of progressive political activism.

Conclusion

These seemingly counter-intuitive ways in which *Star Wars* is made politically meaningful by audiences with opposed politics and worldviews are explained in part by brand culture's fundamental politics of ambivalence, constantly intertwining the "authentic and commodity self" in order to maximize the franchise's commodity value (Banet-Weiser 2012, 14). Thus, whether Princess Leia is being celebrated as a feminist icon or Darth Vader is being referenced as a positive model for political power, the structural ambivalence that underlies the *Star Wars* brand translates in both cases and with equal ease to debates within the political domain.

In this sense, *Star Wars* is not only one of the most visible transmedia franchises, but also one that typifies in many ways the political attitudes, sensibilities, and functions of other similar branded storyworlds. More recent popular transmedia properties like *The Hunger Games*, *Game of Thrones*, *The Walking Dead*, Christopher Nolan's Batman films, and the Marvel Cinematic Universe have similarly inspired wildly divergent political interpretations. For the constantly deferred narratives of these elaborate transmedia properties (Hills 2000, 142), these franchises' structural absence of closure all but precludes a definitive political or ideological textual analysis, while the unchecked proliferation of the storyworld across media makes it equally impossible to base general conclusions on medium-specific formal features or narrative structures. What we do see, however, from Yoda memes to "*Star Wars* is white genocide" boycotts, is that the ways in which these entertainment properties are appropriated in the public sphere vividly illustrate the many ways in which they are made meaningful by their specific association with politics.

References

Althusser, Louis. 2014. *On the Reproduction of Capitalism: Ideology and Ideological State Apparatuses*. Translated by G. M. Goshgarian. London and New York: Verso.

Banet-Weiser, Sarah. 2012. *Authentic™: The Politics of Ambivalence in a Brand Culture*. New York: New York University Press.

Beran, Dale. 2016. "4chan: The Skeleton Key to the Rise of Trump." *Medium*, February 14. Accessed June 2, 2017. https://medium.com/@DaleBeran/4chan-the-skeleton-key-to-the-rise-of-trump-624e7cb798cb.

Booker, M. Keith. 2006. *Alternate Americas: Science Fiction Film and American Culture*. Westport: Praeger.

Brooker, Will. 2002. *Using the Force: Creativity, Community and Star Wars Fans*. New York: Continuum.

Di Placido, Dani. 2016. "Looking Back at the 'Star Wars' Prequel Trilogy." *Forbes*, December 15. Accessed November 21, 2017. www.forbes.com/sites/danidiplacido/2016/12/15/looking-back-at-the-star-wars-prequel-trilogy/#416d57372a9c.

Guynes, Sean A., and Dan Hassler-Forest. 2018. *Star Wars and the History of Transmedia Storytelling*. Amsterdam: Amsterdam University Press.

Hassler-Forest, Dan. 2016. "Politicizing Star Wars: Anti-Fascism vs. Nostalgia in 'Rogue One.'" *Los Angeles Review of Books*, December 2016. Accessed April 26, 2017. https://lareviewofbooks.org/article/politicizing-star-wars-anti-fascism-vs-nostalgia-rogue-one/.

Hills, Matt. 2000. *Fan Cultures*. London: Routledge.

Jameson, Fredric. 1981. *The Political Unconscious: Narrative as a Socially Symbolic Act*. Ithaca: Cornell University Press.

Jenkins, Henry. 1992. *Textual Poachers: Television Fans and Participatory Culture*. London and New York: Routledge.

Jenkins, Henry. 2006. *Convergence Culture: Where Old and New Media Collide*. New York: New York University Press.

Jenkins, Henry, Sam Ford, and Joshua Green. 2013. *Spreadable Media: Creating Value and Meaning in a Networked Culture*. New York: New York University Press.

Johnson, Derek. 2013. *Media Franchising: Creative License and Collaboration in the Culture Industries*. New York: New York University Press.

Loughrey, Clarisse. 2016. "Rogue One: Disney CEO Insists Star Wars Film Contains 'No Political Statements.'" *Independent*, December 13. Accessed May 1, 2017. www.independent.co.uk/arts-entertainment/films/news/rogue-one-disney-ceo-star-wars-political-statements-trump-rebel-facists-a7471446.html.

Murdock, Graham, and Peter Golding. 1973. "For a Political Economy of Mass Communications." *The Socialist Register* 10: 205–234.

Nadel, Alan. 2012. "The Empire Strikes Out: *Star Wars* (IV, V, and VI) and the Advent of Reaganism." In *American Literature and Culture in an Age of Cold War: A Critical Reassessment*, edited by Steven Belletto and Daniel Grausam, 187–208. Iowa City: University of Iowa Press.

Pearson, Roberta. 2013. "*Star Trek*: Serialized Ideology." In *How to Watch Television*, edited by Ethan Thompson and Jason Mittell, 213–222. New York: New York University Press.

Rubey, Dan. 1978. "*Star Wars*: Not So Long Ago, Not So Far Away." *Jump Cut* 18: 9–14.

Scott, Suzanne. 2013. "Fangirls in Refrigerators: The Politics of (In)Visibility in Comic Book Culture." *Transformative Works and Cultures* 13. doi:10.3983/twc.2013.0460.

Siegel, Tatiana. 2016. "'Star Wars' Writers Get Political: Will Anti-Trump Tweets Hurt 'Rogue One'?" *The Hollywood Reporter*, November 21. Accessed April 26, 2017. www.hollywoodreporter.com/heat-vision/star-wars-writers-get-political-will-anti-trump-tweets-hurt-rogue-one-949023.

Srnicek, Nick. 2016. *Platform Capitalism*. Cambridge: Polity Press.

Tryon, Chuck. 2013. *On-Demand Culture: Digital Delivery and the Future of Movies*. New Brunswick: Rutgers University Press.

Turk, Tisha. 2014. "Fan Work: Labor, Worth, and Participation in Fandom's Gift Economy." *Transformative Works and Cultures* 15. http://dx.doi.org/10.3983/twc.2014.0518.

Wasko, Janet. 2001. *Understanding Disney: The Manufacture of Fantasy*. Cambridge: Polity Press.

33

TRANSMEDIA CHARITY

Constructing the Ethos of the BBC's Red Nose Day Across Media

Matthew Freeman

There is tendency inside academic circles to most commonly correlate transmediality with storytelling, particularly with global commercial storytelling scenarios such as large-scale Hollywood franchises. However, the practice of communicating media across multiple platforms is by no means specific to fictional storytelling. This chapter aims to conceptualize transmediality through the lens of charity, discussing the significance of participation, documentary, and community media. To do so, I will use Red Nose Day—a UK-originated charity founded by Richard Curtis in 1985—to trace how the social traditions and sensibilities associated with Red Nose Day have been interlaced with emerging digital technologies to shape and develop this charity campaign across multiple media platforms. In doing so, I will offer an exploration of the relationship between non-fictional iterations of transmediality, charity activism, and social traditions, essentially showing how audiences follow the *ethos* of Red Nose Day across multiple media platforms.

Following an *ethos* across media in fact means questioning the very idea of transmediality, which—at least in fictional contexts—has tended to be closely correlated with seriality. For as Jenkins observes, "transmedia storytelling has taken the notion of breaking up a narrative arc into multiple discrete chunks or instalments within a single medium and instead has spread those disparate ideas or story chunks across multiple media systems" (2009). Serialized media forms such as prequels and sequels are thus adopted in transmedia stories so to build characters across multiple media, guiding the audience from one medium to the next. In that sense, transmediality is a process of expansion, of taking a story and making it bigger. In my book *Historicising Transmedia Storytelling* (2016), I offered the metaphor of the house extension to conceptualize transmediality—suggesting that the core text in a transmedia story is the living room, with all of the subsequent story extensions the rooms that adjoin onto it.

However, when beginning to think about non-fictional cases of transmediality, such as charity campaigns, this analogy is not necessarily appropriate. Non-fiction, in fact, is useful for forcing researchers to re-conceptualize what is meant by transmediality beyond basic notions of storytelling and related concepts of seriality. While much scholarship dwells on the commercial, global industry formations of transmediality, there has been little attempt to track or to understand a more socio-political idea of transmediality. In one sense, examining transmediality from a social or charity perspective means thinking about it as a non-fictional engagement strategy that has ramifications in terms of people, leisure, activism, politics, and society itself. As will be shown, it is possible to trace how social media has informed the way that long-standing charity campaigns are now promoted as transmedia campaigns, with non-fictional forms of narrative engaging people across multiple media platforms. As such, one might question whether it is necessary to theorize a different conceptual model for examining the non-fictional form of transmedia charity campaigns, rather than simply trying to apply its fictional characteristics (such as world-building) to its non-fictional charity counterpart. This chapter explores this question throughout, theorizing the transmedial construction of Red Nose Day as a digital media campaign underpinned by particular non-fictional forms of participation, documentary and community, albeit one where storytelling is still key and where "integral elements ... get dispersed systematically" across media (Jenkins 2007).

Red Nose Day—A Brief Overview

Put simply, the aim of Red Nose Day is to use comedy to raise money and change lives, with a mission statement that seeks to drive positive change through the power of entertainment. But its mantra lies in notions of clean-living, of fun, of joining in, of a common cause, of not taking oneself too seriously, and yet remembering that not taking oneself too seriously is in the name of something quite serious—something that is social, and far bigger than the individual. As a reporter for the *Birmingham Post* once characterized, also, "Red nose people have big hearts and want poverty and starvation to end right now. The political problems which are often the root cause of such suffering interest them less than the thought of doing something good" (*Birmingham Post* 1999, 33). Red Nose Day promotes the message that when you smile, the world smiles with you. How, though, do the themes and money-donating dynamics of Red Nose Day manifest transmedially and evolve alongside the rise of social media platforms? In particular, how do notions of participation, documentary and community each inform Red Nose Day as a transmedia charity? Let's start with participation. What are the parameters of how audiences now participate in Red Nose Day across multiple media?

Participation

Importantly, Kerrigan and Velikovsky argue that non-fictional forms of transmedia bring "unique characteristics and terms to the scholarly debate on non-fiction transmedia, that is: 'non-commercial'" (2016, 250). Indeed, transmediality may have much in common with the structured nature of an expanding commercial storytelling experience, but it is also characteristically epitomized by the participation of audiences. Scolari, Bertetti, and Freeman posit the following formula as a way to conceptualize all participants in transmedia stories: "Media Industry (Canon) + Collaborative Culture (Fandom) = Transmedia Storytelling" (2014, 3). Scolari et al. suggest that there are different levels of participation ranging from the consumer of a single media form, to the "prosumer" who expands the transmedia content by producing new content, which "represents the highest level of transmedia engagement" (2014, 3). Jenkins emphasizes that the rise of new digital technologies and convergence culture have both worked to make this possible, empowering audiences by giving them the "right to participate" (2006, 23). For example, in a non-fictional arena, Grainge and Johnson have discussed how the British Broadcasting Corporation (BBC)'s transmedia coverage of the Olympic Games in 2012 helped the BBC to reinforce its

> "inform, educate and entertain" mantra, since the added coverage of the sporting event available online—alongside its non-fictional books and bonus historical documentaries on DVDs—all served to enrich the ways that audiences were informed, educated and entertained about the Olympics precisely because it occurred across multiple media.
>
> *(2015, 88–90)*

In turn, a heightened sense of engagement across multiple media was seen to encourage more interaction from those audiences via online platforms such as social media, working to shape and re-shape how the cultural meanings of the Olympics circulate.

Regards to transmedia charity, then—again in a BBC context—a key point of Red Nose Day is that its audience participates rather than simply donates money. As Paul Vallely explains,

> Just a third of the money raised comes from telethon donations. Another third comes from the sale of red noses and other merchandise. But the remainder is from activities supporters undertake in the five weeks of the run-up to Red Nose Day, which takes place every other year.
>
> *(2013)*

That, Curtis believes, is what makes Red Nose Day distinctive: "If you involve people in an activity they become more engaged with the issue" (2013). Certainly, the very premise of charity is based on the participation of people; for a charity campaign to be successful, it has to engage people enough so that they feel persuaded to express opinions and, crucially, to be active enough to donate money or resources to a cause. And Red Nose Day's participation has never actually relied on media as such—"from sponsored sweet eating and holding cake sales to fancy dress football matches and paying to dress red for the day, the fundraising possibilities are endless" (*South Wales Echo* 2011, 6).

In this sense, the participatory parameters of Red Nose Day remain not only traditional—that is, based on small local activity or on old media—but share important overlaps with strategies used by more commercial transmedia industry formations. Much of Red Nose Day's brand recognition is based on merchandizing (selling a red nose), which has informed fictional systems of transmediality for many years (Santo 2015; Freeman 2016). Elsewhere, for instance, I have shown how strategies of merchandizing Tarzan in the 1930s enabled the stories of Tarzan to be expanded further across media (Freeman 2015). Jonathan Gray, too, has explored how toys can reshape a transmedia narrative, pointing to examples such as *Star Wars* (1977) and *District 9* (2009) (2010). Similarly, as host Davina McCall proclaimed of Red Nose Day,

> we spend the week leading up to [Red Nose Day] buying every Red Nose Day piece of merchandise that we can get our hands on. We've got T-shirts, noses and things happening on my car. We just basically pimp our lives out to Red Nose Day and do something silly on the day itself, which is really good fun.
>
> *(Holman 2013, 55)*

In another sense, however, and as was hinted earlier, in the contemporary media landscape digital media has been hailed as blurring the lines between producers and consumers, in turn creating a heightened participatory transmedia culture (Jenkins 2006). And this heightened participation has had an impact on Red Nose Day. Krystle Lampshire, a member of Red Nose Day's Digital Marketing team, discussed recently how they are now "developing tactics to inspire their social media audience to take on active fundraising or increase their overall engagement with the campaign on social media" (*Brandwatch* 2015). Part of that heightened engagement now manifests as the ability to follow particular strands of information (or particular threads of plot) from television to the web and back again, granting audiences the choice over which aspect of the Red Nose Day telethon they follow. For example, in 2015 presenter Dermot O'Leary danced for 24 hours without stopping, and comedian Mark Watson embarked on a 24-hour comedy marathon. The way in which audiences can therefore follow ongoing narratives across multiple media is not only indicative of the way in which Jenkins' fictional model of transmedia storytelling continues to pervade multiple sites of media production, distribution and consumption in the twenty-first century, but is also reflective of what Jennifer Holt and Kevin Sanson discuss as "connected viewing"—"a multiplatform entertainment experience [that] … integrates digital technology and socially networked communication with traditional screen media practices" (2014, 1).

However, whereas connected viewing may have led to "the migration of media and our attention from one screen to many," in the case of non-fictional Red Nose Day charity content one can observe how the participation on offer has become increasingly long-form. In fact, even compared to the BBC's transmedia production surrounding the Olympic Games a few years ago, content has evolved from being predominantly short clips to be watched (YouTube clips, web links, etc.) into sites of continuous participatory engagement that both pervade daily life and reflect daily life via a focus on everyday themes such as dancing, comedy and so on. How, though, does this continuous stream of participatory information infuse with documentary aesthetics when Red Nose Day continues to sprawl across multiple media?

Documentary

Bill Nichols argues that "a documentary practitioner should be able to debate social issues such as the effects of pollution and the nature of sexual identity and explore technical concerns such as the authenticity of archival footage and the consequences of digital technology" (2001, 25). But whereas a transmedia documentary project maintains the theoretical underpinning of the documentary form, Siobhan O'Flynn argues that, by definition, "a transmedia documentary distributes a narrative across more than one platform, it can be participatory or not, can invite audience-generated content or not, tend to be open and devolving, though not always" (2012, 144). Nevertheless, a key point to observe is that, somewhat characteristic of how UK media industries are now approaching digital platforms, transmediality does not always equate to storytelling. Elizabeth Evans discusses how "what is most noticeable about the emerging strategies of both Channel 4 and the BBC is their clear attempt to realign the comparative status of broadcasting and digital technologies, with the digital positioned as equal to linear broadcasting" (2015). As Evans elaborates, here:

> Transmedia strategies operate not just at the level of the text; it is also happening at the level of the *channel* or at the level of the broadcaster themselves ... There is increasingly an emphasis on constructing different story forms that may be bound together by an overarching single transmedia estate—guided by a channel or broadcaster's brand identity and, in the case of BBC, a single, coherent online space.
>
> *(2015)*

In the case of Red Nose Day, the digital estate surrounding the charity event seeks similarly to transmedially distribute the meanings and messages of the campaign, and to connect particular platforms with particular themes that create an overarching single transmedia estate. And crucially, those meanings, messages, and themes concern a balance between informing, educating, and entertaining—as per the institutional mantra of the BBC.

However, what is noteworthy is that the Red Nose Day Facebook page is characteristically populated with content that is either entertaining *or* educational in its objectives, but rarely both. Most typically, it is the former that is prioritized in the build-up to Red Nose Day. Short comedic videos, typically fronted by celebrities, including teaser clips of the sketches to be seen in full during the telethon, function much like commercial promos. But in the case of non-fictional transmediality such as Red Nose Day, it is not so much a narrative as it is an *ethos* that is being advanced and developed across platforms, with the meanings of Red Nose Day (fun, clean-living, joining in, not taking oneself too seriously, etc.) not located solely within the broadcasted television event but also extended across multiple promotional forms.

Following the end of Red Nose Day and the television broadcast, for example, it is the role of social media to uphold the more educational aspects of Red Nose Day. Specifically, it is Facebook that works to strengthen the key purpose behind Red Nose Day, and does so by reinforcing the sustained poverty in countries like Kenya and Ethiopia, making use of its status as a perpetual platform—one that more easily pervades audiences' lives every minute of every day—to keep the cause in the public eye long after the telethon has ended. These appeal films are extracted from the television broadcast, but interestingly are often stripped of the comedy trappings that would bookend them on television and re-published online with a phone number to donate clearly emphasized. As such, the "storytelling" of Red Nose Day across television and Facebook thus lies in the juxtaposition between entertainment and education, between comedy and poverty. Audiences are engaged every day with snippets of the former online, watch a mixture of both during the telethon, and then continue to be persuaded online with powerful messages of the latter, each characteristically fronted by a famous British celebrity in a way that continues to hint at a sense of entertainment via the entertainments for the which they are known. Importantly, then, the "storytelling" of Red Nose Day works to always retain its *ethos* of clean-living, of fun, and of a common cause, even when it does not. That is to say that the discourse that is often emphasized in the documentaries about poverty in third-world countries is one of loss, focusing primarily on that which is missing from peoples' lives. We hear of absent parents, for example, of the struggle to find clean water, of the lack of laughter, or of the constant pressure to survive. In other words, it is the presence versus the absence of clean-living and fun that juxtaposes between the scenes of comedy and poverty in the overall Red Nose Day media experience.

Meanwhile, the role of the Red Nose Day Twitter page is partly to document the aftermath of the donations during the television broadcast. Twitter is used to post stories of the local communities that have benefited from Red Nose Day, with web links to interviews and newspaper articles that showcase the good that is now being accomplished. In this sense, at least, one might detect are clear linearity to the narrative of Red Nose Day across platforms: whereas Facebook focuses on the "before" (the lead-up to Red Nose Day itself and the sorts of fundraising under way), and live television broadcast represents the "present" (the main act, as it were), then Twitter showcases the "future" of the story (telling the tales of what happened after the television event). In other words, understanding Red Nose Day—and to some extent other non-fictional transmedia charity campaigns in the UK—is about forging a balance between the spreading of thematically grouped information and the strategic organization of that information in a way that is akin to storytelling. Or as Evans puts it, "transmediality in UK television is as much about distribution as it is about storytelling" (2015). Still, how does this transmedia distribution work alongside themes of community? Let's now trace how the traditions and ways of life in Britain have been interlaced with transmedia platforms to maintain a strong sense of British Red Nose Day community.

Community

Most broadly, community media refers to "locally orientated media access initiatives … dedicated to the principles of free expression and participatory democracy, and committed to enhancing community relations and promoting community solidarity" (*South Wales Echo* 2015, 22–23). Paradoxically, the Red Nose Day telethon continues to hold a status as both "event television" (where audiences are enticed to watch a program live because of its unique cultural importance and sense of rarity) and "tradition" (which speaks of a sense of sameness and ritual). However, what can be argued to unite both concepts of "event television" and "tradition" is the idea of *community* and the notion of bringing together a nation—a theme that has long remained important to the perceived value of Red Nose Day.

For instance, in one sense, Red Nose Day establishes a clear notion of community via its emphasis on distributing money from the Red Nose Day Community Cash fund to small communities across the country. As the *South Wales Echo* reported in 2015, the "Community Cash grants of £500-£1,000 are available to small organisations that are working to help local people living tough lives" (22–23).

But there is often a tension in the discourse of Red Nose Day between its focus on the macro and the micro, between the international and the local. As Victoria Southwell, UK Grants Manager, claimed:

> We have funded some truly fantastic work in the local area and are looking forward to seeing some new groups applying. Thanks to the hard work of these groups, money raised through Red Nose Day can get right to the heart of the local community and make a real difference where it is most needed.
>
> (*South Wales Echo* 2015)

But during the telethon event, community is established at the national level, with the role of celebrities and comedians to prescribe a clear sense of "Britishness" in its look and tone. Memorable and iconic scenes over the years, for example, have included Dawn French kissing Hugh Grant and Billy Connolly dancing naked around Piccadilly Circus.

When conceived of through the lens of transmedia storytelling, moreover, Red Nose Day becomes partly about continuing the extra-textual narratives of films, television programs and their personalities beyond the confines of those films and television programs, essentially extending those narratives onto another platform but in ways that draws attention to the comedic artifice of their construction and thus reinforces themes associated with the British sensibilities of Red Nose Day. For example, the sketches of *Little Britain* (2003–2006) continued over onto a special DVD edition of Red Nose Day, with self-aware guest appearances from Kate Moss, Jonathan Ross, and Patsy Kensit. Previous Red Nose Day telethons have even established themselves as platforms for reviving beloved British sitcoms: 2014 saw a one-off return of *Only Fools and Horses* (1981–2003) starring Sir David Jason and Nicholas Lyndhurst alongside a cameo from David Beckham, while in 2015 *Mr. Bean* (1990–1995) was brought back almost two decades after the television series had ended.

Consider, too, how the story of the judges' relationships seen on *Britain's Got Talent* (2007–) were used to engage Red Nose Day viewers in 2015: "Simon Cowell and David Walliams have taken their bromance up another level after Simon showed off his face-painting skills to make David's face funny for Red Nose Day," reported *South Wales Echo* (2015). "Author and comedian David said: 'I feel absolutely beautiful. Simon's been hard at it all evening and he's even chosen colours that complement my eyes—I suspect this isn't the first time Simon has picked up a makeup brush!'" (2015). Thus, if one were to think of *Britain's Got Talent* and its judges as an ongoing story akin to an amusing soap opera, then one can claim that it was the role of Red Nose Day to extend the non-fictional story of that soap opera, using its non-fictional characters as a way to engage audiences and point them across media platforms.

Importantly, non-fictional narratives such as *Britain's Got Talent* embody the *ethos* defined by Red Nose Day. Like Richard Curtis' charity, Simon Cowell's talent show aims to drive positive change through the power of entertainment, prioritizing notions of clean-living, fun, and joining in—all in the common cause of finding national talent to show to the Queen during the annual Royal Variety Performance. And, in that sense, one can also argue that underpinning the maneuvering of Red Nose Day audiences across multiple media platforms is not necessarily the narrative of the Red Nose Day campaign itself, or the elaboration of its different fundraising skits across platforms in the way that might be theorized of fictional transmedia storytelling. Instead, the transmedia logic of Red Nose Day follows a kind of "transmedia *ethos*" —that is to suggest that audiences are encouraged to migrate to titles and to other pieces of media content that share similar underlying beliefs and values, even if those pieces of media content derive from completely different broadcaster and producer contexts. In this case, the Red Nose Day media experience continues across television, newspapers, websites and social media via other television programs that adhere to the values of life-changing spirit and British fun that have already been ascribed to Red Nose Day. And in so doing, Red Nose Day shares and benefits from the same community built by the likes of *Britain's Got Talent*, *Little Britain*, and a host of similarly themed British television shows.

Conclusion

Since I have shown throughout this chapter that the characteristics, styles, and formats underpinning non-fictional transmedia charity extends across other non-fictional media forms, such as documentary and community media, it might well be necessary to theorize a different conceptual model for examining non-fictional transmedia charity, rather than merely trying to apply its present fictional characteristics to its non-fictional counterpart. In a media landscape marked by consolidation, audience fragmentation, as well as rapid technological change and innovation, something like the Red Nose Day campaign typifies the role of transmediality as a social enterprise right now. Red Nose Day invokes Britain as a long tradition of social activism, and digital platforms work not to transform tradition but to reinforce tradition by using different media to convey different parts of the campaign *ethos*.

The concept of story is therefore still applicable to understanding transmedia charity, but typically in relation only to people, such as following the entertaining activities of television presenters and charity cases across additional platforms to gain more information. Otherwise the concept of *ethos* is perhaps more useful for characterizing the way audiences navigate transmedia charity projects, with people following beliefs, values, themes, philosophies, and meanings (rather than stories) across media. Furthermore, while a term like "prosumer" sounds too commercial to be used in relation to discussions of charity campaigns, I have shown in this chapter how audiences actively expanded the Red Nose Day campaign by contributing to its social impact, suggesting that non-fictional transmedia—itself somewhat synonymous with the non-commercial (Kerrigan and Velikovsky 2016)—inherently represents what Scolari, Bertetti, and Freeman described previously as "the highest level of transmedia engagement," perhaps even more so than with fictional iterations (2014, 3). And in that sense, non-fictional transmedia charity emerges as a means of building continuous daily experiences around audiences; with non-fiction, after all, the story need never end.

Acknowledgements

An earlier version of this chapter was previously published as "Small Change—Big Difference: Tracking the Transmediality of Red Nose Day" in *VIEW: Journal of European Television History and Culture* 5:10 (2016): 87–96. doi: http://dx.doi.org/10.18146/2213-0969.2016.JETHC114.

References

Birmingham Post. 1999. "What A Relief That The Comedy's Over." March 15.

Brandwatch. 2015. "Case Study: Comic Relief." Accessed April 19, 2017. www.brandwatch.com/wp-content/uploads/2015/08/Comic-Relief-Case-Study-2015_v7.pdf.

Evans, Elizabeth. 2015. "Building Digital Estates: Transmedia Television in Industry and Daily Life." Paper presented at the *ECREA TV in the Age of Transnationalisation and Transmediation Conference*, Roehampton University, June 22.

Freeman, Matthew. 2015. "Author-as-Franchise-Product: Edgar Rice Burroughs Inc. and Tarzan and Historical Branded Entertainment." In *Engaging Consumers through Branded Entertainment and Convergent Media*, edited by Jose Martí-Parreño, Carla Ruiz-Mafé, and Lise Scribner, 53–73. Hershey, PA: IGI Global.

Freeman, Matthew. 2016. *Historicising Transmedia Storytelling: Early Twentieth-Century Transmedia Story Worlds*. London and New York: Routledge.

Grainge, Paul, and Catherine Johnson. 2015. *Promotional Screen Industries*. London: Routledge.

Gray, Jonathan. 2010. *Show Sold Separately: Promos, Spoilers, and Other Media Paratexts*. New York: New York University Press.

Holman, Justin. 2013. "Red Nose Day Is A Big Deal." *The Express on Sunday*, March 9.

Holt, Jennifer, and Kevin Sanson (eds.). 2014. *Connected Viewing: Selling, Streaming and Sharing Media in the Digital Age.* London and New York: Routledge.

Jenkins, Henry. 2006. *Convergence Cultures: Where Old and New Media Collide.* New York: New York University Press.

Jenkins, Henry. 2007. "'We Had So Many Stories to Tell': The Heroes Comics as Transmedia Storytelling." *Confessions of an Aca-Fan: The Official Weblog of Henry Jenkins*, December 3. Accessed March 25, 2017. http://henryjenkins.org/2007/12/we_had_so_many_stories_to_tell.html.

Jenkins, Henry. 2009. "The Revenge of the Origami Unicorn: Seven Principles of Transmedia Storytelling." *Confessions of an Aca-Fan: The Official Weblog of Henry Jenkins*, December 12. Accessed February 20, 2017. http://henryjenkins.org/2009/12/the_revenge_of_the_origami_uni.html.

Kerrigan, Susan, and J. T. Velikovsky. 2016. "Examining Documentary Transmedia Narratives through *The Living History of Fort Scratchley* Project." *Convergence: The International Journal of Research into New Media Technologies* 22 (3): 250–268.

Nichols, Bill. 2001. *Introduction to Documentary.* Indiana: Indiana University Press.

O'Flynn, Siobhan. 2012. "Documentary's Metamorphic Form: Webdoc, Interactive, Transmedia, Participatory and Beyond." *Studies in Documentary Film* 6: 45–55.

Santo, Avi. 2015. *Selling the Silver Bullet: The Lone Ranger and Transmedia Brand Licensing.* Texas: University of Texas Press.

Scolari, Carlos, Paolo Bertetti, and Matthew Freeman. 2014. *Transmedia Archaeology: Storytelling in the Borderlines of Science Fiction, Comics and Pulp Magazines.* Basingstoke: Palgrave Pivot.

South Wales Echo. 2011. "Red Nose Day Is Back." March 1.

South Wales Echo. 2015. "A Host of Famous Faces Are Made Up for Red Nose Day." January 26.

Vallely, Paul. 2013. "Richard Curtis: Twenty-five Years Of Laughing In The Face Of Tragedy." *Independent*, February 3. Accessed April 19, 2017. www.independent.co.uk/news/people/profiles/richard-curtis-twenty-five-years-of-laughing-in-the-face-of-tragedy-8478499.html.

34
TRANSMEDIA EDUCATION
Changing the Learning Landscape

Lorena Peret Teixeira Tárcia

The convergent digital age is transforming nearly every aspect of our culture, impacting everything from business to social life to education. As Bass (2012) summarizes:

> A growing appreciation for the porous boundaries between the classroom and life experience, along with the power of social learning, authentic audiences, and integrative contexts, has created not only promising changes in learning but also disruptive moments in teaching. *(1)*

In this scenario, contemporary literacy includes such concepts as Media and Information Literacy (UNESCO 2009), Educommunication (Soares 2011), Transliteracy (Thomas et al. 2007; Frau-Meigs 2012, 2013; Lugo, 2016), New Media Literacies (Jenkins et al. 2009), Media Competencies (Ferrés and Piscitelli 2012), New Literacies, Multiliteracies (Cope and Kalantzis 2009), Transmedia Literacy (Scolari 2016), Ludoliteracy (Zagal 2010), Multimodal Literacy (Jewitt and Kress 2003), and Transmedia Edutainment (Kalogeras 2014).

Although it is important to consider that these proposals coincide with the authors' realities as well as with various theoretical paradigms of communication that influence the way in which the communicative process and its actors are perceived, most views reinforce the belief that education's aim is to provide opportunities for learning, to encourage significant participation at the individual and community levels, and to inspire autonomy and empowerment (Jenkins et al. 2009; Ferrés and Piscitelli 2012; Lugo 2016; Loertscher and Woolls 2014; Soares 2011).

In this chapter, we seek to contextualize and analyze various practices and methodologies related to transmedia education; to this end, we will consider school settings in addition to various learning environments and realities. In our understanding—and based on our experience applying multiplatform communication precepts in support of education in societies with low digital connectivity rates, such as the low-income communities of East Timor and Brazil—we propose to think in terms of transmediality, to look beyond digital environments, and to consider the possibilities of each audience and its available channels.

The Seven Principles of Transmedia Storytelling Applied to Education

Revisiting the seven principles of transmedia storytelling as they apply to education, Jenkins (Jenkins et al. 2009) challenges teachers to involve students and to encourage them to utilize what they see, hear, and read. In such a system, students are urged to seek out content, explore different pieces of information in various contexts, interact easily with other readers, and evaluate ideas across formats. His initial thoughts about the core principles of transmedia teaching/learning include the following:

1. Spreadability vs. Drillability: students can seek out information related to their interests across the broadest possible range of terrains while also drilling deep into what matters to them as individuals. Educators should think more about what motivates students to drill deeper and consider how they can facilitate students' capacity to dig into a topic that they have deemed important.
2. Continuity vs. Multiplicity: by asking "what if" questions, students should be able to think beyond established canons and explore multiple possibilities, various factors, different values, and diverse cultures.
3. Immersion vs. Extraction: the first concept involves a consideration of the potential educational value of virtual learning environments, which rely on notions of immersion and "activities where students build their own virtual worlds—deciding what details need to be included, mapping their relationship to each other, guiding visitors through their worlds, and explaining the significance of what they contain" (Jenkins 2010). The second concept, extractability, encompasses a principle that has long been a part of education—that the elements brought in by students from their home cultures and trips make their way into the classroom.
4. World-building: this concept involves mapping worlds, historical facts, and cultures as integrated systems, thus allowing us to push beyond local insights toward a fuller, richer understanding of a given society. World-building depends on "cultural geography," defined as a "sense of the peoples, their norms and rituals, their dress and speech, their everyday experiences" (Jenkins 2010).
5. Seriality: in light of the high volume of content provided to students at school, seriality refers to the elements that must be present for a story or lesson to take on a satisfying and meaningful form.
6. Subjectivity: this concept involves flipping perspectives and examining how the same events can be understood from multiple points of view.
7. Performance: this concept is about motivation and involves attractors, which draw the attention of students, as well as activators, which give them something to do, thus allowing them to put acquired information into use.

Transmedia Literacy

Transmedia storytelling is consistent with a variety of child development theories and discussions regarding how learning takes place (e.g., the theories of constructivism, Vygotsky, and Piaget). These models hold that learning occurs by playing, exploring, experimenting, remixing, making connections, telling stories in different ways, and communicating visually, orally, or aurally.

To this end, Hovious (2013) proposes the use of transmedia projects to develop seven literacies: (1) multimodal ("Transmedia storytelling facilitates the practice of multimodal literacy because you must make meaning across all the elements in the story to fully understand it"); (2) critical ("Transmedia storytelling requires the 'destructuring' and 'restructuring' of multiple modes of text, a complex task"); (3) digital ("Transmedia storytelling requires navigation of the story and evaluation of the digital elements in the story"); (4) media ("Transmedia storytelling exists across multiple forms of media; each media element must be evaluated separately before multimodal meaning-making can take place"); (5) visual ("Transmedia storytelling is a visually rich experience, and the images play a significant role in the narrative"); (6) information ("The interactivity of transmedia storytelling enhances information literacy skills"); and (7) gaming ("Gaming elements in transmedia storytelling require the use of logical and strategic thinking").

For Lamb (2011), transmedia environments help students to explore, seek out information, and interact with other learners. In this sense, "transmedia is a thoughtful blend of story characters and narrative layered into play (and, in this case, learning experiences); further, the interactions are extended through the use of media channels" (Warren et al. 2015, 69).

The successful implementation of a transmedia-based lesson into a classroom should provide students with an interesting task and with the guidance necessary to select various media platforms from which to embark on an immersive multimodal transmedia learning experience.

The power of the transmedia approach was recognized by the United States Department of Education in 2011. According to the Office of Innovation and Improvement (ED Innovation 2011), "the rich, fictional worlds of transmedia tend to create a greater level of social interaction that can inspire children to create their own stories and media products and to share them with each other."

Between 2010 and 2015, the US Department of Education's Ready to Learn (RTL) Television Program supported three large national projects that attempted to create educational television and digital media products for young children between the ages of two and eight in order to support their development of math and literacy skills. These programs explored whether "a transmedia approach would result in increased educational effectiveness when used with learners from low-income backgrounds" (ED Innovation 2016).

The supposition was that when connected, the strengths of various learning platforms would motivate students to work their way "through a greater variety of content, to encounter a richer variety of learning strategies, and to better link experiences in the home, in school or pre-school, and 'on the go'" (ED Innovation 2016). As grantees experimented with different approaches, the Department encouraged each project to conduct rigorous research into its impact on students' learning. Across these RTL evaluations, researchers generally found positive results when using RTL-produced transmedia products. A report conducted by the Center on Media and Human Development at Northwestern University entitled "The Ready to Learn Program: 2010–2015 Policy Brief" (Wartella, Luricella, and Blackwell 2016) provides a summary overview and independent analysis of the findings of all studies produced as a result of the RTL funding during that time period.

The review findings reveal that the three educational transmedia experiences developed because of the RTL funding were not only effective mechanisms by which young children were taught academic skills pertaining to math and literacy, they also functioned as support systems for parents and educators. The report includes the following major conclusions:

> (1) Positive associations between at-home transmedia engagement and children's math learning; (2) home study intervention parents' increased awareness of and engagement in their children's math learning; and (3) positive associations for at least some students in the intervention groups for the school-based evaluations.
>
> *(Wartella, Luricella, and Blackwell 2016, 1)*

The authors conclude that federal investment in public media and broadband infrastructure is important and should be a national educational priority when seeking to decrease the gap between richer and poorer preschoolers.

In Europe, the European Union's Horizon 2020 program provides funding for the Transmedia Literacy project (EC 2014), which aims "to understand how the young boys and girls are learning skills outside the school." This project involves an interdisciplinary group of 25 researchers who have experience in such fields as media literacy, transmedia storytelling, user-generated content and participatory culture, traditional and virtual ethnography, and pedagogy and innovation in education. This research focuses on teens (12–18 years old) in eight countries across three continents (Australia, Colombia, Finland, Italy, Portugal, Spain, the United Kingdom, and Uruguay) and centers on the development of skills such as transmedia content production and sharing as well as problem-solving in video games (Transmedia Literacy 2017).

Transmedia Educational Projects for Children and Teens

Three iconic transmedia education projects include Inanimate Alice, Robot Heart Stories, and Cosmic Voyager Enterprises.

In 2012, Inanimate Alice was named the Best Website for Teaching and Learning by the American Association of Librarians. Mainly web-based, "the project has inherent transmedia features like a non-linear mode of reading, evolving storytelling and user's interactivity that spans across multiple media platforms ... The reader becomes a part of the story, by participating rather than just consuming the content" (Transmedia Lab 2011).

Set in the early years of the twenty-first century, Inanimate Alice tells the story of Alice and her imaginary digital friend through text, sound, images, music, and games. The story unfolds across multiple platforms and connects various technologies, languages, cultures, generations, and curricula. As the journey progresses, new storylines appear, thus providing additional details and insights. Students are encouraged to develop their own episodes via co-creation, thus filling any gaps by developing new strands. Inanimate Alice was first designed as entertainment and was subsequently adopted by teachers in an effort to connect with students through the media they use and understand. The project is available in English, French, Italian, German, and Spanish and is in use in over one hundred countries.

After using the Inanimate Alice project with her fifth-grade students, Fleming (2013) concluded that the use of transmedia resources not only immerses students in an intense and motivating experience, it also helps educators to develop different modes of educational delivery. "If executed effectively, the curriculum and the technology become one, and at the core is the interaction between technology and story, creating a deep, rich Learning World" (Fleming 2013).

Fleming (2013) thus proposes the creation of a Transmedia Learning World (TLW) that combines transmedia with a pedagogy that is transformative and that allows content to "flow fluidly across the curriculum and from one media to the next." Adopting the pedagogical principles of constructivist and connectivist learning theories, Fleming (2013) believes that educators should be able to "build frameworks for transmedia narratives that enable the learner to take charge of the narrative and then to shape it to their own learning needs." To this end, Fleming (2013) writes that:

> A TLW is a paradigm for learning that combines the capabilities of ubiquitous technologies, real-life experiences, and learner-focused pedagogies, making for profoundly productive and powerful learning experiences. This dynamic ecosystem allows for the creation of a synergy between varieties of learning models and a range of pedagogies that will take students and teachers around the world into new realms.

Another well-known transmedia project is Robot Heart Stories, an experiential learning project that teaches collaboration and creative problem-solving. A student's experience with this project begins when a robot crash lands in Montreal; the robot must then make her way to Los Angeles in order to locate her spacecraft and return home. Students use math, science, history, geography, and creative writing to help the robot make her way across North America. At the same time, Robot Heart Stories extends beyond the classroom by involving a global audience. Each photo or piece of art featuring the robot that is submitted makes the robot stronger and helps her to return home. Robot Heart Stories was the first of three experiential learning projects developed by the storyteller and transmedia pioneer Lance Weiler and creative producer Janine Saunders (Jenkins 2012).

Based on the project analysis, Gambarato and Dabagian (2016, 2) argue that the utilized transmedia strategies "placed the students in the center of the learning and motivated them to learn." The fact that students were characters in the story gave them the opportunity to experience the narrative "instead of just listening or reading it." They also conclude that:

> Moreover, the mediatized environment of Robot Heart Stories developed multiple digital literacies, interpersonal communication skills, and knowledge in various fields, such as geography and arts. In the age of participatory culture, such an approach in education has the potential to contribute to a more effective learning process, which has to be more thoroughly investigated.
>
> *(Gambarato and Dabagian 2016, 2)*

The third educational project classified as transmedia is Cosmic Voyager Enterprises. This project is presented as a simulation of a real-life scenario and is managed by the Conducttr platform, which was created by producer Robert Pratten (2012). Launched in 2012 in Florida in the United States, the project has involved more than 600 students aged 12–17 years.

In this project, Cosmic Voyage Enterprises is a fictional space cargo company. One of the company's rockets crashes into a small town, and the participating students are called upon to formulate a response to the economic and environmental problems that unfold. Various stakeholders are represented on social media, and the Conducttr platform allows the educators behind the project to deliver emails from citizens to unions, families, the mayor, and the environmental protection agency. Each stakeholder offers a unique perspective on the event, and matters are further complicated by business and ethical dilemmas.

During the testing of this project, students across Florida played the game for three weeks, forming teams of five and choosing a role within the company or as legal counsel. "They were then each sent different, unique information pertinent to their role which hence required them to collaborate and share what they knew and decide how the company should respond" (Conducttr 2012).

As a technical platform, Conducttr hosts the content and provides the interactive tools that represent the various stakeholders. Twitter and Facebook posts are responded to automatically; additionally, via email, students can send requests for additional information and receive a reply based on the prior choices made by their team. Developed for such transmedia projects, "Conducttr manages the whole student and teacher registration process and provides the client with detailed metrics" (Conducttr 2012).

According to Gambarato and Dabagian (2016), the transmediality involved in this project contributes to a more realistic experience, engages and challenges the students, and connects them emotionally to the storyworld. The project also teaches students to cooperate during problem-solving activities and offers them a more interesting and meaningful experience.

Transmedia in Higher Education

With regards to older students, Kalogeras (2014) introduces Transmedia Storytelling Edutainment (TmSE) as a model "for developing entertainment media franchises that can be incorporated into pedagogical practice" in higher education, particularly emphasizing the need to integrate images and story into education. To test this concept, Kalogeras wrote a screenplay called *The Goddess Within*, converted it to hypertext, and piloted an online education module using story-based materials. The entire experience is outlined in the book *Transmedia Storytelling and the New Era of Media Convergence in Higher Education*.

Based on this experience, the author suggests "Ten Mandates of Transmedia Edutainment in Higher Education" (Kalogeras 2014): (1) instruct/moderate with passion; (2) incorporate storytelling and have students learn by developing their own stories; (3) aggregate current and interesting content, such as short instructional videos; (4) incorporate multimedia and hypertext; (5) develop personalized creative materials using elements of edutainment; (6) allow room for story-based, student-centered curriculum; (7) define material based on the needs of the learner; (8) design conversations and devise topics for discussion and collaboration; (9) develop coursework with flexibility and provide cases with real-life situations; and (10) design creative assessments that center on different literacies and that accurately measure learning outcomes.

Although this list can be considered somewhat general and familiar to scholars of digital pedagogy as well as to those who work in the digital humanities (Morrissey 2015), Kalogeras emphasizes the need for additional practice with regard to transmedia storytelling edutainment in order to develop a critical evaluation of the processes of learning and teaching.

Tombleson et al. (2016, 340) agree that although the digital media revolution is now more than ten years underway, "a best practice framework to teach transmedia platforms is yet to be developed," as there is little empirical research and few conceptual guidelines "capturing the challenges and opportunities for pedagogy presented by the continuously changing media landscape." The authors thus investigate best practices with regard to the teaching of social media and transmedia, focusing in particular on the communication discipline in higher education. Hence, Tombleson et al. (2016) write that:

> Students may enter learning platforms at any point in time without losing the ability to make sense of learning materials. Information is not replicated on platforms; instead, students are encouraged to engage deeper across the multiple touch points. Using a transmedia approach fosters both a participatory culture, where students see themselves as actively contributing to the curriculum content, and also peer collaboration, as they become part of an active online community of learners.
>
> *(342)*

They therefore proposed a longitudinal action research project to investigate students' engagement with new technologies as well as their learning experiences in a transmedia-focused environment; this study was conducted at Curtin University in Australia and involved students who were pursuing a Public Relations major.

As part of the learning experience, students were encouraged to engage with platforms such as Blackboard, Snapchat, and Twitter, as well as with blogs and online videos. The transmedia material was tailored to each platform, and different styles of engagement were expected from the various students.

The findings ultimately support other studies' (e.g., Flanigan and Babchuk 2015) conclusions that "digital natives" do not know as much about technology as might be commonly assumed. In this study, participating students were not familiar with all available social media platforms and felt overwhelmed by the sheer volume of communication channels and their contained information. "Findings indicate that higher education staff need to dedicate more time to expose learners to (emerging) digital technologies; not only as a means of increasing engagement with students, but also to increase overall digital literacy skills" (Tombleson et al. 2016, 348).

Transmedia Education as a Tool for Citizen Activism

To address inequality in participation opportunities, the McArthur Foundation—with Jenkins at the head (2009)—proposes that schools participate in the development of three cultural competences that serve to confront the challenges of "participatory culture." First, young people should have equal opportunities to actively participate in the contemporary world. Second, they should be able to recognize the ways in which the media influences their perception of the world, thereby developing a critical understanding of the media's lack of transparency. The third competency concerns the prosumer's ethics—young people must have perspective and ethical competence upon becoming content producers and participants on the Internet, as they must be responsible for their messages, their speech, and their influence within the community.

One project that was developed as a new form of literacy education outside the classroom is the HP Alliance. This project adopts the unconventional approach of "making activism accessible through the power of story" (The HP Alliance 2015) by utilizing J. K. Rowling's best-selling *Harry Potter* fantasy novels as a platform for political transformation, thus linking traditional activist groups with new-style social networks and with fan communities. Since 2005, the project has engaged millions of fans via their work in support of equality, human rights, and literacy.

Another example of such an initiative is East Los High (2013–2016), a transmedia edutainment project aimed at young Latino-Americans. Nominated for five Emmy awards, the Hulu original series revolves around the lives of a group of Latino teens who are navigating their final years at a fictional high school in East Los Angeles. Educational messages are embedded in the show's entertainment narratives across various digital platforms in an effort to promote sexual and reproductive health awareness.

As a transmedia education-oriented project, East Los High offers multiple entry points to Latino consumers of dramas and digital entertainment, thus engaging them in the narrative through their preferred platforms (Wang and Singhal 2016).

> [The] transmedia approach was highly strategic because Latinas/os are 40% more likely than is the general population to watch television and videos online or on a smartphone and 3 times more likely to check, via social media, which programs their friends are watching. East Los High capitalized on the digital usage patterns of Latina/o youths to engage them about sexual and reproductive health issues.
>
> *(Wang and Singhal 2016, 1003)*

Analyzing this educational series, Wang and Singhal (2016) conclude that East Los High drew in a wide audience and achieved "strong viewer engagement, and a positive cognitive, emotional, and social impact on sexual and reproductive health communication and education" (1002).

One key aspect of East Los High is the project's sensitive transmedia educational strategy, which functioned in concert with its audience by connecting with the media consumption habits of Latino-Americans and respecting their culture.

Conclusion

During our participation in transmedia educational health projects in low-income communities in East Timor and Brazil, we discovered several problems related to digital connection and the habits of media users. In both communities, because Internet access is still very expensive, we opted to utilize analogue games, theater, and recyclable materials that students could easily find and take home with them rather than promote the use of digital media to engage students, teachers, and their families (Labcon 2017).

In East Timor, although online platforms were used to overcome physical distances, it was still possible to form collaborative networks that resulted in the establishment of connections between people. Approximately 250 students from three universities were involved in this project, which involved various fields of knowledge, including journalism, advertising, graphic design, history, pedagogy, and literature.

The Brazilian project found entrepreneurship to be the communicative bond that was necessary to bring families closer to schools. Through offline actions, a face-to-face connection was proposed as an instrument for problem solving and network expansion. Offline connections reverberated online and allowed re-signification of digital tools, like smartphones, which were used mainly by students for entertainment.

Above all, educational transmedia facilitates human interaction and helps students to participate and create their own learning processes by generating stories, games, and relevant activities. Developing a narrative over multiple platforms, either analogue or digital, while interweaving learning outcomes creates transformational learning experiences. Through the application of transmedia storytelling, students shape their ideas into well-structured narratives and complex storyworlds. Moreover, transmedia provides a platform for students to learn how to identify, understand, and engage different audiences in their stories as well as how to create and link a story across different platforms. As Jenkins (2009) posits, "the best transmedia storytelling serves four key functions. It extends the timeline, maps the world, explores secondary characters, and engages the audience."

In conclusion, offline extensions in transmedia educational projects are extremely important because they have the potential to enhance the learning process regardless of their technology advancement or whether they have all of the perks of digital environments available within their community.

References

Bass, Randall. 2012. "Disrupting Ourselves: The Problem of Learning in Higher Education. Why IT Matters to Higher Education." *Educause Review*. Accessed June 20, 2017. http://er.educause.edu/articles/2012/3/disrupting-ourselves-the-problem-of-learning-in-higher-education.

Conductrr. 2012. "Educational Simulation that Teaches Ethics and Financial Responsibility." Accessed June 2, 2017. www.conducttr.com/success-stories/cosmic-voyage-enterprises/.

Cope, Bill, and Mary Kalantzis. 2009. "Multiliteracies: New Literacies, New Learning". *Pedagogies* 4 (3): 164–195.

EC. 2014. "Communicating Horizon 2020 Projects." *European Commission*. Accessed July 14, 2017. https://ec.europa.eu/easme/sites/easme-site/files/documents/6.Communication-AlexandraRuete.pdf.

ED Innovation. 2011. "Why Use Transmedia in Early Learning?" *Department of Education USA*, Office of Education. Accessed July 3, 2017. https://innovation.ed.gov/2011/04/21/why-use-transmedia-in-early-learning/.

ED Innovation. 2016. "'What does a Cartoon Cat have to do with Learning Math?' New Reports Highlight the Impact of Ready to Learn Television (2010–2015)." *Department of Education USA*, Office of Education. Accessed July 4, 2017. https://innovation.ed.gov/2016/04/13/what-does-a-cartoon-cat-have-to-do-with-learning-math-new-reports-highlight-the-impact-of-ready-to-learn-television-2010–2015/

Ferrés, Joan, and Alejandro Piscitelli. 2012. "La Competencia Mediática: Propuesta Articulada de Dimensiones e Indicadores" [Media Competition: Articulated Proposal of Dimensions and Indicators]. *Comunicar* 38 (XIX): 75–82.

Flanigan, Abraham, and Wayne Babchuk. 2015. "Social Media as Academic Quicksand: A Phenomenological Study of Student Experiences in and out of the Classroom." *Learning and Individual Differences* 44: 40–45.

Fleming, Laura. 2013. "Expanding Learning Opportunities with Transmedia Practices: Inanimate Alice as an Exemplar." *Journal of Media Literacy Education* 5 (2): 370–377.

Frau-Meigs, Divina. 2012. "Transliteracy as the New Research Horizon for Media and Information Literacy." *Medijske Studije* 3 (6): 14–26.

Gambarato, Renira, and Lilit Dabagian. 2016. "Transmedia Dynamics in Education: The Case of Robot Heart Stories." *Educational Media International* 53 (4): 229–243.

Hovious, Amanda. 2013. "The 7 Literacies of Transmedia Storytelling." *Designer Librarian*, November 21. Accessed June 2, 2017. https://designerlibrarian.wordpress.com/2013/11/21/the-7-literacies-of-transmedia-storytelling.

The HP Alliance. 2015. "What We Do." Accessed June 13, 2017. www.thehpalliance.org/what_we_do.

Jenkins, Henry. 2009. "The Revenge of the Origami Unicorn: Seven Principles of Transmedia Storytelling." *Confessions of an Aca-Fan*, December 12. Accessed February 20, 2017. http://henryjenkins.org/2009/12/the_revenge_of_the_origami_uni.html.

Jenkins, Henry. 2010. "Transmedia Education: The 7 Principles Revisited." *Confessions of an Aca-Fan: The Official Weblog of Henry Jenkins*. June 21.

Jenkins, Henry. 2012. "On Transmedia and Education: A Conversation with Robot Heart Stories' Jen Begeal and Inanimate Alice's Laura Fleming (Part One)." *Confessions of an Aca-Fan Blog*, January 27. Accessed July 13, 2017. http://henryjenkins.org/blog/2012/01/on_transmedia_and_education.html?rq=Transmedia%20Education.

Jenkins, Henry, Ravi Purushotma, Margaret Weigel, Katie Clinton, and Alice Robison. 2009. *Confronting the Challenges of Participatory Culture: Media Education for the 21st Century*. Chicago: The MacArthur Foundation.

Jewitt, Carey, and Gunther Kress (eds.). 2003. *Multimodal Literacy*. New York: Peter Lang.

Kalogeras, Stavroula. 2014. *Transmedia Storytelling and the New Era of Media Convergence in Higher Education*. New York: Palgrave Macmillan.

Labcon. 2017. *Projeto de Cooperação Internacional* [International Cooperation Project]. Accessed June 20, 2017. http://labcon.fafich.ufmg.br/projeto-de-cooperacao-internacional/.

Lamb, Annette. 2011. "Reading Redefined for a Transmedia Universe." *Learning & Leading with Technology* 39 (3): 12–17.

Loertscher, David, and Blanche Woolls. 2014. "Transmedia Storytelling as an Education Tool." Paper presented at the annual meeting for IFLA World Library and Information Congress, August 16–22, Lyon, France.

Lugo, Nohemi. 2016. "Diseño de Narrativas Transmedia para la Transalfabetización" [Transmedia Narrative Design for Transiteracy]. Ph.D. diss., Pompeu Fabra University.

Morrissey, Katherine. 2015. "Transmedia Storytelling and the New Era of Media Convergence in Higher Education." Review of the Transmedia Storytelling and the New Era of Media Convergence in Higher Education by Stavroula Kalogeras. *Transformative Works and Cultures* 20.

Scolari, Carlos A. 2016. "Transmedia Literacy." Conference paper at *Communication and Education by Transmedia (CET Congress)*, Girona, Spain, April 22.

Soares, Ismar. 2011. *Educomunicação: O Conceito, o Profissional, a Aplicação* [Educommunication: The Concept, the Professional, the Application]. São Paulo: Paulus.

Thomas, Sue, Chris Joseph, Jess Laccetti, Bruce Mason, Simon Mills, Simon Perril, and Kate Pullinger. 2007. "Transliteracy: Crossing Divides." *First Monday* 12 (12). Accessed July 14, 2017. http://journals.uic.edu/ojs/index.php/fm/article/view/2060/1908.

Tombleson, Bridget, Katharina Wolf, Lydia Gallant, Catherine Archer, and Renae Desai. 2016. "Teaching Transmedia to Millennials: A Critical Reflection on the Embedding of Transmedia Skills in the Communication Curriculum." *Research and Development in Higher Education: The Shape of Higher Education* 39: 340–350.

Transmedia Lab. 2011. "Inanimate Alice: A Cross-platform Educational Project." Accessed June 20, 2017. www.transmedialab.org/en/the-blog-en/case-study-en/inanimate-alice-a-cross-platform-educational-project/.

Transmedia Literacy. 2017. "Exploiting Transmedia Skills and Informal Learning Strategies to Improve Formal Education." Accessed July 3, 2017. https://transmedialiteracy.org/.

UNESCO. 2009. "Media and Information Literacy." Accessed May 23, 2017. www.unesco.org/new/en/communication-and-information/media-development/media-literacy/mil-as-composite-concept/.

Wang, Hua, and Arvind Singhal. 2016. "East Los High: Transmedia Edutainment to Promote the Sexual and Reproductive Health of Young Latina/o Americans." *American Journal of Public Health* 106 (6): 1002–1010.

Warren, Scott, Jenny Wakefield, and Leila Mills. 2015. "Learning and Teaching as Communicative Actions: Transmedia Storytelling." In *Increasing Student Engagement and Retention using Multimedia Technologies: Video Annotation, Multimedia Applications, Videoconferencing and Transmedia Storytelling*, edited by Laura Wankel and Patrick Blessinger, 67–94. Bingley: Emerald Group Publishing Limited.

Wartella, Ellen, Alexis Luricella, and Courtney Blackwell. 2016. *The Ready to Learn Program: 2010–2015 Policy Brief.* Evanston: Northwestern University School of Communication Center on Media and Human Development. Accessed July 9, 2017. http://cmhd.northwestern.edu/wp-content/uploads/2016/04/RTL-Policy-Brief-2010-2015-Wartella-et-al-FINAL-March-2016.pdf.

Zagal, José. 2010. "Ludoliteracy: Defining Understanding and Supporting Games Education." Pittsburgh: ETC Press. Accessed July 29, 2017. http://press.etc.cmu.edu/content/ludoliteracy-defining-understanding-and-supporting-games-education.

35

TRANSMEDIA LITERACY

Rethinking Media Literacy in the New Media Ecology

Carlos A. Scolari

The vast diffusion of digital technologies and new social practices around them has led to the emergence of new concepts in the academic and professional conversations about media literacy. In the last two decades the semantic galaxy around "literacy" has expanded, from "digital literacy" to "new media literacies" or "multimedia literacy." Although each new concept has its own specificities, they all deal with a new set of interactive contents, production skills, and techno-social practices that have resulted from the emergence of the World Wide Web. Most of the concepts focus on how to do things with (new) media at school.

As Meyers, Erickson, and Small (2013) put it, it is necessary to make a "critical turn to the examination of digital literacies, de-emphasizing skills and refocusing attention on diverse contexts of use, and the emergent modes of assessment that are bound by specific circumstances and communities of practice" (360). Livingstone (2004) proposed opening the intervention of media literacy; for example, by analyzing "how the internet mediates the representation of knowledge, the framing of entertainment and the conduct of communication." In tandem with this analysis, research

> must investigate the emerging skills and practices of new media users as the meaningful appropriation of ICT [Information and Communications Technology] into their daily lives ... A top-down definition of media literacy, developed from print and audio-visual media, while a useful initial guide, should not pre-empt learning from users themselves.
>
> *(Livingstone 2004, 11)*

In 2006, Buckingham asked "What do young people need to know about digital media?" In this chapter, however, it is another question that orientates my reflections: What are young people *doing* with digital media? The main objective of this chapter is to propose an alternative and complementary conception to "(new) media literacies" based on informal learning environments (Sefton-Green 2013), bottom-up processes (Livingstone 2004), transmedia storytelling (Jenkins 2003, 2006) and participatory cultures (Jenkins et al. 2006; Jenkins, Ito, and boyd 2016). In this context, the chapter proposes the concept of "transmedia literacy" (TML) to deal with these new practices and processes that have emerged from the new media ecology. The main objective of the chapter is not to propose another fashionable concept to define the "new literacies" but to rethink the same idea of "media literacy" research and practice in the context of the twenty-first-century media ecology.

The New Media Ecology

Since the diffusion of personal computing in the 1980s and the expansion of the web in the 1990s, digital technology has been a catalyst for social change in contemporary societies. From economy to politics, from education to culture, practically all aspects of human life have been transformed due to the different ways of developing and using ICT (Benkler 2006; Rainie and Wellman 2012). The media ecology has mutated from the traditional broadcasting system to a new media ecology, where the old "media species" (radio, cinema, television, books, etc.) must compete with the new ones (YouTube, Twitter, Facebook, mobile devices, etc.) and adapt and change in order to survive (Scolari 2012, 2013).

Even if schools have made great efforts to adapt to the new socio-technical conditions in the past two decades, the general perception is that the social life of children, pre-teens, and teens is built up around a set of digital technologies—from social media to mobile devices—and new practices, which are frequently very different from the educational protocols of schools. According to Castells (2007), there is still a cultural and technological gap between today's youth and a school system that has not evolved along with society and the digital environment: "the idea that today a young person must load a backpack of boring text books set by ministerial bureaucrats, and must remain closed in a classroom to support an irrelevant speech in the name of his/her future, is simply absurd" (Castells 2007, 25).

The transformations in the media ecology include many different phenomena. This chapter focuses on three strongly connected key processes for the redefinition of media literacy: transmedia storytelling, participatory cultures, and informal learning strategies.

Transmedia Storytelling and Participatory Cultures

Together with the expansion of characters and narrative situations through multiple media the creation of user-generated content by fans is the second key element of transmedia storytelling (Jenkins 2003, 2006; Scolari 2009). Jenkins reinforces this vision when he considers that a transmedia franchise relies on the fans to

> Bring the "dispersed pieces together again … The more dispersed they are, the more fans work to assemble them into a meaningful whole." Each piece of information added to the canon opens up new gaps, suggests new kernels, from which new lines of fan speculation and expression can spring.
>
> *(Jenkins 2013, 54)*

Teens are actively involved in these transmedia cultures in which the members "believe their contributions matter, and feel some degree of social connection with one another (at the least they care what other people think about what they have created)" (Jenkins et al. 2006, 3). Jenkins and his team identified different forms of participatory culture, from affiliation (i.e., participation in online communities and social media) to expression (i.e., production of fanfiction, mashups, etc.), collaborative problem-solving (i.e., team organization in video games), and circulation (i.e., shaping the flow of media, such as podcasting, blogging).

Informal Learning Strategies

More broadly, the diffusion of the web and the massive spread of digital technologies in the 1990s repositioned the debates on formal and informal learning processes. As early as 1994, Lemke drew attention to the increasing conflict between two views of education: one based on the traditional curricular model, and the other based on the norms of free access to information. Lemke warned the learning community about the necessity of "articulating this conflict and responding to its implications" (1994, 1). Nowadays informal learning strategies occupy a central position in the everyday life of the new generations and should be put at the center of any new conception of media literacy. The "high-grade interface for multimedia access" pictured by Lemke in the early 1990s, today is called Google/YouTube.

Over recent years, in turn, scholars like Sefton-Green (2006, 2013) have expanded the understanding of informal learning strategies and new educational practices out of the schools. Informal learning is not a new set of skills or a static pedagogical field: it is a bottom-up, dynamic, complex phenomenon that emerges outside of traditional learning institutions. In 2006, research carried out by the Future Lab "mapped out the different approaches to understanding how young people may be learning with ICTs in a range of settings outside the school—especially in contexts not traditionally associated with education" (Sefton-Green 2006, 5). Sefton-Green concluded that "young people's use and interaction with ICTs outside of formal education is a complex 'educational' experience" (2006, 30). According to Ito, in participatory out-of-school environments "learning is a side effect of creative production, collaboration, and community organizing, not the explicit purpose of the activity" (Ito 2016, 93).

It is not easy to analyze these informal learning practices, however. These "wild" informal experiences are often invisible or directly rejected by (adult) researchers:

> So much is projected onto youth that it is often difficult to discuss what they are doing, and why, without observation being obscured by ideas of what they *should* or *shouldn't* be doing. Youth are rarely seen as deserving any agency and, yet, they are also judged based on what they choose to do … people think that they know something about youth either because they were once young or because they are parents to a young person.
>
> *(boyd 2016, 34)*

Transmedia Literacy

Following this brief but necessary description of the state of the new media ecology, the reflection on media literacy can now be understood from a new perspective. Without discarding the identification of new "digital" skills, any new proposal around media literacy should go beyond protective approaches (media literacy as a inoculation of antibodies, as in Potter 2010), putting participatory cultures and bottom-up informal learning processes at the center of its research and intervention program. In this context, instead of asking how do media effects impact youth, a new media literacy approach could start by asking: What are young people doing with digital media?

The changes in the media ecology have altogether unlocked the discussion about the pertinence of traditional definitions of media literacy and the emergence of new literacies. Is it still possible to talk about "media literacy" in a context where the broadcasting (one-to-many) model is being displaced by the network (many-to-many) communication paradigm? As early as 1993, researchers like Buckingham upheld the "need for a new definition of literacy … not tied to particular technologies or practices" but rather one that "allows us to look at the competencies that are developed across the whole range of culture and communication" (1993, 20). Lankshear and Knobel (2007) described this transition from old to new literacies in the following terms:

> The more a literacy practice privileges participation over publishing, distributed expertise over centralized expertise, collective intelligence over individual possessive intelligence, collaboration over individuated authorship, dispersion over scarcity, sharing over ownership, experimentation over "normalization", innovation and evolution over stability and fixity, creative-innovative rule breaking over generic purity and policing, relationship over information broadcast, and so on, the more we should regard it as a "new" literacy.
>
> *(Lankshear and Knobel 2007, 21)*

Toward a Transmedia Literacy

Applying these notions of participation and collective intelligence in an educational arena, it is plain to see that transmedia storytelling practices are challenging researchers and educators alike: media literacy can no longer be limited to the critical analysis of media contents or the acquisition of skills inside the formal education system. The traditional media consumer is now a *prosumer* (a concept introduced by Toffler 1980) or a *participatory creator* (Meyers, Erickson, and Small 2013), an active subject who creates new content and shares it in and across digital networks. In turn, researchers of new media literacy have identified a set of competencies defined as "prosuming skills," which include the skills necessary to produce/create media contents, from the ability to set up an online communicative account to using software to generate digital contents and programming. These skills often work together with distribution, remixing, and participation skills (Lin et al. 2013). It is in this context that the concept of "transmedia literacy" can enrich the concept of traditional media literacies and reposition the theoretical approaches to the new literacies.

At this point, it is useful to return to the participatory cultures described by Jenkins and many other researchers. According to the Pew Research Center, 92 percent of American teens report going online daily, including 24 percent who say they are online "almost constantly"; social media are at the center of this online activity: Facebook is the "most popular and frequently used social media platform among teens; half of teens use Instagram, and nearly as many use Snapchat" (Lenhart 2015). The same report confirms that teens are diversifying their social media site use. A majority of teens (71 percent), for example, report using more than one social media site out of the seven platform options they were asked about. In Europe, too, the situation is comparable:

> The most common online social activities for young people in the EU-28 in 2014 included sending and receiving e-mails (86%) and participating on social networking sites – for example, Facebook or Twitter, by creating a user profile, posting messages or making other contributions – (82%), while close to half (47%) of all young people in the EU-28 uploaded self-created content, such as photos, videos or text to the Internet.
>
> *(Eurostat 2015, 201)*

In this context, TML can be understood as a set of skills, practices, values, priorities, sensibilities, and learning/sharing strategies developed and applied in the context of the new participatory cultures. If traditional literacy was book-centered or, in the case of media literacy, mostly television-centered, then multimodal literacy places digital networks and interactive media experiences at the center of its analytical and practical experience.

This practical experience differs to more traditional forms of literacy, which generally treated the subject as illiterate, while media literacy focused on the consumer as a passive spectator. TML, however, considers the subject to be a prosumer. Another important element of TML is the learning space. The institutional learning environment for traditional forms of literacy is the school, but new generations are now developing their TML skills outside the school (from YouTube to online forums, social media, and blogs). These informal learning spaces will be a key component of TML research.

In traditional literacy the teacher plays the role of a knowledge authority and, at the same time, a mediator between the text and the student. In this context the teacher manages the reading process—as a part of the learning process—and verifies that the text is interpreted correctly. In media literacy, the teacher's role is extended to the inoculation of critical antibodies through the de-automation of the media reception process. In TML the teacher is a knowledge facilitator, an actor who involves learners in a collaborative learning process. In this context, the teacher is a flexible, decentralized actor who promotes bottom-up learning. This role includes another set of activities that could be defined as cultural translation: the teacher as an interface between the educational institution (the classroom, the school) and the external media ecology where the students live and create. Finally, if traditional

Table 35.1 Literacy, media literacy, and transmedia literacy

	Literacy	*Medialiteracy*	*Transmedialiteracy*
Media semiotics (language)	Verbal text (read/write)	Multimodal (audiovisual media)	Multimodal (interactive media and transmedia)
Media supports	Books and printed texts	Broadcasting	Digital networks
Aim of the action	To develop critical readers and writers	To develop critical viewers and, sometimes, critical producers	To develop critical prosumers
Subject interpellation	As an illiterate	As a passive spectator	As a prosumer
Direction of the action	Top-down	Top-down	Bottom-up
Learning environment	Formal (schools)	Formal (schools)	Informal (outside schools)
Role of the teacher	Knowledge authority – Mediator student/text	Knowledge authority – Inoculator of critical antibodies	Knowledge facilitator – Cultural translator
Theoretical references	Linguistics	Media Studies (Theory of media effects)	Cultural Studies / Media Ecology

literacy was inspired by linguistics, and media literacy was strongly anchored in a theory of media effects, TML looks to cultural studies and media ecology as privileged theoretical frameworks.

Table 35.1 summarizes these three conceptions of "literacy": the original conception focused on reading and writing; the second conception, based on the (negative) effects of mass media; and the third conception, inspired by the mutations of the new media ecology and the emergence of transmedia practices.

Transmedia Skills and Informal Learning Strategies

TML therefore focuses on the ever-evolving media practices used by young people. Previous research in this field (such as Jenkins et al. 2006) has identified numerous skills that could be considered as basic competencies of TML, from playing (capacity to experiment with one's surroundings as a form of problem-solving) to performing (ability to adopt alternative identities for the purpose of improvisation and discovery), appropriating (ability to meaningfully sample and remix media content), judging (ability to evaluate the reliability and credibility of different information sources), transmedia navigating (ability to follow the flow of stories and information across multiple modalities), networking (ability to search for, synthesize, and disseminate information), and negotiating (ability to travel across diverse communities, discerning and respecting multiple perspectives, and grasping and following different norms). A first approach to TML, then, should focus on at least three sets of media practices:

Video Game Literacy

In their analysis of young people's digital life in the United States, Ito et al. (2010) identified different kinds of gaming practices inside the ecology of video games. One of the most important outcomes of their research is an understanding of the ways in which young people develop social networks of technical expertise:

> The game has not directly and explicitly taught them technical skills, but game play has embedded young people in a set of practices and a cultural ecology that places a premium

> on technical acumen. This in turn is often tied to an identity as a technical expert that can serve a gamer in domains well beyond specific engagements with games. This is the kind of description of learning and "transfer" that a more ecological approach to gaming suggests.
> *(Ito et al. 2010, 200)*

Many video games involve social practices, from online playing in MMORPGs to player-generated contents like machinima and video-based game walkthroughs (Ito et al. 2010). Scholars consider that video-game players develop and apply different skills from the ability to read; they learn to make decisions and act within a dynamically changing environment (Wagner 2006).

Web/Social Media Literacy

According to Hartley (2009) kids develop an "experimental engagement" with peer-groups and places in the context of both DIY (Do It Yourself) and DIWO (Do It With Others) creative cultures "without the need for institutional filtering or control bureaucracies" (Hartley 2009, 130).

New generations are building their social relationships in digital environments like Facebook or Snapchat, and use the World Wide Web as a huge library for any kind of knowledge necessity (Rainie and Wellman 2012; boyd 2014). Competencies like web navigation, information gathering, taking and sharing pictures, managing different levels of communication, constructing an identity in a virtual environment, watching a web series, or managing the privacy and their personal identity in online platforms are basic skills necessary for navigating the digital environment.

In this area, it is useful to mention the connections between the professional ways of self-learning and young people. Experts in information technology often emphasize that they picked up their skills outside of formal training and instruction. Many members of these technical communities learned how to manipulate code on their own. According to Lange and Ito, young people often reflected these values by describing "how they were largely self-taught, even though they might also describe the help they received from online and offline resources, peers, parents, and even teachers" (Lange and Ito 2010, 262).

Participatory Cultures Literacy

Researchers such as Jenkins et al. (2006) and Gee (2004) have argued that new participatory cultures represent ideal informal learning environments. People can participate in various ways according to their skills and interests because they depend on peer-to-peer teaching with each participant constantly motivated to acquire new knowledge or refine their existing skills. Finally, they allow each participant to feel like an expert while tapping the expertise of others (Jenkins et al. 2006, 9).

As media practices are evolving at such a dizzy speed, TML cannot be reduced to a list of transmedia skills: it also involves many other aspects, from new values like sharing or openness to a complex set of processes and practices around participatory cultures. Further, any literacy should not be limited to cognitive (knowing) and pragmatic (doing) dimensions: the emotional dimension of the subject is also a key element of any literacy acquisition or application process (Ferrés 2014).

Beyond the identification of transmedia skills, the understanding of informal learning strategies is the second key element of TML. According to Jenkins and his team many transmedia skills (playing, performing, navigating, etc.) are being acquired by young people through their participation in the informal learning communities that surround popular culture. Although some teachers and after-school programs are incorporating some of these skills into their teaching and activities, the integration of these important social skills and cultural competencies "remains haphazard at best. Media education is taking place for some youth across a variety of contexts, but it is not a central part of the educational experience of all students" (Jenkins et al. 2006, 56–57). As has been suggested, the new generations are developing and applying informal learning strategies that have been almost ignored by educators and media researchers up until recently. Scholars such as Sefton-Green (2006, 2013) have outlined important research lines in this field that need to be further explored and expanded.

Conclusion

Looking forward, if TML follows the path of traditional media literacy, then it should develop a research agenda and, at the same time, an action program.

In terms of a future research agenda, it is worth remembering that traditional research on media literacy was mostly based on analyzing media effects and reflecting on how to teach critical approaches to media in educational institutions. TML expands and complements this conception by opening up a new territory: the study of transmedia skills developed by children and teens in informal learning environments. The sets of literacies (video game literacy, web and social media literacy, participatory cultures literacy) described earlier in the chapter should be considered as an initial map of a mostly unexplored territory. TML proposes a new set of research questions to researchers: How are teens learning things outside of school? How do teens get started in practices such as video production and editing, web-comics or machinima? How do they learn the basics of audiovisual language? How do they improve their skills? What role do online communities play in this informal learning process? What kind of creative communities and collaborations do youth engage in through the process of producing new media? What kinds of informal learning strategies are they applying? What sharing strategies do online communities put into practice? Research into new transmedia skills has been intensive in the last decade (Jenkins et al. 2006; Ito et al. 2010; boyd 2014), but this kind of study must not stop: the mutations of the media ecology are so fast that any research about teenagers' use of social media, applications, or video games risks becoming outdated within a matter of months. However, the research into teenagers' informal learning strategies is still an unmapped territory that researchers need to explore and define.

With regards to developing a future action program, moreover, the main objective of TML is to improve the life of individuals, as it always was with media literacy (Potter 2010). To do so, TML should aim to create bridges between the new participatory cultures and educational institutions, facilitating the exchange of experiences. In other words, the specific intervention of TML should go beyond academic research and propose ways of exploiting these new skills and learning strategies *inside* formal education institutions. In this context, TML is very close to the spirit of Ito's *connected learning*. According to Ito et al., connected learning can be recognized when "a young person is able to pursue a personal interest or passion with the support of friends and caring adults, and is in turn able to link this learning and interest to academic achievement, career success or civic engagement" (Ito et al. 2013, 4).

Within this context, furthermore, TML posits that educators will play a new non-exclusive role, positioning the teacher as a "cultural translator" who facilitates the recovery and use of transmedia skills (developed in informal learning environments) inside the school. If with conventional media literacy the teacher was considered a knowledge authority and inoculator of critical antibodies, then with TML the educator operates as a knowledge facilitator and translator (or interface) between formal and informal learning environments. Once the informal learning strategies and transmedia skills developed by young people outside of the formal institutions are identified, then TML should work to translate them into a series of didactic activities and proposals that can be implemented inside school settings. This kind of translational activity could ultimately work to materialize Freire's (1970) old dream of constructing a learning environment based on dialogue and people working with each other (rather than one person informing another). In other words, TML does not contradict nor negate traditional media literacy, but instead proposes a complementary approach that moves from formal learning (based on teaching media skills in the school) to informal learning (based on understand how new generations are doing things with media outside of the school). Last but not least, TML may also be helpful for reducing the cultural and technological gaps identified by Castells between today's youth and the school system.

Acknowledgment

This chapter was elaborated with funding from the European Union's Horizon 2020 research and innovation program under the Grant Agreement n. 645238 (TRANSLITERACY research project 2015–18 – URL: https://transmedialiteracy.org/).

References

Benkler, Yochai. 2006. *The Wealth of Networks: How Social Production Transforms Markets and Freedom*. New Haven: Yale University Press.

boyd, danah. 2014. *It's Complicated: The Social Lives of Networked Teens*. New Haven: Yale University Press.

boyd, danah. 2016. "Youth Culture. Youth Practices." In *Participatory Culture in a Networked Era*, edited by Henry Jenkins, Mizuko Ito, and danah boyd, 32–35. Cambridge: Polity.

Buckingham, David. 1993. "Towards New Literacies, Information Technology." *The English and Media Magazine*, Summer, 20–25.

Buckingham, David. 2006. "Defining Digital Literacy. What Do Young People Need to Know about Digital Media?" *Digital Komptanse* 4 (1): 263–276.

Castells, Manuel. 2007. "Estudiar, ¿para qué?" [To Study, for What?]. *La Vanguardia*, November 24. Accessed March 14, 2017. http://egym.bligoo.com/content/view/134411/Manuel-Castells-estudiar-para-que.html#.WiFUCkqWZPY.

Eurostat. 2015. *Being Young in Europe Today*. Luxembourg: European Union. Accessed March 10, 2017. http://ec.europa.eu/eurostat/documents/3217494/6776245/KS-05-14-031-EN-N.pdf/18bee6f0-c181-457d-ba82-d77b314456b9.

Ferrés, Joan. 2014. *Las Pantallas y el Cerebro Emocional* [Screens and the Emotional Brain]. Barcelona: Gedisa.

Freire, Paulo. 1970. *Pedagogy of the Oppressed*. New York: Herder and Herder.

Gee, James Paul. 2004. *Situated Language and Learning: A Critique of Traditional Schooling*. New York: Routledge.

Hartley, John. 2009. "Uses of YouTube: Digital Literacy and the Growth of Knowledge." In *YouTube. Online Video and Participatory Culture*, edited by Jean Burgess and Joshua Green, 126–143. Cambridge: Polity.

Ito, Mizuko. 2016. "Learning and Literacy." In *Participatory Culture in a Networked Era*, edited by Henry Jenkins, Mizuko Ito, and danah boyd, 90–94. Cambridge: Polity.

Ito, Mizuko, Sonja Baumer, Matteo Bittanti, and danah boyd. 2010. *Hanging Out, Messing Around, and Geeking Out: Kids Living and Learning with New Media*. Cambridge, MA: MIT Press.

Jenkins, Henry. 2003. "Transmedia Storytelling." *MIT Technology Review*, January 15. Accessed March 9, 2017. www.technologyreview.com/s/401760/transmedia-storytelling/

Jenkins, Henry. 2006. *Convergence Culture: Where Old and New Media Collide*. New York: New York University Press.

Jenkins, Henry. 2013. *Spreadable Media: Creating Meaning and Value in a Networked Culture*. New York: New York University Press.

Jenkins, Henry, Katie Clinton, Ravi Purushotma, Alice J. Robison, and Margaret Weigel. 2006. *Confronting the Challenges of Participatory Culture: Media Education for the 21st Century*. Chicago: MacArthur Foundation.

Jenkins, Henry, Mizuko Ito, and danah boyd. 2016. *Participatory Culture in a Networked Era*. Cambridge: Polity.

Lange, Patricia, and Mizuko Ito. 2010. "Creative Production." In *Hanging Out, Messing Around, and Geeking Out: Kids Living and Learning with New Media*, edited by Mizuko Ito, Sonja Baumer, Matteo Bittanti, and danah boyd, 243–293. Cambridge, MA: MIT Press.

Lankshear, Colin, and Michele Knobel. 2007. "Sampling 'the New' in New Literacies." In *A New Literacies Samples*, edited by Michele Knobel and Colin Lankshear, 1–24. New York: Peter Lang.

Lemke, Jay. 1994. "The Coming Paradigm Wars in Education: Curriculum vs. Information Access." *Cyberspace Superhighways. Proceedings of the Fourth Conference on Computers, Freedom, and Privacy*, 76–85. Chicago: John Marshall Law School.

Lenhart, Amanda. 2015. *Teen, Social Media and Technology Overview 2015*. Washington, DC: Pew Research Center. Accessed March 12, 2017. www.pewinternet.org/2015/04/09/teens-social-media-technology-2015/.

Lin, Tzu-Bin, Jen-Yi Li, Feng Deng, and Ling Lee. 2013. "Understanding New Media Literacy: An Explorative Theoretical Framework." *Educational Technology & Society* 16 (4): 160–170.

Livingstone, Sonia. 2004. "Media Literacy and the Challenge of New Information and Communication Technologies." *Communication Review* 1 (7): 3–14.

Meyers, Eric, Ingrid Erickson, and Ruth Small. 2013. "Digital Literacy and Informal Learning Environments: An Introduction." *Learning, Media and Technology* 38 (4): 355–367. doi: 10.1080/17439884.2013.783597.

Potter, W. James. 2010. "The State of Media Literacy." *Journal of Broadcasting & Electronic Media* 54 (4): 675–696. doi: 10.1080/08838151.2011.521462.

Rainie, Lee, and Barry Wellman. 2012. *Networked: The New Social Operating System*. Cambridge, MA: MIT Press.

Scolari, Carlos A. 2009. "Transmedia Storytelling: Implicit Consumers, Narrative Worlds, and Branding in Contemporary Media Production." *International Journal of Communication* 3: 586–606. doi: 1932–8036/20090586.

Scolari, Carlos A. 2012. "Media Ecology: Exploring the Metaphor to Expand the Theory." *Communication Theory* 22 (2): 204–225. doi: 10.1111/j.1468-2885.2012.01404.x.

Scolari, Carlos A. 2013. "Media Evolution: Emergence, Dominance, Survival and Extinction in the Media Ecology." *International Journal of Communication* 7: 1418–1441. doi: 1932–8036/20130005.

Sefton-Green, Julian. 2006. *Report 7: Literature Review in Informal Learning with Technology Outside School*. London: Future Media Lab.

Sefton-Green, Julian. 2013. *Learning Not at School*. Cambridge, MA: MIT Press.

Toffler, Alvin. 1980. *The Third Wave: The Classic Study of Tomorrow*. New York: Bantam.

Wagner, Michael. 2006. "Computer Games and the Three Dimensions of Reading Literacy." Proceedings of the 2006 ACM SIGGRAPH Symposium on Videogames. Boston, July 29–30, 139–142.

36

TRANSMEDIA FOR SOCIAL CHANGE

Evolving Approaches to Activism and Representation

Donna Hancox

Since the late 1990s, digital technology has provided the means for stories to be created quickly and shared widely. In the suite of innovations in these new media technologies, transmedia storytelling represents arguably one of the most profound transformations in storytelling. One of the underlying philosophies of contemporary transmedia is a commitment to a de-centralized concept of authorship that does not privilege one voice, one part of the story, or one platform over another. A bricolage approach to transmedia has potential to contribute profoundly to social change projects, particularly those that use personal narratives. This chapter explores the ways in which transmedia storytelling can contribute to social change by both creating and sharing personal narratives, and leveraging the unique qualities of transmedia that allow for immersive representations of complex issues and environments.

Narrative representation and immersive experiences will be examined as two elements of transmedia that can contribute significantly to projects that are centered around agitating for social change. The transmedia project *Hollow* and arts, media and social change company BIGhART and fandoms associated with *Harry Potter* and the social media strategy by the leader of the protest at Standing Rock will be discussed to understand how transmedia narratives and immersive experiences can work toward social change through promoting understanding and empathy.

Activism and Transmedia

The avenues through which communities and community organizations raise awareness about the issues they face and how they agitate for change have developed rapidly in the past ten years; and digital technology has provided community activists with the means to quickly connect with the public and share important information. The practice of transmedia storytelling, in turn, has expanded the possible modes and styles through and in which stories are told, and the opportunities for storytellers to engage audiences.

Activism is commonly seen as a resistant practice and most excursions into any form of online activism can be understood as part of a series of disruptive exercises such as culture jamming. However, back in 2006, Jenkins questioned whether "the old concept of culture jamming has outlived its usefulness. The old rhetoric of opposition and co-option assumed a world where consumers had little direct power to shape the media" (225–226) and "resistance becomes an end in and of itself rather than a tool to ensure cultural diversity and corporate responsibility" (259). The kind of activism illustrated in projects discussed in this chapter are inclusive in their approach rather than combative, but with the exception of The Harry Potter Alliance, exist entirely outside of the corporate sphere. They are also focused on illuminating hitherto unexamined aspects of an issue, particularly the experiences of the people involved to "create alternative media representations and express alternative political imaginaries based on an emerging network ideal" (Juris 2004, 98). This approach echoes Castells' theory of networked power: "the re-programming of communication networks, so becoming able to convey messages that introduce new values to the minds of people and inspire hope for political change" (Castells 2009, 134). If activism such as culture jamming used imagination and appropriation to draw attention to problems, projects such as *Hollow*, organizations like BIGhART and the work of the Oceti Sakowin leaders at Standing Rock attempt to draw attention to potential for change from within the communities, which is an important distinction from earlier forms of transmedia activism.

The projects in this chapter were chosen because they utilize recognizable conventions of transmedia storytelling and borrow elements from other forms of storytelling that pre-date transmedia, such as digital storytelling and documentary filmmaking. And they present a diversity of approaches to both activating change and to transmedia and have re-worked and re-purposed some of the conventions of transmedia storytelling to suit their intentions around raising awareness and facilitating change. All of these examples loosely fall into the category of transmedia activism. The notion of transmedia activism was introduced by Lina Srivastava, who has defined it as "creating social impact by using storytelling by a number of decentralized authors who share assets, create content for distribution across multiple forms of media to raise awareness and influence action" (Srivastava 2009).

Transmedia activism challenges a great deal of what we understand to be transmedia storytelling. Much of what has been identified as transmedia storytelling fetishes mainstream, franchise based stories (and even in the instances where fans have to an extent taken control of the story, it is still always in the interest of the large corporations at the heart of the project) or what James Bridle calls "sleek black box, corporate controlled objects, platforms or services" (Bridle 2013). Without dismissing or diminishing these mainstream projects or the ways in which they are considered, the aim of recognizing evolving approaches to transmedia is to open up the field to encompass other works that instead champion accessible, easy to understand and share alternatives. A key aspect of the following projects and transmedia activism in general is its ability to be spread widely, to be molded by participants and audience and to have low barriers to entry.

Questions about the efficacy of this style of storytelling and activism surround transmedia, but these same doubts were also associated with previous forms of creative and digital activism. One of the persistent criticisms of digital storytelling was that it merely provided an avenue or a form in which people could exercise their voice, but with little effect beyond that.

> Couldry makes clear that serious work on the politics of "voice" requires us to go beyond "a celebration of people speaking or telling stories," but rather must be placed in a larger "political context," one describing the forces that enable or block certain voices from being taken seriously as part of ongoing struggles over power.
>
> (*Jenkins 2016, 30*)

Beyond this notion of voice or amplification, social change requires tactics and strategies to elicit shifts in behavior in the general public, and/or a willingness to advocate for communities outside their own. Two ways that this can be achieved is through effective narrative representations of complex issues and immersive experiences for audiences to gain understanding of unknown environments.

Narrative Representations

In *Narratives, Health and Healing*, Harter, Japp, and Beck (2008) maintained that "narrative is a fundamental way of giving meaning to experience" (3). It is also capable of giving voice and meaning to the experiences of groups and communities who have previously only been represented by others rather than determining their own forms of representation. The belief that stories have an important role to play in social change has an abiding place in many organizations and social movements, and continues to define the philosophy of contemporary activism. In any community, those who have been excluded, whether deliberately or accidently, are often on the bottom rung of the community. Their invisibility has consequences, and

> often their very last asset is their story. It is often valuable, because it acts like a canary in the coalmine. If told in the right way, and placed with the right audiences, these stories can illuminate things we need to know about ourselves and things we need to shift as a society.
> *(Rankin 2016, 35)*

In part, what drives these types of storytelling projects is a belief that having the opportunity to tell their own stories empowers individuals and communities, and that sustainable change occurs from within empowered communities. Nevertheless, stories do not exist in a vacuum and the purpose of many community storytelling or social research projects is to bring those stories into the public discourse and into a dialogue between the storytellers and audience. "The storytelling process, as a social transaction, engages people in communicative relationships. Through identification and co-creation of story, the storytelling and reader/listener create an affective bond and a sense of solidarity: told and re-told 'my story' becomes 'our story'" (Davis 2002, 19). Or as BIGhART state: "it's harder to hurt someone if you know their story" (Rankin 2016).

Polkinghorne (1988) suggested that "narrative is a meaning making structure that organises events and human actions into a whole, thereby attributing significance to individual actions and events according to their effect on the whole" (18) and, as such, it privileges plot structure as a central feature. This emphasis is generally considered integral to narrative and suggests a need for a linear and coherent order of events for a story to be effective in communicating with an audience. The capacity of transmedia to allow makers to tailor stories for specific audiences and platforms means that effective communication moves beyond concepts of plot and structure, and encompass more ephemeral and emotive languages of aesthetics and mood. In this manner, stories and narratives have multiple purposes and one story can be represented through various media and abstract ways to convey the experiences of the storytellers and to engage the audience in a way that "stimulates the audience's creative participation and identification and invites them to supply what is unspecified yet required" (Davis 2002, 16).

The director of *Hollow*, Elaine McMillion, echoes this sentiment when she discusses her decision not to make a linear documentary with an observational approach about McDowell County, in the United States. McMillion stated in 2013 in *Filmmaker* that when she arrived at McDowell County she found "really phenomenal stories of pride and hope" and realized that "she wasn't comfortable editing those into 75minute form and putting a title slide saying 'The End'" (Astle 2013). This shift to more interactive modes and multi-authored stories is more than the inevitable use of technology, it is also a conscious decision to radicalize the nature of the form. *Hollow* is a participatory community project and interactive documentary that explores the social and economic devastation of rural towns in America through the story of McDowell County, in West Virginia. It brings together personal digital stories, photography, sound, interactive data, and grassroots mapping on an HTML5 website which was designed to explore the many stereotypes associated with the area, population loss and potential for the future.

At the center of the project are around 30 stories made about and by the residents of McDowell using video, stills, text and voiceover that are reminiscent of traditional digital stories. In a 2013 interview in *Filmmaker*, McMillion claims that "the stories are encountered within this landscape so that the people featured emerge from a context of place and community. We were really avoiding database storytelling, where you simply sort the videos and watch what you're interested in" (Astle 2013). *Hollow* is interesting in a number of respects: as a dynamic product it looks and feels like the very conscious combination of digital storytelling, documentary filmmaking and transmedia storytelling. The intentions of the creators are also explicit, with McMillion stating:

> I believe it is time that we let the community take control of their identity and allow them to amplify their own voices and ideas. Our hope is that through storytelling and the creation of multidimensional images, the community members will begin to see their environments and neighbors in a new way and begin to work together to preserve the history and make positive contributions to their communities.
>
> *(Astle 2013)*

In *Hollow*, the multiple methods deployed to create the environment in which the project is set are examples of not only the impact that attention to cultural, physical, and economic environment can have on non-fiction projects but also the consideration of how particular media and platforms can be utilized to best portray particular aspects of that world. Perhaps the greatest achievement of *Hollow* is the ways it has captured the "feel" of the McDowell county while also telling a universal story. The stories are specific and detailed, the voices and the landscape are sound and look like nowhere else, but it reveals the same struggle as regional towns all over the world. As with all good storytelling, the specificity shows us the commonality.

BIGhART is an Australian arts media and social change company that works intensively in marginalized communities to create multi-faceted arts events that reflect the stories and the creativity of the participants, and raises awareness about the urgent social issues facing the community. As an organization, BIGhART is focused on collaborating with communities as creative partners rather that as experts arriving to "help" the community and tell their stories for them. Founder and Creative Director Scott Rankin says:

> it became clear that we weren't running with a top-down "tree-shaped" model; we were more like a bamboo plant with a complex and ever- shifting root system that ran the company and resulted in our strong and consistent creative productivity. It was much more a rhizome-based structure.
>
> *(Rankin 2016, 38)*

This networked, flatter structure is suited to transmedia storytelling, which by its very nature is multi-layered and dialogic. As a means of creating and communicating personal narrative, transmedia allows for increasingly horizontal modes of authorship and further dismantling of the binary of the expert practitioners and the amateur community.

One of the most successful BIGhART projects is *NGAPARTJI NGAPARTJI*, which sought to preserve and share traditional Indigenous languages through theater, film, online language courses, and song, and exhibited the outputs via real world performances, online lessons, digital stories, and a feature-length documentary. The two pillars of the project, the Pitjantjatjara language course and the performance piece, existed in tandem during the life of the project, each contributing its own specific perspective and mode of communication in order to generate the strongest possible impact toward achieving the project's goals. This hybrid structure, along with the commitment and generosity with which the story was offered by the Indigenous creators to non-Indigenous audiences resulted in a profoundly moving and affecting series of events and stories. A powerful emotional connection was brokered between makers and audience which sparked widespread interest of various arts and film festival directors along with the mainstream media. "These directors not only supported the presentation of *NGAPARTJI NGAPARTJI* in various venues because of its high production values and aesthetics, but also because it embodied a radically new approach to reconciliation within the Australian nation" (Rankin 2016, 32).

The range of outputs, creative artefacts and opportunities for participants developed through the *NGAPARTJI NGAPARTIJI* project is staggering. A pilot language lesson in Cober Pedy in 2004 motivated participants to join the project and youth workshops were facilitated by local creative producers in towns and remote communities, through which a huge amount of content was created and uploaded to the *ninti* website, and made into DVDs to be distributed for communities without reliable Internet connections. A "Youth Learning" page was created in collaboration with the Centre of Aboriginal Economic Policy to publish the material created through the workshops. The culmination of the project was a stage production with an audience of over 30,000 people and that then formed the basis of a feature-length documentary.

Both *Hollow* and *NGAPARTJI NGAPARTJI* are multi-year projects that relied on multiple media and forms to extensively represent communities and to connect with as larger an audience as possible, and "ensure the work is reaching nodes in multi-layered networks that spark further change. This may mean broader general public audiences, but it also means targeted audiences who can respond to a growing groundswell with shifts in the national story and then shifts in policy" (Rankin 2016, 36). Importantly, they both foreground story as the cohering element, all aspects of the project and media or platform choices are designed to support the story. The stories themselves were made to be a representation of the lived experiences of the communities involved, and to attempt to holistically understand the circumstances of these communities. Though, to do this most effectively, it can be argued that even the most comprehensive narrative representation needs to be supplemented with the opportunity for audience immersion, or to become "part of" the story, even for a small amount of time.

Immersion and Engagement

Hollow and *NGAPARTJI NGAPARTJI* both created spaces for audience immersion—ways to contribute content and to have real-world interactions with the communities—that help to increase the longevity of the worlds depicted. They also understand the increasing need for creators to be able to work in the vocabulary and visual language that is authentic to the community and is meaningful for the audiences. This attention to the kind of engagement that deeply acknowledges the participants was identified by Jenkins first in *Convergence Culture* (2006) and explored further in *Spreadable Media* (Jenkins, Green, and Ford 2013) and *By Any Media Necessary* (Jenkins et al. 2016). It is through the study of fan communities that Jenkins illuminates the ways in which shared stories and worlds—even fictional worlds—are able to rupture the status quo to present possible alternatives.

Brough and Shresthova claim "what is most relevant here however, if that fan communities often form around content worlds, that may not be explicitly political in nature, but can offer resources or spaces for political engagement" (2012, 3.10). This form of activism appeals to fans who feel an affinity with the characters or broad themes of a film/book/television series and the use of the content world serves to encourage fans to either participate directly through events or fundraising promoted explicitly by the creators or to find their own ways to engage with the ideas. It encourages groups of people not generally associated with political action—young people, avid consumers of popular culture—to believe they have a valid voice, that they can find or create familiar spaces from which to express their concerns. These are a diverse set of shared activities and social engagements, ranging from fanfiction writing to gaming, through which people collectively carve out a space for expression and learning. Describing the educational dimensions of participatory culture, Henry Jenkins et al. state "that groups involved in such activities are characterized by 'relatively low' barriers to artistic expression and civic engagement, strong social support for creating and sharing and for the development of 'voice,' informal practices providing mentorship and training for would-be participants, and contributors' sense that what they share matters" (2016, 20).

A well-known example of this type of fan community is The Harry Potter Alliance (HPA). The website for HPA claims it "turns fans into heroes" and that "We're changing the world by making activism accessible through the power of story. Since 2005, we've engaged millions of fans through our work for equality, human rights, and literacy" (The Harry Potter Alliance n.d.). For many young people what counts as "politics" is a distant set of principles and practices that do not align with their own lives, and in most Western democracies the leaders of major political parties and mainstream news services do not even look like the much more diverse world young people inhabit both online and in real life. Thus, it is not surprising that their political action and activism would not resemble the activism of the past. An excellent description of the work of the HPA is given by Jenkins and Shresthova (2016):

> The Harry Potter Alliance's Not in Harry's Name Campaign ... gives us a rich example of how meaningfulness might inspire political action. The group called out Warner Brothers, the studio that produces the Harry Potter movies, because the chocolate manufacturers the studio had contracted to create chocolate frogs and other confections for their theme park attractions were not certified as deploying Fair Trade practices. The HPA cited an independent report produced by Free2Work that gave the involved chocolate companies an F in human rights, suggesting that there were legitimate concerns regarding their labor policies and practices. e group collected hundreds of thousands of signatures on petitions intended to shame the producers into adopting better labor practices.
>
> *(270)*

By congregating around a shared love of *Harry Potter* members of the HPA can feel comfortable communicating with other members and also to take notice of the issues that are identified by the creators of the website. This bond also allows them to feel part of a powerful group rather than a lone voice, and supports sustained engagement in the alliance and continued consideration of the information presented through the HPA.

Also deploying diffuse notions of community and collective action afforded by social media are Indigenous groups fighting for land rights and recognition of the injustices perpetrated against them through continued colonization. A number of these groups have turned to social media and popular culture to raise the profile of their causes. Palestinian activists have dressed as Na'vi tribe members from the film *Avatar* to protest the ongoing occupation of Palestine by Israel, and Indigenous Australians also used the visual tropes of *Avatar* to protest against the opening of a coal mine on their land. The use of instantly recognizable visual cues to communicate long standing exploitation and marginalization allow groups to speak directly in a shared language with a mainstream audience who saw the film but may not know anything about the specific issues being protested. This differs from the HPA in that the activists themselves are not fans, rather they are appropriating the tactics of fan culture to call attention to their own concerns.

A different approach to using multiple types of media in a spreadable way can be seen in the Standing Rock protests of 2016 and 2017. In 2016, the Native American Sioux tribe of Standing Rock fought to stop the Dakota Access Pipeline that would see an oil pipeline run under their reservation and land. Through social media, often utilizing video and live streaming of event, the resistance at Standing Rock garnered international attention. One of the key moments during the protest was when a million people checked in at Standing Rock using Facebook to disrupt and overwhelm police attempts to monitor activity at the site. As previously mentioned with other examples, there was skepticism about the usefulness of the online activity associated with Standing Rock.

> On Tuesday, the hashtag #DAPL trended nationwide for a little while, and then was eclipsed by chatter about the Academy Awards nominations. If social media and live streaming enabled the Standing Rock Sioux to amplify their protest for clean water, its speed and ceaseless flow also allowed the world to forget about them.
>
> *(Dreyfuss 2017)*

The feel-good, no-effort nature of some online activism can result in quick uptake and just as quick forgetting. But, as was the case at Standing Rock, it is rarely considered to be an answer in and of itself, rather a way of broadcasting or amplifying an issue and or the increasing urgency of an issue and to connect people.

> In fact, online connections can help overcome obstacles of space, time, income and knowledge to share stories and information while linking people to each other and the opportunities for action. Indeed, Mic soon revised its article headline: "Checking in at Standing Rock on Facebook is cool—but here's how you can actually help." That signaled an important acknowledgment: While online action alone can't solve a problem, it can be a very useful tool to mobilize people and focus attention on a crucial issue.
>
> *(Torchin 2016)*

It is simple from the outside to consider the protest that occurred at Standing Rock through the lens of social media and to critique as such, but it played out in real life with lasting effects. Author and activist Naomi Klein says, "although stopping the pipeline was crucial, there was something greater at work in this convergence. The camps were now a place where Indigenous and non-Indigenous people alike were learning to live in relationship and community with the land" (Klein 2017, 308). These events were amplified through digital media but they were grounded in the world and signaled a renewed collective power and model of social activism in the United States.

Conclusion

Each new creative form which has emerged at different times in history and has assisted in creating social change—the book, theater, film, photography, music, culture jamming, interactive media—has exploited and radicalized its commercial or mainstream iteration. Transmedia activism is the logical progression of transmedia and reflects the many traditions, philosophies, and practices that precede it. It can be dismissed for its hybrid, ungrounded, playful approach by those working in traditional media or the public sphere but it is impossible to judge the effectiveness of transmedia activism by the merits of older media. In *By Any Media Necessary* (Jenkins et al. 2016), it is noted that

> some critics have dismissed these new forms of activism as attention seeking, yet traditional demonstrations, focused on getting as many bodies as possible into the streets, also seek to render visible their base of support. There's still some tendency to apply standards of broadcast media in looking at social networking practices.
>
> *(266)*

Hollow and BIGhART present examples of deep, committed activism that tells complicated, uncomfortable, but ultimately life-affirming stories from some of the most forgotten communities in otherwise wealthy nations. However, relatively speaking, they remain not especially well known; it is possible to talk to people working in community development who have never heard of the projects. Nevertheless, both have had demonstrable success in regards to policy changes and funding for the communities.

The HPA or AVATAR protests on the other hand can be viewed as short-lived spectacles and the Standing Rock protest is still being evaluated both from within and outside the Sioux community. It is not fair or useful to compare the projects to one another or to judge them by a single set of criteria for success. Each project had a sense of what the outcomes they wanted were or still are (funding, policy change, boycott, coverage) and they made decisions based on capacity, outcome, audience, community to choose the format, the platform, and the approach. One size does not fit all, and we cannot compare contemporary activism to past activism—women's liberation movement to #yesallwomen or the civil rights movement to #blacklivesmatter—rather a new critical language and framework needs to be continually developed to evaluate and improve the application of transmedia for social change based on the intentions of the project, and in the meantime, use any media necessary to resist.

Transmedia activism brings new life and understandings to the role of identity and representation in social change. The past few decades have seen an emphasis being put on organizational structures and the explicit intersection between policy and community as pivotal to successful protest and more recently an infatuation with big data as the holy grail to changing hearts and minds. The social change projects discussed here remind us that people and their stories are at the core of social justice and to achieve that activists need to touch and connect with our shared humanity.

References

Astle, Randy. 2013. "Elaine McMillion and Jeff Soyk on *Hollow*." *Filmmaker*, September 26. Accessed July 28. http://filmmakermagazine.com/72963-elaine-mcmillion-and-jeff-soyk-on-hollow/#.WdS9b_Ykt_Q.

Bridle, James. 2013. "The New Aesthetic and its Politics." *Booktwo.org*, June 12. Accessed July 13, 2017. http://booktwo.org/notebook/new-aesthetic-politics/.

Brough, Melissa, and Sangita Shresthova. 2012. "Fandom Meets Activism: Rethinking Civic and Political Participation." *Transformative Works and Culture* 10: 1–27.

Castells, Manuel. 2009. *Communicative Power.* New York: Oxford University Press.

Davis, Joseph E. 2002. "Narrative and Social Movements: The Power of Stories." In *Stories of Change: Narrative and Social Movements,* edited by Joseph E. Davis, 3–29. Albany: State University of New York Press.

Dreyfuss, Emily. 2017. "Social Media Made the World Care about Standing Rock and Helped Them to Forget It." *Wired,* January 24. Accessed September 14. www.wired.com/2017/01/social-media-made-world-care-standing-rock-helped-forget/.

The Harry Potter Alliance. n.d. Accessed September 2, 2017. www.thehpalliance.org.

Harter, Lyn, Phyllis Japp, and Christina Beck. 2008. *Narratives, Health & Healing: Communication Theory, Research & Practice.* Mahwah: Lawrence Erlbaum Associates Inc.

Jenkins, Henry. 2006. *Convergence Culture: Where Old and New Media Collide.* New York: Routledge.

Jenkins, Henry. 2016. "Youth Voice, Media and Political Engagement." In *By Any Media Necessary: The New Youth Activism,* edited by Henry Jenkins, Sangita Shresthova, Liana Gamber-Thompson, Neta Kligler-Vilenchik, and Arely M. Zimmerman, 1–60. New York: New York University Press.

Jenkins, Henry, and Sangita Shresthova. 2016. "'It's Called Giving a Shit!': What Counts as 'Politics'?" In *By Any Media Necessary: The New Youth Activism,* edited by Henry Jenkins, Sangita Shresthova, Liana Gamber-Thompson, Neta Kligler-Vilenchik, and Arely M. Zimmerman, 255–289. New York: New York University Press.

Jenkins, Henry, Joshua Green, and Sam Ford. 2013. *Spreadable Media: Creating Value and Meaning in a Networked Culture.* New York: New York University Press.

Jenkins, Henry, Sangita Shresthova, Liana Gamber-Thompson, Neta Kligler-Vilenchik, and Arely M. Zimmerman. 2016. *By Any Media Necessary: The New Youth Activism.* New York: New York University Press.

Juris, Jeffrey S. 2004. "Networked Social Movements: Global Movements for Global Justice." In *The Network Society: A Cross-Cultural Perspective,* edited by Manual Castells, 345–376. Northampton, MA: Edward Elgar.

Klein, Naomi. 2017. *No Is Not Enough: Defeating the New Shock Politics.* Penguin: New York City.

Polkinghorne, Donald. 1988. *Narrative Knowing and the Human Sciences.* Albany: State University of New York Press.

Rankin, Scott. 2016. "BIG *h*ART: Beginnings and Trajectories of Development." In *BIGhART: Art, Equity and Community for People, Place and Policy,* 35–64. Sidney: Murdoch University Press.

Srivastava, Lina. 2009. "Transmedia Activism: Telling Your Story across Media Platforms to Create Effective Social Change." Accessed December 11, 2016. http://transmedia-activism.com/.

Torchin, Leshu. 2016. "What Can Mass Check-in at Standing Rock Tell Us about Online Advocacy?" *The Conversation,* November 5. Accessed October 4, 2017. https://theconversation.com/what-can-the-mass-check-in-at-standing-rock-tell-us-about-online-advocacy-68276.

37
TRANSMEDIA IDENTITIES
From Fan Cultures to Liquid Lives

André Jansson and Karin Fast

When Marsha Kinder, in 1991, argued for the relevance of "transmedia" as a label for contemporary modes of world-building, she did so with reference to *Teenage Mutant Ninja Turtles*. When Henry Jenkins (2006), more recently, conceptualized "transmedia storytelling," he did so in relation to *Survivor*, *Star Wars*, and *The Matrix*. What all of these cases of transmediality have in common is that they are enduring media franchises and objects of fandom. Not very surprisingly, these seminal works on transmediality have biased the concept toward popular culture in general and fan culture in particular.

Today, however, the social significance of transmedia stretches far beyond fan cultures and other enclosed communities. The transmediated construction of human identities has gradually made its way into ordinary culture. Anyone with access to a connected media device can sign up for a social media account, start spreading snapshots from his or her life, recommending things to buy or places to go, even setting up a private video channel. As media users, we are also increasingly expected to do this. Social media industries can only survive as long as we are willing to express something online and actively recognize what others are doing. As a consequence, the socio-material prevalence of transmedia in everyday life now affects questions of who to be, who to interact with, and, ultimately, how to go about and feel about one's life, whether connected or not. Even those who are *not* willing to expose themselves online or to put time and energy into their "media image" must reflect upon the prevalence of transmedia.

In this chapter we want to address this broader picture. It means, first, that "transmedia" is understood as a certain type of media ecology and life environment, stretching beyond the confines of popular media franchises and brand identities. The term transmedia then points to how social practices (in addition to texts per se) are molded by and negotiated through different platforms and devices, and interweave with various forms of offline communication (anything from T-shirts to interior design). Second, it means that "identity" is taken in its sociological understanding as a complex and negotiated interface between self and society rather than as a singular entity tied to particular areas of interest, media concepts, or narratives (as in the case of fan cultures). Ultimately, we want to show how the emerging transmedia ecology contributes to the social normalization of "liquid lives" (Bauman 2005).

The chapter is divided into two parts. In the first part we overview the current state of research, identifying three main areas where questions of identity creation have been related to transmedia: *fan cultures*, *social mobilization*, and *play*. We also discuss the theoretical denominators and limitations of these areas. In the second part of the chapter we advance a broader sociological perspective guided by the conceptual framework of Anthony Giddens (see especially Giddens 1991). Our approach is constructed around three of Giddens' key concepts and describes the relationship between self and society as *transmediatized*. The growing indispensability of transmedia spurs the *liquidization of lifestyle sectors*, extended *reliance on abstract systems*, and a state of *ontological vulnerability*. There are of course other theories that could have served as a toolbox for constructing such a broadened view of transmedia identities. We advance the Giddensian approach as a particularly pertinent example of how we should address the social implications of transmedia and connect media change to the structural transformations of modern society.

Main Themes in Research on Identity and Transmedia

Identity generally refers to a sense of wholeness and continuity. Identity is a matter of knowing who one is and *identifying* what something or somebody is. Accordingly, identity in the transmedia literature often refers to the meanings attached to brands and popular media concepts and characters sustained through transmedia marketing and storytelling (e.g., Cooper, Schembri, and Miller 2010; Bertetti 2014). Digital transmedia platforms offer a new potential for "world-building"; a semiotic phenomenon with deep roots in the cultural industries (e.g., Freeman 2015; Fast and Örnebring 2017) that implies that the cross-promotion of a certain brand through different channels may strengthen the audience's engagement and identification with its key themes and values (e.g., Edwards 2012; Granitz and Forman 2015).

When overviewing the literature on transmedia identity it appears that this semiotically oriented perspective, largely emanating from Jenkins' (2003) original thinking around transmedia storytelling (see also Scolari 2009), saturates also most research into the identities of media users/audiences. While we suggest that this field of research contains three main themes—fan cultures, social mobilization, and play—it is also obvious that they overlap, united above all by their proximity to the "Jenkins legacy." First of all, these studies focus on how identities unfold in relation to *particular types of transmedia texts and formats* rather than the broader social implications of transmedia technologies. Second, they implicate more or less *unitary identities*. The transmedia literature tends to highlight particular identity positions related to certain brands, formats, cultural communities, and desired forms of recognition rather than the complexity of everyday identity work. Finally, most research revolves around identities marked by *high levels of expressivity and engagement in the transmedia world*, especially tied to play, popular culture, and leisure interests. Very little has been written about the more banal, indirect, and structural consequences of transmedia in everyday life. As we now continue with a brief overview of the three dominant themes, we also substantiate our call for a more inclusive sociological approach to transmedia identities.

Fan Cultures

The concept of "transmedia storytelling" became popularized with the advent of *The Blair Witch Project*—an independent, low-budget, transmedia production that drew audiences to cinemas worldwide in 1999 (Jenkins 2006). Well before the theatrical release of the film, a mysterious website with—as it would turn out—paratextual content had produced a big, curious, fan base who, upon having seen *The Blair Witch Project* in theaters, could continue their engagement with the "Burkittsville witch" through a comic books series, a pseudo-documentary, and a film soundtrack. If *The Blair Witch Project*, next to other contemporaneous transmedia endeavors like *The Matrix* (1999), raised public awareness about the notion of "transmedia storytelling," Jenkins' (2006) famous conceptualization of this convergent mode of storytelling sparked a plethora of transmedia studies. A majority of these are, like those conducted by Jenkins himself, text-oriented and situated in the realm of popular culture. Consequently, transmedia identities have chiefly been explored in studies of popular franchise texts and related fandoms (see, e.g., Perryman (2008) on *Doctor Who*; Scolari (2009) on *24*; Fast (2012) and Johnson (2013) on *Transformers*; Bourdaa (2013) on *Fringe*; Stork (2014) on *Glee*).

The existent bulk of research identifies deep levels of semiotic, enunciative, and textual productivity as distinguishing features of fan identities in general (Fiske 1992; Sandvoss 2011) and of transmedia identities in particular. At the very least, as Jenkins (2006) explains, the transmedia consumer must be willing to "chase" a story over several content platforms. Rather few studies, however, make serious attempts to discriminate between *types* of transmedia engagement within transmedia franchises. Evans' (2008) focus-group study of *Spooks*—a transmedia drama based on a television series but narrated also in online games—makes an exception. Starting from the premise that *characters* are a "central point" of audience engagement, she investigates the various modes of character engagement enabled by the television and game medium respectively. As for television, emotional engagement is found to be largely connected to feelings of *identification* with the *Spooks* characters. The viewers, Evans finds, admire and empathize with the characters. The gamers, however, experience a different, arguably deeper, type of identification: the game offers a first-person perspective that places the gamer—not a separate, fictional character—*as* the "protagonist" of the storyworld (Evans 2008, 208).

Identification as a driver of transmedia engagement is recognized also by Lemke (2009), although he extends the notion beyond merely character identification: "users identify with a character, or more generally with the ethos, the spirit, the mythos, the 'feel' of the entire franchise world" (147). Identification, he elaborates, is what attracts fans to media franchises in the first place, but also what makes them grow. Engaged fans produce their own paratexts (e.g., fanzines or fan websites), which in turn potentially *reinforces* fan identification. Along these lines, Ellcessor (2012) suggests that prosumer activity can spark identification with transmedia texts by obscuring the lines between text and self. She concludes from her study of the vibrant Twitter activity of actor, writer, and "geek" Felicia Day (*Buffy the Vampire Slayer, The Guild*) that "Stars and others communicate through shared channels in social media, sharing an experience of celebrity and even identity" (Ellcessor 2012, 61). The "others" that Ellcessor write about include Day's showbusiness friends as well as her fans. The "geek," "gamer," and "girl" identities constructed through Day's "authentic" online presence places her work—and audiences—in opposition to the mainstream and hence renders the franchise a potential platform for social mobilization.

Social Mobilization

By social mobilization we refer to the claiming of voice, civil rights and recognition among certain social groups, especially minority groups, and marginalized communities. The boundary between fan cultures and social mobilization is not always a sharp one. Several studies have explored how particular transmedia texts and narratives play into the construction and potential mobilization of identity positions tied to ethnicity, gender, and sexual orientation. For example, Marwick, Gray, and Ananny (2014) analyzed how young adults responded via social media (especially Twitter) to the American TV series *Glee*, known for its portrayal of gay, lesbian, and bisexual high school students. The researchers found that while the viewers (many of them dedicated fans) intervened in the discursive construction of the transmedia production, they also positioned themselves and their identities in relation to the social and cultural themes of the program. Fan engagement thus provided a platform for social mobilization, and vice versa.

Transmedia perspectives have also been used to address the new conditions for social expressivity and recognition. Independent transmedia production and circulation can be used to raise awareness of social inequalities and cultural discrimination (see, e.g., Cheong and Gong 2010), and to strengthen alternative communities. Ramasubramanian (2016) provides an example of how community-oriented transmedia storytelling initiatives can contribute to the mobilization of racial/ethnic minorities and also foster critical literacies for countering the dominant narratives of media conglomerates. Other studies have acknowledged the mobilizing potential of social media. Zimmerman (2016) discusses how undocumented youth activists use "transmedia testimonies" to challenge anti-immigrant policies and claim their rights even if they are not recognized as citizens by the state. These activists use various channels of social media for sharing immigration experiences and documenting their participation in political manifestations. Similar points are made in Fink and Miller's (2014) work on the performance of queer, transsexual, and transgender identities on Tumblr. Transmedia platforms may thus enable disadvantaged groups to establish alternative public spheres. At the same time, however, it should be acknowledged that these types of initiatives are not always successful; they may create a false sense of recognition as well as involving new forms of exploitation and accentuating the gaps between those who are rich and poor with regards to media literacy and cultural capital (see, e.g., Martens 2011; Clark 2016; Soriano 2016).

Some research on transmedia mobilization has claimed the importance of thinking outside the confines of media expressivity and participatory culture. We find such work especially in social movement studies. As Costanza-Chock (2014, 5) notes in his book on the immigrants' rights movement in Mexico, "social movement media-making tends to be cross-platform, participatory, and linked to action." This means that processes of social mobilization are dependent on what Costanza-Chock calls "transmedia organizing" and should be analyzed in relation to "the broader media ecology rather than focusing exclusively on one or a handful of platforms" (2014, 5). A similar approach is found in Gerbaudo's (2012) book *Tweets and the Streets: Social Media and Contemporary Activism*, which assesses how information-sharing and community-building among activist groups occur through digital transmedia technologies as well as more traditional means of communication, such as public posters and flyers. In these cases, the implications of transmedia take us beyond the realm of co-creation and remixing of particular narratives and texts, into a more contextualized view (see also Soriano 2016; Zimmerman 2016; Lin 2017). Transmedia then denotes a media ecology that includes both new *and* old media, thus making justice to the historical legacy of analogue transmediality (see Fast and Örnebring 2017; Freeman 2015), and where it is not just particular texts that are being worked upon but *social communication at large*, e.g., decision-making, social coordination, and self-representation. Together, these continuous transmedia practices shape a "social movement world" (see Costanza-Chock 2013, 100). We return to this broadened approach to transmedia in the second part of the chapter.

Play

Transmedia play is a recurring theme in research on fan cultures and may also tie into the mobilization of cultural identities. Besides these overlaps, however, what marks out research on play is its greater focus on identity experiment, negotiation, and ambiguity. The world of play can provide a space for being somebody else, at least for some time, and rethink one's position in society and the everyday. In games, typically under the shelter of online anonymity, the individual can move beyond pre-existing identity positions and franchised media concepts.

In a study of the Second Life extension of the American teen TV drama *Gossip Girl*, Stein (2009) explores the implications of media audiences stepping into the virtual world of Upper East Side (where the TV series is located), taking on the roles of fictional characters. While this form of commercial game spaces predominantly foster consumerist values (in the case of *Gossip Girl* with an accent on fashion), as Stein argues, one should not overlook the relative agency of users and the possibility for creativity and alternative forms of engagement. In a similar fashion, some research on play has focused on how offline practices interweave with transmedia texts and sometimes challenge the dominant meanings of these texts. Wohlwend (2012), for example, approaches the transmedia products of Disney Princess as "gender identity intertexts" that young audiences use and negotiate collaboratively through play and other forms of interaction. In the school context, which was the site of her fieldwork, different layers of the transmedia texts were articulated and contributing to the modeling of social relations. These were found to not only follow dominant stereotypes but also holding transgressive potential, for instance, when boys were playing the princess. The richness of transmedia material implied that texts and artifacts could be used both as anchors and pivots for the construction of groups, play roles, and identity positions (especially gender).

What is also important to note here is that game worlds may involve more than game-related experiences. Due to their transmedia nature, many games may also function as entrances to new forms of interaction through, for example, chat functions. Some of these may be related to fan cultures, as Willett (2016) found in the case of *Minecraft*, but there are also possibilities for more open-ended social interaction with peers as well as strangers. For young people the virtual game space thus provides a space for pushing social boundaries and claiming greater autonomy in relation to parents. The open-ended nature of transmedia spaces, where conversations and activities may carry over from one platform to another, is of key importance to any understanding of identity work in contemporary culture. As such, it concerns a much broader social realm than the playful transmedia practices of adolescents and young adults.

A Giddensian Approach to Transmedia Identities

The subtitle of this chapter brings to mind sociologist Zygmunt Bauman's work on the escalating liquidity of modern society (see, e.g., Bauman 2000, 2005). One of Bauman's key points is that older distinctions between private and public, work and leisure, home and away, and so forth, are getting more volatile due to overarching transformations like globalization and individualization. The mobile and flexible individual becomes the norm, whereas values like security and continuity lose ground. Liquid life, according to Bauman (2005, 10–11), refers to a state of "constant self-scrutiny, self-critique and self-censure." It is also, we argue, a life of *transmediatized identity work*, which means that modern identities (already individualized) become dependent on transmedia for their maintenance and social recognition. Transmedia here includes the technological systems and material "stuff" of transmedia as well as a variety of texts.

Surprisingly little has been written about the connections between transmedia and liquid lives, however. There has been some research on the liquidity of (media) work in transmedia environments—such as Deuze's (2008) analyses of "liquid journalism." There is also relevant research on "social media," "mobile media," and "connective media" actualizing the increasingly liquid conditions for identity work (e.g., Marwick 2013; Van Dijck 2013; Couldry and Hepp 2016). These studies can be interpreted as accounts of transmediatization. Bauman's own work on liquidity, however, largely preceded the transmedia era. It is also relatively devoid from any systematic analysis of media change as well as analytical categories through which to pinpoint the main aspects of liquidity per se.

Against this background, as we now try to work out the links between transmedia and identity we build not so much on Bauman's work as on the writings of another sociologist, Anthony Giddens. Especially his much acclaimed book *Modernity and Self-Identity* (Giddens 1991) presents a valid framework for thinking through the ways in which transmedia accentuate modernization processes that were in place even before the digital revolution, and which ultimately give shape to the liquid society that Bauman professed. According to Giddens (1991), there are three main elements that explain "the peculiarly dynamic character of modern social life" (16): (1) separation of time and space, (2) disembedding mechanisms (including abstract systems), and (3) institutional reflexivity. In the following three sections, we revisit and recast these developments in light of contemporary transmedia. By way of conclusion, our discussion leads to an elaborated view of liquid lives.

Liquidization of Lifestyle Sectors

The first element in Giddens' characterization of modern social life concerns the gradual loss of spatial and temporal bonds in human affairs. While in pre-modern societies "time and space were connected *through* the situatedness of place" (Giddens 1991, 16, italics in original), that is, mutually constitutive through activities carried out in a particular setting, modernization means that both time and space can be managed. People and their identities are not to the same extent bound to a particular place throughout their lives. Technological and institutional developments such as mechanical clocks, standardized time zones, and new means of communication (including both media and transportation) have made it possible to follow and manage social events at a distance. As Giddens puts it, "modern social organization presumes the precise coordination of the actions of many human beings physically absent from one another" (17). One key example concerns large-scale industrial production and trade, but through modern media like the postal system, print, telephone, radio, and television even the most mundane practices and relationships were connected to other times and places (see also, e.g., Meyrowitz 1985; Scannell 1996).

The shift from mass media to transmedia means two important things. First, information, data, and text are no longer kept within one technology of distribution but circulate between different technologies and platforms. Second, as information, data, and text circulate they can also change through the active involvement of media users. These factors have important implications for the very textures of everyday life, especially the ways in which lifestyles are organized in time and space. As Giddens (1991, 81) points out, the ordering of everyday practices and interests within a coherent lifestyle is essential to modern identity because it gives "material form to a particular narrative" of the self. Under transmediatized life conditions, however, the "lifestyle sectors" that Giddens (83) speaks about become increasingly negotiable and open-ended. It is not just cultural consumption, including music, news, video, and so forth, that are increasingly available in different times and places, ready to be consumed, remixed, shared, and commented upon. Transmedia devices and platforms enable people to plan, coordinate, and represent a steadily growing range of lifestyle practices regardless of where they are. Many work practices as well as social gatherings, events, and identity-defining leisure interests like cooking, traveling, and exercising no longer occur just "here and now" but are also imagined, monitored, and shared with others via transmedia platforms. It means, for example, that the lifestyle sector of "evening running" also holds an online presence (through track records, music playlists, joint "challenges," etc.) and expands into everyday time-spaces beyond the routinized practice itself.

Extended Reliance on Abstract Systems

The second element of modern social life, according to Giddens, is social disembedding. Disembedding is closely related to, and a driver of, the separation of time and space in that it refers to the "'lifting out' of social relations from local contexts and their rearticulation across indefinite tracts of time-space" (Giddens 1991, 18). This "lifting out" of social relations is sustained by so-called *abstract systems*, which have expanded rapidly in modern society. According to Giddens, there are two types of abstract systems; "symbolic tokens" and "expert systems." The former include all kinds of media of exchange that can be used across different contexts, such as money or technological standards. Expert systems include professionalized forms of knowledge that stretches across socio-cultural boundaries and saturates the lifeworld in different parts of the world through technological innovations (affecting anything from food consumption to means of transportation) and advice given by various practitioners as experts (doctors, therapists, scientists, etc.), either directly or through media. A key condition of modern society is that people trust these abstract systems. As people increasingly lead their lives in different time-spaces and get involved in temporally and spatially stretched out processes, they must also rely on abstract systems. Growing autonomy with regards to the modern individual's capacity to expand his/her range of experience (especially through media) and make up life-plans thus brings along a heightened level of dependence (see also Jansson 2013, 2018).

If mass media were a key force of social disembedding, then transmedia implies nothing but an acceleration of these processes. This has to do with the above-discussed liquidization of everyday lives, which also leads to extended reliance on abstract systems. As transmedia practices amalgamate with larger and larger portions of everyday life, and when fundamental elements of life are adapted to transmedia systems (e.g., healthcare, banking, and social communion)—which are the signs of "transmediatization" (Jansson 2013)—then there has to be trust in the functionality and fairness of the systems of mediation. Mobile payment and transferring systems, for example, entail the interconnection of a host of platforms and infrastructures; including the smartphone itself (or any other connected device), the bank account (and associated apps), a secure identification app, credit card, and telephone register. Banking is today entirely transmediatized, and different ways of relating to this social fact becomes constitutive of a person's lifestyle and identity. Furthermore, the transmedia condition has invoked an unprecedented demand on constant upgrading. Anyone who wants to be part of the socially disembedded life promoted by social media and various online services is forced to keep his or her technological equipment and software up to date. This techno-commercial ultimatum is the precondition for liquidity, which, by extension, raises new questions of social inclusion vs. exclusion; autonomy vs. dependence, trust vs. anxiety. Managing transmedia per se becomes an element of identity work.

Ontological Vulnerability

The third theme in Giddens' outline of modernity is reflexivity. Compared to pre-modern life forms, the modern individual (as well as social institutions at large) cannot rely on prescribed social truths and predetermined life trajectories but has to reflect upon information, recommendations, norms, and ideals emanating from a variety of (mediated) sources, many of which are difficult to combine. Social reflexivity may concern relatively banal issues concerning ecological consumption and everyday styles of clothing as well as more life-encompassing decisions. As mentioned above, all such decisions, and their molding into routines, are constitutive of the lifestyle and give shape to identity. It is important to stress that they are not independent from one another but through their successive ordering into lifestyles they foster a sense of meaningful identity and continuity in life. The expansion of modern mass media should be seen as both a source and an object of social reflexivity. The multiplication of mediated experiences, including news, entertainment, artistic expressions, and so forth, has expanded the horizons through which people make sense of the world and reflect upon who they are and who they want to be in the scheme of things. At the same time, decisions concerning which media to use and how to make sense of them are crucial to the formation of communities and the establishment of distinctions in society.

Transmedia entail new avenues for gaining a sense of ontological security, that is, to feel at home in the world and receive positive recognition from others. There are numerous social networking (plat) forms that make it possible for people to stay in touch with one another across time and space and express and document what they think and what they are doing in their lives. Due to the affordances of transmedia, such sharing practices can occur practically instantaneously and involve a widely distributed group of peers. These possibilities for social integration can thus be seen as counterweights to the individualization and disembedding of modern society. At the same time, what is perhaps an even stronger demarcation of the new transmedia ecology is the media-industrial imperative to share, or to "spread," information through networks—and to react upon the messages from others. This is the commercial logic of connective media (see Van Dijck 2013), where "every click counts" and adds to the accumulation of information through which consumer segments can be calculated and targeted with specialized offers. The "click economy" and its algorithms can be seen as a new type of abstract system whose logics shape social life (see Striphas 2015).

The critical side of this development is that the logic of spreadability (Jenkins, Ford, and Green 2013) also reinforces the modern malaise of narcissistic reflexivity—and thus vulnerability. As transmedia expressivity become a key commercial concern, it is also endorsed as a social norm, prescribing constant status updates and ratings of what other people are posting (see also Marwick 2013).

The accumulation of friends, followers, likes, and so forth, is turned into a currency that measures and classifies the social success of individuals, and thus also legitimizes new ways of relating to others. The best example is probably the normalization of the "selfie" as a coercive format of everyday self-representation—a format balancing between empowerment and social risk, between creativity and repression (see Collings 2014; Zhao and Zappavigna 2017). In such a climate, where identities are continuously co-created via the fluctuations of online representations and automated forms of recognition, media users must develop reflexive strategies to handle the insecurities of a connected life. Every new activity on social media is visible (sometimes publicly) and thus part of self-creation. Images and texts can spread via other users into associated networks and far beyond the control of the originator, sometimes leading to unintentional consequences. A life with transmedia is thus a life of continuous doubt, a liquid life where ontological insecurity presents a lingering threat to dreams of self-realization and fulfillment.

Conclusion

In this chapter, we have argued for the broadened relevance of "transmedia identities" as a term that captures not only how popular media franchises and fan identities are constructed but also how transmediatized and *liquidized lives* are constituted more generally. The Giddensian approach was utilized to exemplify how we should address the *social implications* of transmedia and connect media change to the *structural transformations* of modern society. In building our argument, we have proposed two changes to the conventional approaches to transmedia. First, we have suggested that "transmedia" is not confined to popular culture, but rather should refer to a media ecology in which social practices (including, but not only, fan practices) are molded by and negotiated through different media technologies, and interweave with various forms of offline communication. Second, we advise that "identity" is interpreted as a complex and negotiated interface between self and society, rather than as a singular entity tied to particular areas of interest, media concepts or narratives (as in the case of fan cultures). Along these lines, we sympathize with Scolari and Ibrus' (2014) urge for research that acknowledge "transmedia as an important outcome and as a source of contemporary cultural and social complexities—not only as new forms of cultural texts and media institutions or practices but also as new forms of scarcity, inequality, and power struggles" (2193). In calling for such an extension of transmedia (identity) studies, we welcome critical research that addresses questions about, for instance, normalization and legitimization of certain lifestyles/identities, social exclusion, the techno-commercial ultimatum (and its environmental consequences), dependency, ontological insecurity, and other issues of power in the light of accelerating transmediatization.

References

Bauman, Zygmunt. 2000. *Liquid Modernity*. New Jersey: Wiley Blackwell.

Bauman, Zygmunt. 2005. *Liquid Life*. New Jersey: Wiley Blackwell.

Bertetti, Paolo. 2014. "Toward a Typology of Transmedia Characters." *International Journal of Communication* 8: 2344–2361.

Bourdaa, Mélanie. 2013. "'Following the Pattern': The Creation of an Encyclopaedic Universe With Transmedia Storytelling." *Adaptation* 6 (2): 202–214.

Cheong, Pauline H., and Jie Gong. 2010. "Cyber Vigilantism, Transmedia Collective Intelligence, and Civic Participation." *Chinese Journal of Communication* 3 (4): 471–487.

Clark, Lynn Schofield. 2016. "Participant or Zombie? Exploring the Limits of the Participatory Politics Framework through A Failed Youth Participatory Action Project." *The Information Society* 32 (5): 343–353.

Collings, Beccy. 2014. "# selfiecontrol:@ CAZWELLnyc and the Role of the Ironic Selfie in Transmedia Celebrity Self-Promotion." *Celebrity Studies* 5 (4): 511–513.

Cooper, Holly, Sharon Schembri, and Dale Miller. 2010. "Brand-Self Identity Narratives in the James Bond Movies." *Psychology & Marketing* 27 (6): 557–567.

Costanza-Chock, Sasha. 2013. "Transmedia Mobilization in the Popular Association of the Oaxacan Peoples, Los Angeles." In *Mediation and Protest Movements*, edited by Baert Cammaerts, Alice Mattoni, and Patrick McCurdy, 95–114. Bristol: Intellect.

Costanza-Chock, Sasha. 2014. *Out of the Shadows, Into the Streets! Transmedia Organizing and the Immigrant Rights Movement*. Cambridge, MA: MIT Press.

Couldry, Nick, and Andreas Hepp. 2016. *The Mediated Construction of Reality*. Cambridge: Polity Press.

Deuze, Mark. 2008. "The Changing Context of News Work: Liquid Journalism for a Monitorial Citizenry." *International Journal of Communication* 2 (18): 848–865.

Edwards, Leigh H. 2012. "Transmedia Storytelling, Corporate Synergy, and Audience Expression." *Global Media Journal* 12: 1–12.

Ellcessor, Elizabeth. 2012. "Tweeting@ feliciaday: Online Social Media, Convergence, and Subcultural Stardom." *Cinema Journal* 51 (2): 46–66.

Evans, Elizabeth Jane. 2008. "Character, Audience Agency and Transmedia Drama." *Media, Culture & Society* 30 (2): 197–213.

Fast, Karin. 2012. "More than Meets the Eye: Transmedial Entertainment as A Site of Pleasure, Resistance and Exploitation." Ph.D. diss., Karlstad University.

Fast, Karin, and Henrik Örnebring. 2017. "Transmedia World-Building: The Shadow (1931–Present) and Transformers (1984–Present)." *International Journal of Cultural Studies* 20 (6): 636–652.

Fink, Marty, and Quinn Miller. 2014. "Trans Media Moments: Tumblr, 2011–2013." *Television & New Media* 15 (7): 611–626.

Fiske, John 1992. "The Cultural Economy of Fandom." In *The Adoring Audience: Fan Culture and Popular Media*, edited by Lisa A. Lewis, 30–49. London: Routledge.

Freeman, Matthew 2015. "Up, Up and Across: Superman, the Second World War and the Historical Development of Transmedia Storytelling." *Historical Journal of Film, Radio and Television* 35 (2): 215–239.

Gerbaudo, Paolo. 2012. *Tweets and the Streets: Social Media and Contemporary Activism*. London: Pluto Books.

Giddens, Anthony. 1991. *Modernity and Self-identity: Self and Society in the Late Modern Age*. London: Polity Press.

Granitz, Neil, and Howard Forman. 2015. "Building Self-Brand Connections: Exploring Brand Stories through A Transmedia Perspective." *Journal of Brand Management* 22 (1): 38–59.

Jansson, André. 2013. "Mediatization and Social Space: Reconstructing Mediatization for the Transmedia Age." *Communication Theory* 23 (3): 279–296.

Jansson, André. 2018. *Mediatization and Mobile Lives: A Critical Approach*. London: Routledge.

Jenkins, Henry. 2003. "Transmedia Storytelling." *MIT Technology Review*, January 15. Accessed February 4, 2017. www.technologyreview.com/news/401760/ transmedia-storytelling/.

Jenkins, Henry. 2006. *Convergence Culture: Where Old and New Media Collide*. New York: New York University Press.

Jenkins, Henry, Sam Ford, and Joshua Green. 2013. *Spreadable Media: Creating Value and Meaning in a Networked Culture*. New York: New York University Press.

Johnson, Derek. 2013. *Media Franchising: Creative License and Collaboration in the Culture Industries*. New York: New York University Press.

Lemke, Jay. 2009. "Multimodality, Identity, and Time." In *The Routledge Handbook of Multimodal Analysis*, edited by Carey Jewitt, 140–150. London: Routledge.

Lin, Zhongxuan. 2017. "Contextualized Transmedia Mobilization: Media Practices and Mobilizing Structures in the Umbrella Movement." *International Journal of Communication* 11: 48–71.

Martens, Marianne. 2011. "Transmedia Teens: Affect, Immaterial Labor, and User-Generated Content." *Convergence* 17 (1): 49–68.

Marwick, Alice. 2013. *Status Update: Celebrity, Publicity, and Branding in the Social Media Age*. Connecticut: Yale University Press.

Marwick, Alice, Mary L. Gray, and Mike Ananny. 2014. "'Dolphins Are Just Gay Sharks': Glee and the Queer Case of Transmedia As Text and Object." *Television & New Media* 15 (7): 627–647.

Meyrowitz, Joshua. 1985. *No Sense of Place: The Impact of Electronic Media on Social Behavior*. Oxford: Oxford University Press.

Perryman, Neil. 2008. "Doctor Who and the Convergence of Media: A Case Study in Transmedia Storytelling." *Convergence* 14 (1): 21–39.

Ramasubramanian, Srividya. 2016. "Racial/Ethnic Identity, Community-Oriented Media Initiatives, and Transmedia Storytelling." *The Information Society* 32 (5): 333–342.

Sandvoss, Cornel. 2011. "Fans Online: Affective Media Consumption and Production in the Age of Convergence." In *Online Territories: Globalization, Mediated Practice and Social Space*, edited by Miyase Christensen, André Jansson, and Christian Christensen, 49–74. New York: Peter Lang.

Scannell, Paddy. 1996. *Radio, Television, and Modern Life: A Phenomenological Approach*. Oxford: Blackwell.

Scolari, Carlos Alberto. 2009. "Transmedia Storytelling: Implicit Consumers, Narrative Worlds, and Branding in Contemporary Media Production." *International Journal of Communication* 3: 586–606.

Scolari, Carlos Alberto, and Indrek Ibrus. 2014. "Transmedia Critical: Empirical Investigations into Multiplatform and Collaborative Storytelling." *International Journal of Communication* 8: 2191–2200.

Soriano, Cheryll Ruth R. 2016. "Transmedia Mobilization: Agency and Literacy in Minority Productions in the Age of Spreadable Media." *The Information Society* 32 (5): 354–363.

Stein, Louisa. 2009. "Playing Dress-up: Digital Fashion and Gamic Extensions of Televisual Experience in Gossip Girl's Second Life." *Cinema Journal* 48 (3): 116–122.

Stork, Matthias. 2014. "The Cultural Economics of Performance Space: Negotiating Fan, Labor, and Marketing Practice in Glee's Transmedia Geography." *Transformative Works and Cultures* 15. Doi: 10.3983/twc.2014.0490.

Striphas, Ted. 2015. "Algorithmic Culture." *European Journal of Cultural Studies* 18 (4/5): 395–412.

Van Dijck, José. 2013. *The Culture of Connectivity: A Critical History of Social Media*. Oxford: Oxford University Press.

Willett, Rebekah. 2016. "Online Gaming Practices of Preteens: Independent Entertainment Time and Transmedia Game Play." *Children & Society* 30 (6): 467–477.

Wohlwend, Karen E. 2012. "The Boys Who Would Be Princesses: Playing with Gender Identity Intertexts in Disney Princess Transmedia." *Gender and Education* 24 (6): 593–610.

Zhao, Sumin, and Michele Zappavigna. 2017. "Beyond the Self: Intersubjectivity and the Social Semiotic Interpretation of the Selfie." *New Media & Society*. Doi: 10.1177/1461444817706074.

Zimmerman, Arely. 2016. "Transmedia Testimonio: Examining Undocumented Youth's Political Activism in the Digital Age." *International Journal of Communication* 10: 1886–1906.

38

TRANSMEDIA PSYCHOLOGY

Creating Compelling and Immersive Experiences

Pamela Rutledge

Technological developments continually offer storytellers new delivery, connection and sharing options. Each choice raises an equal number of questions, from financial to design considerations, that are both exciting and burdensome. While audience expectations and demands materially change with each shift in technology, the increasing complexity gravitates attention toward the producer.

The wide array of new technologies allows audiences to be more discerning about how they allocate their time as consumption is limited by hours in the day. The psychological implications of "choice" are wreaking havoc leaving media businesses and producers scrambling for a cure. Transmedia storytelling has emerged as a means of addressing the competition for attention and expanding the breadth of access as well as satisfying audience demands for personal control, feedback, interactivity, and social connection integrated into media experience.

Technology changes expectations, but it does not change fundamental needs and motivations. In meaningful ways, people are the same as they have been for thousands of years, driven by instinct and emotion. This includes the drive for social connection, meaningful experience, and sharing stories to be part of something larger than themselves.

Everyone has a story. Stories are how humans construct identity, make sense of the past, anticipate the future, and find the motivation to take action (McAdams 2013). The increased sense of empowerment derived from digital technologies and the proliferation of media options has been accompanied by a desire for increased personal meaning in media experiences. The trend toward transmedia storytelling is well positioned as it can offer richer, more immersive and meaningful experiences by marrying consumption with participation. To satisfy audience demands, transmedia does not have to be a fully built-out, multiplatform, mega-budget storyworld, like the iconic entertainment franchises surrounding *Avatar* or *Star Wars*. Effective transmedia storytelling embraces the storied nature of life where multidimensionality, sociality, and interactivity are core features. Transmedia storytelling replicates human experience in how people naturally interact with their worlds and, at a neural level, in how brains process information and transform sensory input into conscious experience.

Pushing Boundaries

Transmedia continues to push the boundaries of storytelling through the use of multiple platforms and technologies that expand stories from singular linear narratives to multidimensional user experiences. It also, by necessity, disrupts traditional approaches to experience design with the complexity of creating holistic story experiences in danger of overshadowing the need to create sustainable audience engagement across texts.

The psychology of transmedia design is at the root of successful engagement in a storyworld through the experience of immersion and presence across multiple mediated experiences. Presence is an affective-cognitive construct characterized by the sensation of "being there" and describes an audience's perceived connection with a story (Gerrig 1993). Presence lowers cognitive resistance, making the audience more likely to suspend disbelief. In pursuit of presence, producers increasingly integrate the rich technology-enabled content such as augmented and virtual realities. But technology without psychology is a double-edged sword. Presence may come at the cost of decreased usability, narrative confusion, or reluctance to adopt new technologies.

To create a seamless experience across platforms, producers must adapt to new technology guided by psychological implications. They need a way to make judgments about which elements will add value to user experience at a fundamental level and which may detract.

Psychology Shifts the Focus

Findings in neuroscience, cognition, and perception combined with theories of optimal engagement and narrative transportation provide an integrative framework to evaluate the potential for immersion and engagement in and across new technologies. This approach enables creators to shift from a production-focused paradigm to a more user-centric one. This allows the integration of basic subconscious functions such as attention, perception, and cognitive heuristics, with conscious meaning-based processes such as enjoyment, flow, and social connection.

The link between psychology and project execution is especially important for transmedia storytelling because, by definition, the orchestration of elements relies heavily on understanding audience behavior. Success rests on the ability to support audience engagement from the initial connection with the story to migration across media. Transmedia's complexity can easily focus the emphasis on development over consumption. However, as transmedia producer Andrea Phillips (2012) notes, this preoccupation is problematic. To be successful, transmedia cannot assume the audience will appear, but needs a structure that entices, engages, and guides the audience at several junctures.

Producer Decision Points and User Experience

To expand a story into a storyworld, producers face a number of decision points—all with potential psychological consequences and commensurate moments of choice for the audience. While multiple media properties allow a well-crafted storyworld to take on rich dimensions, they require a greater commitment and motivational level from the audience to pursue the story beyond a single linear experience. The producer must decide not just how to tell the story but whether the myriad of potential choices, such as media assets, platforms, activities, and back or side stories, will enhance or disrupt the overall narrative coherence and the ability and desire of the audience to stay engaged.

User experience (UX) has been used as an umbrella term for a wide variety of human–media–technology interactions, yet transmedia migration is seldom one of them. Though the fields of human–computer interaction (HCI) and UX are rooted in cognitive psychology, the qualitative experience of storytelling, meaning-making, identity, and efficacy are uncommon (Petrie and Bevan 2009). While emerging as a usability tool, such as in persona development, they are not the driver of the overall experience. In HCI and UX, users have stories, but users are not in the story. In transmedia, the opposite is true. The audience is in the story, but audience members are rarely acknowledged to have stories of their own. Yet, most transmedia relies on discoverability, ease of use, and the affordances of different technologies, as well as audience preferences and needs. The ABC television show *Lost* had an extended storyworld across multiple platforms, such as mini webisodes, maps, blog, references sites, and a video game. All instances where audience-centric issues would be a factor in access and narrative experience and could ultimately constrain audience reach and, by extension, the user-generated content that played a role in *Lost*'s success.

Mapping transmedia experiences to psychological impact forces an examination of "why" from the perspective of both the creator and the audience. It focuses on how the story, structure, and distribution platforms are intended to engage different psychological processes and forces the articulation of expected outcomes. Deconstruction of transmedia is counter-intuitive in that it is the synergy among the texts that creates the immersive and often magical experience of a good transmedia story. The parts cannot reflect the richness of the whole. The parts do not capture the sensation of breadth in a storyworld like *Game of Thrones*, the shifting attitudes toward teen pregnancy and reproductive health due to Hulu's *East Los High*, or the poignant complexity of teen depression and suicide in Netflix's *13 Reasons Why*. This ability to amplify and broaden experience is transmedia storytelling's strength. However, the effect is reliant on activating individual elements that, taken together, create the sense of immersion that defines the audience's journey.

Psychological Foundations of Transmedia

A critical feature of the psychological theory behind transmedia is the interrelationship of unconscious and conscious processing. The integration of sensory stimuli in the human brain with conscious understanding of experience plays an important role in creating and sustaining engagement. Engagement can have many practical definitions depending upon project goals, from sustained attention to specific actions.

At the most basic level, engagement begins in the unconscious, with attention to sensory stimuli. The brain ultimately translates sensory input into conscious meaning in the context of an individual's previous experiences and beliefs. The multidimensionality of transmedia implies a continued shifting of attention and meaning-making, synthesized into a whole to achieve a level of engagement across texts. Thus, a transmedia experience, and the potential for sustained engagement, continually evolves as stimuli change.

The transmedia experience, therefore, can be tracked from sensory triggers that initiate attention through the conscious and unconscious impact of storytelling to the motivational dynamics of narrative and structure in transmedia migration.

The following are the critical points in this journey:

- Attention precedes audience engagement and is the product of instinctive processing.
- Social connection is a primary human goal and is central to the survival instinct.
- Humans exhibit a biological preference for real over virtual, however, both virtual and physical stimuli activate identical unconscious arousal responses. This response directs attention and is responsible for the ability to respond to an activity or action as relevant, desirable, valuable, pleasurable, or the opposite.
- Instinct and emotion subconsciously dominate the decision-making process.
- The brain processes all information using narrative structure as the sorting device to organize and link multisensory perceptions and meaning for later recall.
- Narrative as a theory of mind makes making storytelling fundamental to all human communication.
- A user's ability to achieve engagement relies on a producer's skill in understanding principles of sensory perception, emotion, and narrative processing.
- Neural networks, information processing, and encoding patterns enable narrative experience with or without overt storylines.
- Narrative is the universal factor enabling the "suspension of disbelief" that underlies psychological immersion.
- The ability to experience presence or to feel transported into a narrative can occur whenever a mediated experience triggers emotion and imagination.
- Presence and narrative transportation allow a narrative to become part of the audience's identity and personal story.

- The narrative zone is a construct describing the experience of being in a narrative.
- Theories of narrative transportation, flow, and presence define the boundary parameters that keep the audience in a narrative zone or storyworld.
- Flow and transportation theories are not interchangeable. They differ in the relative engagement of conscious to unconscious processing.
- Task-based actions that generate flow require higher consciously directed focus compared to narrative-based experiences that place more demand on unconscious processing, such as emotion and visualization, to fuel sense of presence.
- Sustained engagement, as described by both flow and transportation theories, requires the balanced coordination of conscious and subconscious processing, albeit to different degrees.

Attention Starts in the Brain

The critical component for transmedia storytelling is engagement—the ability to attract and keep attention. All physical and psychological experience, including the ability to notice and attend, is first filtered and then constructed by subconscious sensory processing systems, therefore persuasion, as the outcome of attention, starts in the brain.

The brain processes new information based on the survival imperative and gathers multisensory input to evaluate relevance, novelty (movement, newness, unusual behaviors), and pattern comparison (familiarity, sense-making) to determine the potential for threat or reward. Conscious attention is the result of unconscious arousal that occurs in response to the "pain or gain" threshold (Rutledge 2012).

Once information triggers attention mechanisms, cognitive processing continues by comparing new information to previous experience to determine the level of reward or threat. Content that is perceived as a reward will engage conscious processing to evaluate the positive potential. Research demonstrates that information that is both relevant to the user's goal and consistent with or enhancing the user's sense of self heightens the perception of value and motivates further attention. Continued attention creates concentration which increases the probability of liking. The ability to self-reference and self-identify promotes a favorable evaluation no matter what the quality of content logic or information (Escalas 2007).

Using the Triune Brain Heuristic in Transmedia Design

Insights drawn from neuroscience provide a valuable lens for transmedia design decisions and post mortem analysis of previous projects. A modular concept of the human brain was proposed by neuroscientist Paul D. MacLean in the 1960s. It continues to provide a simple approach to conceptualize and anticipate the impulsive and often unpredictable nature of human behavior and is widely applied in fields of neuromarketing and neuroleadership.

In the Triune Brain theory, MacLean proposed that people process information in three ways: instinctively, emotionally, and cognitively via separate brain centers that map to the acknowledged stages of evolutionary development of the human brain. These are the primitive or instinctive brain, the limbic system or emotional brain, and, most recent from an evolutionary perspective, the neocortex or rational brain (MacLean 1990).

The order of dominance in message processing has a profound impact on interpretation and response. Emotions are the functional directors of the neural interactions that govern attention and meaning, from perception to inference and goal choice (Reiner 1990). Therefore, instinct and emotion frequently drive behaviors. Unconscious instincts and emotions not only control reactive responses, but evidence suggests that as much as 95 percent of decision-making is based on instinct and emotion and is later rationalized by the conscious brain to align desires with conscious meaning. This predisposition is frequently exploited by marketers who link primal drivers of emotional arousal such as sex and rejection with congruent products.

Table 38.1 The Triune Brain theory highlights the three pathways of experience processing: lower level processing such as instinct and emotion and higher level, conscious processing that creates directed attention and meaning. Narrative is uniquely able to bridge all three, facilitating immersion and persuasion

Instinct and Emotions	*Conscious Attention*
Emotion-dominant	Achieving mastery bolsters self-esteem
Visual images have instant impact	Responsive feedback increases agency and validation
Gain and rewards create approach attitude	Participation creates a sense of ownership and affiliation
Pain and threat trigger avoidance response	
User-centric, personal relevance	Social collaboration meets connection needs
Finding answers and solving mysteries creates perceptions of safety	Empathy builds intimacy
← Narrative →	

The Triune Brain theory is a practical hierarchical framework for anticipating how content and structural choices will be received and what behavioral responses can be anticipated. Great content will not offset consumer frustration over poor user interface design because negative emotions trigger the instinctive responses of fight or flight, resulting in avoidance and distancing behaviors. Conversely, offering rewards of prizes or appreciation can increase a consumer's tolerance of difficulty because positive emotions enhance trust and liking. Content that is difficult to understand or find or that activates self-consciousness or self-doubt increases demands on conscious cognitive processing and impedes hedonic enjoyment (Barasch, Diehl, and Zauberman 2014). As indicated in Table 38.1, different attributes activate different brain functions. However, the critical rule of thumb is that instinct and emotion take precedence.

The Psychology of Story

Transmedia producers are well familiar with the nuts and bolts of storytelling since the foundation of transmedia is a good story. In cases such as entertainment properties that are anchored around a movie franchise like *Pirates of the Caribbean*, the story is well developed and articulated with accuracy and consistency across a number of media, such as books, video games, and even amusement park rides. In other cases, the storyworld is defined largely within the consumer's brain. This is most common in brand stories. Apple, for example, has created a clearly defined brand storyworld based on the celebration of individualism, creativity, and innovation that is demonstrated and reinforced emotionally by all customer touchpoints from package design, advertising campaigns, and product functionality to store design, sales personnel training, and the Genius bar. Chipotle Mexican Grill has also developed a brand story that is far more than that of a "Mexican fast food restaurant." The food genre is almost incidental to the brand story of social responsibility emphasized in the "food with integrity" tag line. Chipotle maintains story coherence across the restaurant experience and advertisements, and demonstrates their larger purpose through support of community fundraisers, sponsoring local farms, and the slow food movement. Chipotle has successfully amplified their brand story and expanded their reach, urging the audience to "cultivate a better world" with animated advocacy videos such as *Back to the Start* and *Chipotle the Scarecrow*, an accompanying free arcade-style adventure game for mobile devices, and even a satire Hulu TV series *Farmed and Dangerous*. In Apple's and Chipotle's storyworlds, consumers are the hero. As in any powerful story, the hero is transformed enabling customers to rewrite their personal stories and embrace aspirational identities.

The Brain Fills in Gaps

Transmedia storytelling campaigns benefit from the predisposition of the human brain to use narrative to facilitate meaning-making. As illustrated by Apple and Chipotle, a storyworld does not need to be fully articulated. Narrative is the native language of the brain and people are unconsciously motivated to make sense of information by organizing it as a story (Haven 2007). When stories are incomplete or information unavailable, the audience will infer connections, motives, and significance to achieve a sense of narrative logic (Bruner 1991). The "gaps" in brand stories make room for consumers to insert themselves, increasing brand salience. By definition, transmedia relies on the audience to insert themselves through sharing and user-generated content (UGC). The psychological UGC, the invisible by-product of the interaction between the consumer and the mental process of meaning making, goes unacknowledged.

The gestalt principles of perception demonstrate how the brain attributes meaning to information, assigning hierarchy, movement, dimension, and intentionality to visual objects based on their colors, relative sizes, location, and perceptions of movement (O'Connor 2015). This attribution of meaning is not restricted to visual information. The brain fills in the gaps and assigns relationships to all information using a narrative organizing structure that enables sequencing and extrapolations of causality. This ability to complete patterns, make judgments, and infer actions is an innate tendency essential to human survival. It not only drives the propensity to anticipate story endings, it is at the root of a number of cognitive heuristics that inform decision-making.

Decisions are the product of needs moving toward goals. Contrary to popular beliefs, human needs are not hierarchical (Rutledge 2012). Social neuroscience has confirmed earlier work on attachment and motivation and shown that social connection is fundamental to human physical and emotional survival (Lieberman 2013). Thus, storytelling, as a social activity, is not just an entertaining pastime. It defines what it means to be human. It is a core skill that moves a need toward a goal—human connection. In fact, storytelling is such a critical skill that the ability to tell stories is used as a measure of cognitive development.

Stories serve many functions critical to human survival. They:

- provide the primary mode of meaning construction and transfer of understanding;
- enable the exchange abstract concepts;
- are a conduit for social norms and cultural transmission;
- allow assignations of significance and intention that are essential to achieving interpersonal connection and emotional intimacy.

All Stories Start as Chemical Signals

All information comes through the five senses—sight, touch, smell, taste, and sound. Narrative structures enable the brain to translate, interpret, and encode sensory input as conscious meaning. Internally, narratives provide the sequencing and multisensory links that facilitate understanding and recall and become the foundation of beliefs, schemas, and mental models. The internalization of culturally shared mental models and schemas defines salient individual and social identities and provides the textual, visual, and auditory metaphors that storytellers can employ to effectively engage and motivate their listeners.

The implications of information processing suggest that storytellers should begin as listeners. The audience's stories contain the beliefs, assumptions, and metaphors that operate as filters and magnets, unconsciously influencing meaning and perceptions of relevance, authenticity, and liking. Internal cognitive structures and schema create a myriad of expectations, from the practical, such as where content should be found or how to navigate a site, to the emotional, such as motivating the audience to migrate across texts.

Metaphors and Archetypes: Cognitive Patterns of Shared Meaning

The psychologist Carl Jung used the construct of archetypes to conceptualize personality development and motivation. To Jung, archetypes represented universally shared primal forces in what he described as a collective unconscious. Jungian archetypes are patterns of human experience that create a common language for understanding life, people, and behavior. In contrast to personality traits, archetypes are a constellation of profound qualities that are instantly recognizable across cultures and evoke deep emotions (Read, Fordham, and Adler 1959/2014). The resurgence of interest in storytelling has created an accompanying appreciation for the power of archetypes and their ability to encapsulate core concepts, social roles, and behavioral attributes. Despite cultural variations, archetypal themes, settings, and characters embody essential elements of universal human experience that are identifiable across myth, literature, and rituals. Archetypal characters and themes feature prominently in the stories of well-known transmedia franchises, such as *Star Wars*, *The Matrix*, *Avatar*, and *The Hunger Games*. They are equally powerful communication tools in brand stories. A clear and consistent archetype sends an instantly understandable signal about a company's brand promise and correlates financially through strong brand equity and relationally through customer commitment fueled by a desire to belong (Mark and Pearson 2001). They are equally effective in focusing brand stories. Companies like Harley Davidson, Coca Cola and Disney have all benefited from the application of brand archetypes.

In storytelling, archetypes function as heuristics that deliver a large amount of meaning with relatively little information and effort. Archetypes tap into the audience's pre-existing knowledge and emotion and clearly communicate information about the characters. This instantly establishes expectations about plot, genre, and action, increases identification with characters and enhances audience commitment through meaning construction (Woodside, Sood, and Miller 2008). Identification with characters has been shown to predict the emotional impact and enjoyment of media entertainment and to increase the persuasive effects of messaging (Zillmann and Vorderer 2000; Green and Brock 2002). In addition, archetypes allow the story to bridge the cognitive gaps between texts during migration because the audience implicitly understands the roles and intentionality of the characters, keeping the audience in the narrative zone.

Stories Link the Parts of the Brain

Stories evoke emotion and images, activating the instinctive brain while actively engaging cognition and meaning. In the Triune Brain framework, stories have the unique ability to bridge all levels of the brain, linking instinct, emotion with rational thought.

Case Study: The Lizzie Bennet Diaries

The *Lizzie Bennet Diaries*, the transmedia adaptation of Jane Austen's classic story *Pride and Prejudice*, was hailed as an unqualified success due to its creativity and innovative approach. It is also a good case study to explore the psychological dynamics that can benefit transmedia.

The *Lizzie Bennet Diaries* drew on a known and loved story and storyworld reinforced by archetypal characters. The ready fan base fueled expectations, familiarity, and anticipation, increasing liking. Sequels and storyworld expansions of popular properties, such as Marvel Universe extensions, benefit similarly. The casting was consistent with known intentions of the characters; variations in side narratives aligned with current cultural metaphors, such as transforming Lizzie Bennet's best friend, the industrious, practical Charlotte Lucas, into an entrepreneur.

Multiple platforms took advantage of affordances to enhance relational connections between audience and characters. For example, the vlog format allowed actors to speak directly to the audience. Conversational speech and using the camera to simulate the experience of eye contact triggers mirror neurons which collapses the distinction between real and virtual. This personalization was reinforced by adopting the within-family nickname for Elizabeth (Lizzie), a social convention that signals access and intimacy.

The characters were available to engage with the audience on multiple social media channels. Twitter, in particular, was successful at breaking the fourth wall and further amplifying the sense of unmediated and authentic connection. The lack of perceived mediation on Twitter facilitates the development of parasocial relationships where the audience begins to feel they actually know the characters and actors, increasing emotional commitment (Horton and Wohl 1956).

Audience participation enhanced ownership and buy-in to the project's success, motivating people to share and encourage others to participate to maintain ego consonance. Comments, shares, and unearned media provided social proof and validated the *Lizzie Bennett Diaries*, creating opportunities for affiliation as a fan or arbiter of something new and special (Cialdini 2007).

To celebrate the fifth anniversary of the original launch, the creators re-released all of the content in real time on Facebook; this not only extended the fan base but rewarded existing fans by letting them re-experience the series as it was originally launched. This type of event triggers the neural networks storing the multisensory memories and amplifying current experience with previous emotions and meanings.

Case Study: Simplemente Maria

The 1969–1970s Peruvian telenovela, *Simplemente Maria* [*Simply Maria*], is an earlier example of how psychological dynamics can help propel a narrative. One of the longest running and most popular telenovelas in Latin America, it told the story of a young country girl who came to the city to find work and was seduced and abandoned. The audience lived Maria's story over 448 episodes. They watched her struggle with the social stigma of having a child out of wedlock, overcome economic hardships by learning to read and sew and, ultimately, emerge triumphant in business and love. While the goal of the program was to provide commercially viable entertainment, unintended social consequences included a spike in the sales of Singer sewing machines and young housemaids across Peru emulating Maria's behavior by enrolling in literacy programs (Singhal, Obregon, and Rogers 1995).

Like the *Lizzie Bennett Diaries*, *Simplemente Maria* drew on clear archetypal characters and themes. Maria was hardworking and idealistic despite challenges. Her Orphan (or Cinderella) archetype was placed in a culturally relevant story, reflecting the social struggles of the pressures from rural-to-urban migration in Peru at the time. Maria was relatable and aspirational; her success modeled a behavioral path and validated the hopes and efforts of many, reflecting, ultimately, the "just world" cognitive bias that good things should happen to good people (Dalbert 2004).

Simplemente Maria was also a harbinger of transmedia storytelling. Strong parasocial relationships developed in response to well-known and appealing characters and the frequency of television and concurrent radio broadcasts. Multiple distribution channels extended reach and access points and took advantage of the cultural habits of shared viewing and verbal discourse, increasing the retelling and psychological appropriation of content. Involvement was amplified by framing social and political issues in personal and familial terms around archetypal themes, increasing both salience and emotional participation.

Recognizing the power of subconscious drivers in programs like *Simplemente Maria*, Televisa executive Miguel Sabido was inspired to develop a theory-based approach to telenovela storytelling that would intentionally target social change. Drawing on experts in behavior change and communication, Sabido created a distinct methodology that was the basis for his productions, making him a pioneer in the Education-Entertainment genre. For over 25 years, he successfully produced media structured on psychological theory and achieved significant social impact, tackling topics from birth control and literacy to HIV testing (Papa et al. 2000).

The Sabido Methodology remains a model for the incorporation of psychological theory into media construction. It integrates: (1) dramatic theory to extend emotional range and increase attention and direct focus; (2) circular communication theories to capture the impact of messages and the renorming of beliefs reaffirmed by offline discussion and social sharing; (3) universals drawing on Jungian archetypes to trigger recognizable and relatable characters and plots and to enhance buy-in; (4) transitional characters to model a pathway for behavior change based on Albert Bandura's social cognitive theory; (5) instinctive, emotional, and cognitive triggers drawing on MacLean's Triune Brain theory; (6) enhanced personalization to promote parasocial relationships; and (7) Sabido's theory of tone which used sound variations to activate targeted somatic responses. The Sabido Methodology informs many successful serialized entertainment-education programs, notably the award-winning transmedia production *East Los High*.

The Sabido Methodology effectively targets both subconscious and conscious processing, creating multiple psychological access points to enter and emotionally engage with the story. Unlike today's producers, however, Sabido did not have to contend with a complex transmedia structure and the challenge of maintaining narrative immersion across media.

Structuring the Narrative Zone

Several factors enable or dissuade audience engagement and migration. The theories of flow and narrative transportation can be used in conjunction with the Triune Brain theory to conceptualize a narrative zone. The narrative zone serves as a guide to strategic decision making and navigating the tensions to achieve immersion and motivation.

Optimal Engagement: The Theory of Flow

Flow is the state of optimal engagement where the challenge of an activity matches the skills of the user. It is a continual balancing act between effort and concentration within the boundaries of the user's capabilities (Csikszentmihalyi 1991). Achieving flow results in a sense of deep enjoyment. Flow is defined as focused concentration on the activity with clear goals and feedback, with a sense of control and a loss of self-consciousness and time. Maintaining flow requires continually offsetting challenge with the requisite skill level. If the challenge is greater than the skill, anxiety results; if the skill is greater than the challenge, the individual becomes bored and disinterested. Optimal experience is not a steady state, but an evolving process of skill matching challenge through increasing and decreasing difficulty levels and opportunities for skill-building and mastery. The flow theory is applied to the development and analysis of gameplay, user experience, creative endeavors, and other activities with intentional concentration (see Figure 38.1).

In a transmedia strategy, the Triune Brain model can be used to operationalize the theory of optimal experience where the flow zone must extend across media. Independent of the narrative experience, migration requires energy expenditure, whether following a link or looking for hidden clues. The levels of challenge and arousal in the necessary activities must maintain attention and interest. This cognitive cost must be commensurate with the audience member's ability to achieve the gain. The actions necessary to reach the next entry point must not only promise visible reward, but they must support the unconscious perceptions of self-efficacy and positive emotion that fuel motivation. The Triune Brain model translates flow into instinct: a balance between threat and reward, avoid and approach. An activity must be challenging enough to achieve arousal and get the attention of the instinctive brain, but it must not surpass ability, creating threat. When the skill and challenge equilibrium work within the zone, it enhances self-efficacy and triggers the dopaminergic reward system which bolsters identity and self-esteem at the conscious level. Challenges that overwhelm skills threaten identity and self-competence, and are translated physically into a threat response triggering the hypothalamic-pituitary adrenal axis resulting in cortisol release. This physical manifestation is consciously interpreted as anxiety, dislike, anger, or frustration (Gregory and Rutledge 2016).

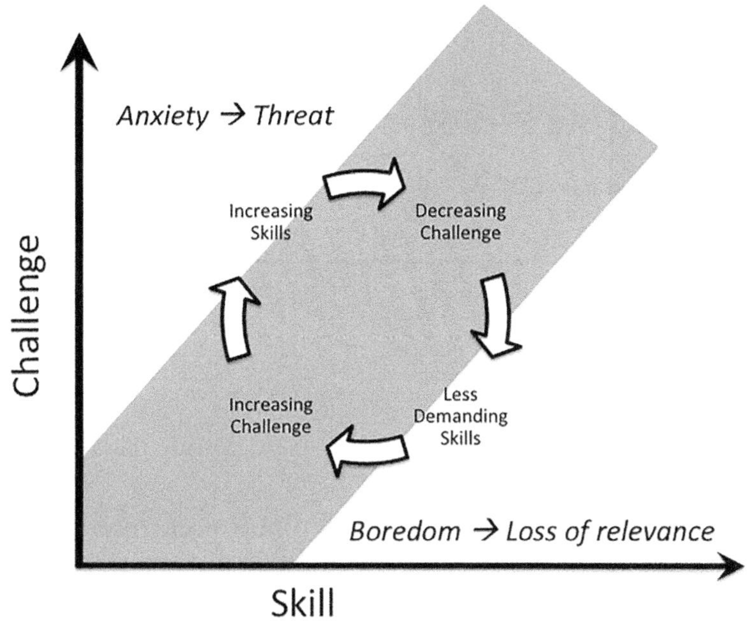

Figure 38.1 Flow theory combined with Triune Brain theory provides a useful heuristic to evaluate user experience as an ebb and flow of challenge and emotion that sustains engagement.

Research shows that staying in the flow zone increases positive affect; positive emotions increase motivation by inhibiting cognitive dissonance and increasing optimism and resilience (Fredrickson 2004). Negative experiences create a cognitive challenge to individual identity and can have a halo-effect, influencing the global opinion of a media property. Feelings of incompetence create cognitive dissonance and trigger the need to preserve or restore a positive sense of identity. To restore self-esteem and ego consonance, a person frequently attributes the negative affect to some aspect of the related experience (Elliot et al. 1994). In transmedia, challenges and cognitive dissonance that disrupt the ease of movement during migration create flow exit points, resulting in the potential loss of interest, the falling "out" of the story, and a lapse of motivation to continue exploring the storyworld (see Figure 38.2).

The greater the amount of cognitive and emotional investment in successfully meeting a challenge or task, the more absorbing it becomes. In flow theory, optimal engagement occurs when all available energy and skills are devoted to an activity (Nakamura and Csikszentmihalyi 2002). This implies that a flow-inducing activity must be responsive to and designed to account for player cognition and incorporate a range of emotional and perceptual limits.

Narrative Transportation

Transporting into a narrative is phenomenologically similar to flow in that it results in complete absorption. Where flow speaks to the emotional experience of being within a transmedia structure so that movement across texts is seamless, the narrative zone addresses a larger, more meaning-based and subjective experience. Flow involves smooth, uninterrupted focus on an activity, where narrative immersion is the fluency of imagining which results in the ease of processing and a redirection of attention away from conscious awareness of surroundings or tasks and into the narrative (Busselle and Bilandzic 2008).

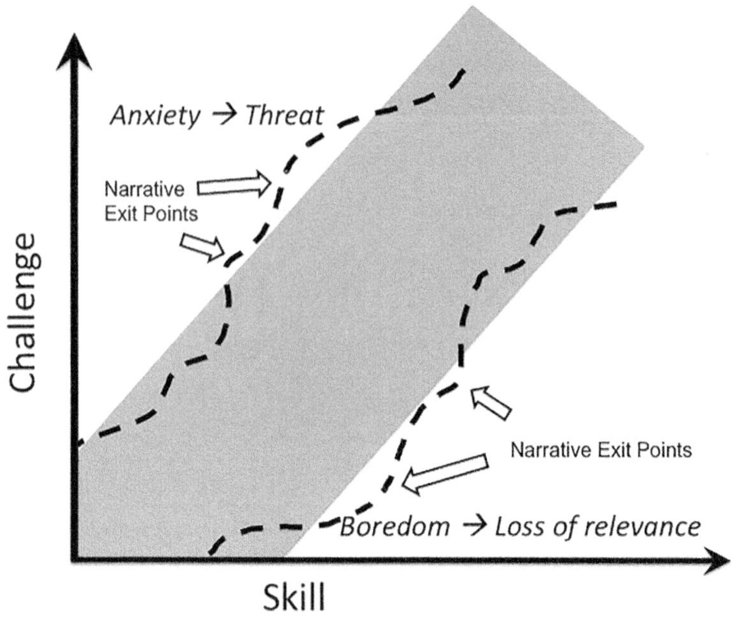

Figure 38.2 Transmedia experiences require effort on the part of the audience to follow the story. Therefore, exit points undermine the commitment to continue and can result in complete loss of audience engagement.

The flow state is often equated to narrative immersion with similar impact on subjective experience. However, in Triune Brain framework, flow is a new brain activity undertaken with intention, such as seeking out a companion social media site to a video series or following a link from one text to the next. Narrative actively engages all levels of the brain, but relies primarily on the engagement of old brain responses of emotion, visualization, and instinct prior to processing and meaning-making in the new brain.

Stories are holistic and, as such, replicate authentic human experiences. The brain's instinctive ability to project and visualize create the sense of presence and transportation. Narratives activate neural mechanisms creating genuine physical responses and emotions. These nonverbal impressions are translated into conscious meaning.

Similar to the flow zone, immersion in a story creates a narrative zone that is a balance of tension and release through the relational investments in characters and plot. It also inhibits cognitive challenge, thereby increasing the motivation to pursue further opportunities for relationships and being part of the characters' world. Crossing platforms breaks the fourth wall and challenges traditional mental schemas of entertainment narratives. These moments are vulnerable to narrative breaks where the audience loses the sense of immersion and psychologically falls out of the story. When moving to social media platforms, the powerful allure of personal relationships creates an emotional thread that can bridge the gap as long as signposts make the transition clear and the gap is cognitively narrow. Social media maintains the commitment to the narrative by moving the story into the audience's personal space, engaging and impacting identity, emotion, and beliefs.

The narrative zone defines the storyworld. There are three main components necessary for traveling in a storyworld: (1) cognitive engagement; (2) emotional engagement; and (3) mental imagery. These correspond to the functional breakdown of the three processing centers of the Triune Brain theory. It is the mutually reinforcing combination of cognition, emotion, and visualization that produces the experience of narrative transportation. Every within-narrative input, or audience interaction within the storyworld, amplifies the experience, whether it is exchanging tweets with Gigi from *Lizzie Bennet Diaries* or posting comments to the pregnant teen Cici's vlogs in *East Los High*. For example, applying the Triune Brain theory explains how the physical interaction with characters increases the drive to maintain connection due to the instinctive motivation to protect against the emotional loss of social disconnection.

In a transmedia story, the audience becomes the traveler, moving into and across a storyworld. While the narrative determines the traveler's role and identity, the platforms, with their distinct social norms and affordances, function as both context and content. Each platform creates a distinctive embodied experience and frames cognitive meaning.

The brain's power of projection and imaging can overcome any emotional distancing from migration as long as the content and platform are appropriately matched so that context and content are true to the canon psychologically as well as factually. When the story maintains internal integrity, active identification and imagination allow the audience to psychologically adapt to the logic and laws of the narrative world across media. While the narrative zone becomes more porous during migration, transmedia uniquely enables transformation from audience to stakeholder when the traveler takes ownership of the narrative and the journey, making it no longer the property of the storyteller alone.

Ownership shifts the audience mindset, amplifying the commitment to immersion as the narrative becomes intertwined with identity and self-efficacy. This enhances persuasion through the disinhibition of critical thinking functions accompanied by heightened emotional responses and amplified self-focus. Research shows that audience members who imagined themselves as the "star" of a narrative had stronger emotional responses and reported a more positive experience (Escalas 2007).

Media and Platform Choices Matter

Platform brand and message congruence is an opportunity to maximize message impact and story continuity. The differential impact of a common message across various media shows that the communication medium frames message perception. People have medium-specific cognitive processing strategies and social expectations, focusing on content type related to the source. Identical content is perceived differently depending on the platform. For example, content on Facebook was seen as having more personality (Walsh et al. 2013). Political messages on Twitter increased the consumer's sense of connection to the politician and the consumer's ability to become immersed in the narrative compared to identical messages delivered in a newspaper (Lee and Shin 2014).

Unconscious expectations and beliefs about any given medium affect how people manage their social contacts and communications strategies. This includes content-creation and interaction, independent of the message. Viewer-held assumptions about source influence the perceptions of information credibility, relevance, and potential for enjoyment (Coe et al. 2008). Consequently, to avoid narrative exit points and maximize experience, content must align with the platform brand personality and social norms.

Cost–Benefit Analysis

People run a cognitive cost–benefit analysis in the face of any decision or behavior change, including migrating across media. The costs and benefits can be broken down into categories of subjective evaluation and unconscious responses that manifest in observed behaviors and attitudes. These include beliefs and expectations about:

- Success, such as feelings about one's abilities, competence, and potential for attainment.
- The intrinsic value of the action, such as the enjoyment from the balance between stimulation and control, the act of performing a deed or the opportunity to play.
- Usefulness, such as longer-term goal achievement, skill-building, or learning.

- Perceived or actual costs, such as investment of time, money, identity, or the loss of other opportunities.
- Potential for social connection and affiliation with desired or undesired groups or identities.

These factors can influence motivation for migration across transmedia texts independent of story quality or creativity. The audience wants as few costs as possible to achieve their goals. An audience-centric perspective looks through the audience's eyes to understand their goals and see the potential obstacles that may hinder experience or conflict with outcome.

Conclusion

The continuing shifts in the media landscape and the increasing menu of media tools and platforms demand a better understanding of the psychological implications of content, structural choices, and their interaction. This is particularly true of the complex mix of story and strategic decisions necessary to create a larger, more meaningful and impactful experience in transmedia storytelling. Integrating a psychological framework enables an audience-centric approach to experience building. It integrates the audience's beliefs, values, and goals and highlights the underpinnings of individual preferences, meaning-making, and decision-making processes. Theories translated into actionable, audience-centric guidelines give storytellers, designers, and producers another dimension in which to conceptualize their decisions and keep the audience engaged in the narrative zone. The competition for the audience's attention will continue to grow as technology evolves. Being an early adopter of the "next new thing" is not a sustainable strategy. Applying psychology is, however, because fundamental human goals and needs remain constant. This knowledge allows producers to get in front of trends and anticipate how new technologies and approaches can satisfy the audience while minimizing resistance. Integrating psychology will enable a more purposeful intersection of story, structure, and audience to unleash the creativity and innovation that fuels a successful transmedia practice.

References

Barasch, Alixandra, Kristin Diehl, and Gal Zauberman. 2014. "When Happiness Shared Is Happiness Halved: How Taking Photos to Share With Others Affects Experiences and Memories." *Advances in Consumer Research* 42: 101–105.

Bruner, Jerome. 1991. "The Narrative Construction of Reality." *Critical Inquiry* 18 (1): 1–21.

Busselle, Rick, and Helena Bilandzic. 2008. "Fictionality and Perceived Realism in Experiencing Stories: A Model of Narrative Comprehension and Engagement." *Communication Theory* 18 (2): 255–280.

Cialdini, Robert B. 2007. *Influence: The Psychology of Persuasion*. Revised ed. New York: HarperCollins.

Coe, Kevin, David Tewksbury, Bradley J. Bond, Kristin L. Drogos, Robert W. Porter, Ashley Yahn, and Yuanyuan Zhang. 2008. "Hostile News: Partisan Use and Perceptions of Cable News Programming." *Journal of Communication* 58 (2): 201–219.

Csikszentmihalyi, Mihaly. 1991. *Flow: The Psychology of Optimal Experience*. New York: HarperCollins.

Dalbert, Claudia. 2004. "The World is More Just for Me than Generally: About the Personal Belief in a Just World Scale's Validity." *Social Justice Research* 12 (2): 79–98.

Elliot, Andrew J., Holly A. McGregor, Shelly Gable, and Patricia G. Devine. 1994. "On the Motivational Nature of Cognitive Dissonance: Dissonance as Psychological Comfort." *Journal of Personality and Social Psychology* 67 (3): 382–394.

Escalas, Jennifer Edson. 2007. "Self-Referencing and Persuasion: Narrative Transportation versus Analytical Elaboration." *Journal of Consumer Research* 33: 421–429.

Fredrickson, Barbara L. 2004. "The Broaden-and-Build Theory of Positive Emotions." *Philosophical Transactions Royal Society London* 359: 1367–1377.

Gerrig, Richard J. 1993. *Experiencing Narrative Worlds: On the Psychological Activities of Reading*. New Haven: Westview.

Green, Melanie C., and Timothy C. Brock. 2002. "In the Mind's Eye: Transportation-imagery Model of Narrative Persuasion." In *Narrative Impact: Social and Cognitive Foundations*, edited by Melanie C. Green, Jeffrey J. Strange, and Timothy C. Brock, 315–342. Mahwah: Lawrence Erlbaum Associates Publishers.

Gregory, Erik M., and Pamela Rutledge. 2016. *Exploring Positive Psychology: The Science of Happiness and Well-Being*. Santa Barbara: ABC-Clio Praeger.
Haven, Kendall. 2007. *Story Proof: The Science Behind the Startlng Power of Story*. Westport: Libraries Unlimited.
Horton, Donald, and R. Richard Wohl. 1956. "Mass Communication and Para-social Interaction." *Psychiatry* 19: 215–219.
Lee, Eun Ju, and Soo Yun Shin. 2014. "When the Medium Is the Message: How Transportability Moderates the Effects of Politicians' Twitter Communication." *Communication Research* 41 (8): 1088–1110.
Lieberman, Matthew D. 2013. *Social: Why Our Brains are Wired to Connect*. New York: Crown Publishers.
McAdams, Dan P. 2013. "The Psychological Self as Actor, Agent, and Author." *Perspectives on Psychological Science* 8 (3): 272–295.
MacLean, Paul D. 1990. *The Triune Brain in Evolution*. New York: Plenum Press.
Mark, Margaret, and Carol S. Pearson. 2001. *The Hero and the Outlaw: Building Extraordinary Brands Through the Power of Archetypes*. New York: McGraw-Hill.
Nakamura, Jeanne, and Mihaly Csikszentmihalyi. 2002. "The Construction of Meaning through Vital Engagement." In *Flourishing:Positive Psychology and the Live Well-Lived*, edited by Corey L. M. Keyes and Jonathan Haidt, 83–104. Washington, DC: American Psychological Association.
O'Connor, Zena. 2015. "Colour, Contrast and Gestalt Theories of Perception: The Impact in Contemporary Visual Communications Design." *Color Research & Application* 40 (1): 85–92. doi: 10.1002/col.21858.
Papa, Michael J, Arvind Singhal, Sweety Law, Saumya Pant, Suruchi Sood, Everett M. Rogers, and Corinne L. Shefner-Rogers. 2000. "Entertainment-education and Social Change: An Analysis of Parasocial Interaction, Social Learning, Collective Efficacy, and Paradoxical Communication." *Journal of Communication* 50 (4): 31–55.
Petrie, Helen, and Nigel Bevan. 2009. "The Evaluation of Accessibility, Usability and User Experience." In *The Universal Access Handbook*, edited by Constantine Stepanidis, 1–30. Boca Raton: CRC Press.
Phillips, Andrea. 2012. *A Creator's Guide to Transmedia Storytelling*. New York: McGraw-Hill.
Read, Herbert, Michael Fordham, and Gerhard Adler (eds.). 1959/2014. *C.G. Jung: The Collected Works—The Archetypes and the Collective Unconscious*. Second ed. Vol. 9. New York: Routledge.
Reiner, Anton. 1990. "An Explanation of Behavior: The Triune Brain in Evolution." *Science* 250 (4978): 303–305.
Rutledge, Pamela. 2012. "Augmented Reality: Brain-based Persuasion Model." 2012 EEE International Conference on e-Learning, e-Business, Enterprise Information Systems, and e-Government, Las Vegas, NV, July 16–19.
Singhal, Arvind, Rafael Obregon, and Everett M. Rogers. 1995. "Reconstructing the Story of Simplemente Maria, the Most Popular Telenovela in Latin America of All Time." *International Communication Gazette* 54 (1): 1–15.
Walsh, Patrick, Galen Clavio, M. David Lovell, and Matthew Blaszka. 2013. "Differences in Event Brand Personality Between Social Media Users and Non-users." *Sport Marketing Quarterly* 22 (4): 214–223.
Woodside, Arch G., Suresh Sood, and Kenneth E. Miller. 2008. "When Consumers and Brands Talk: Storytelling Theory and Research in Psychology and Marketing." *Psychology & Marketing* 25 (2): 103–145.
Zillmann, Dolf, and Peter Vorderer. 2000. *Media Entertainment: The Psychology of its Appeal*. Mahwah: Lawrence Erlbaum.

39

TRANSMEDIA RELIGION

From Representations to Propaganda Strategy

Marie-Eve Carignan

Closely linked with the human condition, the culture and the communication system of the world in which it takes place, the concept of religion is difficult to define and often described "as the belief in or the worship of a god or gods" (Croucher et al. 2017). In this chapter, I will explore the transmediality of religion by looking at how it evolves from its first manifestations to the coordinated use of mass media and socio-digital tools by religious organizations. As a place to start, Croucher et al. (2017) observe that:

> Researchers are beginning to explore the influence of mediation on religion and culture, how our globalized world affects the communication of religions and cultures, and how interreligious communication is misunderstood; and researchers are recognizing the need to extend studies into non-Christian religious cultures.

This relative absence of research into media in/as religion also extends to the field of transmedia studies, as there are very few studies about transmediality and religion despite it being a relevant area of analysis. As Wagner explains, we should "think of transmedia as religion, and religion as transmedia in the way that they both provide ways of negotiating with the mediums of life with the goal of unification and actualization of ultimate reality" (2012, 208 in Ringlestein 2013, 376). Perhaps one of the few ways in which transmedia scholarship has considered the religious examples of life is when it has chosen to mobilize them by comparing their actual influence on popular culture, such as fandom, as will be hinted at briefly below.

This chapter, however, presents a deeper examination of the comprehensive transmedia propaganda strategies used by extremist organizations, mainly Islamic State (IS). Certainly, we know that there is a major difference between religion, extremists, radicalized individuals, and terrorism (a concept that shares no singularly agreed upon definition in the scientific community). These concepts are all closely linked though, as Seib (2017) evokes, and this chapter's subsequent analysis of the use of transmedia propaganda strategies by terrorist organizations serves as a good example of how transmedia storytelling strategies can be mobilized by religious-motivated organizations.

Religion and Transmediality Over the Years

As noted, the application of the transmedia storytelling phenomenon to religious groups has received little consideration by scholars to date. Such relative absence is especially surprising given the way that some scholars have pointed to the transmediality of religion in their accounts of the histories of transmediality, which dates back to before the emergence of mass media. As showcased by other authors in this book, the use of transmedia storytelling can be seen across time, and is in no way a new phenomenon. Jenkins (2006), for example, recalls "the story of Jesus as told in the Middle Ages":

> Unless you were literate, Jesus was not rooted in a book but was something you encountered at multiple levels in your culture. Each representation (a stained-glass window, a tapestry, a psalm, a sermon, a live performance) assumed that you already knew the character and his story from someplace else.
>
> *(Jenkins 2006, 119)*

In this same vein, Burroughs (2013) also gives some examples of various transmediated religious practices in the context of ancient religions:

> A myriad of ancient religious practices might be conceptualized as incorporating multiple media platforms into worship. Egyptian ceremonies that integrate murals, smoke, and smell are inherently transmediated experiences designed to ingrain religious practice into the everyday. Murals were public reminders of Egyptian duty to deities, which seamlessly studded religious ritual into daily life. The burning of incense and smoke animated the gods in ritual ceremonies (Wise 2009). The utilization of bells, flags, ram horns, clocks, and pillars (Peters 2013) can all be thought of as historical religious media.
>
> *(Burroughs 2013, 73–74)*

Indeed, religions have always used a variety of support structures to convey their ideals and share their beliefs. This can include writing, painting, sculpture, songs, dance, music, singing, and architecture, either orally or in the collection of relics. Religions have placed a priority on different geographic locations to share their story with different people in different contexts at different times. Examples include ceremonial and religious sites, and through symbols, iconography, and so on. Wagner (2012), too, has shown how traditional religions have been effective at honoring different components or mini-stories that contribute to a much larger narrative.

The rise of transmedia storytelling on a grand scale is closely linked with industrialization, and with the emergence of consumerism and the development of mass media (Freeman 2016a). With the power to reach mass audiences and continued technological advancement, some researchers have compared the impact and strategies employed by mass media to those used by religion to gain audience reach. For example, Gerbner et al. (2002) compared religion and television:

> Television provides, perhaps for the first time since the preindustrial religion, a daily ritual that elites share with many other publics. The heart of the analogy of television and religion, and the similarity of their social functions, lies in the continual repetition of patterns (myths, ideologies, "facts," relationships, etc.) that serve to define the world and legitimize the social order.
>
> *(194)*

In the same way, the consumption of content from large cinema franchises such as *Star Wars, Harry Potter, The Twilight Saga, The Matrix* (see Wagner 2012) or *The Hunger Games* (see Ringlestein 2013) has been compared to a new form of religiosity closely linked with the transmedial consumption of content, with the aforementioned scholars citing various religious-like practices, as diverse sets of fans "have accumulated rituals, myths, codes of morals, clothing, texts, holidays, mass celebrations at [Comic-Con] and other ritual gatherings, and devoted followers who look to the worlds they present as motivation for how to live a good life" (Wagner 2012).

In turn, Burroughs (2013, 73) evokes how transmedia storytelling applied as a theoretical framework "can help portray how technological shifts in the digital era have impacted religious expression and can be expanded beyond mere popular culture," with the latter far more frequently serving as the object of transmedia studies. These industrial developments have allowed religions to assume a strategic transmedia approach in a way that allows them to make use of new and different media so to develop a more complex and complete message. As an example, Burroughs (2013, 74) demonstrates how the Mormon media systems use a religious transmedia framework "augmented by the use of social media and technology." Technological advances are creating a context that opens up new channels of communication to the public. This context is particularly relevant to the study of transmedia religion considering the use of different channels by some religious organizations and the impact of the new convergent media system on the appropriation of that religion. With this idea in mind, indeed, Burroughs uses the concept of techno-faith, explaining that:

> Techno-faith is the practice of inscribing mediated spaces with religiosity but also how religious institutions work to structure that practice. This is the role of digital rituals. In a transmediated religious ecology where multiple media platforms and texts are open to construct notions of faith, digital rituals intervene in the struggle between institutions and individuals and mediate face-to-face and digital disjunctures.
>
> *(2013, 77–78)*

Internet and social media both work to facilitate the interaction with a global audience and the implementation of innovative strategies, which in turn facilitate the capacities of all religious groups to disseminate their message across the planet. Socio-digital tools are an important manifestation of how the media ecosystem evolves in a more convergent way and change the religious groups' transmedia strategies. To demonstrate all of this, I will now focus the next sections on the use of these particular strategies by extremist organizations.

Transmedia Propaganda Strategies Used by the Islamic State Terrorist Organization

As mentioned at the beginning of this chapter, religion (understood here as the belief in a god or gods) and extremist movements are closely linked. Seib (2017) draws several additionally relevant links between the two concepts that are worth outlining in detail here:

> This history underscores the importance of understanding religion as part of the foundation of some forms of violent extremism, and of recognizing the role of religion in countering that extremism …
>
> But in practice, even the word of God means different things to different people, as do to particular terms. Consider *jihad*, which can be translated as "sustained struggle". According to some, this means a struggle within oneself to live according to the precepts of Islam, but others would say that it refers to an aggressive struggle against the enemies of Islam, a "holy war". Particularly in the West, the latter has been widely accepted as *jihad's* meaning, and *jihadist* is often used loosely as a synonym for "terrorist". Further, *jihad* can be viewed as both an individual and a collective duty, and so, if the community of Islam is attacked, all Muslims are required to join in *jihad* in response. But what is an "attack": combat such as that occurred during the Iraq War, or nonmilitary cultural encroachment by the West? […]
>
> In her study of IS's use of religion in its pronouncements, Theresa Ford noted that "87 percent of jihadist propaganda included justifications from the Qur'an, Hadith [words and actions

of the Prophet Mohammed], or scholarship," and some of the group's statements cite the Qur'an dozens of times ...

> Whatever their beliefs, and whether driven by reasoned malice or unreasoning insanity, many violent extremists wrap themselves in religion to disguise their destructiveness as something noble. Religion can never justify criminal irresponsibility on the part of individuals, organizations, or even nations. Nevertheless, religion and other aspects of culture cast broad shadows. Beyond individual acts are ramifications related to broad culture-based conflict, a "clash of civilizations."
>
> *(Seib 2017, 47–50)*

If religion and extremism are closely linked, then, so too is IS and the rise of the radicalization phenomenon that is itself closely associated with violent Islamism, as featured by news agencies around the world. Radicalization, particularly in the context of religion, refers to a phenomenon by which "a person is initiated to an ideological message and encouraged to replace his moderate beliefs with extreme opinions. In this regard, radicalization means adopting beliefs that most people do not hold" (Bélanger et al. 2015, our translation). Sometimes, radicalized individuals adopt "extremist belief systems—including the willingness to use, encourage or facilitate violence—with the aim of promoting an ideology, political project or cause as a means of social transformation" (Center for the Prevention of Radicalization Leading to Violence (CPRLV) 2016). This latter phenomenon can be defined as "radicalization leading to violence," which entails the "belief in the use of violent means to promote a cause" (CPRLV 2016). Transmediality, meanwhile, "is powerful in the way it constructs worlds for people seeking meaning, order, purpose and a sense of belonging" (Ringlestein 2013). In this sense, the use of a comprehensive transmedia propaganda strategy by an extremist group can be understood as a good way to convert new capability around the globe.

Indeed, through its widely developed and cleverly executed communication strategies, IS has repeatedly been successful in radicalizing (and enlisting) young Westerners, who should be very reluctant to respond to such group calls, as it is far from their realities. But the use of new communication tools, such as social media and video games, enables the terrorist organization to interest them and to raise awareness about their cause. The Internet is therefore a key element in this process of radicalization, and it plays a crucial role in all the phases of radicalization up until the deployment of young radicals until they are on the ground with formal organization (Precht 2007). However, Pauwels et al. (2014) cite the need to distinguish between the active exposure to content of extremist groups on digital networks, the deliberate search for certain information, and passive exposure. This would include discovering select content by accident, namely by doing other things online. The results of the interviews conducted by Pauwels et al. (2014) demonstrates that digital networks are mainly used as a useful resource or tool in order to then pursue offline interests that were developed online, or to keep up to date in terms of the movement and organization of networks. According to Pauwels et al., it "is very unlikely that an individual can radicalize only by using social networks" (2014, 30; our translation); instead, the social networks need to be a part of a bigger organized system or "is a facilitator who will be involved when the process of radicalization is already under way" (2014, 28; our translation). These new and advanced communication strategies, of which we have already seen the first manifestations by groups such as Al-Qaeda, have led researchers to focus more on the circulation strategies of these groups' radicalization messages within its various communication platforms. The synergy of different media is now enabling these groups to refine their communication strategies and to use a diverse array of platforms so to broadcast more complex messages (Preston et al. 2011).

As Monaci also demonstrates, IS uses transmediality to reach targets and by exploiting "different media platforms, audiovisual contents, and synergies among various media assets to enhance its messages" (2017, 2845). This particular strategy echoes Jenkins' original definition of transmedia storytelling: "integrate multiple texts to create a narrative so large it cannot be contained within a single medium" (2006, 95). Monaci (2017, 2845) goes on to explain how

> IS pursues a comprehensive transmedia propaganda strategy aimed at amplifying the reach of messages presented in *Dabiq* [a high-quality magazine produce by IS, published in several languages from 2014 to 2016 and recently replaced by the magazine *Rumiyah*] through synergic relations with multiple online contents, such as videos released on various websites, *nasheed* (vocal music that carries Islamic beliefs), Twitter hashtags, and so on.

In fact, the IS transmedia propaganda strategy is so well organized that:

> No other terrorist organizations—and not many governments—have matched the volume and sophistication of IS media content. This kind of productivity does not ensure ultimate victory (however IS might define that), but it does guarantee a prominent place in the global public's consciousness and establishes a precedent for other terrorist groups, present and future.
>
> *(Seib 2017, 123)*

This strategy is not only based on the synergistic use of the organization's own media, then, but also on the constant use of the mass media to reach wider populations and governments.

IS Strategies for Reaching Mass Media

While developing this particular communication strategy that takes advantage of the complementary use of the media, IS may also try to reach mass media to make their message heard. But since their extremist position rarely gets redistribution by said mass media, they try to provoke spectacular events that will capture the media's attention. By doing so, IS expect that the media will cover the events and, at the same time, talk about their espoused cause and, by extension, the media will further their public and political reach beyond what they usually get. This process is what Brigitte Nacos (2002) calls "media-oriented terrorism." By doing spectacular attacks that target a large public, the group's intended affect is to gain media attention and to use the reciprocal communication process between media, public, and government that Nacos called the "Triangle of Political Communication." This is why in recent years we have seen many concurrent attacks intended to increase the number of victims and the impressive effects of these attacks, as well as the reach of the organization. The series of attacks on November 13, 2015 in Paris and the suburbs of Paris, France, including shootings and suicide attacks, is an example of this strategy. In a short time, the terrorists attacked the Stade de France, the terraces and cafes in the city center, as well as the theater of the Bataclan, during the concert of the American group Eagles of Death Metal. The suicide bombings of March 22, 2016 at Brussels airport and the metro train near the Maelbeek station in Belgium are another example. Some terrorist organizations are also willing to claim attacks they have not committed themselves in order to gain the media's attention. In recent years, in fact, many attacks claimed by IS have not been clearly linked with the organization.

Some terrorist organizations have also started to adopt advanced communication and public relations strategies in order to better engage mass media. Said groups have started to use journalistic techniques on their media platforms, hoping that this information will more easily reach the journalists and can be used "as is" in the traditional media (Linera Rivera 2016). To quickly reach the media, they will often choose to disseminate this information in English first, rather than in Arabic. One good example of these communication strategies put in place by terrorist organizations are videos of beheading hostages distributed by IS. About these videos, Seib explain:

> Its execution images are best known and even though they constitute only a relatively small percentage of IS online content, they illustrate sophisticated media manipulation skills. As Judith Tinnes has written, "Many mass media outlets—most notably tabloids—show unashamed voyeuristic fascination with execution footage, often ignoring the full (much broader and complex) spectrum of ideological messaging." The mass media, wrote Tinnes, serve IS "as free-of-charge multipliers for their propaganda". She added that during 2015 alone, IS executed more than 1,000 captives in front of still or video cameras.
>
> *(Seib 2017, 122)*

Even if the use of beheading for a religious purpose, or as a symbol of war success, is not a new phenomenon (see Baquedano and Graulich 1993 in Zech and Kelly 2015), "contemporary jihadist beheading is not only an extension of hostage-taking, it is also an independently evolving terrorist tactic" (Lentini and Bakashmar 2007). These tactics are employed for many different reasons, including religious motives, follower recruitment, securing tangible assets (like acquiring funds, an exchange of prisoners or publicity), cycle of violence (take revenge, provoke or respond to other governments for example) and deterrence (as using the public display of video to terrify the audience) (Tures 2015). The concrete example of these motivations could include "obtaining ransom payments, hampering foreign investment, discrediting transitional states, and recruiting supporters" (Lentini and Bakashmar 2007, 303). Zech and Kelly (2015, 85) suggest that IS' goal by distributing these videos—other than to react to external threats and to obtain collaboration from foreign governments or as internal sanctions—is "to provoke Western powers and raise IS' profile abroad." As Zech and Kelly (2015, 88) explain further: "The actions of Jund al Khilafa (Soldiers of the Caliphate) in Algeria also suggest that beheading is perceived as a strong brand-building activity that other groups might use to cement their position on the side of the Islamic State."

This kind of distribution of carefully staged beheading videos that show a jihadist speaking in a perfectly mastered western language expressing demands to foreign governments proceeding a beheading of a hostage placed before him, and on his knees in an orange jumpsuit, this is reminiscent of American inmates (think US prisoners in the Guantanamo jail), and thus the IS group is seeking to create a strong visual that will leave a mark on the public imagination. In many of these videos "the victims are seen to 'repent and recant.' The confessional aspect is part of IS' attempt to control the narrative and legitimize the murders in the eyes of viewers. Surely only the guilty would feel compelled to confess their sins?" (Zech and Kelly 2015, 87).

> One crucial aspect of beheadings by IS militants is [also] the visceral symbolism captured in decapitation videos with the ritualistic placement of the head back onto the body. Politically, beheading has run the gamut from special punishment for traitors or "enemies of the state," to a standardized method of carrying out the death penalty, as in revolutionary France.
>
> *(Zech and Kelly 2015, 87)*

From a transmedia perspective, these videos of hostage beheadings are seen on various web and social media platforms, before being featured on traditional media, including print media forms that use diffused screenshots of the videos. Whether each media form uses this imagery or not is often quite varied. Some media present only the first images of the videos, whereas others stop the video before the beheading (sometimes showing images taken after the decapitation with the head in the background), while others consider that it is of public interest to present the video in its entirety, albeit offering viewers a warning before the start of the broadcast. Other media, though, have chosen not to (or no longer) broadcast these videos at all so as not to participate in the game of the terrorist groups; such media are now opting to broadcast images of the hostages before they are captured instead. In other words, the transmedial construction of these images and narratives is rather fragmented and distorted, with the regulatory guidelines of some media companies meaning that parts of the content are available on some platforms but not others. Despite the fact that very little analysis has been offered to the influence of these beheadings on the public and political attitudes, the terrorists groups "assume that beheadings work because they receive so much coverage in the Western press. And perhaps those in the media assume they work because people are watching" (Tures 2014, 129). For as Friis (2015, 726) puts it, "the extensive attention devoted to ISIS's beheading videos [by the governments, the media and the citizens], and the significance ascribed to them, are highly illustrative of the ever more apparent importance of visual imagery and visual media in contemporary warfare."

Is the Mass Media Being Caught in the Strategies of Terrorist Organizations?

Given the example outlined above, this question to do with the unintended role of mass media in the strategies of terrorist organizations becomes a pertinent one, especially given the broader collapsing of traditional media business models in favor of free information being available to audiences online. This contemporary journalistic context has brought foreword an emphasis on "infotainment," a concept that Freeman (2016b) describes as the concrete manifestation of transmediality for informational and non-fictional content. The shift to more entertaining journalism stems from even broader industrial shifts concerning the ways that journalists are now required to do their job more quickly, reducing opportunities for detailed analysis of events. Indeed, journalists now have less time to approach multiple sources of information, particularly in crisis situations, such as armed attacks, simply as a result of the need to publish information faster than competitors. Chaouch (2016) and Seib (2017) both argue that the temporal and economic pressures faced by journalists today benefits those terrorists that take advantage of compressed timelines and reduced analysis to impose media images that they have staged themselves.

By way of example, elsewhere I have examined the Quebec Press Coverage of how radicalization leading to violence in Canada, show that the mass media rarely dwells on defining or explaining this phenomenon, instead tending to focus on the events themselves so to get the attention of the public—which is exactly what the terrorist group are expecting by provoking spectacular events (Carignan and Marcil-Morin 2018). In this previous work, I also looked at the media coverage of the October 2014 attacks that occurred in Canada (St-Jean-sur-Richelieu and Ottawa), highlighting the media practice of focusing attention on the perpetrators of the attacks, the radicalized individuals, and their private lives. By doing so, the media tend to present personalized images of these individuals and speculate about the "potential" motivation behind such attacks. At the same time, the mass media will also share different information made available by the terrorist organization deployed to gain media attention. And by doing so, the media further legitimize the propaganda messages that have been orchestrated by these organizations. For example, in this instance, various mass media shared a video created by Michael Zehaf-Bibeau, the perpetrator of the Ottawa attacks, in his car mere moments before taking action, a video in which he partly cites his motives for committing the violent acts (Carignan and Marcil-Morin 2018).

Echoing the workings of transmediality, such extremist messages released by the perpetrators of Saint-Jean-sur-Richelieu and Ottawa attacks were extensively distributed across their own social media platforms as well as via more traditional broadcast channels, with both parties relaying these messages back and forth. Considering that a number of researchers are pointing to the potential dangers of giving "too much space" and importance to those responsible for violent events, and spreading their propaganda (Hénin 2015; Berthomet 2015), which would in turn glorify them and spur copycats to reproduce similar acts of violence, this analysis raises a lot of questions, namely: is the mass media playing the game of terrorists without knowing it? And is the transmediation of today's media content making the acts of terrorism more prolific and accessible? As Augé (2016) has also suggested:

> Even though social networks have given more autonomy to terrorist groups, studies on the influence of the mass media show that television and the print media remain at the

> forefront of information tools, so [IS] is still far from free himself from the mass media that "credibility" its danger and shows the extent. Nevertheless, the more the noise is resounding on the Internet, the more it is inevitable to talk about it on television. In the face of the monopoly that [IS] exercises on its territory, journalistic independence becomes a daily challenge.
>
> *(216, our translation)*

Conclusion

The objective of this chapter was first to demonstrate that the implementation of transmedia strategies by religious movements is not a new phenomenon. From the earliest religious communications and in the first forms of representations of gods or divinities throughout the ages, we can see the formation of transmedia representations that give meaning and greater effect to the story. Consider, for example, religious representations combining text, orality, music, and singing allowing their author to reach a wide audience regardless of age or reading ability (Pearson 2009).

What does appear to be new in the transmediality of religion, however, is the introduction of new technologies that have enabled the implementation of even more advanced communication strategies, which have expanded the reach of followers to distant territories, creating a new form of "techno-faith." This chapter's example of IS and the many tactics that it adopts to ensure complementarity of its messages across various media platforms to reach future Western followers, and the cleverly orchestrated communication strategies it puts in place to reach the mass media, are all illustrative of how a terrorist organization can make use—and are making use—of comprehensive transmedia propaganda strategies in the twenty-first century. In academic circles, this land of transmedia religion seems particularly fertile and I invite more research to be done on the ways in which various religious practices are being developed as transmedia practices. It would be interesting, not only to look at the strategies put in place by religious groups, but also to consider the social impacts of these strategies and at their efficiency in communicating different sets of beliefs.

Acknowledgements

I would like to thank Mikaëlle Tourigny, from Université de Sherbrooke, for her significant contribution to the theoretical framework of this chapter.

References

Augé, Rawaa. 2016. "Daech et les Médias: Coulisses d'un Mariage Forcé" [ISIS and the Media: Behind the Scenes of a Forced Marriage]. *Hérodote* 1 (160/161): 209–222.

Baquedano, Elizabeth, and Michel Graulich. 1993. "Decapitation Among the Aztecs: Mythology, Agriculture and Politics, and Hunting." *Estudios De Cultura Nahuatl* 23: 163–178.

Bélanger, Jocelyn J., Noëmie Nociti, P. E. Chamberland, V. Paquette, D. Gagnon, A. Mahmoud, and C. Eising. 2015. *Bâtir une Communauté Résiliente dans un Canada Multiculturel: Trousse de Renseignements sur l'Extrémisme Violent* [Build a Resilient Community in a Multicultural Canada: Information Kit on the Violent Extremism]. Montreal: Université du Québec à Montréal.

Berthomet, Stéphane. 2015. *La Fabrique du Djihad: Radicalisation et Terrorisme au Canada* [The Jihad Factory: Radicalization and Terrorism in Canada]. Montreal: EDITO.

Burroughs, Benjamin. 2013. "And I'm a (Social Media) Mormon: Digital Ritual, Techno-faith, and Religious Transmedia." *Open and Interdisciplinary Journal of Technology, Culture and Education* 8 (2): 71–81.

Carignan, Marie-Eve, and Sara Marcil-Morin. 2018. "Canada: Transmediality as News Media and Religious Radicalization." In *Global Convergence Cultures: Transmedia Earth*, edited by Matthew Freeman and William Proctor, 121–139. London and New York: Routledge.

Center for the Prevention of Radicalization Leading to Violence (CPRLV). 2016. *What is Radicalization Leading to Violence?* Accessed March 9, 2017. https://info-radical.org/en/radicalization/definition/.

Chaouch, Maxime. 2016. "La Représentation Médiatique du Terrorisme: Analyse de la Construction Médiatique de l'Événement Terroriste et Approche des Implications Psychosociales" [The Media Portrayal of Terrorism: Analysis of the Media Construction of the Terrorist Event and the Approach of the Psychosocial Implications]. M.A. thesis, Université de Caen Basse Normandie.

Croucher, Stephen M., Cheng Zeng, Diyako Rahmani, and Mélodie Sommier. 2017. "Religion, Culture, and Communication." *Oxford Research Encyclopedia of Communication*. Doi: 10.1093/acrefore/9780190228613.013.166.

Freeman, Matthew. 2016a. *Historicising Transmedia Storytelling: Early Twentieth-Century Transmedia Story Worlds*. London and New York: Routledge.

Freeman, Matthew. 2016b. "Small Change–Big Difference: Tracking the Transmediality of Red Nose Day." *VIEW Journal of European Television History and Culture* 5 (10): 87–96.

Friis, Simone Molin. 2015. "Beyond Anything We Have Ever Seen: Beheading Videos and the Visibility in the War Against ISIS." *International Affairs* 91 (4): 725–746.

Gerbner, George, Larry Gross, Michael Morgan, and Nancy Signorielli. 2002. "Growing Up with Television, The Cultivation Perspective." In *Against the Mainstream: The Selected Works of George Gerbner*, edited by Michael Morgan, 193–213. New York: Perter Lang.

Hénin, Nicolas. 2015. *Jihad Academy: Nos Erreurs Face à l'État Islamique* [Jihad Academy: Our Mistakes in Facing the Islamic State]. Paris: Fayard.

Jenkins, Henry. 2006. *Convergence Culture: Where Old and New Media Collide*. New York: New York University Press.

Lentini, Pete, and Muhamad Bakashmar. 2007. "Jihadist Beheading: A Convergence of Technology, Theology, and Teleology?" *Studies in Conflict & Terrorism* 30 (4): 303–325.

Linera Rivera, Rafael E. 2016. "Social Representation of Threat in Extended Media Ecology: Sochi 2014 Olympics, Jihadist Deeds, and Online Propaganda." Ph.D. diss., Fielding Graduate University.

Monaci, Sara. 2017. "Explaining the Islamic State's Online Media Strategy: A Transmedia Approach." *International Journal of Communication* 11: 2842–2860.

Nacos, Brigitte L. 2002. *Mass-Mediated Terrorism: The Central Role of the Media in Terrorism and Counterterrorism*. New York: Rowman & Littlefield.

Pauwels, Lieven, Fabienne Brion, Brice De Ruyver, Marleen Easton, Nele Schils, and Julianne Laffineur. 2014. *Comprendre et Expliquer le Rôle des Nouveaux Médias Sociaux dans la Formation de l'Extrémisme Violent. Une Recherche Qualitative et Quantitative* [Understanding and Explaining the Role of New Social Media in the Formation of Violent Extremism. A Qualitative and Quantitative Research]. Accessed March 20, 2017. www.belspo.be/belspo/fedra/TA/synTA043_fr.pdf.

Pearson, Roberta. 2009. "Transmedia Storytelling in Historical and Theoretical Perspectives." Paper presented at The Ends of Television Conference, Amsterdam, June 29–July 1.

Precht, Tomas. 2007. *Home Grown Terrorism and Islamist Radicalisation in Europe: From Conversion to Terrorism*. Copenhagen: Danish Ministry of Justice.

Preston, John, Jane Binner, Layla Branicki, Maria Angela Ferrario, and Magnalini Kolokitha. 2011. *Multiple Attacks on Transport Infrastructure: An Inter-Disciplinary Exploration of the Impact of Social Networking Technologies upon Real Time Information Sharing, Response and Recovery*. London: University of East London.

Ringlestein, Yonah. 2013. "Real or Not Real: The Hunger Games as Transmediated Religion." *Journal of Religion and Popular Culture* 25 (3): 372–387.

Seib, Philip. 2017. *As Terrorism Evolves: Media, Religion and Governance*. New York: Cambridge University Press.

Tures, John A. 2014. "Have Beheadings Intimidated or Angered the American Public." *Seton Hall Journal of Diplomacy and International Relations* 2: 117–134.

Wagner, Rachel. 2012. *Godwired: Religion, Ritual and Virtual Reality*. New York: Routledge.

Zech, Steven T., and Zane M. Kelly. 2015. "Off with Their Heads: The Islamic State and Civilian Beheadings." *Journal of Terrorism Research* 6 (2). doi: 10.15664/jtr.1157.

PART V

Methodologies of Transmediality

40

A NARRATOLOGICAL APPROACH TO TRANSMEDIAL STORYWORLDS AND TRANSMEDIAL UNIVERSES

Jan-Noël Thon

Despite their manifold differences, transmedial franchises such as *The Lord of the Rings*, *A Song of Ice and Fire*, or *Harry Potter*, *Batman*, *X-Men*, or *The Walking Dead*, *Star Wars*, *Indiana Jones*, or *The Matrix*, *Doctor Who*, *Star Trek*, or *Lost*, and *Tomb Raider*, *Warcraft*, or *Halo* tend to be most visibly defined by their narrative functions. Accordingly, a number of influential approaches to the study of transmedial franchises such as Henry Jenkins' concept of "transmedia storytelling" (see Jenkins 2004, 2006) or Lisbeth Klastrup and Susana Tosca's concept of "transmedial worlds" (see Klastrup and Tosca 2004, 2018) focus on the representation of characters, stories, and worlds across media such as novels, comics, films, television series, and video games. The significant heuristic value of Jenkins' as well as Klastrup and Tosca's approaches notwithstanding, however, it seems that both the concept of "transmedia storytelling" and the concept of "transmedial worlds" suffer from a largely unexamined commitment to what one could call the model of the "single world."

While such a model would, of course, be entirely appropriate in those cases in which transmedial franchises not just aim at but actually succeed in representing a "single world" as part of what Jenkins describes as the ideal of "a unified and coordinated entertainment experience" (Jenkins 2007), on closer inspection, these cases appear to be rarer than one might expect. One way or another, it seems somewhat unsatisfying to exclusively base the analysis of transmedial franchises' narrative functions on the assumption that they ideally should represent a "single world," independently of whether we call that world a "story world" (Jenkins 2004, 124; see also Harvey 2015), a "transmedial world" (Klastrup and Tosca 2004; see also Wolf 2012), or something else altogether. Instead, it may be helpful to move from the model of the "single world" to a model of "multiple worlds," thus more explicitly acknowledging transmedial franchises' specific brand of narrative complexity.

Against this background, the following presents a narratological approach to the analysis of transmedial storyworlds and transmedial universes, using the theoretical framework of transmedial narratology to further contribute to the development of a nuanced account of the complex arrangements of narrative works that characterize the above-mentioned transmedial franchises. Aiming to complement rather than contradict existing approaches within current transmedia studies, transmedial narratology offers a theoretically and methodologically refined foundation upon which we can build in order to extend the existing tools for the analysis of transmedial franchises' narrative functions beyond the model of the "single (story)world" (see also Thon 2016 for an in-depth discussion of transmedial narratology as a theoretical frame and method of analysis; as well as Thon 2015 for a more detailed version of the argument presented here in abbreviated form).

The Concept of the Storyworld and Representational Correspondence across Media

While the concept of storyworld has recently begun to gain currency within transmedia studies, as well, it has been primarily developed within the field of narratology. Indeed, the history of the concept can be traced from Gérard Genette's "diegesis" (1988, 17) and Seymour Chatman's "world of potential plot details" (1978, 29), via the "fictional worlds" of possible worlds theorists such as Thomas Pavel (1989), Ruth Ronen (1994), or Lubomír Doležel (1998), to cognitive narratologists such as Marie-Laure Ryan (1991), Richard Gerrig (1993), or David Herman (2002). Despite a common conceptual core, then, the various approaches to storyworlds that are located within different strands of narratological practice not only use a variety of terms to refer to them but also conceptualize them rather differently. Herman, for example, understands storyworlds as "mental models of who did what to and with whom, when, where, why, and in what fashion in the world to which recipients relocate" (2002, 9) or as "global mental representations" of "the world[s] evoked implicitly as well as explicitly by a narrative" (2009, 107).

Yet, equaling storyworlds with their mental representations unhappily ignores that we usually presuppose some kind of intersubjective plausibility when we talk about what is represented by narrative works across media. We can, in other words, construct more or less accurate or appropriate mental models of a given narrative work's storyworld. Of course, this does not at all mean that the ways we imagine the storyworld based on a given narrative work will be entirely alike, nor that they should be—but equaling the storyworld that a narrative work represents with the way in which individual recipients imagine it rather obviously runs counter to how we usually talk about storyworlds, both as "mere" recipients and as academics. One way to solve this terminological problem within the context of transmedial narratology would be to distinguish between the medial representation of a storyworld as part of a narrative work, the mental representations of that storyworld that recipients build during the reception process, and the storyworld itself as an intersubjective communicative construct (see, e.g., Eder 2008; as well as, once more, the more detailed discussion in Thon 2016).

One of the advantages of a narratological perspective on transmedial franchises is that it provides a better understanding of the processes that underlie narrative representation and comprehension. While the present chapter cannot discuss these processes in too much detail, it still seems helpful to mention two complementary principles that are generally taken to be relevant in this context. According to Marie-Laure Ryan, there is a *principle of minimal departure* at work during narrative meaning making that allows the recipients to "project upon these worlds everything [they] know about reality, [making] only the adjustments dictated by the text" (Ryan 1991, 51). It is worth stressing, though, that recipients do not "fill in the gaps" from the actual world itself but from their actual world knowledge, and that, moreover, "[t]he frame of reference invoked by the principle of minimal departure is not the sole product of unmediated personal experience" but may include various forms of medial and generic knowledge, or even a specific "textual universe as frame of reference" (Ryan 1991, 54).

Clearly, the principle of minimal departure is no less relevant in transmedia(l) contexts, as recipients will usually draw on previously established specific "fictional world knowledge" when trying to comprehend a work that is part of a given transmedial franchise. Nevertheless, works that are part of a transmedial franchise can and usually will "depart" from what was previously established to be the case in that franchise's world(s), as well, ranging from "sloppy contradictions" (Jenkins 2006, 105) and minor changes to an established "worldness" (Klastrup and Tosca 2004) to more substantial cases of "retconning" (Wolf 2012, 213) and "reboots" (Wolf 2012, 215). While the latter in particular may change the storyworlds established by previous entries in a transmedial franchise quite substantially—for example by casting the black actor Idris Elba as Norse god Heimdall in *Thor* (2011) or by reimagining the previously male team of parapsychologists as female in *Ghostbusters* (2016)—"reboots" will never change a storyworld "beyond recognition." Accordingly, even those (few) Marvel or *Ghostbusters* fans that did not appreciate the above-mentioned changes will have had no trouble "filling in" quite a few "gaps" based on their previous fictional world knowledge.

Yet, despite the importance of the principle of minimal departure and the "filling in" of the "gaps" of narrative works for which it allows, some "gaps" can never be "filled" in an intersubjectively valid manner. The main reason for this is that recipients' world knowledge—whether historical or contemporary, nonfictional or fictional, universal or particular—can provide only comparatively general additional information so that they cannot conclusively infer the answer to specific questions such as "Does character X have a birthmark on his or her back?" if the narrative representation does not provide it. While recipients may *pretend* that storyworlds are complete in the process that Marie-Laure Ryan calls "fictional recentering" (1991, 24), most theorists of fictional worlds agree that represented worlds are *actually* incomplete. This still holds with regard to the worlds represented by transmedial franchises, though "filling in the gaps" of a previously established storyworld is one of the core functions that new works being added to a transmedial franchise may fulfill.

Even with regard to the less extensive question of how storyworlds may be represented within the confines of a single work in a single medium, however, it is also important to note that recipients routinely "ignore" some aspects of narrative representations in order to intersubjectively construct the storyworlds thus represented. Narrative meaning making is based on an acute awareness of the intricacies of what Gregory Currie calls *representational correspondence*, a term designed to capture the general observation that, "[f]or a given representational work, only certain features of the representation serve to represent features of the things represented" (Currie 2010, 59). Particularly in cases where the assumption of representational correspondence becomes problematic, recipients will look for alternative *external explanations* related to hypothetical authorial intentions or established representational conventions before trying to imagine contradictory or otherwise problematic storyworlds based on a rigid insistence on *internal explanations*.

Kendall Walton pointedly describes this aspect of narrative meaning-making in terms of a *principle of charity*, stressing that, "if there is another ready explanation for the artist's inclusion of a feature that appears to generate a given fictional truth, it may not seem that he meant especially to have it generated. And *this* may argue against recognizing that it is generated" (Walton 1990, 183, original emphasis). Beyond readily apparent "sloppy contradictions" (Jenkins 2006, 105), these external explanations will often refer to medium-specific representational conventions, as the "distance" between a given narrative representation and what that representation represents can indeed be quite pronounced (see also Walton 1990, 181–182, for a famous discussion of the "distance" between the words of Shakespeare's Othello and the words of the actors that play him; as well as Thon 2017 for a more in-depth examination of the application of charity across media).

Take, for example, the audiovisual representation of Yoda in *Star Wars Episode V: The Empire Strikes Back* (1980), *Star Wars Episode VI: Return of the Jedi* (1983), and the original theatrical release of *Star Wars Episode I: The Phantom Menace* (1999), which evidently differs from the audiovisual representation of Yoda in *Star Wars Episode II: Attack of the Clones* (2002), *Star Wars Episode III: Revenge of the Sith* (2005), and *Star Wars Episode VIII: The Last Jedi* (2017). It amounts to a "silly question" (Walton 1990, 176), however, to ask for an internal explanation of these differences, as the film series' switch from using a puppet to using computer-generated imagery (CGI) in order to present Yoda readily provides an external explanation (which was further emphasized by the implementation of a CGI representation of Yoda in the Blu-ray release of *The Phantom Menace* [2011]).

Likewise, it would be "silly" to ask for an internal explanation of the differences between the audiovisual representation of Albus Dumbledore in the first two feature films of the *Harry Potter* series (2001, 2002) and the corresponding audiovisual representation in its third through eighth installment (2004–2011), as the use of actors to represent characters is a core representational convention of narrative films and the untimely death of actor Richard Harris, who was then replaced by Michael Gambon, provides a ready external explanation of these differences. In both the case of *Star Wars* and *Harry Potter*, then, the intersubjective construction of the serial storyworlds represented by the films would entail applying charity in order to "ignore" at least some aspects of their audiovisual representation.

It is, of course, also entirely possible for (series of) narrative works to provide internal explanations for these (as well as other) kinds of apparent contradictions. One of the more salient cases would be the BBC's long-running television series *Doctor Who* (1963–), which internally explains the regularly occurring changes of the actor used to represent its protagonist as the result of all Time Lords' capability for "regeneration," a process that transforms both their physical form and some aspects of their personality—including, for example, the possibility of a sex change that has recently caused a bit of a stir among some of the more entitled male fans of the franchise when Jodie Whittaker took over from Peter Capaldi as the first female doctor in 2017 (see Hills 2017 for some context; as well as Evans 2011; Harvey 2015; Hills 2015 for additional discussion of the franchise).

In the absence of this kind of internal explanation, though, some degree of apparently contradictory difference in the audiovisual representation of a character tends to be "charitably ignored"—and this arguably also applies to transmedial representations such as those of Yoda in the *Star Wars* movies, the animated television series *Star Wars: Clone Wars* (2003–2005), the CGI feature film *Star Wars: The Clone Wars* (2008) and the CGI television series of the same title (2008–2014), the more recent CGI television series *Star Wars Rebels* (2014–) and the animated YouTube series *Star Wars Forces of Destiny* (2017), the various *Star Wars* comics, and the no less numerous *Star Wars* video games—or to those of Albus Dumbledore in J. K. Rowling's seven *Harry Potter* novels (1997–2007), their eight feature film adaptations, and the various *Harry Potter* video games that, in turn, tend to base their audiovisual representations of Dumbledore on the way he is represented in the films (for further discussion of the *Star Wars* and *Harry Potter* franchises, see, e.g., the contributions in Guynes and Hassler-Forest 2017; and Brenner 2015).

Just as the principle of minimal departure does not allow recipients to "fill in" every last "gap" that a given narrative work leaves in the storyworld it represents, however, so does the principle of charity not allow recipients to generally "ignore" all kinds of contradictions that these narrative works may "appear to generate." Indeed, it is quite frequently the case that narrative works successively represent situations that do not immediately "add up" to a noncontradictory storyworld without providing a plausible external explanation for these apparent contradictions, leading to various contemporary storyworlds being best described as compounds of two (or more) noncontradictory storyworlds. While it seems generally helpful to conceptualize storyworlds as noncontradictory "by default" and analyze contradictory storyworlds as compounds of noncontradictory sub-worlds, then, we are evidently confronted with "compounding" on a significantly larger scale in many transmedial franchises.

Work-Specific Storyworlds, Transmedial Storyworlds, and Transmedial Universes

Whether a given franchise is conceived as transmedial from the beginning (as is, for example, the case with *The Matrix*) or expanded across media after an initial commercial success (as is, for example, the case with *Star Wars*), the narratological approach sketched above postulates that every narrative work that is part of the franchise represents a storyworld of its own, but, at the same time, establishes a relation between that storyworld and the storyworlds represented by the other narrative works that are part of the franchise. Instead of assuming that transmedial franchises generally represent a "single world," then, this perspective allows for a systematic distinction between the local medium-specific storyworlds of single narrative works, the glocal but noncontradictory transmedial (or, in many cases, merely trans*textual*) storyworlds that may be constructed out of local work-specific storyworlds, and the global and often quite contradictory transmedial storyworld compounds that may, for lack of a better term, be called transmedial universes (for further discussion of the rather contested terminology, see also, e.g., Harvey 2015; Packard 2015; Wolf 2012, 216–220).

Drawing on Henry Jenkins's discussion of "adaptation and extension" (2011), Mark J. P. Wolf's discussion of "adaptation" and "growth" (2012, 245), and Marie-Laure Ryan's discussion of "expansion" and "modification" (2008, 385), one can then further ask to what extent two single narrative works within a transmedial franchise are defined, first, by a relation of *redundancy*, when one is aiming to represent the same elements of a storyworld that the other represents; second, by a relation of *expansion*, when one is aiming to represent the same storyworld that the other represents, but adds previously unrepresented elements; or, third, by a relation of *modification*, when one is aiming to represent elements of the storyworld represented by the other, but adds previously unrepresented elements that make it impossible to comprehend what is represented as part of a single, noncontradictory storyworld. While this rather basic distinction will not capture all the intricacies of transmedial universes, it still seems to be a helpful heuristic for the in-depth analysis of the interrelations between a given franchise's work-specific storyworlds.

Take, for example, the transmedial franchise *A Song of Ice and Fire*, which is based on George R. R. Martin's novel series *A Song of Ice and Fire* (1996–), but has since been developed to include not only the wildly successful HBO television series *Game of Thrones* (2011–) but also several comics and graphic novels, a collectible card game, two board games with several expansions, two pen-and-paper role-playing games, and various video games (as well as a series of prequel novellas by Martin himself, an extensive companion book that he co-authored, a book of maps, and two collections of artwork). At first glance, the strong presence of Martin as the author of the franchise's "ur-text" (Jenkins 2007) might suggest that its subsequent entries should generally be expected to be redundant adaptations of (specific parts of) the transtextual world over which Martin has exerted direct authorial control. Even if we accept that, say, Daniel Abraham and Tommy Patterson's comics series *A Game of Thrones* (2011–2014) is a largely redundant adaptation of the first novel (from which it has also borrowed its title), though, this evidently does not apply to many of the other entries in the franchise.

While there are obvious differences in the ways that the novel and the comics series present their respective work-specific storyworlds, these differences might arguably be "ignored" by reference to the diverging representational affordances and limitations of literary texts and graphic narratives. In the case of HBO's television series *Game of Thrones*, however, it would seem that no amount of charity will allow for the intersubjective construction of a storyworld sufficiently similar to that represented by Martin's series of novels in order to speak of a primarily redundant adaptation. While the television series does not simply "re-tell" the story originally told by the novels, then, it also quite clearly does not expand that storyworld in a noncontradictory way. Rather, it takes certain elements and leaves others, changing and re-arranging them to an extent that it seems more appropriate to speak of a modification of the novel series' transtextual storyworld than of its redundant adaptation or noncontradictory expansion (even though the television series certainly also "re-tells" and expands on elements of the storyworld that is represented by the "ur-texts" of the novel series).

Much more could be said on the *A Song of Ice and Fire* franchise, of course (see, e.g., the contributions in Gjelsvik and Schubart 2016), but the main point here is that transmedial franchises are often not appropriately described as representing a single storyworld, even though the work-specific storyworlds of the various works that are part of a given franchise may, to various degrees, add up to noncontradictory transmedial storyworlds. Indeed, while one can also find franchises such as *The Matrix* or *Halo*, which appear to be primarily defined by an attempt at the noncontradictory expansion of a single transmedial storyworld (see, e.g., Jenkins 2006; Harvey 2015), it seems not uncommon even for comparatively small-scale franchises (in terms of storyworld "spread," not in terms of commercial success) to establish two clearly distinct storyworlds via a high-profile modifying adaptation, while still aiming at a further expansion of each of these storyworlds via works in other media (see also Parody 2011 for additional discussion of the role that adaptations play in transmedial franchises).

An even clearer example of this would be the comics-based franchise of *The Walking Dead*, which, similarly to the novel-based franchise of *A Song of Ice and Fire*, has recently experienced a surge in both commercial and critical success. At first glance, AMC's *The Walking Dead* television series (2010–) may appear to be a straightforward adaptation of Robert Kirkman's *The Walking Dead* comics series (2003–) and Telltale's *The Walking Dead* adventure game series (2012–2014) may, in turn, appear to be a straightforward adaptation of the television series, but the differences of the stories these series tell turn out to be quite striking and do, in fact, make a description of the television series' work-specific storyworld as a modification of the comics series' work-specific storyworld appear more appropriate. In contrast, the adventure game series takes even more liberties—including a change in the main protagonist—but does so in a way that makes it appear as a largely noncontradictory expansion of the comics series' work-specific storyworld rather than as either a redundant adaptation or a contradictory modification of that storyworld (see also, e.g., Beil and Schmidt 2015). And, last but not least, Terminal Reality's first-person shooter *The Walking Dead: Survival Instinct* (2013) is, perhaps, not a very well-made video game, but still manages to convey its aim to expand the work-specific storyworld of the television series rather than that of the comics series with sufficient clarity.

Again, there would be more to say about the web series, the board games, the card game, and the various other video games that have been published since the commercial and critical success of AMC's *The Walking Dead* television series, but the general principle of storyworld interrelation should have already become clear. Indeed, it seems that both Telltale's *The Walking Dead* and Terminal Reality's *The Walking Dead: Survival Instinct* are authorized to expand the storyworlds established by the comics series and the television series, respectively. They are, in other words, what is commonly described as *canonical* expansions, which sets them aside from quite a large number of other works that could, at first glance, be comprehended as noncontradictory expansions of a previously established storyworld based on what they represent, but are not authorized to expand the storyworld in question, remaining *apocryphal*. The current proliferation of author collectives and licensing practices certainly plays an important role here (see, e.g., Evans 2011; Johnson 2013; Wolf 2012; and the contributions in Gray and Johnson 2013), yet it should also be noted that the question of canonicity does not entirely coincide with the question of authorship.

As has already been mentioned, the franchises of *The Matrix* and *Halo* appear to be primarily defined by an attempt at the noncontradictory expansion of a single transmedial storyworld, despite the fact that their individual works are still created by different author collectives. Now, it seems clear that the authors or author collectives of officially licensed works such as the video game *Enter the Matrix* (2003) or Eric Nylund's novel *Halo: The Fall of Reach* (2001) are more likely to have the authority to expand the transmedial storyworlds that are at the "canonical core" of the *The Matrix* and *Halo* franchises than, say, the authors or author collectives of the various *The Matrix* and *Halo* fan fictions that can, for example, be found at *An Archive of Our Own*. However, one can also find licensed works such as the video game *The Matrix: The Path of Neo* (2005) or the last of the seven short anime films collected in *Halo Legends* (2010) that are explicitly marked as apocryphal despite being created by officially licensed author collectives.

As Mark J. P. Wolf remarks, then, "for a work to be canonical requires that it be declared as such by someone with the authority to do so" (2012, 271)—but the importance of the canonical/apocryphal distinction for the analysis of transmedial franchises' narrative functions should not be overemphasized. Indeed, it would seem that the relevance of the question of a given licensed work's canonicity remains largely limited to those cases where that work's storyworld could, at least in principle, be comprehended as a noncontradictory expansion of a previously represented storyworld (whether work-specific, transtextual, or transmedial). Yet, in the case of the *A Song of Ice and Fire* and *The Walking Dead* franchises, it is obvious that the *Game of Thrones* and *The Walking Dead* television series constitute modifying rather than redundant adaptations of George R. R. Martin's novel series and Robert Kirkman's comics series, respectively. In those cases, asking which of the works is "more canonical" seems to miss the point—which is that they do not contribute to the representation of a "single world," to begin with.

However, while there seems to be neither an immediate need nor theoretically solid ground to exclude any work—whether canonical or apocryphal, licensed work or fan creation—from consideration as contributing to the representation of a franchise's transmedial universe, the fact remains that the canonical/apocryphal distinction often plays an important role in the intersubjective construction of these universes. Recently, this has been forcefully illustrated by the discussions surrounding the acquisition and subsequent re-ordering of the *Star Wars* franchise by Disney in 2012 (see, once more, the contributions in Guynes and Hassler-Forest 2017). Indeed, the *Star Wars* franchise's 2014 move from the six "canonical levels" of the "Holocron continuity database" to the simpler binary distinction between canonical *Star Wars* content and apocryphal *Star Wars Legends* content (see Wookieepedia 2017) not only reaffirms the necessity to go beyond the model of the "single world" in analyzing transmedial franchises' narrative functions but also exemplifies rather well that the relations between a transmedial universe's storyworlds are in constant flux.

Conclusion

This chapter has presented a narratological perspective on work-specific storyworlds, transmedial storyworlds, and transmedial universes as a productive way of transcending the model of the "single world" in the analysis of transmedial franchises' narrative functions. As has further become clear, the "toolbox" of transmedial narratology can be fruitfully applied to a variety of different cases, ranging from the gradual expansion of a single transmedial storyworld in comparatively small-scale franchises such as *The Matrix* or *Halo*, via the more complex combination of redundancy, expansion, and modification that defines the transmedial universes of franchises such as *A Song of Ice and Fire* or *The Walking Dead*, to the synchronically complex and diachronically variable transmedial universes that long-running franchises such as *Star Wars* generate. One way or another, though, the narratological perspective presented here is meant to complement rather than contradict the wide range of existing approaches to the study of transmedial franchises, thus leading to a richer understanding of these franchises specific brand of narrative complexity without in any way dismissing other aspects of their production, aesthetics, and reception.

Acknowledgments

This is a shortened and revised version of Jan-Noël Thon's previous publication: "Converging Worlds: From Transmedial Storyworlds to Transmedial Universes." *Storyworlds: A Journal of Narrative Studies* 7 (2) (2015): 21–53. (c) University of Nebraska Press.

References

Beil, Benjamin, and Hanns Christian Schmidt. 2015. "The World of *The Walking Dead*—Transmediality and Transmedial Intermediality." *Acta Universitatis Sapientiae* 10: 73–88.

Brenner, Lisa S. (ed.). 2015. *Playing Harry Potter: Essays and Interviews on Fandom and Performance*. Jefferson: McFarland.

Chatman, Seymour. 1978. *Story and Discourse: Narrative Structure in Fiction and Film*. Ithaca: Cornell University Press.

Currie, Gregory. 2010. *Narratives and Narrators: A Philosophy of Stories*. Oxford: Oxford University Press.

Doležel, Lubomír. 1998. *Heterocosmica: Fiction and Possible Worlds*. Baltimore: Johns Hopkins University Press.

Eder, Jens. 2008. *Die Figur im Film: Grundlagen der Figurenanalyse* [The Character in Film: Foundations of Character Analysis]. Marburg: Schüren.

Evans, Elizabeth. 2011. *Transmedia Television: Audiences, New Media and Daily Life*. London: Routledge.

Genette, Gérard. 1988. *Narrative Discourse Revisited*. Translated by Jane E. Lewin. Ithaca: Cornell University Press.

Gerrig, Richard J. 1993. *Experiencing Narrative Worlds: On the Psychological Activities of Reading*. New Haven: Yale University Press.

Gjelsvik, Anne, and Rikke Schubart (eds.). 2016. *Women of Ice and Fire: Gender,* Game of Thrones *and Multiple Media Engagements*. London: Bloomsbury.

Gray, Jonathan, and Derek Johnson (eds.). 2013. *A Companion to Media Authorship*. Chichester: Wiley-Blackwell.

Guynes, Sean, and Dan Hassler-Forest (eds.). 2017. *Star Wars and the History of Transmedia Storytelling*. Amsterdam: Amsterdam University Press.

Harvey, Colin B. 2015. *Fantastic Transmedia: Narrative, Play and Memory across Science Fiction and Fantasy Storyworlds*. Basingstoke: Palgrave Macmillan.

Herman, David. 2002. *Story Logic: Problems and Possibilities of Narrative*. Lincoln: University of Nebraska Press.

Herman, David. 2009. *Basic Elements of Narrative*. Chichester: Wiley-Blackwell.

Hills, Matt. 2015. *Doctor* Who*: The Unfolding Event—Marketing, Merchandising and Mediatizing a Brand Anniversary*. Basingstoke: Palgrave Macmillan.

Hills, Matt. 2017. "Casting a Female Doctor Who Wasn't So Bold—Choosing Another White Male Would Have Been Really Risky." *The Conversation*, July 26. Accessed July 3, 2017. http://theconversation.com/casting-a-female-doctor-who-wasnt-so-bold-choosing-another-white-male-would-have-been-really-risky-81410.

Jenkins, Henry. 2004. "Game Design as Narrative Architecture." In *FirstPerson: New Media as Story, Performance, and Game*, edited by Noah Wardrip-Fruin and Pat Harrigan, 118–130. Cambridge, MA: MIT Press.

Jenkins, Henry. 2006. *Convergence Culture: Where Old and New Media Collide*. New York: New York University Press.

Jenkins, Henry. 2007. "Transmedia Storytelling 101." *Confessions of an Aca/Fan*, March 22. Accessed July 30, 2017. http://henryjenkins.org/2007/03/transmedia_storytelling_101.html.

Jenkins, Henry. 2011. "Transmedia 202: Further Reflections." *Confessions of an Aca/Fan*, August 1. Accessed July 22, 2017. http://henryjenkins.org/2011/08/defining_transmedia_further_re.html.

Johnson, Derek. 2013. *Media Franchising: Creative License and Collaboration in the Culture Industries*. New York: New York University Press.

Klastrup, Lisbeth, and Susana P. Tosca. 2004. "Transmedial Worlds—Rethinking Cyberworld Design." *Proceedings of the International Conference on Cyberworlds 2004*. Accessed July 3, 2017. www.itu.dk/people/klastrup/klastruptosca_transworlds.pdf.

Klastrup, Lisbeth, and Susana P. Tosca. 2018. *Transmedial Worlds and Everyday Life: Networked Reception, Social Media, and Fictional Worlds*. New York: Routledge.

Packard, Stephan. 2015. "Closing the Open Signification: Forms of Transmedial Storyworlds and Chronotopoi in Comics." *Storyworlds: A Journal of Narrative Studies* 7 (2): 55–74.

Parody, Clare. 2011. "Franchising/Adaptation." *Adaptation* 4 (2): 210–218.

Pavel, Thomas. 1989. *Fictional Worlds*. Cambridge, MA: Harvard University Press.

Ronen, Ruth. 1994. *Possible Worlds in Literary Theory*. Cambridge: Cambridge University Press.

Ryan, Marie-Laure. 1991. *Possible Worlds, Artificial Intelligence, and Narrative Theory*. Bloomington: Indiana University Press.

Ryan, Marie-Laure 2008. "Transfictionality across Media." In *Theorizing Narrativity*, edited by John Pier and José Á. García Landa, 385–417. Berlin: De Gruyter.

Thon, Jan-Noël. 2015. "Converging Worlds: From Transmedial Storyworlds to Transmedial Universes." *Storyworlds: A Journal of Narrative Studies* 7 (2): 21–53.

Thon, Jan-Noël. 2016. *Transmedial Narratology and Contemporary Media Culture*. Lincoln: University of Nebraska Press.

Thon, Jan-Noël. 2017. "Transmedial Narratology Revisited: On the Intersubjective Construction of Storyworlds and the Problem of Representational Correspondence in Films, Comics, and Video Games." *Narrative* 25 (3): 286–320.

Walton, Kendall L. 1990. *Mimesis as Make-Believe: On the Foundations of the Representational Arts*. Cambridge, MA: Harvard University Press.

Wolf, Mark J. P. 2012. *Building Imaginary Worlds: The Theory and History of Subcreation*. London: Routledge.

Wookieepedia. 2017. "Canon." *Wookieepedia*. Accessed July 4, 2017. http://starwars.wikia.com/wiki/Canon.

41

AN ONTOLOGICAL APPROACH TO TRANSMEDIA WORLDS

Frank Branch and Rebekah Phillips

Use of ontology as a means of scholarly discovery has a long history in the sciences and is a valuable tool for examining transmedia fictional worlds. Merriam-Webster defines ontology as "a particular theory about the nature of being or the kinds of things that have existence" (Merriam-Webster.com 2017b). Scholars working in the hard sciences, such as chemistry, biology, and geology, have a long history of using ontological approaches to structure their inquiries into the natural world. They classify compounds as proteins, describe species as insects, or categorize rocks as igneous. These ontologies are used to create theories of existence that structure basic lines of scientific inquiry. For example, to answer, "What are the characteristics of rocks that remain unaltered after the extrusion of lava?" geologists created the ontological category of "igneous rocks." This classification allows scholars to communicate new knowledge about igneous rocks without introducing ambiguity into the discussion. In addition, the classification can be used to frame new lines of inquiry, such as, "How does radioactive decay progress in igneous rock formations?"

There is considerable debate about the usefulness of such classification systems in the arts and humanities. Concepts such as aesthetics and genre are substantially more subjective and lack the clearly defined physical characteristics found in hard sciences (Degani-Raz 2003, 2005; Eco 2009; Thomasson 2005). For instance, defining an "igneous rock" as "rock formed by solidification of a molten magma" (Merriam-Webster.com 2017a) is a far more objective definition than can be achieved when defining more fuzzy concepts like "science fiction" or "supervillain."

Though transmedia studies are more subjective than hard sciences, ontological approaches are still helpful with communication and framing of inquiry. For example, definition of terms like science fiction or supervillain tell as much about the beliefs and practices of the persons defining them as it does about the objects themselves. Ontologies assist scholars in understanding what is knowable within the aesthetics of transmedia versus what is not, while also setting these boundaries through their ability to define frames of reference (Thomasson 2005). A scholar cannot answer the question, "How are female supervillains portrayed in the Marvel Cinematic Universe?" without a universally accepted frame of reference as to what constitutes a supervillain. This supposition is true even if the understanding of supervillain is somewhat subjective and nebulous around the edges. Doležel (1995) has even stated that you cannot truly interpret a fictional world without grounding it to terms and concepts found in the real world. A well-structured ontology becomes an encyclopedia that grounds frames of reference within a transmedia fictional world to the world we live in. This grounding allows a scholar to traverse these worlds using a common theory of existence.

In the rest of this chapter, we will explore three specific frames of inquiry where the use of an ontology is particularly helpful. These queries are not the only ones facilitated by an ontological approach to transmedia studies, but illustrate how ontology is used to create a mutual understanding of existence within a transmedia world. These frames of inquiry are: understanding how transmedia works interconnect to form various canonical works; exploring the internal dynamics of the things found within transmedia fictional worlds (e.g., people, places, laws, events); and understanding the interaction and information-seeking behavior of various user groups (e.g., fans, brand managers, authors) with transmedia properties.

Ontology to Study the Structure of Interconnected Works

The very nature of transmedia storytelling creates a complex interconnection of independent stories that make up a larger narrative. Prominent examples of these stories include the multiple universes in Marvel, *Star Wars*, *Aliens*, *Harry Potter*, and *Lord of the Rings*. The Marvel transmedia property alone has multiple different and interlinked storylines. For instance, the Civil War storyline unfolds over multiple series of comic books within the Marvel Ultimates Universe (Miller and McNiven 2006; Straczynski, Lee, and Garney 2006) and separately, in a movie, as part of the Marvel Cinematic Universe. These storylines have different narratives in each universe, even though both are representative of the Marvel Civil War. Another occurrence of differing narratives can be found in the comic book series when Spider-Man reveals his secret identity to the world in a press conference. This critical narrative is left out of the movie version of the same story (Errico 2016; Miller and McNiven 2006; Markus and McFeely 2016; Straczynski, Lee, and Garney 2006).

These differences are more than cosmetic and are essential to the structure of the narrative itself. In one universe within the Marvel transmedia property, Spider-Man's identity is secret, while in the other it is a matter of public record. This singular change has a significant impact on the character's narrative within each world. Spider-Man's actions in the cinematic universe come with anonymity, while in the Ultimates universe, they influence Peter Parker's civilian life. Scholars interested in understanding how the consequences of fame and anonymity are portrayed in transmedia can directly compare the portrayals of the famous and anonymous Peter Parkers by linking them to a single storyline. An ontology facilitates this linkage because it creates a common frame of reference for the idea of a storyline that can describe the nature of this relationship.

In addition, the question of what is considered to be the canon of a transmedia property is complicated by narrative variation of the same story elements in different universes. The issue of canon is important to fans who debate it endlessly on the internet (Booth 2009), and is also significant to scholars. For example, scholars trying to understand questions, such as, "What is the literary theme of the character of Spider-Man?" need to determine which Spider-Man (Bondi 2011) they are talking about. Is anonymous Peter Parker, or famous Peter Parker the authoritative version? Or, are they both authoritative in their own versions of the universe?

Understanding what is the canon of a transmedia property becomes even more complex when you incorporate the business priorities of transmedia property owners to the question: What is authoritative? For example, when Disney purchased *Star Wars* from Lucasfilm they declared over 20 years of narratives from the Star Wars: Expanded Universe to no longer be canon. This change in canon was done so Disney would not be constrained by the stories in the Expanded Universe in their own future storytelling (Lucasfilm 2014). Disney, in this one decision, changed the entire "canonical" universe of *Star Wars*, including elements as valuable to scholarly research as the Skywalker family tree.

These types of transformations are not always so grand as the deliberate change in corporate direction. Decisions as simple as changing a property's showrunner can alter the canon. For instance, the lifecycle of the Xenomorph creature from the Aliens franchise changed when the property was managed by James Cameron, versus Ridley Scott. Each director changed the Xenomorphs narrative to suit their own storytelling objectives (Aliens Wikia 2017). Scholars studying these directors need to understand how the canon of the Xenomorph reflects each director's storytelling style.

To obtain this knowledge, scholars often need to review dozens of original works or rely on fan created wikis to understand the complex web of storylines found within a single transmedia property. These forms of media do not make this information easily discoverable, maintained, or able to be commonly communicated. This difficulty is simply because transmedia is a dense information format and wikis are constantly evolving based on their crowdsourced nature (Booth 2009; Gray, Sandvoss, and Harrington 2007; Stephens 2006).

An ontology is the critical tool used to create the common frame of reference required to understand the complex relationships between storylines found in transmedia. Ontologies allow for individual works within a large transmedia property to be categorized and related to each other systematically. In the simple ontology found in Figure 41.1, you can see how different narratives and their inter-relationships can be systematically categorized into a shared understanding. For example, Marvel itself is a Transmedia Property. That property has multiple Story Worlds, such as the Marvel Ultimates and the Marvel Cinematic Universe. Civil War is a single Storyline that exists simultaneously in both Story Worlds. The ontology can represent this relationship accurately by relating Storylines to Story Worlds through a specific property called Has World. Individual Transmedia Creative Works, such as, *Civil War #1* and *Captain America: Civil War*, can then be related to the appropriate Storyline via a Has Storyline property. This relationship is separate from the Has World property that is used to relate the same Transmedia Creative Works to the appropriate world. This ontological structure allows the scholar to discover, navigate, and communicate the full web of these relationships using common ubiquitous language (Branch et al. 2017). The use of ubiquitous language is crucial for scholars wanting to test hypotheses and communicate conclusions.

Other tools, like bibliographic series, fail to adequately represent the interconnected nature of the narratives, as they traditionally exist outside of a well-structured ontology. For example, the Civil War

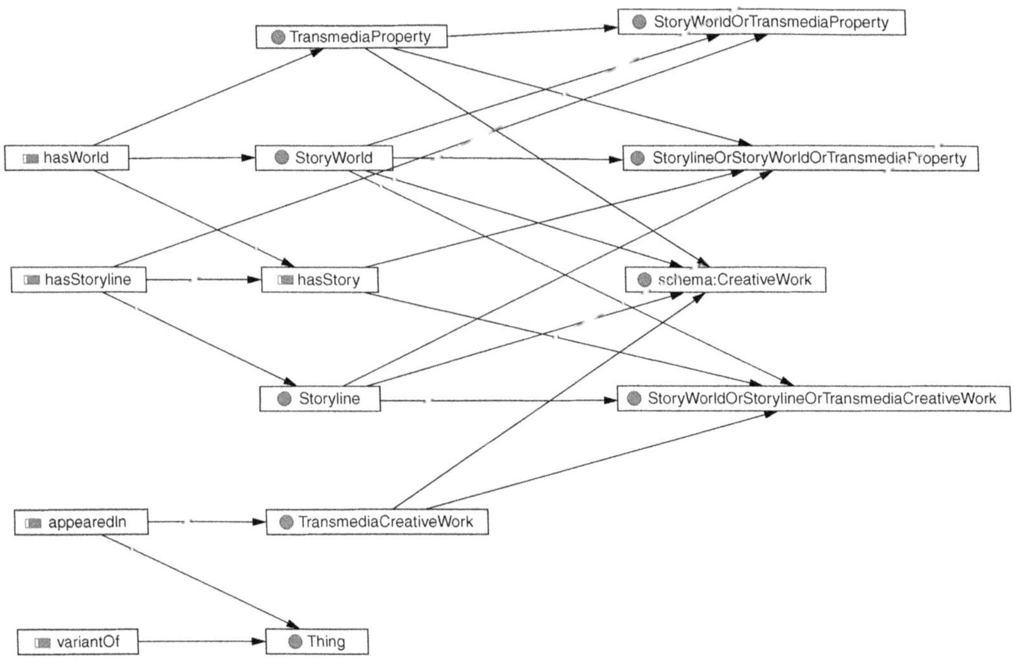

Figure 41.1 Ontological approach to transmedia works.
Source: Branch et al. (2017).

storyline is found in more than just one bibliographic series within Marvel Ultimates. It can be found in both the Civil War comic book series (Miller and McNiven 2006) and the Amazing Spider-Man comic book series, issues 532–534 (Straczynski, Lee, and Garney 2006). There is no established bibliographic relationship between the two series, and Civil War content is only found in three issues of the Amazing Spider-Man series. It is impossible to find the connection between the Amazing Spider-Man and the Civil War series using the bibliographic structure as traditionally cataloged. Scholars and fans would need to resort to scouring internet wikis to find this connection (Branch et al. 2017). An ontology, such as the one in Figure 41.1, has the power to overcome the missing bibliographic data by exposing the relationships between these works. It does so by explicitly revealing them via named properties like Has Story, Has World, and Has Storyline. Scholars can then navigate these named relationships either manually or by using modern semantic search engines such as Falcon, Swoogle, and Watson.

Ontology to Study the Internals of Transmedia Worlds

Ontology is a critical tool to help scholars explore the contents within transmedia fictional worlds. An ontological approach to structuring the contents of these worlds goes beyond just understanding the relationships between each of the individual works. The very nature of a transmedia fictional world means that there are complex story elements, such as characters, events, places, and tropes layered across multiple independent narratives. For example, the story of the Skywalker family unfolds across eight movies, two television series, and dozens of books. A thorough analysis of the evolution of this one family requires a scholar to develop a descriptive framework to talk about these elements.

All scholars must develop a semiotic framework to understand a story element such as the Skywalker family. This personal framework is required to construct meaning and derive conclusions about what those narratives tell us about our culture (Eco 2009; Degani-Raz 2003, 2005). Doležel (1995) even states that without a structured encyclopedia containing the contents of a fictional world, it is difficult, if not impossible, to talk about them in a consistent manner. Consequently, scholars, either explicitly or implicitly, create such an encyclopedia when conducting their own research. Without such a tool, there is no way for them to describe elements of the narrative.

Case in point, a scholar is likely to create a Skywalker family tree when examining the narrative evolution of those characters. This would include more nuance than one finds in a genealogy containing simply marriages and parentage. An ontology's power to draw clear distinctions between various related descriptive properties clarifies the differences between these kinds of familial relationships more concretely. The ontology in Figure 41.2 subdivides the property of spouse by describing

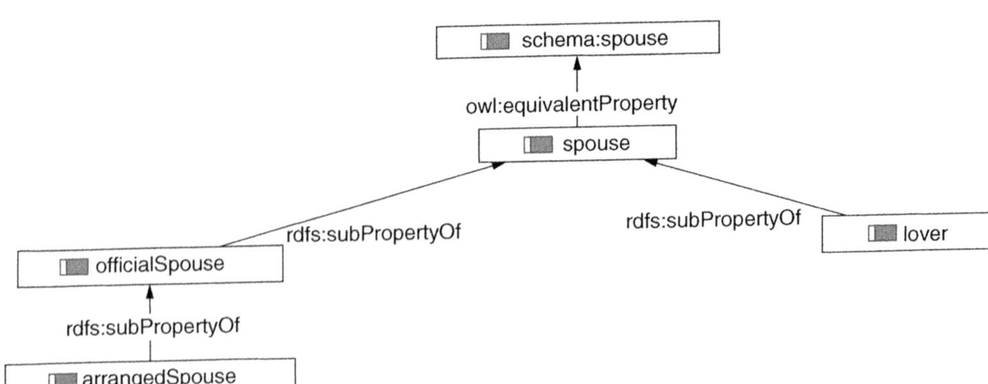

Figure 41.2 Ontology of spousal relationships.
Source: Branch et al. (2017).

additional nuances such as a "lover," an "arranged spouse," or even simply who is the "official spouse." You can see how this formal ontology goes further than simple genealogy by defining these variations in spousal relationships. This type of ontological approach is critical for scholars to have a common way to discuss issues like infidelity in transmedia properties. The nature of marriage as portrayed in a transmedia property like *Game of Thrones* can be compared to the portrayal found in *Star Wars* by the very act of labeling the kind of spousal relationships found in these narratives.

Ontology can be used to structure the information of almost any kind of narrative element found in a transmedia property. Ontological approaches to the content within a transmedia story are not just limited to traditional hierarchical elements like family trees. For example, Branch et al. (2017) found that fans place a high value on the evolution of characters throughout the story arc of a transmedia property. One such example is the metamorphosis of Anakin Skywalker into Darth Vader in the world of *Star Wars*. This evolution of character is so important that fans of *Star Wars* consider Darth Vader and Anakin Skywalker to be different characters even though they are the same person.

In Figure 41.3 you can see an ontological representation of narrative change. In this ontology, Darth Vader and Anakin Skywalker are simply two different things (of type: character) that are linked together by metamorphosis. This metamorphosis can be given a rich set of properties that allow you to describe the changes in Anakin's character completely. For example, a metamorphosis can have a catalyst that causes it to happen, such as the death of Anakin's mother. In addition, it can have a location, a time, even actors that participated in the change. This descriptive power comes from the ontological definition of a metamorphosis being "a kind of event that profoundly changes a narrative element" (Branch et al. 2016). Because of this definition, metamorphosis contains all the other properties of a real-world event. The metamorphosis can even be given a change type property that describes the fictional trope represented by this change in character. Anakin "Falls" into the Dark Side and becomes Darth Vader. Then, Darth Vader is "Redeemed" in the end and returns to being Anakin.

All researchers, including fans, will create their own ontologies to describe the complex elements within transmedia fictional worlds. Humans need to generate some semiotic context in order to think about any topic and especially topics more prone to subjectivity like the arts and humanities (Eco 2009; Degani-Raz 2003, 2005; Doležel 1995). Each person creates their own ontology of a transmedia world in their head as they explore its contents. This representation is created if only to allow them to understand the subjective information found within the narrative, in their own way.

However, without a common ontology for transmedia studies, scholars will continue to communicate using their own representations of these fictional worlds. This individualized communication will by its nature inject noise into scholarly communications around the topic. These noisy

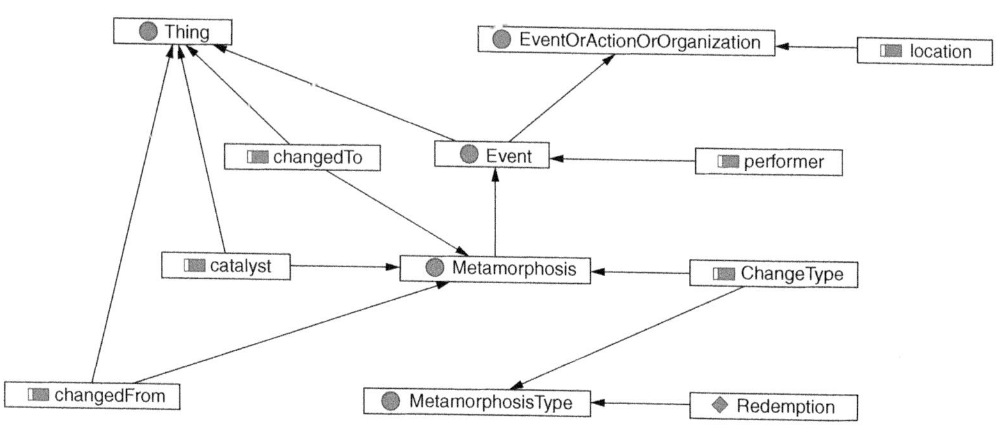

Figure 41.3 Ontology of narrative change.
Source: Branch et al. (2017).

communications happen in all fields and are among the reasons why hard sciences have been adopting ontologies and taxonomies for centuries. There is a convincing argument to be made that the more subjective a field of study is, the more important it is to have a common ontology in place to facilitate common ubiquitous language around the subject. In the world of speculative fiction, this common language becomes increasingly important because fantastical elements of the narrative must be connected to things in the real world to be understood (Doležel 1995). For example, a light saber is a fantastical element from the *Star Wars* universe that can be described as a sword, a ritualized weapon, and a symbol of a warrior class. This ontological description connects a light saber to the samurai sword found in the real world in Japan. Labeling a light saber in this way creates a mutual understanding of what this fictional thing is and how it relates to the real world.

Creating a formal ontology provides a way for transmedia scholars to communicate about narrative elements that are shared across the field. This common ubiquitous language works in the same way that the taxonomy of species has allowed biologists to communicate about biological evolution to advance the field of biology. The creation of transmedia ontologies represents the maturing of transmedia studies into a more formal discipline that will lead to an accelerated advancement of knowledge, just as it has in the hard sciences.

Ontology to Study Fan Information-Seeking Behavior

While public perception of groups who identify as fans of science-fiction and fantasy based transmedia universes has improved over the last few decades, the stereotype of fans being socially awkward loners who speak fluent Klingon and live in their parents' basement persists. However, several research studies indicate that this stereotype tends to be false. Individuals who self-identify as fans are generally successful and socialize a considerable amount with other fans (Bondi 2011; Booth 2009; Gray, Sandvoss, and Harrington 2007). We can see how perceptions have shifted, in part, by the prevalence of new transmedia fictional works based in these science-fiction and fantasy universes that have been recently released to film, print, and gaming mediums.

While there is something to be said for appealing to a large audience (rather than catering to a more specific, smaller group of fans) and creating works that do not require extensive research or background knowledge, the success of these works can be largely attributed to the groups who devour anything related to the universes they adore. Fandom is akin to religion for some (Chidester 2005; Wagner 2012). They spend vast sums of money on replica merchandise; travel great distances to attend conventions such as Comic Con in San Diego; and invest their time and attention to understanding every nuance of these universes. Due to the significant investment that fans are willing to make to support their favorite works, scholars need to understand their information-seeking behavior (Branch et al. 2017; Lewis 1992).

Branch et al. (2017) used an ontological approach to study how fans interact with information using a test of their ontology created from domain analyses and conducting interviews with an assortment of fans from around the country. The statistics from that study's data archive (Branch et al. 2016) provide additional insight into fans' information-seeking behavior.

They found that the 19 fans interviewed were looking for meaningful connections between events, people, places, and items when conducting routine searches for transmedia fictional elements. This was determined through the initial interview, and confirmed by a subsequent card sort study. Based on the input received from these fans, the team used the previously discussed ontological technique to establish a way to represent not only the basic connections between the beings, locations, powers, etc., but also the tone of those relationships. For example, interviewees agreed with the initial assessment that a forced alliance or arranged marriage should be represented differently than when a character chooses a mentor or spouse for themselves. This finding was discovered using an ontological approach but has significant importance for any scholar studying the representation of gender roles and relationships in popular fiction.

Branch et al. (2017) also inquired about other aspects of fan behavior as part of an integrated approach to developing an ontology. For example, nearly all the interviewees answered additional questions regarding why and how they conduct their fandom research. A total of 14 participants reported spending anywhere from 20 minutes to eight hours per week on both active and passive information-seeking behavior. The additional five people expressed difficulty in quantifying their weekly research because it was largely passive and happened throughout the course of the day as they would check various social media sites. This behavior is critical in understanding both the usage requirements of a fan-based ontology and informs scholars about fans' media consumption habits that are critical to other fields of research.

The development of a solid ontology requires the information scientist to dig deep into user behavior so that the ontology is useful to its target audience. Insights into fan psychology are uncovered as part of that process. One user (StarWars_Jack) in the Branch et al. (2016) study, stated that his research each week is based on: "Conversations will bring something up. Looking up information, clarifying, reading an article on the web will spark an, 'oh I want to go check this.' Just the normal, 'oh I really want to know the answer to that.' What comes up in a normal day" (2), indicating that this is a component of his everyday routine. Another explained that active research occurs when they are about to watch or read a creative work and are interested in discovering prior origin stories to determine variances between old and new storylines. In addition, responses indicated that these fans appreciate the convenience of being able to search for relevant information about their favorite universe online, but also use primary sources (e.g., poring over printed works or watching a video, ready to press pause and examine a scene) to identify a specific detail in the storyline. It was also found that many of the interviewees would search relentlessly for information to win an argument. These findings, from an ontological study of transmedia, demonstrate the kind of fan psychology that is of interest to scholars digging into the sociology of popular culture, as well as understanding audience behavior from an information consumption perspective (Branch et al. 2017).

Conclusion

The creation of an ontological structure is a labor-intensive process of a qualitative nature. These two specific aspects present significant obstacles to the successful implementation of an ontology. The amount of time necessary for adequately representing each element that can be found in a single transmedia fictional universe is certainly a daunting task and may be considered cost-prohibitive. There is also the need to maintain cohesion of the various elements within an ontology so that there is no loss of informational integrity. This is where the qualitative nature of domain analysis and ontological development can cause problems.

One of the challenges faced is how to represent fictional versions of real things (e.g., calendars or cities). Additional challenges are found in the representation of characters and relationships that change over time. As stated previously, there must be a widely agreed upon definition of types, properties, and interrelationships used within an ontology. Otherwise, the structure loses stability and the connections between elements lose some of their meaning and veracity.

Future research on use of ontologies for these universes should expand to include scholars, brand managers, and creators. This approach will allow researchers to finely tune the ontological structure in such a way that it is intuitive for a wider spectrum of users and revealing of popular culture to the scholar. Card sort studies, a common ontological technique, can be conducted with small groups of users, the subsequent qualitative data analyzed, and insight acquired from these processes incorporated into an ontology that reflects the groups' information-seeking behavior.

Potential future research on the ontological approach is vast for those interested in pursuing it. Since this is a relatively new field of scrutiny, there is much yet to be discovered regarding the information-seeking behavior of various groups. For instance, the aesthetic intent behind a specific storyline can hold value for scholars. By paying attention to the psychological and sociological context brought to the work by the creator, research of the transmedia works themselves may have greater value and depth. Creators must find a delicate equilibrium when using ontology to maintain their narrative across a variety of mediums while also appealing to a large audience. A successful implementation of transmedia caters to the different needs of fans and leaves enough room within the storyline for fans to create their own creative works that flesh out the perceived gaps in the narrative (e.g., where did an artifact come from) (Hills 2012). Scholars can further research on the ontological approach by studying how creators use ontology to maintain this balance and assist creators of future transmedia works by identifying specific techniques that should be avoided.

Although it is difficult to organize the vast swathe of information found in a single transmedia fictional universe, there are significant long-term rewards for creators, scholars, and fans of these fictional havens. Creators of canonical works will be better equipped to provide new works across a variety of media platforms while maintaining the continuity and depth that fans seek. Scholars benefit because any source of information that is more easily searchable, lends itself to the advancement of research. Fans benefit from this approach because, not only would their favorite fictional universe provide more of the coherence and unity they want in a narrative, but non-canonical works would be able to be more easily located, shared, and consumed (Branch et al. 2017).

Hills' (2012) research states that transmedia storytelling can become "an immensely rich space, capable of sustaining a great volume of fan engagement over time" (411). There is an art to building a transmedia fictional universe and it takes a multitude of resources to provide fans with an environment that compels them to explore beyond the experience of watching a single movie or reading a single book. To be truly successful at the endeavor of studying these worlds built over various media platforms, one must first have all the information contained in the fictional universe organized and easily accessible. An ontological approach provides scholars with the means to understand stories and how they serve their fans' needs.

References

Aliens Wikia. 2017. *Xenomorph.* Accessed May 21, 2017. http://aliens.wikia.com/wiki/Xenomorph.

Bondi, Gail A. 2011. *Close Encounters of a Different Kind: A Study of Science Fiction an Culture and its Interactions with Multiple Literacies.* Indiana: Indiana University of Pennsylvania.

Booth, Paul. 2009. *Fandom Studies: Fan Studies, Re-written, Re-read, Re-produced.* New York: Rensselaer Polytechnic Institute.

Branch, Frank, Theresa Arias, Jolene Kennah, Rebekah Phillips, Travis Windleharth, and Jin Ha Lee. 2016. *Knowledge Organization in Transmedia Fictional Worlds: A Study of Harry Potter, Lord of the Rings, Marvel Universe, and Star Wars.* Accessed June 10, 2017. https://digital.lib.washington.edu/researchworks/handle/1773/36214.

Branch, Frank, Theresa Arias, Jolene Kennah, Rebekah Phillips, Travis Windleharth, and Jin Ha Lee. 2017. "Representing Transmedia Fictional Worlds through Ontology." *Journal of the Association of Information Science and Technology*. doi: 10.1002/asi.23886.

Captain America: Civil War. 2016. Directed by Anthony Russo and Joe Russo.

Chidester, David. 2005. *Authentic Fakes: Religion and American Popular Culture.* Berkeley: University of California.

Degani-Raz, Irit. 2003. "Possible Worlds and the Concept of 'Reference' in the Semiotics of Theater." *Semiotica* 147 (1/4): 307–329.

Degani-Raz, Irit. 2005. "Theatrical Fictional Worlds, Counterfactuals, and Scientific Thought Experiments." *Semiotica* 157 (1/4): 353–375.

Doležel, Lubomír. 1995. "Fictional Worlds: Density, Gaps, and Inference." *Style* 29 (2): 201–214.

Eco, Umberto. 2009. "On the Ontology of Fictional Characters: A Semiotic Approach." *Sign Systems Studies* 37 (1/2): 82–98.

Errico, Marcus. 2016. "Captain America: Civil War' Comics vs. Movies: The 10 Biggest Differences (SPOILERS!)." *Yahoo*, May 10. Accessed May 21, 2017. www.yahoo.com/movies/captain-america-civil-war-comics-vs-movies-011546407.html.

Gray, Jonathan, Cornel Sandvoss, and C. Lee Harrington. 2007. *Fandom: Identities and Communities in a Mediated World.* New York: New York University Press.

Hills, Matt. 2012. "Torchwood's Trans-transmedia: Media Tie-ins and Brand 'Fanagement'." *Participations* 9 (2): 409–428.

Lewis, Lisa A. (ed.). 1992. *The Adoring Audience: Fan Culture and Popular Media.* New York: Routledge.
Lucasfilm. 2014. "The Legendary Star Wars Expanded Universe Turns a New Page." *Star Wars*, April 25. Accessed May 21, 2017. www.starwars.com/news/the-legendary-star-wars-expanded-universe-turns-a-new-page.
Merriam-Webster. 2017a. *Definition of Igneous Rock by Merriam-Webster.* Accessed May 13, 2017. www.merriam-webster.com/dictionary/igneous%20rock.
Merriam-Webster. 2017b. *Definition of Ontology by Merriam-Webster.* Accessed May 13, 2017. www.merriam-webster.com/dictionary/ontology.
Miller, Mark (w), and Steve McNiven (p). 2006. *Civil War.* New York: Marvel Entertainment.
Stephens, Michael. 2006. "Wikis: Usage and Services." *Library Technology Reports* 52–57.
Straczynski, J. Michael (w), Stan Lee (w), and Ron Garney (p). 2006. "Amazing Spider-Man." New York: Marvel Entertainment.
Thomasson, Amie L. 2005. "The Ontology of Art and Knowledge in Aesthetics." *Journal of Aesthetics and Art Critism* 63 (3): 221–229.
Wagner, Rachel. 2012. *Godwired: Religion, Ritual, and Virtual Reality.* New York: Routledge.

42

AN EXPERIENCE APPROACH TO TRANSMEDIA FICTIONS

Susana Tosca and Lisbeth Klastrup

Frodo is badly hurt, his shoulder pierced by a Morgul blade. His life his ebbing away as his face turns pale, then nearly blue. The Nazgul are closing in and Aragorn cannot get the hobbits out of danger quickly enough. But then, help arrives. Arwen, the elven princess, saves Frodo from the galloping Nazgul in an amazing horse chase across the forest. I let out a long-held breath when they get to safety. I am awed, relieved, and puzzled. This is December 2001, and I am watching *The Lord of the Rings* in the cinema in its premiere day. I am an avid reader of Tolkien's books, and a veteran player of the *Middle Earth* role-playing game. Arwen should not have appeared in this part of the narrative, according to canon, so her presence bugs me a little, but how important is literal adaptation of the novels? It was actually a nice scene, and it is good that she gets more screen time. I remember how frustrated I was as a teenager that there were no females in the fellowship of the ring. In fact, in my role-playing troupe, I have played a character like her, who gets to do all sorts of exciting things. I make a mental note of checking the Tolkien fora as soon as I get home to my computer. I bet the hardcore fans are not satisfied with this change and I need to get into that discussion.

The transmedial experience of watching a movie like *Lord of the Rings*, which belongs to a vast transmedial world (Klastrup and Tosca 2004), is a complex affair, where we interpret, get emotionally involved, and access our memories or repertoire of knowledge about the transmedial universe. We not only react to the scenes we are watching, but also to how they compare to everything else we know about the world, and to what others know and think about the world. Like any other experience, a transmedial experience affects our senses, engages our intellect and is embedded in our everyday life in various ways. This chapter looks at transmedia fictions through an experiential lens, and proposes a model to understand and analyze transmedial experiences, taking their specificity into account. We illustrate our model by applying it to a case, the Jane Austen inspired game *Regency Love*.

The Transmedial Experience

From the beginning of our transmedial research, we have been interested in transmedial worlds (TMWs) as acts of the imagination, that is, as mental images gradually built from diverse encounters with fictions that share the same universe, and which become into existence in acts of aesthetic reception:

> Transmedial worlds are abstract content systems from which a repertoire of fictional stories and characters can be actualized or derived across a variety of media forms. What

> characterises a transmedial world is that audience and designers share a mental image of the "worldness" (a number of distinguishing features of its universe).
>
> *(Klastrup and Tosca 2004, 1)*

This *worldness* can be further described by using the concepts of mythos, topos, and ethos, which respectively refer to the backstories that explain the TMW, its settings (places and peoples) and the philosophy and ethics that make sense in that world (Klastrup and Tosca 2004).

Our transmedial world framework is heavily indebted to classic reception theory and its focus is on the incompleteness of fictions (Iser 1978; Eco 1979; Jauss 1982). Users (readers/viewers/players) need to perform interpretive and emotional work in order to aesthetically experience the narrative they interact with. Works of representational art inhabit a gray ontological zone in that they both exist beforehand and become through the art of aesthetic consumption. Reception theory is a phenomenological approach that builds upon the concept of experience as investigated by philosophers such as Husserl, Heidegger, Sartre, Merleau-Ponty, and Ingarden, among others (Armstrong 2005). We are also inspired by an interaction design approach to experience, in particular by the emphasis of this field on the bodily situatedness of any experience (Dourish 2001) and the idea of experience as felt-life made of several threads: sensual, emotional, compositional, and spatial and temporal (McCarthy and Wright 2004).

We have earlier argued that any encounter with a transmedial product will evoke all the knowledge and affect that the user associates with the particular TMW, which will then be rearranged in the light of the new incorporation (Klastrup and Tosca 2011). The transmedial experience is thus the actualization (by a user) of a TMW's *worldness* as manifested in a concrete platform with its corresponding affordances (Tosca and Klastrup 2016). Thus, the transmedial experience is always situated, both in time (in relation to previous and future encounters with the TMW), in space (the materiality of the medium also plays a part), and in the body of the user, who is involved sensorially, intellectually, and emotionally. Expectations, nostalgia, and various forms of transmedial desire (Tosca 2015) play a part in motivating and directing the interaction, that we have previously illustrated as seen in Figure 42.1 (Tosca and Klastrup 2016).

The vignette we used at the beginning of this chapter would be a good illustration of this situatedness. Our encounter with the *Lord of the Rings* first movie prompted hermeneutic work, not

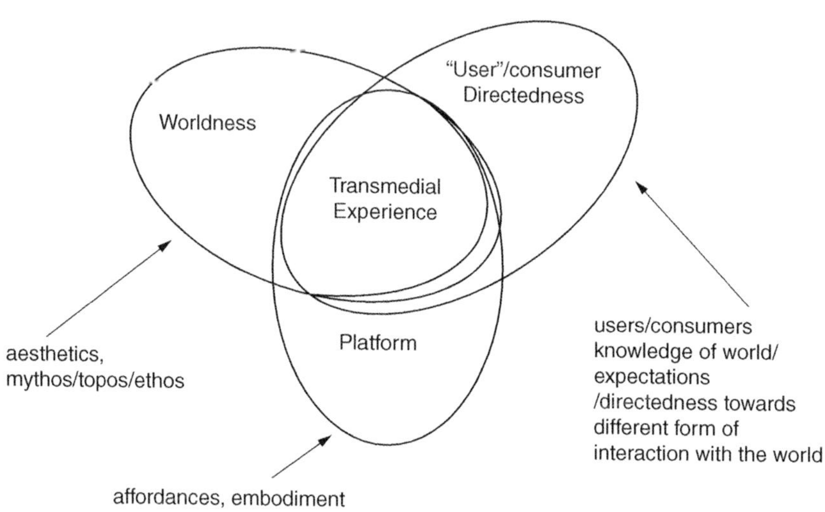

Figure 42.1 A model of transmedial experience.

only at the diegetic interpretive level (to understand what is being shown on screen) but also activating our memory and the repertoire of the Tolkien TMW (to evaluate how what we are seeing fits or does not fit with the canon, in other words, complies to the *worldness*). This process is not merely intellectual, but embedded in a myriad of emotions, again related to different times. That is, what we feel about what we are experiencing now is colored by the expectations we had before entering the movie theater, and the emotional attachment to this particular world that we have developed over time. Each new experience of the TMW is fueled by a nostalgic desire that brings back the thrill of the first encounter, so that our very life story also becomes a part of the current experience. Interpretation, emotion, and action are thus entwined in a circular dance not dissimilar to the hermeneutic circle in which each transmedial instantiation would add to the whole TMW universe, whose meaning gets revised and expanded with each encounter. Transmedial experiences thus transcend the individual fictions and integrate a whole network of perceptions and meanings that has been building up in the user's imagination, sometimes for many years.

How to Observe

But how to capture the transmedial experience? We propose to combine an aesthetic-analytic approach with a qualitative empirical investigation, a hybrid method which we have often used in the past (Klastrup and Tosca 2011, 2014). That is, we are interested in investigating each fiction as an individual work (what are its formal properties, how it is experienced according to the medium, what are its aesthetic qualities), and as a part of a TMW universe that gets activated by the user's imaginative interaction. It is this latter part that requires some methodological creativity, usually a combination of different qualitative approaches.

The first step is to document our own encounter with the work of fiction, which we do by systematically keeping field notes while reading/viewing and/or playing diaries (if it is a game). These notes will often be messy, including at once interpretive evaluations, gut reactions, aesthetic appraisals, or musings on how the work relates to the TMW in question. Our field notes can be used both for a rigorous analysis of the work and for guiding the compilation of questions that we would like to pose to other users.

Once the work has been experienced, we are ready to produce an aesthetic/formal analysis, that will be certainly different, according to the medium and platform we are dealing with. For instance, the analysis of a novel could deal with how changes in focalization present different characters life worlds, while the analysis of a video game could show how the fighting rhythm conveys a particular meaning. Each medium has its strengths, for instance a novel excels at portraying inner life, while a three-dimensional computer world can give an unparalleled sense of immersion. The critic shall judge each medium by its own premises, and not reject modes of expression just because they are different from whichever work that initiated the TMW. This is quite commonsensical, but still, it is surprising how many discussions about adaptation or media migration get entangled in medium-ranking battles (the book is always better than the movie), where some modes of expression are always automatically superior to others.

Once we have formed our own idea of the fictional work as an aesthetic object and how its properties actualize the TMW at hand, we are ready to reach out to other users in order to attain a wider perspective of reception. In fact, we would go as far as to arguing that without reaching to other users, a reception study remains incomplete, trapped in the individual mind of the researcher, who, no matter how insightful, can never account for a more general experience of the work. It is therefore that we always make an effort to investigate the fan communities attached to each TMW. This can involve several methods, for example, participating in discussion fora online and closely following the blogging and social media productions of fan communities, using an digital ethnography approach (Pink et al. 2015). We can also approach individual users and carry out in-depth interviews or focus groups about their media consumption experience, where the ways in which the TMW intertwine with their personal stories are explored in an open-ended way. Other methods for qualitative data elicitation can also be applied, such as asking users to narrate stories of experience as we, for example, did in order to investigate how the death of their characters was significant for players of *World of Warcraft* (Klastrup 2008). Their candid storytelling revealed many of the qualities of the TMW, which we would not have been able to grasp otherwise. In general, we consider methods of narrative inquiry (Kim 2016) to be productive when investigating TMW experiences, because it is not always easy for people to verbalize their thoughts, and the effort of narrating an aesthetic encounter with a work (or an aspect of it) can reveal things that they themselves were not consciously aware of.

In the context of this chapter, we have played *Regency Love* as an individual work of fiction, and considered it in relation to the TMW of Jane Austen, which we know well due to our background in European literature studies and our sustained consumption over the years of all sorts of products related to this TMW, such as films, other games, tourist tours, and even merchandising related to Jane Austen. We have also located all the fans' reviews and player conversations we could find online, viewed playing diary videos by fans in YouTube, as well as carefully perused the extensive materials of the *Regency Love* wiki (Tea for Three Studios 2015), where game developers and fans discuss many aspects of the fiction and the interaction in detail. Moreover, we have conducted two interviews with players of the game. All these materials will be combined in our illustration of a transmedial experience. But let us introduce the game and explain how it relates to the TMW of Jane Austen.

Case Sample: *Regency Love* and the Jane Austen TMW

Regency Love is a romance game (also known as dating sims or Otome games) for iOS devices set in the English 1800s as depicted in the novels of Jane Austen. It was launched in 2013 by the independent game developer Tea for Three, and has enjoyed continued (if niche) popularity and very positive player reviews. The player's role is that of a young woman, whose father has been dead for a year and now lives alone with her mother in the fictional English town of Darlington. Her goal is to get married to one of the eligible bachelors in the village. The gameplay is based on conversational interaction (through the choice of dialogue options) and the obtention of "motivation points" through interacting and playing small puzzle games (answering questions about the period or playing hangman games filling in sentences). One playthrough takes a couple of hours, but the game can be replayed many times to try different suitor routes or different outcomes.

Regency Love belongs to what we can call the Jane Austen TMW. This universe, emerged from the worlds that the author conjured in her eighteenth-century novels, is interesting because, unlike the heavily franchised worlds of, for instance, *Game of Thrones* or *Lord of the Rings*, the stories and products adopting an "Austenesque" mythology, ethos, and topos have emerged independently of each other. *Regency Love* shares thus the TMW of other very different products like, for instance, the film *Clueless* (Amy Heckerling, 1995) or the book *Pride and Prejudice and Zombies* (Grahame Smith and Jane Austen, 2009). In a Jane Austen TMW, place and storylines often seem to matter less than character and the central plot in an Austen-inspired story: the coming of age of a young woman, based on her social blunders and encounters with various prospective love interests/husbands. Following, what is central in the Austen universe is therefore not particular characters (though some, such as Emma or Elizabeth Bennett, remain highly recognizable, and feature in many screen adaptations of Austen's novels), but rather their archetypical traits: a wish for independence; a strong will; good but often misdirected intentions; and the inability to recognize (at first) which man is the right one for her. There are doubts, self-discovery, and usually a great deal of witty dialogue and prose. This demonstrates that a transmedial universe as mental construct can also center around archetypical characters and the particular worldview (ethos) that they embody, rather than on specific narratives and specific characters.

As for other video games in the Jane Austen TMW, there are several visual novels/Otome games, like *The Lady's Choice* (Seraphinite, 2016) or *Northganger Abbey* (Spiral Atlas, 2016), with similar approaches to gameplay as *Regency Love*. There are also games in other genres, such as the multiplayer role-playing game *Ever Jane* (3Turn Productions, 2013), where reputation, social status, and the acquisition of property are central elements. The player advances, not by fighting, but by participating in social events, improving selected character traits, such as "duty" or "kindness," and building their fortune. There are even less plot-heavy genres, like the two-dimensional platform side scroller *Stride and Prejudice* (No Crusts Interactive, 2014), an endless runner where you jump over the text of the novel. However, the "get a husband/survive in society" seems to be the main motivation around which gameplay is constructed in games belonging to the Jane Austen TMW. This is hardly surprising given the social structure of the period. What is maybe more interesting is that, through their emphasis on strategic conversation, these games let players perform the constriction and uncertainty that eighteenth-century social exchanges had for women of the English (mostly) middle class. We will later return to this important aspect.

A Transmedial Playing Experience

In this section, we will deal with the three areas described in our model above (Figure 42.1): the user, the world, and the platform, as the transmedial experience occurs when they conflate during the reading/playing experience. Surely, it is rather artificial to separate the three factors since they are affected and constituted by each other in the experience of playing, but it is necessary for clarity.

The User: Expectations, Motivations, Desires

"It is a truth universally acknowledged, that a single man in possession of a good fortune, must be in want of a wife" (Austen 1813, 1). If you have read *Pride and Prejudice*, chances are that you recognized the quote, and if you appreciated it, the words will also hold some kind of power over you. We are shameless fans of Jane Austen. For us, these are not only words. They are also the faraway thrill of opening a new black paperback, the pleasure of taking the plunge into another world far from our siblings' squabble, the freshness of the newly cut grass, the light touch of a white summer dress, and the butterflies in our stomach as a young man we like asks us if we want to go with him to the harvest fair. That is why we picked this game from the app store, to play in a world that we have known and loved for decades. And all these memories are with us the first time we open the game, certainly shaping the expectations we have upon what we are to encounter. We want to get immersed into a world of country houses, tea invitations, walks, candlelight, and quiet gossiping. Everybody will be polite, witty, and "know their place."

Not everybody's motivation is literary, though, as players have approached the game for other reasons as well. For example, there is a thriving community of people (mostly women), who play dating sims/Otome games, the genre which *Regency Love* could be said to belong to. For them, *Regency Love* is just a "setting" among many. As one of the players we interviewed said, "I just thought that it looked like an exciting new theme. I haven't read Jane Austen but I have seen films from this period and have always like the dresses and the houses. I figured this has to be really romantic." This player was later disappointed that the romance was less explicit than in the other games she is used to playing, although she acknowledged that she had been positively surprised by the quality of the writing and the dialogue options.

It is very common to play a game because someone, whose taste we trust, has recommended it. This was obvious in the online conversations about the game, where players advice each other on which aspects of the game might be appealing, or tag each other in social media thinking that this is something for someone they like: "I just LOVE it! If you're into Jane Austen or the Brontë sisters' literature, then you are going to love this game for sure. You have the possibility to fall in love with two totally different gentlemen and can also change the fate of those around you."

In fact, there is a very clear social aspect to user motivation. Affective consumption is often integrated into a culture of sharing, where the affects that the user has both for the TMW and the person she wants to connect it with, come together. For instance, the other player we interviewed, said she would try to introduce her 20-year-old sister to the game, because she studies literature at university, so "maybe this is something we could have together."

The World: Jane Austen Mythos, Topos, and Ethos

Our game starts with a picture of a quaint village and the following sentence: "All delightful stories revolve around others, and so preoccupied you are with theirs that you do not notice stumbling upon your very own tale." From this opening screen, the game immerses us, communicating the appropriate *worldness* through a well-framed topos. Here we have the promise of an enjoyable tale, an interesting cast of characters, and a "clueless" heroine that is so attuned to everybody else's thoughts and actions that she might not be as self-aware as would be desirable, although we are also made to understand that this will change.

The topos is an idyllic village, hand-painted in beautiful watercolour-like shades. The women are wearing empire-cut gowns in light colors, and the men are dressed in white shirts with a high collar, tailcoats, and waistcoats. They are "delighted to make our acquaintance," they "daresay," or ask us to "do them the honour of," so the language also feels right. Our mother in the game is very keen on us improving our skills and trying to find a husband, and also our late father's friend, Mr. Worthington, who instructs us very explicitly in his first visit: "My dear Miss B. I am sure you know that the way to secure happiness is by marrying a kind gentleman with the means to provide for you." Thus, starts the real game of finding a husband.

However, and as opposed to other kind of games where player action is unambiguous (shoot all that moves), the player does not really know how to go about this. The most "gamey" element is arguably the obtention of motivation points in order to raise our character skills (riding, reading, music, dancing, drawing, and needlework), but we cannot see which one is important, or even how many points are desirable, so a typical strategy is just to fill them up randomly, dividing points equally between the skills, just to be on the safe side. The motivation points are earned by answering questions about *Regency Life*, for example about the appropriate time that a dead spouse should be mourned, or as to what clothing could be indecent, as well as playing hangman games to fill in words belonging to the Jane Austen TMW. These games and questions invariably begin to repeat themselves after half an hour of play, so it is not a very strenuous challenge to harvest many points. In a way, the questions about the proper mores of the time are a literal simulation of Regency England's ethos; while the hangman games that dwell on Jane Austen quotes bring the mythos to the foreground, as it is her works that *Regency Love* is inspired by.

From a gameplay perspective, the motivation points mechanics is of much less importance than the plot; *Regency Love* is mainly a narrative game. In accordance with the rest of the TMW, the important thing is the story unfolding before our eyes. The player spends her time going to parties, attending tea invitations, going for walks in the woods, or running errands in town; but all this is a means to the ends of getting her into conversation with the non-playing characters, both the desirable bachelors and the supporting cast.

The conversation options allow the player to mold the personality of her character very differently, as she can answer in different ways: modest, bold, witty, kind, arrogant … For example, when the dashing Mr. Ashcroft compliments her at a party: "But how terrible of me, to have neglected you. Miss Westfall, you are quite beautiful," the player has three possible answers:

1. "Thank you. I admit to have partaken in vanity tonight"
2. "You had almost missed your opportunity to pay such a compliment, but alas, you have made it just in time"
3. [Looks down].

Over time, our choices will determine which character traits we develop, which in turn will have an influence in how the non-playing characters react to us and the endings that we can achieve in the story. For example, Mr. Ashcroft will react negatively if we are "vulgar" rather than "compassionate," while other suitors might value other traits. When playing this, we often felt constrained by the lack of immediate feedback to our actions or, worse still, by a feeling that only the most submissive of attitudes will secure us the husband. One solution is to save at different points during the game and then experiment with the kinds of answers we use in our interactions with the desired bachelors. But again, in this, there is a conflict between the requirements of efficient gameplay, and the entertaining exploration of different personalities, as the following note from our playing diary can illustrate:

> I want Mr. Ashcroft, but why does it all have to be about his wimpy sister? I want to tell her to pull her act together and stop being a cry-baby but if I am nasty to her as I feel like to, he will stop liking me! I have to pretend to like her so that he likes me. I am so fake!

One way around this, suggested by several of the fans, whose reviews we documented, might be to choose a bachelor who actually prefers us being witty and non-conventional: Mr. Curtis. Leala Turkey, in her YouTube review of the game says that he is "an aloof, grumpy, broody type … he doesn't really like people or social events … He is kind of snobby which I think is fun" (Turkey 2015). But the player is not completely free here either. Like Leala, she is also wondering if "he will like that," when she picks the sarcastic sentences in her dialogue options.

Ultimately, it might be about the different suitors embodying different attitudes toward romance, all of which are possible within the Jane Austen TMW, with its fair share of pretentious heroes, brooding young gentlemen, flirty officers, or grumpy bachelors that turn out to be good people. In fact, perhaps the most salient characteristic of the Jane Austen TMW, as previously noted, is the strength of her archetypical characters, both the heroines and their potential suitors.

The Platform: Affordances, What Can a Game Do

In an interview with *Regency Love* developer, Samantha Lin, she explains what the aim of the game was, and how it relates to other games:

> why not have a fantasy RPG [role-playing game] with all the romance and none of the fighting? So basically, a dating sim—only, not quite, because we wanted to have all the dialogues and conversations from those RPGs we loved, as well as the ability to make an actual difference with your conversation choices.
>
> *(Lien 2013)*

Here, she is pointing at the affordances of the medium of the video game, where fight mechanics would appear easier to implement than love mechanics, at least if we attend to the critical mass of the medium. However, conversation can also be a worthwhile avenue to walk down in game design. *Regency Love* allows us to think of visits and balls as wartime campaigns, and social chitchat as a weapon. Indeed, the guides and walkthroughs about the game reveal the steps we must follow if we want to "conquer" a particular bachelor, as if he was a besieged city in a strategy game. The dating sims genre is also characterized by a certain degree of neuroticism, as players usually try to get all the possible "romance routes" in one single game through successive replays. This is not so much about looking for the optimal experience as about making sure that we have enjoyed all the possible worlds that the game contains, by virtue of letting us experience different love stories that let us perform multiple sides of our personality and know other characters (to some extent also archetypes), all in harmony with the TMW.

As for the abilities, we must make an effort to improve (our needlework, our dancing ...), Lin is adamant: the domestic sphere can be as challenging and shape character as much as any jungle adventure. Why should women chores and pursuit be less worthy of being gamified that men's? "A well-darned sock shows resourcefulness, skill, financial awareness and a respect for one's belongings. So we wanted to focus on those types of stories, to allow for a kind of gentle introspection" (Lien 2013). This is, indeed, as much in the spirit of the Jane Austen TMW as any adaptation we have seen. A computer game can simulate and immerse, provide the illusion of inhabiting this world.

For how is it to be a young middle-class English lady in Regency England? We will never know, but a game like this can, more than any other medium, convey the difficulty of deciding to favor one suitor over another, and the maddening uncertainty of choosing the right social event and saying the right word. And it is not only about interpretation, because we feel this tension in our body, as we impatiently run through the dialogue options, frustrated at the opaqueness of all human conversation. The game ultimately and perfectly reproduces the social claustrophobia of a very small world where happiness depends on betting on the right horse, so to speak, and then being able to ride it to the finish line. As a young lady in Darlington, we are trapped, with very limited possibilities for action. We despair, spending points randomly in our skills to improve our chances without knowing how this will happen. We are afraid of speaking our own mind because that might close some possible romance route. We want to make our suitor understand that we want him, but not too clearly should our virtue be compromised. And what if we picked the wrong man? One that is gentle on the outside and turned out to be a bad husband? We will never be closer to inhabiting Mansfield Park than here.

Conclusion

We have in this chapter argued that researching transmedial experiences must combine an aesthetic/formal analysis with the qualitative investigation of user reception in order to get the full picture of the TMW experience, thus avoiding exclusive focus on the text's form or its fandom. We have proposed a transmedial experience analysis model and illustrated our holistic ambition through a case, the video game *Regency Love*, set in the Jane Austen TMW. This is a novel addition to our body of empirical work, as it deals with a non-franchised TMW, whose strength lies in introspection and the appeal of its archetype-characters.

References

Armstrong, Paul B. 2005. "Phenomenology." In *The Johns Hopkins Guide to Literary Theory and Criticism*, edited by Michael Groden, Martin Kreiswirth, and Imre Szeman, 2nd ed., 562–566. Baltimore: Johns Hopkins University Press.

Austen, Jane. 2012 [1813]. *Pride and Prejudice*. Lanham: Start Publishing LLC.

Dourish, Paul. 2001. *Where the Action Is: The Foundations of Embodied Interaction*. Cambridge, MA: MIT Press.

Eco, Umberto. 1979. *The Role of the Reader: Explorations in the Semiotics of Texts*. Bloomington: Indiana University Press.

Iser, Wolfgang. 1978. *The Act of Reading: A Theory of Aesthetic Response*. Baltimore: Johns Hopkins University Press.

Jauss, Hans Robert. 1982. *Aesthetic Experience and Literary Hermeneutics*. Minneapolis: University of Minnesota Press.

Kim, Jeong-Hee. 2016. *Understanding Narrative Inquiry: The Crafting and Analysis of Stories as Research*. Los Angeles: Sage.

Klastrup, Lisbeth. 2008. "What Makes World of Warcraft a World? A Note on Death and Dying." In *Digital Culture, Play, and Identity: A World of Warcraft Reader*, edited by Hilde Corneliussen and Jill Rettberg, 143–166. Cambridge: MIT Press.

Klastrup, Lisbeth, and Susana Tosca. 2004. "Transmedial Worlds – Rethinking Cyberworld Design." In *Proceedings – 2004 International Conference on Cyberworlds, CW 2004*, 409–416.

Klastrup, Lisbeth, and Susana Tosca. 2011. "When Fans Become Players: The Lord of the Rings Online in a Transmedial World Perspective." In *Ring Bearers, The Lord of the Rings Online as Intertextual Narrative*, edited by Tanya Krzywinska, Esther MacCallum-Stewart, and Justin Parsler, 46–69. Manchester: Manchester University Press.

Klastrup, Lisbeth, and Susana Tosca. 2014. "*Game of Thrones*: Transmedial Worlds, Fandom, and Social Gaming." In *Storyworlds across Media: Toward a Media-Conscious Narratology*, edited by Marie-Laure Ryan and Jan-Noël Thon, 295–314. Lincoln: University of Nebraska Press.

Lien, Tracy. 2013. "*Regency Love*: A Game about 19th Century Courtship Inspired by Baldur's Gate." Accessed August 1, 2017. www.polygon.com/2013/12/25/5231746/regency-love.

McCarthy, John, and Peter Wright. 2004. *Technology as Experience*. Cambridge, MA: MIT Press.

Pink, Sarah, Heather Horst, John Postill, Larissa Hjorth, Tania Lewis, and Jo Tacchi. 2015. *Digital Ethnography: Principles and Practice*. London: SAGE.

Tea for Three Studios. 2015. *Regency Love Wikia*. Accessed August 1, 2017. http://regencylove.wikia.com/wiki/Regency_Love_Wikia.

Tosca, Susana. 2015. "We Have Always Wanted More." *International Journal of Transmedia Literacy* 1 (1.1): 35–43. doi:10.7358/ijtl-2015-001-tosc.

Tosca, Susana, and Lisbeth Klastrup. 2016. "The Networked Reception of Transmedial Universes: An Experience-Centered Approach." *MedieKultur: Journal of Media and Communication Research* 32 (60): 16. doi:10.7146/mediekultur.v32i60.23362.

Turkey, Leala. 2015. "Lets Play Regency Love." Accessed August 1, 2017. www.youtube.com/watch?v=nyhryQTLqzs.

43

A DESIGN APPROACH TO TRANSMEDIA PROJECTS

Renira Rampazzo Gambarato

The design approach to transmedia projects implies the design thinking process behind the ideation, building, and execution of transmedia stories. This vast domain has yet to be further explored by scholars, researchers, and practitioners alike. This chapter reviews recent relevant developments in transmedia project design and focuses on the intertwined relations of transmedia design, systems, and models. The theoretical framework refers to the design thinking process (Dubberly, Evenson, and Robinson 2008; Liestøl 2003; Mendel 2012; Moles and Caude 1970) and transmedia stories as systems (Bunge 1979; Gambarato 2012; Weingartner and Dorn 1990). The empirical approach addresses transmedia design analytical and operational models (Ciancia 2015; Lovato 2017; Moloney 2018; Srivastava 2013; Von Stackelberg 2011), highlighting Gambarato's (2013) transmedia project design model.

Design Process and Transmedia Projects

The design thinking process is a methodological stance that provides a solution-based approach to problems. Design thinking, as a "meta-disciplinary methodology" (Lindberg, Noweski, and Meinel 2010, 35), embraces the planning and implementation, ideation, and execution of solutions. It is not just about creating human-centered products, services, stories, and experiences, but about the process for tackling complex systemic challenges. The design thinking process is composed of stages or phases that vary according to the author, as can be seen in Brown (2008) and Hoffman (2016). Although there are multiple definitions and models related to design thinking, to a certain extent, the variations revolve around the classical creative process framework, for instance, proposed by Moles and Caude (1970). The process, immersed in a feedback loop, generally integrates phases, such as (1) apprehension, (2) preparation, (3) incubation, (4) illumination, (5) verification, and (6) communication. These phases describe problematization, ideation, representation, and implementation stages that ideally lead to more effective designs.

Moloney (2018) intertwines design thinking and transmedia storytelling, arguing that the design of a systemically complex transmedia project configures an act of implementation and management. Thus, Moloney (2018) proposes, in the realm of transmedia journalism, a decision flow model that poses questions to be asked in the initial stages of the design process of transmedia stories. The list is composed of 12 questions, addressing the starting points and continuity of storyworlds, the interconnectivity of stories and media platforms, the structure, the business model, and the timeline. Moloney's management treatment is also found in Lovato's (2017) design of non-fiction transmedia narratives spreadsheets and Srivastava's (2013) narrative design canvas. Lovato presents 11 spreadsheets that handle the narrative, experience, media platforms, and execution dimensions of a transmedia project. Srivastava proposes a simpler canvas, involving a narrative statement, activities, audience engagement, distribution channels, resources, costs, and revenue streams.

Ciancia (2015) developed a transmedia design framework as a "conceptual and operational tool for designing engaging narrative environments" (132). Her analytical approach is based on Chow and Jonas' (2008) model: analysis, synthesis, projection, and communication. Ciancia's design-oriented research process is as follows: (1) analysis, the phase dedicated to discovering valuable data; (2) projection, the phase oriented to the case-study method; and (3) synthesis, the phase of shared practices for designing narrative environments. During the analytical phase, Ciancia (2015, 135) proposes a model that focuses on the storytelling, structure, and business and production dimensions involved in the transmedia design process.

Von Stackelberg (2011) offers "an ontology for transmedia narrative design" (115), identifying three key design phases (user engagement, narrative, and interaction design) and four-level design tasks (transmedia project, storyworld, story, and scene/sequence levels). A set of questions accompanies his four-level process. The aim is to standardize the design approach to transmedia projects. Although Von Stackelberg's key design phases, tasks, and questions provide an extremely detailed framework of the intricacies of writing transmedia stories and designing audience engagement, they are not practical for implementation compared with the models by Moloney (2018), Lovato (2017), Srivastava (2013), and Ciancia (2015).

Transmedia Supersystem

In order to address the complexity of transmedia projects and the implications in transmedia design, the systemic nature of transmedia storytelling is explored. The concept of a system is used extensively *urbi et orbi*. It refers to an assemblage of parts forming a unitary whole, but since Aristotle, the notion has been the whole is something over and above its parts. The set of correlated parts is not simply the sum of them due to changeable interactions that form the integrated whole (Gambarato 2012). Mario Bunge's (1979) scientific approach to general systems theory focuses on structural characteristics of systems and defines them as complex objects, whose components are more interrelated than loose. This aspect corresponds to the essence of transmedia projects. The formal definition of system developed by Bunge (1979) represents the relationship of the components to each other and between the components and the environment (*Umwelt*). A system (σ) is an ordered triple:

$$\sigma = < C, E, S >$$

Where:

C = composition (set of components)
E = environment (milieu)
S = structure (set of relations on the union of C and E)

The composition of a system is the set of its components: the environment, the set of items to which the composition is connected, and the structure (the relations among the components and between them and the environment). Bunge (1979) gave the following example: The composition of a school is the group formed by the staff and pupils; the environment is the natural and social milieu, and the structure consists of the relations between teaching and learning, managed and being managed, and others.

The relations among components are called the internal structure, and the relations between the components and the elements of the environment are designated the external structure. When the external structure is empty, the system is called closed, and when a system presents an external structure, the system is considered open (Weingartner and Dorn 1990). In addition, the properties of a system can vary with time. Therefore, systems may be open in some respects and closed in others. Furthermore, parts of a system can also be systems in their turn, that is, a subsystem. Conversely, the whole surroundings of a system can be a system as well, that is, a supersystem. The definition of a system should consider characterizations of the supersystem that include the system and the subsystems included in it (Weingartner and Dorn 1990).

Considering Bunge's (1979) view of "a system of nested systems, i.e. a collection of systems each of which is a subsystem of a larger system (or supersystem)" (12), Gambarato (2012) argues that "a transmedia story could be seem as a supersystem composed of nested systems and subsystem like Russian dolls or Chinese boxes" (73). A transmedia project is a supersystem composed of systems, such as a story, platforms, audiences, etc. The story system, for instance, has subsystems, such as a plot, characters, time, location, genre, and so forth. Gambarato (2012) transposes Bunge's (1979) system equation $\sigma = < C, E, S >$ to the realm of transmedia storytelling (TS) as follows: "TS = < (story, experience, platforms, audience, business model, etc.), (community of people who share common interests related to the storyworld), (interaction, participation) >" (73).

The relationship (S) between the components (C) and the environment (E) is of particular interest within transmedia projects. Participatory transmedia stories are open systems, allowing participants to collaborate, co-create, and influence the project. Participation (S) occurs when the community of people who share common interests (E) can, with respect at least to a certain aspect, influence the set of components (C), such as the story. The Great British Property Scandal is an example of an open system transmedia project. The project was created by Channel 4 to address the British housing problem: Although many livable homes were empty or being demolished, a significant number of families needed social housing. The core aim of the project was to force the issue on the government agenda, seeking an actual policy, such as the creation of a low-cost loan fund to support owners of properties standing empty who would like to renovate their houses and put them back in the market. Furthermore, the project focused on taking action, through audience engagement, to change government policies to promote the revitalization of abandoned properties. The Great British Property Scandal exemplifies grassroots circulation practices in which "a media text becomes material that drives active community discussion and debate at the intersection between popular culture and civic discourse—conversations that might lead to community activism or social change" (Jenkins, Ford, and Green 2013, 168). The transmedia story spread across a TV series, a website, a mobile application, and social media profiles. Audiences were encouraged to participate by doing the following: (a) signing a petition on the website, (b) reporting empty homes they were aware of via the mobile application, and (c) volunteering to help with the renovations needed to put empty properties back into use. The mobile application offered the opportunity for audiences to easily take pictures of the properties they wanted to report. The Global Positioning System (GPS) built into the app facilitated the precise identification of the location and automatically submitted the details to the empty homes officer of the correct local council. According to Channel 4 (2013), the project succeeded in the following ways: (a) More than 118,200 petition signatures were collected (100,000 within a week after the project was launched), (b) 10,000 empty homes were reported, (c) £17 million was allocated for the new low-cost loan fund in the United Kingdom, and (d) architect George Clarke, the transmedia project's host, was appointed the independent empty home advisor to the government. Thus, the participatory nature (S) of The Great British Property Scandal allowed the environment (E) to influence the components (C) of the open system.

Participation can be differentiated from access and interaction in relation to media content, according to Carpentier (2015), considering: (1) Access is the presence of previously produced content, (2) interaction is the selection and interpretation of content, and (3) participation is the co-decision between producers and consumers on/with content.

When participation is not allowed in transmedia projects, the systems can be considered closed: Audiences can act, react, or interact but cannot interfere with the narrative. Closed system transmedia stories involve interaction but not participation, "in the sense that audience can decide the path to experiencing it, can click here or there, can react to social media entries, but it is not able to collaborate and co-create" (Gambarato 2012, 76). Alma: A Tale of Violence is a transmedia example of a closed system. Produced by the Franco-German channel ARTE, this non-fictional project tells the life story of Alma, a young woman, who shares her memories about her criminal experience as a former gang member in Guatemala. The story unfolds across an interactive tablet application, a web documentary, a TV documentary, a photography exhibition, and two printed books. Accessing the interactive documentary on the tablet application, audiences can swipe back and forth between two image streams, presented above one another. The bottom screen introduces Alma in front of a black background, telling her violent confessions and difficulties leaving the gang called Maras. The second screen, accompanied by an immersive sound mix and voice-over, contextualizes Alma's ordeal through photographs and drawings. In addition, the interactive documentary features four modules with more information on Guatemala, the Maras, the country's violence, and crime prevention. Audiences can choose how they will experience the story but cannot influence the project's design or change the results. The interaction mechanism (S) of Alma: A Tale of Violence does not allow the environment (E) to transform the components (C) of the closed system.

Transmedia Design Project

Drawing on the notion of transmedia projects as complex supersystems, the design process of this kind of project also entails the analysis and synthesis phases, or preparation and inspiration stages in a dynamic process, in which synthesis is preceded by analysis or vice versa (Liestøl 2003). "Analysis begins as thoughtful reflection on the present and continues as conversation with the possible" (Dubberly, Evenson, and Robinson 2008, 57), while synthesis represents the possible outcome, solution, future, or response. Investigating how designers move from analysis to synthesis, Dubberly, Evenson, and Robinson (2008) developed the analysis–synthesis bridge model based on classic models, such as the Beer (1966) model and the Alexander (1964) model. The binominal analysis–synthesis bridge model "enables designers to develop larger and more complex systems and makes the process of working with larger and more complex organizations easier" (Dubberly, Evenson, and Robinson 2008, 61). Mendel (2012) reassures:

> Models are increasingly important in design—as design, in collaboration with other disciplines, increasingly deals with systems and services. Many aspects of customer experience unfold over time and location, and thus are intangible. With their ability to visualize and abstract various aspects of a given situation, models become tools for exploring relationships in ways that aren't otherwise possible. To this end, models are able to synthesize different types of data (qualitative and quantitative), as well as inputs from various perspectives to provide visibility into issues occurring at the boundaries of disciplines. Where differences in discipline language, practices, and approaches can get in the way of problem solving, models can provide insights and frame discussions that must take place.
>
> *(81)*

In this context, the transmedia design project model by Gambarato (2013) is proposed. It is an analytical and synthetical model that contributes to a qualitative understanding of the design process of complex projects that unfold across multiple media platforms. The model was elaborated as a tool for facilitating the analysis of transmedia projects in fictional and non-fictional realms, because analysis is a crucial aspect of the design process that can lead toward synthesis (Dubberly, Evenson, and Robinson 2008; Liestøl 2003). The model aims at outlining the essential features of transmedia stories as supersystems in order to support the analytic and synthetic needs of transmedia designers and the applied research in the interest of the media industry. The purpose of the transmedia design project model is to assist practitioners to better organize their approach to complex transmedia experiences. The model comprises ten specific dimensions that are guided by a series of practical questions: (1) premise and purpose, (2) narrative, (3) world-building, (4) characters, (5) extensions, (6) media platforms and genres, (7) audience and market, (8) engagement, (9) structure, and (10) aesthetics. The structure of the model is based on Strickler's (2012), Jenkins' (2010), and Long's (2007) considerations of transmedia storytelling. A brief description of the model is presented in Table 43.1. This perspective is objective but not restrictive. Other questions and layers of understanding can be considered and added to the analysis as well. Qualitative and quantitative methods can be used according to the nature of the question and the availability of data (Gambarato 2013). "Models integrate information from a wide range of practice domains and data types to reveal patterns hidden in data and identify relationships. They can integrate with a wide range of methods and tools for problem solving" (Mendel 2012, 85).

To briefly exemplify and illustrate the application of the transmedia design project model, the case of the Fish Fight campaign is presented. Produced by Channel 4 in the United Kingdom, the Fish Fight transmedia campaign dealt with the wasteful discarding of healthy fish at sea under European fishing policies: (1) Premise and purpose: This transmedia campaign was designed to draw the public's attention to the discarding of caught fish because of the quota system intended to conserve fish stocks within the European Union (EU) and to pressure the authorities to change the EU Common Fisheries Policy (CFP). The project was a sustainable fishing campaign focused on stopping what could be considered environmental crime and on eliminating waste of food. (2) Narrative: The storyline unfolded toward the popularization of bycatch species among consumers and active support for banning discarding practices. (3) World-building: Fish Fight developed between 2010 and 2013 as a local British project but soon became an international endeavor with worldwide participation. The project impacted people involved in the marine product production and consumption process, including final consumers, supermarkets, fish suppliers, fisheries, and the European Commission. (4) Characters: The campaign's protagonist was Hugh Fearnley-Whittingstall, a celebrity chef, food writer, and television host, but he was not alone. The audience can be considered a primary character of the project as well, because people played a crucial role in the campaign. Celebrities, such as Prince Charles and Stephen Fry, and renowned chefs like Jamie Oliver and Gordon Ramsay also joined the fight. In addition, Fish Fight had spin-offs, such as the What Are Your Prawns Eating? campaign and Hugh's Fish Fight: Save Our Seas. (5) Extensions: The project had multiple extensions, revolving around the Channel 4 TV series. Extensions such as YouTube videos, a website, social media profiles, and a mobile application, enabled audiences to express their opinion and make an impact on the campaign. (6) Media platforms and genres: The media platforms encompassed television, the Internet, DVD, mobile, print, and social media. The rollout strategy started with the release of a viral video in 2010, followed by the website, social media profiles, TV series, and mobile application. The application displayed useful information about several fish species, offering 50 recipes with sustainable fish choices, and revealing a list of restaurants committed to sustainability. The devices required to access the project were a computer, a television set, and a smartphone or tablet to use the mobile application. Documentary is the genre that characterizes Fish Fight. Overall, the synergy between various media platforms resulted in multiple options for audiences to access the campaign and motivated them to join the fight to reform policies for the social good. (7) Audience and market: The campaign reached international audiences because of the direct involvement of the EU in the issue, Channel 4 penetration, and the spreadability of social media networks. Access to the transmedia extensions of Fish Fight was free of charge, collaborating to hook diverse prospective audiences inside and outside the EU. (8) Engagement: The campaign turned audiences into activists by supplying them with mechanisms for influencing government decisions related to the issue. In 2011, when the new CFP was being discussed in the European Parliament, the Fish Fight webpage offered the public the

Table 43.1 Concise description of the transmedia project design analytical model

Topic	*Practicable questions*
1. Premise and purpose	What is the project about? Is it a fiction, non-fiction, or mixed project? What is its fundamental purpose? Is it to entertain, to teach, or to inform? Is it to market a product?
2. Narrative	What is the summary of its storyline? What is the time frame of the story? What are the strategies for expanding the narrative? Are negative capability and migratory cues included? Is it possible to identify intermedial texts in the story?
3. World-building	Which is the central world where the project is set? Is it a fictional world, the real world, or a mixture of both? How is it presented geographically? Is the storyworld large enough to support expansions?
4. Characters	Who are the primary and secondary characters of the story? Does the project have any spin-offs? Can the storyworld be considered a primary character on its own? Can the audience be considered a character as well?
5. Extensions	How many extensions does the project have? Are the extensions adaptations or expansions of the narrative through various media? Is each extension canonical? Does it enrich the story? Are the extensions able to spread the content and provide the possibility to explore the narrative in-depth?
6. Media platforms and genres	What kind of media platforms (film, television, Internet, etc.) are involved in the project? Which devices (computer, game console, tablet, mobile phone, etc.) are required by the project? What is the rollout strategy to release the platforms? Which genres (action, adventure, detective, science fiction, fantasy, etc.) are present in the project?
7. Audience and market	Who is the target audience of the project? What kind of "viewers" (real-time, reflective, and navigational) does the project attract? Do similar projects exist? Do they succeed in achieving their purpose? What is the project's business model? Was the project successful revenue-wise? Why?
8. Engagement	Through what point of view (PoV) do audiences experience this world: first person, second person, third person, or a mixture of them? What role do audiences play in this project? What are the mechanisms of interaction in this project? Is there any participation involved in the project? Does the project work as a cultural attractor/activator? Is there user-generated content (UGC) related to the story (parodies, recaps, mashups, fan communities, etc.)? Does the project offer audiences the possibility of immersion into the storyworld? Does the project offer audiences the possibility to take elements of the story and incorporate them into everyday life?

Table 43.1 (Cont.)

Topic	*Practicable questions*
9. Structure	When did the transmediation begin? Is it a proactive or retroactive project? Is this project closer to a transmedia franchise, a portmanteau transmedia story, or a complex transmedia experience? Can each extension work as an independent entry point to the story? What are/were possible endpoints of the project? How is the project structured?
10. Aesthetics	What kinds of visuals are used (animation, video, graphics, etc.) in the project? Is the overall appearance realistic or a fantasy environment? Is it possible to identify specific design styles in the project? How does audio work in this project? Are there ambient sounds, sound effects, music, and so forth?

opportunity to sign a petition and send emails to members of the Parliament to directly push for a policy change. In 2012, the campaign developed a tool for audiences to tweet every fisheries minister across Europe in their own language. The transmedia campaign led to concrete results. Maria Damanaki, the EU fisheries commissioner, praised the work of the Fish Fight campaign in calling attention to the subject and mobilizing citizens to sign the petition and pressure the European Parliament (Harvey 2013). The demersal discard ban was implemented in phases beginning in January 2015, to help fishermen adapt to the new regulations, and took effect in January 2016. "A complete ban on the discarding of all quota species will be in place by 2019" (Holland 2016). (9) Structure: Fish Fight is a proactive transmedia project that was planned to be transmediatic since the beginning. The transmedia strategy involved launching the website, social media profiles, and a YouTube video before the TV series premiered and later releasing the mobile application and the second season of the TV series. The different extensions of Fish Fight functioned as independent entry points to the storyworld. (10) Aesthetics: The campaign had a high-quality production of audiovisual content (documentary style), and the digital extensions, especially the website, largely invested in information graphics—infographics. The Fish Fight design focused on a simple, clean, and clear layout, relying on blue palettes to communicate and disseminate data and information about discard issues, and to call audiences to action (Gambarato and Medvedev 2015a). For other analyses of fictional and non-fictional transmedia stories that have applied this transmedia design project model, see Gambarato (2014, 2016), Gambarato and Dabagian (2016), and Gambarato and Medvedev (2015b).

Conclusion

The complexity of transmedia stories embodies intricate entanglements between all the constituent elements of [super][sub]systems, that is, the set of components, the environment, and the set of relations. The design of transmedia projects is not isolated from the interlaced social, cultural, economic, and political constructs, including the complexity of the dispositions of audiences to interpret and identify with (or dis-identify from) semiotic elements of the story (e.g., characters, themes, environments, events, and outcomes). The complexity of transmedia projects extends further to the complexity of the identity markets that help shape our dispositions, and to the social networks within which we conduct our interactions with transmedia storytelling (Gambarato 2012).

The transmedia design process models discussed in this chapter depict, in a simpler or deeper manner, the essential structure of transmedia projects and the relevance of each element of what we called transmedia supersystems. In the case of the transmedia design project according to Gambarato (2013), premise and purpose, narrative, world-building, characters, extensions, media platforms and genres, audience and market, engagement, structure, and aesthetics are indispensable for designing a transmedia project and analyzing its pertinence.

Reflecting upon design philosophy, Christy Dena (2017) states: "I disagree about the design philosophy being about ontology (things) than about epistemology (knowledge). Actually, isn't it more a bit of both, and the user, so phenomenology?" Drawing on Dena's consideration and bringing this argument to the transmedia realm, phenomenology (experience) perfectly relates to the fact that transmedia practice can be identified as a "possible procedure to address the issue of contemporary complexity through a phenomenological approach to the coeval reality" (Ciancia 2015, 133).

References

Alexander, Christopher. 1964. *Notes on the Synthesis of Form*. Cambridge, MA: Harvard University Press.

Beer, Stafford. 1966. *Decision and Control: The Meaning of Operational Research and Management Cybernetics*. New York: John Wiley & Sons.

Brown, Tim. 2008. "Design Thinking." *Harvard Business Review* 86 (6): 84–92.

Bunge, Mario. 1979. *Treatise on Basic Philosophy (Volume IV). Ontology: A World of Systems*. Amsterdam: Reidel.

Carpentier, Nico. 2015. "Differentiating between Access, Interaction and Participation." *Conjunctions: Transdisciplinary Journal of Cultural Participation* 2 (2): 7–28.

Channel 4. 2013. "Campaign Results." *The Great British Property Scandal*, November 3. Accessed July 14, 2017. www.channel4.com/programmes/the-great-british-property-scandal/articles/all/campaign-results/1094.

Chow, Rosan, and Wolfgang Jonas. 2008. "Beyond Dualisms in Methodology: An Integrative Design Research Medium 'MAPS' and Some Reflections." Paper presented at Undisciplined! Design Research Society Conference, Sheffield, England, July 16–19.

Ciancia, Mariana. 2015. "Transmedia Design Framework: Design-Oriented Approach to Transmedia Research." *International Journal of Transmedia Literacy* 1 (1): 131–45. Accessed July 21, 2017. doi:10.7358/ijtl-2015-001-cian.

Dena, Christy. 2017. *Facebook*. July 21. Accessed June 3, 2017. www.facebook.com/christydena/posts/10159015648645654?pnref=story.

Dubberly, Hugh, Shelley Evenson, and Rick Robinson. 2008. "The Analysis-Synthesis Bridge Model." *Interactions* 15 (2): 57–61.

Gambarato, Renira R. 2012. "Signs, Systems and Complexity of Transmedia Storytelling." *Communication Studies* 12: 69–83.

Gambarato, Renira R. 2013. "Transmedia Project Design: Theoretical and Analytical Considerations." *Baltic Screen Media Review* 1: 80–100.

Gambarato, Renira R. 2014. "Transmedia Storytelling in Analysis: The Case of Final Punishment." *Journal of Print and Media Technology Research* 3 (2): 95–106.

Gambarato, Renira R. 2016. "The Sochi Project: Slow Journalism within Transmedia Space." *Digital Journalism* 4 (4): 445–461.

Gambarato, Renira R., and Lilit Dabagian. 2016. "Transmedia Dynamics in Education: The Case of Robot Heart Stories." *Educational Media International*. doi:10.1080/09523987.2016.1254874.

Gambarato, Renira R., and Sergei Medvedev. 2015a. "Fish Fight: Transmedia Storytelling Strategies for Food Policy Change." *International Journal of E-Politics* 6 (3): 43–59.

Gambarato, Renira R., and Sergei Medvedev. 2015b. "Grassroots Political Campaign in Russia: Alexei Navalny and Transmedia Strategies for Democratic Development." In *Promoting Social Change and Democracy through Information Technology*, edited by Jakob Svensson, and Vikas Kumar, 165–192. Hershey, PA: IGI Global.

Harvey, Fiona. 2013. "Fish Fight Renews Campaign to Protect the Seabed." *Guardian*, February 14. Accessed July 13, 2017. www.theguardian.com/environment/2013/feb/14/fish-fight-campaign-marine-life.

Hoffman, Libby. 2016. "10 Models for Design Thinking." *Medium*, July 29. Accessed July 12, 2017. https://medium.com/@elizabeth7hoffman/10-models-for-design-thinking-f6943e4ee068.

Holland, Jason. 2016. "New EU Fisheries Discard Ban Gets Underway." *SeaFood Source*, January 4. Accessed June 2, 2017. www.seafoodsource.com/news/environment-sustainability/new-eu-fisheries-discard-ban-gets-underway.

Jenkins, Henry. 2010. "Transmedia Storytelling and Entertainment: An Annotated Syllabus." *Continuum: Journal of Media & Cultural Studies* 24 (6): 943–958.

Jenkins, Henry, Sam Ford, and Joshua Green. 2013. *Spreadable Media: Creating Value and Meaning in a Networked Culture*. New York: New York University Press.

Liestøl, Gunnar. 2003. "'Gameplay': From Synthesis to Analysis (and Vice Versa): Topics of Conceptualization and Construction in Digital Media." In *Digital Media Revisited*, edited by Gunnar Liestøl, Andrew Morrison, and Terje Rasmussen, 389–413. Cambridge, MA: MIT Press.

Lindberg, Tilmann, Christine Noweski, and Christoph Meinel. 2010. "Evolving Discourses on Design Thinking: How Design Cognition Inspires Meta-Disciplinary Creative Collaboration." *Technoetic Arts: A Journal of Speculative Research* 8 (1): 31–37.

Long, Geoffrey. 2007. "Transmedia Storytelling: Business, Aesthetics and Production at the Jim Henson Company." Master's thesis, Massachusetts Institute of Technology.

Lovato, Anahí. 2017. "Diseño de Narrativas Transmedia de no Ficción: Hacia un Modelo Posible. Caso: De Barrio Somos" [Design of Nonfiction Transmedia Narratives: Towards a Possible Model. Case: De Barrio Somos]. Paper presented at the 9th Foro Internacional de Periodismo Digital and 4th Encuentro de Narrativas Transmedia, Rosario, Argentina, April 19–20.

Mendel, Joanne. 2012. "A Taxonomy of Models Used in the Design Process." *Interactions* 19 (1): 81–85.

Moles, Abraham, and Roland Caude. 1970. *Créativité et Méthodes d'Innovation dans l'Entreprise* [Creativity and Methods of Innovation in Enterprises]. Paris: Fayard-Mame.

Moloney, Kevin. 2018. "Designing Transmedia Journalism Projects." In *Exploring Transmedia Journalism in the Digital Age*, edited by Renira R. Gambarato and Geane Alzamora, 83–103. Hershey, PA: IGI Global.

Srivastava, Lina. 2013. "The Narrative Design Canvas." *Transmedia Activism*, November 12. Accessed July 27, 2017. www.transmedia-activism.com/.

Strickler, Elizabeth. 2012. "10 Questions." *Cross Media Design*, July 23. Accessed June 3, 2017. http://transmediadesign.wordpress.com/10-questions/.

Von Stackelberg, Peter. 2011. "Creating Transmedia Narratives: The Structure and Design of Stories Told Across Multiple Media." Master's thesis, State University of New York.

Weingartner, Paul, and George Dorn (eds.). 1990. *Studies on Mario Bunge's Treatise*. Amsterdam: Editions Rodopi B. V.

44

A MANAGEMENT APPROACH TO TRANSMEDIA ENTERPRISES

Ulrike Rohn and Indrek Ibrus

There are two ways to look at the phenomenon of transmedia. First, one perspective on transmedia can be about ways of storytelling (Jenkins 2003, 2010; Scolari 2009) or world-building (Lisbeth and Tosca 2004; Ryan and Thon 2014; Saldre and Torop 2012). Much of the writing on transmedia employs this broadly humanities based perspective. The other perspective sees transmedia as a business model by media providers, responding to the realities of convergent and saturated markets and to new opportunities for market potentials. The business and management perspective on transmedia sees the transmedia approach as a response to changing market conditions that marks an end to the defense of old business models and modes of value creation and the beginning of new ones (Hartley et al. 2013). Where media providers follow a transmedia approach, this affects their business models and strategic management. The term "transmedia," however, is not common among media management scholars who aim to understand decision-making and behavior by media firms. Scholars in this field usually speak of cross-media, cross-promotion, cross-platform, or multiplatform strategies (Doyle 2015a). Media management scholars have a tradition of speaking about cross-media instead of transmedia, as their focus is on the business side of the phenomenon and less on forms of storytelling. The difference between transmedia and cross-media concepts, as Ibrus and Scolari (2012) explain, is that "crossmedia" denotes broader strategies where a coherent narrative told across multiple media may be only one way to establish connections between the different media. Cross-media strategy as a diversification strategy (Chan-Olmsted and Chang 2003) may be about dispersing semantically related forms of content across multiple media or platforms.

Compared to this, transmedia storytelling could be understood as a narrower term, referring to attempts to use the means of a gradually unfolding narrative to connect different media in the era of user participation. Yet, it does not mean that it is less complex. Instead, it may demand significant management skills and efforts to develop complex narratives across platforms. This chapter will introduce the most important notions, aspects and dimensions of managing transmedia storytelling/world-building efforts. It will introduce various aspects of how to study firms involved in transmedia management in order to understand their decision-making and behavior, as well as their strategies' influence on the general media ecology. It will point to the challenges of researching transmedia management and the need for an interdisciplinary approach for studying transmedia enterprises, projects, and products.

From Corporate Strategies to Network Management

Davidson (2010) proposes dividing transmedia projects into proactive and retroactive projects. The first refers to projects designed to be transmedia from the beginning and the second to those that become transmediatic down the line, in evolutionary or ad hoc ways. Yet, in both cases there is always at least some striving to secure certain consistency in the "story-universe" (Lisbeth and Tosca 2004; Ryan and Thon 2014; Saldre and Torop 2012). Hence, transmedia management needs to be, in effect, a form of strategic management; one that cannot be dealt with on the operational and product- or platform-specific level alone. A management approach to transmedia involves the decision-making process and the strategies a firm undertakes in order to create and sustain competitive advantage via a transmedia approach. Transmedia management is less about meeting single product- or platform-specific targets than it is about corporate targets and, therefore, it needs a corporate view. Much more than product-specific or platform-specific choices, a transmedia approach to managing media presumes strategic planning that starts with corporate goals that derive from the corporate mission and vision. Transmedia strategies need to be based on a situation analysis that regards both external analysis of the environment and competitors as well as internal analysis of the competencies and resources available. The implementation of the strategies then includes allocation of the required resources, formulation of the behavior guidelines and controlling of the realization process (Wirtz 2011).

Managing transmedia strategies very often entails the management of a network that spans various enterprises, subsidiaries, platforms, and brands. As such, transmedia projects can also be viewed in terms of network strategies, where a network cooperates on a value creation process. Enterprises involved in such a network may maintain their legal independence and be merely connected by their common goal (Wirtz 2011). The contemporary complexity of media markets and the increasing rate of innovations in the media sector make cooperation between enterprises not only attractive but very often also unavoidable (Shapiro and Varian 2013). Fundamentally, a transmedia strategy means moving out of the comfort zone of individual products and platforms, leaving the philosophy of defending one's properties, but engaging in new modes of value chains across various products and platforms even in corporation with other firms as well as audiences that together form the network of transmedia storytelling.

Transmedia Human Resource Management

In most media firms, managers deal with two different cultures: the culture of the administrative and managerial people in the firm, and the culture of the creative and content-producing people in the firm. Very often, there is a "Chinese wall" between those two areas in a company. Managers are wary of interfering in the creative process, for fear of demotivating the creatives (Deslandes 2016). While the managers are accountable for economic performance, the brand value and editorial reputation that some of the creative talents enjoy may be difficult to replace. At the same time, creatives often resist commercial interests (Aris and Bughin 2005) and may therefore be reluctant to accept the imposition of external authority or controls (Aris and Bughin 2005; Deslandes 2016; Perez-Lattre and Sanchez-Taberno 2003). Deuze (2016) points to the powerful link between work and self-realization that is generally evident in the motivation to pursue creative careers in the media as well as to the individualization in work, all of which may make it difficult to manage creative workers.

While personnel management in media firms is complicated, managing personnel in transmedia projects and enterprises deals with an extra portion of complexity. The management of transmedia projects not only involves the management of the business and creative people involved; it also involves managing people from media types and platforms. People who work with distinct media or with different modalities may have rather different motivations, and their differences in training and production routines have conditioned different work cultures. The managerial challenge is to make these cultures, expectations, and routines complementary and to make people with different media sector-specific backgrounds cooperative with each other. It is not uncommon for enterprises that start a transmedia project first to have to overcome problems that are a result of cultural clashes between departments or firms (Erdal 2009). In terms of management, therefore, working on a transmedia project is effectively, as already suggested above, network management that may involve not only the knowledge of a variety of distinct media, but oftentimes also negotiating distinct medium-specific creative practices, professions, cultures, and ethoses (Dena 2009, 68). A transmedia specialist may need to have skills across various types of the media, and be able to design for various specific media, use protocols that are specific to the devices, develop relationships with its gatekeepers, and understand the discourses and processes that are specific to the related industry. Since this is oftentimes too challenging for creative practitioners with backgrounds in only one of the included media, a new breed of practitioners, the transmedia producers (with a focus on intermediation between media and platforms), has been on the rise in recent years in both Northern America and Europe (Ibrus 2016). According to Creative Skillset, a British council for skills in cultural industries, skills to develop content for multiple platforms were perceived as the second most important skills gap in the sector in 2011, behind only technical skills.

Personnel management in transmedia projects needs to take an integrative and holistic approach not only because the boundaries between media types and platforms are blurring, but also because the traditional distinctions between management and creativity are blurring (Bilton 2007). Successful transmedia management crosses the boundaries between commercial acumen and creative enterprise as well as the boundaries between different disciplines within an organization, including editorial, marketing, public relations, sales, and advertising. The challenge of managing a networked team for transmedia strategies that includes people with different backgrounds, expertise, and employment status is to ensure that such a constellation is not encouraging looseness and non-commitment, but that everyone involved is committed to a common understanding of the goals.

Transmedia Audience Management

Where content reaches and engages audiences across platforms, including platforms that enable a two-way connectivity, transmedia management increasingly is about building and sustaining relationships with audiences. In the area of the Internet and the seemingly endless media offers for entertainment and information, audience demand has become very fickle and increasingly fragmented. Audiences are increasingly used to being in control of their own media menus, to participating in or influencing content creation for some media services, and they often expect to receive some of the services or content for free. Hence, there are three main challenges for transmedia audience management: (1) guiding audiences across different media and platforms; (2) managing participatory practices and two-way communication with audiences; and (3) managing complex business models where access to some media content is free and some is provided for a fee, and where some of the user activity may be either monetized or exploited for further marketing or product/service/content development purposes.

Media enterprises have developed a variety of ways to deal with these intertwined challenges. Deploying a so-called blockbuster strategy is one of them. The blockbuster strategy is conditioned by the understanding that in saturated markets only salient brands and narratives that enjoy high visibility can attract fragmented audiences. Hence, a blockbuster strategy focuses on a few high-impact ideas—i.e., "blockbusters" (Doyle 2010). As such, it is a risk reduction strategy, in which a few media products generate the bulk of revenues for a media organization. Instead of increasing the chances for company success by increasing the number of attempts with small margins, a blockbuster strategy devotes large budgets to a small number of blockbusters (Küng 2008). The idea is that a large budget for talent, production, and promotion will attract large numbers of audiences. Critically viewed, blockbuster strategies decrease output diversity and neglect the demands of a niche audience. Blockbuster strategies target a mass audience to avoid the risk that particular audience groups or niches will reject the project. Hence, the narratives of blockbuster projects are never extremely sophisticated, though their special effects and the technology they use may well be (Küng 2008; Vogel 1999). While effectively it is only the larger media firms that are able to follow a blockbuster strategy as only they have the means to invest in a blockbuster project, blockbuster projects may also encourage the cooperation of firms and the development of new kinds of value networks between organizations that are associated with different media and modalities (Küng 2008).

Closely connected to the blockbuster strategy and also a way to address audience fragmentation and market saturation is the approach of building and maintaining strong brands. In fact, a large part of transmedia management is brand management. Owning and operating visible media brands helps to achieve cross-platform visibility to secure a foothold with audiences across platforms. Especially as the quality of a media product emerges only upon consumption, having a strong brand is crucial to support audiences with consumption choices (Rohn and Baumann 2015). Hence, companies spend increasing amounts of resources on communicating brand identity and promise with the aim of creating a desired brand image with audiences (Chan-Olmsted 2006; Siegert et al. 2015). Where they have successfully established a recognizable transmedia brand, this also increases the value that their media product or narrative universe may have for advertisers and other corporate partners along value chains. A large part of the management of brands and, especially, transmedia brands includes assuring that all brand-related activities and content are consistent with each other and aim to create the same brand image with audiences. If this is not the case, single activities and content pieces may, in fact, harm the brand. Assuring that brand content and activities are coherent is especially important where audience input and activities are involved.

Transmedia is by definition participatory, and an important strategy for transmedia audience management is to engage with audiences directly. As Wikström (2014) suggests, enabling audience engagement should follow the principles of providing audiences with the tools to generate their own content, giving them building blocks that facilitate their creative activities, and recognizing and awarding them. Transmedia enterprises tend to rely on the increasingly intimate relationships with audiences, and audience participation in the production and development processes is often seen as key to success in transmedia management. The aim is not only to reduce costs through free user-created content, but also to embrace "user innovation" to generate market demand for particular products. Such open innovation processes are also expected to increase brand loyalty through the engagement of dedicated users and the formation of fan communities.

Building and nurturing communities of fans and dedicated users is very important in the era of abundance of media offers and saturated markets. An industry consensus suggests that to facilitate such communities, an appropriate combination of free and paid-for content needs to be offered across platforms. Such "freemium" business models tend to be applied so that free offers on selected platforms promote premium offerings on other platforms.

Especially independent content producers may experience the intimate relationships between media producers and audiences as empowering (Ibrus 2012). Micro-companies "normally" have sales limited by a small number of distributors and other intermediaries. But having started to experiment with transmedia strategies and brands and additionally developing direct relationships with increasingly platform-agnostic users, they may experience an increase in bargaining power. If they also develop close relationships with fans or other followers, they may be better positioned to negotiate terms with, for instance, intermediaries, distributors, and channels. Hence, especially independent and micro-businesses may be motivated to build direct relationships with their audiences and to offer various and innovative kinds of transmedia extensions.

Transmedia audience management may also have to deal with the potential conflicts between transmedia audiences and producers. Despite industry discourses that celebrate user empowerment and participation, the reality is that users' possibilities for meaningful contribution may be limited. There is evidence (for instance, Bolin 2010; Perryman 2008) that shows that viewers' options in terms of creativity are often restricted in the interest of what is perceived as professional production quality or narrative coherence. In addition to professional insecurities, the economic rationale may serve to encourage content owners to limit fan contributions because of the risk of diluting intellectual property value—as a result of losing legal control over content or by undermining its brand value as, for instance, in the case of pornographic renditions of their popular narratives.

Another challenge is that the interests of dedicated fans of a transmedia product or brand may differ from those of the general mainstream audience that creates crucial revenues for transmedia blockbusters. As such, casual viewers may lose interest if the content is too complex and full of references to storylines on other platforms. As most of the audience members are mainstream viewers, this segment influences the business model bottom-line. However, as fans have high conversion rates, the challenge is not to exclude audience segments with varying degrees of dedication by experimenting with the product design principle known as tiering. Through tiering, media providers address different audiences with different content in different media and environments (Dena 2009).

The Economics of Transmedia Investment

Transmedia projects seem financially attractive due to the high economies of scope and scale they enable. In transmedia projects, costs, such as for content ideas, marketing, advertising, and personnel, do not occur for every platform individually. Instead, costs may be shared among platforms, making it more cost-effective compared to if a single platform had to operate individually (economies of scope). Especially the costs related to intellectual property rights (IPR) may be shared among and fully exploited by platforms. The more output produced and sold from a single IPR, the more the average costs for each of these goods decrease as their share of the fixed cost, i.e., the costs for the IPR, decreases (economies of scale) (see also Doyle 2013).

Financial savings due to economies of scope and scale, however, are only one side of the coin for transmedia projects and enterprises. A firm's journey from a single-sector focus to a transmedia content provider in which previous industry processes and output are altered constitutes an investment. Reallocation of resources, change of business practices and any process of "creative destruction" (Caballero 2006) come with a cost attached. Furthermore, the opportunities of transmedia encourage an increase in the volume of media content produced and delivered at a time when budgets of media organizations are often tightly constrained (Doyle 2015b), and many firms face an unknown future. Some of the extra costs that occur include costs for transmedia training and workshops, new technologies and equipment, software solutions for multiplatform content management and social media monitoring tools, and new personnel, such as community or social media managers (Rohn 2015; Rohn and Baumann 2015).

Return on investment in transmedia projects is usually slow, and it may take time before cost-efficiency in terms of economies of scale and scope outweighs the investment made and/or any profit is made. Audience revenues as well as participation and engagement depend not only on the quality of the content provided but also on whether the transmedia content enjoys network externalities, which only occurs when content reaches a critical mass of audiences. The more people are watching or engaging with certain content, the higher its value for each audience member, particularly if they were to miss out on a big blockbuster or media hype otherwise.

It is due to these extra costs that media companies have been growing wary of transmedia projects where returns from online and new outlets are not immediate or substantial enough. Yet, Picard suggests that one should apply a strategic management perspective by "focusing on the long-term growth of firms, not only on immediate financial goals (that) ignore the future" (2006, 33). Furthermore, Küng (2008) and Oliver (2014) suggest that the ability of media companies to act on the changing market demand in the digital and transmedia age is, indeed, a source of competitive and strategic advantage.

It needs to be recognized, however, that different types of organizations face different financial opportunities and challenges with their transmedia projects. Transmedia innovations are more attractive for larger firms than for smaller firms who may not have the same opportunities to cross-subsidize their own operations within the organization. The more platforms a company operates, the more opportunities it has for transmedia activities and the more easily and cost-efficiently they can be conducted. Furthermore, although start-up enterprises may have the advantage of not having to restructure, retrain, and reform their organization for transmedia purposes, they cannot—in contrast to older players—use profit from ongoing legacy media to invest into new endeavors. Hence, they lack this comfort of financial backup when experimenting with platforms, channels, and new forms of storytelling. Given the economics of transmedia projects, transmedia projects may, in fact, support the power of existing conglomerates and a further consolidation of the industry.

The Challenges of Studying Transmedia Enterprises

Studying the firms involved in transmedia projects and networks, in terms of understanding their strategic decision-making, behavior, and output, is a complex endeavor. In order to fully understand the processes, practices, and outcomes of transmedia strategies, a holistic approach is needed that takes into consideration various perspectives and applies insights and methods from different approaches. No matter what research question, being in close dialogue with representatives from the industry is of crucial importance to the study of transmedia enterprises. Though research should be independent and ready to criticize conditions where needed, the dialogue with the industry helps to identify relevant research questions and to receive industry feedback to ensure relevance and reliability of research findings. What is more, it helps to keep up to date on industry insights and practices, as the industry is usually much more in touch with the latest developments than academia.

The phenomenon of transmedia is about innovative artistic expression as much as it is about new forms of business model. Researchers who focus on understanding transmedia narratives and storytelling tend to have different backgrounds from researchers who are more interested in the management and business aspects of it. Where transmedia enterprises are studied from a holistic perspective, research faces both the challenges as well as the opportunities that any interdisciplinary research entails, especially where members of a research team have different backgrounds, approaches, assumptions, and priorities. Such an interdisciplinary approach is not uncommon in the study of media, where media management research, for instance, aims at bridging media studies with management studies.

The particularity of studying the transmedia phenomenon, however, is that it needs not only to cross various academic disciplines and approaches, but also to follow a multi-sector approach to the media. The study of transmedia enterprises and their strategies requires knowledge about practices in various sectors of the media, such as the television and film or the book and magazine publishing industries.

Researching questions regarding transmedia enterprises can be manifold. We distinguish three different levels of analysis that are crucial for understanding transmedia enterprises and their strategies:

(1) Product level: Transmedia enterprises as producers of transmedia products

Transmedia products are the outcome of transmedia management within and among organizations and other involved parties, such as freelancers or audiences. Business approaches to transmedia products and enterprises study the composition and changes in the transmedia product portfolios that firms own and operate. Related brand management studies may study the characteristics of transmedia brands and the value that branded transmedia products create for their target group. Transmedia products may also be studied in terms of their innovative character and new production formats.

(2) Firm level: Transmedia enterprises and their decision-making and strategies

On a firm level, studying transmedia enterprises aims at providing an answer as to why firms do what they do and how they do this. Hence, of interest are decision-making processes as well as the formulation and nature of transmedia strategies. This is usually in the realm of media management studies, but it may also include media economics for understanding the competitive and regulatory environment in which firms operate and which influences their decision-making. Also understanding the value that transmedia products create for audiences, advertisers, intermediaries, and investors is crucial for understanding what influences decision-making by organizations. Research may include analysis of best-case examples and identification of success factors for transmedia projects and products. The academic domain known as media production studies (e.g., Gawer 2011) focuses generally on the complex (power) relationships among the people and institutions involved in media production processes that may shape the face of the media in one way or another. With an organizational studies approach, research on transmedia enterprises may be concerned with questions regarding organizational structures and processes of restructuring, team building, cooperation as well as leadership in transmedia projects and transformations.

(3) Industry and society level: Transmedia enterprises as part of industries and the general society

Transmedia enterprises may also be studied in terms of their positioning in or contribution to the creative and media industries as well as their value for and influence on society at large. In terms of industries, transmedia enterprises that target certain markets are most likely not alone in doing so, but find themselves in a competitive environment where they compete with other firms and media providers for audience attention and advertising revenues. Research questions of relevance may include the understanding of the entry barriers for new players in the industry, the evolution of value chains or networks, clustering, standardization or codification processes, the contribution of the media industry to the general economy, or the internationalization of the industry. Studying transmedia enterprises on an industry level also includes questions regarding professional training and education as well as employment opportunities or working conditions. Innovation studies research the dynamics and evolution of emergent technologies and other innovations in an industry, including the necessary conditions for these innovations to emerge. The platform studies (e.g., Gawer 2011) look at media platforms as increasingly central intermediaries in transmedia value chains. A critical study of the transmedia industry can be found within the media industry studies tradition (Havens, Lotz, and Tinic 2009), which has adapted and combined the critical approaches of cultural studies and political economy to media industries. A critical perspective on transmedia enterprises may discuss the value of transmedia products and innovations for society at large as well as the conditions of productions.

A holistic approach to studying transmedia enterprises and management includes research questions, considerations, and discussions on all of these levels (product level, company level, and industry and society level), though some of these overlap. In order to tackle the complexity of the subject matter, much of the studies are case-study driven, with certain companies or transmedia products and brands serving as cases. This is particularly true for the business and management studies approach to the transmedia phenomenon. It is important, however, that the findings of these studies are discussed in terms of their implications on the industry and societal level. Of increasing importance for future research is also an international perspective on the evolution of transmedia strategies, forms and practices (see Freeman and Proctor 2018) as well as processes of international standardization or local adaptation. Though international format trade, for instance, has received a lot of attention in terms of research into the variations of local adaptations (e.g., Esser, Bernal-Merino, and Smith 2016; Rohn 2014), research, so far, lacks an understanding of globally standardized or locally specific processes and products of transmedia strategies and what conditions them.

Acknowledgements

This work was supported by two grants funded by Estonian Research Council – PUT1176 and PUT1674.

References

Aris, Annet, and Jacques Bughin. 2005. *Managing Media Companies: Harnessing Creative Value*. Chichester: John Wiley & Sons.

Bilton, Chris. 2007. *Management and Creativity: From Creative Industries to Creative Management*. Malden: Blackwell.

Bolin, Göran. 2010. "Digitization, Multiplatform Texts, and Audience Reception." *Popular Communication* 8: 72–83.

Caballero, Ricardo. 2006. *The Macroeconomics of Specificity and Restructuring*. Cambridge, MA: MIT Press.

Chan-Olmsted, Sylvia M. 2006. *Competitive Strategy for Media Firms: Strategic and Brand Management in Changing Media Markets*. Mahwah: Lawrence Erlbaum Associated.

Chan-Olmsted, Sylvia M., and Byeng-Hee Chang. 2003. "Diversification Strategy of Global Media Conglomerates: Examining its Patterns and Determinants." *Journal of Media Economics* 16 (4): 213–233.

Davidson, Drew. 2010. *Cross-Media Communications: An Introduction to the Art of Creating Integrated Media Experiences*. Pittsburgh: ETC Press.

Dena, Christy. 2009. "Transmedia Practice: Theorising the Practice of Expressing a Fictional World across Distinct Media and Environments." Ph.D. diss., University of Sydney, Sydney.

Deslandes, Ghislain. 2016. "Leadership in Media Organisations: Past Trends and Challenges Ahead." In *Managing Media Firms and Industries*, edited by Gregory Ferrell Lowe and Charles Brown, 311–327. London: Springer.

Deuze, Mark. 2016. "Managing Media Workers." In *Managing Media Firms and Industries*, edited by Gregory Ferrell Lowe and Charles Brown, 329–341. London: Springer.

Doyle, Gillian. 2010. "From Television to Multi-platform: Less from More or More from Less?" *Convergence: The International Journal of Research into New Media Technologies* 16 (4): 431–449.

Doyle, Gillian. 2013. *Understanding Media Economics*. London: Sage.

Doyle, Gillian. 2015a. "Guest Editor's Introduction to the Special Issue: Multi-platform Strategies." *Journal of Media Business Studies* 12 (1): 3–6.

Doyle, Gillian. 2015b. "Multi-platform Media and the Miracle of the Loaves and Fishes." *Journal of Media Business Studies* 12 (1): 49–65.

Erdal, Ivar John. 2009. "Cross-Media (Re)Production Cultures." *Convergence: The International Journal of Research into New Media Technologies* 15 (2): 215–231.

Esser, Andrea, Miguel Á. Bernal-Merino, and Iain Robert Smith (eds.). 2016. *Media Across Borders: Localizing TV, Film and Video Games*. New York: Routledge.

Freeman, Matthew, and William Proctor (eds.). 2018. *Global Convergence Cultures: Transmedia Earth*. London: Routledge.

Gawer, Annabelle. 2011. *Platforms, Markets and Innovation*. Cheltenham: Edward Elgar Publishing Inc.

Hartley, John, Jason Potts, Stuart Cunningham, Terry Flew, Michael Keane, and John Banks. 2013. *Key Concepts in Creative Industries*. Los Angeles, London, New Delhi, Singapore, and Washington, DC: Sage.

Havens, Timothy, Amand D. Lotz, and Serra Tinic. 2009. "Critical Media Industry Studies: A Research Approach." *Communication, Culture & Critique* 2: 234–253.

Ibrus, Indrek. 2012. "The AV Industry's Microcompanies Encounter Multiplatform Production." In *Crossmedia Innovations: Texts, Markets, Institutions*, edited by Indrek Ibrus and Carlos A. Scolari, 44–68. Frankfurt: Peter Lang.

Ibrus, Indrek. 2016. "Micro-studios Meet Convergence Culture: Crossmedia, Clustering, Dialogues, Auto-communication." In *Media Convergence Handbook – Vol. 2: Firms and User Perspectives*, edited by Arthur Lugmayr and Cinzia Dal Zotto, 155–173. Berlin, Heidelberg: Springer.

Ibrus, Indrek, and Carlos A. Scolari. 2012. "Introduction: Crossmedia Innovation?" In *Crossmedia Innovations: Texts, Markets, Institutions*, edited by Indrek Ibrus and Carlos A. Scolari, 7–21. Frankfurt: Peter Lang.

Jenkins, Henry. 2003. "Transmedia Storytelling: Moving Characters from Books to Films to Video-games Can Make Them Stronger and More Compelling." *Technology Review*, January 15. Accessed November 17, 2017. www.technologyreview.com/biotech/13052.

Jenkins, Henry. 2010. "Transmedia Storytelling and Entertainment: An Annotated Syllabus." *Continuum: Journal of Media and Cultural Studies* 24 (6): 943–958.

Küng, Lucy. 2008. *Strategic Management in the Media: Theory to Practice*. London Sage.

Lisbeth, Klastrup, and Susana Tosca. 2004. "Transmedial Worlds: Rethinking Cyberworld Design." *Procedings of the International Conference on Cyberworlds*. Los Alamitos.

Oliver, John. 2014. "Dynamic Capabilities and Superior Firm Performance in the UK Media Industry." *Journal of Media Business Studies* 11 (2): 55–77.

Perez-Lattre, Francisco, and Alfonso Sanchez-Taberno. 2003. "Leadership, an Essential Requirement for Effecting Change in Media Companies: An Analysis of the Spanish Market." *The International Journal on Media Management* 5 (3): 199–208.

Perryman, Neil. 2008. "Doctor Who and the Convergence of Media: A Case Study in 'Transmedia Storytelling'." *Convergence: The International Journal of Research into New Media Technologies* 14 (1): 21–39.

Picard, Robert G. 2006. *Journalism, Value Creation and the Future of News Organizations*. Cambridge, MA: Joan Shorenstein Center of the Press, Harvard University.

Rohn, Ulrike. 2014. "Small Market, Big Format: Idols in Estonia." *Baltic Screen Media Review* 2: 122–137.

Rohn, Ulrike. 2015. "Social Media Business Models." In *International Encyclopedia of Digital Communication & Society*, edited by Robin Mansell and Peng Hwa Ang, 1–12. New Jersey: Wiley-Blackwell.

Rohn, Ulrike, and Sabine Baumann. 2015. "Media Brands in Social Network Sites: Problems German Media Companies Have Faced and Lessons They Have Learned." *Journal of Brand Strategy* 4 (1): 70–82.

Ryan, Marie-Laure, and Jan-Noël Thon. 2014. *Storyworlds across Media: Toward a Media-Conscious Narratology*. Lincoln: University of Nebraska Press.

Saldre, Maarja, and Peeter Torop. 2012. "Transmedia Space." In *Crossmedia Innovations: Texts, Markets, Institutions*, edited by Indrek Ibrus and Carlos A. Scolari, 14–28. Frankfurt: Peter Lang.

Scolari, Carlos A. 2009. "Transmedia Storytelling: Implicit Consumers, Narrative Worlds and Branding in Contemporary Media Production." *International Journal of Communication* 3: 586–606.

Shapiro, Carl, and Hal R. Varian. 2013. *Information Rules: A Strategic Guide to the Network Economy*. Boston: Harvard Business Press.

Siegert, Gabriele, Kati Förster, Sylvia Chan-Olmsted, and Mart Ots. 2015. *Handbook of Media Branding*. London: Springer International.

Vogel, Harold L. 1999. *Entertainment Industry Economics: A Guide for Financial Analysis, 4th edition*. New York: Cambridge University Press

Wikström, Patrik. 2014. "Tools, Building Blocks, and Rewards: Traditional Media Organizations Learn to Engage with Productive Audiences." *Journal of Media Business Studies* 11 (4): 67–89.

Wirtz, Bernd W. 2011. *Media and Internet Management*. Wiesbaden: Gabler Verlag.

45

A MICRO-BUDGET APPROACH TO TRANSMEDIA IN SMALL NATIONS

Kyle Barrett

This chapter explores forms of low-budget transmedia production in Scotland. It does so as a means of articulating a micro-budget approach to understanding the workings of transmediality from the perspective of small nations as opposed to the mass-audience franchise contexts most commonly interrogated in transmedia studies. First, I will discuss Scottish cinema, its dominant styles and the role that micro- and low-budget production plays in attempting to diversify the types of projects produced in Scotland. The limited output in terms of Scottish film production has seen little development over the past few decades despite alternative avenues for filmmaking that have become readily available for emerging practitioners. I will provide an example of alternative avenues for film production through the use of online crowdfunding, an increasingly used method of financing that offers a participatory element for fans/potential funders: "By seeking various contributions from a large group of people via online platforms, screen producers can independently create, market and distribute their own feature-film, short-film, documentary, multi-platform or digital-media projects" (Laycock 2016, 112).

The purpose of this analysis is to contextualize the need for further development of significant media projects that challenge and subvert the dominant styles and themes, particularly within Scottish cinema, as "there has been ongoing discussion about what kinds of films should be made in Scotland with the—still limited—resources available" (Hutchinson 2015, 22).

To illustrate this "discussion," I will then present an example of micro-budget transmedia production through the Lock-Up Your Daughters (LUYD) collaborative group. Based in Glasgow, the group have embraced multiple media platforms for their content. LUYD have employed magazines, short films, and club nights to create a variety of projects and have embraced the restrictions of working within micro-budgets. I will argue that these various platforms feed into one another, particularly magazine content which mirrors many narratives in the short films they have produced. Whilst not producing conventional transmedia narratives, LUYD have benefited from deploying their content across a variety of media:

> Not every project has to be a transmedia project of course, or is suitable to be presented in such a way, but almost every project can benefit from some aspect of it—whether the goal is to attract an audience, market or brand the content, or expand its popularity, or as a new storytelling methodology. This holds particularly true for independent film and media-makers.
>
> *(Nikolic 2016, 139)*

Through group collaboration and LUYD's aims of expression, I will also argue that this form of transmediality, however unconsciously these platforms have fed into one another, has resulted in some of the most creative, low-budget media projects in Scotland.

Scotland's New Talent

Emerging filmmakers working in Scotland utilize affordable technology to produce a wide variety of projects, as "new digital technology and micro-budget/DIY production approaches all suggest the possibility for resilient productivity even as Scottish film-makers continue to require stronger sources of base funding" (Nowlan 2015, 182). Collaborative groups, such as LUYD, continuously produce work within a local network of filmmakers. This cultivates talent despite working within low-budgets. Helen Wright, filmmaker and a founder of the group, comments:

> the community base that we're building—strong creative friendships, camaraderie, shared aesthetic and political interests—will benefit us if we develop towards bigger projects [be] cause we'll be able to still keep costs low by helping each other out, working on each other's films for a small level of pay, etc.
>
> *(Personal Communication, 2013)*

Sharing the responsibilities of various production roles and the drive from their passion of filmmaking, the open community indicates a sense of participatory culture, which I will discuss in detail later in this chapter. The group also demonstrates a progression of grassroots filmmakers that are gravitating toward larger projects, as it is "most sensible to start with low-cost content creation first, build a fan following, and only then try to fundraise and move to larger, more ambitious formats" (Nikolic 2016, 198). Additionally, the transmedia avenues pursued by the collaborators in LUYD exhibits a shift from the traditional cinematic representation of Scotland through social realism. LUYD utilizes narratives that stem from personal experience, as witnessed through their magazine articles. The group also adopt various genres within their short films, including science fiction, black comedy, and thriller, whilst never abandoning the character-driven narratives central to each project, which I will discuss later.

Scottish Cinema: Themes, Styles, and Micro-Budget Approaches

Scotland has for many years been associated with "gritty," "kitchen sink" realist dramas from directors such as Ken Loach and Peter Mullan. Emerging filmmaker Scott Graham has also adopted a similar approach, particularly with his feature-debut *Shell* (2012). Scotland's association with social realism typically relies on presenting the grimness of working-class life:

> many of the most notable Scottish fictional feature films of the past 40 years have engaged with "social realism," which "British cinema," especially from the 1950s through the 2000s, has been closely associated—or, perhaps, better put, which has often been touted as the mode of film-making in which British cinema has seen its greatest accomplishments, at least in terms of "serious drama."
>
> *(Nowlan 2015, 182)*

While funders have been sympathetic to these films, Scottish audiences have been resistant to this style of filmmaking and instead flock to the latest blockbuster releases. Creative Scotland have identified in a report published in 2014 which states that "annual audience share for indigenous Scottish productions in Scotland has varied from as much as 7% to as little as 1%" (Creative Scotland 2014, 11). This is in comparison to 84.2 percent of the box office taken by US-backed films (Creative Scotland

2014, 11). Practitioners often work within the confines of minimal resources and create their own movements and collaborations, particularly in a small nation context, as "so much work in the creative industries takes place in project networks—bringing together the talent and efforts of a diverse group of people and organizations—most creative workers rely on their local contacts" (Davies and Sigthorsson 2013, 66).

We can consider the result of the adoption of digital technologies in media production to have created a "post-film" era where "we are now witnessing the emergence of a different media regime, and indeed a different mode of production, than those that dominated the twentieth century" (Shaviro 2010, 2). This has created a paradigm shift since the replacement of analogue technologies. Characteristics of this new paradigm include alternatives to traditional media production in terms of funding, production practice, and distribution:

> On one hand, low- and micro-budget production models were symptomatic of significant change within local thinking about the best ways to achieve long-term film industrial sustainability. But on the other, many movies emerging from that shift proved surprisingly amenable to traditional critical agendas that see artistic exploration and representation of national identity as Scottish cinema's primary *raison d'être*.
>
> *(Murray 2015, 147)*

This shift has not only expanded the opportunities for emerging media practitioners but has also enabled experimentation with traditional forms of production. This includes transmedia storytelling that at first may have seen to be an anomaly but now can be seen to be part of an innovative approach to creating projects that have a life beyond one platform. As Ethem (2015, 32) reflects, "if an innovation or novelty like transmedia emerges within a society, society first perceives it as abnormal. If it is recognized and accepted, it becomes normalized within the society." Not only has transmedia storytelling been embraced by both filmmakers and audiences, the line between consumer and producer becomes blurred further as this digital shift has enabled anyone to become a filmmaker: "Transmedia is a process precisely because the old top–down industry model of creator–spectator, producer–consumer is giving way to a nonlinear, immersive, and dialogical model of participatory co-creation" (Elwell 2014, 240).

Transmedia storytelling provides an opportunity for creators to engage closely with their audience. However, as noted above, audiences are somewhat reluctant to engage with Scottish-produced content, which has resulted in a limited industry. Regarding its cinema, moreover, Scottish film production over the past 20 years has been relatively small compared to the 1990s boom, with notable features such as *Shallow Grave* (Danny Boyle, 1994) and *Ratcatcher* (Lynne Ramsay, 1999). This rise in the 1990s launched the careers of director Boyle, actor Ewan McGregor, director and actor Peter Mullan, and many others, leading Duncan Petrie to comment that the "emergence of a distinctive Scottish cinema has been just one element of the rich and diverse fermentation of creativity and cultural expression in Scotland ... that has done so much to forge a new culture of possibility" (2000, 226). This "culture of possibility" saw a decline in filmmaking in Scotland during the 2000s. This was due to a variety of reasons, not least because of the continuous restructure and restriction of funding agencies, including Scottish Screen and the Scottish Arts Council—later merged into Creative Scotland in 2010. However, during the 2000s there were "increasing numbers of local artists produc[ing] work partly (and on occasion, wholly) outside of established channels of state support. Nowhere was that phenomenon more visible than in the post-millennial mushrooming of micro- and low-budget filmmaking practices in Scotland" (Murray 2015, 143). This "mushrooming" saw projects such as *The Inheritance* (Charles-Henri Belleville, 2007) come to fruition, which was made on a budget of £5,000 and shot on mini-DV. The film later went on to win the Raindance award at the British Independent Film Awards (Martin-Jones 2009, 63).

Creative Scotland have been sympathetic toward films that follow the social realist pattern, rather than projects that slant toward genre filmmaking: "Certainly, some Scottish genre films have received ample attention, and respected Scottish film-makers have made important genre films. But a bias remains in favour of the art film over the genre film" (Finch 2015, 266). Financing has also contributed to works that have an identifiable "Scottishness," such as golfing drama *Tommy's Honour* (Jason Connery, 2016) which received £400,000 through Creative Scotland's Screen Production fund (Miller 2016). Creative Scotland also financed *T2: Trainspotting* (Danny Boyle, 2017), supporting a significant franchise within Scottish cinema. It falls on those working within minimal resources and micro-budgets to create genre films, which "often travel well, finding broad audiences in a way idiosyncratic or art films may not. For instance, horror film fans are more likely to seek out horror films regardless of their national origin" (Finch 2015, 267). Films such as the above mentioned *The Inheritance* utilize their low-budgets and available resources to produce films that are moving away from grim social realism.

Micro-Budget Production

The Inheritance, for instance, employs the "road movie" within its narrative. The road movie has been a staple of Scottish cinema for decades, with films such as *Soft Top, Hard Shoulder* (Stefan Schwartz, 1992) and David Mackenzie's low-budget debut feature *The Last Great Wilderness* (2002) using the travel film as a means of exploring Scotland. It also offers the opportunity for filmmakers to merge other genres into their narrative, including comedy and drama for the former title, and horror and surrealism in the latter. Characters are typically located outside of their homeland and must return home to reconnect with their past. This approach is particularly useful for Scottish-based features on limited budgets as "a low-budget road movie offers the opportunity for an exploration (no matter how minimal) of the border-crossing potential of the travel film for examining Scottishness and Scottish identity" (Martin-Jones 2009, 64). *The Inheritance* offers a wider view of Scotland, and Scottish culture, by moving away from the urban grimness featured heavily in the works of Ken Loach and writing partner Paul Laverty in films such as *Carla's Song* (1996), *My Name is Joe* (1998), and *Sweet Sixteen* (2002). This break from tradition is a refreshing approach to demonstrate the capabilities of a diverse Scottish cinema. Not only this, it provides opportunity to explore the unseen, or under-represented aspects of Scotland and its culture that does not necessarily fall into the social realist category:

> Scotland, in fact, encompasses "many Scotlands" with their own considerably distinct characteristics in turn—and these "Scotlands" can be distinguished along lines of region, locale, class, ethnicity, gender, sexuality, generation, subcultural affiliation, political identification, language, religion and yet many more lines of demarcation.
>
> *(Nowlan and Finch 2015, 7)*

If we place Scottish films within the context of urban versus rural landscapes, we can note varying differences of representations of region, locale, and class. The edgy, working-class characters featured in the above-mentioned Loach films struggle to survive the mean city streets, whereas the characters featured in *The Inheritance*, two brothers—foils for one another, adding a "buddy" element to the narrative—try to reconcile their differences within the landscape of the Highlands, a return to nature, away from urban society.

Crowdfunding: Participatory Culture and "Everyone is a Producer"

Crowdfunding can be described as "a way of fundraising by asking for financial contributions from a large number of individuals through an Internet platform" (Nikolic 2016, 75). Alternative methods of funding have become increasingly prolific over the past few years, given its relatively short life span

(Tomczak and Brem 2013, 335). Independent filmmakers can develop their projects without the traditional avenues of funding and "no longer have to wait for government bodies to green-light their project; they can source funds prior to, during and/or after production to guarantee their film comes to fruition" (Laycock 2016, 113). Practitioners can actively engage with the public to fund their projects to "pitch ideas and ask for small up-front investments in the project, in return for getting the eventual product at a lower price—in some cases making it exclusively available to the participants" (Davies and Sigthorsson 2013, 213). Websites such as Kickstarter and Indiegogo encourage and support independent practitioners to acquire investment, advertising their plans "by featuring the project's trailer, pitch and pledge to produce/distribute the film, if they raise set sum, decided at the outset, in one month" (Sørensen 2012, 736). However, there are issues with this funding model, particularly when considering the campaigns that filmmakers have to develop in order to grab the public's attention. In short, there are two types of fundraising—direct and indirect:

> Direct crowdfunding is when the fundraiser makes a direct appeal to a specific audience via their own fundraising platform (for example, the fundraiser's own website) or to their own supporters (for example, a band raising money from its fans). Indirect crowdfunding is, on the other hand, a general appeal for funding to the unknown general public or "crowd."
>
> *(Tomczak and Brem 2013, 342)*

"Indirect" crowdfunding is generally the approach many filmmakers adopt for their projects, attempting to appeal to as large an audience as possible. However, whilst online platforms may support low-budget filmmakers, it is still difficult to pinpoint a specific audience. For instance, art house and experimental cinema tend to have a particular audience and may only appeal to fans of that specific type of film. This is problematic for campaigning, as "the potential profitability of a movie is unpredictable, it is hard to obtain financing, especially for art house movies or specific documentaries that have limited market potential" (Braet, Spek, and Pauwels 2013, 2). However, the possibility for successful funding relies on how best the filmmaker "sells" their idea to potential investors. In this instance, a short video is created and displayed on the webpage, which explains the idea and how the money will be utilised to bring the proposal to fruition. It is here where we can observe the notion of participatory culture in crowdfunding, through Henry Jenkins et al.'s definition as: "one with ... strong support for creating and sharing creations with others ... members who believe their contributions matter and members who feel some degree of social connection with one another (at the least, they care what other people think about what they created)" (2009, 5–6).

Additionally, Paul Booth argues that crowdfunding can be understood as "Spimatic Fandom." Building from Bruce Sterling's (2005) work, Booth defines a "spime" as "the lifecycle of an object from initial design through physical substantiation to final digital trace" (Booth 2015, 150). "Spimatic Fandom," particularly when focusing on the participatory aspects of crowdfunding, then, "integrates production and reception as coexistent paradigms of crowdfunding" (Booth 2015, 150).

In response to any donation made, investors will receive a reward, which can range from "Thank you" at the end of the project, to not only a copy of the film but a credit as an executive producer for large sums donated. By contributing, engaging, and participating in the development of a project, everyone can become a producer. Participation through crowdfunding, then, is "no longer simply an opening up, an expansion, a liberation, it is now also a principle of improvement, an instrument of change, a creative force" (Kelty 2013, 24). However, there are problems with the reward-based system, for not only can there be poor incentives, there is a heavy responsibility for the creators and faith must be placed within them:

> an independent filmmaker who is collecting funds to produce a movie usually offers a copy of the movie as a reward for pledging. Sometimes it isn't emphasized enough, but it

> is important to understand: The filmmaker is not "selling movies". After all, the movie isn't even created yet; it is just an idea in the filmmaker's mind.
>
> *(Young 2013, 52)*

While incentives may be enticing for those who wish a project to be realized, it is still problematic for funders for various reasons, particularly when investing in a film as, "the movie was in the filmmaker's mind, and it may change during the creation process. As a matter of fact, this is almost expected" (Young 2013, 52). The creative process varies and there are no reassurances that the final film that people separately envision will be the final product.

Riptide: A Scottish Crowdfunded Feature

Utilizing Indiegogo, Glasgow filmmaker Tim Barrow was successful in raising funds to complete his new feature film *Riptide* (2017). Previously, Barrow wrote and produced *The Inheritance* through his company Lyre Productions and issues the film on DVD through their website, www.lyreproductions.wordpress.com. The self-distribution method was in response to watching "distribution companies go bust through traditional models," as Barrow explains himself (Personal Communication 2017). This self-distribution method not only allows for complete control of the distribution of the film but also an accessible platform for anyone to see the film.

Despite the limitations of its budget, *The Inheritance* demonstrates the level of control filmmakers maintain through each stage of production. *The Inheritance* was also the result of "the rapid evolution of digital production and post-production technologies helped to facilitate low- and micro-budget Scottish filmmaking's contemporary rise" (Murray 2015, 146). As Barrow notes, it is also vital to the development of the Scottish film industry:

> It's tough making films on tiny budgets, and takes a lot of faith as well as hard work. But it's vital that people living in Scotland make films here. There's plenty of outside production companies coming to shoot here (*World War Z, Avengers: Infinity War*, etc.), which is terrific but I want Scottish audiences seeing films made by people living in these communities. And supported by distributors and cinemas. We all want Scottish audiences to see our work.
>
> *(Personal Communication 2017)*

Whilst there has been an increase in using Scotland as a location for outside production companies, it is important for Scotland to present its own stories that reflect its society and culture, as these Hollywood productions "de-emphasise Scottish uniqueness, depicting Scotland as a place that is largely indistinct from, or indeed interchangeable with, other locations around the globe" (Marmysz 2014, 28). Lyre Productions followed *The Inheritance* with another low-budget feature, Barrow's directorial debut *The Space Between* (2012) with a budget of £15,000 (The Space Between the Movie n.d.). The acclaim and touring of *The Inheritance* with question and answer sessions enabled the filmmakers to self-distribute *The Space Between* on a wider basis (Personal Communication 2017). By generating contacts, the film premiered at the Filmhouse Edinburgh in 2011. The film subsequently completed 57 cinema screenings before, again, being distributed on DVD through Lyre Productions' website, which "gave it a much longer life than had we signed with a distributor. We still sell DVDs of both features, posting them around the world," states Barrow (Personal Communication 2017). This success also refers to Booth's notion of spimatic fandom, where fans of the film, specifically those who purchased the DVD, could contribute to future projects, especially through online crowdfunding.

Riptide, listed as *UNF* on www.indiegogo.com, sought £4,000 for completion funds with the remainder of the budget from private funds, the overall budget being £10,000, as Barrow confirms (Personal Communication 2017). Attaining £4,088 at the end of the campaign (Indiegogo 2014), the film is currently in postproduction (Lyre Productions 2017). Much like *The Inheritance*, *Riptide* is billed as "A road movie love story" (Indiegogo 2014) that is focusing on two characters that collide with one another and explores the psychiatric disorder schizophrenia.

According to Barrow, the experience of using crowdfunding was "a good way to build interest in your work" (Personal Communication 2017). However, the drawback can be that filmmakers do not understand their audiences or engage with potential investors and that "crowd-based communication and fundraising used by independent film and media-makers has to be—or at least appear to be— levelled and genuine" (Nikolic 2016, 78). Returning to Jenkins et al.'s notion of participatory culture, the perks of this campaign reinforces the notion of community. Raising £4,000 for completion funds and offering a variety of rewards for investment, each with their own "intimate" attributes, the participation from the community "must believe they are free to contribute and that what they contribute will be appropriately valued" (Jenkins et al. 2009, 5). The campaign features five contribution amounts. Starting at £5, this gives the contributor a "Bear Hug" as a thank you. £10 will get an investor a ticket to the screening of the film. £25 will provide a ticket for the screening as well as a copy of the film on DVD. £50 provides all the previous incentives as well as copies of Lyre Productions' other films, *The Inheritance* and *The Space Between*, on DVD. The highest amount one can contribute is £100, which will credit the investor as an Executive Producer, include all the perks of the previous donations and a dinner with director Barrow.

The £25 contribution was the most claimed, with 29 investors and £100 being sold out, which only allows for a maximum of five investors. The outreach of the filmmakers, whether it be the "Bear Hug" or the Executive Producer dinner reinforces Barrow's own sense of appealing to "Scottish audiences seeing films made by people living in these communities" (Personal Communication 2017). This in turn demonstrates an openness within the community, inviting the public to participate in the creation of a project and contributing to their national cinema.

LUYD: Personal Transmedia Narratives

I will now focus the remainder of this chapter analyzing LUYD's work. This analysis will explore the pilot issue of the magazine created by the group in spring 2008 and the short film *Clt. Alt.* (Helen Wright, 2011). Whilst the film is not a direct adaptation of the publication, it spreads some of the prevalent themes, which "serve to be tasted and, if the taster likes it, he or she can go to the source and continue with it to completion" (Ethem 2015, 36).

The magazine is presented as "your new alt queer rag" (LUYD 2008, 3) and contains several articles that explore personal perspectives about dating, the LGBTQ+ scene in Scotland as well as reviews of films and music released at that time. The pilot issue is described as: "a concentrated version of LUYD, intended as a little taster of what is to come. Designed to flaunt all our regular features and we've given our lovely contributors free reign to share their most precious memories. This pilot is all about style and experiences" (LUYD 2008, 3).

It is these "experiences" that translate into the short film noted above and provide the film with a personal perspective and richness that is often lacking in micro- or low-budget efforts. LUYD is a group collaboration, the aim of which, Wright notes, "is to give individuals the chance to express themselves, so that means we come out with a variety of different types of films" (Personal Communication 2013). The content created for each media platform and its fluidity to be expanded upon in another format, enables the filmmakers to explore, as Wright comments, a variety of different films, thus lending itself a rich, transmedial dimension. The traditional approach to developing transmedia projects would have the author:

> identify the most suitable model for the publication and technological aspects of the project. The most important difference between these models is the way in which the story is managed through multiple media within a project, a condition on which both the dramatic

> universe of the project and the variety of experiences presented to the audiences clearly depend.
>
> *(Giovagnoli 2011, 48)*

Whilst the magazine offers personal accounts of dating, LUYD's use of genre makes their film projects a site of interest. *Clt. Alt.*, for instance, blends relationship drama with a science-fiction concept. Arguably, the magazine article in the pilot issue written by Du Prez, in which the author discusses Internet dating and their experience, offers a pathway into the film. Prez shares: "Faced with a long, hot and above all, lonely summer in London I decided to entrust my search for romance to the internet …" (2008, 6). Later, the author notes the importance of friendship when venturing into the realm of online dating: "You may have chosen them for yourself but without friends sifting through the slop and providing references you're playing a terrifying kind of dating roulette" (2008, 6). The personal expression of the apprehensiveness of dating from this article is somewhat refashioned into a different context in *Clt. Alt.*

The film opens on two co-workers, Cate (Zara Badman) and Alex (Claudie Qumsieh), having a drink. It transpires they have a secret relationship that is hidden from everyone else. Cate's boyfriend phones her during the scene, irritating Alex. Alex, frustrated that the evening plans have suddenly changed, receives a phone call herself. Everyone in the bar, except Alex, freezes. The voice on the phone, parodying a sales call, informs Alex that she can "delete" Cate from her life. She agrees and after providing payment details, gives Cate one final embrace. After a quick conversation, Cate vanishes, and everything resumes as normal. The blend of relationship drama and sci-fi genre tropes create a fascinating insight into the dating scene. The phone, acting as a subverted symbol of *disconnection* rather than communication, has a running audible motif throughout the film through the noise of phone interference. Cate gladly has a conversation with her boyfriend on the phone, whilst the silent Alex is unable to express her anger clearly. However, Alex is later chastised for having an innocent conversation with the bartender. Cate is treating her relationship with Alex as a casual occurrence, keeping it secret from her heterosexual partner. Spreading the content, or rather its themes, from the magazine enables the filmmakers to present narratives that "shift and drift, progress and regress, grow and flow across film, television, print, digital artifacts, the web and interpretive communities" (Ruppel 2009, 282). Prez's article is the clear starting point for the theme of "romance in the 21st century." The strength of both the article and the short film is the creators' personal perspectives on the subject matter. Whilst it is not a straight translation of the article to the screen, this is another form of transmedia storytelling as we "ordinarily think of adaptation as retelling existing stories, whereas transmedia storytelling tends to be characterised as telling new stories in different media but set within a consistent diegetic world" (Harvey 2015, 3). Looking at the "world" of LUYD there is a subversive notion of transmediality by building on one another's work. By using the genre of sci-fi, Wright fictionalizes the experience of dating, something that audiences everywhere can relate to. However, as this is an iteration, or rather reinterpretation of this theme, both platforms have their own parameters in how they explore this theme. This is common in transmedia storyworlds as they are: "constructed as a series of self-contained story events and objects that exist as independent yet interconnected narrative arcs or artifacts … with their own beginnings, middles, ends and narrative emplottment, which is entwined with and constitutive of the architecture of the story-world as a whole" (Elwell 2014, 241).

The storyworld "as a whole" is LUYD itself, i.e., the community it has created. Audiences that tasted the magazine, as noted by Ethem above, can then explore the magazine, the film and the club nights, which offered opportunities of networking, for a fuller integrated experience of the community. The club nights and events from 2011–2013 listed on the LUYD website (www.lockupyourdaughtersmagazine.co.uk) demonstrate a variety of pursuits from film screenings to engaging with creating articles for the magazine. This not only further demonstrates the participatory culture theorized by Jenkins but also a form of Booth's spimatic fandom, where fans of the magazine can contribute to the creation itself. *Clt. Alt.* screened to both the Cut! Filmmaking Gala and the LUYD Queer Screenings in Glasgow's GFT cinema (LUYD 2014).

As a contrast to the works of Barrow, LUYD's does not purposefully set out to make their projects as a distinctly Scottish venture, meaning that their films *happen* to be set in Scotland rather than drawing overt attention to this fact regarding locations. Wright comments on this topic: "I think that as a group, we are not particularly interested in Scottishness or telling Scottish stories. However, because all our films are set in Scotland, I would say that this is never not a factor, if that makes sense" (Personal Communication 2013). Whilst Barrow utilized the road movie for a number of projects, and wanted to explore the Scottish landscape for a Scottish audience, LUYD made use of the limited resources they had at hand to explore personal narratives.

Conclusion

Micro- and low-budget filmmaking in Scotland has seen an unprecedented shift over the past ten years. The accessibility and affordability of new technologies enable anyone to become a filmmaker. Barrow and LUYD represent two interesting filmmakers, each with their own aims of expression. Barrow wishes to explore Scotland and present Scottish narratives that speak to not only locals but international audiences, as seen through their DVD distribution worldwide. The role of crowdfunding as a participatory element for Barrow's projects reinforces Jenkins et al.'s notion of participatory culture as well as engagement through community. Online platforms, such as Indiegogo, open the process of film production, allowing potential funders to have a "piece" of the project, whether it is a "Thank You" credit or a copy of the final film. The notion of spimatic fandom as a participatory/transmedial process in the creation of a project is the result of digital convergence, and the outreach required on the part of the creator of a project is essential in ensuring its development:

> Fans' engagement at multiple nodes within the production process portends a consequential role in the production of the media object. Crowdfunding campaigns that successfully engage their fans in a more participatory manner—acknowledging previous fan work, noting the saliency of fan activities in the past, appealing to fan attention in the future—highlight the temporal existence of a fandom.
>
> *(Booth 2015, 151)*

This form of fandom requires a community both on and offline. LUYD's sense of community, again, reinforces Jenkins et al.'s notion of participatory culture, as it "provides strong incentives for creative expression and active participation" (2009, 6). This facilitation of self-expression within the group enables the filmmakers to present personal narratives that expand upon content generated for the magazine. The strength of LUYD's approach to transmediality is characterized by the spread of themes across platforms, with direct adaptations being largely redundant due to the nature of the articles in the magazines, i.e., personal expressions, and the visions of the filmmakers. Instead, LUYD borrow, adapt, and spread their content, as it is "not about offering the same content in different media platforms, but it is the world-building experience, unfolding content and generating the possibilities for the story to evolve with new and pertinent content (Gambarato 2012, 72–73). Barrow and LUYD both make strong cases that regardless of available resources, varied and personal projects can emerge, revealing more of Scotland's diverse culture.

References

Barrow, Tim. Email message to author, May 8, 2017.

Booth, Paul. 2015. "Crowdfunding: A Spimatic Application of Digital Fandom." *New Media & Society* 17: 149–166.

Braet, Oliver, Sander Spek, and Caroline Pauwels. 2013. "Crowdfunding the Movies: A Business Analysis of Crowdfinanced Moviemaking in Small Geographical Markets." *Journal of Media Business Studies* 10: 1–23.

Creative Scotland. 2014. "Enjoying Film." *Review of the Film Sector in Scotland*, January 11–12. Accessed January 13, 2017. www.creativescotland.com/__data/assets/pdf_file/0018/25245/Review_of_the_Film_Sector_in_Scotland_-_Jan_2014.pdf.

Davies, Rosamund, and Gauti Sigthorsson. 2013. *Introducing the Creative Industries: From Theory to Practice*. London: SAGE Publications Ltd.

Elwell, J. Sage. 2014. "The Transmediated Self: Life between the Digital and the Analog." *Convergence: The International Journal of Research into New Media Technologies* 20 (2): 233–249.

Ethem, Ibrahim. 2015. "Transmedia Storytelling and Transforming Human Imagination." *AJIT-e; Istanbul* 7 (23): 31–40.

Finch, Zach. 2015. "Comedy, Fantasy and Horror." In *Directory of World Cinema: Scotland*, edited by Bob Nowlan and Zach Finch, 266–271. Bristol: Intellect.

Gambarato, Renira Rampazzo. 2012. "Signs, Systems and Complexity of Transmedia Storytelling." *Estudos em Comunicação* 12: 69–83.

Giovagnoli, Max. 2011. *Transmedia Storytelling: Imagery, Shapes and Techniques*. Pittsburgh: ETC Press Ltd.

Harvey, Colin. 2015. *Fantastic Transmedia: Narrative, Play and Memory Across Science Fiction and Fantasy Storyworlds*. Basingstoke: Palgrave Macmillan.

Hutchinson, David. 2015. *The Media in Scotland*. Edinburgh: Edinburgh University Press.

Indiegogo. 2014. "Waiting for the Riptide – Short Film." Indiegogo.com, August 2. Accessed May 30, 2017. www.indiegogo.com/projects/waiting-for-the-riptide-short-film-family-drama#/.

Jenkins, Henry, Kevin Clinton, Robin Purushatma, Andrew Robinson, and Matthew Weigel. 2009. *Confronting the Challenges of Participatory Culture: Media Education for the 21st Century*. Cambridge, MA: MIT Press.

Kelty, Christopher. 2013. "From Participation to Power." In *The Participatory Cultures Handbook*, edited by Alan Delwiche and Jennifer Jacobs Henderson, 22–32. New York: Routledge.

Laycock, Rebecca A. 2016. "The Audience's Worth: Crowdfunding as a Source of Film Finance." *Metro* 188: 112–117.

Lock Up Your Daughters. 2008. "Pilot Issue Introduction." Accessed March 3, 2017. https://issuu.com/luydmagazine/docs/luydpilot.

Lock Up Your Daughters. 2014. "*Clt.Alt.*" Accessed March 22, 2017. www.lockupyourdaughtersmagazine.co.uk/content/?portfolio_item=ctrl-alt.

Lyre Productions. 2017. Accessed January 10, 2017. https://lyreproductions.wordpress.com/.

Marmysz, John. 2014. "The Myth of Scotland as Nowhere in Particular." *International Journal of Scottish Theatre and Screen* 7: 28–44.

Martin-Jones, David. 2009. *Scotland: Global Cinema—Genres, Modes and Identities*. Edinburgh: Edinburgh University Press.

Miller, Phil. 2016. "Golfing Drama Directed by Jason Connery to Open Edinburgh International Film Festival in 2016." *The Herald*, April 27. Accessed January 11, 2017. www.heraldscotland.com/news/14455241.Golfing_drama_directed_by_Jason_Connery_to_open_Edinburgh_International_Film_Festival_in_2016/.

Murray, Jonathan. 2015. *The New Scottish Cinema*. London: I. B. Tauris & Co Ltd.

Nikolic, Vladan. 2016. *Independent Filmmaking and Digital Convergence: Transmedia and Beyond*. New York: Routledge.

Nowlan, Bob. 2015. "The Angels' Share." In *Directory of World Cinema: Scotland*, edited by Bob Nowlan and Zach Finch, 118–124. Bristol: Intellect.

Nowlan, Bob, and Zach Finch. 2015. *Directory of World Cinema: Scotland*. Chicago: University of Chicago Press.

Petrie, Duncan. 2000. *Screening Scotland*. London: BFI Publishing.

Prez, Du. 2008. "Internet Dating – More Freaks than Saturday in Lidll?" *Lock Up Your Daughters*. Spring. https://issuu.com/luydmagazine/docs/luydpilot.

Ruppel, Marc. 2009. "Narrative Convergence, Cross-Sited Productions and the Archival Dilemma." *Convergence: The International Journal of Research into New Media Technologies* 15: 281–298.

Shaviro, Steven. 2010. *Post-Cinematic Affect*. Ropley: O-Books.

Sørensen, Inge. 2012. "Crowdsourcing and Outsourcing: The Impact of Online Funding and Distribution on the Documentary Film Industry in the UK." *Media, Culture & Society* 34 (6): 726–743.

The Space Between the Movie. n.d. Accessed January 8, 2017. https://thespacebetweenthemovie.wordpress.com/.

Sterling, Bruce. 2005. *Shaping Things*. Cambridge, MA: MIT Press.

Tomczak, Alan, and Alexander Brem. 2013 "A Conceptualized Investment Model of Crowdfunding." *International Journal of Entrepreneurial Finance* 15: 335–359.

Wright, Helen. E-mail message to author, August 30, 2013.

Young, Thomas Elliott. 2013. *The Everything Guide to Crowdfunding: Learn How to Use Social Media for Small-business Funding*. Avon, MA: Adams Media.

46

A GENETTIAN APPROACH TO TRANSMEDIA (PARA) TEXTUALITY

Raúl Rodríguez Ferrándiz

In the 1980s, Gérard Genette constructed a complete and extraordinarily solid theory around "textual transcendence of the text" or "all that sets the text in relationship, whether obvious or concealed, with other texts" (1992, 81; 1997a, 1). After some vacillations over terminology, he called this textual transcendence of text "transtextuality." The common term used for this phenomenon up to that period was "intertextuality," which Kristeva (1980, 64–91) had translated from Mikhail Bakhtin's "dialogism." However, Genette discovered the complexity and extension of this phenomenon and preferred to reserve "intertextuality" for one of the five types of transtextuality he identified, described, and illustrated with examples.

Despite the fact his work was translated into English (Genette and Maclean 1991; Genette 1992, 1997a, 1997b), many pioneering studies on transmedia storytelling (TS) did not allude to this enlightening precedent. Although it is true that Genette's work addressed transtextual relations between literary or at least written texts, and barely mentioned transtextual relations within narratives of different media (between theater and cinema or between novel and film, for instance).

Screen studies and other media studies were, however, receptive very early on to Genette's work. They inspired, for example, reflections on (trans)textuality in cinema (Stam, Burgoyne, and Flitterman-Lewis 1992, 206–209; Klecker 2015), television (Hansen 1999), film trailers (Kernan 2004), video games (Consalvo 2007, 2017), or multimodal interactive fiction (Stewart 2010).

It is reassuring to find that a second wave of researchers in the midst of the transmedia turn, have recuperated Genette's work and recognized it as precursor to their own. Scholars such as Gray (2010), Grainge (2011), McCracken (2013), Desrochers and Apollon (2014), Booth (2015), Geraghty (2015), Pesce and Noto (2016), and Dehry Kurtz and Bourdaa (2017) evaluate the Genettian speculation, and in particular the "paratext" category as a theoretical framework capable of explaining the diachronic genesis and synchronic functionality of transmedia textual proliferation. Furthermore, the journal *Critical Studies in Media Communication* has recently devoted a special issue (34(2), 2017) to "Paratexts, Promos, and Publicity." Thus, we could say that the transmedia turn has become genetically Genettian, playing on words. However, it is important to explain that paratextuality has adopted new characteristics regarding the theory with which it emerged and gained meaning.

Genettian Paratextuality

Genette spoke of five different types of transtextuality. Arranged in ascending order of abstraction, implication and globality, these are (1) *intertextuality*, (2) *paratextuality*, (3) *metatextuality*, (4) *hypertextuality*, and (5) *architextuality*. He devoted a volume to the fifth ([1979], 1982), the fourth ([1982], 1997a) and the second, which is the fundamental reference for this work ([1987], 1997b; Genette and Maclean, 1991), entitled *Seuils* in its original French edition, and *Paratext: Thresholds of Interpretation* in the North American edition. Paratext is everything that introduces a text that comes into our hands. As Genette notes, a paratext is not a limit or monitored border, but a place of passage, a lobby, which invites us to learn more, and also provides us with what may eventually make us decide not to read the actual text. The paratext is an undecided space between the inside and the outside of the text, between strictly literary, rhetorical, stylistic, and genre regulations, and social, economic, industrial, and advertising regulations; "a zone not only of transition but also of *transaction*" (1997b, 2), "a kind of canal lock between the ideal and relatively immutable identity of the text and the empirical (sociohistorical) reality of the text's public … the lock permitting the two to remain 'level'" (1997b, 408).

Genette distinguishes between what he calls the *peritext* (titles and subtitles, pseudonyms, forewords, dedications, epigraphs, prefaces, intertitles, notes, epilogues, and afterwords, which obviously all refer to the literary text and its privileged vehicle, the book) and the *epitext*, which can be both public (interviews, author's comments) and private (authorial correspondence, oral confidences, diaries, and pretexts). Genette also takes into consideration the epitexts created by the editor (advertisements, promotions, synopses). While the peritext is sewn to the text, the epitext is not published alongside it: it is placed in another *transmedia* dimension—newspapers, magazines, radio or television programs, lectures and colloquia, and all public performances perhaps preserved on recordings or in printed collections.

This spatial disjunction may also be accompanied by temporal disjunction: the epitexts may precede the text they refer to (*prior paratexts*: testimonies of the author about their ongoing publishing projects, prospectuses, and announcements of forthcoming publications), may appear at the same time as the text (*original paratexts*: interviews or press conferences during the launch of the text), or appear after the text (*later* or *delayed paratexts*: second or successive editions, comments, notes, memories of the writing of a past text) (1997b, 5–6).

Genette dismisses to some degree what he calls the editorial epitext (in comparison to the author's text, whether public or private) because its advertising and promotional functions do not directly involve the participation of the author, who is not responsible for the exaggerations that this epitext can make about the text. For the French author, paratextuality "is characterized by an authorial intention and assumption of responsibility" (1997b, 3).

Updating Paratextuality

Genette stated that

> defining a paratextual element consists of determining its location (the question *where?*); the date of its appearance and, if need be, its disappearance (*when*); its mode of existence, verbal or other (*how?*); the characteristics of its situation of communication—its sender and addressee (*from whom? to whom?*); and the functions that its message aims to fulfill (*to do what?*).
>
> *(1997a, 4)*

He answered these questions in the field of literature, above all French realism (Stendhal, Balzac, Flaubert, Zola, Proust). However, unstoppable transmedia text expansion in the era of telematic digital convergence requires that his theory to be adapted to new dimensions of "textual transcendence" of the text, which was for him transtextuality.

In effect, paratextuality can be seen in a flattering new light because the Internet is, in a sense, paratext paradise: everything is together in some way, everything has its surroundings, and we just have to find out how many clicks separate us from them, and how much, if anything, we have to pay to access them and, last but not least, which product will be the text and which will be the paratext for each visitor, and even for each visit. Some authors claim that the difference between text and paratext is in fact diluted in the digital world as textual hierarchies are not so important and user-centered design makes the focus of interest unpredictable and consequently the center and periphery of the story (Pellizzi 2006, 5). Nevertheless, it is evident that digital paratexts not only frame and present the text, but also operate on it and from it. This can even be demonstrated in the world of literature, which was Genette's original field of research, but now consumed through reading devices such as eReaders, tablets, or hybrids like vook (McCracken 2013). In effect, an infinite number of paratexts are adhered to every text on the Internet, which allows continuous transactions between different parts of the text itself and between the text and its periphery; that is to say in a centripetal or centrifugal way. However, this enhanced paratextuality does not only pertain to the digital world, although this is an exceptional tool for paratext production. It also interacts with the physical world of objects and the meaning they bear as paratextual commodities are designed to have a textual (symbolic and indeed narrative) appeal as part of their own selling operation (Rodríguez-Ferrándiz 2014).

Paratextuality and Screen Studies

With the implicit purpose of extending Genette's speculations from the literary field to film and television texts, Jonathan Gray (2010) has offered an aggiornamento of the term "paratext." For him, a film or television text is a small portion of a large textual universe that surrounds it: previews, teasers, trailers, sneak peaks, promos, interviews with creators, online discussion forums, entertainment news, reviews, podcasts, merchandising, guerrilla marketing campaigns, bonus materials, spoilers, fan creations, posters, video games, alternate reality games, DVDs and CDs with the show's soundtrack, prequels and sequels, the "previously on" and the "on the next episode." Today, both film and television studies cannot be understood without taking into account these paratextual proliferations, as they can, to a great extent, determine the meaning of a film or television series (2010, 22).

When Gray was writing his thesis on the intertextuality of *The Simpsons*, he perceived a surprising phenomenon: no mainstream text in American television history has been more continuously critical of consumerism and commercialism than *The Simpsons* and yet, at the same time, few television shows have sold more merchandise of all types: keychains, T-shirts, notebooks, bedding, toys, etc. The reason for this contradiction or ambivalence could not be found in the television text itself, consequently, Gray had to look for an explanation which transcended the show. Thus, Gray analyzed some dimensions of the paratextual constellation around the show: the opening credits of *The Simpsons*, and also *The Simpsons Game*, a set of online ads for *The Simpsons Game*, the political controversies arising from the TV series, which derived in cultural wars between conservative and liberal America, etc. From the moment the White House, through George H. W. Bush and his first lady, publicly protested against the series, *The Simpsons* succeeded in selling its products, especially T-shirts, not so much as a powerful (and paradoxical) anti-consumerism brand, but more as a rejection of the neoliberal politics of the early 1990s. Gray's book sheds light on some dimensions of the paratextual constellation around many other film and television productions: the DVDs of *The Lord of the Rings* saga, the promotion campaign for Christopher Nolan's *Batman* trilogy, the starkly different trailers (one American, one Canadian) for Atom Egoyan's film *The Sweet Hereafter* (1997), the fan spoilers for *Lost*, and the opening credits of *The Sopranos* or *Dexter*. He even looked at off-screen paratexts, like the toys and games created around *Star Wars*.

In Gray's opinion, the term "paratext," apart from its prestigious academic background, was more appropriate than other, perhaps more fashionable terms, used in media studies, such as hype, promotion, synergy, or peripherals. These all have exaggerating and bombastic connotations (hype) or are strategies designed to create profits, not primarily to create meaning (2010, 4–6). However, the term *paratext* does have other problems. The prefix "para" suggests what is parallel and attached, but also evokes, contemptuously, what is subsidiary, auxiliary, clandestine, unregulated, and even usurping.

Gray, however, gives a certain dignity and flexibility to paratext, which was not in Genette's theory. For Genette, paratexts were always in subservience to the text.

Genette did also predict, however, that paratexts had functions that had not yet been exploited during his time and which broadened horizons and did not fiercely delimit them. When he reflected on the paratextual value of notes, very common in discursive texts (history, essays, and so forth), but rare in fiction, he suggested:

> One could imagine a more emancipated regime in which the note would no longer come under the heading of the documentary type of discourse but would be narrative in type and would – in itself and for its own account – pursue some momentary fork in the narrative.
>
> *(1997b, 335–336)*

Furthermore, in the epilogue to his book, Genette noted that it would be interesting to study three paratextual phenomena that he had to skip: *translation*, *illustrations*, and *serial publication*. The latter seems particularly inspiring for the study of television narratives. Undoubtedly, TV series offer in advance, even before their transmedia expansion, a great affinity with paratextuality. It is precisely seriality which guarantees, on the one hand, the possibility of complex narrative plots, and, on the other, long-term viewer engagement. This implies establishing strategies which ensure narrative continuity between deliveries, and these strategies are in many cases paratextual.

Why Do We Want Paratexts? The Multiple Functions of Paratexts

Theoretically, each medium involved in a TS should contain a text which offers something new, narratively speaking, so that its contribution simultaneously enriches and widens the narrative as a whole (Jenkins 2006, 95–96; Straumann 2015: 256–257). The literature on TS debates whether Jenkins' concept applies better to the construction of the narrative itself or rather to the furnishing of fictional worlds, rightly called *transmedia worlds* (Klastrup and Tosca 2004). In other words, crafting a transmedia narrative is about crafting the story or about crafting the world in which this story exists (Long 2007, 60). It has also been debated whether the term should only be applied to texts that are transmedia from birth (native transmedia), extended to be transmedial after initial success or regarded as transmedia *post factum* in the cultural memory (Saldre and Torop 2012).

In any case, it is a complicated task to distinguish between those paratextual features which directly contribute to narrative progression (undoubtedly qualified for being components in TS) and those paratextual features that point to the work, but contribute nothing or very little to the narrative progression within it (although they contribute to the production of meaning in reception, for example, by anchoring generic expectations). In other words, it seems that the concept of paratext, at least in Gray's work, is wider than Jenkins's concept of TS (as well as being wider than Genette's concept incidentally). In order to conciliate both perspectives, the notion of *paratextual orientation* formulated by Jason Mittell (2013, 2015) can be inspiring and act as a bridge between the strictly narrative and hype, promotional paratextuality.

Mittell reflects on the concept of paratext applied to fictional television series (2013, 165–181; 2015, 261–291). For him, "in the digital era, a television programme is suffused within and constituted by an intertextual web that pushes textual boundaries outward, blurring the experiential borders between watching a programme and engaging with its paratexts" (2015, 7). The complexity of contemporary television narratives is largely due to the overflow of the television text across other platforms. This, undoubtedly, challenges "the ease with which casual viewers might make sense of a programme, inviting temporary disorientation and confusion" (2015, 261). As Mittell points out, in order to prevent this effect, there are *orienting paratexts*, which belong to a third category alongside *transmedia paratexts* (which are the narrative extensions described by Jenkins and Mittell himself (2015, 292–318) and *promotional paratexts*, coming soon formats that aim to hype, promote, or introduce a program (Gray 2010,

47–79). Orienting paratexts aim "to create a layer atop the program to help figure out how the pieces fit together or to propose alternative ways of seeing the story" (Gray 2010, 261–262).

It could be argued that transmedia (storytelling) paratexts underpin the story, and ultimately merge with it. Promotional paratexts, for their part, recommend or prescribe the story, on behalf of the TV production company or network, which explicitly communicates with its target audience. Orienting paratexts occupy an intermediary and instrumental position.

Orienting paratexts "reside outside the diegetic storyworld, providing a perspective for viewers to help make sense of a narrative world by looking at it from a distance" (Gray 2010). Their role is to intervene in four basic storytelling facets that might require orientation: time, events, characters, and space. Mittell includes in this category wikis, guides, timelines, mapping chronologies, graphics, family trees, recaps—which summarize narrative material in a straightforward manner—split-screen synchronizing, and even reedited versions of series in chronological order.

This category of paratext also includes pieces that look beyond to connect the series with other textual, intertextual or extratextual realms outside the core program. In other words, on the one hand, there are figures and even synoptic schemes about the transmedia paratextual universe around the text: the network of associated products that compose the intertextual matrix of the franchise. And on the other hand, the series is connected with another fictional series or texts or aspects of the real world, intertextually interwoven with the mothership. In the case of the TV series *Lost*, for instance, Lostpedia provides information about video games (*Lost: Via Domus*), ARG (*The Lost Experience*, *Find 815*), mobisodes (*Lost: Missing Pieces*), novels (Gary Troup, *Bad Twin*), websites (Oceanic Airlines, Dharma Initiative, Janelle Granger Dairy …), podcasts, among others. In addition, Lostpedia highlights the intertextual references of the series: the novels read by some of the characters (Dickens, Dostoyevsky, Joyce, Nabokov, Agatha Christie, and Stephen King, among others); the biographies of the philosophers and scientists after whom some characters are named (Locke, Hume, Rousseau, Faraday) and their connection with the characters' personality and role; and the cultural matrixes referenced by the series: novels about shipwreck survivors in an apparently desert island (Defoe, Verne, H.G. Wells, Huxley, William Golding …); films and series about that same subject, whether original or adaptations, techno-scientific conspiracy theories, supernatural or mystical explanations of the events, etc. (Jones 2007; Johnson 2009).

Both the network and fans themselves offer these compasses so people are able to navigate through the transmedia narrative universe. This produces a very interesting tension between the official canon and the fanon (fan-produced content), between the actual story and the theoretical, parodic and promotional meta-discourse. The mapping consists in faithfully describing the territory, but the temptation to increase it (to expand the territory or to fill it with new information: *auctor* is one who *augments*), hype it up, summarize or excerpt highlights for fast viewing, or even falsify, ridicule, or caricature is very strong.

Paratextuality and Authorization: Who Can Produce Paratexts?

Regarding the question "from whom, to whom?" it is obvious that Genette was interested in the response to a notion of strong authorship and silent reception. For him, the first part of this question only had two possible answers, the author and the publisher, acting as the delegate of the author (1997b, 408). The second part also had two possible answers: public or private paratexts, depending on whether the paratexts constitute public communications (although with potentially different reach: the general public, the effective readers of the book, the critics, the distributors, or the retailers) or private communications (intimate diaries, letters, private notes). Genette never conceived of the possibility that there would be a reversal of roles: paratextually active public, and authors and editors turned into receptors, and maybe beneficiaries, of amateur production.

John Fiske (1987, 108–112) suggested distinguishing between "horizontal forms" and "vertical forms" of intertextuality. The former being those produced between "primary texts"—television programs or series—focused around concepts such as gender and character, and the latter between primary texts and other texts which are not primary: on the one hand, advertising, station identification journalistic articles, and criticism ("secondary texts") and on the other, discussions and commentaries in the audience, generated by television productions ("tertiary texts"). Fiske could be said to be preserving the centrality of the show as opposed to the proliferation of by-products around it, both amateur and corporative, the analytical consolation prize or bronze medal of textuality. But he was at least recognizing the possibility of active and productive reception.

In fact, Fiske himself devoted a specific study to this fan productivity, distinguishing between semiotic, enunciative, and textual production (1992, Hills 2013). While semiotic productivity was "essentially interior," that is to say, how audiences understand media texts, enunciative productivity concerns meanings shared or spoken in face-to-face culture, those which "occur only within immediate social relationships" (Fiske 1992, 39). On the other hand, textual productivity is when "fans produce and circulate among each other texts which are often crafted with production values as high as any in official culture" (Fiske 1992), but without financial interest, which is absent from fan production.

Obviously, Fiske's theory, like that of Genette, corresponds to a pre-Web 2.0 era. In today's world, it is clear that fandom not only blurs the limits between enunciative and textual productivity (informal conversations have become online forums, postings, reviews, and commentaries on the Internet), but also makes it easier for "interior" semiotic productivity to feed social networks in real time. Even the reveries of the solitary websurfer are poured onto the Internet and pass directly from semiotic to textual productions.

There is now no restriction at source on the reach of fan productivity, neither textual, nor enunciative or semiotic (the latter two being textualized—paratextualized), as they all circulate on the Web. Furthermore, they do not only reach other fans, but can also reach an unpredictable and potentially huge audience, and have very different results. They may be promotional, for example, prescribe or recommend products to people who do not necessarily belong to a closed fan community. It is also possible that they do not merely praise or strongly criticize; they may aspire to be a part of the story itself, expanding the narrative universe or delving deeper inside it, even influencing the official creations of the franchise, which owns the rights to the story.

Much of the theoretical work on TS, starting with Jenkins' formulation, has turned precisely towards the benefits of fan productivity, while the most critical authors have wondered whether the very digital media that have been said to blur lines between producers and consumers and create a more participatory culture, instead reinforce cultural hierarchies and free-labor exploitation (Andrejevic 2009). This is what Roberta Pearson (2008) calls the Jekyll and Hyde of TS.

The current panorama is in fact even more complex. In the opinion of John T. Caldwell (2011, 175–194), paratextual agency has been simplistically divided into two types: corporative or official and amateur or fan-made, i.e., between a "top-down" corporate ephemera on the one hand (authorized by the industry) and "bottom-up" user-generated content (UGC), on the other. However, a third instance deserves to be taken into account: a worker-generated content (WGC), i.e., "professional worlds that operate in the shadow of both the multimedia conglomerates and the celebratory, ostensibly unruly fans" (Caldwell 2011, 183).

In any case, paratextual productivity by fans has become increasingly the focus of attention: the spoilers of *Survivor* (Jenkins 2006, 25–58) or *Lost* (Gray and Mittell 2007), the recaps of *The Sopranos* or the spoof trailers of *The Shining* (Dusi 2015), the webpages, Weblogs, and wikis created by *24*'s fans (Scolari 2009), the twittersodes that are built around fictional characters in dialogue with fans of *Mad Men* (Jenkins, Green, and Ford 2013, 30–37), the mashups of *Titanic* (Jensen 2013) or the videos of *Star Trek* (Coppa 2008), the fan activism around *Harry Potter* (Jenkins 2015, 206–229), the fan fiction around *Game of Thrones* (Tosca and Klastrup 2014; Fathallah 2016) or the fandom paratextual memory around *Doctor Who* (Hills and Garde-Hansen 2017), to give only a few examples.

Conclusion

The role of paratexts in narrative construction demands our attention; this role is not merely peripheral or ancillary (Gray 2010). Neither the temporal limitation derived from its qualification as *ephemeral* (Uricchio 2011) nor the spatial limitation suggested by *interstitial* (Ellis 2011) seem to fit their current functionality. Paratext suggests an interesting coalescence between the center and periphery of the story, between a transitional zone and a transactional space. On this threshold, the paratext efficiently deploys its functions—promotional, guiding, and narrative—which should not be conceived as mutually exclusive, but as effects which are updated in all paratexts to varying degrees. Recent updating of paratextuality encompasses the whole range of phenomena and formats involved in transmedia studies, and allow them to be ordered and classified effectively (Rodríguez-Ferrándiz 2017).

However, paratextual environmentalism must be encouraged. It is important to watch over paratexts, take care of them (and not simply watch out for them, as Genette recommended), not only as essential pieces in the construction and circulation of the social sense of the texts, but also for their physical conservation, their filing and classification. Just as television scholars prior to the emergence of the VCR (videocassette recorder) suffered from the anguish provoked by the irreparable loss of emissions from the flow of the Hertzian waves, today there is paratextuality, both digital and analogue, which runs the risk of vanishing, when it has undoubtedly contributed to the meaning of the texts we enjoy.

References

Andrejevic, Mark. 2009. "Exploiting YouTube: Contradictions of User-generated Labour." In *The YouTube Reader*, edited by Pelle Snickers, and Patrick Vonderau, 406–423. Stockholm: National Library of Sweden.

Booth, Paul. 2015. *Game Play: Paratextuality in Contemporary Board Games*. New York: Bloomsbury.

Caldwell, John T. 2011. "Corporate and Worker Ephemera: The Industrial Promotional Surround, Paratexts and Workers Blowback." In *Ephemeral Media: Transitory Screen Culture from Television to YouTube*, edited by Paul Grainge, 175–194. London: Palgrave.

Consalvo, Mia. 2007. *Cheating: Gaining Advantage in Video Games*. Cambridge, MA: MIT Press.

Consalvo, Mia. 2017. "When Paratexts Become Texts: De-centering the Game-as-text." *Critical Studies in Media Communication* 34 (2): 177–183.

Coppa, Francesca. 2008. "Women, Star Trek, and the Early Development of Fannish Vidding." *Transformative Works and Cultures* 1. doi:10.3983/twc.2008.0044.

Dehry Kurtz, Benjamin W.L., and Mélanie Bourdaa. 2017. *The Rise of Transtexts: Challenges and Opportunities*. London: Routledge.

Desrochers, Nadine, and Daniel Apollon (eds.). 2014. *Examining Paratextual Theory and its Applications in Digital Culture*. Hershey, PA: IGI Global.

Dusi, Nicola M. 2015. "Remixing Movies and Trailers Before and After the Digital Age." In *The Routledge Companion to Remix Studies*, edited by Eduardo Navas, Owen Gallagher, and xtine burrough, 154–165. London: Routledge.

Ellis, John. 2011. "Interstitials: How the 'Bits in Between' Define the Programmes." In *Ephemeral Media: Transitory Screen Culture from Television to YouTube*, edited by Paul Grainge, 59–69. London: Palgrave.

Fathallah, Judith. 2016. "Statements and Silence: Fanfic Paratexts for *ASOIAF/Game of Thrones*." *Continuum* 30 (1): 75–88. doi:10.1080/10304312.2015.1099150.

Fiske, John. 1987. *Television Culture*. London and New York: Routledge.

Fiske, John. 1992. "The Cultural Economy of Fandom." In *The Adoring Audience: Fan Culture and Popular Media*, edited by Lisa A. Lewis, 30–49. London: Routledge.

Genette, Gérard. 1992 [1979]. *The Architext: An Introduction*. Berkeley: University of California Press.

Genette, Gérard. 1997a [1982]. *Palimpsests: Literature in the Second Degree*. Lincoln: University of Nebraska Press.

Genette, Gérard. 1997b [1987]. *Paratexts: Thresholds of Interpretation*. New York: Cambridge University Press.

Genette, Gérard, and Marie Maclean. 1991. "Introduction to the Paratext." *New Literary History* 22 (2): 261–172. Accessed February 3, 2017. https://goo.gl/8FFPzF.

Geraghty, Lincoln (ed.). 2015. *Popular Media Cultures: Fans, Audiences and Paratexts*. London: Palgrave.

Grainge, Paul. 2011. *Ephemeral Media: Transitory Screen Culture from Television to YouTube*. London: Palgrave.

Gray, Jonathan. 2010. *Show Sold Separately: Promos, Spoilers and Other Media Paratexts*. New York: New York University Press.

Gray, Jonathan, and Jason Mittell. 2007. "Speculation on Spoiler: Lost Fandom, Narrative Consumption, and Rethinking Textuality." *Participations* 4 (1). Accessed January 19, 2017. www.participations.org/Volume%204/Issue%201/4_01_graymittell.htm.

Hansen, Ole E. 1999. "Television Stations and the Internet: Paratext, Intratext or Hypertext." In *Intertextuality & Visual Media*, edited by Ib Bondebjerg and Helle K. Haastrup, 195–217. Copenhagen: University of Copenhagen.

Hills, Matt. 2013. "Fiske's 'Textual Productivity' and Digital Fandom: Web 2.0 Democratization Versus Fan Distinction?" *Participations* 10 (1): 130–153. Accessed February 14, 2017. www.participations.org/Volume%2010/Issue%201/9%20Hills%2010.1.pdf.

Hills, Matt, and Joanne Garde-Hansen. 2017. "Fandom's Paratextual Memory: Remembering, Reconstructing, and Repatriating Lost Doctor Who." *Critical Studies in Media Communication* 34 (2): 158–167.

Jenkins, Henry. 2006. *Convergence Culture: Where New and Old Media Collide*. New York: New York University Press.

Jenkins, Henry. 2015. "'Cultural Acupuncture': Fan Activism and the Harry Potter Alliance." In *Popular Media Cultures: Fans, Audiences and Paratexts,* edited by Lincoln Geraghty, 206–229. London: Palgrave.

Jenkins, Henry, Sam Ford, and Joshua Green. 2013. *Spreadable Media: Creating Value and Meaning in a Networked Culture*. New York: New York University Press.

Jensen, Peter K. 2013. "Clever Mashups: Online Parodies and the Contingency of Meaning." *Continuum* 27 (2): 283–293. doi: 10.1080/10304312.2013.766312.

Johnson, Derek. 2009. "The Fictional Institutions of Lost: World Building, Reality, and the Economic Possibilities of Narrative Divergence." In *Reading Lost. Perspectives on a Hit Television Show*, edited by Roberta Pearson, 29–52. London: I. B. Tauris.

Jones, Steve. 2007. "Dickens on *Lost:* Text, Paratext, and Fan-Based Media." *The Wordsworth Circle* 38 (2): 71–77.

Kernan, Laura. 2004. *Coming Attractions: Reading American Movie Trailers*. Austin: University of Texas Press.

Klastrup, Lisbeth, and Susana Tosca. 2004. "Transmedia World. Rethinking Cyberworld Design." In *Proceedings International Conference on Cyberworlds 2004*, 409–413. Los Alamitos: IEEE Computer Society. Accessed February 4, 2017. http://doi.ieeecomputersociety.org/10.1109/CW.2004.67.

Klecker, Cornelia. 2015. "The Other Kind of Film Frames: A Research Report on Paratexts in Film." *Word & Image* 31 (4): 402–413. doi:10.1080/02666286.2015.1053035.

Kristeva, Julia. 1980. "Word, Dialogue, Novel." In *Desire in Language: A Semiotic Approach to Literature and Art*, edited by Leon S. Roudiez, 64–91. New York: Columbia University Press.

Long, Geoffrey. 2007. "Transmedia Storytelling: Business, Aesthetics and Production at the Jim Henson Company." Master's thesis, Massachusetts Institute of Technology.

McCracken, Ellen. 2013. "Expanding Genette's Epitext/Peritext Model for Transitional Electronic Literature: Centrifugal and Centripetal Vector son Kindles and iPads." *Narrative* 21 (1): 103–123. Accessed January 2017, 2017. https://goo.gl/WZ3KjX.

Mittell, Jason. 2013. "Serial Orientations: Paratexts and Contemporary Complex Television". In *(Dis)Orienting Media and Narrative Mazes*, edited by Julia Eckel, Bernd Leiendecker, Daniela Olek, and Christine Pieporka, 165–182. Bielefeld: Transcript.

Mittell, Jason. 2015. *Complex TV: The Poetics of Contemporary Television Storytelling*. New York: New York University Press.

Pearson, Roberta. 2008. "The Jekyll and Hyde of Transmedia Storytelling." Paper presented to the conference Television and the Digital Public Sphere, Paris, France, October 22–24.

Pellizzi, Federico. 2006. "Dialogism, Intermediality and Digital Textuality." *IasOnline Netzkommunikation in ihren Folgen*. Accessed February 27, 2017. www.iasl.uni-muenchen.de/discuss/lisforen/pellizzi_dialogism.pdf.

Pesce, Sara, and Paolo Noto. 2016. *The Politics of Ephemeral Digital Media: Permanence and Obsolescence in Paratexts*. New York: Routledge.

Rodríguez-Ferrándiz, Raúl. 2014. "Cultural Industries in a Post-industrial Age: Entertainment, Leisure, Creativity, Design." *Critical Studies in Media Communication* 31 (4): 327–341. doi:10.1080/15295036.2013.840388.

Rodríguez-Ferrándiz, Raúl. 2017. "Paratextual Activity: Updating the Genettian Approach with the Transmedia Turn." *Communication & Society* 30 (1): 165–182. doi:10.15581/003.30.1.165-182.

Saldre, Maarja, and Peeter Torop. 2012. "Transmedia Space." In *Crossmedia Innovations: Texts, Markets, Institutions*, edited by Indrek Ibrus and Carlos A. Scolari, 25–44. Frankfurt: Peter Lang.

Scolari, Carlos A. 2009. "Transmedia Storytelling: Implicit Consumers, Narrative Worlds, and Branding in Contemporary Media Production." *International Journal of Communication* 3: 586–606. Accessed January 19, 2017. http://ijoc.org/ojs/index.php/ijoc/article/view/477/336.

Stam, Robert, Robert Burgoyne, and Sandy Flitterman-Lewis. 1992. *New Vocabularies in Film Semiotics: Structuralism, Post-structuralism and Beyond*. London: Routledge.

Stewart, Gavin. 2010. "The Paratexts of *Inanimate Alice*: Thresholds, Genre Expectations and Status." *Convergence* 16 (1): 57–74. doi:10.1177/1354856509347709.

Straumann, Barbara. 2015. "Adaptation-Remediation-Transmediality." In *Handbook of Intermediality*, edited by Gabriele Rippl, 249–267. Berlin: De Gruyter.

Tosca, Sara, and Lisbeth Klastrup. 2014. "A *Game of Thrones*: Transmedial Worlds, Fandom, and Social Gaming." In *Storyworlds across Media: Toward a Media-conscious Narratology*, edited by Marie-Laure Ryan and Jan-Noel Thon, 295–314. Lincoln: University of Nebraska Press.

Uricchio, William. 2011. "The Recurrent, the Recombinatory and the Ephemeral." In *Ephemeral Media: Transitory Screen Culture from Television to YouTube*, edited by Paul Grainge, 23–36. London: Palgrave.

47

A SEMIOTIC APPROACH TO TRANSMEDIA STORYTELLING

Geane Carvalho Alzamora

Semiotics, the "science of signification and all kinds of signs" (Santaella and Nöth 2004, 7), offers relevant contributions for understanding transmedia dynamics, which is characterized by crossing different mediatic environments and the involvement of audience participation in the expansion of the narrative universe. Theoretical and methodological frameworks derived from semiotics promote, among other aspects, an accurate understanding of the textual modalities that permeate each mediatic environment and its intersections, the processes of signification possible in transmedia dynamics, and the socio-symbolic interactions involved.

According to Jenkins (2006), a transmedia story unfolds across multiple media platforms, with each new text making a distinctive and valuable contribution to the whole. The author posits that the economic logic of a horizontally integrated entertainment industry dictates the flow of content across media. From a semiotic point of view, this process is sign related and, thus, must be examined as a signification system.

The term transmedia storytelling, as it has been used by Jenkins since 2003, reveals the textual-narrative nature of transmedia dynamics, which stimulates the development of analyses based on semiotic approaches known as textual-semiotic and narrative-semiotic. The first approach examines the communicative relation built around sets of textual practices (Wolf 1995), while the second approach demonstrates the exchange of semiotic values involved in figurative representations that permeate the narrative discourse (Santaella and Nöth 1995). Both approaches distance themselves from the informational-semiotic current of thought that identifies the first semiotic studies on communication (Wolf 1995) and interact with one another, since both belong to the field of discursive semiotics, which derives from the studies of Swiss linguist and philosopher Ferdinand de Saussure (1857–1913).

Various theoretical-methodological perspectives originated from discursive semiotics. One is connected to social semiotics, a current of thought that prioritizes the social context of the production and circulation of signs taking into account each of the communication situations analyzed, such as the topic of conversation, the interpersonal relationship involved, and the communication channels used (Lemke 2009). Discursive semiotics has also defined studies on transmedia storytelling, especially its multimodal focus, which ascertains the coordination between semiotic resources involved in the elaboration and circulation of texts.

According to Nöth (1995), the field of text semiotics covers related disciplines, such as literature, sociology, philosophy, and aesthetics. He emphasizes concepts that are important to this approach, such as textuality, which might be circumscribed as a coherent whole, and intertextuality, which pertains to the influence of one set of texts on another. Both concepts are used extensively in transmedia studies.

Although less frequently applied to transmedia studies, Peircean semiotic theory, which originated from studies by North American mathematician, physicist, chemist, and philosopher Charles Sanders Peirce (1839–1914), is suitable for investigating various aspects of transmedia dynamics. Among the possible contributions of this kind of semiotics to studies of transmedia storytelling, we emphasize its potential to describe the pragmatic improvement of transmedia dynamics and its communicational logic. Using this perspective, we analyze the notion of transmedia television. The argument is built on the Peircean notion of semiosis, as will be demonstrated below.

The Main Approaches

According to Scolari (2009), semiotics is useful for describing the instruments that create meaning in transmedia storytelling. Based on the notion that every text constructs its reader (Eco 1979), Scolari (2009) considers that the same text may often create different implicit consumers. Therefore, he states that to understand the complexity of transmedia textualities, it could be useful to compare them with traditional strategies for constructing implicit readers. He emphasizes that transmedia storytelling proposes a new narrative model, inspired by different media and languages, thus making an analysis of it from a narratological-semiotic point of view pertinent. This perspective, according to Scolari (2009), considers a new dimension of multimodal discourse and perceives the narrative as a device that produces meaning.

In Herman's (2004) view, transmedial narratology presupposes the study of linguistic features of the means involved because although narratives in different media exploit a common stock of narrative design principles, narratives exploit the principles among certain ways determined by the properties of each medium. Herman considers that the narrative is medium-independent but is also radically dependent on the medium.

Ryan (2004) clarifies that the study of narrative across media is not the same as the interdisciplinary study of narratives. Transmedial narrative theory focuses, she says, on the particular semiotic substance and on the narrative's technological mode of transmission. "Whether we call it 'narrative media studies' or 'transmedial narratology,' the study of narratives across media is a project from which the understating of both media and narrative should benefit" (Ryan 2004, 35).

Media studies is an area of investigation that dates to the 1930s and which has been progressively permeated by theoretical and methodological assumptions that originated in semiotics. An area of investigation, thus, has been outlined, which Danesi (2010) names media semiotics. Danesi highlights the prominent position occupied by the Internet in contemporary studies of media semiotics and that seems to confirm the relevance of a rapprochement between media and narrative studies, as suggested by Ryan (2004) to characterize what she calls "transmedial narratology." This area is defined by the strong presence of approaches stemming from textual semiotics.

Bertetti (2014), for instance, argues that textual semiotics offers relevant resources for understanding the logic of transmedia. In a study about transmedia characters, he asserts that transmediality and transtextuality pose some problems to the status of the characters, in particular to their identity. Based on a textual-semiotic approach, Bertetti (2014) argues that a different logic of transmedia construction is centered on shared fictional characters, including possible variations at each textual level.

Richards (2017) also views transtextual and transmedial storytelling as correlated concepts but clarifies the conceptual comprehensiveness of each notion: "Transtextual storytelling refers to a narrative created through accretion across multiple texts rather than within a single work, while transmedial storytelling refers to a narrative unfolding across more than one medium" (Bertetti 2014, 29). He argues that transtextuality does not derive from convergence culture (Jenkins 2006) and exhibits early examples of folk, canonical, and branding logics related to contemporary transmedia within cultural narratives and discourses in a wider perspective. Following Marsha Kinder (1991, cited by Richards 2017, 16), Richards opts for the expression "transmedia intertextuality."

The importance of intertextuality in the definition of transmedia is endorsed by Jenkins (2009, cited by Delwiche 2017, 35), for whom "a work needs to combine radical intertextuality and multimodality for the purposes of additive comprehension to be at transmedia story." Delwiche (2017) considers that multimodality is a core component of transmedia storytelling, an aspect often addressed by social semiotics.

According to Lemke (2009), social semiotics develops this core model in a particular way that is quite useful for the practical analysis of multimodal texts. In the context of traversal media analysis which represents meanings created across institutional boundaries, across media, genres, settings, and contexts what is particularly interesting for social semiotics is that all these various media and genres create coordinated meanings; that is, each one is an intertext for all the others, and none of their meanings is entirely independent from the others' meanings.

The notion of intertextuality is frequently related to the idea of intermediality. According to Herkman (2012), the intermediatic flux is intertextual and grounded on the continuity and variation of arrangements in a specific context of meaning. Therefore, the term intermedia refers to the context of mutual affectation in which a mediatic environment shares elements of its language with another.

Herkman (2012) highlights that the terms intermedia and convergence were coined in the 1960s. The first was conceived in the technology realm and the latter in the artistic sphere. According to Herkman (2012), in the 1990s the term convergence acquired a connotation of technological and industrial synergy, and began to relate more to the concept of hypermedia. Meanwhile, the term intermedia started to emphasize the intertextual perspective, and was frequently related to studies of media art. More recently, convergence and intermedia began to designate, almost indistinctly, the socio-communicative processes in digital connections. Thus, these terms integrate the semantic universe of transmedia and convergence culture, as established by Jenkins (2006).

The relationship between transmedia storytelling and convergence culture (Jenkins 2006) underlies the idea that transmedia is a contemporary cultural phenomenon. Based on the cultural semiotics approach, Ibrus and Ojamaa (2014) state that the practices of transmedia storytelling are becoming mainstream, which could facilitate intensified and widespread processes of intersemiotic translation at all cultural levels. As a result, those practices effectively broaden the possible variety of perceptions, meanings, and texts/representations available in a culture.

Peircean Semiotics Contributions

In contrast to discursive semiotics, which is based on the notion of linguistic sign, Peircean semiotics is founded on phenomenology and conceives the sign as the coordination, in constant movement, of three elements: the sign (representamen), object, and interpretant. Peirce named semiosis, or mediation, the continuous and auto-corrective transformation of the semiotic triad.

In a brief but dense account, Merrell (2010) explains how signs occur in accordance with the Peircean point of view: "What I mean by mediation is that a sign component acts as an intermediary between two other sign components" (Merrell 2010, 29). It is a reticular mediation process that involves the determination from a preceding sign, its object, and the representation by association (collateral experience) of a posterior sign, its interpretant, to create another semiotic triad and continue successively.

According to Parmentier (1985), mediation combines the semiotic process of representation, deriving from the interpretant as it is perceived in relation to the object through the mediation of the sign, and the semiotic process of determination, arising from the object as it is perceived in relation to the interpretant through the mediation of the sign. Representation, therefore, is an aspect of mediation. In 1906, Peirce elaborated two major trichotomies of interpretants: (1) One is related to the subdivision of interpretants into the immediate interpretant (related to the quality of the impression that a sign is to produce), the dynamical interpretant (which is whatever interpretation any mind actually makes of a sign), and the final interpretant (which arises if the ideal ultimate opinion is reached). (2) The other subdivision of the interpretant is into the emotional (related to feeling as the first effect of a sign), the energetic (related to interpretation efforts), and the logic, considered the essential effect caused on the interpreter (Alzamora and Gambarato 2014). The subdivisions of the interpretant define the course to be taken by meaning and are related to the forms of engagement in semiosis (Colapietro 2004), thus constituting specificities in the representation process.

The varied semiotic process of representation associates new signs with semiosis through collateral experience, thus constructing a network of signs that is constantly expanding in such a way that the reference of an object is never lost, and the creative capacity of semiosis is never compromised. Therefore, the process of semiotic mediation occurs on two levels: between the object and the interpretant (through the mediation of the sign) and between a chain of signs (sign/object/interpretant) and another, through the mediation of the interpretant (Alzamora and Bicalho 2016).

From this viewpoint, transmedia dynamics may be understood as a mediation process that involves successive semiotic processes of determination and representation linked together on two levels: in a media environment (through intertextual references) and in the intersection between two or more media (through transtextual representations).

On both levels, the semiotic process of determination appertains to the narrative universe of reference, and the semiotic process of representation emphasizes the creative capability of transmedia storytelling in relation to the referential determination. Thus, transmedia dynamics operates on drillability (determination) and spreadability (representation), as implied by Jenkins (2009) in one of the seven principles of transmedia storytelling: spreadability/drillability, continuity/multiplicity, immersion/extractability, world-building, seriality, subjectivity, and performance.

According to Jenkins (2010), the media industry often talks about continuity in terms of canons, that is, the information accepted as part of the definitive version of a particular story. In contrast, multiplicity is related to the multiple alternative versions compared to the established canon.

According to the approach taken by Peircean semiotics, continuity is related to the endless process of semiosis. "For Peirce, there is an essential continuity between utterance and interpretation, between the process in which an interpretant is grasped as such" (Colapietro 1989, 20). In this perspective, continuity incorporates multiplicity by the variability of interpretants generated in the process of semiosis. Continuity, therefore, delineates the expansion of the transmedia storytelling based on references from the canon story associated, by collateral experience, with variations of the narrative driven by citizen participation. The other principles proposed by Jenkins (2009, 2010) operate under the same semiotic logic.

In Jenkins' (2010, 2017) view, transmedia storytelling is one of a range of transmedia logics, which might also include transmedia branding, transmedia performance, and transmedia learning. Although there may be some overlap between these different logics, each transmedia dynamic can trigger specific regulatory principles.

Jenkins (2017) introduced two related concepts, transmedia logics and locations, suggesting how this framework might allow us to build more models of transmedia production. "By transmedia logics, I mean two interrelated things: the goals a transmedia production is intended to serve and the assumptions made about the desired relationship among transmedia consumers, producers, and texts" (Jenkins 2017, 222). He suggests that transmedia locations stem from the premise that different forms of transmedia are apt to emerge within different social, cultural, and industrial contexts. "Many of these logics also assume a strong connection between transmedia production practices and some ideal form of audience participation" (Jenkins 2017).

In Peirce's view, logic is another name for semiotics (Colapietro 1989). According to Gambarato and Nani (2016), in the realm of Peircean logic or semiotics, the action of the sign results in interpretants that are neither ultimate nor static but are continuously generated. "Through the variability of signification, connection and integration specific to transmedia stories, new articulations are made between the narrative instalments dispersed across diverse media outlets" (Gambarato and Nani 2016, 154).

From the point of view of Peircean semiotics, transmedia dynamics is a sign process in permanent reticular expansion, which involves the diverse proliferation of interpretants. The interpretative potential of the sign (immediate interpretant) may be understood in transmedia semiotics as a kind of invitation to participation that will become a new sign (dynamic interpretant) only through the associative action of signs (collateral experience) outlined by the semiotic operation of representation. In transmedia dynamics, the phenomenological variety of interpretants (emotional, energetic, and logical) is translated into different forms of representation.

In a study of media transformation, Elleström (2014) characterizes transmediation based on the Peircean differentiation between the concepts of mediation and representation. According to Elleström (2014), all transmediation involves some degree of transformation, and media representation is a kind of transmission of media characteristics. "Media representation means that the representamen of the target medium conjures up in the mind of the perceiver both the representamen and the object of the source medium" (Elleström 2014, 17). He differentiates simple transmediation of qualified media—"elementary form being transmediated from one qualified medium to another" (Elleström 2014, 21)—from complex transmediation of qualified media—"a target medium triggers representations of multifaceted media traits similar to those of a source medium" (Elleström 2014, 22).

In Elleström's (2014) view, the notion of transmediation approaches adaptation. However, Jenkins (2009) distinguishes transmedia expansion from adaptation, by arguing that the adaptation process does not introduce in the narrative universe elements that are significantly different from the original that serves as a reference. On this topic, Scolari (2011) highlights that adaptation, or intersemiotic translation, is a creative process established between one language system and another. He claims that although different from transmediatic expansion, intersemiotic translation can be a possible transmedia storytelling strategy.

Based on the semiotics of Charles Sanders Peirce and the system theory of Mario Bunge, Gambarato (2012) discusses the process of transmediatic expansion of the narrative between language systems. Gambarato correlates the concepts of sign, system, and complexity in order to better understand the intricate nature of transmedia storytelling. According to her, a transmedia project is more complex because it allows user participation, which leads to unpredictable developments in the story and enhances the whole experience, thus contributing to its permanence. "Complexity of Transmedia Storytelling definitely embodies intricate involvement between all the constituent elements of [super] [sub] system, i.e., set of components, environment and set of relations" (Gambarato 2012, 81). This perspective distances itself from the notion of adaptation.

Transmedia dynamics is characterized by Alzamora and Gambarato (2014) as a pragmatic offshoot of semiosis in media. In this perspective, the productive incompleteness of the interpretant is taken as a conceptual parameter for understanding how media consumption regulates habits and delineates the transmedia narrative in the semiotic process of network associations. The interpretant generated by the sign is also another sign. Thus, the interpretant necessarily generates another sign that acts as its interpretant, and so forth. In this approach, the incompleteness of semiosis itself, and consequently, of the interpretants generated in this open-ended process, corresponds to the richness of the variability that transmedia storytelling can evoke.

According to Colapietro (2004), from 1903 onward, under the influence of Charles Darwin's evolutionism, Peirce's pragmatism, or *pragmaticism*, emphasizes notions of historicity and temporality in the process of the logical enhancement of meaning. Therefore, the pragmatic perspective of semiotics presupposes that mediation operates on evolutionary continuity, resulting in an improvement in action habits—which outline interpretation modes in semiosis, be it human or not.

From this viewpoint, transmedia dynamics may be considered a type of logical improvement of mediatic semiosis, as argued below for the notion of transmedia television. According to this perspective, transmedia television is a form of mediatic semiosis outlined temporally and based on historical language references (Alzamora 2017).

Transmedia Television According to Peircean Semiotics

The reticular configuration of contemporary television affects its expressive form, or televisuality, historically established around specific notions such as programming and serial repetition (Sarlo 1997). The current context, as stated by Jost (2010), is characterized by the multiplication of screens and the subsequent difficulty in knowing which of these media should be called television. Televisual syntax becomes diffuse (Jost 2010), composed by transmissions defined by the present time, institutional identification of the broadcaster, and references to program schedules, the last formed notably by content related to information, entertainment, and instruction.

According to Evans (2011), the technical and socio-cultural conditions of contemporaneity bring forth unprecedented consequences for the configuration of television. In contrast to the mediatic perspective derived from mass communication, which historically defined it, Evans (2011) claims that contemporary television is cross-media, characterized by digital connections.

From the viewpoint of semiosis, according to its pragmatic angle, television is a mediation process referenced by a variety of languages and interactive practices under regulation of determination. These references are partially incorporated by television dynamics in a persistent manner, although not necessarily an enduring one. Therefore, the dynamics of television alter inasmuch as interaction practices arising from other mediatic dynamics affect, in an intense and cumulative manner, the uses and appropriations of television.

Semiotic regulation of determination and representation configures semiotic mediation under the auspices of historicity and temporality. In the first situation, television semiosis recursively indicates the languages used as references in the domain of intermedia dynamics. In the second, television semiosis is expanded in the domain of transmedia dynamics, which functions in varied temporalities according to the diversity of interaction practices that have been engaged in the form of collateral experience.

For Elleström (2010), intermediality must be understood in accordance with the complex network of materially tangible qualities that constitute the fundamental conditions of each medium and act in synchronicity with perceptive and interpretive operations undertaken in the context of reception. The intermedia approach presupposes the conditions of mediatic production, circulation, and consumption, as well as the material interfaces associated with social, cultural, historical, communicative, and aesthetical circumstances.

From this viewpoint, television semiosis acts in control of intermedia dynamics when it updates and individualizes its social, cultural, aesthetic, and discursive references in a certain mediatic pattern according to the historically defined conditions of production, circulation, and consumption. Representative capability is delineated by this set of references but also expanded in accordance with the current communication scenario. The characteristics of intermediality, therefore, are altered throughout time, although they configure a stable set of references. Such features, here related to intermediality, have notably been mentioned in the sphere of transmedia television.

Based on Jenkins (2006), Evans (2011) distinguishes *transmedia storytelling distribution* from *transmedia storytelling engagement* in order to define transmedia television. Evans' first proposition corresponds to the idea of utterances from transmedia television based on the inclusion of multiple texts, scattered across several mediatic environments. Meanwhile, the second proposition concerns the changes resulting from the practices of the reception of television content distributed simultaneously across platforms but consumed differently.

In a similar perspective, Orozco (2012) differentiates the transmediality of transmission from the transmediality of reception to characterize contemporary television. The author argues that the growing ubiquity and connectivity of the audience strengthen the impression that the consumption of television content in digital connections is a decision of the connected audience, in spite of transmediatic transmission strategies.

From the viewpoint of semiosis, this ambivalence defines transmedia dynamics because the latter is constructed in the tension between reference and creativity. Television semiosis in transmedia dynamics is intermedia by reference and is characterized by complementarity and drillability. However, transmedia dynamics acts pragmatically in television semiosis by expanding its referential nucleus of meaning by diversion and incorporation, in sync with contemporary habits of mediatic consumption.

According to Jansson (2013), transmedia dynamics configures the preferable mode of contemporary mediatization in what he calls the Transmedia Age, in contrast to the Age of the Masses. In his view, contemporaneity is characterized by mediatic textures, that is, amalgams that combine interaction practices mediated by mediatic arrangements. Mediatic textures configure plurimediatic forms, integrated and flexible and that normalize interaction practices both online and offline, thus producing an impact on the circulation of mediatic textures that originated in the Age of the Masses, of which television is an example. Therefore, as television becomes more pervasive and diffuse, the more polycentric its mediatic texture tends to become, and the bigger its potential of integration with contemporary interaction practices should become.

Transmedia is a logic, not just a discursive mode (Jenkins 2010). In accordance with this viewpoint, we uphold the idea that transmedia dynamics configures a logical rapprochement with intermedia dynamics by acting in television semiosis in order to improve it, in compliance with contemporary mediation processes and the emerging habits that regulate interaction practices in digital connections.

The use of hashtags in transmedia dynamics on television is an indication of this scenario (Alzamora 2017). Hashtags are seen here as mediation processes that function as a reference to the proposal of connection to the premise of multiscreen television that determines them, as well as related to the numerous contexts of interaction practices associated through collateral experience that evoke new meanings in dynamics derived from social engagement. Thus, hashtags configure semiotic processes exerting a mediating function on two levels. First, hashtags operate as mediating signs between the determination originating from the communicational proposal of the broadcaster—its object—and the desired effect of association in the form of interaction practices engaged in on online social media—its interpretant. Second, given the interpretant's productive diversity, always incomplete, the hashtag associates varied stages of meaning in keeping with the social, technical, and discursive nature of each mediatic environment in which the hashtag is inscribed and with each communication scenario the hashtag outlines.

Thus, the hashtag operates as a symbolic link that associates similar positions materially, albeit not necessarily consensually. Therefore, social engagement mediated by hashtags might endorse referential transmedia planning, in the case of television, but may also reconfigure it, as well as eventually generate a transmedia dynamic of its own. The social use of hashtags interferes with the reticular configuration of the television audience and has become a relevant feature in the transmediatic circulation of television content today.

Conclusion

According to Harrington, Highfield, and Bruns (2013), the relationship between television and social media is highly complex, but the authors argue that the relationship also improves the engagement between the audiences and television content. "Transmedia content does not need to be live to make use of a wide range of media channels and platforms, and Twitter can play a role also in the anticipation and follow-up discussion of television shows" (Harrington, Highfield, and Bruns 2013, 407). In this case, the authors (2013) claim that Twitter should be used more frequently to sustain a community of enthusiasts and facilitate their interactions with program makers, rather than for (or in addition to) live interaction during the broadcast itself.

The relationship between television and Twitter is usually mediated by hashtags. The purpose is to improve television engagement by connecting it to mediated social conversation. This scenario is a transmedia dynamic because it associates interactive habits on Twitter with the television audience of reference. Thus, this scenario favors the pragmatic enhancement of the habits of the television audience in accordance with contemporary parameters of mediated social interaction on online social media connections. Therefore, transmedia television becomes a constituent part—never autonomous—of digital media connections in a broader sense.

References

Alzamora, Geane. 2017. "Televisão em Semiose: Mídia, Intermídia, Transmídia." Proceedings of the XXVI Encontro Anual da Associação Nacional dos Programas de Pós-Graduação em Comunicação, 1–18. Accessed August 20, 2017. www.compos.org.br/data/arquivos_2017/trabalhos_arquivo_FG1AZ7KPGK43U2SK4OT2_26_5335_21_02_2017_13_09_09.pdf.

Alzamora, Geane, and Luciana Andrade Gomes Bicalho. 2016. "A representação do Impeachment Day mediada por hashtags no Twitter e no Facebook: semiose em redes híbridas." *Interin* 21 (2): 100–121.

Alzamora, Geane, and Renira Gambarato. 2014. "Peircean Semiotics and Transmedia Dynamics: Communicational Potentiality of the Model of Semiosis." *Ocula -Occhio Semiotico Sui Medi* 15: 1–15. Accessed August 20, 2017. www.ocula.it/files/OCULA-15-PEIRCE-Alzamora_Gambarato_[270,532Kb].pdf.

Bertetti, Paolo. 2014. "Toward a Typology of Transmedia Characters." *International Journal of Communication* 8: 2344–2361. Accessed August 20, 2017. http://ijoc.org/index.php/ijoc/article/view/2597.

Colapietro, Vincent. 1989. *Peirce's Approach to the Self: A Semiotic Perspective on Human Subjectivity*. Albany: State University of New York Press.

Colapietro, Vincent. 2004. "The Routes of Significance: Reflections on Peirce's Theory of Interpretants." *Cognitio* 5: 11–27. Accessed November 20, 2017. https://revistas.pucsp.br/index.php/cognitiofilosofia/article/view/13206.

Danesi, Marcel. 2010. "Semiotics of Media and Culture." In *The Routledge Companion to Semiotics*, edited by Paul Cobley, 135–149. New York: Routledge.

Delwiche, Aaron. 2017. "Transmedia Storytelling and the Audience Question." In *The Rise of Transtexts: Challenges and Opportunities*, edited by Benjamin W. L. Derhy Kurtz and Mélainie Bourdaa, 33–48. New York: Routledge.

Eco, Umberto. 1979. *Lector in Fabula*. Milan: Bompiani.

Elleström, Lars. 2010. "The Modalities of Media: A Model for Understanding Intermedial Relations." In *Media Borders, Multimodality and Intermediality*, edited by Lars Elleström, 11–50. New York: Palgrave Macmillian.

Elleström, Lars. 2014. *Media Transformation: The Transfer of Media Characteristics Among Media*. New York: Palgrave Macmillan.

Evans, Elizabeth. 2011. *Transmedia Television: Audiences, New Media, and Daily Life*. New York: Routledge.

Gambarato, Renira R. 2012. "Signs, Systems and Complexity of Transmedia Storytelling." *Estudos em Comunicação* 12: 69–83. Accessed August 20, 2017. www.ec.ubi.pt/ec/12/pdf/EC12-2012Dez-4.pdf.

Gambarato, Renira R., and Alessandro Nani. 2016. "Blurring Boundaries, Transmedia Storytelling and Ethics of C. S. Peirce." In *Ethics in Screenwriting: New Perspectives*, edited by Steven Maras, 147–175. Melbourne: Palgrave Macmillan.

Harrington, Stephen, Tim Highfield, and Axel Bruns. 2013. "More than a Backchannel: Twitter and Television." *Participation: Journal of Audience and Reception Studies* 10 (1): 405–409. Accessed August 23, 2017. www.participations.org/Volume%2010/Issue%201/30%20Harrington%20et%20al%2010.1.pdf.

Herkman, Juha. 2012. "Convergence or Intermediality? Finnish Political Communication in the New Media." *Convergence: The International Journal of Research into New Media Technologies* 18 (4): 369–384.

Herman, David. 2004. "Toward a Transmedial Narratology." In *Narrative Across Media: The Languages of Storytelling*, edited by Marie-Laure Ryan, 41–46. Lincoln: University of Nebraska Press.

Ibrus, Indrek, and Maarja Ojamaa. 2014. "What Is the Cultural Function and Value of European Transmedia Independents?" *International Journal of Communication* 8: 2283–2300. Accessed November 23, 2017. http://ijoc.org/index.php/ijoc/article/view/2650/1204.

Jansson, André. 2013. "Mediatization and Social Space: Reconstructing Mediatization for the Transmedia Age." *Communication Theory* 23: 279–296.

Jenkins, Henry. 2006. *Convergence Culture: Where Old and New Media Collide*. New York: New York University Press.

Jenkins, Henry. 2009. "The Revenge of the Origami Unicorn: Seven Principles of Transmedia Storytelling." *Confessions of an Aca-Fan*, December 12. Accessed August 20, 2017. http://henryjenkins.org/2009/12/the_revenge_of_the_origami_uni.html.

Jenkins, Henry. 2010. "Transmedia Education: The 7 Principles Revisited." *Confessions of an Aca-Fan*, June 21. Accessed August 23, 2017. http://henryjenkins.org/2010/06/transmedia_education_the_7_pri.html.

Jenkins, Henry. 2017. "Transmedia Logics and Locations." In *The Rise of Transtexts: Challenges and Opportunities*, edited by Benjamin W. L. Derhy Kurtz and Mélanie Bourdaa, 220–240. New York: Routledge.

Jost, François. 2010. *Compreender a Televisão* [Understanding Television]. Porto Alegre: Sulina.

Lemke, Jay. 2009. "Multimodal Genres and Transmedia Traversals: Social Semiotics and the Political Economy of the Sign." *Semiotica* 173 (1/4): 283–297. Accessed August 20, 2017. www.academia.edu/3033681/Multimodal_genres_and_transmedia_traversals_Social_semiotics_and_the_political_economy_of_the_sign.

Merrell, Floyd. 2010. "Charles Sanders Peirce's Concept of the Sign." In *The Routledge Companion to Semiotics*, edited by Paul Cobley, 28–39. New York: Routledge.

Nöth, Winfried. 1995. *Handbook of Semiotics*. Bloomington and Indianapolis: Indiana University Press.

Orozco, Guillermo. 2012. "Audiencias Conectadas y Desconectadas. Dos Modos de Estar Frente a la Pantalla Televisiva y Buscar la Interlocución" [Connected and Disconnected Audiences. Two Ways to Stand in Front of the Television Screen and Search for Interlocution]. In *TVMorfosis: La Television Abierta Hacia la Sociedad de Redes* [TVMorfosis: Open Television Towards Network Society], edited by Guillermo Orozco, 187–200. Mexico City: Procutora de Contenidos Culturales Sagahón.

Parmentier, Richard J. 1985. *Signs and Society: Further Studies in Semiotic Anthropology*. Bloomington: Indiana University Press.

Richards, Denzell. 2017. "Historcizing Transtexts and Transmedia." In *The Rise of Transtexts: Challenges and Opportunities*, edited by Benjamin W. L. Derhy Kurtz and Mélainie Bourdaa, 15–32. New York: Routledge.

Ryan, Marie-Laure. 2004. "Introduction." In *Narrative Across Media: The Languages of Storytelling*, edited by Marie-Laure Ryan, 1–40. Lincoln: University of Nebraska Press.

Santaella, Lúcia, and Winfried Nöth. 1995. *Handbook of Semiotics*. Bloomington: Indiana University Press.

Santaella, Lúcia, and Winfried Nöth. 2004. *Comunicação e Semiótica* [Communication and Semiotics]. São Paulo: Hacker Editores.

Sarlo, Beatriz. 1997. *Cenas da Vida Pós-Moderna: Intelectuais, Arte e Video-Cultura na Argentina* [Scenes of Postmodern Life: Intellectuals, Art and Video-Culture in Argentina]. Translated by Sérgio Alcides. Rio de Janeiro: Editora UFRJ.

Scolari, Carlos. 2009. "Implicit Consumers, Narrative Worlds, and Branding in Contemporary Media Production." *International Journal of Communication* 3: 586–606. Accessed August 20, 2017. http://ijoc.org/index.php/ijoc/article/view/477/336.

Scolari, Carlos A. 2011. "Demoliendo sartenes: La Nova Cuina de Ferran Adria." *Designis* 17.

Wolf, Mauro. 1995. *Teorias da Comunicação* [Communication Theories]. Lisbon: Presença.

48

A MYTHOLOGICAL APPROACH TO TRANSMEDIA STORYTELLING

Nicoleta Popa Blanariu and Dan Popa

A cognitive pattern based on archetypal images articulated in an exemplary story, myth works in the history of culture as a "hypotext" (Genette 1982; Eliade 1963). The texts that resume, re-contextualize, and reinterpret (elements from) the mythical scheme may be regarded as "hypertexts" in relation to it. The relationships between the myth and its echoes in literature or other modes and media of expression—verbalized or not, static or dynamic, printed or digital, virtual or *live* performances, etc.—may be described in terms of "inter-" and "transtextuality" (Genette 1982) or, as the case may be, "inter-" and "transmediality." Aiming to obtain a formal model of the narrative, the structuralists have deconstructed and correlated the versions of the same myth. They thus identified a set of minimal units, which they called "mythemes," the different combinations of which generate the versions of the respective myth (Lévi Strauss 1958). Myth can therefore be reduced to a "system of motifs" (Tomashevsky 1973, 247–285) or "mythologems" (Jung 2014). Lévi-Strauss (1958) refines such an approach: the myth is not just a set of "motifs," but also the very "set of its variants." This means, first, that the mythical story is a whole class of (re)tellings obtained by combining the same structural motifs, which makes the myth radically different from other narratives. Second, in my opinion, Lévi-Strauss's definition finds today a confirmation and possible reformulation as a result of the proliferation of transmedia phenomena, especially transmedia storytelling. What do we mean? Myth is indeed confused with "all its variants," as the father of structural anthropology observed. But, these variants are circulated through multiple media, which, in turn, condition the form that mythical content acquires. Therefore, keeping and completing the formula of Lévi-Strauss, we could say that a myth is *the inter- and transmedial corpus "of its variants."* We shall further operate with this meaning.

Despite the distinctive features, the versions of the same myth exhibit, as a stable nucleus, a number of structural constants, which Lévi-Strauss (1958) calls "mythemes," Jean Rousset (1978) "invariants," and Carl Gustav Jung (2014) "mythologems." The myth involves the existence of a "mythical scenario," a narrative "scheme" composed of such "invariants" (agents, attributes, circumstances, actions). Through the intervention of the variable elements, the general scheme is incorporated into a particular story. The analysis of the myth can, therefore, approach two complementary aspects: the continuity of a scenario and its variability. If the myth may be understood as the "set of its variants," as per Lévi-Strauss, then it also subsumes the "set of transformations of a scenario," under the influence of the practices and representations specific to the contexts in which it is upgraded (Pageaux 2000, 131–132).

A generation or an author may rediscover a myth, reinterpreting it, rendering its latent values. During the German occupation, the French public saw, for example, in Antigone, a symbol of Resistance (Brunel 1988). Beyond what separates them, between Prometheus, Antigone, Oedipus,

Faust, Don Juan, and Tristan and Isolde, is a family resemblance that transpires in the different versions of these myths: everyone is associated with the same "mytheme": transgression of a limit, violation of an interdiction. This is what triggers the hero's whole journey: fault, (down)fall, and greatness. Similarly, Parsifal is recommended by another fundamental "mytheme": the *search* (for the Grail, in this case) doubled by a fatal error caused by a lack of spiritual ability (along with which the former heroic performance becomes unnecessary).

It is probably symptomatic for the contemporary imaginary the recurrence of "mythologems" such as *the child, the old man* (*the wise man, the magician*), *the trickster*, in some of the most loved transmedia stories: *Harry Potter, Lord of the Rings, Trigun, Kamichu!*, and *Arc the Lad*, for instance. Without embarking on a thorough analysis of these mythologems, we here nevertheless mention some of the valences that one of them has: "the *infans* archetype," of "great significance" and "frequency" "throughout the world" (Jung 2014, 173). It is associated with the "maturing of personality" (Jung 2014, 182), the need to restore the "connection with an original condition" of innocence, vitality, hope. It prefigures a "future transformation of personality," through the "synthesis of conscious and unconscious elements" (Jung 2014). In short, the child's motive is "a symbol that unites opposites, a mediator, a *saviour*, that is, someone who completes" and who is able "to overcome the monster of darkness" (Jung 2014, 188). Harry Potter and the Hobbit Frodo save the world from such evil forces. The "schoolgirl goddess" Kamichu is, in turn, a mediator between worlds, men and spirits, between the visible and invisible, *corpus et anima*, between what *seems* and what *is*. At the level of symbolic value, Jung distinguishes between two closely related mythologems: the "child god" and the "child hero" (Jung 2014, 189). The former is "completely supernatural," the latter "has a human nature," however located in the vicinity of the "supernatural," which makes it a "semi-god" (Jung 2014). Harry Potter, Frodo, and Kamichu have, each in his/her own way, an inner *daimon* that gives them this exceptional character, located in the uppermost part of the human being and yet within its limits. The *infans* archetype is often associated in myths, fairy tales, and transmedia stories, with the "mysterious and miraculous" birth, "abandonment, exposure" and "endangering of the child" (Jung 2014). Harry Potter's biography confirms it.

Another complex psychological phenomenon is also revealed in *Lord of the Rings* and *Harry Potter*: the link between the "*trickster* and the individual *shadow*," the latter being a "summation" of the "inferior character traits of the individual" (Jung 2014, 304). In the former, the phenomenon is manifested through the ambiguous relationship between Frodo and Gollum, and in the second, through the relationship between the hero and Voldemort. Voldemort's evil potential is accentuated by a linguistic taboo: avoiding the direct designation of the character, which is replaced by a periphrasis, "You-Know-Who." The denominative substitution relies on an ancient magic principle, according to which the name concentrates the essence of the reality it designates. Little Frodo is bound to wear the ring of power until it is thrown into the volcano and the end of the mission. Frodo's guide is the cursed hobbit Gollum, a sort of Cain whom the fascination of power pushes into murder and who ultimately turns out to be disloyal toward his master. But Frodo is betrayed not only by Gollum, but also by himself: in a moment of weakness, Frodo is reluctant to destroy the fatal ring, being mastered by the mirage of the absolute power that it offers. Thus, the *trickster* (the opponent) is only the reflection of the shadow being—the dark but hidden side of the hero himself. Despite the previous hesitation, Frodo finally accomplishes his salutary gesture, convinced that he must destroy the ring for saving the world. His final decision somehow reminds us of that of a Shakespearean character, Prospero, who at the end of *The Tempest* destroys his magic book after he has punished the usurpers and restored the moral order of the world. At the end of the mission they have undertaken, both little Frodo and the old dethroned Duke Prospero voluntarily give up the source of their superhuman power. Likewise, Professor Dumbledore from *Harry Potter* destroys the Philosopher's Stone, to prevent it from being used by evil forces.

In fact, in fairy tales, dreams, in some transmedia stories, the "archetype of the spirit" often takes the appearance of an old man who offers "advice and help" in "stalemate" situations (Jung 2014, 246) such as Albus Dumbledore from *Harry Potter*, Gandalf from *Lord of the Rings*, Getafix from

Asterix, etc. Sometimes, the image of the old man is "ambivalent," doubly claimed by both good and evil, even ultimately serving the latter, as it is the case with the magician Sauron from *Lord of the Rings*. Sauron's perversion leads to an ample deployment of "theriomorphic symbols of the spirit" (Jung 2014, 256–259). Numerous mythologems from transmedia stories lend themselves to some other similar annotations.

The "Bardic Function" in Transmedia Storytelling

If "mythological motifs" are "structural elements of the psyche," and myths are, originally, "involuntary statements about unconscious soul events" (Jung 2014, 174–177), a question arises: what could the mythical substrate of so many transmedial products say about today's man? There is not enough space to try to give an exhaustive answer here. But, it would also be an explanation—not the only one—for the extreme popularity of transmedia storytelling. In any case, even "understanding the history" remains tributary to "mythical structures." The historical fact is often assimilated to a mythical *pattern*, without which the event would be but an "ordinary fact" (Durand 1998, 29). Here are just two examples of the many possible ones. "Without waiting for the Messiah—which is mythical—there is no Jesus Christ" (Durand 1998). Similarly, Napoleon's mythization takes place against the backdrop of "mythical messianism" enveloping (pre)romantic Europe at the end of the eighteenth century and the beginning of the nineteenth century. In such a context, the "historical incarnation" of the promethean myth in the person of Napoleon Bonaparte becomes possible. It is precisely in this sense that "the destiny of the West" is a "mythological warp in which heroes, titans and gods clash" (Durand 1998). In a way, the myth "shares the roles in history and decides what exactly 'constitutes' the historical moment, the soul of an age" (Durand 1998, 29–30).

Thus, the "mythological process" does not have, by far, only a "religious meaning," but also a "general" one, comprising both "historical truth" and "physical truth" (Cassirer 1972, 24) and—we may add by relying on Freud, Jung, and their followers—the psych(olog)ic truth. This readiness of the myth to be relevant to a multitude of aspects of human experience, whether individual or social, validates the interpretation that Roland Barthes (1973) gives to the myth from a semiological perspective. Barthes assimilates the cognitive and axiological function of the myth with the question of the meaning, of some socially shared symbolic values. More specifically, he identifies two levels of signification: the first is that of "denotation" and the second is that of "connotation and myth." In this context, the myth corresponds to a "chain of concepts" widely accepted within a culture, through which community members "conceptualize and understand" a particular subject or part of their social experience (O'Sullivan et al. 2001, 299–300). The good functioning of the community depends (also) on this consensus of interpretations, which ensures the synergy of social behavior around mythically circulated force-images. That is why the loss of mythological heritage—and reference—represents "a moral catastrophe, always and everywhere, even for the civilized man" (Jung 2014, 177).

Such a shortcoming may, to a certain extent, be controlled by activating the "bardic function" (Fiske and Hartley 1978) of the broadband media, including digital media, video games, and transmedia storytelling. By their "bardic" function, these media may articulate "the main lines of the established cultural consensus about the nature of reality (and therefore the reality of nature)" (Fiske and Hartley 1978). Inevitably, such a function uses "representations and myths" (O'Sullivan et al. 2001, 146), in a Barthesian sense. The episode "Crossing the River of Time" from the *Kamichu!* series illustrates the manifestation of the "bardic" dimension. Relying on a (psycho)drama inspired by modern Japanese history and formulated in terms of animistic mythology, the bardic function aims, in this case, to overcome a collective trauma and to reconcile the two Japans, the two mentalities. One of them, conservative, rooted in the aristocratic code of honor of the samurai, is incapable of accepting the defeat in World War II and, especially, the decision to surrender broadcasted by radio by the emperor. It feeds a deep sense of guilt and self-discredit because of the seeming abdication from an implacable moral duty. The latter, on the contrary, valorizes the heroic behavior of the Japanese Army, recovering, with dignity and recognition, the vestiges of a difficult past. The mythical symbol of this historical experience is, in *Kamichu!*, the famous Yamato, the largest battleship ever built, formerly the admiral ship of the Japanese fleet. To slow the advance of the allied troops, Yamato was sent to Okinawa in April 1945 with the order to defend the island and fight to self-destruction. However, before being able to carry out his mission, it was sunk off the coast. For Japanese culture, the Yamato battleship symbolizes the splendor and collapse of the Empire. In *Kamichu!*, in an animistic parable, there is the attempt to bring Yamato's *spirit* to the surface, under the close supervision of the "schoolgirl goddess." The initiative is, nevertheless, struck by the resistance of wreck: "It would be shameful for me to go back there looking like this," "the people of Kure put a lot of effort into building me and they sent me out with high hopes." "I can't limp into port like the shipwreck I am. It's disgraceful." "Maybe it would be better for everyone if I stay put and slowly rust away … into nothing" (see "Kamichu! Episode 09: Crossing the River of Time"). In reply, Yamato is praised and rehabilitated: "It's not disgraceful!" "Nobody thinks badly of you. You're a hero to all of us, Yamato! We're proud of you!" "There are so many people here in Japan who have never forgotten about you." "Coming home wouldn't just make you happy, Mr. Yamato. It would make everyone else happy, too" (see "Kamichu! Episode 09: Crossing the River of Time"). The episode evokes a way to collectively metabolize a difficult historical experience, illustrated by Yukio Mishima's biography, among others.

In a parodic stance, hammed here and there, the *Asterix* series demystifies the bardic function through one of the characters, Cacofonix. The initial meaning—historical and etymological—of the function is distorted. The bard Cacofonix—an appellative which is inconsistent with the social role of the designated person—is tied during parties, so that he cannot sing, therefore, to hinder him from performing an emblematic attribution. The bard is here an antiphrase replica to the old function of the poet and singer (*aoidos*, bard, minstrel, troubadour, etc.) in traditional societies: that of vehicle of the exemplary word (around which the consensus of the community is formed) and also of *melos* and harmony. In the Aesopic manner from *Asterix*, the blocked word and cacophony are the symptoms of a disturbance of the bardic function. Such a (highlighting through) parody of the bardic function is complicated in *Asterix* for being taken over by the atypical bard of some marks of the controversial, original, and inconvenient artist: sometimes admired and often unwanted, listened to and finally censored. In the eyes of some, Cacofonix has yet a solution—exile: his incomprehensible talent could find recognition in Lutetia, where, more than 15 centuries later, a great European artistic and intellectual capital, Paris, would rise. The *Asterix* series recovers, however, without caricature distortions, another component of the bard's function, namely that of transmitting a heritage of collective knowledge. Thus, Cacofonix reactivates the memory of the community, helps the Druid Getafix—the religious leader of the group and the guardian of traditional wisdom—to heal his amnesia. Together, the two—the bard and the druid—occasionally exercise a pedagogic function, that of teachers of the village.

Mytho(do)logy: Myth, Mythanalysis, Mythocriticism

By relying on previous approaches to myth—especially psychoanalysis and structuralism—Gilbert Durand (1998) proposes two complementary ways of investigating a mythological corpus: "mythocritique" and "mythanalysis." The first highlights "the directing myths and their significant transformations." The other "extracts" from myths "not only the psychological meaning," but also the "sociological" one. More specifically, "sociological mythanalysis" "attempts to outline the great guiding myths of historical moments and types of groups and social relationships." By what does Durand's mythanalysis differ from Jungian analysis? The French anthropologist responds by way of example: while Jung "generalizes and uniformalises" the "animate archetype," "mythanalysis deals with finding different types of *anima* according to the typologies of ancient mythology: Venus,

Demeter, Junona, Diana etc." (1998, 308–309). Mythocritique allows for the discovery of both the "wear, disappearance" and the reappearance of a myth, as well as its "transformation," depending on how a particular era favors a particular group of mythemes. Thus, "by the amputation of a group of mythemes, a myth changes its meaning," also changing the "soul of an age" (1998, 313).

Crystallized in the Renaissance, in the German folk culture, amid rich mythological, historical, religious, and philosophical sources, Faust's myth may be seen as the result of the "transformation" of a much older Greek myth: Faust is an "amputated Promethean myth" which lacks the "altruism of the titan" (Durand 1998, 315). The way old myths have been reinterpreted over the last few decades, through an alliance of mythology with technology, is also relevant to such permutations and combinations of mythemes. Numerous video games exploit—sometimes only for the exotic or archaic atmosphere—allusions to the mythological Buddhist, Celtic, Chinese, Japanese, Egyptian, Greek, Biblical, Gnostic, and Slavic mythological imaginary, not once by blending and hybridizing different traditions. In the *Too Human* video game, Scandinavian mythological motifs are intertwined with emblematic elements for the (post)modern imagery, the latter functioning as Barthesian-like mythologems. Ymir, Loki, Baldur, Midgard find themselves among people, advanced machines, *cyborgs*, in a life-and-death conflict, between defenders and opponents of humanity. The Aesir gods have become a kind of *cyborg*, their bodies have been perfected by cybernetic means. Conflict—a classic mytheme—no longer opposes people and gods, divine families (Vanir and Aesir, as in the Nordic mythology) or generations of immortals (Titans and Olympians at the Greeks), but the mankind and Aesir *cyborgs* on the one hand, and the killing machines, "Children of Ymir," on the other. We will also further indicate some aspects of the "transformation" of the Promethean myth into older and more recent inter- and transmedial productions.

Civilizing and *philanthropos*, Prometheus is also the mythical figure of revolt and (self)assertion even at the cost of (self)sacrifice. Otherwise, Prometheus and Faust are "the major figures of the Renaissance," according to Ernst Bloch (1974; also Brunel 1988; Bonnardel 1993; Kerényi 1963). The (pre)romantics associate the Titan with other mythical-religious figures such as Jesus (Maistre 1922; Shelley 1820), Lucifer (Shelley 1820), Ixion, Tantalus, Sisyphus (Goethe 1904). Shelley (1820) oxymoronically associates Prometheus with both Christ and Satan. Prometheus and Satan are similar in their rebellion; Prometheus and Christ, in their sacrifice. André Gide (1953) proposes a parodic and deheroing version of the myth, *Prometheus Misbound* (*Prométhée mal enchaîné*). His Prometheus is a Nietzschean immoralist, a follower of the "great style" and the "will to power," the latter being the "will to have will," unbroken by self-blaming (Popa Blanariu 2016, 2017).

From the Renaissance to the twentieth century, there are many transpositions of Prometheus in painting. Beyond them, Ridley Scott's movie *Prometheus* (2012) is a surprising rebranding of the myth, and transmedia campaign for promoting it is a masterpiece of marketing. The campaign was conceived as a "mix of social, traditional, and transmedia storytelling" (see "Prometheus Transmedia Campaign"). In a mixture of horror, religion, and science fiction, the movie associates anthropogenesis and eschatology, artificial intelligence, the quest for God, for knowledge about immortality, about evil and faith. Scott's movie also evokes some invariants of the Greek myth of Prometheus: rebellious Titan, anthropogenesis (creation of the android David), humanity's relationship with the gods, and destruction of mankind. The film interweaves motifs from several mythological traditions: Greek-Latin mythology, Jewish and Christian beliefs on creation, some Gnostic allusions, etc. The transmedia campaign is an interactive one, and continues the film plot. It transmedially integrates several viral videos, online game, websites, television through which the "line between the reality and imagination" is "blurred" (Arnašiūtė 2012; Przegalińska 2015; Jenkins 2011; López-Varela and Saavedra 2017; Popa Blanariu 2017).

Actantial Structure and Narrative Design

Vladimir Propp's (1970) formalization of the fairy tale is based on the identification of universal type-functions, namely structural invariants. The function is an action of the character, defined from the point of view of the meaning it achieves in the course of the action. By analyzing 100 Russian fairy tales, Propp discovers 31 universal "functions." The distinction between "actants" and "actors," drawn by Greimas, allows for the separation of "two autonomous levels" (1973, 161). The actants are related to "a narrative syntax," and the actors "can be recognized in the particular discourses where they manifest themselves" (Greimas 1973). The two categories are realized syncretically: an actant may be manifested by several actors, just like a single actor may manifest several actants (Greimas 1973). Greimas identifies three pairs of actants: subject/object, sender/receiver, adjuvant/opponent. The hero may be the receiver of his own action, when he proposes the mission to himself, just like he becomes its receiver when the adventure of the search is to his advantage. The game of relationships can be transformed: an opponent may help; a friend may become a rival. For example, in *Lord of the Rings*, the magician Sauron turns from Gandalf's ally into his rival. By exploiting the three actantial pairs, Greimas formulates an actantial model based on the object of desire, targeted by the subject (hero). Obtaining the object is facilitated or, on the contrary, prevented by the action of adjuvants and opponents. Usually, the narrative is reducible to the "search" of such an object (Greimas 1966, 177). Greimas identifies three actantial axes. The subject/object relationship formalizes the search for the object of desire (the golden fleece that, in the Greek myth, the Argonauts led by Jason are off to find, the secret of immortality that Gilgamesh is longing for, saving the world as in *Lord of the Rings* or *Harry Potter*, etc.) (Greimas 1966). The sender/receiver axis is that of knowledge and power, corresponding to the distribution of values among the characters. The adjuvant/opponent axis describes how the action of the characters is facilitated or prevented. The categories adjuvant and opponent are not necessarily represented by characters as such. Thus, the two categories sometimes cover the inner states of the subject, projections or suppressions of the desire to act (Greimas 1966, 190). For example, Frodo's temporary weakness who, in *Lord of the Rings*, hesitates for a moment to get rid of the evil ring.

The canonical scheme of the story contains three fundamental types of evidence: the skill test (the hero is endowed with magic instruments, ubiquity, invisibility, etc.); the decisive test (the hero carries out the actions for which he has been invested: the destruction of the forces of evil, the release of their prisoners, etc.); the glorification (the hero is recognized for what he is and what he has done). Jean-Marie Floch describes the invariants of the story in Chomsky's terms of *competence* (what the subject wants/knows/can do) and *performance* (what the subject actually does) (Floch cited by Rovenţa-Frumuşani 1999, 114). The narrative scheme thus consists of four sequences: contract, competence, performance, and sanction. The *contract* consists of the proposal by the sender and acceptance by the subject of a future program of heroic actions. The sender embodies the *ethos* of the community and has a persuasive effect on the subject: he determines the subject to accept the contract after other actants have refused or missed the qualification test. Hobbit Bilbo Baggins abandons the custody of the ring of power, and magician Gandalf, in the name of the conjuration of Good, convinces young Frodo to accept the difficult mission of secret bearer of the ring until he succeeds in destroying it. *Competence* corresponds to the "skill test": the hero acquires the qualities necessary to achieve the program proposed by the sender. Before entering the dangerous land, the Queen of Elves equips Frodo and his comrades with magical objects useful in the encounter that awaits them. The *performance* coincides with achieving the program or the "decisive test." The hero now applies what he has learned, doing what is expected of him and what he himself wants, as a trustee of values and collective will. With the latest powers, Frodo reaches the volcano and throws the ring in which the essence of evil lies, thus releasing the world from dark forces. *Sanction* (reward) involves the hero's "glorification" and the sender's acknowledging his success. After the end of the mission, Frodo is invited by Gandalf to join the immortal people of the elves and leave with them in a land beyond the world's borders, death and human weakness, in order to find peace. The reward that the hero receives is a half-god condition. Like in fairy tales, the "sanction" initiates here an axiological approach based on the compensatory revenge of fiction upon the real, of good upon evil.

Conclusion

Beyond the elements of mythological inspiration, there is still a connection between mythical narrative and transmedia storytelling: their *performative* dimension. In traditional communities, the myth—an exemplary story telling the founding events of the group—is remembered periodically, in ritual circumstances, by telling or dramatizing it in front of a circle of initiates (Eliade 1963). The ritual reiteration of the myth has a role of community solidarity around common values, embodied by mythical representations and transposed into norms of collective behavior. Hence, the paradigmatic component of the myth (Eliade 1963; Cassirer 1972). The function of solidarity traverses the *identification* of the assistance with mythical archetypes by overcoming the observer condition and *participation* in the mythical drama, narrated or played on the stage of the rite. Transmedia storytelling responds to the same requirements for *participation* and *identification*. Henry Jenkins integrates the two categories into a quintet of "logics of commitment" that contributes to the emergence of transmedia storytelling: the "logic of entertainment," "of social connection," "of expert" (manifested by the creation of fan sites), that of the "immersion" (which facilitates "participation") and, finally, the "logic of identification" (which allows fans to "perform an identity according to what they see") (Gallarino 2012).

On the other hand, transmedia storytelling is a much older practice than digital technologies (Jenkins 2014; Freeman 2017). Numerous examples may be invoked here, including a prehistoric sample of transmedia storytelling: "The first 'cartoons' are to be found in the … stone age" (Muscan 1984, 36–37) in Bessov Noss, Russia. The static images engraved there, on the shores of the lake—the hunter, the wizard, the boats, a rich fauna—are enlivened and combined in cinematic sequences at a certain moment of the day. The optical effect is due to the water refraction properties (Muscan 1984). The intuition of the physical and aesthetic phenomenon is exceptional at the Neolithic engraver. Transmedia storytelling has, therefore, a long (pre)history and, apparently, a promising future. It responds to a fundamental human need, that of (the organization of the world's image through) the story. The myth is the archetypal, primordial, paradigmatic story.

References

Arnašiūtė, Živilė. 2012. "Prometheus." *Transmedia Storytelling*, December 5. Accessed April 22, 2017. https://zivilearnasiute.wordpress.com/2012/12/05/prometheus.

Barthes, Roland. 1973. *Mythologies*. St. Albans: Paladin.

Bloch, Ernst. 1974. *La Philosophie de la Renaissance* [Philosophy of the Renaissance]. Paris: Payot.

Bonnardel, Françoise. 1993. *Philosophie de l'Alchimie. Grand Oeuvre et Modernité* [Philosophy of Alchemy]. Paris: Presses Universitaires de France.

Brunel, Pierre (ed.). 1988. *Dictionnaire des Mythes Littéraires* [Dictionary of Literary Myths]. Monaco: Éditions du Rocher.

Cassirer, Ernst. 1972. *La Philosophie des Formes Symboliques. La Pensée Mythique* [The Philosophy of Symbolic Forms. Mythical Thought]. Translated by Jean Lacoste. Paris: Minuit.

Durand, Gilbert. 1992. *Figures Mythiques et Visages de l'Oeuvre. De la Mythocritique à la Mytanalyse* [Mythic Figures and Faces of the Work. From Mythocriticism to Mythanalysis]. Paris: Dunod.

Durand, Gilbert. 1998. *Figuri Mitice şi Chipuri ale Operei. De la Mitocritică la Mitanaliză* [Mythic Figures and Faces of the Work. From Mythocriticism to Mythanalysis]. Translated by Irina Bădescu. Bucharest: Nemira.

Eliade, Mircea. 1963. *Aspects du Mythe* [Aspects of Myth]. Paris: Gallimard.

Fiske, John, and John Hartley. 1978. *Reading Television*. London: Methuen.

Freeman, Matthew. 2017. *Historicising Transmedia Storytelling: Early Twentieth-century Transmedia Story Worlds*. New York: Routledge.

Gallarino, Aurore. 2012. "Henry Jenkins Explique sa Vision du Transmedia et de l'Engagement des Publics" [Henry Jenkins Explains His Vision of Transmedia and Audience Engagement]. *Transmedia Lab*. Accessed January 19, 2017. www.transmedialab.org/autre/henry-jenkins-explique-sa-vision-du-transmedia-et-de-lengagement-des-publics.

Genette, Gérard.1982. *Palimpsestes. La Littérature au Second Degré* [Palimpsests: Literature in the Second Degree]. Paris: Seuil.

Gide, André. 1953. *Marshlands and Prometheus Misbound: Two Satires*. Translated by George D. Painter. New York: New Directions.

Goethe, Johann Wolfgang von. 1904. "Prometeu" [Prometheus]. Translated by Panait Cerna. *Sămănătorul*, III, 51.

Greimas, Algirdas Julien. 1966. *Sémantique Structurale* [Structural Semantics]. Paris: PUF.

Greimas, Algirdas Julien. 1973. "Les Actants, les Acteurs et les Figures" [Actants, Actors, and Figures]. In *Semiotique Narrative et Textuelle* [Narrative Semiotics], edited by Claude Chabrol, 161–176. Paris: Librairie Larousse.

Jenkins, Henry. 2011. "Transmedia 202: Further Reflections." *Confessions of an AcaFan*, July 31. Accessed April 4, 2017. http://henryjenkins.org/blog/2011/08/defining_transmedia_further_re.html.

Jenkins, Henry. 2014. "'All Over the Map': Building (and Rebuilding) Oz." *Acta University Sapientiae, Film and Media Studies* 9: 7–29.

Jung, Carl Gustav. 2014. *Opere Complete*. Vol. 9. Partea I: *Arhetipurile Inconştientului Colectiv* [The Archetypes and the Collective Unconscious]. Translation by Vasile Dem Zamfirescu and Daniela Ştefănescu. Bucharest: Editura Trei.

"Kamichu! Episode 09: Crossing the River of Time." *Anime transcripts*. Accessed December 10, 2017. Accessed February 12, 2017. http://animetranscripts.wikispaces.com/Kamichu!%3EEpisode+09.

Kerényi, Carl. 1963. *Prometheus: Archetypal Image of Human Existence*. Translated by Ralph Manheim. New York: Bollingen Foundation.

Lévi-Strauss, Claude. 1958. *Anthropologie Structurale* [Structural Anthropology]. Paris: Plon.

López-Varela Azcarate, Asuncion, and Estefania Saavedra. 2017. "The Metamorphosis of the Myth of Alquemy in the Romantic Imagination of Mary and Percy B. Shelley." *Icono 14* 15(1): 108–127.

Maistre, Comte J. de. 1922. *Les Soirees de Saint-Petersbourg* [St. Petersburg Dialogues]. Paris: Garnier.

Muscan, Catinca. 1984. *Explorări in Enigmatic* [Explorations into the Enigmatical]. Bucharest: Albatros.

O'Sullivan, Tim, John Hartley, Danny Saunders, Martin Montgomery, and John Fiske. 2001. *Concepte Fundamentale din Ştiinţele Comunicării şi Studiile Culturale* [Key Concepts in Communication and Cultural Studies]. Translated by Monica Mitarcă. Iaşi: Polirom.

Pageaux, Daniel-Henri. 2000. *Literatura Generală şi Comparată* [Comparative and General Literature]. Translated by Lidia Bodea. Iaşi: Polirom.

Popa Blanariu, Nicoleta. 2016. *Cand Literatura Comparată Pretinde că Se Destramă. Studii şi Eseuri*. Vol. I: "Invarianţii, un Fir al Ariadnei?" vol. II: "(Inter)text şi (Meta)spectacol" [When Comparative Literature Pretends To Be Falling Apart. Studies and Essays. Vol. I: "Invariants, an Ariadne's Thread?", vol. II: "(Inter)text and (Meta) performance"]. Bucharest: Eikon.

Popa Blanariu, Nicoleta. 2017. "Transmedial Prometheus: From the Greek Myth to Contemporary Interpretations." *Icono 14* 15 (1): 88–107. doi: 10.7195/ri14.v15i1.1040.

Propp, Vladimir. 1970. *Morphology of the Folktale*. Texas: University of Texas Press.

Przegalińska, Aleksandra. 2015. "Prometheus: A Transmedia Campaign." In *Management in Virtual Environments*, edited by Grzegorz Mazurek, 79–89. Warsaw: Kozminski University.

Rovenţa-Frumuşani, Daniela. 1999. *Semiotică, Societate, Cultură* [Semiotics, Society, Culture]. Iaşi: Institutul European.

Rousset, Jean. 1978. *Le Mythe de Don Juan* [The Myth of Don Juan]. Paris: Armand Colin.

Shelley, Percy Bysshe. 1820. *Prometheus Unbound, A Lyrical Drama in Four Acts, With Other Poems*. London: C. and J. Ollier.

Todorov, Tzvetan. 1975. *Poetica. Gramatica Decameronului* [Poetics. The Grammar of Decameron]. Translated by Paul Miclău. Bucharest: Univers.

Tomashevsky, Boris. 1973. *Teoria Literaturii. Poetica* [Theory of Literature. Poetics]. Translated by Leonida Teodorescu. Bucharest: Univers.

49

A QUALITATIVE NETWORK APPROACH TO TRANSMEDIA COMMUNICATION

Matthias Berg and Andreas Hepp

Network is a concept that in essence describes the linkages between different entities. In media and communication research, these entities typically are human beings. The analysis of human social networks and the role of media for building up the social relations of these networks has a long tradition, which goes back to the origins of empirical media and communication research (see Katz and Lazarsfeld 1955). This long-term history corresponds with social sciences in general, where since the 1950s "network" became a more and more established concept to describe social relations (for example Bott 1957), typically referring back to Alfred R. Radcliffe-Brown's (1940) theorizing of social structure. However, with the spread of the global infrastructure of the Internet, "network" became a far more common concept. Whole books were written about the idea that society as such can be described as a "network society" (Castells 1996; Dijk 1999). From this point of view, "network" became a very general metaphor for describing the linkages of various forms of human (and non-human) connectivity (Hepp et al. 2008).

With this extended perspective, it also became usual to use the concept of the network in a much wider sense within media and communication research. Quite obvious is the use to describe the linkages between "artefacts" such as different websites, for example, as they are captured in a crawler analysis (Rogers 2010). Other examples are the references between tweets, something that also can be understood as a certain kind of network of references (Bruns and Stieglitz 2013). Similarly, Paul C. Adams and André Jansson refer to Henry Jenkins' (2006) work on convergence, defining it as "the erosion of boundaries between media, which disassociates texts from particular mediated contexts and transforms them into nodes within a network of many different media" (Adams and Jansson 2012, 303). Actor network theory (ANT) even went one step further in using this metaphor: its main interest lies in the "networks" being built by humans and non-humans within a certain assemblage (Latour 2007). Methodologically speaking, these networks become reconstructed by "following" the human and non-human actors within the linkages of the respective assemblage (Marcus and Saka 2006). We can notice at this point the diverse use of the concept of network. This corresponds with very different kinds of network analysis: while all are using "network" as a metaphor to describe linkages between different entities, they differ highly in their methods. The reconstruction of an actor network in ANT has, for example, nothing to do with a crawler analysis of online references in webpages.

Social Networks and Communication Networks

In qualitative network analysis within media and communication research the term "network" typically captures social relations between humans. At this point, it is helpful to analytically distinguish two forms of networks: social networks and communication networks. A *social network analysis* reconstructs the social relations between different human actors. These social relations can be described by their character, for example as "strong" or "weak" (Granovetter 1983). A *communication network analysis* focuses not on social relations in general but on practices of communication in particular: it reconstructs the practices of communication that "link" different human actors (Hepp, Roitsch, and Berg 2016). While social networks and networks of communication are closely related, this distinction is nevertheless beneficial. First, it is analytically useful, because social networks and communication networks partly can fall apart. It is possible to have a "strong tie" to someone (a very old friend) with whom one communicates only infrequently (but in important situations such as a crisis). Therefore, the pure frequency of communication does not reflect one to one the intensity of social networks. Second, the distinction is useful as it allows us to classify different kinds of qualitative network analysis: such that are directed to social relations, and such that are directed to communicative relations.

Nowadays, both social network analysis and communication network analysis have a strong transmedia or cross-media orientation: in the perspective of media and communication research, neither social networks nor networks of communication are built up by a single medium but typically transmedia: the "strong" or "weak ties" one person might have to another rely on various media—not only the media used for communicating with each other (telephone, messenger, social media, etc.), but also the media shared as important resources in life (films, books, series, etc.) (Baym 2015). Therefore, grasping social networks and communication networks typically means not to focus on one single medium but on the variety of (digital) media that are important for the respective social relation. Evidently, the same is true for communicative networks.

However, this transmedia character of networks does not mean that the "affordances" (Hutchby 2001) of certain media and their infrastructures would not be of importance for the character of the respective social relation. This is demonstrated quite well by the discussion about "networked individualism" (Castells 2001, 116–136; Rainie and Wellman 2012; Wellman et al. 2003). This idea is related to the argument that present forms of individualism rely on the Internet (and mobile technologies) as its "material basis": social network sites support, for example, the building of social relations which are rather dedicated to certain topics and interests instead of a localization in time and space (here and now). Such topics, among others, may refer to politics, religion, or popular culture (e.g., sports, music, computer games, fictional media texts), which are potential resources for communities "held together through the mutual production and reciprocal exchange of knowledge" (Jenkins 2006, 27). While social relations exist across a variety of media, the Internet together with mobile communications supports a certain "shaping" or "molding" of these relations. This is one main issue of mediatization research: investigating how social relations transform with the change of media and communication (Couldry and Hepp 2017, 53–56).

Qualitative Communication Network Analysis

Such assumptions are also the background for qualitative network analysis. In general, qualitative approaches of network analysis are gaining relevance in social science, especially in the context of mixed method designs (see Hollstein 2014). The strength of a qualitative network analysis is to integrate the analysis of structural relations (linkages) with the analysis of practices (meaning). Starting with this, we can define qualitative network analysis as a form of network analysis that uses qualitative instruments to make network structures and practices visible. Therefore, qualitative network research is not just the interpretation of network visualizations (Freeman 2000; Lima 2013, 21–72) or of standardized network data (Scott 2013). It is an independent method, being part of the more far-reaching qualitative methodology of social sciences. This said, a relation to network visualizations and standardized network analysis remains in qualitative network analysis: also for qualitative network analysis, it is a relevant question as to which visual methods are appropriate for collecting data (for example network maps or drawings) as well as for presenting data (network diagrams). A relation to the standardized network analysis is given on the one hand by a shared history (Scott 2013). On the other hand, both approaches are attached to the idea of bringing society back into audience research; that is, to overcome the dominance of psychological and causal models in media use (see Rogers and Kincaid 1981, 38–39). In such a sense, early (also qualitative) network approaches like those adopted by Elihu Katz and Paul Lazarsfeld have much in common with later reflection on media appropriation as we know it from cultural studies (Morley 1992).

At this point, it is worth referring back to the distinction of social networks and communication networks as we did at the beginning of this chapter. The reason is that qualitative analysis of social networks typically reflects transmediality in a rather indirect way. One example for this is a study by Maria Löblich and Senta Pfaff-Rüdiger (2011). In their qualitative network analysis, "expert interviews" and "network cards" (Baxter and Babbie 2004, 334) were combined to reconstruct the networks of political actors. The study did not focus on (mediatized) communication networks as such but on the networks of political actors and their influence, considering a special topic of media and communication research: political actors in the field of media politics of youth protection. This relates to questions of transmediality insofar as young people use a variety of media, and as the networks are built across face-to-face communication and various forms of mediated communication. However, transmediality in the sense of how the variety of these media qualitatively influences the character of the network is not a focus of this kind of research.

This is a different case with many studies of communication networks. Their actors can vary, starting with everyday people, such as members of fan communities, and ending with experts and elite actors of specific fields. At the center of such analyses are the communication networks of these actors and how they are mediated by various media. Therefore, a special research interest is the communicative mediation of these networks—a mediation which typically takes place in transmedia—and the influence of various media in this. This is also a frame of our network analysis, for example on the transmedia networks of migrants (Hepp, Bozdag, and Suna 2012) or our recent research on the communicative networking and community building of different media generations (Hepp, Berg, and Roitsch 2014; Hepp, Roitsch, and Berg 2016). This research also integrates insights of ANT (Latour 2007; Couldry 2008), as the materiality of media technologies is fundamental for such networks. In the following, we want to outline our approach as an example of a qualitative network analysis that reflects the transmedia character of communication networks.

Three Levels of Transmedia Qualitative Network Research: A Case Study

Our approach of a qualitative network analysis addresses the role of media for communicative networking and community building. The aim was to develop a cross-generational typology of people's "horizons of mediatized communitization" which are defined by the aggregates of communities to which individuals feel they belong. Following a qualitative approach that can be best described as media-ethnographic "miniatures" (Bachmann and Wittel 2006, 191), we conducted what we call a "contextualized communication network analysis" (Hepp, Roitsch, and Berg 2016). Within three waves of data collection from October 2010 until July 2017, we gathered 161 media-ethnographic miniatures (61 persons of younger age (16 to 30 years), 41 persons of middle age (31 to 59 years), as well as 59 elderly and retired persons (60 to 88 years)) in the two German cities of Bremen and Leipzig as well as their rural surroundings. The sampling process followed the concept of "theoretical sampling" as developed by Glaser and Strauss (1999, 45–78).

At the core, our study comprises three instruments of data collection. First, we carried out qualitative interviews with an average duration of 127 minutes to capture the *meaning dimension* with topics including the interviewees' media appropriation, the communitizations relevant to them, as well as corresponding communicative practices. During the course of the interviews, the participants were asked to draw the subjective perception of their communicative networks on two unstructured cards and to comment on them. In this way, we aimed to capture the *structural dimension* of their communicative networking. In order to describe the *process dimension* of their egocentric communication networks and communitizations, we employed semi-standardized media diaries, which the interviewees filled out using printed notebooks or a smartphone app. Finally, the data were analyzed according to a coding procedure oriented toward a "grounded theory" approach (Glaser and Strauss 1999).

To clarify the relevance of meanings, structures, and processes of communicative networking as three dimensions of a transmedia approach to qualitative network analysis, we now refer to the example case of Berit Haller, a 58-year-old school teacher. Berit Haller was born and raised in southern Germany before she moved to Bremen with her husband and two adolescent sons at the beginning of the 1990s. Her horizon of communitization is what we consider "multilocal," which becomes apparent when looking at the three dimensions in detail.

Level of Meaning: Qualitative Interviews

The multilocal type first and foremost implies a significant translocal extension of relevant communities in the individual's horizon. In Berit Haller's case, the multilocal horizon of communitization is based on the significance of certain localities, which derive from the close social relations located there. Most important in this respect are her family and friends, which are both nationally and internationally dispersed. When it comes to her family, Berit Haller talks about her parents living close to her hometown in southern Germany and a brother who is located in eastern Germany. Of the two grown-up sons, one lives in Hamburg and the other is enrolled in a master's program in Colombia. Next to people in Bremen, Berit Haller's circle of friends includes contacts going back to her childhood days and her best friend living in Switzerland.

Altogether, Berit Haller reflects that she is "used to being international," which on the one hand "would not be possible without [media]." On the other hand, the teacher also stresses the importance local mobility bears for a multilocal horizon of communitization. This refers to situative mobility such as visiting friends and family or traveling to foreign countries a lot already in her childhood, as well as biographical forms of mobility such as moving to Bremen or a semester abroad that she spent in Great Britain during her university studies. As a consequence, she feels less connected to a specific region, nationality, or Europe as a supra-national construct, but rather sees herself as a world citizen.

A multilocal horizon of communitization, however, does not mean that local community life can be disregarded. In Berit Haller's case, this becomes apparent when looking at the city of Bremen as her everyday urban environment. Having grown up in a village of some 3,000 inhabitants she says she experienced "very strong social control" so that her moving to Bremen amounted to "a leap into freedom." The meaning of this city, however, derives on the one hand from an urban structure that allows her to do as she pleases and, on the other hand, from the social relationships, especially colleagues and even more her local friends.

Level of Structure: Network Maps

Nowadays mediatized horizons of communitization massively depend on media communication, as already mentioned in a citation above. This becomes especially apparent when turning to the structural dimension—and thus to the network maps that Berit Haller sketched out during the interview. Those drawings instantly reveal the transmedia character of communicative networking within the horizon of communitization represented in her comprehensive media repertoire. To begin with, Berit Haller's subjective visualization of her reciprocal media communication (Figure 49.1a) shows that the landline telephone and emails are relevant for more or less all communitizations which are present in the drawing. Furthermore, she also writes and receives letters (*Post* and *Briefkontakt*) and postcards

(*Postkarten*) in order to maintain relationships with her parents, remotely living friends, and sons. By contrast, newer digital technologies such as Facebook and Skype are used in the specific contexts of a few friends from her hometown and her son in South America, respectively. This is hardly surprising since her children are the primary motivation for engaging with digital media in order to stay connected, despite the fact that at the age of 58 she feels rather "incompatible concerning new media." Before he went to South America, for example, her older son told her "either you get a smartphone with WhatsApp … or you join Facebook." At the same time, her sons are the major resources of digital media literacy, teaching her how to operate the computer and various programs and platforms. Similarly, the mobile phone and SMS (Short Message Service) are exclusively relevant for reciprocal transmedia networking with the two sons and her brother.

The second network map (Figure 49.1b) represents Berit Haller's view of her produced media communication. In contrast to the reciprocal media communication, this drawing is organized according to the media types which are important to her: Internet, television (*Fernseher*), newspapers (*WeserKurier, Fernsehzeitung, Frauenzeitschrift*), books (*Bücher*), DVDs and videos, theater, cinema (*Kino*), and music (*Musik*). It is important to note that produced media communication also fundamentally refers to an individual's mediatized horizon of communitization. When it comes to watching television, in addition to the news, Berit Haller is especially interested in documentaries about South America (*Reportagen: Südamerika*). Thus, her multilocal orientation is not only related to her son living there, but also becomes expressed in the consumption of specific mass-media content. Furthermore, the network map bears witness to her passion for animated films as well as stories by Agatha Christie and Alfred Hitchcock, which can be regarded as specific forms of popular cultural communitizations.

From another angle, it becomes clear that Berit Haller's major fields of interest in terms of produced media content are connected to what Maren Würfel (2014) calls "transmedia appropriation": she consumes news via Internet, television, and newspapers, informs herself about South America using the Internet as well as television, and listens to music on CDs as well as in theatrical performances. The relevance of the thriller and crime genre for her produced media networking is

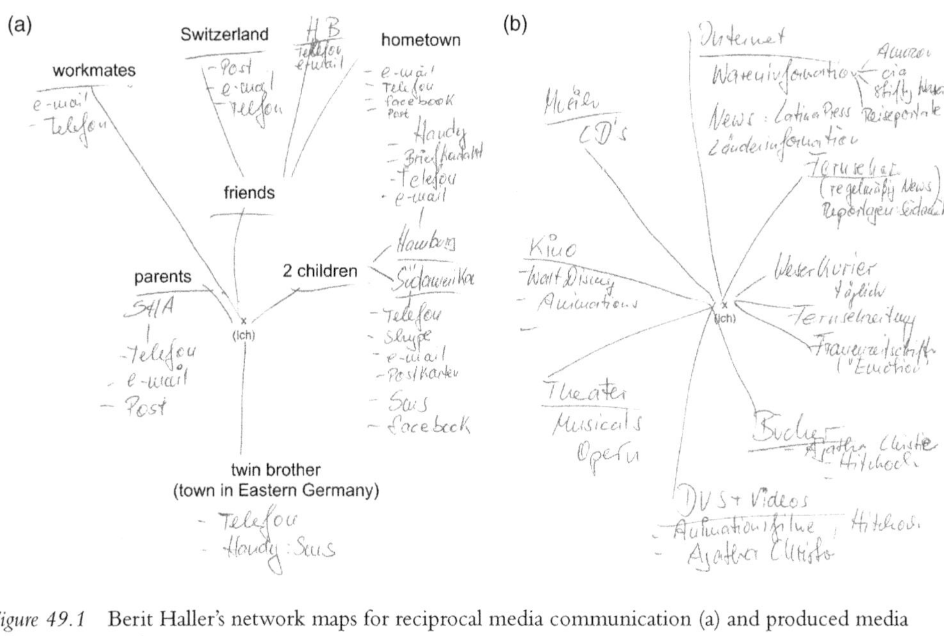

Figure 49.1 Berit Haller's network maps for reciprocal media communication (a) and produced media communication (b).

mirrored in Berit Haller's consumption of fictional content in the form of movies (*DVDs + Videos*) as well as books (*Bücher*) and thus can be understood as appropriation of transmedia storytelling.

Level of Process: Media Diaries

The process level, finally, allows for an analysis of communicative practices connected to the communicative networking within a person's horizon of communitization. This is realized through the interpretation and visualization of the data gathered by means of media diaries. Accordingly, Figure 49.2 shows the visualization of Berit Haller's media diary, which on one page (left-hand column) contains groups of particular media while, on the other, lists the social groupings (legend) which are relevant to Berit Haller. The numbers in the columns indicate the days of the week, while those with a grey background indicate the weekend. By use of this graph, we now can visually follow the temporal progress and processual patterns of transmedia networking through which Berit Haller's social groupings are reinforced and reproduced.

Berit Haller's multilocal orientation on the one hand is represented in the translocal communicative connections to friends (xxxxxx) and family (xxxxxx). On the other hand, the media diary visualization indicates the importance of popular culture and the world in general as reference points of her feelings of belonging. In the former case, personal connections are predominantly realized through emails and telephone conversations, which among other things includes calling her parents at least "every other day." In addition, her morning routine becomes visible, which she describes as follows: "I get up, then I turn on the Internet and check my mails, check Facebook for pictures posted by my older son." Also, a single incident of using Skype to video-chat with her son in South America is reported in the afternoon of day 6.

The extent to which popular cultural products and a feeling of belonging to a world community are expressed in her communicative practices becomes much more transparent in the diary data as compared to the interview material and network drawings. For one thing, the diary entries bear witness to the reception of animated and cartoon movies as an instance of popular cultural communitization. Apart from that, the intensity of watching television in the evenings is connected to the broadcasts and coverage of the soccer world championship that took place in Brazil at that point in time. In addition, we can assume that Berit Haller also dealt with this event reading the newspaper which she regularly reads "all over," as she says. Not only does her vivid enthusiasm for sports resonate in following this global popular cultural event via the media. It also becomes amplified by the fact that it took place in Brazil, which coincides with her interest in South America. Finally, this example indicates how produced and reciprocal media communication overlap in the sense that she also receives pictures via Facebook from her son in South America, attending matches of the German national soccer team. Hence, this example refers to transmediality in two ways: on the one hand,

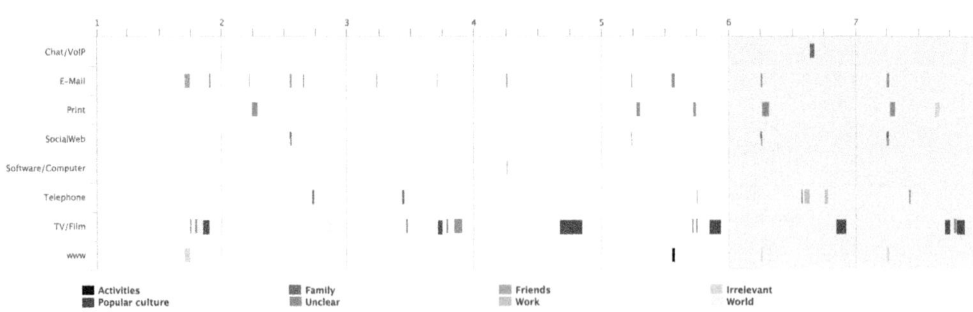

Figure 49.2 Visualization of Berit Haller's media diary.

Berit Haller follows what can be regarded an instance of "transmedia news coverage" (Gambarato, Alzamora, and Tárcia 2016, 1450). On the other hand, the same topic is taken up in the reciprocal transmedia networking within the communitization of her family.

Conclusion

As this chapter demonstrates, qualitative network analysis is an emerging field in media and communication research. Irrespective of whether this research is focused on social networks or on communication networks, the transmedia character of these networks is at one and the same time an empirical fact (people build up their networks across various media) as well as an empirical challenge (how to analyze this transmedia character in a proper way). We understand the contextualized network analysis as introduced by us as one possible answer to this challenge: the drawing of network maps makes it possible to analyze structural links across a variety of media. Qualitative interviews give access to the practices of transmedia networking and their meaning. And (app-based) media diaries offer the chance to reconstruct communicative networking across different media as a process.

Drawing on an example case from our empirical work on mediatized horizons of communitization, we have shown that transmedia networking refers to reciprocal as well fictional and non-fictional produced media communication. More precisely, those two forms of communicative networking are interrelated, and they frequently overlap within ego-centered communication networks. Thus, it has been possible to connect transmedia networking to other forms of transmediality such as "transmedia appropriation" (Würfel 2014) and "transmedia news coverage" (Gambarato, Alzamora, and Tárcia 2016). This suggests that qualitative communication network analysis might be a useful approach to research not only concerning the reception or production side of media communication but also with regard to their intersections. The latter have become especially prominent in times of convergence which, as Jenkins states, "occurs within the brains of individual consumers and through their social interactions with others" (2006, 3).

This said, a fundamental challenge remains for all approaches to qualitative network research that try to address the transmedia character of networking, and that is the increasing level of complexity: the media manifold. By "manifold," we refer not just to the plurality of today's media channels and interfaces (Madianou 2014), but to their interlinked nature and the multi-dimensional order that results and that encompasses our whole media environment (Couldry and Hepp 2017, 11). Therefore, the challenge for any qualitative network analysis which takes a transmedia point of view is to investigate the various "affordances" and "shaping" or "molding" forces of the interrelated variety of media that are part of social and communicative networking. While it is a practical question as to which approach of qualitative network analysis is taken in a specific study, they are all confronted with the necessity to address this overall challenge.

Acknowledgements

This chapter is based on research that was funded by the German Research Foundation (DFG) within the Priority Program 1505 "Mediatized Worlds."

References

Adams, Paul. C., and André Jansson. 2012. "Communication Geography: A Bridge Between Disciplines." *Communication Theory* 3: 299–318.

Bachmann, Götz, and Andreas Wittel. 2006. "Medienethnografie." In *Qualitative Methoden Der Medienforschung*, edited by Ruth Ayaß, and Jörg Bergmann, 183–219. Reinbeck b. Hamburg: Rowohlt.

Baxter, Leslie A., and Earl R. Babbie. 2004. *The Basics of Communication Research*. Toronto: Thomson Wadsworth.

Baym, Nancy K. 2015. *Personal Connections in the Digital Age. Second Edition*. Cambridge: Polity.

Bott, Elizabeth. 1957. *Family and Social Network*. London: Tavistock.

Bruns, Axel, and Stefan Stieglitz. 2013. "Towards More Systematic Twitter Analysis: Metrics for Tweeting Activities." *International Journal of Social Research Methodology* 16 (2): 91–108.

Castells, Manuel. 1996. *The Rise of the Network Society: The Information Age: Economy, Society and Culture. Vol. 1*. Oxford: Blackwell.

Castells, Manuel. 2001. *The Internet Galaxy: Reflections on the Internet, Business, and Society*. Oxford: Oxford University Press.

Couldry, Nick. 2008. "Actor Network Theory and Media: Do They Connect and on What Terms?" In *Connectivity, Networks and Flows: Key Concepts for Contemporary Communications*, edited by Andreas Hepp, Friedrich Krotz, Shaun Moores, and Carsten Winter, 93–110. Cresskill: Hampton Press.

Couldry, Nick, and Andreas Hepp. 2017. *The Mediated Construction of Reality*. Cambridge: Polity Press.

Dijk, Jan van. 1999. *The Network Society*. London: Sage.

Freeman, Linton C. 2000. "Visualizing Social Networks." *Journal of Social Structure* 1 (1): 1–30.

Gambarato, Renira R., Geane C. Alzamora, and Lorena P. T. Tárcia. 2016. "Russian News Coverage of the 2014 Sochi Winter Olympic Games: A Transmedia Analysis." *International Journal of Communication* 10: 1446–1469.

Glaser, Barney G., and Anselm L. Strauss. 1999. *Discovery of Grounded Theory: Strategies for Qualitative Research*. New Brunswick: Aldine Transaction.

Granovetter, Mark. 1983. "The Strength of Weak Ties: A Network Theory Revisited." *Sociological Theory* 1: 203–233.

Hepp, Andreas, Matthias Berg, and Cindy Roitsch. 2014. "Mediatized Worlds of Communitization: Young People as Localists, Centrists, Multi-Localists and Pluralists." In *Mediatized Worlds: Culture and Society in a Media Age*, edited by Andreas Hepp, and Friedrich Krotz, 174–203. London: Palgrave.

Hepp, Andreas, Cindy Roitsch, and Matthias Berg. 2016. "Investigating Communication Networks Contextually: Qualitative Network Analysis as Cross-Media Research." *MedieKultur* 32 (60): 87–106.

Hepp, Andreas, Cigdem Bozdag, and Laura Suna. 2012. "Mediatized Migrants: Media Cultures and Communicative Networking in the Diaspora." In *Migrations, Diaspora, and Information Technology in Global Societies*, edited by Leopoldina Fortunati, Raul Pertierra, and Jane Vincent, 172–188. London: Routledge.

Hepp, Andreas, Friedrich Krotz, Shaun Moores, and Carsten Winter. 2008. "Connectivity, Network and Flow." In *Connectivity, Networks and Flows: Conceptualizing Contemporary Communications*, edited by Andreas Hepp, Friedrich Krotz, Shaun Moores, and Carsten Winter, 1–12. Cresskill: Hampton Press.

Hollstein, Betina. 2014. "Mixed Methods Social Networks Research: An Introduction." In *Mixed Methods Social Networks Research: Design and Applications*, edited by Silvia Dominguez and Betina Hollstein, 3–35. Cambridge: Cambridge University Press.

Hutchby, Ian. 2001. "Technologies, Texts and Affordances." *Sociology* 35 (2): 441–456.

Jenkins, Henry. 2006. *Convergence Culture: Where Old and New Media Collide*. New York: New York University Press.

Katz, Elihu, and Paul F. Lazarsfeld. 1955. *Personal Influence: The Part Played by People in Mass Communication*. New York: Free Press.

Latour, Bruno. 2007. *Reassembling the Social: An Introduction to Actor-Network-Theory*. Oxford: Oxford University Press.

Lima, Manuel. 2013. *Visual Complexity: Mapping Patterns of Information*. Princeton: Princeton Architectural Press.

Löblich, Maria, and Senta Pfaff-Rüdiger. 2011. "Network Analysis: A Qualitative Approach to Empirical Studies on Communication Policy." *International Communication Gazette* 73 (7): 630–648.

Madianou, Mirca. 2014. "Polymedia Communication and Mediatized Migration: An Ethnographic Approach." In *Mediatization of Communication*, edited by Knut Lundby, 323–348. Berlin and New York: de Gruyter.

Marcus, George E., and Erkan Saka. 2006. "Assemblage." *Theory, Culture & Society* 23 (2–3): 101–106.

Morley, David. 1992. *Television Audiences and Cultural Studies*. London and New York: Routledge.

Radcliffe-Brown, Alfred R. 1940. "On Social Structure." *Journal of the Royal Anthropological Society of Great Britain and Ireland* 70: 1–12.

Rainie, Harrison, and Barry Wellman. 2012. *Networked: The New Social Operating System*. Cambridge, MA: MIT Press.

Rogers, Everett M., and Lawrence D. Kincaid. 1981. *Communication Networks. Towards a New Paradigm of Research*. New York: Free Press.

Rogers, Richard. 2010. "Mapping Public Web Space With the Issuecrawler." In *Digital Cognitive Technologies: Epistemology and the Knowledge Society*, edited by Bernard Reber and Claire Broussard, 89–99. London: ISTE.

Scott, John. 2013. *Social Network Analysis*. 3rd ed. London, Thousand Oaks, and New Delhi: Sage.

Wellman, Barry, Anabel Quan-Haase, Jeffrey Boase, Wenhong Chen, Keith Hampton, Isabel Díaz, and Kakuko Miyata. 2003. "The Social Affordances of the Internet for Networked Individualism." *Journal of Computer-Mediated Communication* 8 (3).

Würfel, Maren. 2014. "Transmedia Appropriation and Socialization Processes Among German Adolescents." *International Journal of Communication* 8: 2240–2258.

50

A METRICS MODEL FOR MEASURING TRANSMEDIA ENGAGEMENT

Eefje Op den Buysch and Hille van der Kaa

How do you measure transmedia? What types of metrics will help transmedia producers to better understand and subsequently compare and contrast the impact of a transmedia story? These are million-dollar questions begging for answers. In some cases, even literally: when transmedia productions are being deployed, the stakes are often high to gain the interest of audiences, and thereby their spending.

Measuring engagement is important because it contextualizes what an audience size means and how a transmedia production is performing. It helps producers and storytellers to see what actually gets fans to engage. Measuring engagement, however, can be desirable from different perspectives. A producer or marketer will be interested in metrics to evaluate the campaign in terms of return on investment. A storyteller specifically wants to know how the audience responds to the narrative.

We believe that measuring engagement means placing audience size into the broader context of how the transmedia production as a whole is actually performing. Stakeholders in the production can see where, how, and when fans engage, which allows them to make refinements. Transmedia metrics provide transmedia storytellers tools to design, create, and execute productions so that they have impact.

In 2014, we presented a transmedia metrics model that can be integrated into the daily activities of a transmedia storyteller. Inspired by current developments in marketing, we take this model one step further and make it applicable for use in the age of extreme personalization.

One Step Back: Developing a Transmedia Metrics Model

Working with students in our various transmedia storytelling courses inspired our research on transmedia metrics. Our students learn to design, develop, and execute transmedia productions. In this process, the students experience all facets of transmedia production. We discovered that the production teams needed meaningful measuring tools to determine whether their campaigns performed as expected. They needed insights on audience behavior to give direction to their campaign and tap into the user need. To learn more about what and how to measure, we started with an exploration of existing insights and models of engagement. We made an analysis of possible missing elements and, subsequently, we proposed a new model.

Overview on Engagement Models and Insights

In their work, Attfield, Kazai, and Lalmas (2011) define user engagement from a broad perspective. Their definition reads: "User engagement is the emotional, cognitive and behavioral connection that exists, at any point in time and possibly over time, between a user and a resource." This approach emphasizes the emotional, cognitive, and behavioral factors that, overall, define user engagement. All three factors are open to measurement.

According to Forrester Research (2007), emotional engagement attempts to measure the affection or aversion of a user. Intimacy is considered an emotional engagement factor and is measured by satisfaction rating or by sentiment analysis in blogs, comments, surveys, and questionnaires. Cognitive engagement can be measured through awareness, interest, and intention. It is important to measure the likelihood of a user advocating, for instance by forwarding content or inviting friends. Besides measuring influence, Forrester Research stresses involvement and interaction. This "behavioral" engagement focuses on the presence and action of the user, which can be measured by the number of visitors or the amount of time spent on a website. Additionally, one could look at the click-through ratio, the number of online transactions, or the number of uploaded videos. In transmedia productions, on- and offline media go hand in hand. Measuring behavior or sentiment is not limited to online media. The use of offline media can be measured just as well, as long as an appropriate method is used. Methods like questionnaires, observations, or interviews can serve this purpose, for instance. For transmedia storytellers it is also customary to connect the physical world with the digital world through text messaging (Conducttr 2016), mobile applications, QR-codes, sensors, RFID-chips (Radio-frequency identification) and location-based services for making interactions with physical objects and locations digitally measurable.

For Robert Bole, director of Broadcasting Board of Governors in the United Stated, it is clear that creators of "the new narratives" need analytics. He notes:

> Rather than the old way of 'more is better' as a proxy for quality (more audience = bigger success), the digital world allows storytellers to hone their craft through nearly real-time feedback across multiple channels that can result in quick shifts, additions and changes to narratives depending on pockets of audience behavior and interests.
>
> *(Bole 2013)*

In his talk at the DC Digital Analytics Association Symposium in 2013, Bole defined six points of interest that need to be addressed to help understand engagement. The metrics and analytic tools that help understand the impact of stories that unfold across multiple platforms are, according to Bole: (1) narrative attribution, (2) conversion and pathways, (3) segmentation format feedback, (4) quantifiable impacts, (5) engaged loyalty, and (6) audience-created enhancement. By measuring the narrative attribution, a storyteller can identify which story element (or media mix) converted the audience member into an engaged, multi-channel user. Measuring the conversion and pathways helps determine "optimal" paths for constructing narrative arcs. Segmentation format feedback measures the success of the format in engaging and propelling a narrative arc by audience type. Quantifiable impact is needed to measure audience participation across the narrative arc. Engaged loyalty means finding out if additional context and information leads to more informed, loyal, and frequent users. Audience-created enhancement involves measuring how audiences add value through contributions to the narrative arc.

Although we found several works on the needs of a storyteller, we found less information about transforming those needs into practical, actionable insights. We did, however, find a variety of models on the desired outcome of transmedia analysis. Mayfield's (2006) "Power Law of Participation model" is such an example. The model shows 12 stages of participation, ranging from the rather inactive "reader" to the "lead," a highly engaged and active audience member.

Mayfield's work relates to the "roles" described by Jenkins, Green, and Ford (2013). There are at least two significant roles that users can play to push a campaign forward. Jenkins, Ford, and Green mention two of those roles: lurking and spoiling. Lurking could be defined as a stage where users look at the content produced but are not actually interacting with it or sharing it. Lurking plays an important part in the evaluation phase. "Lurkers" are those who are interested but not ready (or they will never be ready) to start interacting with the transmedia campaign. Another important role is the spoiler. "Spoilers" are a very active group of people that seek to know everything that can be known about a campaign. Often, they group together on forums or Facebook groups, out of sight of the campaign creators. They also create blogs related to a campaign.

Pratten (2011) defines five levels of increasing engagement: (1) attention, (2) evaluation, (3) affection, (4) advocacy, and (5) contribution. In his work, Pratten proposes measurements aiming specifically at measuring engagement in a transmedia world. His five levels of increasing engagement expand Forrester Research's earlier-mentioned measures of engagement, by adding "affection," meaning "affinity towards the world." Having this kind of information about the segmentation of the participants in a transmedia production is very useful for storytellers. They will be able to measure and to adjust the story accordingly, based on these insights.

Engagement Models: What is Missing?

Our students found various models and methods to measure engagement (not all of which were mentioned here). A range of models is available to either classify levels of engagement or to create typologies of "the engaged." We need to fill the gap between the existing models to find a way of gathering meaningful insights that can be used to give direction to a transmedia production in a practical way.

Based on the research performed by our students on existing models, we want to make sure that our new transmedia metrics model:

- measures engagement and classifies the user;
- transforms "raw" data into meaningful insights;
- tracks the user journey;
- tracks the connection between different channels (on- and offline).

The Need for Actionable Insights: Introducing a New Transmedia Metrics Model

When students learn to develop transmedia productions, they start off with the notion that three aspects need attention throughout the development process: (1) there is a storyworld that is to be discovered; (2) there is an audience that is involved; and (3) there is a period in which the story unfolds. These three principles form the basis of the new transmedia metrics model.

Audience members who interact with the world are considered to be engaged users. By tracking the behavior of individuals, we can map how they discover the world and how they interact with it over time (see Figure 50.1). We record each time a user "touches" something in the storyworld. By listing to all of these *points of interaction* and structuring them into chapters, scenes, and beats, we can track the journey and the hotspots of engagement for individual users as they progress through the story.

Points of interaction are predefined by the creator (the storyteller) of a transmedia production in the form of acts, scenes, plot points, beats, and so forth. An audience member, referred to as a user, progresses in the storyworld by discovering these scenes and beats, interacting with characters, solving puzzles, sharing content with friends, or viewing content on specific channels like YouTube or blogs. By tracking this individual user journey through the storyworld, the storyteller receives insights into

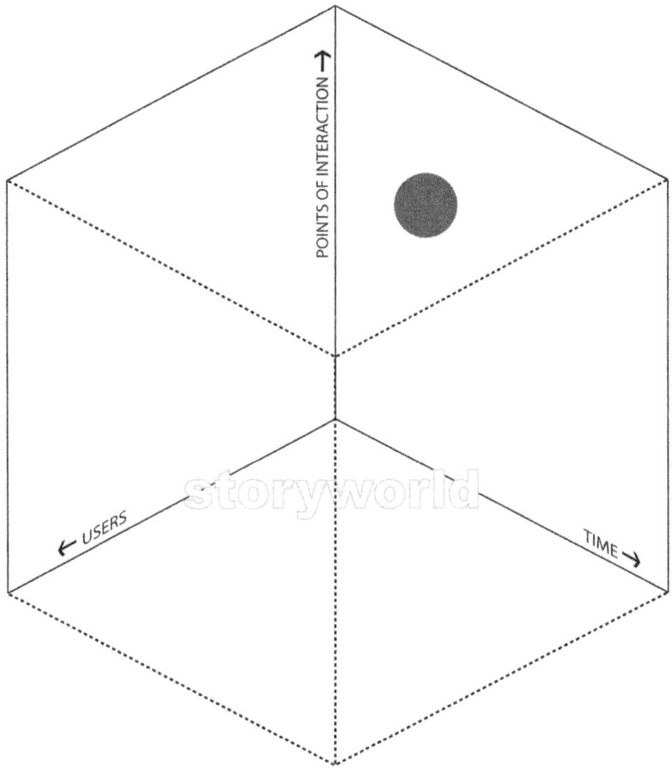

Figure 50.1 The new transmedia metrics model: users interact with the storyworld over a period.

the behavior of that individual. With this information, the storyteller is able to analyze whether the transmedia production is performing the way it should. Is all the content discovered? Does the user progress? Does the user need help? Now, actionable insights are at the storyteller's disposal.

Interactions are tracked using the concept of "toggle switches." If a user interacts with a certain point, the toggle switch changes from OFF to ON. Picture taking a tour through a house and switching on the light when entering a room, leaving a trail of lights in all the rooms visited.

The key benefit of our conceptual model is that the behavior of an individual audience member can be compared to the behavior of others. In doing so, we can interpret the relative engagement of individuals compared to some or all others at certain points of interaction as a ratio. Tracking all the behaviors of all individuals allows us to group them on different levels.

Examples of relative engagement measurement:

- Compare the amount of interactions of User A in [Scene 5] to all interactions in [Scene 5].
- Show the progress of User B compared to the most progressed user.
- List the number of users that interacted with [Scene 7, Beat 2] compared to all users.

The gathered data about the user's journey through the world (points of interest touched by a user at specific moments in time) is combined with that of other users. This also opens the possibility for ways of segmentation and classification that the reviewed models offer. Segmentation is based on actual data of individual users, which is an asset over previous models. Now, users can be classified as "lurkers," "contributors," or "cognitively engaged" relative to all the users in the storyworld.

Expert Evaluation of Our Transmedia Metrics Model

We asked leading experts to evaluate the clarity, completeness, effectivity, applicability, and benefit of our model. Among them are Sam Ford (Peppercomm, New York, US), Pamela Rutledge (Media Psychology Research Center, Boston, US), Bart Robben (Elastique, Hilversum, Netherlands), Egbert van Wyngaarden (Transmedia Desk, Munich, Germany), and Soraia Ferreira (University of Texas at Austin, Porto, Portugal).

Since we focus on content engagement (story) rather than on return on investment (ROI) in the development of our conceptual model, we first wanted to know what respondents sought in a new transmedia metrics model for storytellers. As a cross-section of the various models, we selected: (1) loyalty measurement, (2) participation ratio, (3) target characteristics, (4) media channel effectiveness, (5) content effectiveness, and also (6) ROI measurement as conventional metrics used in defining the success of transmedia productions. Therefore, we asked our respondents to rank the importance of these metrics.

Our respondents chose content effectiveness as the most important focus of a storyteller's measurement model. The second most important was the measurement of ROI. Loyalty measurement, participation ratio, target characteristics, and media channel effectiveness seemed to be less important. These results are in line with our focus on the storyteller, but they show the relative importance of ROI as well.

We also asked the respondents to evaluate our model. First, we asked them to rate the following items on a scale from 1 to 10: applicability, effectiveness, completeness, and clarity.

Next, we asked our respondents to explain this numerical evaluation using semi-structured interviews. In general, we found that our respondents were positive about the applicability and clarity of the model. "Most models process the use, engagement or participation of one medium, but this model combines all different media and their results" (Welhuis). "It measures the journey, and also the overall journeys" (Ferreira). "The model would be helpful for producers who are interested in understanding how each element of their story was successful, compared to the others, or to find out in what way each audience member wants to travel through their story and engage with some parts of the story" (Ford). In addition, they liked the focus on the ability to measure user journeys. At the same time, they were skeptical about the effectiveness and completeness of our model. For example, Vegt and others mentioned the missing ability to respond to something unexpected. Related to the completeness of the model, our respondents highlighted missed opportunities like no focus on the financial side of transmedia measurement (mentioned by Tan) and no monitoring of the soft side, e.g., emotions (mentioned by Wijngaard and others).

The Transmedia Metrics Model Put into Practice

Based on the feedback of the experts, we think the model could be put into practice. Looking at today's developments in the context of our model we can find that the notion of measuring (all) points of interaction has entered the world of consumer analytics in a big way. In 2017, Gartner states that Customer Journey Analytics (CJA) can bring "business-critical insights to understand the customer journey and improve customer experiences" (Gartner 2017). Evaluating the transmedia experts' opinions of our research, we learn that they too would like to measure everything, even "the unexpected." This follows recent developments that show a need for measuring the customer journey to offer customers (extremely) personalized experiences. Comparable to Gartner's claim, Walker Information (Walker 2017), a B-to-B customer experience consulting firm, shares its insights on future consumer behavior in its report *Customers 2020*. Walker Information distills three elements that will influence companies in the upcoming years: (1) personalization of experience, (2) anticipation on customers' future needs, and (3) multichannel savviness. Walker Information finds the key future metric to be engagement. Gartner (2017) analyzed and reviewed a number of tools in their "Market Guide for Customer Journey Analytics." Transmedia storytellers benefit from the findings and can consider incorporating mentioned tools in their production.

Transmedia storytellers have frequently placed the user journey at the heart of creating a transmedia production. In addition, we like to think that, in the field of transmedia storytelling, producers are focused on customer need from the start. Transmedia storytellers have a competitive advantage in comparison to their marketer counterparts. Transmedia minded professionals master the art of engaging audiences with stories next to putting to practice their knowledge of telling stories "systematically across multiple delivery channels" (Jenkins 2007).

Designing Stories that Engage

A point of discussion when working with students often is whether transmedia storytelling is a marketing "tool" for big production companies or a way for creatives like filmmakers, writers, documentary makers, for instance, to express their views in a different format than before. When showing examples of successful transmedia productions, the blockbuster introductory campaigns often come to mind first ("Why So Serious?", *Westworld*, *Game of Thrones*). This connects to the ongoing discourse on East Coast and West Coast schools of transmedia that Andrea Phillips (2012) codified several years ago.

The distinction Philips made was not meant to become a dichotomy, she explains in an interview with Henry Jenkins (2012) on his blog: "I find it extremely frustrating that a perception exists that big franchise-style intertextual works and very personalized independent works can't both be transmedia at the same time." In our view, each transmedia production, big or small, marketing campaign or stand-alone production, can gain from a "transmedia metrics way of thinking." We illustrate the use of our model by a reasonably straightforward transmedia production, created by students.

Case Study: The Curse of the Diamonds

What started with a photo from a collection at the regional heritage center in the Dutch city Tilburg, developed into a story called The Curse of the Diamonds. The Curse of the Diamonds is a transmedia storytelling production developed by students of the Fontys Transmedia Storytelling Lab at the Academy for Creative Industries (Tilburg, the Netherlands). The student team consisted of: Javi Bus, Brenda Pattipleohy, Massimo Carolus Adrianus America, Tamara Scholten, Helena Comninos, Lisette Vlassak, and Maria Herholdt Engermann. This transmedia project tells the story of refugees during World War I in Belgium and the Netherlands. The group of students approached the heritage center to ask for their approval to combine facts and fiction into a transmedia story. It is a story about doctor Otten and his three daughters who fled from Belgium to Tilburg.

The main fictional character in the transmedia production is the graduate student Amy Heerkens. Based on stories told by her grandmother about refugees in World War I, she is creating a documentary together with her classmate and cameraman Rob. Amy has a blog, Twitter and Facebook accounts, and a YouTube channel. Amy uses these channels to tell the information she discovers about the Otten family that lived at her grandmother Heerkens' house during the war. There appear to be rumors about curses and disappearances, so Amy embarks on an exciting investigation.

The audience could progress in the story through several videos, posts, puzzles, and interactive 360-degree images. At the end, Amy called for help from the audience by texting the address to a location in the city. In a spooky hall near the railroad tracks, audience members could step into a virtual reality (VR) experience by putting on goggles to find out the story ending.

Looking at this production from the point of view of our transmedia metrics model, we can distill:

- A storyworld that is to be discovered: the story about Amy in search of the curse of the Otten family.
- An audience that is involved: the people that follow Amy on her journey.
- The period: the timeframe between the first encounter with Amy and the end-event with VR experience.

These three elements feed the design of the user journey, a diagram of channels and story elements plotted on a timeline. The story of Amy Heerkens starts on Facebook. She posts a blog about the situation of refugees in the world nowadays. Several calls-to-action take users from Facebook to other channels and thereby to subplots in the storyworld: her friend Rob's page, a link to a survey, national news outlets, etc.

Measuring the Performance

From a transmedia storyteller's standpoint, it is crucial to make sure that every plot point involves some kind of interaction. This is the way to record how audience members evolve through the transmedia production. Here we have the key to our transmedia metrics model: every step in the story should be accompanied by a designed interaction. Why? Because it is a way to listen to your audience. More specifically: your individual audience member. In transmedia storytelling, the design of interaction is just as important as the design of a strong story. A point of interaction means a behavior that can be recorded, for instance, a click, a post, an email. It starts with audience members making themselves known to you as the producer, and have them interact with the story through (measurable!) points of interaction. This takes effort, but remember that audiences *want* to be engaged. They want to talk to you. They want to be involved. Transmedia forefather Jeff Gomez describes this phenomenon as follows:

> When we encounter a narrative, whatever the form, if it appeals in specific ways to our most powerful emotions, our fundamental desires, the values ingrained in us as children by our families, we respond. We like it, because it feels right. We love it, because it reinforces our sense of rightness. We become loyal to it. We're not going to leave it.
>
> *(Gomez 2017)*

Gomez continues describing how fan bases are formed, brands are built, and movements grow: people want to share their emotional connection to a story with others. In this sense, now that we have this arsenal of tools and channels, storytellers can put them to their advantage by crafting them into ways of listening to audiences. When following Gomez' line of thinking, audiences give back response to what is provided. In a constant flow of listening and responding, stories become engaging and meaningful. Thinking of points of interaction as a way of sharing, instead of merely recording then, a transmedia production can be brought to a higher level.

When the team of students created The Curse of the Diamonds, they measured the performance of their production by keeping track of all views, clicks, and likes in a dashboard. By tracking how many audience members progressed to the first video by Amy Heerkens by clicking the link on her Facebook page, the team got a rough idea about the effectiveness of this designed interaction. Compare this to a sales funnel for online stores: for every 1,000 people that visit a web shop, only a fraction will place a product in the shopping basket, and even less will actually buy a product. Thus, this metric merely provided the team with insights on the number of people that progressed, not on who those people were, and certainly not on overall engagement.

Conclusion

Looking at transmedia metrics as a way to listen and give back to audiences makes creating transmedia productions increasingly exciting. As an example, in our classroom, students learn by following Robert Pratten's (2015) guide "Getting Started in Transmedia Storytelling." In designing transmedia productions, Pratten uses a method of beats (within scenes) to connect the production's narrative layer (story) to the interactive layer (actions) and the presentation layer (experience). Since in transmedia storytelling time is not the only way to progress in a story, beats with triggers make it possible for the audience to advance in a transmedia production.

Imagine how the team of students would incorporate the transmedia metrics model together with the insight that analytics can be a way of listening to audiences. All elements of the model (storyworld, timeframe, user) would dictate the building blocks of the production: scene, beat, trigger, condition, and action. In every scene there should be a story-element to be discovered and it would have to be triggered by an action. In other words: de user would need to do something to trigger progression to the next step.

In the online story, an audience member could want to have a look at some files stored in the heritage center. Often in transmedia production a user would have to find a code of some sort to get access to a room. The scene would look something like that illustrated in Table 50.1.

Through incorporating the transmedia metrics model, the triggers would be "recorded" in a database (see Table 50.2).

With recent technological developments like chatbots, voice-controlled input, or face recognition, points of interaction can get more and more seamless and unnoted. Introducing a chatbot as a character having a conversation with an audience member can be an applicable way of both gaining knowledge about the person and, at the same time, provide a way to the storyteller to disclose new story elements. Upcoming techniques make it possible to tailor the disclosure of story elements to a single, unique user. A trigger can be conditional to the recognition of a face, a touch identification (ID), a biosensor, and so forth. Opening a door at the heritage center could then be triggered by a face recognition application. Not only would this open the door for the user, but it would give the transmedia production team the opportunity to store this face scan and use the information it provides later on in the story. In this example: when users want to talk to Amy Heerkens' friend Rob, the character can then say he does not want to talk to users he has not seen before.

From a story design point of view, storytellers can now make smart use of all the personal information they gather from their users. In the field of marketing, personal information feeds profiled marketing communication. Think of Facebook using a person's likes or location to show advertising on a timeline. In a transmedia production—given the precondition that audience members choose

Table 50.1 Schematic view of a scene, beat, trigger, and action

Scene	*Beat*	*Trigger*	*(Condition)*	*Action*
Visiting the heritage center to look at files of the Otten family	Enter the heritage center	Click on the door		Door opens
	Open a file cabinet	Enter code	Right code	Cabinet opens
			Wrong code	Error message

Table 50.2 Schematic view of database entry for user Robin

User name	*User email*	*Scene*	*Beat*	*Trigger*	*Condition*	*Action*	*Time*
Robin	robin@gmail.com	Visiting the heritage center to look at files of the Otten family	Open a file cabinet	Enter code	Right code	Cabinet opens	11:00 pm, 12-10-17

to step into the storyworld by signing up—users can be offered personalized storylines by incorporating what they have told about themselves. Transmedia productions could even go as far as to use the information for personalized branding. But this, of course, depends on the production team's standpoint, since profiling of consumers within the fields of marketing and communication is becoming an increasing point of discussion in society. Transmedia storytellers and producers might want to make sure that they are aware of ethical consequences when developing projects. Being open about collecting data, and providing the necessary information on security, forms the basis from which users can make a choice to step into your designed experience. As we have addressed in this chapter: users want to be engaged, they want to talk to you (Gomez 2017), but this will only work if that conversation is built on (mutual) trust.

References

Attfield, Simon, Gabriella Kazai, and Mounia Lalmas. 2011. "Towards a Science of User Engagement." WSDM Workshop on User Modelling for Web Applications, Hong Kong, February 9–12. Accessed May 14, 2017. www.dcs.gla.ac.uk/~mounia/Papers/engagement.pdf.

Bole, Rob. 2013. "Going Old School: Telling Digital Stories in Analog Media." Accessed May 10, 2017. www.innovation-series.com/2013/06/14/going-old-school-telling-digital-stories-in-analog-media/.

Conducttr. 2016. "Location-based Manga Quest." Accessed May 19, 2017. www.conducttr.com/success-stories/location-based-manga-quests/.

Forrester Research. 2007. "Marketing's New Key Metric: Engagement." Accessed May 8, 2017. www.forrester.com/report/Marketings+New+Key+Metric+Engagement/-/E-RES42124.

Gartner. 2017. "Market Guide for Customer Journey Analytics." August 30. Accessed November 12, 2017. www.gartner.com/doc/3794964/market-guide-customer-journey-analytics.

Gomez, Jeff. 2017. "The Collective Journey Series: Regenerative Listening." *Collective Journey*, March 8. Accessed May 11, 2017. https://blog.collectivejourney.com/the-secret-to-new-storytelling-regenerative-listening-5250c65b6391.

Jenkins, Henry. 2007. "Transmedia 101." *Confessions of an Aca-Fan*, March 21. Accessed May 14, 2017. http://henryjenkins.org/blog/2007/03/transmedia_storytelling_101.html.

Jenkins, Henry. 2012. "Creating Transmedia: An Interview with Andrea Phillips (Part One)." *Confessions of an Aca-Fan*, November 1. Accessed May 12, 2017. http://henryjenkins.org/blog/2012/11/creating-transmedia-an-interview-with-andrea-phillips-part-one.html.

Jenkins, Henry, Joshua B. Green, and Sam Ford. 2013. *Spreadable Media: Creating Value and Meaning in a Networked Culture*. New York: New York University Press.

Mayfield, Ross. 2006. "Power Law of Participation." *Ross Mayfield's Weblog*, April 27. Accessed May 3, 2017. http://ross.typepad.com/blog/2006/04/power_law_of_pa.html.

Phillips, Andrea. 2012. *A Creator's Guide to Transmedia Storytelling: How to Captivate and Engage Audiences across Multiple Platforms*. New York: McGraw-Hill.

Pratten, Robert. 2011. *Getting Started with Transmedia Storytelling*. London: CreateSpace Independent Publishing Platform.

Pratten, Robert. 2015. *Getting Started in Transmedia Storytelling, 2nd Edition – A Practical Guide for Beginners*. London: CreateSpace Independent Publishing Platform.

Walker. 2017. "Customers 2020: The Future of B-to-B Customer Experience." Accessed May 15, 2017. www.walkerinfo.com/knowledge-center/featured-research-reports/customers2020-1.

AFTERWORD: THE PRESENT AND FUTURE OF TRANSMEDIA PRACTICES—A CONVERSATION

Alison Norrington, Kate Pullinger, Nataly Rios Gioco, and Kate Fitzpatrick

With the intent of both echoing and elaborating on the themes, sentiments, and arguments explored throughout this entire volume, this final chapter—a conversational interview between the editors and four transmedia practitioners, each based in a different media sector in the United Kingdom—uses as its jumping-off point the idea that transmediality continues to mean different things to different people in different ways, albeit with some overarching conceptual overlaps that continue to make the term useful. As has been articulated throughout this book, the creative strategies employed to produce content across multiple media platforms are largely informed and characterized by the contexts in which they operate. This Afterword, then, while reiterating this same importance of contexts of specificity, will build on the work of the book's Introduction and all of its subsequent chapters by further teasing out the overarching industries, arts, practices, and cultures of what transmedia really is and where it operates today via the views of four well-placed practitioners working across four creative industries.

Revisiting our initial question of whether transmediality should—or can—be conceptualized as one clearly defined phenomenon, or whether it is instead a series of separate, albeit related, phenomena, this Afterword brings together four creative voices: one based in digital writing, one who works primarily in television, one in digital marketing, and one in online gaming. The following interviews are with a digital author as well as with creative personnel from StoryCentral Ltd., Great State, and Conducttr. Specifically, the interviewees comprise: Kate Pullinger, writer of multimedia interactive novels; Alison Norrington, Creative Director at StoryCentral Ltd.; Kate Fitzpatrick, Senior Strategist at Great State; and Nataly Rios Goico, Creative Consultant at Conducttr. Kate Pullinger has won numerous prizes and accolades for her novels, short stories, and digital fiction. Her most recent work, *Breathe*, is a ghost story that is personalized for every reader created in collaboration with publisher Visual Editions and Google Creative Lab Sydney. In 2017 she published *Jellybone*, a media-rich novel for smartphones. StoryCentral Ltd. is a London-based transmedia consultancy firm that has worked globally with such companies as Walt Disney Imagineering, SundanceTV, AMC Networks, and FOX International to incubate new media franchises, participatory experiences, story architecture, and audience engagement strategies—notably with a focus on television-based transmedia stories. Great State is one of the UK's most awarded independent digital marketing agencies, helping its partners to develop a great brand, make it real through a great experience, and deliver it at scale through great technology. And Conducttr, led by its London headquarters, is a world-renowned audience engagement platform that integrates storytelling with gaming automation to build immersive, personalized gaming experiences. All four of these figures and companies are working at the cutting-edge of their respective sectors, lending distinctive insights into the meaning of transmediality across digital writing, television, digital marketing, and gaming.

Editors: How did you come to discover transmedia?

Alison Norrington (AN):

"I came to transmedia through literature, producing 'chick-lit' kind of books back in 2003–2005. And because they were chick-lit books, which was quite a disrespected genre, I tried with each of my books to add some culture. In every one of my books, I had a European city, I had an art form, and I had a form of music that ran through each of the books. And I wished even back then that I could have let my readers see the impressionist paintings or hear the classical music. But it was 2003 and the only way that would happen was through a CD in the back of the book, and the publisher was not going to pay for that. So I was looking at ways of adding a richness to my storyworlds that went beyond just one platform."

Kate Pullinger (KP):

"I was looking at emerging platforms, writers, fragmented interaction and pervasive media—which of course comes under the 'transmedia' title."

Editors: Broadly, what does "transmedia" mean in each of your industries?

KP:

"Speaking as an author, it is much more of a cottage industry for most writing projects, compared to Hollywood projects, like a Marvel film or something like that. There are lots of interesting examples of transmedia writing that have print and digital iterations; they might have theater iterations linked to them for example. That is quite common now, but often in an indie kind of way. And this is because the publishing industry simply has not recognized transmedia, really, as a revenue stream: they are focused on the book as the primary format."

Natalie Rios Gioco (NRG):

"For me, at Conducttr, our model is to make everyone's life an adventure, and the way we do this is by creating immersive experiences, specifically by combing powerful storytelling with the digital and the physical world. And stories are important because stories are the way that we understand, remember and experience the world. So what we do is use our technology at Conducttr to transform the world into a storytelling and gaming canvas. For me, there are three key aspects of transmedia: engagement, blended digital and physical worlds, and experiential learning. With the first of these, transmedia allows you to engage audiences, to make connections, to create an emotion with the audience. With companies, it can create differentiation, and with brands, and it can increase the brand sentiment. For the second aspect, a blend of the physical and digital might spur ideas of wearable technologies, or the Internet of things, but it is not really that in this case. Transmedia allows you to create digital, interactive experiences that enable discussion, that allow you to connect to other people in real physical spaces, and to connect those spaces as well. And the other key aspect is experiential learning, or learning by experiencing. But this does not mean necessarily educational experiences. Rather, even if it is a piece of art, it is about the way that we want to deliver a message, some content, some knowledge, in some way—and transmedia allows you to do that by making the audience to be active instead of passive."

Kate Fitzpatrick (KF):

"Okay, I am going to start with the 'marketing' bit of this question, rather than the 'transmedia' bit. Most of us know that essentially the basic activity of marketing revolves around your customer, your buyer, your audience, your user. And it is about satisfying their needs with that product, service or experience. And marketing is about trying to get someone (i.e., the consumer) to do something, in our case via digital channels, for a business benefit. And that is where the commercial side comes in, for more often than not there is some line to revenue down the road. What we find is that you either get the direct correlation, which is the purchasing of a product, or the indirect, which might be someone signing up to start the journey of becoming a giver for a charity, for example. So they start off with a one-off donation but we actually want to turn them into someone who donates on a regular basis. And this is where the transmedia element comes in for us. We see the application of transmedia as an effective way to get our consumer to do something, which might well be purchasing a product, signing up for a newsletter, requesting more information, applying for a job, and so on. Today, in the context of digital marketing, the concept of transmedia itself means creating a journey or experience that uses the most relevant mix of channels and platforms for your intended audience."

AN:

"My approach to transmedia, from a story perspective, is: I want to define the big idea; I want to know the theme of the story. Because, for me, the theme of the story—the big idea—ties into your audience and the experience you want to deliver. These three things go hand in hand. For me, the experience design and the emotional design of any project is informed by the theme. And I feel that is what is lacking in many attempts at transmedia storytelling. There is often no reason, or why. It is done as marketing, there is no heart and soul behind it, there is no emotion. What is the theme? What is the message? Why are you telling this story? What do you want to evoke? That said, there are ad campaigns that have killed the term 'transmedia.' When I work in LA or New York, there are people who reject that term, mainly because the buzzword-iness of it has been so overused that it is now almost as if it is a solution to a problem. 'Oh just transmediate it ...' We are not actually 'transmediating' anything. The word has been used so incorrectly and so often that it has done it a disservice as an approach. Because, importantly, transmedia is an approach to storytelling. There are advertising campaigns that have attempted to tell a story across platforms, and the audience has been interested and followed that story, but only to find a dead end. [In these cases], it is all about them wanting your money and them wanting you to buy their thing. We need to treat people the same way in the online environment as we do in real life; you would not gather a group of people in real life, get them to go on a journey with you, and then let them find out that it was only to serve you."

Editors: What, then, typifies transmedia as a practice across each of your industries?

KP:

"The indie narrative game sector is very interesting currently. It is incredibly story-led, with small companies that are one- or two-person led. People who are interested in games seem to be much more comfortable in the digital sphere than people who are interested in books. These games are often highly experimental but also accessible, in ways that experimental literature quite often is not. So it really depends on genre I think. There is the TV chef phenomenon in the world of writing, which supports a lot of publishing. There has always been books that support television series. But the ability to think about, say the novel, and its potential to be a transmedia event, is beyond the scope of publishing, a narrowly focused industry without much R&D [research and development]. However, all these industries are competing for attention; evidence shows that reading is declining, and publishing needs to do what it needs in order to survive."

AN

"Yes, genre and platform do come into this. Usually, though, clients come to me and say, 'well, we want to tell a story on Facebook, Twitter and YouTube.' And I say, 'well, okay, but hang on a minute …' Because nine times out of ten, yes, those platforms play a part in the story, but they may not be the primary ones. Always, platform is the first consideration, I find. Which is absolutely the wrong thing. For me, it is about themed storytelling: how can we pull the real theme of the story out? Because if you can pull the theme out, then you transcend looking at transmedia in terms of it being about divides between marketing, storytelling, communication, and so on. If you get the theme in there, then you tick all of those boxes as you go round, rather than putting it into separate segments. In five years' time, too, Facebook could be dead; Twitter could be dead. And my point is: understanding your audience, your experience, and your emotional design means essentially that you could walk into any commissioner's office and pitch your work, saying 'this is a story about this, my audience are these kinds of people who will behave in this way.' That way, it does not matter if Facebook goes down; there will always be another platform. Crucially, it is the behavior of audiences that underpins transmedia, not the actual technology of the platform."

NRG:

"For me, the practice of transmedia is about immersion, and allowing audiences to be immersed. Sometimes we think of immersion as filling the audience with everything, with a lot of events and all of the content across all of the platforms. But this only succeeds in blocking the view, as it were, and drowning the audience. Instead, we need to remember that immersion starts with emotion. We need to think about what emotions we want to create for the audience; that, and only that, will lead to immersion. Because I can then choose the platforms, choose the interactive components, and choose the narrative, that will create the desired emotion and, in turn, the desired immersion for that audience."

KF:

"What I am experiencing at the moment in digital marketing is that we simply talk about the practice of transmedia as developing experiences. We have all become channel-agnostic in a way; it does not really matter where [a given piece of content] sits; what matters is what the audience wants to be doing. The most successful digital marketing campaigns are those that understanding this, and get their channel mix just right so that audiences are interested in doing something. And transmedia is indeed a perfect way to get customers to do something. And yet audiences do not really think in terms of things being campaigns for this or that; they think about it in terms of an interaction with a brand or with a shop—they have gone online and just looked at it on their phone or laptop. The point is that audiences do not really consider all of these channels and platforms themselves, and as marketers we often lose sight of that."

Editors: All of you seem to be strongly highlighting the importance of audience in terms of defining the cultures of transmedia …

AN:

"Yes, there is a major misconception, I think, when clients talk about a 'build it and they will come' mantra, as if producing great transmedia content will naturally lead to big audiences. It is not about 'build it and they will come'; the correct mantra is 'go where they are': find out where the target audience is hanging out, and deliver your content to them directly."

NRG:

"Agreed, but transmedia is really about conversation, between storytellers and audiences, between audiences and other audiences, and between digital and virtual worlds. I see the function of Conducttr to be to listen to events, and those events then create a response. For example, something happens that triggers a response that may be in the form of sending a tweet, publishing a video, sending an email, making a bracelet, turning off and on a light switch, and so on. Transmedia becomes a system of cause and effect—a distribution of information (cause) that triggers an integrated, expansive response (effect)."

KF:

"Absolutely. Our entire approach to transmedia campaigns at Great State can really be boiled down to two characteristics—user led and experience driven. Your audience is always central to everything; if you do not know who your audience is then it becomes incredibly difficult to produce a successful transmedial work or campaign. If you understand their wants and needs, then you can develop an effective narrative that will engage them. By 'experience driven' I mean developing a frictionless, multilinear journey. The idea here is that audiences do not start at A and end at Z; instead, they go all over the place, conducting different tasks in different areas. People are using different devices in different areas to complete different tasks. And those cases you need your design and all your different places to be intuitive. Remembering that where one user's story begins might be where another's end can be quite important. Hence one of the main outputs of our user led/experience driven approach is what is called an experience map, which shows the user's journey through a particular element."

KP:

"Not to buck the trend here, but audience is not quite as important for me as an author. This is actually a small frustration, and one that is typical of the digital writing sector right now, I think. There are no mechanisms for obtaining this kind of review data, at least in the way that it would be in traditional book or film culture. With print books, you can track reviews and see whether those reviews are any good. As a digital writer, though, experimenting with form, it is not something I am all that interested in personally. I am more interested in the next project."

Editors: That leads us nicely to the art of transmedia. How would you each go about characterizing the creativity of this kind of multiplatform work?

NRG:

"One of the things about transmedia is that it allows you to associate virtual spaces and real spaces with a story. Everything about a project can be real—through transmedia we can expand the experience of the real world to the digital world, and vice versa. So it is about delivering information by experiencing, and that is the art of transmedia for me."

KP:

"One thing that I set out to do as a writer was to create a storyworld that exists across different media. That, for me, is a key part of what I do; even though it does not resemble a Hollywood-style transmedia project. For me, storyworld means identifying a clearly defined character that exists in a clear world around them. It is then about trying to figure out what that character could get their platform to do, and what they could not get their platform to do. You always start with a blue-sky approach, and then narrow the perimeters. I do think about participation carefully—for example, there are points in the story of *Jellybone* where the reader could create their own story within the story, which is an idea worth exploring further, namely because there are a lot of readers who are also writers."

AN:

"We are now at a point where storytelling is so sophisticated; as a storyteller, that is both really exciting and incredibly terrifying, because we have got a high bar to reach. But the bottom line is this: we have audiences at the forefront of modern storytelling; they are vocal and active on social media, so you cannot hide away. Storytellers need to understand relevant and credible processes for building storyworlds, to understand the fundamentals of thematic storytelling, and also the levels of immersion and experience design. With all that in mind, for me there are three things that circles story: we have the narrative design of the story, which is the pace, the flow, the rhythm, the genre, the way that it is written, and so on. This narrative design of a story is something that we already get. But two other things that I think are often ignored with story are, first, the experience design, so what rollercoaster do you want to take your audience on? Even in terms of if you are writing just a single-platform story, whether it is a book or a television series or whatever, what is the experience you want to deliver? And the third thing that circles story for me is emotional design: how do you want your audience to feel at the end? As storytellers we are puppet-masters of people's emotions, and we need them to be clear how we want them to feel at the end of our story."

Editors: What do you each see to be the main problems with transmedia right now?

KF:

"Well, it can be somewhat challenging, and that is partly because a narrative is not actually owned by one entity, whether that is a brand, an author, an agency, and so on. I myself have worked with brands who have a particular product—they own a product—but then you have a creative agency who have developed the idea to market that product. And then you have another agency who executes the media to then market that product. You might have someone else who is really great with virtual reality and is working on part of it over here, and someone else working on display advertising over there. So it becomes rather like a proverbial fight with everyone trying to get their piece into the puzzle, and I think what we are facing is quite a big challenge in terms of trying to get people to understand why we are doing what we are doing in the first place."

AN:

"Yes, I still feel that we have problems with silos in terms of how we make the teams come together across storytelling, marketing, and research and development, and social media, to really push these ideas through. Some of my experiences of transmedia in the UK is that many of the projects have fallen down because the structure of the companies are not equipped to roll out the project. They like the idea, but when it actually comes to the point of making, campaigning and stewarding it through, they do not have the structure."

Editors: And, finally, as you all see it, where next for transmedia?

KP:

"Next, I am personally involved in a number of commissions that explore transmedia as a kind of temporal engagement reading form. One commission, for example, is a collaboration with publisher Visual Editions and Google Creative Lab Sydney. The story I have developed is called *Breath*, a web-based mobile reading experience. The story personalizes for every reader through the use of APIs [application processing interfaces], which draw in data about place, weather, season, according to where and when the reader accesses the story. The project resides in the same storyworld as *Jellybone*, mentioned earlier, but for me the way it uses data-flows that are all around us begins to hint at possible futures for transmedia storytelling."

KF:

"From my perspective, thinking about transmedia as an audience-led approach transcends sector and possibly even time, too. Any given campaign, be it produced now or tomorrow, might well be completely different depending on where it has been produced, but that is really only because the user need is different. In summary, the meaning of transmedia in the context of digital marketing is about communicating with the consumer in a way that understands and supports their needs. And I cannot see that changing. It is about discipline, and thinking about whether a given platform or technology is really needed as part of the campaign in terms of the needs and wants of the user. And it is also about patience in the sense that building up successful transmedia campaigns takes a lot of time, energy, research, and resources."

AN:

"Well, in terms of where we are going next, virtual reality, escape rooms, and augmented reality are all fun new ways of communicating a transmedia story. Platforms will come and go, though, and for me what will always remain key is the thematic heartbeat of the story, and understanding who your audiences are as well as focusing on the experience or the rollercoaster that we want to deliver to those audiences. Content can be shareable and spreadable, but making people excited enough to take the leap means remembering the heart and soul of the story in the first place. You want to make your audience feel like heroes, to let them have a voice, and to instil feelings of being change-makers, so that these transmedia strategies and approaches are being used for something that is ultimately for the good."

INDEX

Note: Page numbers in **bold** refer to tables.